Second Edition

AUTOMATIC TRANSMISSIONS
Fundamentals and Service

William L. Husselbee

A RESTON BOOK
Prentice-Hall
Englewood Cliffs, New Jersey 07632

To my wife, Laurie,
whose patience, understanding,
and clerical assistance
made this project possible.

Library of Congress Cataloging-in-Publication Data

Husselbee, William L.
  Automatic transmissions.

  Rev. ed. of: Automatic transmission service. c1981.
  "A Reston book."
  Includes index.
  1. Automobiles—Transmission devices, Automatic.
2. Automobiles—Transmission devices, Automatic—
Maintenance and repair.   I. Title.
TL263.H88   1986   629.2'446   85-25704

ISBN 0-8359-9373-6 (pbk.)

Editorial/production supervision and interior design
by Barbara J. Gardetto

A Reston Book
Published by Prentice-Hall
A Division of Simon & Schuster, Inc.
Englewood Cliffs, New Jersey 07632

© 1986 by Prentice-Hall
Englewood Cliffs, New Jersey 07632

10   9   8   7   6   5   4

PRINTED IN THE UNITED STATES OF AMERICA

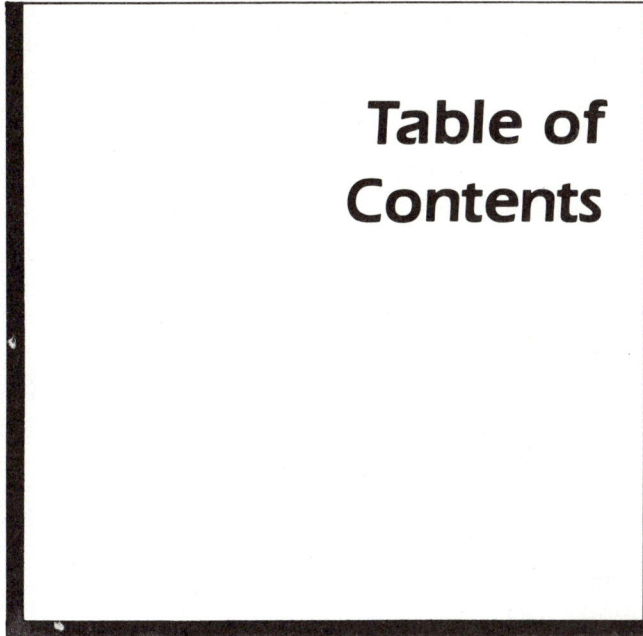

Table of Contents

Chapter 3
Automatic Transmission Fluids 38

Chapter 4
Fluid Couplings and Torque
Converters 51

Chapter 5
Clutches Used in Automatic
Transmissions 83

Chapter 6
Bands and Servos 98

Chapter 10
Pressure Regulating Devices   158

Chapter 11
Hydraulic Circuit Relay Valves
and Flow Control Devices   171

Preface

In the last decade, there have been many changes in automatic transmission designs and with these, obviously, are alterations in operating characteristics. The majority of these changes are, in effect, new transmission types or the addition of lock-up converters, (either hydraulically or electronically operated) to existing automatic units. But regardless of the modification, the main reason for it is that it increases the fuel economy of a vehicle, whether it is achieved by the production of a compact car with a smaller engine or by providing some form of overdrive to decrease engine operating speeds.

In any case, these new transmission types as well as the older ones require periodic maintenance by skilled technicians. These people work in the thousands of repair facilities throughout the United States and abroad.

Due to the large increase in the number of new vehicles produced and sold each year and the holding on of vast numbers of reliable older vehicles by the public, along with the retirement of skilled workers, there is always a need for new employees to service and repair automatic transmissions. However, these new people require some extension training before they can enter the trade, and many skilled technicians need updating on the newest equipment on the market.

Automatic Transmissions: Fundamentals and Service can be of immense help in this technical training. This text provides up-to-date information on the subject matter in a straightforward and concise manner and provides many tricks-of-the-trade to assist the new as well as the skilled mechanic in reducing his labor time on difficult jobs.

To accomplish this task, the book is divided into two sections, one on fundamentals and the other on service.

FUNDAMENTALS

In a typical chapter on theory, the purpose, construction, and operation of a component is explained in easy-to-understand language. These chapters also point out many of the differences in basic component or system design used by the various manufacturers in their respective transmissions. Once the reader understands the basic principles under which a given transmission component or system operates, the knowledge is easily applied to master the differences encountered on the many types of automatic transmissions in use today.

Along with explaining the function, design, and operation of each of the major transmission components, Section I also points out clearly how these parts interact with one another in order to make the whole transmission operational. The last chapter in the section puts all this information to good use in explaining the mechanical and hydraulic operation of a three- and four-speed transmission.

The mechanical component covered first is the planetary gear train. Chapter 2 reviews gear design and how the gear uses the principle of the lever to control both torque and speed of the engine. It then describes the design, function, and ratios achieved by simple planetary gearsets, a Simpson gear train, and the Ravigneaux system.

Chapter 3 supplies important information regarding transmission fluids. For example, it describes the purpose, types, and structure of the fluid. Also, the chapter provides information on the two main types of factory systems, along with several kinds of after-market devices used to cool the fluid.

Chapter 4 is devoted to torque converters and deals with the purpose, construction, and operation of these units. It describes in detail all the converter components and how they function, along with how the impeller produces vortex and rotary flows within the converter. Finally, the chapter explains the operation of the standard, variable-pitch, lock-up, overdrive, and splitter-gear converters.

Two types of clutches are covered in Chapter 5, the

multiple-disc and the overruning. This description points out the purpose and design of both units and how the transmission uses both clutch designs to control planetary operation.

In Chapter 6, the reader will find a thorough description of the different types of bands and their operating servos. These explanations contain the design, function, and operation of these units and relate how they are necessary to control planetary gear train operation.

Chapters 7 through 13 deal with the subject of hydraulics. Without the effects of hydraulics, the transmission could not function. Therefore, the subject matter contained in these chapters is very important to understanding automatic transmission operation. The reader should study it carefully.

Hydraulic system sealing devices are covered in Chapter 7. Various problems encountered in attempting to seal the hydraulic system against excess fluid leakage are described. The chapter also covers the different types of sealing devices used in an automatic transmission, how the mechanic should install certain types, and explains why the manufacturers use certain seal designs in a given location.

Hydraulic fundamentals are dealt with in Chapter 8, including hydraulic levers, force, pressure, and "Pascal's Law." Further, the chapter describes the basic components of a simple hydraulic system and then explains how these parts operate together to make the system transmit pressure and motion.

Chapter 9 begins the explanation of the actual hydraulic components of the transmission by describing the various types and designs of oil pumps and how they operate, including the gear, rotor, and vane units. Chapter 10 covers the relief valves, pressure-regulating, and converter-control valve circuits. Chapter 11 explains the design, function, and operation of the manual, shift, relay, and check valves. There are descriptions in Chapter 12 of the three types of hydraulic signaling devices, i.e., the governor, throttle, and kickdown valves.

Chapter 13 covers circuit or systems diagrams. This chapter describes the special symbols and codes used on the schematics, and it also explains how to interpret these charts and use them to troubleshoot a malfunction within the transmission. Knowledge of how to use these charts is an invaluable aid to the technician when trying to locate the cause of a problem.

Chapter 14 explains the operation of two transmissions, the TorqueFlite and the automatic overdrive (AOD) as well as an electronically controlled lock-up converter. This description includes the major hydraulic and mechanical components of each unit and covers the hydraulic operation and mechanical powerflow of each transmission in the Drive and Reverse ranges.

SERVICE

The service section of the text is basically a shop or lab manual. Its function is to provide the reader with the "how to" information necessary to inspect, service, troubleshoot, or rebuild the modern automatic transmission.

This section covers every aspect of shop training necessary to enter the field of automatic transmission repair. The section has nine well-thought-out and complete chapters which cover the following: shop equipment and tools; measuring devices and fasteners; automatic transmission problem diagnosis; changing the transmission fluid and filter; band and linkage adjustments; transmission and seal removal and installation; torque converter and hydraulic pump inspection, testing, and service; subassembly cleaning, inspection, and service; and transmission overhaul.

The material contained in the service section represents an overview of the many types of tools and equipment along with the diagnosis, service, and inspection procedures commonly used by the industry. It does not cover the overhaul procedures for every automatic transmission on the market because this information would fill many volumes. Instead, the service chapters present those concepts that apply to every unit or can be applied to the majority of transmissions the mechanic will encounter in the field today.

ORGANIZATION OF THE BOOK

Each chapter of the text contains various types of training aids to assist the reader in learning the material. For instance, each chapter contains a generous number of illustrations. Moreover, each illustration has a brief explanation that ties it to the text material. The hydraulic schematics in Chapters 13 and 14 are coded to indicate the fluid flow in a given circuit. This assists the reader in following the fluid flow as described in the text material. Furthermore, important steps or precautions are placed in bold for emphasis.

At the end of each chapter are Review Questions, and each chapter on fundamentals also has a Summary. Both assist the reader in determining how well he remembers the material contained in the chapter. The answers to the Review Questions are found in the Appendix located in the back of the book.

The text also provides two methods of quickly locating a topic. One is the index located in the back of the text; the other is an expanded Table of Contents. This table contains both the main and subheadings of all subject matter within all the chapters.

The author has 30 years' experience in the industry in the following areas: automotive machinist, automobile mechanic, service advisor, instructor, Auto/Diesel Department Director, and technical writer. Five years of his teaching experience has been in Automatic Transmissions at the college level. This experience is reflected in the practical approach to the subject matter and the content and design of the text.

ACKNOWLEDGMENTS

The author wishes to thank Chrysler Corporation, Ford Motor Company, General Motors Corporation, and Reston Publishing Company for the use of materials and illustrations in this text. The author would be grateful for information from readers on errors or omissions from the text.

William Husselbee

Section 1

FUNDAMENTALS

Since 1940, the automotive industry has designed and manufactured many types of fully automatic transmissions. The differences that exist among them lie in component structure, size, and some variations in hydraulic system operation. As for their basic operation, they all operate in much the same manner. Therefore, the person who understands the operation of one automatic transmission can apply this knowledge to other transmission designs.

Before studying the various chapters on the construction and operation of a typical automatic transmission, the reader should understand a few basic facts relating to the transmission of power from the engine to the drive wheels. These include the functions of the power train and automatic transmission, characteristics of piston engines that make a transmission necessary, speed and torque ratios, and the advantages of an automatic transmission over the manual shift type.

CHAPTER 1

Introduction to Automatic Transmissions

Power Train

Located between the engine and drive wheels, the automatic transmission is a component of the power train. The **power train** consists of a number of parts responsible for carrying the rotary motion developed in the engine to the drive wheels. The number and type of components needed by the power train depend on the type of transmission used, the location of the drive wheels, and the location of the engine. For example, the power train of a vehicle with a standard (manual-shift) transmission consists of a friction clutch, transmission, drive shaft, differential, and the rear axles (Fig. 1-1). On the other hand, vehicles with an automatic transmission use a torque converter instead of the foot-operated clutch as part of the power train. Also, motor vehicles that use either front-wheel drive or rear-mounted engines do not require drive shafts to complete their power trains. In these situations, the transmission, final drive, and differential are commonly called a transaxle (Fig. 1-2).

AUTOMATIC TRANSMISSION FUNCTIONS

The **automatic transmission** is a torque transferring and multiplying device that can allow the engine to act as a braking device and can change the direction of the drive wheels. **Torque** is a twisting effort produced by the engine; this torque is responsible for turning the drive wheels (Fig. 1-3).

Multiplies Torque

If the engine is operating and the driver places the transmission's shift lever into a forward or reverse driving range, torque passes through the automatic transmission. Moreover, if the engine becomes overloaded during acceleration or from pulling a vehicle up an incline, the transmission multiplies torque without the operator moving the gear selector into a lower driving range. However, most automatic transmission designs also give the driver partial to full manual control of all driving ranges.

Assists Braking Action

To assist the brakes in slowing a vehicle, most automatic transmissions provide a lower gear range. This gear range, when activated under certain conditions, forces the engine to act like a brake to slow the vehicle down. This braking action is especially useful when a vehicle is moving down a long mountain grade to prevent the brakes from overheating.

Reverses Direction

The automatic transmission also provides a means to reverse the direction of the drive wheels so the vehicle can go backwards. This ratio provides a speed reduction and

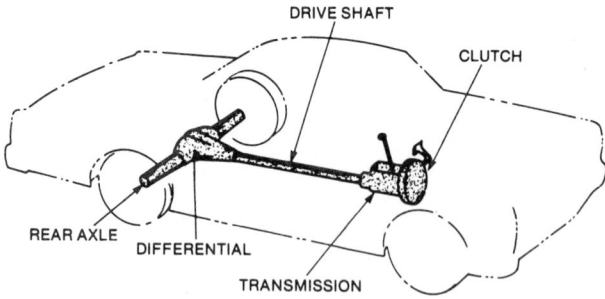

FIGURE 1-1 The power train of an automobile with a conventional transmission.

torque increase in the same manner as the lower gear range mentioned above.

Increases Fuel Economy

A recent trend within the automotive industry has been the development and production of overdrive automatic transmissions. The main reason for this is to increase a vehicle's fuel economy by permitting the engine to operate at a reduced rpm at cruising speed. Another, but less advertised advantage to the use of an overdrive is an extension of normal engine life, which also results from lowering its rpm during high road speeds.

TORQUE

The Engine Needs Help

The function of an engine is to produce torque that turns the drive wheels in order to set the vehicle in motion and to maintain certain road speeds. However, engine torque is not sufficient to propel the vehicle under all driving conditions, due to certain engine operating characteristics (Fig. 1-4). First, if an engine is under a load and operating at low speeds, torque output is low but reaches maximum at about 40 to 50 percent of engine speed, producing the highest horsepower. Above this 40 to 50 percent range,

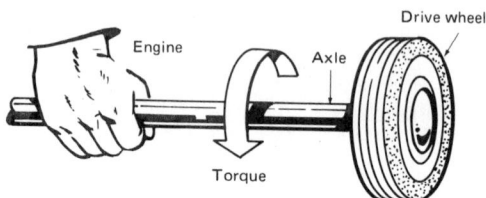

FIGURE 1-3 Engine torque is responsible for turning the drive wheels.

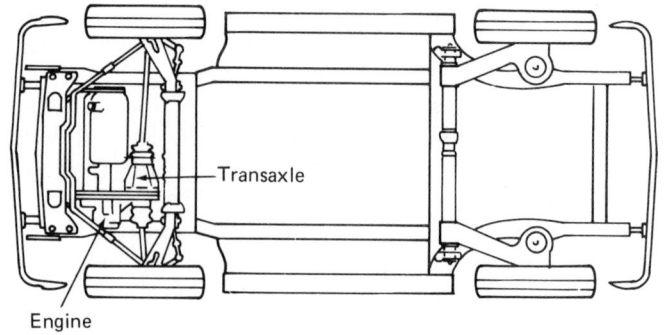

FIGURE 1-2 The drive train of a front-wheel-drive vehicle with a transverse engine and transaxle.

torque begins to taper off. Second, if an engine is operating with no load applied, torque output will be lower than that of a loaded engine operating at the same speed. Third, the engine will not produce any additional torque than is necessary to rotate the drive wheels. Finally, the engine will not operate under an excessive load condition.

In order to clarify the reasons for these engine characteristics, a description of torque production by a running engine is necessary. For simplicity, the discussion will focus on the operation of a single cylinder engine. Just keep in mind that final engine torque output is the result of the total number of engine cylinders all working together.

Force

Torque is the tendency of a **force,** a pushing or pulling action, to produce rotation of an object on its axis; the unit of measurement for torque is foot-pounds. In the case of the gasoline-piston engine, the force is the pressure caused by the burning air/fuel mixture pressing downward on the piston head (Fig. 1-5). Since the connecting

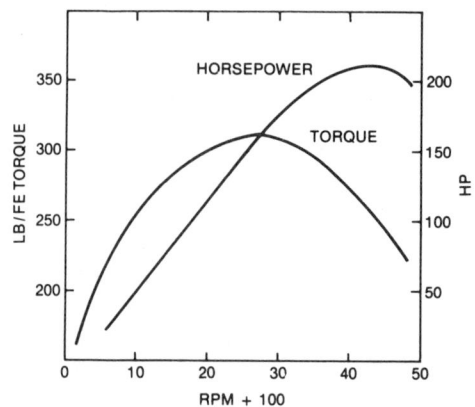

FIGURE 1-4 Torque and horsepower output curves of a typical piston engine.

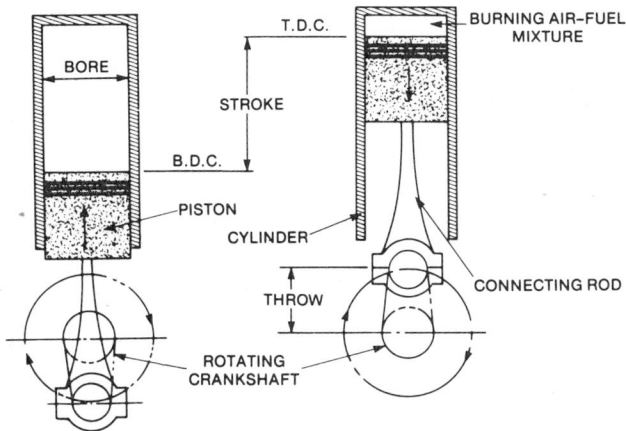

FIGURE 1-5 Torque production within the piston engine.

rod attaches the piston to the crankshaft, the downward piston movement forces the crankshaft (the object) to rotate.

The factors that determine the extent of the force produced by a piston on the crankshaft are the bore of the cylinder, the stroke of the piston, the compression ratio, and the quantity of air/fuel mixture admitted into the cylinder and combustion chamber. The **bore** of an engine is the diameter of the cylinder; the **stroke** is the distance the piston travels from the top to the bottom of the cylinder or from the bottom to the top; and the **compression ratio** is the volume of the cylinder and combustion chamber when the piston is at the bottom of its stroke divided by the volume when the piston is at the top.

All the above-mentioned factors are design dimensions of an engine. In order to change them, a complete reworking of the engine is necessary.

However, the main factor used to raise or lower torque output of an operating engine is the amount of air/fuel charge allowed to enter the cylinder. The quantity of the air/fuel mixture entering the cylinder varies with throttle valve position. The throttle valve of a carburetor operates by mechanical linkage connected to the gas pedal. As the driver presses down on the gas pedal, the throttle valve, a disk-shaped valve, opens to allow additional air/fuel mixture into the cylinder (Fig. 1-6). When the driver releases foot pressure on the gas pedal, the throttle valve closes to restrict the amount of mixture entering the cylinder.

Torque Varies With the Engine's Operating Mode

The amount of torque produced by a loaded engine varies with speed. For example, when an engine is operating at idle speed (measured in revolutions per minute [rpm]) with the automatic transmission in gear, the throttle valve allows a relatively small air/fuel charge to enter the cylinders (Fig. 1-6). Engine vacuum at this time will be relatively high. Now, since the quantity of the air/fuel charge determines the force applied to the pistons, total engine torque output at this point will be too low to set the vehicle in motion.

However, as the driver opens the throttle valve to accelerate the vehicle, the vacuum drops and additional air/fuel mixture enters the cylinders. As a result, engine torque rapidly increases until it reaches a maximum figure near the midpoint in the engine's safe operating range (Fig. 1-7). To assist the engine, the automatic transmission and torque converter multiply engine torque from the moment the driver releases the brakes until the engine reaches sufficient torque to propel the vehicle.

An unloaded engine operating at the same rpm as a loaded engine of the same size produces less torque than the loaded one. This engine characteristic is also the result of throttle valve action on the air/fuel mixture flow. If

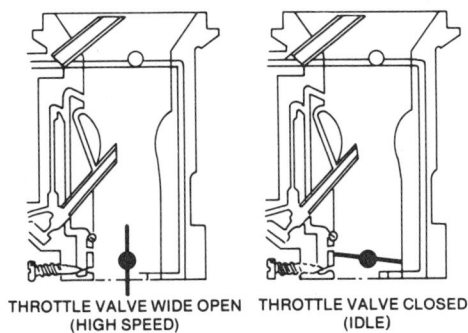

FIGURE 1-6 The carburetor throttle valve controls engine speed and torque by regulating the amount of air/fuel mixture entering the cylinders.

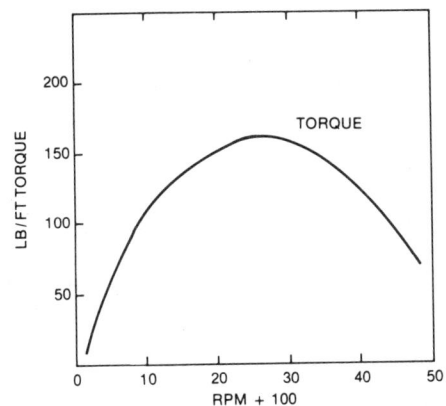

FIGURE 1-7 Torque curve of a loaded piston engine.

both engines (loaded and unloaded) are accelerated to a given rpm, for instance, the unloaded engine requires a smaller throttle valve opening than the loaded engine. This results in a higher engine vacuum and a smaller quantity of air/fuel mixture entering the cylinder of the unloaded engine. This reduced charge, when ignited, produces less push on each of the piston heads, resulting in a decrease in torque output.

Since the rate of air/fuel flow controls torque output, the engine never produces any more additional torque than is necessary. For example, to maintain a vehicle's given road speed, an engine must develop a torque of 150 foot-pounds. The driver depresses the gas pedal until the vehicle reaches the desired road speed. This action opens the throttle valve sufficiently to permit enough of an air/fuel charge to enter the cylinders to produce the 150 foot-pounds of engine torque—and no more. In other words, the engine produces just enough torque to bring the vehicle up to a certain road speed.

The engine will not operate if excessively overloaded. An engine of a certain bore, stroke, and compression ratio will produce, with a sufficient charge of air/fuel mixture, an output of torque that is enough to rotate the drive wheels of most moving vehicles on a level road. If the torque needed to rotate the drive wheels exceeds the maximum torque production of the engine, the transmission or some other torque multiplying device must increase the torque coming from the engine, or the engine will eventually stall.

Even if the engine was operating at a point of maximum torque output and the amount of torque was high enough to set the vehicle in motion, the acceleration would be slow, noisy, and sometimes uncomfortable for the occupants of the vehicle. Furthermore, the engine speed would be too high, placing a strain on the engine as well as the other power train components of the vehicle. With the knowledge of the engine's torque characteristics in mind, it should be apparent why the automatic transmission and other power train components are necessary to multiply torque and control engine speed.

SPEED OR GEAR RATIOS

As previously mentioned, the automatic transmission provides the vehicle with various driving ranges, or speed ratios. When related to any form of gear set, a **speed ratio** is the relative speed in number of revolutions that the input (driver) gear makes to one revolution of the output (driven) gear. As an equation,

$$\text{speed ratio} = \frac{\text{number of turns of (output) driven gear}}{\text{number of turns of (input) driver gear}}$$

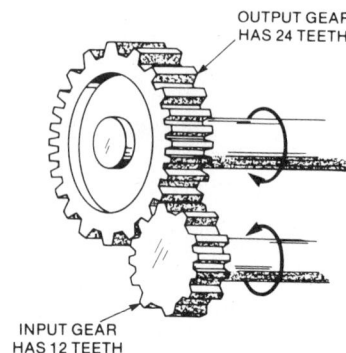

FIGURE 1-8 *Gear ratio is determined by dividing the number of teeth of the output gear by the number of teeth found on the input gear.*

To determine a typical speed ratio, sometimes known as gear ratio, count the number of teeth of each gear and then divide the number of teeth on the output gear by the number of teeth on the input. For instance, in Fig. 1-8, the drive (input) gear has 12 teeth, and it meshes with the driven (output) gear having 24 teeth. The ratio for this set is 24/12 or 2:1. This means that the input gear must make 2 revolutions in order for the output gear to make one revolution.

PLANETARY GEAR SPEED RATIOS

The component within the automatic transmission that produces speed ratios through gearing is the planetary gear set (Fig. 1-9). The actual ratio, in this case, is the relative speed of the transmission's input shaft (which brings torque into the transmission) to the speed of the transmission's output shaft (which delivers torque to the drive shaft or directly to the ring and pinion). For instance, a certain planetary gear set provides a ratio of 2.50:1 in low ratio. This 2.50:1 ratio simply means that the input shaft is making two and a half revolutions to each complete revolution of the output shaft.

Planetary gear trains provide two or more forward ratios and a reverse. A lower gear ratio, such as the 2.50:1 mentioned above, allows the engine to operate at a higher speed while the drive wheels turn at a lower speed. A high-gear ratio, on the other hand, permits the input and output transmission shafts to revolve at the same speed, 1:1 (i.e., one to one). In the case of the 1:1 ratio, the engine will still be operating faster than the drive wheels because of the gear ratio provided by the ring and pinion along with the speed ratio provided by the torque converter, in most cases. An overdrive ratio, approximately 1:1.50, permits the engine and input shaft to revolve slower than

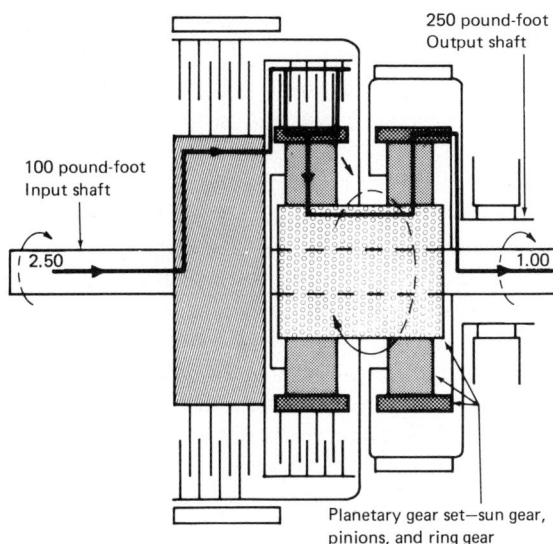

FIGURE 1-9   The tear ratio between the automatic transmission's input and output shafts.

the transmission's output shaft. Finally, the reverse gear ratio provides not only a speed differential between the input and output shafts but also a change of direction.

Torque Converter Speed Ratios

The torque converter produces an unlimited number of speed ratios. The **torque converter** is a device that uses the impact energy of moving fluid to transmit torque from the engine's crankshaft to the automatic transmission's input shaft. To transmit torque, the converter has two main moving parts: the impeller and the turbine (Fig. 1-10).

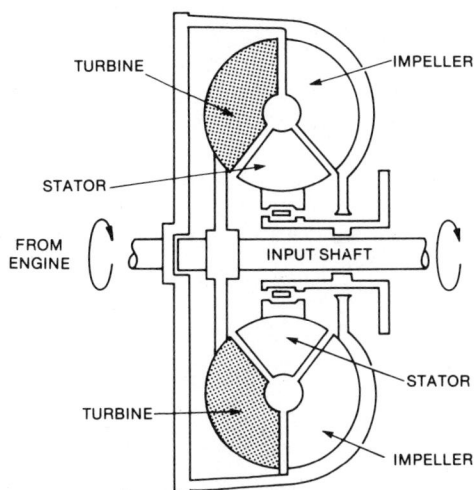

FIGURE 1-10   The impeller and turbine of the torque converter transmit torque between the engine and the transmission's input shaft.

These components operate within a sealed housing that is full of fluid.

The speed ratio of the torque converter is the number of revolutions the impeller rotates compared to a single rotation of the turbine and the input shaft. For example, with the vehicle stationary and the engine operating at idle with the transmission in gear, the impeller rotates at engine speed, but the turbine is stationary. The speed ratio is zero with the converter producing zero percent efficiency. The engine will be turning the impeller at about 600 rpm, but the turbine is still. On the other hand, if the impeller is turning at 1,000 rpm and the turbine at 900 rpm, the converter efficiency now is about 90 percent. Between the zero speed ratio and a ratio of about 1:0.9, the converter produces a wide range of constantly changing speed ratios. This text covers the design and operation of the torque converter in more detail in another chapter.

Ring and Pinion Gear Speed Ratios

The ring and pinion gears that are part of the drive axle assembly produce a fixed speed or gear ratio between the transmission's output shaft and the drive axles (Fig. 1-11). This gear ratio varies between about 2.5 and 3.5 for most passenger car applications and serves to control engine rpm along with its torque output during all phases of vehicle operation. For instance, when the automatic transmission is operating in low gear, the transmission's ratio is multiplied by the ring and pinion ratio to produce the total speed ratio between the input shaft and the drive axles. Therefore, if the transmission's low gear ratio is 2.5:1 and the ring and pinion ratio is also 2.5:1, the total speed or gear ratio is 5:1.

The converter ratios are also effective during this time. However, since these ratios are changing very rapidly, it is difficult to figure them into the total speed reduction between the engine and drive wheels until the trans-

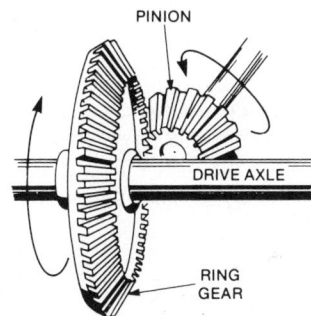

FIGURE 1-11   The ring and pinion gear of the drive axle assembly provides a fixed gear ratio between the transmission's output shaft and the drive axles.

mission shifts into high gear and the converter reaches its 1:0.9 ratio.

When the automatic transmission upshifts to high gear, the ring and pinion ratio remains effective in controlling engine rpm and torque output. At vehicle cruising speed, the engine should be nearly producing maximum torque, but the ring and pinion ratio has the effect of reducing engine load by increasing its torque. This leaves the engine with some reserve torque, and the automatic transmission will not have to downshift into a lower gear ratio whenever the engine receives additional demands for torque necessary for acceleration.

The disadvantage in utilizing a ring and pinion ratio of more than 1:1 (direct drive) is a loss of fuel economy. A 1:1 ratio is more economical because the engine would be operating at a more efficient speed, but when additional engine torque is necessary to accelerate the vehicle, the transmission would have to downshift into a ratio that provides the necessary torque increase.

GEAR RATIOS ARE ALSO TORQUE RATIOS

Whenever a set of gears provides a speed reduction between two objects, the gears also multiply torque or provide a torque increase ratio between the two. For example, the low gear of an automatic transmission produces a gear ratio of 2.5:1 between the input and output shafts (see Fig. 1-9). This same ratio provides a torque increase of 1:2.5. Therefore, if the input shaft's torque is 100 pounds-foot, the output shaft's would be 100 pounds-foot times 2.5, or 250 pounds-foot.

The same rule applies for the ring and pinion gear ratio except that the torque increase would be between the transmission's output shaft and the drive wheels. As in the case of the speed ratios, the total vehicle torque increase ratio is equal to the transmission's ratio multiplied by the ring and pinion ratio. The important idea to remember is that speed reduction also means torque increase, and the

speed reduction allows the engine to operate at a more efficient torque producing rpm while the gears themselves are increasing engine torque.

The torque converter also has the ability to multiply as well as transmit torque. In order to accomplish a torque increase, the converter requires a third component called a stator (see Fig. 1-10). The stator fits between the impeller and turbine and begins to multiply engine torque whenever there is an impeller-to-turbine speed ratio greater than about 1:1. Since the speed ratios of the torque vary, the torque ratios do also. The maximum torque increase ratio of most automotive converters is about 1:2.25 and varies during acceleration until it reaches 1:0.9 during the converter's coupling phase. At this point, there is a small torque loss to the drive wheels due to slippage within the converter between the impeller and turbine.

ADVANTAGES OF USING AUTOMATIC TRANSMISSIONS

In any type of motor vehicle, the automatic transmission offers several advantages over the manual-shift type. The automatic transmission provides the vehicle with a smooth start, proper upshift pattern, and the best possible gear ratios for all driving conditions. The automatic transmission also responds to the will of the driver as well as the requirements of the various highway and vehicle speed conditions. An automatic transmission makes a motor vehicle more comfortable and easier to drive because this unit requires no foot-operated clutch to set the vehicle in motion or to shift gears. Consequently, the driving of a vehicle in heavy traffic or up a steep, twisting mountain road requires little driver effort. Finally, the automatic transmission, with its simple-to-operate controls, requires little skill on the part of the driver to start the vehicle in motion or to select the proper gear ratio at the correct time.

SUMMARY

1. The automatic transmission is a member of the power train team.

2. The automatic transmission is a torque transferring and multiplying device that can also change the direction of the drive wheels and act as a braking device.

3. Under certain driving conditions, the engine will not produce enough torque to propel a vehicle.

4. High gas pressure, caused by a burning air/fuel charge, forces the piston and rod to rotate the crankshaft.

5. In an operating engine, the quantity of air/fuel mixture admitted to each cylinder determines torque output.

6. At engine idle, the torque output is low; engine torque reaches a maximum around the midpoint in the engine's operating range.

7. Engines will not produce any more torque than needed and will not operate under an overload condition.

8. Gear ratios control engine rpm and multiply torque.

9. The automatic transmission, torque converter, and ring and pinion provide speed ratios and torque multiplication.

10. The automatic transmission offers several advantages over the manual-shift type.

REVIEW

This section will assist you in determining how well you remember the material contained in this chapter. Read each item carefully. If you can't complete the statement, review the section in the chapter that covers the material.

1. Torque defined is a _____.

   a. twisting effort

   b. push-pull effort

   c. up and down effort

   d. law of inertia

2. The engine, when it is operating, produces _____.

   a. inertia

   b. energy

   c. torque

   d. force

3. Engine torque output is low during _____ speeds.

   a. idle

   b. high

   c. intermediate

   d. both b and c

4. The factor that alters torque production of an operating engine is the _____.

   a. bore

   b. stroke

   c. size of the valves

   d. quantity of the air/fuel mixture

5. Automatic transmission speed ratios are between the _____ and _____.

   a. output shaft, differential

    b. input shaft, output shaft

    c. engine, input shaft

    d. engine, differential

6. A ring and pinion ratio of 3:1 means that the transmission's output shaft will be rotating _____ faster than the drive wheels.

    a. one time

    b. two times

    c. three times

    d. both a and c

7. If an engine produces a torque of 100 pounds-foot and the transmission has a 3:1 gear ratio, torque increases to _____ pounds-foot before leaving the transmission.

    a. 50

    b. 150

    c. 225

    d. 300

8. If the torque converter is operating at 90 percent efficiency, the impeller will be turning at about 1,000, but the turbine will be rotating at _____ rpm.

    a. 700

    b. 900

    c. 1,000

    d. 1,200

9. The ring and pinion ratio increases _____.

    a. torque

    b. fuel economy

    c. force

    d. energy

10. The part of the torque converter that multiplies torque is the _____.

    a. impeller

    b. turbine

    c. housing

    d. stator

For the answers, turn to the Appendix.

In any type of automotive transmission, there are a number of different sized gears that mesh together to form the various speed and torque ratios. Before discussing how manufacturers arrange gears to construct the various types of planetary systems as found in automatic transmissions, it is appropriate to explain the construction and design of gears in general along with how they use the principle of the lever to multiply torque.

Gear Construction

A **gear** is nothing more than a wheel, which has small extensions called teeth around its circumference (Fig. 2-1). In order for a gear to perform its functions of changing speed, torque, and direction, it must be very strong. For this reason, manufacturers use a high quality material such as steel alloy, nickel, and chromium in the production of gears.

These materials undergo several machine processes and a hardening procedure before becoming a transmission gear. For instance, a machine hammers the materials, while they are red hot, into the general shape of a gear. Then other precision machinery cuts the teeth and other critical areas. Finally, a special heat-treating process produces a smooth, hard surface of the gear teeth, combined with a somewhat softer but tough gear body.

Gear Design

Designed and then machined into each gear are features such as teeth clearances, circle, pitch, and root diameters, in addition to gear teeth angle and location (Fig. 2-2). Before two gears can mesh properly, some tip and back-lash clearances must exist between the mating teeth. These clearances, which are several thousandths of an inch each, allow for lubrication expansion, possible size irregularities, and the heat expansion of the gear itself.

Circle diameter is a measurement taken across the circumference (the full distance around a gear). Pitch diameter is taken at a point about halfway from the bottom to the tip of the teeth. Pitch diameter is the true measurement utilized in calculating exact gear ratios. Lastly, minimum diameter of a gear is the measurement taken across it at the root, or base, of the teeth.

GEAR TYPES

Spur

The automotive industry identifies gears by the angle as well as the location of its teeth. For instance, a **spur-type** gear, the simplest type of gear, has teeth that are parallel

CHAPTER 2

Planetary Gear Systems

to the centerline, or axis of the gear (see Fig. 2-1). **Axis** is the centerline around which a gear or a shaft rotates.

Helical

A **helical gear** is a unit that has teeth machined at an angle to the axis of rotation (Fig. 2-3). There are several advantages and one disadvantage of using this gear shape. The first and most important advantage of using a helical gear is that it is stronger because more than one tooth is carrying the torque loads at any given time. This is not the case with the spur gear.

The second advantage of using a helical gear is its quieter operation. This is due to the gear's natural wiping action as it engages and disengages the teeth of another gear.

There is one disadvantage to the use of helical gears, which are in mesh and turning. That is, they tend to slide apart in an endwise direction. This is a direct result of the angle cut of the gears. In automatic transmission usage, helical gears are held in a given position, so they cannot move far enough apart to become demeshed. However, the tendency for endwise movement does cause undue stress on the ends of the gears, thrust washers, or the component that supports the gears themselves. As a result, these areas tend to wear out faster than those found on spur gears that are operating under the same conditions.

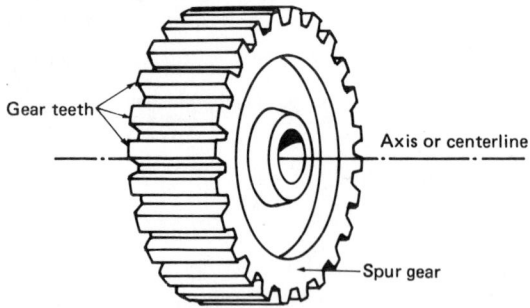

FIGURE 2-1 A gear is a wheel that has extensions called teeth around its circumference.

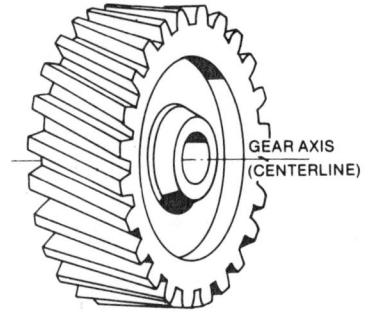

FIGURE 2-3 The teeth of a helical gear are cut at an angle to its axis.

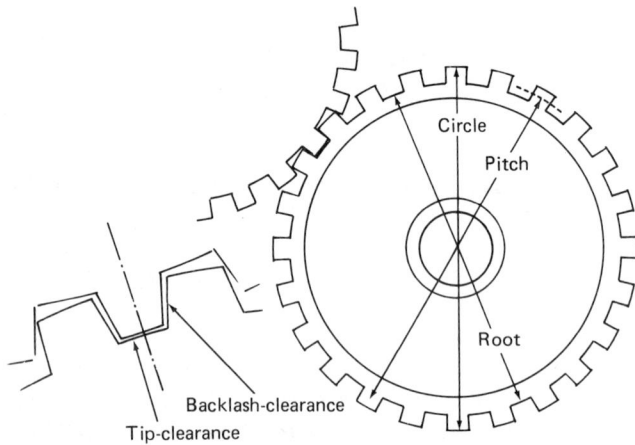

FIGURE 2-2 The circle, pitch, and root diameters in addition to the backlash and tip clearance of a typical gear.

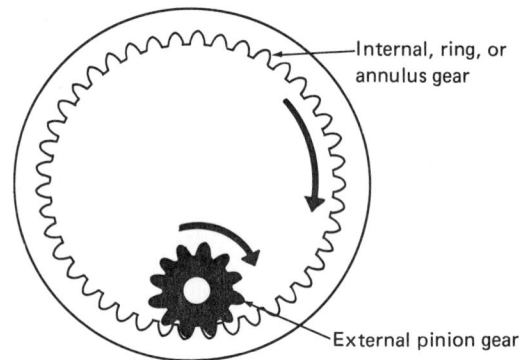

FIGURE 2-4 An internal gear has teeth cut into its inner circumference.

External Gears

Gears are also typed as to the location of their teeth. An **external gear,** for example, has its teeth machined into the outside circumference of the gear body (see Figs. 2-1 and 2-3). This type of gear is the most common one used in transmission and drive axle assemblies.

Internal Gears

In a planetary gear system, there is a gear that has teeth machined into its inside circumference (Fig. 2-4). This gear is known as either the **internal, ring,** or **annulus.**

The external gear meshed into the internal gear is known as a **pinion.** The pinion rotates on a pin while traveling around the inside of the internal gear. Note also in Fig. 2-4 that when an external gear meshes with an internal gear, they both rotate in the same direction.

TORQUE MULTIPLICATION USING GEARING

Lever and Fulcrum

To increase torque, transmission gears use the principle of the lever and the fulcrum. A **lever** is a simple machine. It is nothing more than a rigid bar, which rotates around a fixed point called a **fulcrum** (Fig. 2-5).

The combination of a lever and a fulcrum can serve two

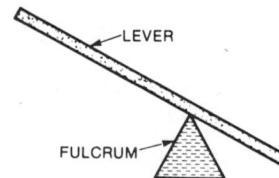

FIGURE 2-5 A simple lever and fulcrum

purposes. First, it can provide a mechanical advantage. Second, the lever and fulcrum can change the direction of applied force.

Mechanical Advantage

By applying a small force over a greater distance, the lever provides a mechanical advantage by using a small input force to move a larger output force. For example, if a person cannot lift a heavy box, he could position a crowbar or lever (Fig. 2-6) and lift the box with half the effort that would normally be necessary to raise it.

Suppose that in order to raise the end of the box, 200 pounds of force is necessary. By placing the lever under the box at a location one foot from the fulcrum and pushing downward on the opposite end of the lever at a point two feet from the fulcrum, only 100 pounds of downward force is necessary to move the box upward.

Change of Direction

The use of a lever and fulcrum to multiply torque as shown in Fig. 2-6 causes a change of direction of the

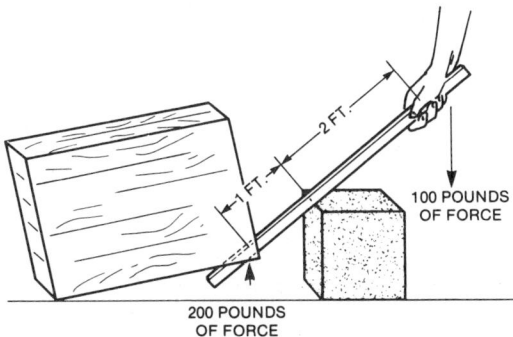

FIGURE 2-6   A lever used to lift a heavy box.

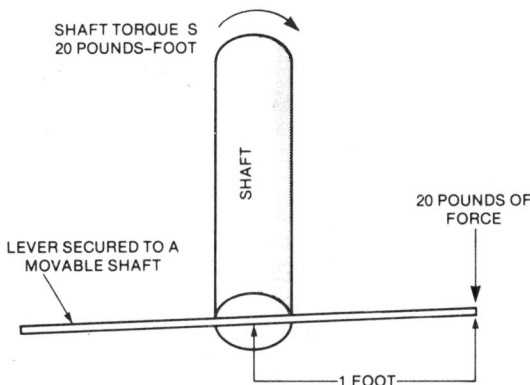

FIGURE 2-7   The movement of a shaft mounted lever causes the shaft to twist or torque.

forces involved. Notice that in this situation the box (output force) and the downward force (input force) move in opposite directions.

Mechanical Advantage Ratios

The term given to the relationship, or ratio, between the input and output forces is **mechanical advantage.** In the example above, the ratio is between the input force of 100 pounds and the output force of 200 pounds, or 1:2. If a ratio is 1:1, the lever provides no mechanical advantage, but it does create a change in direction.

Mechanical Advantage Alters Distance Traveled

In gaining a mechanical advantage between the input and output forces through the use of leverage, the distance each force travels is different. In Fig. 2-6, the input force travels two inches downward to raise the output force one inch. In other words, the person lifting the box must apply the downward input force over a longer distance in order to raise the box. The ratio for the distance traveled by the input and output forces is two inches to one inch, 2:1. Therefore, whenever a lever produces a mechanical advantage, watch for an accompanying loss in the distance traveled by the output force.

Fulcrum or Shaft Torque

Whenever a lever rotates on its fulcrum point, it is subjected to a given amount of torque. For instance, if a simple lever, at its fulcrum point, is secured to a movable pipe or shaft and someone applied a force to one end of the lever that caused the lever to move either up or down, the shaft would rotate with a given amount of twist or torque (Fig. 2-7). This shaft rotation is known as **shaft torque.**

The amount of shaft torque is equal to the applied force on the lever arm multiplied by the distance the force is applied from the center of the fulcrum or shaft. For example, if a 20-pound force pushes down on the lever at a point 1 foot from the fulcrum point, the attached shaft turns with a torque of 20 pounds times 1 foot, or 20 pounds-foot.

If the shaft torque is known, the output force on the ends of the lever is equal to the torque divided by the lever's radius. **Radius,** in this case, is the distance from the center of the lever to the point where the force is applied. In Fig. 2-8, the shaft delivers 30 pounds-foot torque to a lever arm with a radius of one foot. The output force on the ends of the lever is therefore

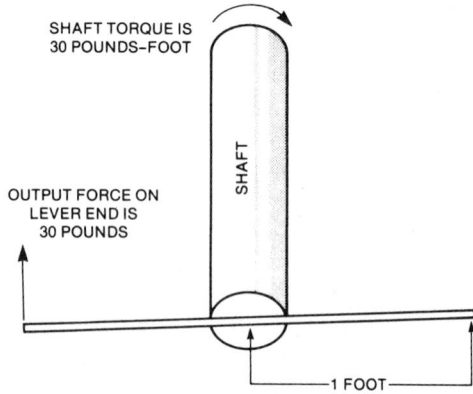

FIGURE 2-8   Rotation of the shaft causes the ends of the lever to move with a given amount of force.

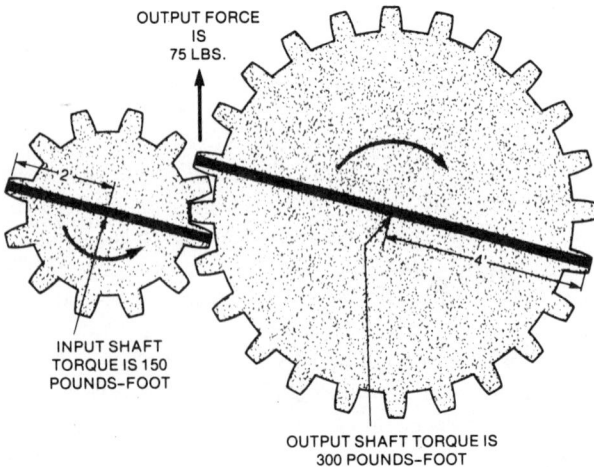

FIGURE 2-10   Torque multiplication along with a speed reduction occurs when two meshed gears have different number of teeth.

$$\frac{30 \text{ pounds-foot}}{1 \text{ foot}} \text{ or 30 pounds}$$

If the end of one simple lever contacts the end of a second lever, force can move from one lever to the other. The same situation occurs anytime two gears are in mesh. The gear teeth act as lever ends that extend all around each gear (Fig. 2-9). The teeth of the small gear (its lever ends) apply a push (force) to the teeth (lever ends) of the second gear.

Each of these gears rotates around its own center, which is the fulcrum point, but in opposite directions. As shown in Fig. 2-9, the smaller gear is the input (drive) in that the input shaft drives the gear at its fulcrum point. The larger gear is the output (driven gear) because the shaft attached to its fulcrum point is the output shaft.

Torque Multiplication Between Gear Driven Shafts

In the gear arrangement shown in Fig. 2-9, anytime the input shaft turns, torque multiplication occurs in the output shaft. To understand how this torque increase takes place, an explanation of each step in the flow of torque (which many people refer to as powerflow) between the two shafts is necessary. If the input shaft applies a torque of 150 pounds-foot to the fulcrum point of a gear which has a radius of 2 feet, the gear teeth have an output force of

$$\frac{150 \text{ pounds-foot}}{2 \text{ feet}} \text{ or 70 pounds}$$

Since the output gear has a radius of 4 feet, the torque on the output shaft is 75 pounds times 4 feet or 300 pounds-foot. The resulting mechanical advantage (torque increase) between the input and output shaft is 150 pounds-foot to 300 pounds-foot, or 1:2.

An even easier way to figure the torque multiplication accomplished through gearing is to count the number of teeth located on the input and output gears. For instance, Fig. 2-10 shows two spur gears. The input gear has 12 teeth, and the output has 24 teeth. In this case, the ratio of

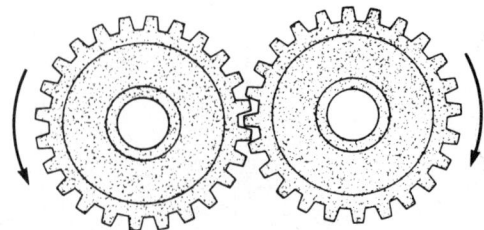

FIGURE 2-9   The gear teeth act as lever ends that extend all around each gear.

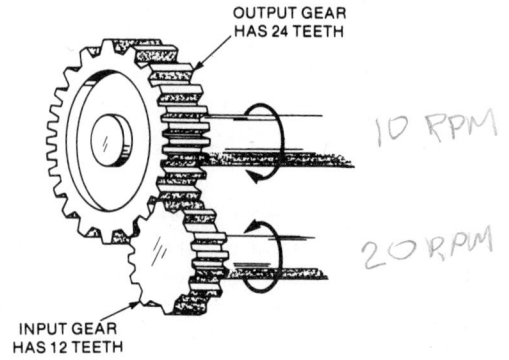

FIGURE 2-11   No torque multiplication or speed reduction occurs if two meshing gears have the same number of teeth.

torque increase between the two shaft mounted gears is 12 to 24 (12/24), or 1:2. Note also in the illustration that both gears are turning in opposite directions.

Idler Gears Transfer Torque

Figure 2-11 illustrates two gears having the same diameter and number of teeth. Each gear has 24 teeth; therefore, the torque multiplication is 24 to 24, 1:1. In other words, the two gears are just transmitting torque from one shaft to another while providing a change of direction.

It is also not uncommon in transmissions to have an arrangement like this one. By the addition of a third gear, the center gear acts as an idler to obtain a change in direction between the input and output gears, without changing their torque. For example, suppose a 48-tooth gear was meshed to the right of the center gear. The torque increase ratio would be input gear (24) to idler gear (24), 1:1 times the idler gear (24) to output gear (48) or 1:2

$$\frac{\text{input gear 24}}{\text{idler gear 24}} = \frac{1}{1} \times \frac{\text{idler gear 24}}{\text{output gear 48}} = \frac{1}{2} \text{ or 1:2}$$

In this case, the idler gear did not change the torque ratio between the input gear and the output gear. The idler gear, in this situation, is only used to turn the output gear in the same direction as the input, which is necessary in some gear train arrangements.

Torque Increase Causes a Speed Reduction

Increasing torque through the use of gears does bring about a speed reduction. To figure the speed reduction that gears produce, compare the number of teeth on the output gear to the number of teeth on the input gear. For example, in Fig. 2-10, the output gear has 24 teeth, while the input gear has 12 teeth. The speed ratio between the two gears, with the smaller one driving the larger, is

$$\frac{\text{output gear 24}}{\text{input gear 12}} = \frac{2}{1} \text{ or 2:1}$$

In other words, the smaller input gear turns twice as fast as the output gear that has double the number of teeth.

In Fig. 2-11, both gears have the same number of teeth. Consequently, the speed ratio is

$$\frac{\text{output gear 24}}{\text{input gear 24}} = \frac{1}{1} \text{ or 1:1}$$

In other words, they both rotate at the same speed.

It should be obvious that there is a definite relationship between torque and speed ratios. For example, utilizing two gears to double the torque output cuts the output speed to half the input. Expressing this in ratio form: Torque ratio 1:2; speed ratio 2:1. Ignoring the relatively minor friction losses of gears and shafts, this means that a 1:2 torque ratio doubles the output shaft torque, while permitting the input shaft to turn at twice the speed.

PLANETARY GEAR TRAINS

The **planetary gear train** is the heart of any automatic transmission (Fig. 2-12). This mechanical device is responsible for all forward-driving ratios and reverse. The complete gear train may consist of one or more simple planetary gearsets, two simple planetary gearsets connected by a common sun gear, or a compound planetary gearset.

Simple Planetary Gearset Design

A simple planetary gearset consists of a sun gear, a carrier with three or more planet pinions, and an internal gear, also called the ring or annulus gear (Fig. 2-13). The gear located in the center of the unit is the sun gear. The teeth

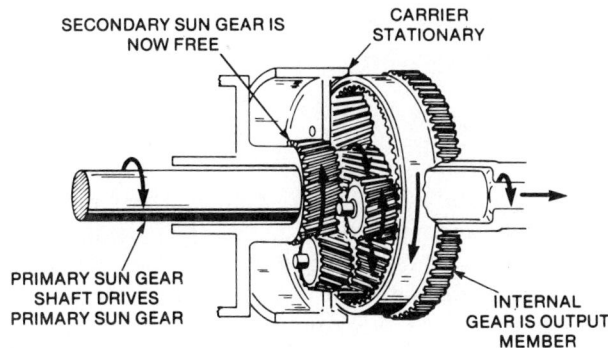

FIGURE 2-12   A typical compound planetary gear train.

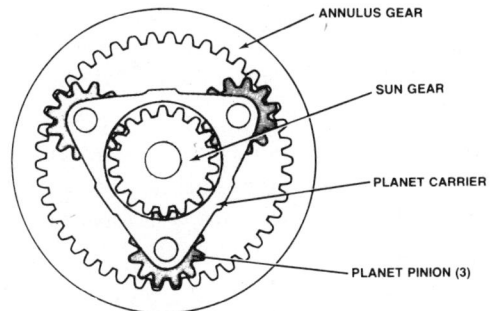

FIGURE 2-13   The design of a simple planetary gearset. (Courtesy of Chrysler Corp.)

of the sun gear are in constant mesh with the teeth of the planet pinions.

The **carrier** supports the pinion gears on rigid pins, and each gear is free to run on these fixed shafts. The **planet pinions** are in mesh with the sun gear and the internal gear, which has teeth cut into its inside circumference. Since the pinions have the ability to turn on their centers and at the same time revolve with the carrier around the sun gear, the unit's activity is similar to the planet's movement in our solar system; therefore, the entire unit is commonly known as a planetary.

Planetary Gear Train
Advantages of Using a

The planetary gear train offers several advantages over the sliding gear arrangement found in manual-shift transmissions. First, the device is smaller in size because all planetary gear train components revolve around a common axis, the sun gear. Second, all planetary gears are in constant mesh, which eliminates the possibility of gear teeth damage from gear clashing caused by partial engagement or improper shifting techniques. Moreover, the constant mesh characteristic permits quick gear ratio changes without the loss of torque flow through the transmission. Third, planetary gear trains are stronger because the distribution of torque load is over more gears; consequently, more teeth are in contact to carry the loads. Fourth, the location of the various planetary gear train components makes the task of holding or locking the components together for ratio changes relatively easy. Finally, planetary gearsets are very versatile because the gearset can produce a wide variety of gear ratio combinations. However, it is more common for manufacturers to use only six different combinations and a direct drive from a simple gearset.

General Rules Pertaining to
the Action of a Planetary

Before studying some of the actual gear ratios produced by a simple planetary gearset, it is important to understand certain rules that pertain to the action of planetary gears.

1. Whenever the carrier is stationary with either the ring gear or the sun gear driving, the planet pinions rotate, "idle" on their pins, and force the remaining unused gear to turn in a reverse direction.

2. If either the sun or ring gear is stationary and the free sun or ring gear acts as the driving unit, the planet pinions "walk" the carrier, output unit, around the stationary gear.

3. The sun and planet pinion gears always rotate in opposite directions.

4. The planet pinions and the ring gear turn in the same direction.

5. In order to achieve a reduction of either speed or torque through the planetary, one member of the gearset, either the sun gear, carrier, or ring gear, is stationary or locked to the transmission; and the input shaft drives a second member. The third remaining member is output and transmits torque to the output shaft.

6. To achieve direct drive, high gear, the planetary must have two of its three members driven at the same speed, which will lock up the gearset. As a result, the input shaft's speed is the same as that of the output shaft.

7. When there is only one drive member and no held member, the gearset is in neutral.

8. Overdrive results whenever the sun gear is the output member.

9. If the sun gear is the input member, the gearset provides a forward overdrive ratio.

10. If the carrier is the input (drive) member, the gearset produces a forward overdrive ratio.

11. Whenever the carrier is the output member, the gearset produces a lower gear ratio in a forward direction.

RATIOS ACHIEVED BY SIMPLE PLANETARY GEARSETS

As mentioned, a simple planetary gearset usually produces six different ratios and a direct drive, depending upon what conditions it is operating under. The conditions, in this case, refer to which gearset member is input (driving), which is held stationary, and which is the output (driven) member.

Ratio 1

A simple planetary operating in ratio number 1 produces a speed increase (overdrive) and a torque reduction (Fig. 2-14). In this case, the input shaft drives the carrier, the sun gear is stationary, and the ring gear is the output, driven member. As the input shaft rotates the carrier clockwise, the planet pinions "walk" around a stationary sun gear in a clockwise direction. The walking pinions also cause the ring gear to turn in the same direction as the

SPEED INCREASE   FORWARD ROTATION   TORQUE REDUCTION

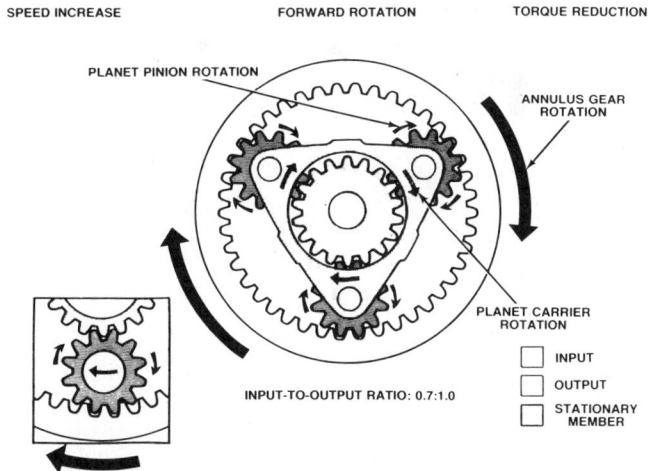

INPUT-TO-OUTPUT RATIO: 0.7:1.0

☐ INPUT
☐ OUTPUT
☐ STATIONARY MEMBER

FIGURE 2-14   *A simple planetary gearset operating in ratio number 1. (Courtesy of Chrysler Corp.).*

SPEED INCREASE   REVERSE ROTATION   TORQUE REDUCTION

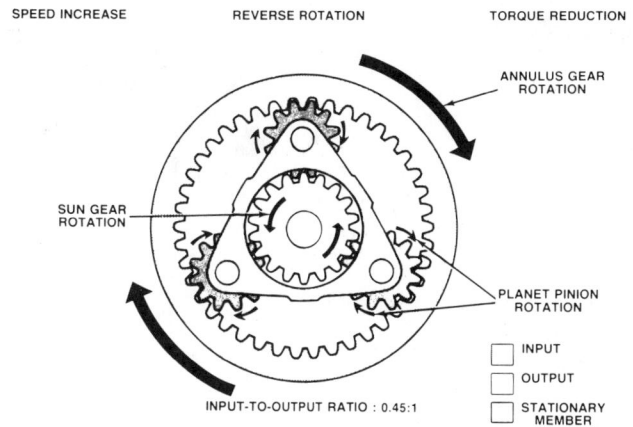

INPUT-TO-OUTPUT RATIO : 0.45:1

☐ INPUT
☐ OUTPUT
☐ STATIONARY MEMBER

FIGURE 2-16   *A simple planetary gearset operating in ratio number 3. (Courtesy of Chrysler Corp.)*

pinions. With this arrangement, the input shaft is rotating slower than the output at a ratio of 0.7:1. In other words, the gearset is producing an overdrive gear ratio that reduces torque but increases output speed in comparison to that of the input.

To calculate the actual gear ratio of a simple planetary operating in gear ratio number 1, overdrive, divide the sum of the number of teeth on the sun gear (S) and the ring gear (R) into the number of teeth on the driven gear (D). For example, in Fig. 2-14, the sun gear has 18 teeth and the ring gear has 42; the sum of which is 60. Also, the driven gear is the ring gear, which has 42 teeth. Since the formula states that this gear ratio is equal to D/(S + R), or 42/60, the input to output ratio is 0.7 to 1 (0.7:1).

Ratio 2

A simple planetary operating in gear ratio number 2 also produces a torque reduction and speed increase (Fig. 2-

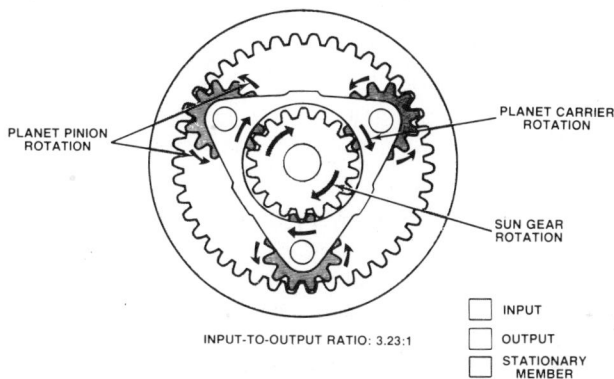

INPUT-TO-OUTPUT RATIO: 3.23:1

☐ INPUT
☐ OUTPUT
☐ STATIONARY MEMBER

FIGURE 2-15   *A simple planetary gearset operating in ratio number 2. (Courtesy of Chrysler Corp.)*

15). In this situation, the input shaft drives the planet carrier, the ring gear is stationary, and the output member is the sun gear. As the input shaft drives the carrier clockwise, inside the ring gear, the planet pinions turn counterclockwise. Since the pinions are in mesh with the sun gear, the sun gear has to rotate in a clockwise direction. The input to output ratio is 0.3:1, which is the best gear ratio for a forward speed increase, but the ratio also achieves a greater reduction in torque than ratio number 1.

To calculate the actual gear ratio of any simple planetary operating in ratio number 2, overdrive, use the number 1 gear ratio formula, D/(S + R) to find the answer. Since the sum of both the sun gear teeth (18) and the number of ring gear teeth (42) is 60 and the driven sun gear has 18 teeth, the ratio is 18/60, or 0.3 to 1 (0.3:1).

Ratio 3

Gear ratio number 3 produces an overdrive and a reverse relationship between the input and output shafts (Fig. 2-16). In this case, the planet carrier is the stationary member and the output member is the sun gear. The input shaft drives the ring gear clockwise. As the ring gear turns, it causes the planet pinions to "idle" clockwise. Due to this pinion rotation, the sun gear and output shaft turn in a direction opposite to that of the input member, the ring gear. The ratio is 0.43:1; that is, the output shaft is spinning faster in a reverse direction than the input shaft, but with reduced torque.

To calculate the actual gear ratio of a planetary operating in ratio number 3, divide the number of teeth on the driving gear (D1) into the number of teeth on the driven gear (D). In Fig. 2-16, the driven gear (sun) has 18

SPEED REDUCTION   FORWARD ROTATION   TORQUE INCREASE

ANNULUS GEAR
ROTATION

PLANET CARRIER
ROTATION

PLANET PINION
ROTATION

INPUT-TO-OUTPUT RATIO: 1.45:1

☐ INPUT
☐ OUTPUT
☐ STATIONARY
   MEMBER

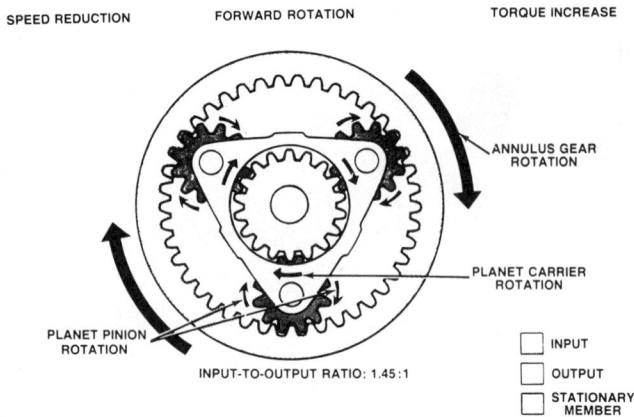

FIGURE 2-17   A simple planetary gearset operating in ratio number 4. (Courtesy of Chrysler Corp.)

teeth and the driving ring gear has 42. The ratio is D/D1, or 18/42, that equals a rounded-off figure of 0.43 to 1 (0.43:1).

Ratio 4

Gear ratio number 4 provides a speed reduction and a torque increase between the input and output shafts of the planetary (Fig. 2-17). During this ratio, the input shaft drives the ring gear in a clockwise direction with the sun gear held stationary. The output member is the carrier. As the ring gear turns clockwise, it forces the planet pinions to also rotate in a clockwise direction. The pinions, in turn, "walk" the carrier around the held sun gear in a clockwise direction. The ratio between the input and output shafts is 1.43:1, which indicates a speed reduction between the two but an increase in torque.

To determine the actual gear ratio of a simple

planetary functioning in ratio 4, divide the number of teeth on the driving gear (D1) into the sum of the teeth on both the sun and ring gears. In Fig. 2-17, the driving member, the ring gear, has 42 teeth. The sun gear (18 teeth) and the ring gear (42 teeth) combined to have 60 teeth. Therefore, the gear ratio is equal to $(S + R)/D1$, or 60/42, or 1.43:1.

Ratio 5

Simple planetary gear ratio number 5 also produces a speed reduction and a torque increase between the input and output shafts (Fig. 2-18). In this situation, the ring gear is stationary while the input shaft drives the sun gear. The output member is the planet carrier. With the sun gear rotating clockwise, the pinions turn counterclockwise. Since the ring gear is held, the pinions "walk" the carrier in a direction opposite to their rotation—or clockwise—inside the ring gear. This action results in an input-to-output shaft gear ratio of 3.33 to 1 (3.33:1).

The same formula that provided the number 4 gear ratio of a simple planetary will also provide the number 5 ratio. Since the driving gear is now the sun gear, which has 18 teeth, and the sum of the sun and ring gear teeth is 60, the gear ratio is equal to $(S + R)/D1$, or 60/18, or 3.33 to 1 (3.33:1).

Ratio 6

Gear ratio number 6 is the most common method of providing the automatic transmission with a reverse gear ratio that produces a speed reduction and a torque increase (Fig. 2-19). In this ratio, the sun gear is the input, driving member, and the output (driven) member is the ring gear. The carrier is the held stationary member. As

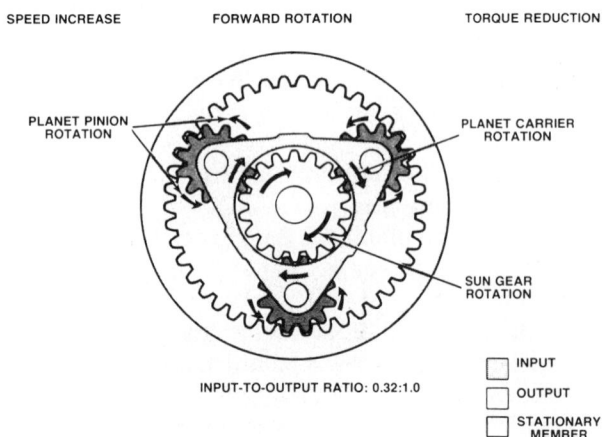

SPEED INCREASE   FORWARD ROTATION   TORQUE REDUCTION

PLANET PINION
ROTATION

PLANET CARRIER
ROTATION

SUN GEAR
ROTATION

INPUT-TO-OUTPUT RATIO: 0.32:1.0

☐ INPUT
☐ OUTPUT
☐ STATIONARY
   MEMBER

FIGURE 2-18   A simple planetary gearset operating in ratio number 5. (Courtesy of Chrysler Corp.)

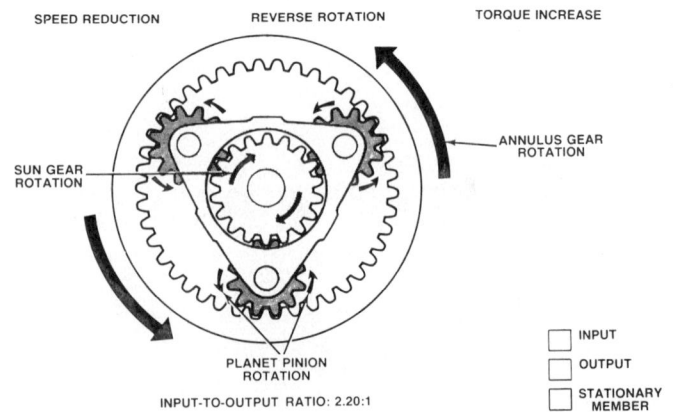

SPEED REDUCTION   REVERSE ROTATION   TORQUE INCREASE

SUN GEAR
ROTATION

ANNULUS GEAR
ROTATION

PLANET PINION
ROTATION

INPUT-TO-OUTPUT RATIO: 2.20:1

☐ INPUT
☐ OUTPUT
☐ STATIONARY
   MEMBER

FIGURE 2-19   A simple planetary gearset operating in ratio number 6. (Courtesy of Chrysler Corp.)

the input shaft drives the sun gear in a clockwise direction, the planet pinions "idle" in a counterclockwise direction because the carrier is held. The idling pinions force the ring gear to turn the output shaft counterclockwise, a direction opposite to that of the input. The input-to-output ratio is 2.33 to 1 (2.33:1).

To calculate the actual gear ratio of any planetary gearset functioning in ratio 6, divide the number of teeth on the driving gear (D1) into the number of teeth on the driven gear (D). In Fig. 2-19, the driving sun gear as 18 teeth and the driven ring gear 42. Therefore, the ratio is equal to D/D1, or 42/18, or 2.33 to 1 (2.33:1).

Restrictions to the Use of Simple Planetary Gearsets

A simple planetary gearset that is operating alone in an automatic transmission will only produce a limited number of the above-mentioned gear ratios. As pointed out, a simple planetary can function in six different ways to produce a speed reduction, an overdrive, or a reverse. Also, by driving any two members at the same speed, the gearset will operate in direct drive.

But whenever the manufacturer permanently fastens one member of the gearset to the output shaft, this action limits the number of ratios achievable to either two forward or one forward and a reverse. Consequently, in order to provide a vehicle with additional forward ratios, an automatic transmission must have several individual simple planetaries, two simple planetary gearsets connected together by a common sun gear (a Simpson gear train), or a compound planetary.

SIMPSON PLANETARY GEAR TRAIN

In place of using several individual simple planetaries, an alternative method commonly used now to provide a transmission with three forward gear ratios is the Simpson gear train, invented, developed, and patented by Howard W. Simpson. This gear train is extensively used by domestic automobile manufacturers such as Chrysler in its TorqueFlite series; Ford in its C-3, C-4, C-5, and C-6; and General Motors in its 125, 200, 250, 325, and 350 and 400 series transmissions. Furthermore, a number of European and Japanese manufacturers use the Simpson gear train in their automatic transmissions.

Design

The Simpson gear train system (Fig. 2-20) **consists of two simple planetary gearsets,** referred to in this chapter as the front and rear units, which Simpson ingeniously con-

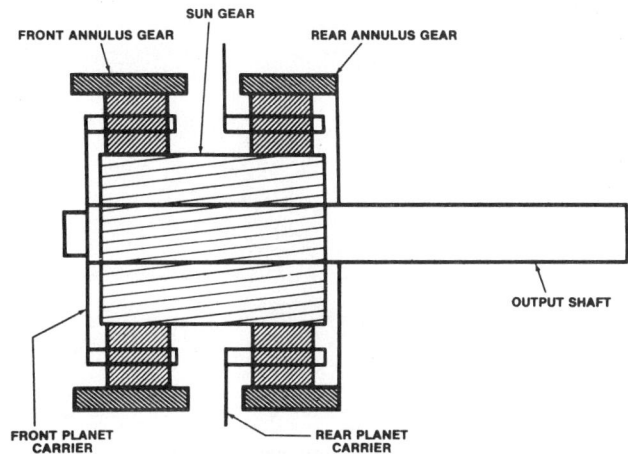

FIGURE 2-20   The design of a Simpson planetary gear train. (Courtesy of Chrysler Corp.)

nected together with a common sun gear. The gear train has two identical ring gears, six identical planet pinions, and the single long sun gear serving both gearsets. The use of identical gears allows a minimum tooling because only three different gears and two slightly varied carriers are necessary to complete the entire gear train.

Gear Train Buildup

In order to make the flow or torque (powerflow) through the Simpson gear train easier for the reader to understand, let's examine the buildup of this unit, piece by piece, as it is found in an actual automatic transmission. While studying this buildup, consider the construction of each piece and its possible connection with other components.

An input shaft that has splines on both ends brings torque into the gear train (Fig. 2-21). **Splines** are slots or grooves cut into the shaft; they mate with similar splines cut into a bore located in the center of the turbine and the rear clutch drum. These splines ensure that the turbine

FIGURE 2-21   The input shaft splines into the turbine and rear clutch drum.

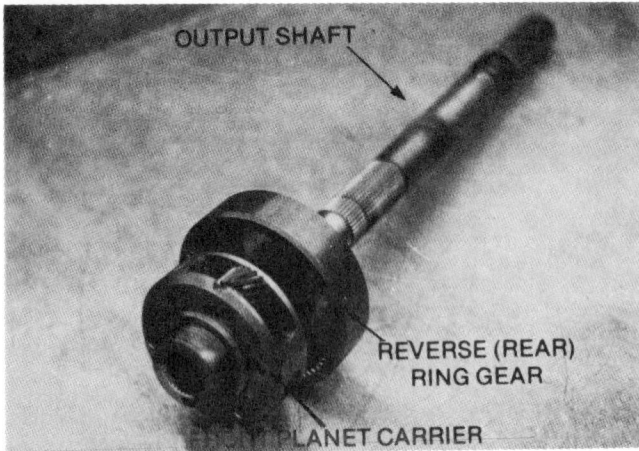

FIGURE 2-22   Splined into the output shaft are the front unit planet carrier and the rear unit ring (annulus) gear.

drives the input shaft which, in turn, turns the rear clutch drum and clutch assembly.

The rear clutch assembly connects and disconnects the front unit ring (annulus) gear to the input shaft. For the moment, this is all the reader needs to know about this unit. Further on in the text is a detailed explanation of the construction and operation of the different types of clutch assemblies.

Splined to the left end of the output shaft (Fig. 2-22) is the front unit carrier with its planet pinions. Since the front unit carrier attaches to the output shaft, it is the output planetary member for the front unit gearset.

The long sun gear can be a driving or stationary planetary member. In order to accomplish this task, the sun gear attaches to a driving shell by means of splines cut into its outer circumference at the center of the gear (Fig.

FIGURE 2-23   The driving shell and sun gear, front clutch cylinder, and brake band of a typical Simpson gear train.

2-23). The driving shell itself has a series of large teeth that mate into slots machined into the front clutch drum (cylinder). The front cylinder and clutch assembly, when activated, connect the driving shell and sun gear to the input shaft. On the other hand, if the front clutch cylinder is stationary, the driving shell and sun gear will not rotate.

The brake band shown in Fig. 2-23 is responsible for stopping the rotation of the front clutch cylinder and driving shell. For now, this is all the reader needs know about brake bands. Further on in this text is a detailed explanation of brake bands and their operating units, called servos.

If the sun gear is rotating, it delivers torque into the reverse unit planet pinions (see Fig. 2-20). These pinions are part of the rear unit carrier, which is the rear unit's stationary member. The rear unit carrier attaches into the low and reverse drum. By holding this drum stationary, the rear unit carrier will not rotate and the rear planet pinions will "idle" on their support pins. As a result, torque moves from the sun gear, to the pinions which "idle," and finally to the rear unit ring (annulus) gear.

The rear unit ring gear splines to a flange. This flange also has internal splines, which mate with external splines machined into the output shaft (see Fig. 2-22). The rear unit ring gear serves as the output member for the rear unit gearset.

In summation, the two driving planetary members of the Simpson gear train discussed here are the front unit ring gear and the long sun gear. The holding members are the long sun gear and the rear unit carrier. The output members are the front unit carrier along with the rear unit ring gear. Finally, by using both the front and rear units or each unit by itself, the Simpson gear train can produce three forward gear ratios and reverse.

GEAR RATIOS PRODUCED BY A SIMPSON GEAR TRAIN

First Ratio

In first ratio, low gear, the front unit ring gear is the driving member for both gearsets; the rear unit carrier is the stationary member for both (Fig. 2-24). With the engine operating, torque flows through the converter and drives the input shaft in a clockwise direction. Since the input shaft drives the now activated rear clutch assembly, the front unit ring gear turns in a clockwise direction. During this time, the rear carrier is held stationary by a brake band or one-way clutch. A **one-way clutch** is a device that permits a unit to turn in one direction but stops its rotation in the other.

The front unit ring gear also rotates the front planet

pinions in a clockwise direction. The front planet carrier, which splines to the output shaft, has the weight of the vehicle imposed on it. Consequently, the carrier has a tendency to remain stationary. This action causes the front planet pinions to temporarily "idle" on their support pins and turn the long sun gear counterclockwise.

Counterclockwise rotation of the long sun gear causes clockwise movement of the rear unit planet pinions. Since the rear unit carrier is the permanently held, stationary member, its pinions "idle" and drive the rear unit ring gear and output shaft in a clockwise direction but at a slower speed than the input shaft.

This output shaft rotation also forces the front unit carrier to turn at the same speed and in the same direction as the output shaft. Consequently, both the front unit ring gear and planet carrier are rotating clockwise, but the front carrier is spinning at a slower speed than the ring gear.

The actual gear ratio achieved by a Simpson gear train in low or first ratio is somewhat more difficult to calculate due to the front gearset having two output torque members—the carrier and sun gear. However, the total ratio is still the sum of those produced by both the front and rear gearsets.

To calculate the ratio of the front set, use this formula: 1 plus the number of teeth on the driven gear (D) divided by number of teeth on the drive gear (D1) minus D/D1 (1 + D/D1 - D/D1). If the driven sun gear has 28 teeth and the drive ring gear has 62, the ratio of the front unit is 1 + 28/62 - 28/62, or 1.

The ratio of the rear unit under these circumstances is 1 plus the number of teeth on the drive gear (D1) divided by the number of teeth on the driven gear (D). In the rear unit, the drive sun gear has 28 teeth and the driven ring gear has 62, so the ratio is 1 + 28/62, or 1.45. The sum of the two gearsets is 1 + 1.45, or 2.45. Thus, the input to output shaft ratio in first gear is 2.45:1 (Fig. 2-24).

Second Ratio

When the planetary gear train is in second ratio, the front unit gearset alone produces the speed reduction and torque increase. In this case, the front unit ring (annulus) gear is the driving member. The long sun gear is held stationary by the kickdown brake band, and the front unit carrier is the output member (Fig. 2-25).

Engine torque passes from the torque converter turbine to the input shaft, to the reach clutch drum, through the activated (applied) clutch, and to the front unit ring gear. Therefore, the front unit ring gear turns clockwise, which also forces the front unit planet pinions to spin clockwise. Since the long sun gear is stationary, the spinning front unit planet pinions "walk" the front carrier and output shaft around the held sun gear at a speed which is slower than the driving front unit ring gear.

To determine the actual gear ratio of the Simpson gear train functioning in second, divide the number of teeth on the driving gear (D1) into the sum of the teeth located on both the sun and ring gears. In the gear train shown in Fig. 2-25, the driving ring or annulus gear has 62 teeth. The sun gear has 28 teeth, and the sum of the teeth on both the sun and ring is 90. Therefore, the gear ratio between the input and output shafts is equal to (S + R)/D1, or 90/62, or 1.45:1.

Third Ratio

During the third gear ratio, direct drive, the front unit ring gear and the long sun gear are the driving members. In this case, the front clutch connects the long sun gear to

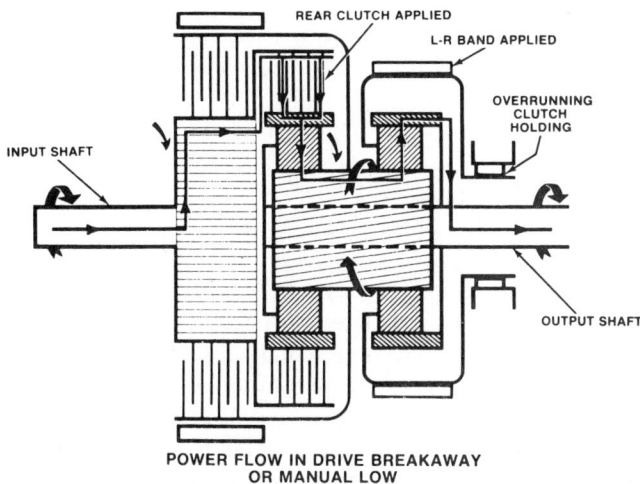

FIGURE 2-24 The Simpson gear train operating in first ratio. (Courtesy of Chrysler Corp.)

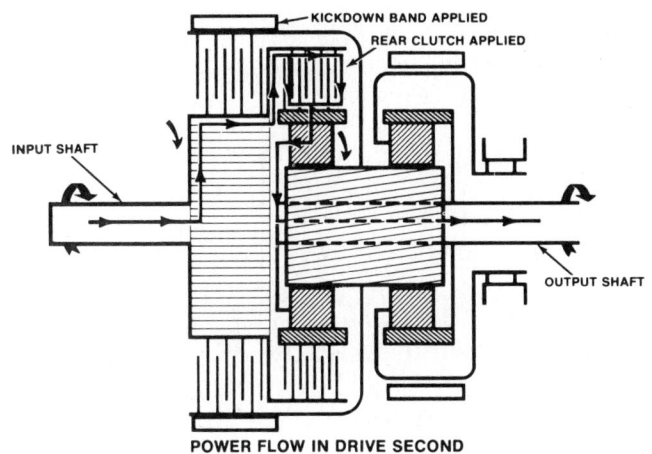

FIGURE 2-25 The Simpson gear train operating in second ratio. (Courtesy of Chrysler Corp.)

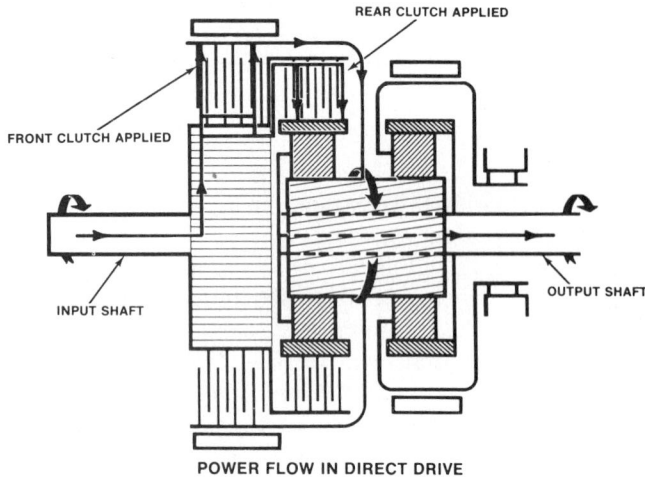

FIGURE 2-26   The Simpson gear train operating in third ratio. (Courtesy of Chrysler Corp.)

the input shaft. At the same time, the rear clutch connects the front unit ring gear to the input shaft (Fig. 2-26). Notice that during direct drive the front unit gearset is again responsible for providing the ratio.

During third ratio, engine torque flows through the torque converter to the transmission's input shaft, through the rear clutch assembly, and to the front unit ring (annulus) gear. Therefore, the front unit ring gear turns clockwise at input shaft speed. At the same time, input shaft torque passes through the front clutch and turns the driving shell and sun gear in a clockwise direction at input speed. Since the front unit ring gear and the sun gear both turn at the same speed and in the same direction, all gear rotation in the gearset stops and it spins as one piece.

This lock-up condition results from the inability of the

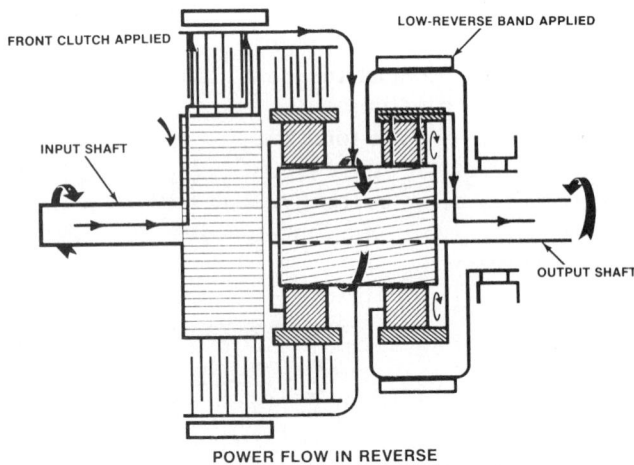

FIGURE 2-27   The Simpson gear train operating in reverse ratio. (Courtesy of Chrysler Corp.)

front unit pinions to turn in two directions at the same time. For instance, when the front unit ring gear spins clockwise, it should turn the front unit carrier pinions clockwise. But at the same time, the long sun gear, which is also turning clockwise, attempts to drive these same pinions counterclockwise. These front unit pinions cannot rotate in two directions at the same time; as a result, the entire planetary gear train locks up and the output shaft and the input shaft spin at the same speed. Although the front unit gearset started the process, the rear unit must, due to its interaction with the front, lock up also.

Reverse Ratio

In reverse ratio, the long sun gear is the driving member, the rear unit carrier is the stationary member, the output member is the rear unit ring gear. Since none of the front unit gearset members are held, the rear unit has the total responsibility for this ratio (Fig. 2-27).

With the engine running, torque flows through the torque converter into the input shaft, to the rear clutch assembly, across the applied clutch, and to the driving shell and sun gear. This action causes the sun gear to turn clockwise at input shaft speed. The sun gear turns the rear unit planet pinions in a counterclockwise direction that, in turn, spins the rear unit ring gear and output shaft in a direction opposite to the sun gear input shaft.

This reversal of the ring gear's direction occurs because the rear unit planet carrier is stationary; therefore, the planet pinions "idle" and drive the ring gear backwards (counterclockwise). The low and reverse brake band that wraps around the low reverse drum holds the reverse drum and rear carrier stationary during this ratio.

To calculate the actual gear ratio of a Simpson gear train in reverse, divide the number of teeth on the driving gear (D1) into the number of teeth on the driven gear (D). In the gear train illustrated in Fig. 2-27, the driven ring gear has 62 teeth, while the driving sun gear has 28. So the gear ratio between the input and output shaft is equal to D/D1, or 62/28, or 2.2:1.

Neutral

In neutral, the input shaft drives none of the planetary gear train members and the brake bands hold no members stationary. However, engine torque does pass through the converter to drive the input shaft clockwise. The input shaft turns the rear clutch drum clockwise, but at this point, torque flow stops.

NOTE: When Simpson gear trains appear in different makes of automatic transmissions, the names given to the various components may be different from those chosen

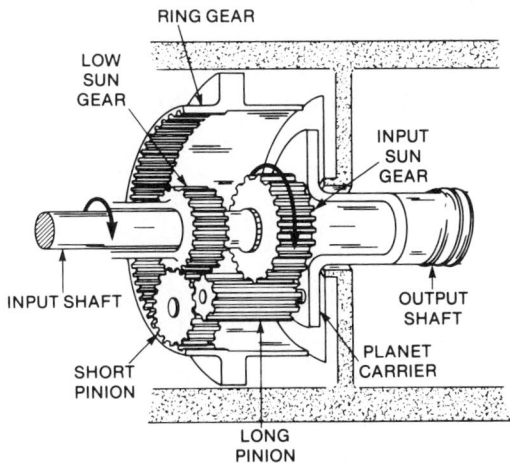

FIGURE 2-28 The design of a two-speed compound planetary gear train.

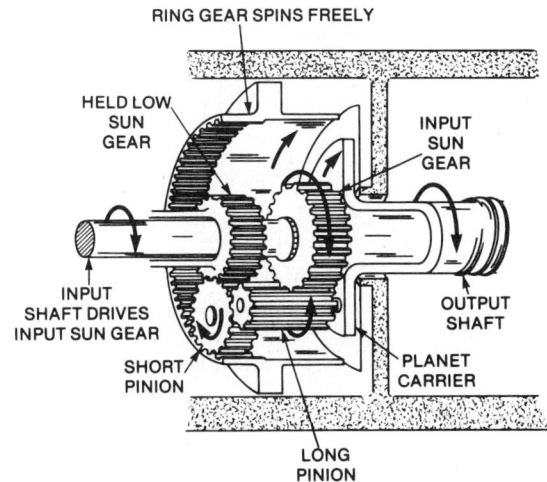

FIGURE 2-29 A two-speed compound planetary gear train operating in low ratio. A brake band (not shown) holds the low sun gear stationary.

for use in this chapter. Also, the size and position of components may vary slightly in each brand of transmission. But all these gear trains operate in the same manner to produce three forward gear ratios and a reverse.

COMPOUND PLANETARY GEAR TRAINS

The compound (Ravigneau) planetary has been in use in domestic-built automatic transmissions since 1949, about 11 years before the Simpson design went into production. Although this form of gear train can produce either two, three, or four forward speed ratios and a reverse, it is found only in a few current transmission models. The reasons for this is that the Ravigneau gear train is somewhat more expensive to produce than the Simpson and its unique construction places some design limitations on the remaining components of the transmission, which make the unit difficult to use in many vehicle configurations.

Two-Speed Planetary Design

Although the two-speed transmissions are no longer in current production, there are a great number of these units still in service in older vehicles. Typical automatic transmissions that incorporate the two-speed version of the compound planetary are the General Motors Powerglide and T-300 units; Ford used the same style of planetary for a short time in its two-speed Fordomatic transmission. However, manufacturers have not used the two-speed design of this planetary since the early 1970s, due to the popularity and versatility of a three-speed transmission.

A typical version of the two-speed compound planetary consists of an input sun gear and shaft, a set of long planet pinions that mesh with the input sun gear, and a set of short planet pinions, which mesh with the long pinions but not the input sun gear (Fig. 2-28).

In addition, the short pinions mesh with the low sun gear that the input shaft drives during one ratio and that a brake band holds stationary during another ratio. All the planet pinions mount on a single planet carrier, which is the output member for the gear train. Finally, a single ring gear meshes with the short planet pinions. With this arrangement, the gear train can produce low, direct, and reverse ratios.

Powerflow: Low

In low ratio, the input shaft drives the input sun gear clockwise (Fig. 2-29); a brake band prevents the low sun gear from turning. Since the input sun gear spins clockwise, the long planet pinions turn counterclockwise. The long planet pinions also force the short planet pinions to rotate clockwise. The short pinions, in turn, "walk" the planet carrier and output shaft around the low (held) sun gear in a clockwise direction but at a slower speed than the input shaft is turning.

Powerflow: Direct

When this planetary gear train upshifts to high, direct drive (Fig. 2-30), the input shaft drives both the input and low sun gears in a clockwise direction. A clutch assembly connects the low sun gear to the input shaft during this ratio.

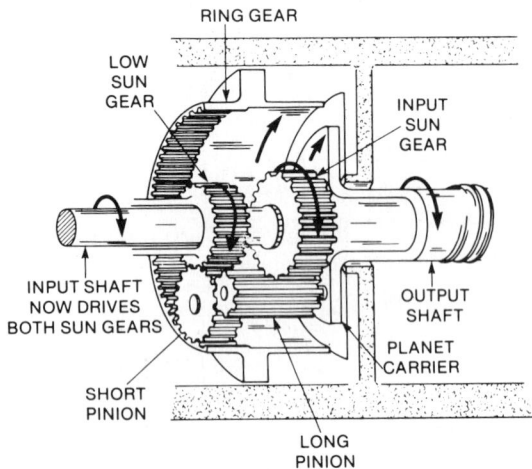

FIGURE 2-30 A two-speed compound planetary gear train operating in high, direct drive. A clutch assembly (not shown) holds the low sun gear stationary.

By observing the rotation of both sets of planet pinions in low ratio and again in direct drive, the reason why the unit turns at a 1:1 ratio will be obvious. During low ratio, the long pinions rotated counterclockwise and the short pinions turned clockwise. Now in direct drive, the input sun gear still causes the long pinions to spin counterclockwise, but the low sun gear also drives the short pinions counterclockwise. Remember that both the long and short pinions are in mesh and must turn in opposite directions. Since the sun gears attempt to spin the long and short pinions in the same direction, they lock up and the entire gear train rotates as a solid unit. In other words, the input and output shafts turn at the same speeds.

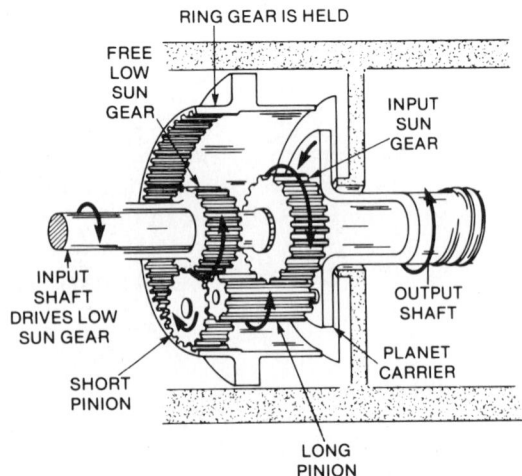

FIGURE 2-31 A two-speed compound planetary gear train operating in reverse ratio. A clutch assembly (not shown) holds the ring gear stationary.

Powerflow: Reverse

During a reverse gear ratio, the input shaft drives the input sun gear clockwise; the ring gear is held stationary by a clutch assembly; and the carrier acts as the output member (Fig. 2-31).

Since the sun gear turns clockwise, the long pinions rotate counterclockwise. This action causes the short pinions to turn clockwise. Because the ring gear is stationary, the short pinions "walk" around inside the ring gear pulling the planet carrier with them. This action imparts a reverse motion to the output shaft and a reduction in speed between the input and output shafts.

THREE-SPEED COMPOUND PLANETARY

Unlike its two-speed counterpart, the three- or four-speed compound (Ravigneau) planetary is still in current production by some auto manufacturers. For instance, Ford uses this design in its FMX, automatic overdrive (AOD), and ATX transaxle units. General Motors also utilizes this planetary in its (MD3) 180 as well as several foreign-built vehicles equipped with a Borg-Warner transmission.

Design

The compound planetary gear train develops three forward ratios by arranging its members in a slightly different manner than they were in the two-speed unit (Fig. 2-32). The center, inside gear of this unit is the primary sun gear, which the primary sun gear shaft drives. In mesh with the primary sun gear are three short primary

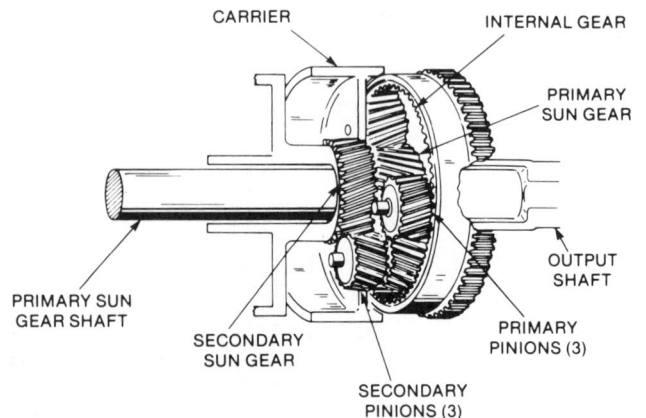

FIGURE 2-32 The design of a three-speed compound planetary gear train.

planet pinions. These planet pinions rotate in an opposite direction to the primary sun gear. Note also that the primary sun gear is smaller in diameter than the secondary sun gear. This is opposite to what is found in the two-speed unit.

Meshed with the primary pinions and also the secondary sun gear are three long secondary planet pinions. The secondary pinions rotate in a direction opposite to the primary pinions and mesh with the internal (ring) gear, which attaches to the output shaft. Notice also, in the three-speed compound planetary, that the long and short pinions are reversed, relative to their position on a two-speed unit. However, both sets of pinions still mount on a single carrier, which can serve this planetary as either a driving or stationary member.

Powerflow: First Ratio

With this gear train arrangement, the unit can produce first, second, third, and reverse ratios. In first ratio, the turbine (input) shaft drives the primary sun gear shaft, which, in turn, drives the primary sun gear clockwise. The planet carrier is the stationary member (Fig. 2-33), and the internal gear is the output.

The clockwise direction of the primary sun gear forces the primary pinions to turn counterclockwise. Since the primary pinions are in mesh with the secondary pinions, they rotate clockwise. The carrier is stationary; therefore, the secondary pinions "idle" and cause the internal gear and output shaft to rotate in a clockwise direction at a slower speed than the turbine (input) shaft. During this gear ratio, the secondary sun gear turns freely in a reverse direction and has no effect on the gear train.

Powerflow: Second Ratio

During second ratio, the turbine shaft again drives the primary sun gear shaft, which, in turn, drives the primary

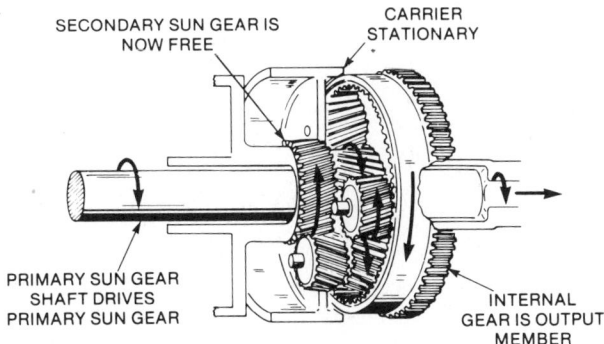

FIGURE 2-33   The three-speed compound planetary gear train operating in first ratio. The turbine shaft (not shown) turns the primary sun gear shaft.

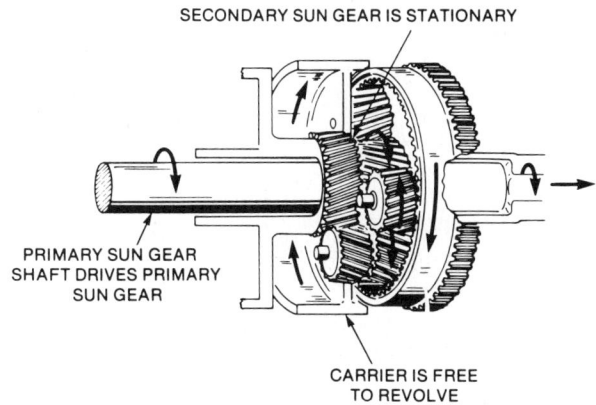

FIGURE 2-34   The three-speed compound planetary gear train operating in second ratio. The turbine shaft (not shown) turns the primary sun gear shaft.

sun gear clockwise. The secondary sun gear is the stationary member (Fig. 2-34). Since the primary sun gear turns clockwise, the primary pinions spin counterclockwise and drive the secondary pinions clockwise. The secondary pinions "walk" the carrier around the held secondary sun gear. At the same time as the secondary pinions "walk" the carrier, they drive the internal gear and output shaft in a clockwise direction but at a slower rate than the primary sun gear.

Powerflow: Third Ratio

In third ratio, direct drive, the turbine shaft drives both the primary and secondary sun gears in a clockwise direction (Fig. 2-35). The primary sun gear attempts to drive the primary pinions counterclockwise, and the secondary sun gear tries to turn the secondary planet pinions counterclockwise. The primary and secondary

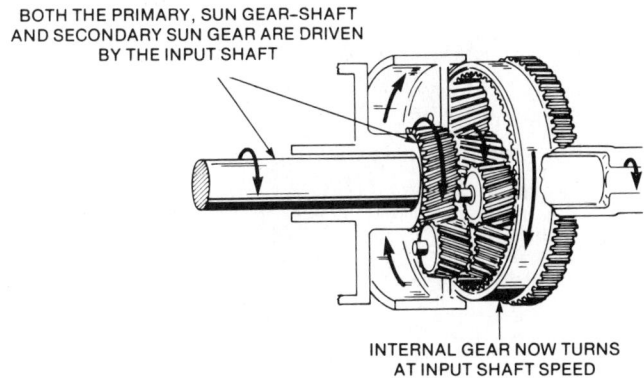

FIGURE 2-35   The three-speed compound planetary gear train operating in third ratio. The turbine (input) shaft (not shown) drives both the secondary sun gear and the primary sun gear shaft.

TURBINE (INPUT) SHAFT
DRIVES SECONDARY SUN GEAR

CARRIER IS STATIONARY

PRIMARY SUN GEAR SHAFT
IS FREE TO REVOLVE

RING GEAR TURNS BACKWARDS

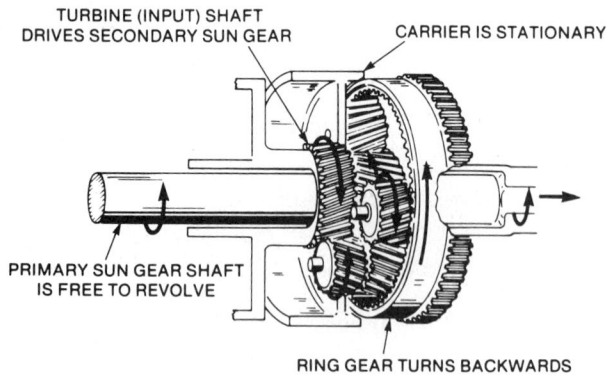

FIGURE 2-36 The three-speed compound planetary operating in reverse ratio. The turbine shaft (not shown) drives the secondary sun gear.

planet pinions are in mesh with each other and must therefore rotate in opposite directions. Consequently, the planet pinions cannot turn, and the entire gear train revolves as a unit. In other words, the output shaft spins at the same speed as the turbine (input) shaft.

Powerflow: Reverse

To produce reverse, the turbine shaft drives the secondary sun gear clockwise and the planet carrier is the stationary member (Fig. 2-36). Since the secondary sun gear is turning clockwise, the secondary pinions are rotating counterclockwise. The planet carrier is stationary; consequently, the secondary pinions "idle" and turn the internal gear in a reverse (counterclockwise) direction. In this situation, the internal gear (output member) spins slower than the secondary sun gear, the input member. The primary pinions and sun gear spin freely and have no effect on the gear train.

FOUR-SPEED COMPOUND PLANETARY

Probably the one aspect of a compound (Ravigneau) planetary that has kept it from being phased out completely in favor of the Simpson design is its ability to

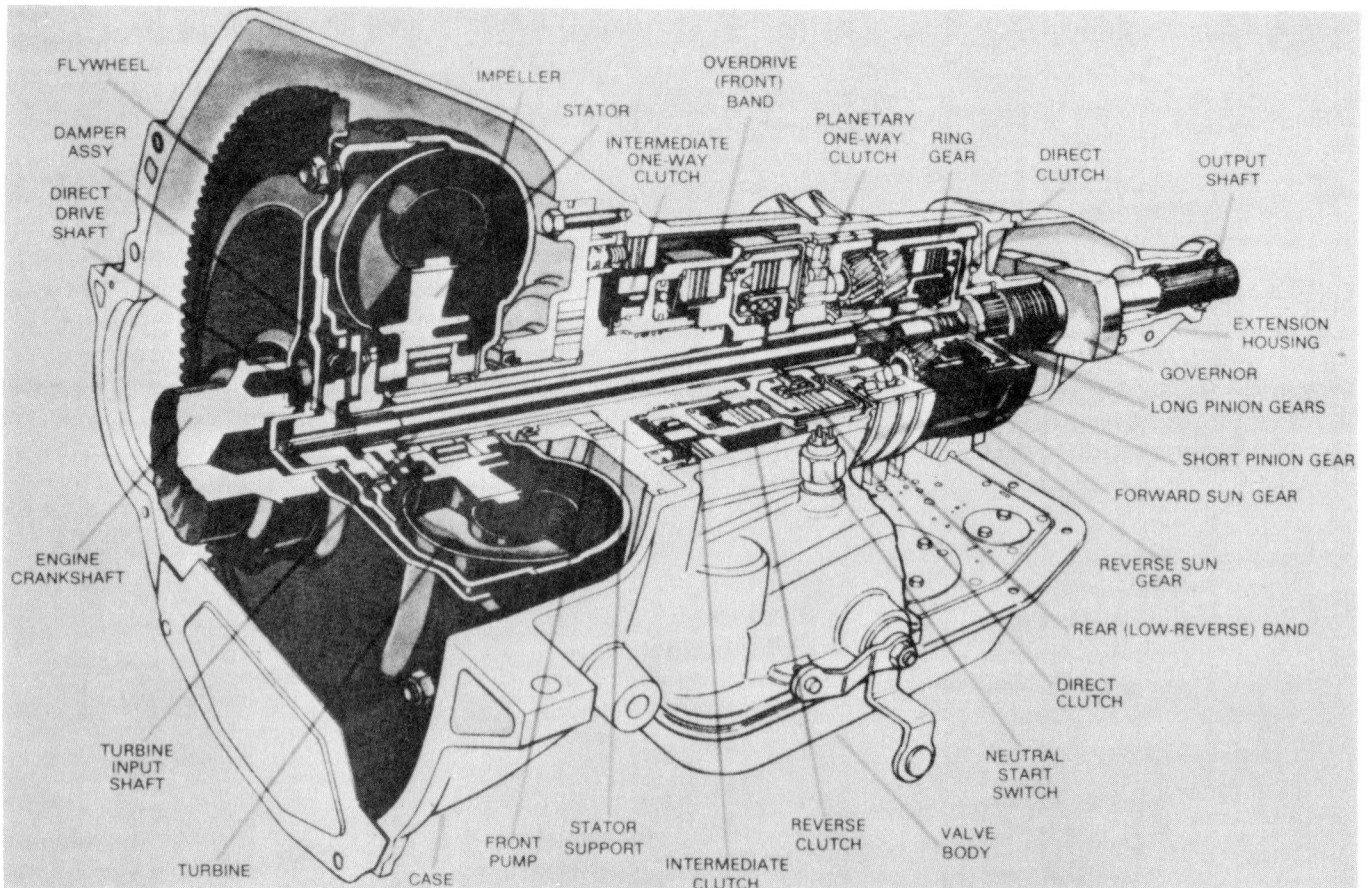

FIGURE 2-37 Cutaway view of a Ford four-speed automatic overdrive transmission with the Ravigneau gear train. (Courtesy of Ford Motor Company)

produce four forward speeds and a reverse. The fourth forward speed ratio is, in this case, an overdrive. Moreover, the Ravigneau gear train can produce the overdrive without any additional planetary members. This is not the case with the pure Simpson gear train.

Ford Motor Company put this positive quality of the Ravigneau planetary to good use in the development of its Automatic Overdrive (AOD) transmission (Fig. 2-37). This unit produces three forward speeds and a reverse in a similar manner as the three-speed compound planetary previously described. However, at a given speed with the transmission in drive range, the unit also provides an overdrive between the direct-drive shaft and the output shaft.

To understand how the gear train provides the overdrive ratio, let's examine the components of the transmission in greater detail. However, before doing so, just remember that although the planetary has the same basic design, its appearance may be slightly different from the three-speed unit previously described and the names of the parts are changed. Also, there are some additional parts added to the transmission in order to make the planetary produce the four gear ratios.

Design

The Ford AOD transmission has four clutch assemblies, two one-way clutches, two bands, and two input shafts to control the gear train (Fig. 2-38). The forward clutch connects the forward sun gear to the turbine shaft in the first three forward ratios. (NOTE: In the three-speed compound planetary discussed earlier, the forward sun gear was called the primary sun gear. In the AOD transmission, it is still the smaller of the two sun gears.)

The reverse clutch assembly connects the reverse sun gear, which is the front or larger sun gear, to the turbine shaft. This gear was called the secondary sun gear in the three-speed compound planetary described earlier. However, unlike the three-speed unit, the reverse sun gear is not used as a driving member in any forward ratio. In other words, the reverse clutch only connects the front sun gear to the turbine shaft in reverse.

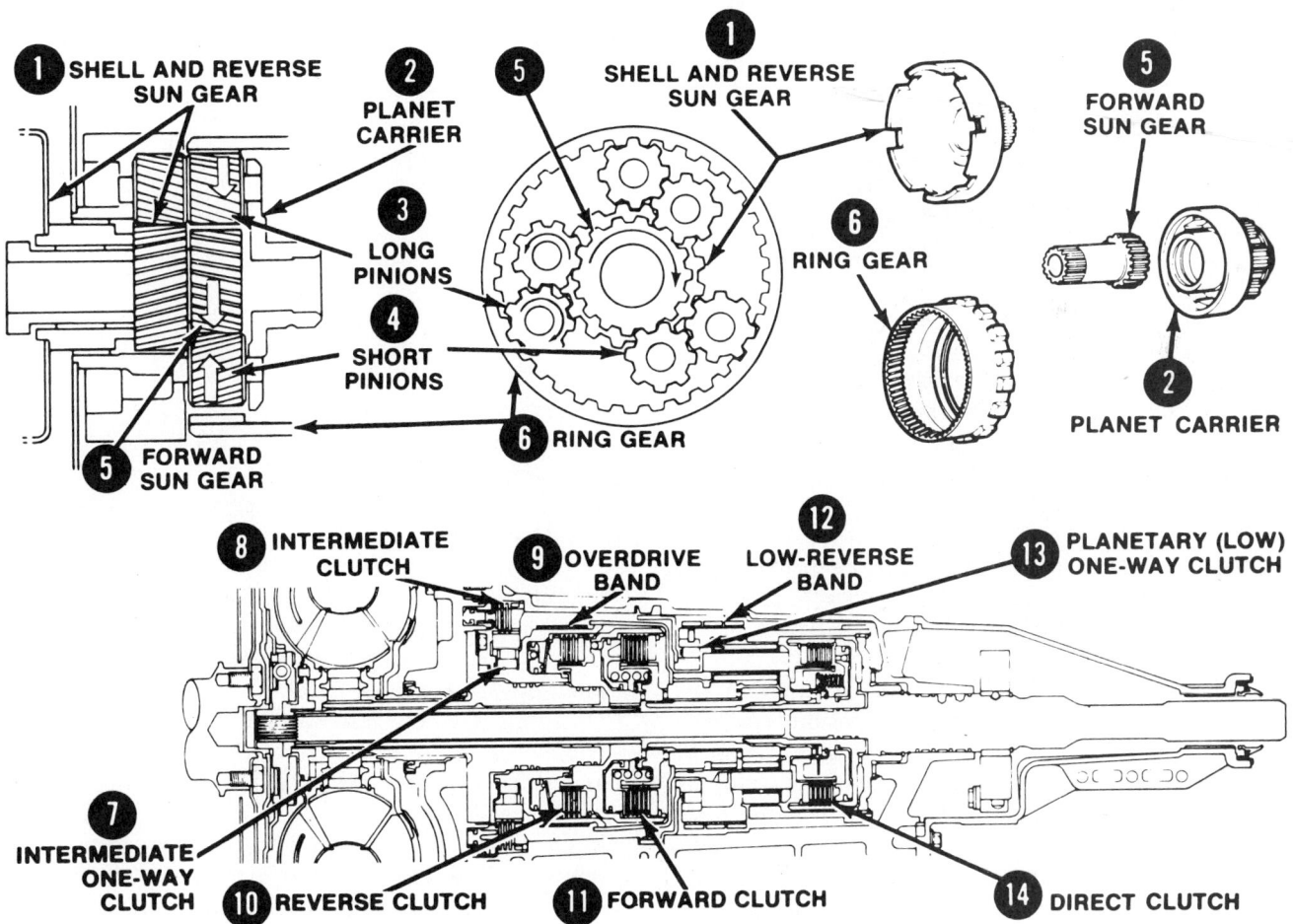

FIGURE 2-38   The gear train components of the Ford AOD transmission. (Courtesy of Ford Motor Company)

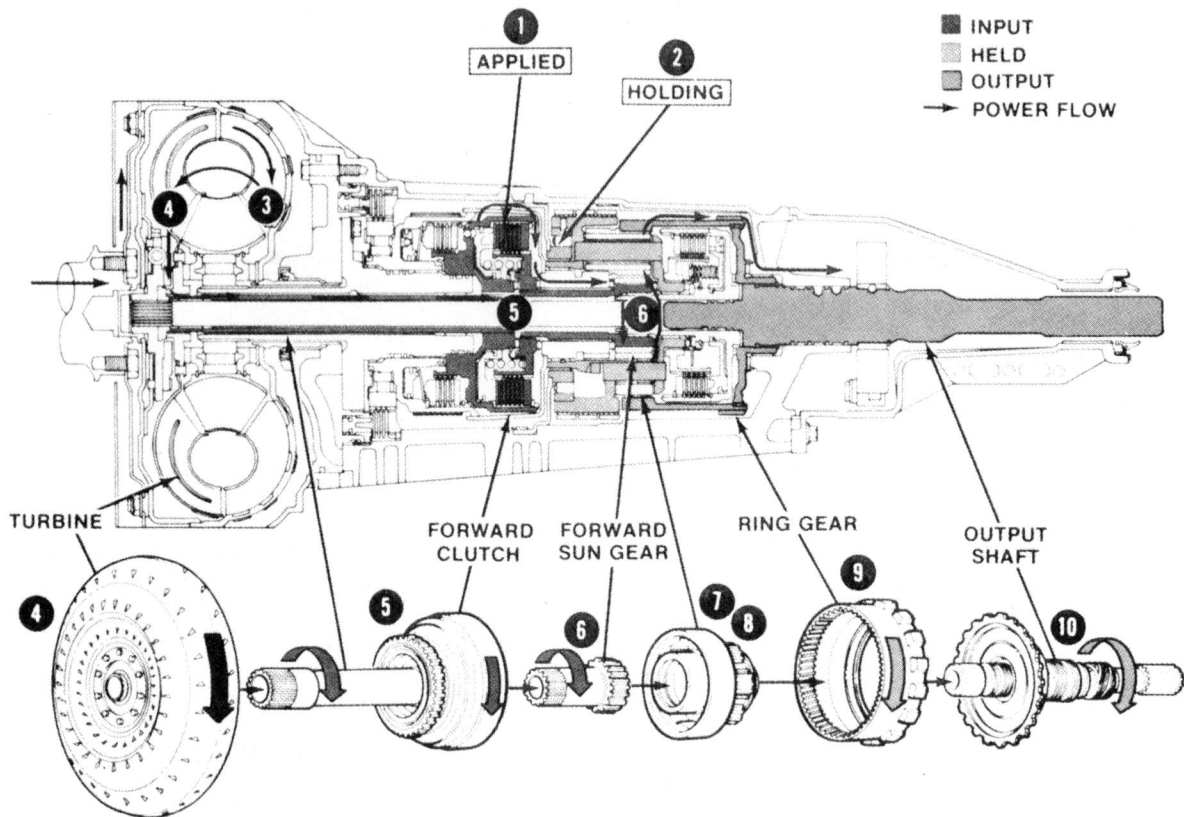

FIGURE 2-39   *Powerflow within an AOD transmission in first ratio. (Courtesy of Ford Motor Company)*

The intermediate clutch assembly is responsible for locking the outer race of the intermediate one-way clutch to the transmission case. This action occurs in second gear to prevent the reverse sun gear from turning counterclockwise. However, the one-way clutch will permit the gear to rotate clockwise.

The direct clutch connects the carrier to the direct-drive shaft (not the turbine shaft). The direct-drive shaft fits inside the tubular turbine shaft and is driven by the torque converter itself.

There are two mechanical one-way clutches in the AOD transmission: the intermediate and the low. The intermediate prevents the counterclockwise rotation or the reverse sun gear in second ratio. The low clutch stops the counterclockwise rotation of the carrier in low (first) ratio.

The AOD transmission also has two brake bands: the overdrive and low-reverse. The overdrive band, when applied, holds the reverse sun gear stationary. The low-reverse band is responsible for holding the planet carrier.

Powerflow: First Ratio

In the first ratio, the forward clutch is applied; this causes the turbine shaft to drive the forward sun gear in a clockwise direction. The planet carrier is the holding member of the gear train (Fig. 2-39).

With the forward sun gear turning clockwise, the short pinions spin counterclockwise. The short pinions, in turn, drive the long pinions clockwise. Resistance to motion of the ring gear creates a reaction in the carrier, which attempts to turn it backwards (counterclockwise). However, the low one-way clutch prevents this from happening. As a result, the carrier's long pinions "idle" and drive the ring gear and output shaft forward or clockwise.

During vehicle deceleration, the low one-way clutch overruns. This allows the carrier to rotate, which prevents engine braking through the transmission. However, in manual low driving range, the low-reverse band holds the carrier in both directions to permit engine braking.

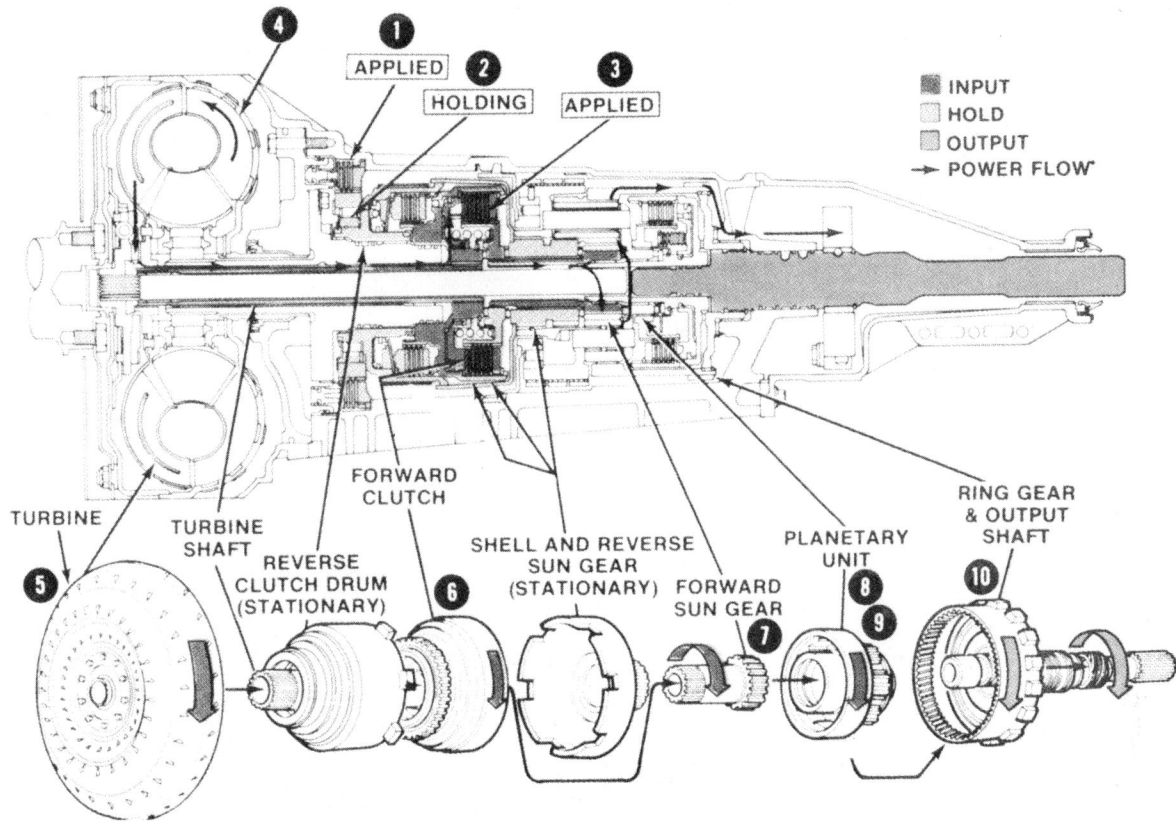

FIGURE 2-40   *Powerflow within an AOD transmission in second ratio. (Courtesy of Ford Motor Company)*

Powerflow: Second Ratio

During the second ratio, the forward clutch assembly remains applied to connect the forward sun gear to the turbine shaft. The intermediate clutch assembly and one-way clutch are responsible for holding the reverse sun gear from counterclockwise rotation (Fig. 2-40).

With the forward sun gear spinning clockwise, the short pinions turn counterclockwise. The short pinions, in turn, drive the long pinions in a clockwise direction. The long pinions "walk" the carrier around the stationary reverse sun gear and, at the same time, drive the ring gear and output shaft clockwise. With this arrangement, the input-to-output speed in second ratio is increased as compared to low gear.

Powerflow: Third Ratio

In third ratio, the direct clutch applies to connect the carrier to the direct-drive shaft. The forward clutch assembly remains activated to connect the forward sun gear to the turbine shaft (Fig. 2-41). NOTE: During third and fourth ratios, the intermediate clutch assembly remains applied. However, this action has no effect on the ratios because the entire gear train is turning in a clockwise direction anyway. The reason the clutch remains applied is to provide a smoother 3-2 downshift.

Since the forward sun gear and the carrier are both turning clockwise, the planet pinions, for all practical purposes, cannot turn. Therefore, the entire gear train is forced to rotate at engine speed, or a 1:1 ratio. About 60 percent of the input torque comes into the gear train through the direct clutch, while the remaining 40 percent transmits via the forward clutch.

Powerflow: Fourth Ratio

When the AOD transmission automatically upshifts to fourth ratio, the forward clutch assembly releases, but the direct clutch assembly stays applied. Also, the overdrive brake band holds the reverse sun gear stationary (Fig. 2-42). The planetary gear train is now operated by the

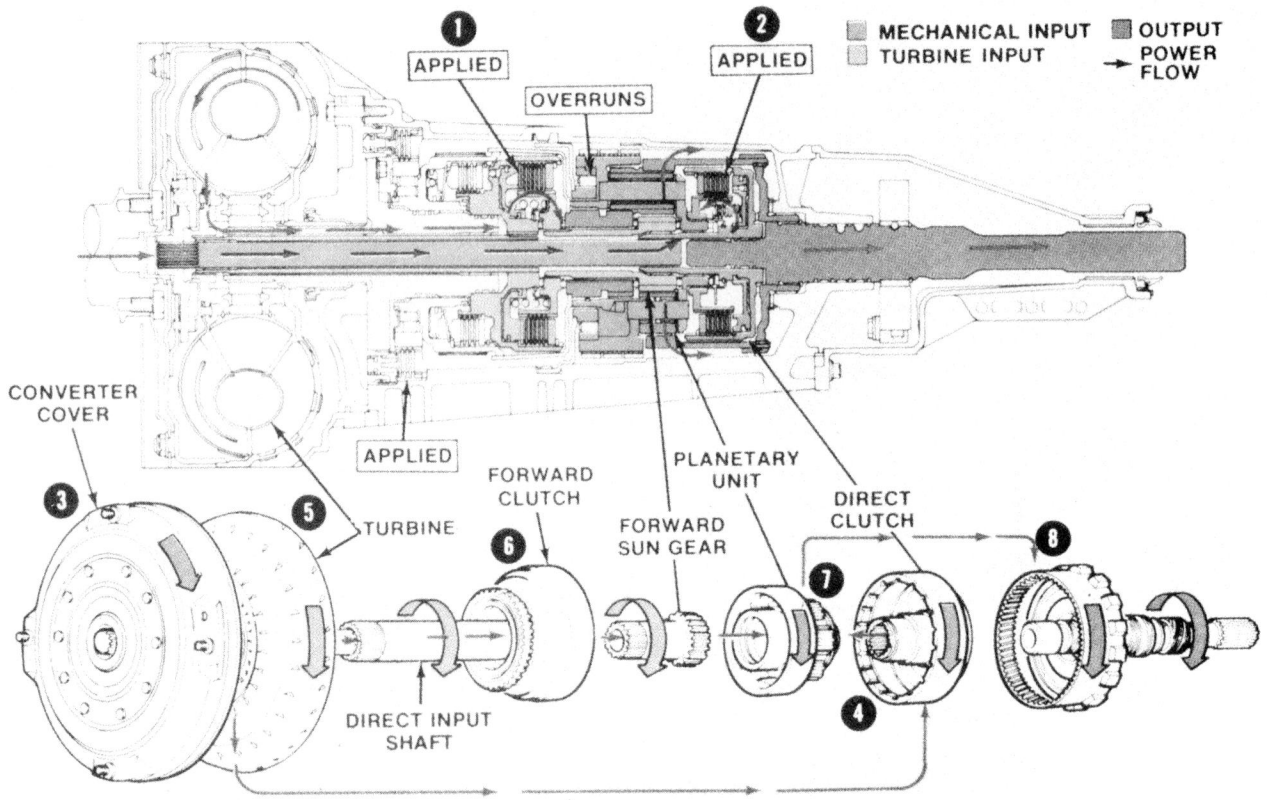

FIGURE 2-41   Powerflow within an AOD transmission in third ratio. (Courtesy of Ford Motor Company)

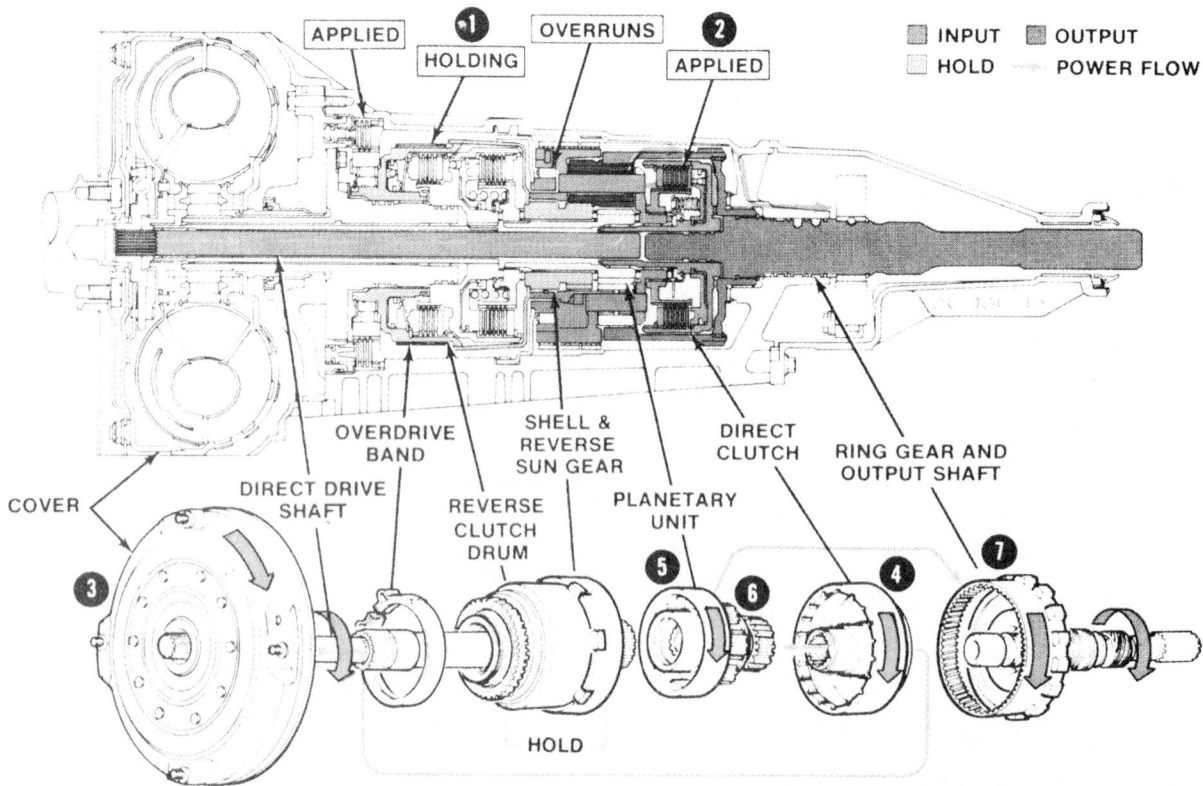

FIGURE 2-42   Powerflow within an AOD transmission in fourth ratio. (Courtesy of Ford Motor Company)

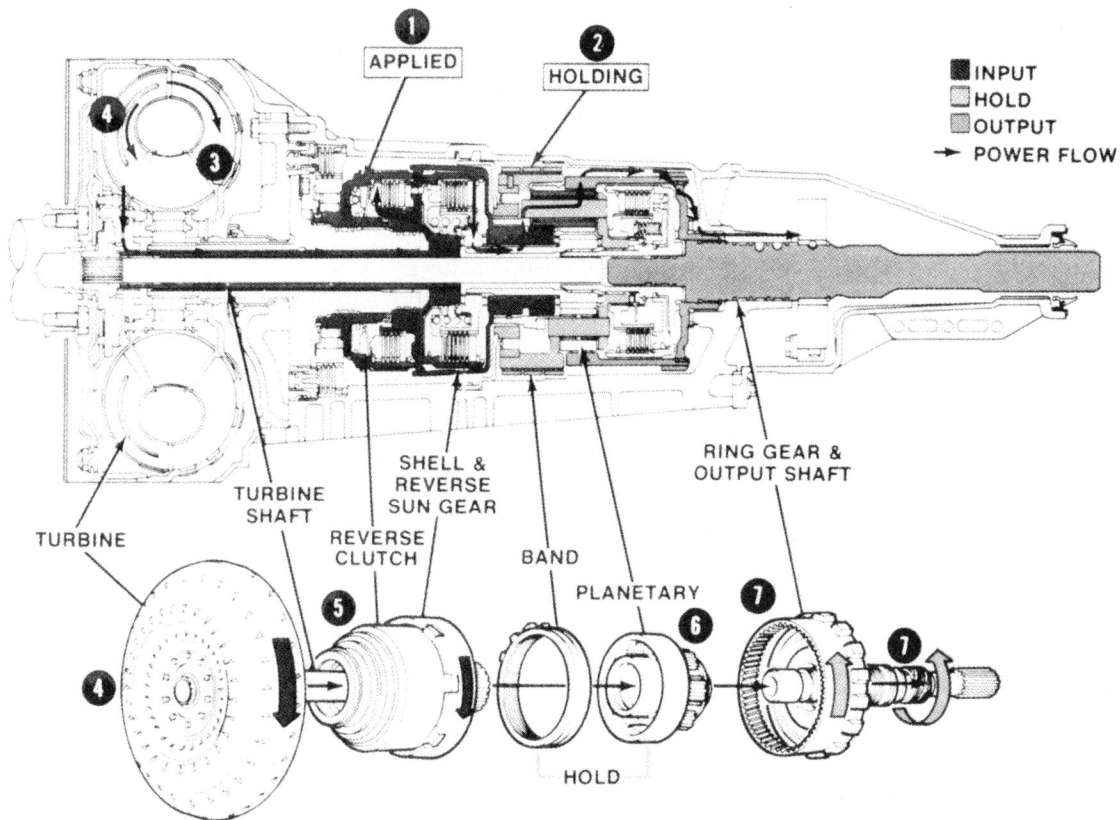

FIGURE 2-43 Powerflow within an AOD transmission in reverse ratio. (Courtesy of Ford Motor Company)

direct-drive shaft and receives no input torque from the turbine shaft.

In fourth ratio, torque moves from the direct-drive shaft, through the applied direct-clutch assembly, to the planet carrier. This drives the carrier in a clockwise direction. Since the reverse sun gear is held stationary by the band, this carrier rotation causes the long pinions to spin on their support pins and drive the ring gear in a clockwise direction. In this case, the ring gear and output shaft turn faster than the carrier and direct-drive shaft— an overdrive condition.

Powerflow: Reverse

In reverse ratio, the reverse clutch assembly transmits torque from the turbine shaft to the reverse sun gear. The low and reverse brake bands hold the planet carrier stationary in both directions (Fig. 2-43).

The turbine shaft, through the reverse clutch, drives the front (reverse) sun gear clockwise. This gear, in turn, rotates the long pinions counterclockwise, which drive the ring gear and output shaft counterclockwise, the reverse of input shaft direction.

SUMMARY

1. To be able to carry the torque loads, gears must be very strong.

2. Designed into the construction of each gear are features such as teeth clearance, circle, pitch, and root diameters, in addition to gear teeth angle.

3. To alter torque, transmission gears use the principle of the lever and fulcrum.

4. By applying a small force over a greater distance, the lever can use a small input force to move a larger output force.

5. The term given to the relationship between the input and output forces is mechanical advantage.

6. In gaining mechanical advantage between the input and output forces, the distance each force travels is different.

7. Shaft torque is equal to the force on the gear teeth times the gear's radius.

8. Gear teeth force is equal to shaft torque divided by the gear's radius.

9. If the end of one simple lever contacts the end of a second lever, force can move from one lever to the other; the same situation occurs anytime two gears are in mesh.

10. An easy way to figure the torque multiplication accomplished through gearing is to count the number of teeth located on the input and output gears.

11. Torque multiplication results in a speed reduction.

12. The planetary gear train is the heart of an automatic transmission.

13. A planetary gear train offers several advantages over the sliding gear arrangement found in manual-shift transmissions.

14. A simple planetary gearset consists of a sun gear, a carrier with three or more planet pinions, and an internal (ring) gear.

15. Simple planetary gearsets provide six different gear ratio combinations plus direct drive.

16. During gear ratio 1, the input shaft drives the carrier; the sun gear is stationary; the ring gear is the output member.

17. During gear ratio 2, the input shaft drives the planet carrier; the ring gear is stationary; the output member is the sun gear.

18. During gear ratio 3, the input shaft drives the ring gear; the carrier is the stationary member; the sun gear is the output member.

19. During gear ratio 4, the input shaft drives the ring gear; the sun gear is stationary; the output member is the carrier.

20. During gear ratio 5, the input shaft drives the sun gear; the ring gear is stationary; the carrier is the output member.

21. In gear ratio 6, the input shaft drives the sun gear; the carrier is stationary; the ring gear is the output member.

22. During direct drive, two planetary members are driven by the input shaft.

23. A simple planetary gearset that is operating alone in an automatic transmission will produce a limited number of gear ratios.

24. A Simpson gear train produces three forward speed ratios and a reverse.

25. The Simpson gear train consists of two simple planetary gearsets connected together by a common sun gear.

26. The two driving members of the Simpson gear train are the front unit ring gear and the long sun gear.

27. The holding members of the Simpson gear train are the long sun gear and the rear unit carrier.

28. The output members of the Simpson gear train are the front unit carrier and the rear unit annulus gear.

29. When a Simpson gear train is operating in first ratio, the front unit ring gear is the driving member; the rear unit carrier is the stationary member; the front unit carrier and the rear unit ring gear are the output members.

30. When a Simpson gear train is operating in second ratio, the front unit ring gear is driving; the long sun gear is stationary; the front unit carrier is the output member.

31. When the Simpson gear train is operating in direct drive, the front unit ring gear and the long sun gear are the driving members.

32. When a Simpson gear train is operating in reverse, the long sun gear is the driving member; the rear unit carrier is the stationary member; the rear unit ring gear is the output member.

33. A compound planetary gear train can produce two, three, or four forward gear ratios and a reverse.

34. When the two-speed compound planetary gear train is in forward reduction, the input sun gear is the driving member; the low sun gear is stationary; the carrier is the output member.

35. When the two-speed compound planetary gear train is in direct drive, both the sun gears are driving members.

36. When the two-speed compound planetary is in reverse, the input shaft drives the input sun gear; the ring gear is stationary; the carrier is the output member.

37. A compound planetary gear train can develop three forward gear ratios by arranging the members in a slightly different manner than they were in the two-speed units.

38. With a three-speed compound planetary gear train in first ratio, the primary sun gear is the driving member; the carrier is the stationary member; the ring gear is the output member.

39. With a three-speed compound planetary gear train in second ratio, the primary sun gear is the driving member; the secondary sun gear is the stationary member; the ring gear is the output member.

40. With a three-speed compound planetary gear train in direct drive, both the primary and secondary sun gears are the driving members.

41. With a three-speed compound planetary gear train in reverse, the secondary sun gear is the driving member; the carrier is the stationary member; the ring gear is output.

42. When the Ford AOD transmission is in second ratio, the forward sun gear is the driving member; the reverse sun gear is the stationary member; and the ring gear is the output member.

43. When the Ford AOD transmission is in third ratio, the carrier and forward sun gear are the driving members; the ring gear is the output member.

44. When the Ford AOD transmission is in fourth ratio, the carrier is the driving member; the forward sun gear is stationary; the ring gear is the output member.

45. When the Ford AOD transmission is in reverse ratio, the reverse sun gear is the driving member; the carrier is the stationary member; the ring gear is the output member.

REVIEW

This section will assist you in determining how well you remember the material contained in this chapter. Read each item carefully. If you can't complete the statement, review the section in the chapter that covers the material.

1. That portion of the gear that acts like a lever end is the _____.

    a. root

    b. center

    c. teeth

    d. clearance

2. The term given to the relationship between the input and output forces is _____.

   a. mechanical advantage

   b. mechanical advance

   c. work

   d. inertia

3. If the shaft torque is 100 pounds-foot and the gear's radius is 2 feet, the output force of the teeth is _____.

   a. 50 pounds-foot

   b. 200 pounds-foot

   c. 100 pounds-foot

   d. 150 pounds-foot

4. The torque multiplication ratio of two gears in mesh which have 12 teeth and 36 teeth is _____.

   a. 1:2

   b. 1:3

   c. 1:4

   d. 1:5

5. What is the speed ratio of the gear train in the above question if the smaller gear is driving?

   a. 0.5:1

   b. 1:1

   c. 2:1

   d. 3:1

6. The mechanical heart of the automatic transmission is the _____.

   a. planetary gear train

   b. clutches

   c. bands

   d. torque converter

7. The outer gear of a simple planetary gearset is the _____.

   a. sun gear

   b. pinions

   c. ring gear

   d. carrier

8. The two planetary gears that always rotate in opposite directions are the _____.

   a. sun and ring gears

   b. sun and planet pinion gears

   c. planet pinions and ring gear

   d. carrier and ring gear

9. If the sun gear rotates clockwise, with the carrier held, the ring gear will _____ .

    a. not turn

    b. turn counterclockwise

    c. turn clockwise

    d. lock up

10. A simple planetary usually provides, excluding direct drive, how many ratios?

    a. six

    b. five

    c. four

    d. three

11. When installed in an automatic transmission, a simple planetary will produce a maximum of _____ forward ratios.

    a. five

    b. four

    c. three

    d. two

12. The compound planetary gear train has several single members. They are the _____ .

    a. pinions and the sun gear

    b. carrier and the pinions

    c. carrier and the ring gear

    d. ring gear and the sun gear

13. A compound planetary that has the ring gear as the output member can produce _____ forward speeds.

    a. one

    b. four

    c. two

    d. three

14. The planetary gear train that has two simple planetary gearsets connected together by a long sun gear is the _____ .

    a. Simpson   c. Ravigneau

    b. compounds   d. complex

15. When a planetary gear train is in direct drive, _____ .

    a. one member is driving and one is stationary

    b. two members are stationary

    c. no members are stationary or are driving

    d. two members are driving

16. In order for a planetary gear train to be in a speed or torque reduction, _____.

    a. one member is driving and one is stationary

    b. two members are stationary

    c. no members are stationary or are driving

    d. two members are driving

17. The gearset that produces second ratio in a transmission using a Simpson gear train is the _____ unit.

    a. forward

    b. rear

    c. front

    d. reverse

18. If the carrier is stationary and the sun gear rotates, the planet pinions are said to _____.

    a. walk

    b. idle

    c. slide

    d. run

19. The unit that normally drives the transmission's input shaft is the _____.

    a. impeller

    b. stator

    c. turbine

    d. clutch

20. The component which the pinions always "walk" is the _____.

    a. carrier

    b. ring gear

    c. sun gear

    d. planet pinions

21. In the AOD transmission, which clutch connects the front or large sun gear to the turbine shaft?

    a. intermediate

    b. reverse

    c. direct

    d. forward

22. In first ratio within the AOD transmission, the holding member within the planetary is the _____.

    a. carrier

    c. low sun gear

    b. reverse sun gear

    d. ring gear

23. In fourth ratio with the AOD transmission, the driving planetary member is the _____ .

    a. ring gear

    b. low sun gear

    c. reverse sun gear

    d. carrier

For answers, turn to the Appendix.

CHAPTER 3

Automatic Transmission Fluids

Up to this point, this text has covered in detail the construction and operation of various planetary gear trains and, briefly, the functions of the torque converter, clutches, and brake bands. These parts by themselves will not produce an automatic transmission that is operational. The automatic transmission, like any other **hydraulic** device, must have fluid to make it function. **Automatic transmission fluid** (ATF) is therefore of utmost importance to the operation of the unit and plays an important role in extending the life of the transmission as well.

Fluid is the name given to the lubricant used in automatic transmissions and other hydraulic devices. Over the years, manufacturers have used the term "fluid" along with a special reddish dye to prevent confusion of this substance with other lubricating oils. Also, since engine oil and ATF have different colors, the reddish color of ATF assists the mechanic in determining the location of leaks from either unit.

FLUID FUNCTIONS

ATF must be superior to other lubricating oil because it has to operate under more severe conditions and has more functions to perform. The functions include transmitting torque and hydraulic pressure, acting as a coolant, lubricating components, cleaning and sealing components, and controlling friction.

Transmits Torque

One of the more difficult jobs that ATF performs is transmitting engine torque from the crankshaft to the transmission's input shaft. To accomplish this task, the fluid circulates in a sealed torque converter (Fig. 3-1). The converter has an impeller, which delivers quantities of fluid at high speeds into the turbine that connects to the transmission's input shaft. When the impact energy of the moving fluid is great enough to overcome the load on the turbine and input shaft, torque transfers from the crankshaft to the input shaft.

Transmits Hydraulic Pressure

Another function of ATF is to transmit hydraulic pressure in an operating transmission. The fluid under pressure is the medium used to operate the various clutches and brake bands. If you remember, the clutches connect and disconnect various planetary gear train members to the input shaft and sometimes hold a member stationary. Brake bands, on the other hand, just hold a planetary member stationary.

Fluid, under pressure, also serves as hydraulic signals. These fluid signals are responsible for altering hydraulic pressure and determining upshift and downshift patterns. In order for the automatic transmission to function smoothly, hydraulic pressure and shift points must relate properly to engine load and vehicle speed.

Acts as a Coolant

Fluid acts as a coolant in an operating transmission to carry off excess heat that builds up in the unit. During the operation of the transmission, the fluid activity within the torque converter generates a great deal of heat. Also, during an upshift, the operation of clutches or bands produces excessive temperatures. This heat passes directly into the fluid.

The fluid, in turn, acts as a medium to carry the heat to a cooling device. This device may be either an air-cooled or water-cooled transmission cooler. In either case, the device lowers the temperature of the fluid, thereby preventing the otherwise excessive temperatures from ruining the fluid and transmission components.

Lubricates, Cleans, and Seals Components

ATF lubricates, cleans, and seals all the moving parts of an automatic transmission. As a lubricant, fluid reduces friction and its negative effects (heat and wear) when introduced as a film between moving parts. If the fluid does not maintain this lubricating cushion between the

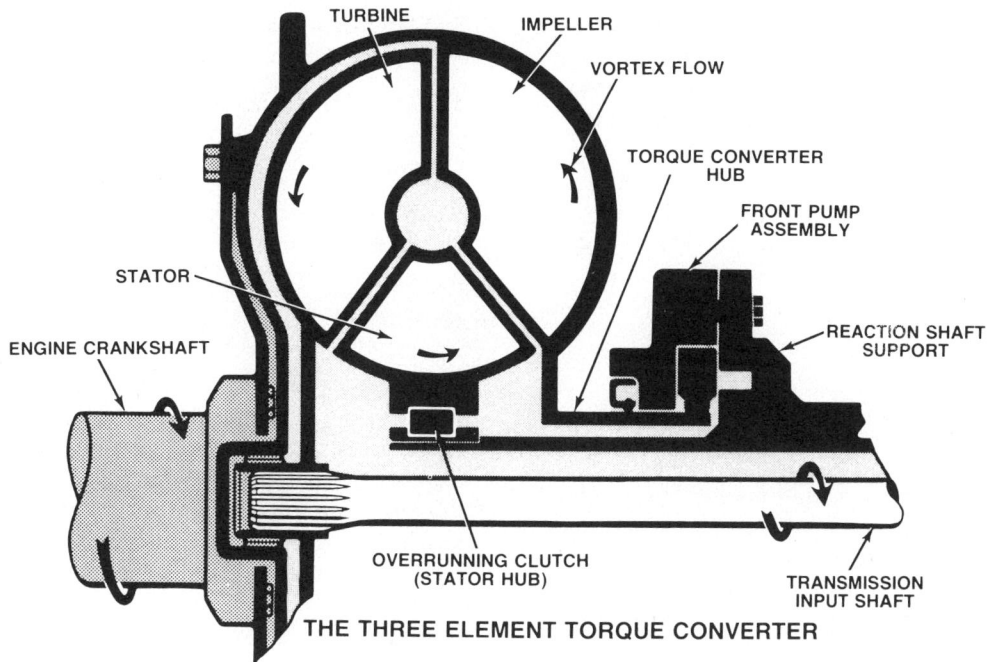

TURBINE

IMPELLER

VORTEX FLOW

TORQUE CONVERTER HUB

FRONT PUMP ASSEMBLY

REACTION SHAFT SUPPORT

STATOR

ENGINE CRANKSHAFT

OVERRUNNING CLUTCH (STATOR HUB)

TRANSMISSION INPUT SHAFT

THE THREE ELEMENT TORQUE CONVERTER

FIGURE 3-1   The impeller and turbine of the torque converter. (Courtesy of Chrysler Corp.)

moving parts, friction will cause premature failure of the moving parts.

Automatic transmission fluid cleans the parts and valves within the unit. The fluid has a number of additives that reduce the buildup of varnish and sludge in order to prevent these deposits from forming into clots that could plug the pickup screen or jam a shift valve. Furthermore, the fluid carries any dislodged clutch or band material, as well as any metal particles, down into the fluid reservoir, the pan. At this point, the foreign matter will either sink to the bottom of the pan, or as the particles attempt to move through the filter, it traps and holds them.

It is a natural characteristic of ATF to also act as a sealing agent for moving parts. In the automatic transmission, the fluid prevents excessive leakage around valves, rotating shafts, and the servo and clutch pistons.

Controls Type of Friction

The final function of ATF is to control the type of friction between all moving parts in the transmission. **Friction** is the resistance to motion between two objects in contact with each other. In the automatic transmission, direct or dry friction should never exist even between the clutch plates (Fig. 3-2) or between the band and drum. If dry friction existed between these units, the shifts would be very harsh and the units would wear out very rapidly.

To avoid this problem, automatic transmissions use lubricated "wet" clutches and bands to change ratios. In this situation, a constant supply of fluid passes between the clutch plates or between the band and drum as well as all other moving parts. As a result, friction occurs not between dry units but between a number of fluid layers and between the fluid layers and components themselves. This results in a type of friction known as **viscous friction** —resistance to motion between adjacent layers of fluid.

In the automatic transmission, the fluid layers cling to each of the components and act as wedges to keep all the parts separated. In the case of the clutch plates or band and drum, the fluid layers lubricate and separate the units and little or no friction exists between them.

When a clutch assembly or band activates, the fluid continues to form a wedge between the units to prevent

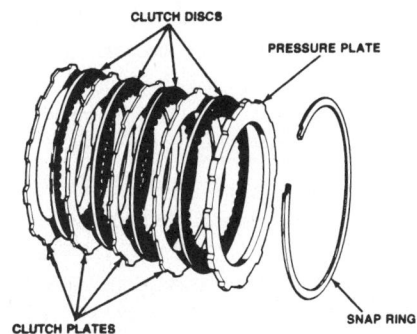

CLUTCH DISCS

PRESSURE PLATE

CLUTCH PLATES

SNAP RING

FIGURE 3-2   Fluid prevents dry friction from occurring between the clutch plates and discs. (Courtesy of Chrysler Corp.)

actual contact while most of the fluid "squeezes" out from between the components. During this time, the viscous friction between the fluid layers will either increase or decrease due to additives in the fluid. However, when the clutch plates or band and drum come together, a film of fluid still remains on these parts. This remaining fluid, under pressure resulting from the force applied to it by the clutch or servo piston, prevents direct contact and forms a "liquid lock" that helps to control slippage between the contacting surfaces.

FLUID STRUCTURE AND ADDITIVES

In order to meet the above demands, manufacturers blend a high quality mineral oil base with a number of additives. An **additive** is a small amount of a material added to another substance to provide it with special properties. The performance qualities that the additives give the fluid are oxidation stability, antifoam suppression, viscosity improver, compatibility, wear resistance, pour point depressant, corrosion and rust inhibitor, detergent-dispersant, friction modifier, and seal swell agent.

Oxidation Stability

Additives are necessary to stabilize oxidation because of the relatively high temperatures that the automatic transmission may encounter while operating. **Oxidation** is the union of oxygen with the fluid under high temperature conditions. For instance, when a vehicle is pulling a trailer or operating under conditions encountered in hot weather or in congested city traffic, the temperature within the automatic transmission may exceed 300° F. Furthermore, during an upshift, local temperatures at the clutches or bands may reach as high as 600° F.

For long life, ATF should operate at about 175° F. At the temperatures mentioned above, along with the introduction of air through normal transmission breathing, the fluid begins to oxidize, which changes its normal characteristics and causes a number of detrimental effects.

Fluid oxidation can produce five abnormal effects on automatic transmission operation which reduce its overall life.

1. Oxidation can alter friction characteristics of the fluid that will result in excessive clutch and band slippage. Excessive slippage produces high clutch or band temperatures, which, in turn, make the oxidizing condition more severe.

2. Acids or peroxides, formed during fluid oxidation, may corrode bushing and thrust washer materials, damage seals, and harm the composition clutch plates.

3. Oxidation increases viscosity (rate of flow) enough to lower effective transmission operation.

4. Oxidation forms a brown substance, varnish, which can lead to control valve or governor sticking and ultimate failure of the automatic transmission.

5. Byproducts of oxidation reduce the effectiveness of the antifoam additives.

Antifoam Suppression

The suppression, by use of additives, of the foaming tendency of fluid in an automatic transmission is essential to the proper operation of the unit. **Foam** is a light, frothy mass of fine bubbles formed in or on the surface of the fluid caused by the "egg-beater" action of internal transmission components violently agitating the fluid. However, a high or low fluid level can also cause foaming in the fluid.

Foam creates several problems in the transmission. For example, foam can produce erratic pump, torque converter, clutch, band, and hydraulic system operation due to the presence of air, which is compressible. In addition, foaming can result in the loss of fluid through the breather or filler tube.

Viscosity Improver

Using additives to improve fluid viscosity has improved the torque converter's and the automatic transmission's efficiency under all types of temperature conditions. **Viscosity** is the resistance to flow exhibited by the fluid, and heat has a definite influence on viscosity. High temperature will normally thin the fluid and permit it to flow easier. A fluid viscosity that is lower than normal can cause internal leakage in pumps, valves, and servos, which will reduce hydraulic pressure.

Cold weather, on the other hand, makes fluid heavier and flow much more slowly. High viscosity fluid produces sluggish transmission performance, improper shift patterns, in addition to possible bushing and clutch plate failures from lack of proper lubrication. To reduce the above-mentioned problems, manufacturers add viscosity improvers to their fluids to stabilize fluid viscosity to a nearly constant level during changes in temperature.

Compatibility

There are additives in the fluid to make it compatible to the various materials used in the transmission. This simply means that the additives protect the various types of metals such as steel, bronze, and aluminum as well as the seals and friction materials used in clutches and bands from chemical activities that would deteriorate them.

Wear Resistance

ATF also contains additives that work chemically to form a strong, tough surface film between moving parts. This film helps to stop any metal to metal contact, which causes wear and scoring on contacting surfaces. **Scoring** is a scratching or grooving effect caused by foreign materials caught between two contacting surfaces. In many cases, the foreign material is small particles of metal lost from transmission components due to wear. This metal would cause severe scoring on the surfaces of components such as a drum, where a brake band rides, if it were not for additives in the fluid.

Pour-Point Depressant

A **pour-point depressant** is an additive that improves the ability of ATF to flow (its fluidity) when cold. **Fluidity** is the ease in which the ATF flows through the transmission's hydraulic passages and spreads out over all the bearing surfaces. At very low temperatures, a fluid can become so thick that it would not flow at all. A pour-point depressant lowers the temperature point at which the fluid will be too thick to flow. As a result, the additives keep ATF more mobile at low temperature for adequate transmission lubrication during cold weather operation and initial warm-up.

Corrosion and Rust Inhibitors

Corrosion and rust inhibitors resist the formation of harmful acids in the fluid and keep water from damaging the transmission parts. High transmission temperatures cause acids and water to form in the fluid. These contaminants, while circulating through the transmission in the fluid, could etch, corrode, or rust the unit's metallic components. However, the corrosion inhibitor attacks the acid-forming ingredients and neutralizes their effect, while the rust inhibitor keeps the water from damaging all metal surfaces.

Detergent-Dispersant

A **detergent-dispersant** is an additive in the fluid that loosens and detaches any deposits of varnish, sludge, friction material, or metal particles. The fluid then carries these loosened materials to the pan, where the heavier particles sink to the bottom. The smaller, lighter material tends to remain suspended in the fluid until caught in the pickup screen or filter.

Friction Modifier ATF contains additive substances that modify the friction characteristics of the fluid. The additives alter a fluid's **coefficient of friction,** which is a number representing the amount of friction between the fluid and other contacting surfaces. For example, fluid along with clutch assemblies and brake bands require a certain coefficient of friction in order for them to perform their functions of either transmitting torque to or holding planetary members stationary.

However, since manufacturers use different types of friction materials for clutch and band linings, no single fluid satisfies the best friction requirements of all transmission types now in service. Therefore, a particular fluid has an additive that gives it a proven friction characteristic, so it reacts favorably with certain friction materials. This modification is responsible for two of the basic fluid types in service—Type F and Dexron.

ATF Type F has a coefficient of friction that increases as the sliding speed between drive and driven clutch plates or between the brake band and the drum decreases (Fig. 3-3). Sliding speed, in feet per minute, refers to the difference in speeds between the drive and driven clutch plates or between the band and drum. At the start of application of these units, the speed difference is great, but as the clutch or band steadily applies, the speed of the drive and driven members reduces as the clutch assembly or band picks up the load. With Type F fluid, the friction level increases as sliding speeds equalize. This action gives the transmission a firm, aggressive shift.

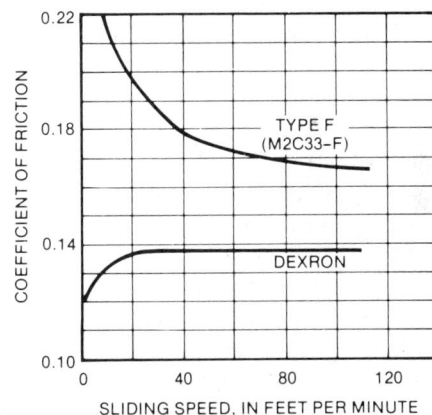

FIGURE 3-3 *The coefficient or amount of friction between sliding components varies with the type of fluid utilized.*

Dexron fluid, on the other hand, provides a lower friction level as the sliding speeds equalize (Fig. 3-3). This provides a transmission with a soft, smooth shift. Finally, it is important to remember that the mating of the fluid's friction level with the transmission's clutch and band material is necessary to obtain the desired shift characteristics and prevent premature failure of the clutch plates and bands.

Seal Swell Agent

Fluid producers add a seal swell agent to their product to keep seals within the transmission soft and pliable. High temperatures dry out seals and make them hard, which reduces the seal's ability to control leakage. The seal swell agent reduces this problem by assisting the fluid's passage into the seals, thus keeping them pliable and soft.

FLUID TYPES AND USAGE

Since the introduction of the automatic transmission, auto manufacturers have developed and used a number of different fluid types. In some cases, a newly developed fluid type was nothing more than the incorporation of more additives into the basic mineral oil base. This became necessary when manufacturers began to install engines with high horsepower output into vehicles, vehicles became factory equipped with air conditioning, and the motoring public started using their automobiles and light trucks to pull trailers and boats. Finally, as mentioned earlier, different fluid types are available that actually change the shift feel by altering the fluid's coefficient of friction.

Fluids Used in Ford Transmissions

Since 1951, the Ford Motor Company has used a number of fluid types in its automatic transmissions. From 1951 to 1961, Ford automatic transmissions used Type A Suffix A fluid (Fig. 3-4). This fluid was, at that time, the only one utilized by vehicle manufacturers for automatic transmissions. Several fluid suppliers marketed this product after receiving approval for their product from the Armour Research Foundation, Illinois Institute of Technology, now I.I.T. Research Institute. If a fluid met all the specifications of Type A Suffix A fluid, it received an AQ (Armour Qualification) number.

In 1961, Ford vehicles equipped with automatic transmissions came from the factory filled with a new fluid, Type F. Ford Motor Company developed this fluid and gave it a Ford Specification M-2C33D. By 1964, Ford recommended this fluid for all their automatic transmissions; however, pre-1961 units could use either Type F or Type A Suffix A fluids. In 1968, Ford upgraded Type F into a long-life fluid. The new fluid had a specification M-2C33F (Fig. 3-5) and replaced the old Type F (M-2C33D) for use in all Ford automatic transmissions but a few.

Since the introduction of Type F, Ford has developed three other fluids, C-J, Type H, and MV. In 1976, Ford redesigned its C-6 automatic transmission for the 1977 model year; this and later C-6 and other Ford transmissions use C-J fluid, Ford Specification M-2C138-CJ. The C-J fluid is red with an orange tint compared to the darker red color of Type F and is only available at Ford Service Centers. However, because C-J fluid is quite similar to Dexron II in friction characteristics, Ford announced in 1978 that Dexron II is a suitable alternate to C-J for use in 1977 and later C-6 and other Ford transmissions, which use C-J fluid.

Beginning in 1982, Ford began to use another new fluid, Type H. This fluid has a Ford Specification of ESP-M2C166-H and is required for service in the C-5 select-shift automatic transmission with locking select torque converter.

To correct some cold weather operating problems with its automatic transaxle, Ford introduced a multi-viscosity (MV) fluid in 1981. This automatic transaxle fluid has a Ford specification of ESP-M2C164-A, and it is only used in ATX transaxle vehicles exposed to subfreezing weather conditions to improve cold starting and shift operations. Otherwise, the transaxle used C-J or Dexron II.

Since many petroleum suppliers market Type F fluid, make sure the Ford specification number is on the can, or do not use it because the product may not meet Ford's standards for ATF. Also, now that Ford automatic transmissions can use either Type F, H, MV, C-J, or the alternate for C-J-Dexron II — *be sure to check the owners or the Ford Service manual before adding or changing the fluid in a Ford automatic transmission for the correct type to use.* See Table 3-1 for Ford's recommendations through 1984.

FIGURE 3-4   Type A Suffix A fluid.

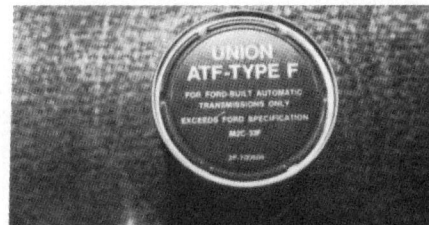

FIGURE 3-5   Type F fluid.

TABLE 3-1   Fluid Recommendations for Ford Vehicles Through 1984

TRANSMISSION	FLUID TYPE
AOD	CJ or Dexron II-D
ATX	CJ, Dexron II-D, or in cold climates, MV
C3	Before 1980, Type F; 1981 and on, CJ or Dexron II-D
C4	March 8, 1979, and earlier models, Type F March 8, 1979, and on, CJ or Dexron II-D
C5	Type H
C6	1976 and earlier models, Type F 1977 and on, CJ or Dexron II-D
JATCO	Passenger car, CJ or Dexron II-D Courier, Type F
FMX	Type F
ZF	Dexron II-D

Fluids Used in American Motors, Chrysler, and General Motors Transmissions

Over the years, American Motors, Chrysler, and General Motors have used two types of fluid in their automatic transmissions. Before 1968, they all used Type A Suffix A fluid in all their units. In 1968, General Motors developed a new fluid—Dexron—(Fig. 3-6) for all its automatic transmissions. This fluid had a qualification number, B-XXXXX, on its containers. The X's represented the five-digit number that each fluid manufacturer assigned to the fluid. This new fluid functioned better than Type A fluid at low temperature; it resisted fluid oxidation more efficiently; and it contained properties needed for smooth shifting that Type A Suffix A did not have. Lastly, American Motors and Chrysler adopted Dexron for their

FIGURE 3-6   Dexron fluid.

units and followed the same policy as General Motors for using either Dexron or Type A Suffix A for all their pre-1968 transmissions. (NOTE: With the advent of Dexron and Type F fluids, I.I.T. Research Foundation no longer certifies automatic transmission fluids marketed by major oil companies. General Motors and Ford Motor Company are their own authorizing agents for Dexron and Type F.)

Since 1968, General Motors has changed its formula for Dexron. For example, in 1973, GM introduced Dexron II with a C-XXXXX on its container. This new fluid became necessary due to the United States ban on killing whales in late 1971. The fluid additive used as the friction modifier in the early Dexron fluid was supplied from sperm whales.

A chemical was substituted for the whale oil and Dexron II-C was developed as a long-life fluid. This new formula also had a higher resistance to oxidation and higher frictional stability than Dexron-B.

However, the new additives in Dexron II-C caused a corrosive action between the solder used by General Motors in their tubular-type transmission coolers at the inlet fittings on some models. As a result, in 1976, GM developed Dexron II with a D-XXXXX. This new fluid performed as well as Dexron II-C but did not cause cooler, inlet-fitting corrosion.

Dexron II-D now replaces both Dexron-B and Dexron II-C fluids in all General Motors, American Motors, Chrysler, and some Ford automatic transmissions. However, before adding this fluid to any GM, AMC, Chrysler, or Ford transmission, always make sure which formula of Dexron is recommended; and never use a product if it does not have the trademark on the can of "Dexron" followed by the proper number or letter designation.

Fluids Used in Foreign-Built Transmissions

Foreign automobile manufacturers have extensively used either Dexron II-D or Type F in their respective automatic transmissions. Table 3-2 shows an overview of the fluid types each manufacturer is using. But always refer to the owner's or service manual for the correct fluid type to use in a vehicle because the factory can change which type it recommends from one year to the next.

FLUID COOLING

The purpose behind providing a cooling system for ATF is to prolong the fluid's useful life and thereby directly affect the life of the transmission. As previously men-

TABLE 3-2   Fluid Recommendations for Foreign-Built Vehicles

MAKE OF VEHICLE	FLUID TYPE
Audi	Dexron II-D
BMW	Dexron II-D
Datsun	1971 and earlier models, Type F 1972 and later, Dexron II-D
Fiat	Dexron II-D
Honda	Dexron II-D
Mazda	Type F
Mercedes-Benz	Dexron II-D
Opel	Dexron II-D
Peugeot	Dexron II-D
Porsche	Dexron II-D
Saab	Type F
Subaru	Dexron II-D
Toyota	Type F
Triumph	Dexron II-D
Volkswagen	Dexron II-D
Volvo	Type F

tioned, the constant action of the recirculating fluid in the torque converter and the application of a clutch or band increase the fluid's temperature a great deal. This system does not "cool" the fluid but attempts to keep the fluid at a normal operating temperature of around 175° F. At this temperature, the fluid has a potential of lasting about 100,000 miles or longer.

Water-Cooled System

Vehicle manufacturers have used two different systems, the water-cooled and the air-cooled, to reduce the operating temperature of ATF. The most popular method of reducing fluid temperature is the **water-cooled** system, which consists of a water-tight cooler that is located in the radiator's lower or side tank, two lines that carry the fluid to and from the transmission, and valving to control fluid pressure and flow (Fig. 3-7).

When the transmission is operating, fluid under pressure flows from the transmission to the radiator cooler and returns to the transmission pan or to the transmission's lubricating system. As the fluid moves through the cooler, the excess heat in the fluid passes into the engine's coolant, which surrounds the tubing of the transmission cooler.

Engine coolant can cool as well as heat the fluid. If the

engine coolant has a lower temperature than the fluid, fluid temperature goes down. But, if the engine coolant's temperature is much hotter than that of the fluid, the fluid warms up as it passes through the cooler. In extremely cold climates, this action helps to warm up the fluid. However, in very hot climates, this same action can raise the fluid's temperature high enough to shorten its useful life.

Air-Cooled System

One **air-cooled system** used on some early model automobiles and trucks consisted of a torque converter that had a series of air fins around its outer circumference, an inlet baffle or cooling shroud that was fixed to the converter's housing and directed the incoming air over the entire converter, along with an outlet duct (Fig. 3-8). As the torque converter rotated with the engine, its attached fins acted like a centrifugal air pump. This action created a low-pressure area at the air inlet, which pulled in outside air. The air flow passed over the converter, removed excess heat from the fluid contained inside the converter, and expelled the heated air by centrifugal force through the outlet duct into the atmosphere.

A more recent version of a system using the same principle has a cooling shroud cover, tacked over the rear half of the converter, in place of air fins. Also, the converter housing incorporates a series of rounded ventilation ports instead of inlet and out ports.

As the converter with this system design rotates, it also acts as a centrifugal pump. However, in this case, the low-

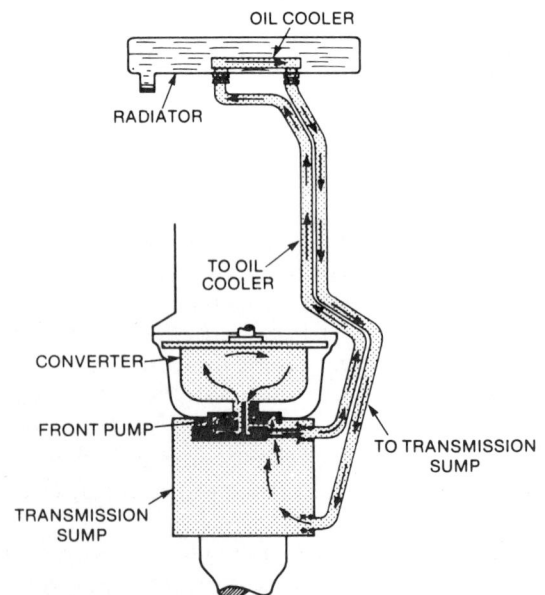

FIGURE 3-7   A schematic of a typical water-cooled, fluid cooling system.

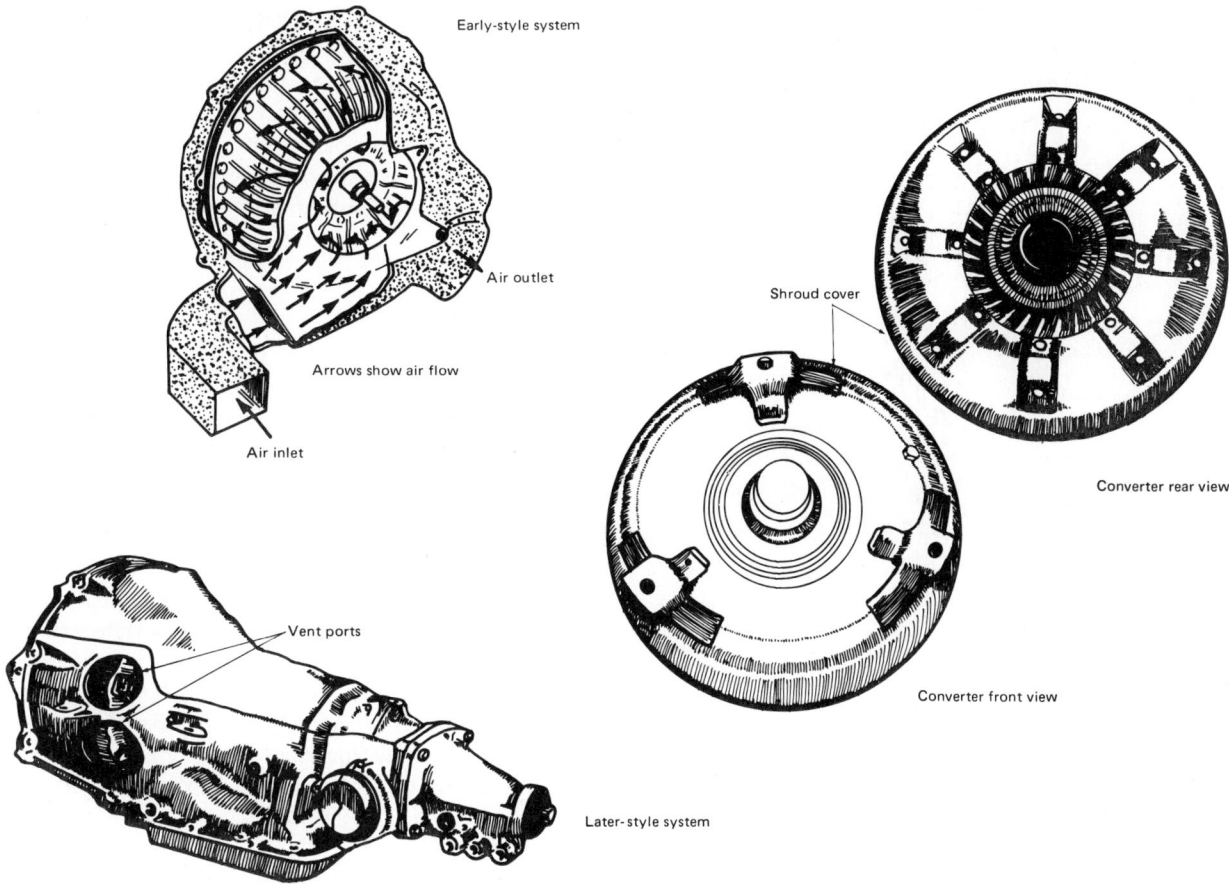

FIGURE 3-8   A typical air-cooled, fluid cooling system.

pressure area forms between the attached shroud and the converter shell. This causes outside air to enter the space and absorb converter heat. After absorbing the heat, the air is expelled out of the case housing through the ventilation ports due to centrifugal force.

After-Market Cooling Devices

Even with the cooling systems mentioned above, certain driving conditions could cause the fluid to overheat. For instance, if a vehicle becomes stuck in sand or mud and the driver rocks the vehicle back and forth, the fluid temperature will increase faster than the standard system can cool it. Also, when a vehicle operates in very hot climate, pulls excessive loads such as a trailer or camper, or has a restricted or blocked fluid flow to the cooler in the radiator, the transmission fluid will overheat to a point that its useful life is cut short.

The fluid's life is less because fluid temperatures above normal cause rapid oxidation. Oxidation, remember, is the union of fluid with oxygen in the presence of high temperature and results in failure of the fluid to lubricate properly. The rate of oxidation in some fluids can double

for each temperature increase of 20° F above normal, which is 175° F. Furthermore, if the oxidation rate doubles, useful life is cut in half (Fig. 3-9). Although much has been done to reduce oxidation in long-life fluids through the use of additives, the effect of excessive heat on any fluid, over a period of time, is the same—its service life is cut short.

Temperature above normal — Rate of oxidation to double for each temperature increase of 20 degrees above normal. As oxidation rate doubles, useful life is cut in half.

Now;

at 175 degrees F, life is 100,000 miles
at 195 degrees F, life is 50,000 miles
at 212 degrees F, life is 25,000 miles
at 235 degrees F, life is 12,000 miles
at 255 degrees F, life is   6,250 miles
at 275 degrees F, life is   3,000 miles (approx.)
at 295 degrees F, life is   1,500 miles
at 315 degrees F, life is   750 miles

FIGURE 3-9   The rate of oxidation of a certain fluid at a given temperature above normal.

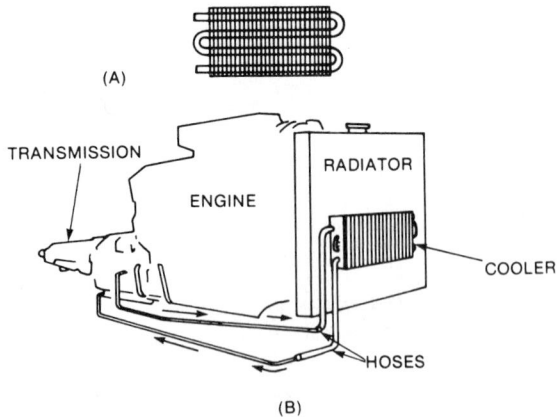

FIGURE 3-10   Auxiliary cooler shown at (a). Part (b) shows the cooler connected directly to the transmission's cooler lines.

Auxiliary Cooler

To assist the standard transmission cooling system in maintaining a reasonable fluid temperature, several after-market devices are available that help reduce fluid temperature and oxidation buildup. The first device is an auxiliary cooler mounted in front of the radiator or anyplace where air can pass through it (Fig. 3-10). The cooler itself has a tube or series of tubes that run the full length of the unit, and layers of air fins surround the tubes.

If the auxiliary cooler is to handle all the fluid cooling, both the transmission lines connect to the open ends of the tubes (Fig. 3-10). This type of installation is sufficient for normal cooling demands, where no other type of cooler is available or for assisting the factory installed air-cooled system in reducing fluid temperature.

When the auxiliary cooler is to assist the water-cooled system, its open tube ends connect in series with the transmission cooler return line (Fig. 3-11). This installation allows all the fluid leaving the radiator cooler to pass through the auxiliary cooler; consequently, it reduces the

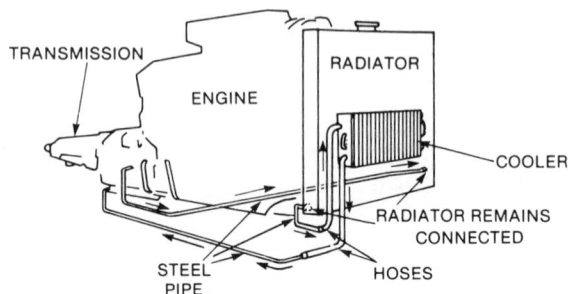

FIGURE 3-11   An auxiliary cooler connected in series with the radiator cooler in order to provide extra fluid cooling.

temperature even more before the fluid returns to the transmission.

With either type of installation, the auxiliary cooler reduces fluid temperature in the same way. As the fluid passes through the cooler's tubes, it gives off heat to the tubes. The tubes, in turn, release their heat, via the air fins, to the air passing around and through the tubes. The engine cooling fan or the forward movement of the vehicle assures adequate air flow through the auxiliary cooler.

Larger Capacity Fluid Pan

The second after-market device that assists in cooling the fluid is a specially designed transmission pan. This pan (Fig. 3-12) has about a one-quart larger capacity than the standard unit, and running the length of the pan are a series of air fins.

Fluid in the transmission will normally give off heat to the pan, and the air passing over its base removes some of this heat. With the new type of pan, the fins provide additional radiating surfaces for heat transfer and, at the same time, direct the cooling air over the base of the pan. In addition, the extra quart of fluid tends to reduce the overall temperature of the fluid in the transmission.

FLUID ENERGY

A fluid in motion is said to possess kinetic energy. For example, water flowing down a mountain can turn a water wheel because of this form of energy. **Kinetic energy** is the stored capacity for performing work possessed by the moving fluid by virtue of its momentum. **Fluid momentum** is the effect that the moving fluid has on other objects which determines the length of time required to bring the moving fluid to rest when the fluid is under action of a constant force. In other words, kinetic fluid

FIGURE 3-12   Special air-cooled fluid pan with additional capacity for fluid storage.

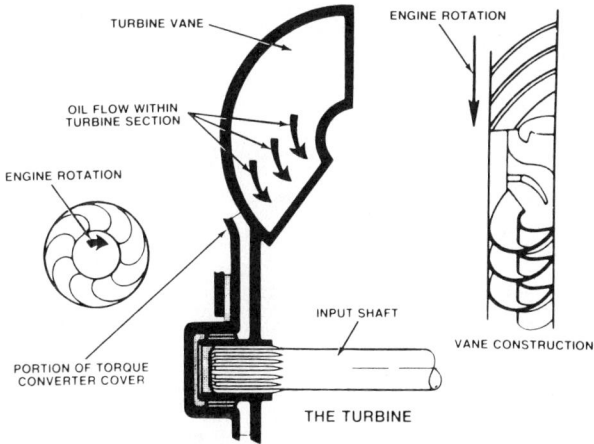

FIGURE 3-13 *The turbine blades (vanes) attempt to slow or stop the moving fluid. (Courtesy of Chrysler Corp.)*

energy represents the work necessary to bring the fluid from its actual velocity (rate of flow) to a state of rest. This energy is what drives the turbine within the torque converter (Fig. 3-13).

The engine supplies the needed constant rotary force (torque) necessary to drive the impeller inside the torque converter (see Fig. 3-1). In turn, the impeller blades set the fluid in motion. With increases in engine speed, fluid motion (velocity) intensifies, along with kinetic energy.

The device that attempts to slow or stop the moving fluid is the turbine blades. The turbine, which is also part of the converter, attaches to the transmission's input shaft, and if the transmission is in gear, the input shaft and turbine will have the vehicle's load imposed on them (Fig. 3-13).

As the fluid flows through the turbine, the fluid decelerates or slows down as it presses against the edges of all the blades. During this process, part of the kinetic energy, originally possessed by the moving fluid on

entering the turbine, expends itself against the turbine blades. If sufficient kinetic energy is available in the fluid, the turbine will rotate and the vehicle will move. In other words, the moving fluid has performed work.

FLUID FORCE AND PRESSURE

In studying the laws of fluids and other liquids, the reader must understand the differences between "force" and "pressure." **Force** means to push or pull an object. The unit of measurement for force is pounds. **Pressure,** on the other hand, is the result of a force applied to a fluid trapped in a sealed container system. Since a liquid is not compressible, the end result is fluid pressure. The unit of measurement for pressure is pounds per square inch (psi).

To determine the amount of pressure in the system shown in Fig. 3-14, divide the area of the input piston into the force applied to the fluid. In this case, the area of the input piston is 10 square inches and the force is 50 pounds. Therefore, the fluid pressure is 50 pounds/10 square inches, or 5 pounds per square inch, 5 psi. (NOTE: Attempting to figure out the actual pressure in an automatic transmission is much more difficult because a rotating pump supplies a varying force to the fluid, and the system itself contains valve bores and passages of different areas. These factors will be covered later in the text.)

Fluid pressure can apply a force (push) to an output piston (Fig. 3-15). An output piston, placed in a hydraulic system, will then be able to deliver this force to another object. The output force is equal to the piston's area multiplied by the pressure within the system. In Fig. 3-15, the output piston has an area of 20 square inches and the system pressure is 5 psi. Therefore, the output force is 20 square inches times 5 psi, or 100 pounds. In automatic transmission hydraulic systems, the output pistons apply force to the clutch plates or bands.

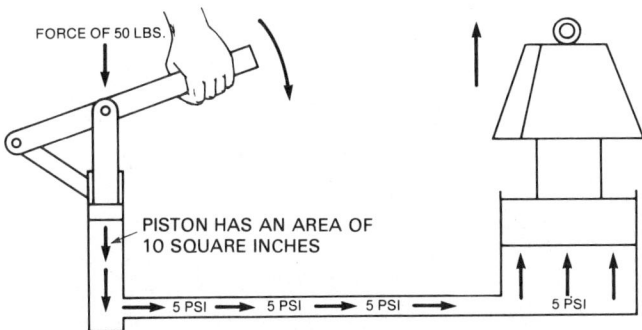

FIGURE 3-14 *Force is a push or pull measured in pounds; pressure is the result of force applied to a fluid in a sealed system.*

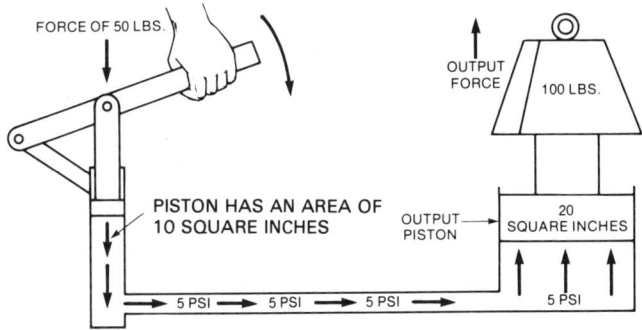

FIGURE 3-15 *Fluid pressure can apply a force to an output piston.*

SUMMARY

1. Automatic transmissions will not function without ATF.

2. Fluid is the name given to the lubricant used in automatic transmissions.

3. Fluid transmits torque between the engine and the transmission.

4. Fluid transmits pressure in an operating transmission.

5. Fluid carries heat to the transmission or auxiliary cooling system.

6. Fluid controls the type of friction within a transmission.

7. Fluid lubricates, cleans, and seals moving transmission parts.

8. Fluid acts as a friction modifier.

9. ATF has a mineral oil base plus a number of additives.

10. Fluid additives retard oxidation.

11. Additives reduce the foaming tendency of fluid.

12. Fluid additives stabilize viscosity.

13. Additives make ATF compatible with other transmission materials.

14. Fluid additives reduce wear and scoring.

15. Additives act as pour-point depressants.

16. Fluid contains rust and corrosion inhibitors.

17. Fluid additives act as a detergent-dispersant to keep the transmission clean and prevent sludge from blocking passages.

18. Additives modify the friction characteristic of the fluid.

19. ATF contains seal-swell agents.

20. Type F fluid has a high friction level as sliding speed equalizes.

21. Dexron provides a lower friction level as sliding speed equalizes.

22. Ford automatic transmissions now use Type F, C-J, H, or MV fluids.

23. American Motors, Chrysler, and General Motors all use Dexron II-D in all their automatic transmissions.

24. A fluid cooling system prolongs the useful life of the fluid.

25. The two basic fluid cooling systems are the water-cooled and the air-cooled.

26. Two after-market devices, the auxiliary cooler and a specially designed pan, can assist the regular fluid cooling system in reducing high fluid temperatures.

27. Moving fluid possesses kinetic energy.

28. Kinetic fluid energy drives the turbine within the converter.

29. Fluid transmits pressure and force.

REVIEW

This section will assist you in determining how well you remember the material contained in this chapter. Read each item carefully. If you can't complete the statement, review the section in the chapter that covers the material.

1. What is the term used to designate the substance that lubricates an automatic transmission?

    a. oil

    b. lubricant

    c. fluid

    d. Type A

2. Fluid acts in the _____ to transmit torque between the engine and transmission.

    a. torque converter

    b. clutch

    c. band

    d. pump

3. What type of friction exists in an automatic transmission?

    a. dry

    b. greasy

    c. viscous

    d. kinetic

4. Fluid has a _____ oil base.

    a. caster

    b. mineral

    c. synthetic

    d. fish

5. Oxidation of the fluid occurs in the presence of _____.

    a. additives and oxygen

    b. cold and additives

    c. cold and oxygen

    d. heat and oxygen

6. What is a frothy mass of fine bubbles called?

    a. oxidation   c. foam

    b. viscosity   d. compatibility

7. What is the main cause of varnish in the transmission?

    a. friction

    b. heat

    c. oxidation

    d. cold

8. Which fluid has a high friction level as sliding speeds equalize?

    a. Type F

    b. Dexron

    c. Type A

    d. C-J

9. Chrysler automatic transmissions use _____ fluid.

    a. Type F

    b. C-J

    c. Dexron

    d. Type A or C-J

10. Normal operating fluid temperatures should be around _____.

    a. 300° F

    b. 250° F

    c. 200° F

    d. 175° F

11. Which type cooling system uses a converter with fins?

    a. air-cooled

    b. water-cooled

    c. after-market

    d. both a and c

12. If a vehicle is pulling a trailer, fluid temperature may reach an average temperature of _____.

    a. 175°

    b. 225°

    c. 300° F

    d. 500° F

For the answers, turn to the Appendix.

Every motor vehicle, whether it is an automobile, truck, or motorcycle, requires some form of coupling between the engine and transmission. A **coupling** is a device, in this case, that connects or disconnects the engine from the transmission. When a manual-shift transmission is used in a vehicle, the coupling is in the form of a friction clutch assembly (Fig. 4-1). However, if the vehicle has an automatic transmission, a fluid coupling or torque converter takes the place of the friction clutch. The only exception to this is in the case of a few vehicles such as race cars, tractors, trucks, and some recreational vehicles that use some form of friction clutch to couple the engine to an automatic transmission.

FUNCTIONS OF A FLUID COUPLING AND TORQUE CONVERTER

In order for a vehicle to operate properly, the fluid coupling or torque converter must perform a number of important functions. These include transmitting torque, multiplying torque, acting as an automatic clutch, dampening out torsional vibrations, acting as a flywheel, providing a means of mounting the starter ring gear, and driving the transmission's hydraulic pump. (NOTE: The design of a fluid coupling is such that it cannot multiply torque, but the unit can still perform many of the other functions mentioned above and explained more in detail in the following paragraphs.)

Transmits Torque

The fluid coupling or torque converter transmits the torque developed by the engine to the transmission's input shafts. In the converter illustrated in Fig. 4-2, engine torque (the twisting effort) passes from the crankshaft to the flexplate. In turn, the flexplate drives the attached converter housing with its integral bladed impeller. But without additional components, engine torque transfer would cease at the impeller.

To complete the transfer process, a coupling or torque converter requires a bladed turbine and a quantity of fluid. The turbine is free to rotate on support bearings inside the converter or coupling housing and attaches to the input shaft by means of splines. There is no physical connection between the turbine and the impeller.

The actual medium used to transmit engine torque from the impeller blades to those of the turbine is the fluid. As mentioned in the last chapter, fluid in motion possesses kinetic energy—has the ability to perform work. The work, in this situation, would be to move the turbine and attached input shaft against the load of the vehicle, imposed on the shaft.

CHAPTER 4

Fluid Couplings and Torque Converters

The fluid is set in motion by the engine driving the bladed impeller, which is a form of pump. If the resulting impact energy of moving fluid striking the turbine blades is great enough, the turbine and input shaft will begin turning. It should be obvious then that the moving fluid is the actual medium that is transmitting engine torque between the engine and transmission and that the amount of torque transfer depends on the amount of kinetic energy contained in the fluid by virtue of its motion.

Multiplies Torque

As mentioned in Chapter 1, the engine does not produce sufficient torque to start a vehicle in motion or keep it in motion under certain load conditions. Consequently, the vehicle must have a number of torque multiplying devices such as the transmission, ring and pinion gears, or—in the case of vehicles with automatic transmission—the torque converter. The transmission along with the ring and pinion gears multiply engine torque and control its speed mechanically through the use of gears. These gears act as a series of rotating levers that multiply engine torque while permitting an engine to operate at its most efficient speed.

The fluid torque converter also acts like a transmission by providing, up to its design capabilities, varying

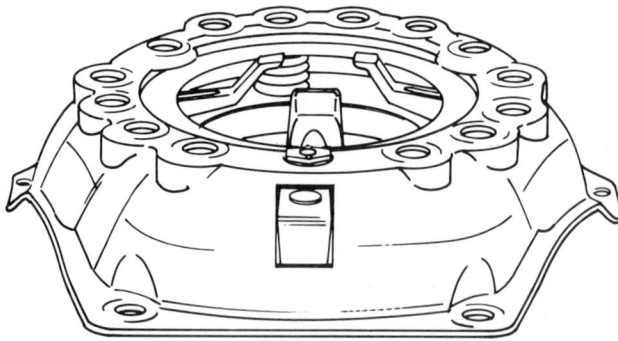

Friction clutch assembly

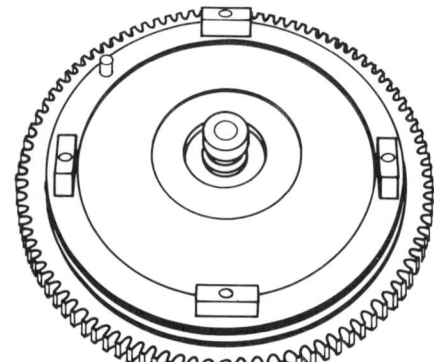

Torque converter

FIGURE 4-1   Two types of coupling devices, the friction clutch and the torque converter.

amounts of torque multiplication while at the same time controlling engine speed. However, the torque converter performs these functions not through gears but by controlling the fluid flow returning from the turbine to the impeller. In the torque converter, this action is accomplished through the use of a third component, called the stator (see Fig. 4-2).

The **stator** is the vaned reaction member placed between the impeller and turbine. The purpose of the stator is to change the direction of the fluid returning from the turbine to the impeller, which without such assistance, would strike the impeller blades in such a way as to hinder its rotation. By redirecting the fluid flow back into the same direction as the impeller turns, the returning fluid assists the engine in driving the impeller, thus increasing its ability to initially accelerate the confined fluid within the housing.

Acts as an Automatic Clutch

As in the case of any other form of transmission, the automatic transmission does require a device for smoothly disconnecting and connecting the flow of power from the engine. This action is necessary so a vehicle can operate in gear at curb idle while, for example, waiting at a corner for the traffic light to change. Also, the device must reconnect the engine power to the transmission smoothly to reduce the wear and tear on other drive train components, while increasing the comfort of the vehicle's occupants.

In a motor vehicle with an automatic transmission, a torque converter or fluid coupling controls the transfer of power between the engine and transmission. **Power,** as used here, is the rate at which work is accomplished and is a factor of applied force times the distance traveled by an object. **Work** is nothing more than the moving of an

object against an opposing force. In the case of a converter or coupling, the object is the turbine and input shaft.

Power is sometimes confused with torque. **Torque** is a twisting effort, which may or may not result in motion. For instance, when the engine is driving the impeller inside a converter or coupling, the resulting fluid flow may transmit a given amount of torque (twist) to the turbine and input shaft. However, unless the turbine actually moves, no work is done; therefore, there is no transfer of power.

There are two reasons for this interruption of power flow inside a converter or fluid coupling when a vehicle is at curb idle with the transmission in gear. First, the engine is idling and the impeller is therefore pumping only small quantities of fluid into the turbine. Second, the load of the

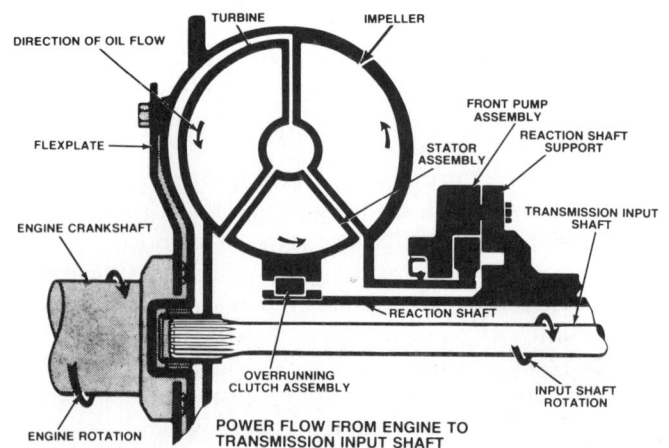

FIGURE 4-2   Torque converter components: The coupling members are the impeller and turbine; the torque multiplying device is the stator. (Courtesy of Chrysler Corp.)

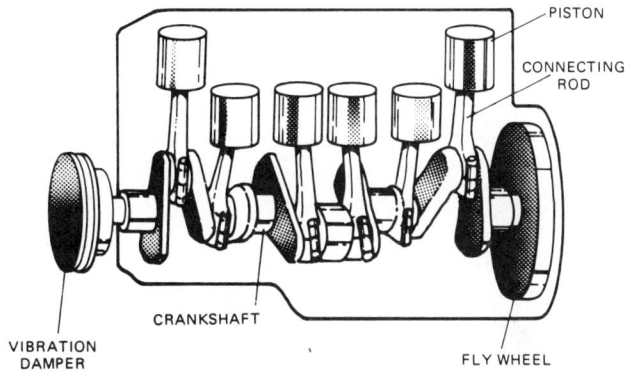

FIGURE 4-3 Torsional vibrations are set up in the crankshaft as a result of power impulses attempting to twist the crank pins ahead of the rest of the crankshaft.

vehicle (its weight and the tire-to-road friction) are imposed on the turbine through the input shaft.

The only way power can transfer through to the input shaft is by increasing fluid motion. As the driver accelerates the engine, fluid velocity increases until its resulting impact energy is great enough to overcome the load on the input shaft and turbine. At this point, the turbine begins to revolve, resulting in a smooth transfer of power into the transmission.

Dampens Out Torsional Vibrations

A negative characteristic of the piston engine is that it produces a torsional vibration while running. **Torsional vibration** is the twist-untwist motion produced in the crankshaft as each cylinder in turn produces power. When the spark plug fires the air-fuel charge in a cylinder, the resulting pressure pushes the piston down in the cylinder. This motion transmits through the connecting rod to the crankpin, where the rod attaches (Fig. 4-3).

The resulting forces tend to twist, or drive, the crankpin ahead of the rest of the crankshaft. Then, in a moment, the termination of the power impulse relieves the force on the crankpin. It now tends to untwist, or snap back, into its original relationship with the rest of the crankshaft. This twist-untwist tendency, repeated with every power impulse, sets up an oscillating motion in the crankshaft, known commonly as torsional vibration. If not controlled, these oscillations can build up so much that a crankshaft may actually break at certain speeds or damage gears or other components of the drive train.

An engine has a vibration damper installed on the end of the crankshaft opposite to the flywheel that reduces the oscillations to protect the crankshaft. However, the damper does not relieve the entire problem, so the crank-

shaft does transmit some vibration into the transmission via the coupling device.

The coupling device itself therefore must be designed to stop these vibrations. The standard torque converter (without direct-drive capability) and the coupling both use their fluid to dampen out the remaining vibrations so they do not affect the remaining drive train components. The direct-drive converter also uses the fluid as a dampening device during a portion of its operating phase but must also have torsional springs, which absorb the oscillations during direct-drive operation.

Acts as a Flywheel

The engine also requires a flywheel. The **flywheel** is a heavy wheel that attaches to the end of the crankshaft opposite to the vibration damper and keeps the engine turning smoothly between successive power strokes (see Fig. 4-3). Each cylinder of the engine operates on a cycle that has only one power stroke, where the piston is actually driving the crankshaft. During the other three strokes, the rotating crankshaft is moving the piston up or down in its cylinder. Thus, during the power stoke, the crankshaft tends to speed up; during the other three strokes, it tends to slow down.

The heavy flywheel resists any effort to change its speed of rotation. When the crankshaft attempts to speed up or slow down, flywheel inertia resists it. In effect then, the flywheel absorbs power from the crankshaft during the power stroke and returns it to the crankshaft during the remaining three piston strokes of the piston's cycle.

In a vehicle with a manual-shift transmission, the flywheel is a heavy one-piece assembly that also performs two other functions. First, the flywheel provides a smooth driving surface for the disc of the friction clutch assembly. Second, the flywheel supplies the means for mounting the starter ring gear, which is needed to turn the engine over for starting.

In a vehicle with an automatic transmission, the torque converter or coupling, filled with fluid, along with the flex or drive plate provide the weight necessary to smooth out engine operation. Also, the manufacturer welds the starter ring gear directly to the converter (Fig. 4-4) or to the flex plate.

Drives the Transmission's Hydraulic Pump

For lubrication and the fluid pressure necessary to operate the clutches and bands, the automatic transmission must have a hydraulic pump. This pump must supply hydraulic pressure whether the vehicle is stationary or in motion. In other words, the pump must produce

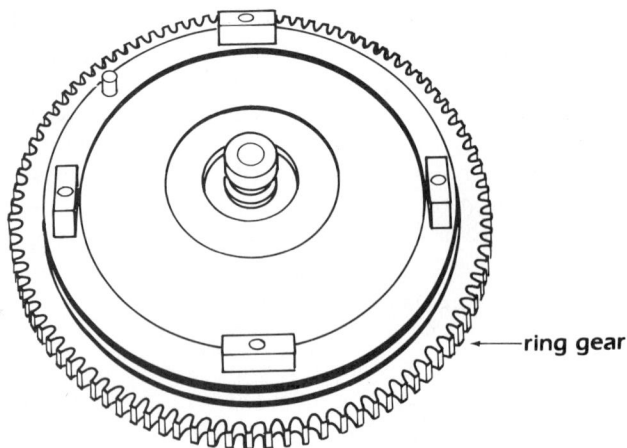

FIGURE 4-4   A starter ring gear welded to the converter housing.

this pressure whenever the engine is operating even if the vehicle is stopped.

Manufacturers accomplish this by having the converter mechanically drive the hydraulic pump. The torque converter attaches to a flex or drive plate, which is bolted to the crankshaft. The hub, located on the backside of the converter housing, has either two flats or two slots machined into its end. These mate with similarly machined areas located on the drive gear or rotor of the hydraulic pump (Fig. 4-5). With this arrangement, the converter housing rotates with the engine, and the hub drives the gear or rotor which, in turn, causes the pump to produce flow and pressure.

THE FLUID COUPLING

The coupling is the simplest form of fluid clutch. Therefore, a discussion of its design and operation will serve as an ideal introduction to how the impact energy of moving

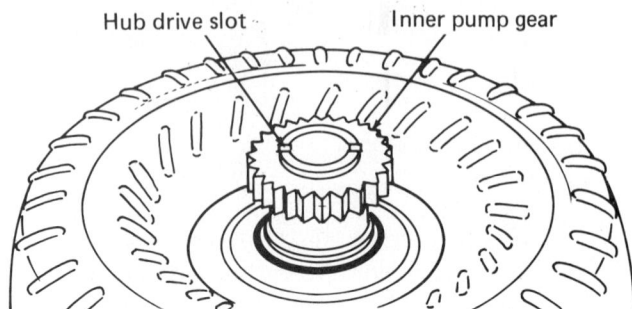

FIGURE 4-5   Slots in the converter's hub drive the inner, pump gear, or rotor.

fluid can place a solid object into motion. Moreover, some of the principles presented here will definitely apply later to the discussion of torque converters.

However, keep in mind that although some early automatic transmission types used the fluid coupling as the device for connecting or disconnecting engine power to the transmission, the unit did not multiply torque. Consequently, since 1964, the torque converter has been the only type of fluid clutch used by the various auto manufacturers in their automatic transmissions.

A Simple Coupling

To illustrate a simple fluid coupling, consider the action of two electric fans facing each other (Fig. 4-6). Fan number 1 is plugged in and permitted to run, while fan number 2 has no power and is immobile before fan number 1 is turned on. As fan number 1 begins to rotate, its blade pulls air through the back of the fan cage and blows it out through the front. These air currents strike the blades of fan number 2 causing it to turn also. In this situation, the air acts as the fluid or medium of power transfer between blades of the two fans.

In the illustration, each fan represents half of a fluid coupling, with fan number 1 being the impeller or pump. Fan number 2 is the turbine section of the coupling.

Note also in Fig. 4-6 the air leakage around the number 2 fan cage. This results in an inefficient coupling and is one of the reasons why fan number 2 never reaches the operating speed of fan number 1. A shroud, built around both fans in order to contain the air, would make this form of coupling more efficient.

FLUID COUPLING DESIGN

As in the simple example of the two fans mentioned above, a functioning fluid coupling, as used in an automatic transmission, also consists of two main rotating members (Fig. 4-7). The **impeller** is the driving member or pump, which corresponds to the power-driven fan. The **turbine** is the driven member that serves a similar function as the fan without power.

Housing

However, to improve the operating performance of the transmission coupling, its design is quite a bit different from the two-fan example. For instance, to ensure that there is no loss of directional fluid flow to the driven member, the turbine, both the impeller and turbine are enclosed in a sealed stamped steel or (in a few cases) aluminum housing. The housing attaches to the engine

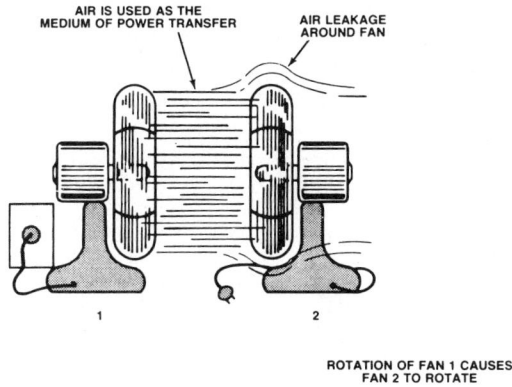

FIGURE 4-6   Two electric fans illustrate a simple fluid coupling. (Courtesy of Chrysler Corp.)

crankshaft usually through a flex or drive plate. This design ensures that the impeller will rotate with the housing whenever the engine is running.

To improve torque and power transfer between the impeller and turbine, the coupling housing is full of fluid (ATF) instead of air. Obviously, due to its additional mass and incompressibility, ATF will provide a great deal more impact energy than an equal quantity of air. The transmission's hydraulic system supplies the fluid necessary to keep the housing full whenever the engine and coupling are in operation. A low fluid level in the coupling would reduce its operating efficiency.

Impeller and Turbine

The design of the impeller and turbine is also much different from the two electric fans illustrated in Fig. 4-6. In a fluid coupling, like the one shown in Fig. 4-7, the

impeller and turbine both have a shape similar to a donut that is cut in half. One half, which represents the impeller for the coupling, attaches permanently to the housing. Therefore, it rotates with the housing at engine speed.

The other half forms the turbine. The turbine is free to rotate on bearings in close proximity to the impeller; but it is in no way physically connected to the impeller. Moreover, a set of splines connect the turbine to the transmission's input shaft. The **input shaft** transmits engine torque and power into the transmission whenever the turbine is set in motion.

Notice in Fig. 4-8 how the blades are set into the two hollowed-out halves. The blades or vanes in each half are positioned radially around the inside section of each half. The placement and number of blades determine the coupling's efficiency under operating conditions.

Note also in Fig. 4-8 how both the impeller and turbine blades are **flat** from their leading to tailing edges. The **leading edge of a turbine blade** is the initial contact point of the fluid as it enters the unit near the rim; whereas, the **leading edge of an impeller blade** is the beginning point of the vane at the center of the impeller, where its metal skin covering begins to curve around the blade.

The **tailing edge of the turbine blade** is its portion that is adjacent to or across from the leading edge of the impeller vane; whereas, the **tailing edge of an impeller blade** is the section that is directly across from the leading edge of the turbine blade.

In regard to flow, fluid always leaves a given blade from its tailing edge. Fluid enters either the turbine or impeller by first striking the leading edges of their blades. This information will be helpful during the upcoming discussion of fluid flow patterns within the coupling.

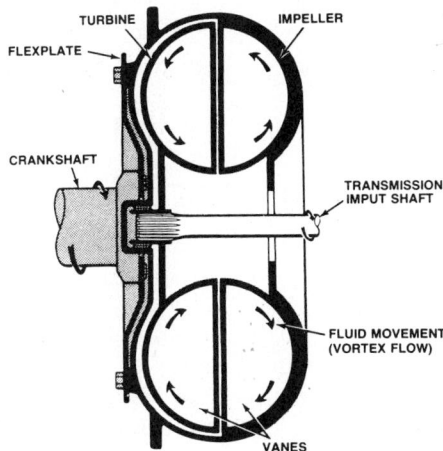

FIGURE 4-7   The design of a fluid coupling. (Courtesy of Chrysler Corp.)

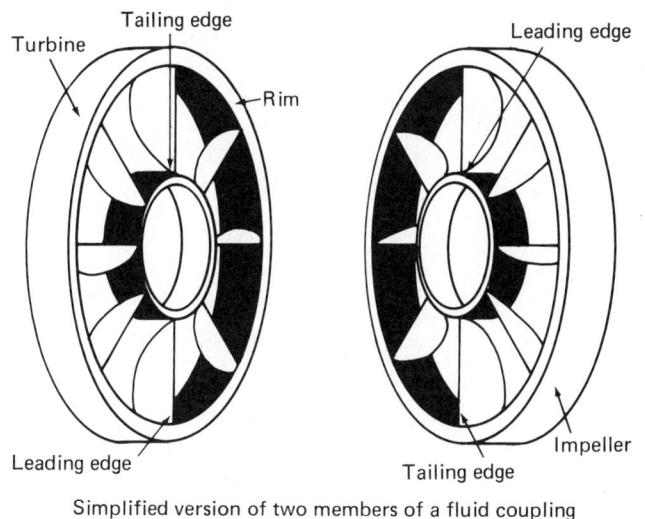

Simplified version of two members of a fluid coupling

FIGURE 4-8   The blades or vanes of a coupling are flat and are set radially into the impeller and turbine.

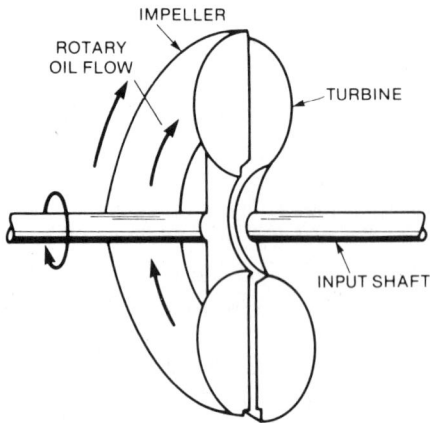

FIGURE 4-9 Fluid trapped between the spinning impeller blades causes rotary flow.

FLUID MOTIONS PRODUCED INSIDE A FLUID COUPLING

Before attempting to explain the operating phases of a coupling, it is an ideal time to point out the two types of fluid motions produced by the impeller and turbine as they turn. These fluid motions are responsible for the transfer of torque and power within both the fluid coupling and the torque converter.

Rotary Flow

The first type of flow produced by the impeller as it turns at engine speed is rotary. **Rotary flow** is the result of the impeller blades pushing on the trapped fluid within the metal housing (Fig. 4-9). The blades force the fluid to follow the clockwise direction of the impeller as it spins. This action creates a vertical, spinning mass of fluid, a vertical fluid whirlpool.

Vortex Flow

The second fluid motion is **vortex,** which is produced by centrifugal force acting upon the fluid trapped between the blades of the impeller. **Centrifugal force** is a force that tends to impel the fluid outward from the center of rotation. In this case, the fluid moves from the center of the impeller toward the rim. The exact same thing happens to a marble placed at the center of a spinning phonograph turntable. But since the rim of the impeller has a curvature, the fluid leaves it at a right angle to the spinning fluid mass produced by rotary motion (Fig. 4-10). This vortex flow is responsible for moving the fluid into the turbine, and the resulting impact energy contained in the fluid transmits torque and sets the turbine and input shaft in motion.

Fluid Exit Angle

The relative strengths of the rotary and vortex fluid motions determine the actual exit angle of the oil as it leaves the impeller's rim and enters the turbine (Fig. 4-11). As previously mentioned, vortex flow is at right angles to

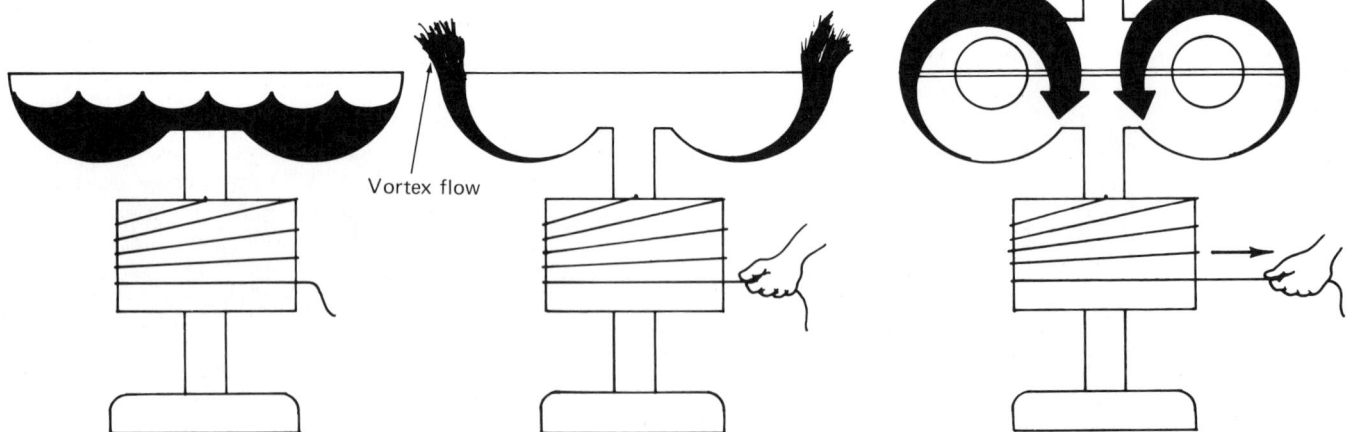

(a) Fluid lies level in stationary impeller

(b) Impeller is spun and fluid spills out of its rim due to centrifugal force

(c) Fluid passes into turbine where it transmits torque to its blades

FIGURE 4-10 Centrifugal force, acting on the fluid trapped between the rotating impeller blades, produces vortex flow.

that of rotary. But rotary flow deflects the vortex flow from the perpendicular to form the actual exit angle.

If, for example, rotary and vortex flows are weak due to slow impeller rotation, the combined motions produce an exit angle that causes the fluid to strike the turbine blades with only a glancing blow (Fig. 4-12). In other words, the entering fluid tends to enter the turbine between the blades instead of striking them straight on where its impact energy could be expanded in a useful manner in attempting to drive the turbine.

This condition occurs within the coupling whenever the engine is idling. During this time, the turbine blades do not absorb much energy from the moving fluid; consequently, the input shaft cannot deliver power into the transmission and the vehicle cannot move.

As engine speed increases, the turbine will begin to turn, and the fluid's exit angle from the impeller gradually changes until it reaches its most efficient point as the

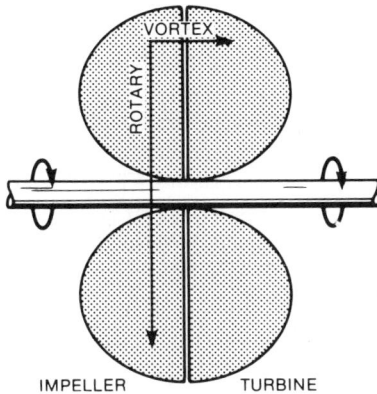

FIGURE 4-13   As the turbine begins to rotate, it begins to produce rotary and vortex flow.

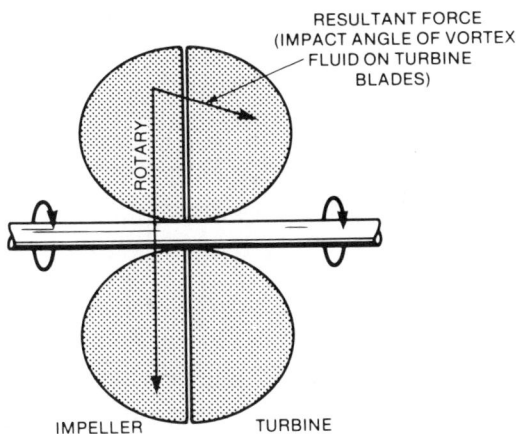

FIGURE 4-11   The normal vortex flow is at a right angle to that of rotary flow.

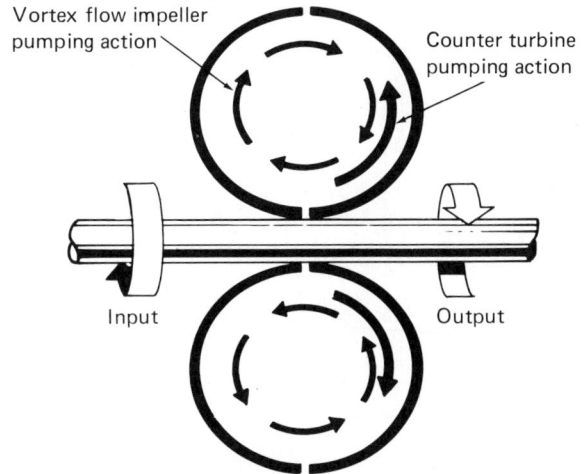

FIGURE 4-12   Vortex flow exit angle with the engine at idle speed.

turbine approaches impeller speed. With an increase in impeller speed as the engine accelerates but low turbine rpm, rotary and vortex flows are high. However, the exit angle they produce is about the same as the one shown in Fig. 4-12. In other words, an increase in both flows from the impeller has little effect on the exit angle because they both grow in intensity at the same rate.

But as the turbine starts to move, the impeller's vortex flow begins to slow down and its exit angle begins to change. This action results from the counterpumping action of the turbine; the turbine itself begins to produce rotary and vortex flows (Fig. 4-13). Both rotary flows are in the same direction so the rotary flow from the turbine just combines with that from the impeller. However, the vortex flow from the turbine is in an opposite direction to that from the impeller. Thus, the weaker turbine vortex flow slows the one from the impeller by bucking its free movement.

This action continues as the turbine picks up more speed until it approaches the rpm of the impeller. All during this period, the impeller's exit angle changes even more from the perpendicular to form a more effective impact angle on the turbine blades until it reaches a point similar to the one shown in Fig. 4-14. Now the fluid leaving the impeller has the best angle to impact the turbine blades in a more direct manner. Consequently, torque and power transfer is very efficient at this time.

FLUID COUPLING OPERATION

With these facts on fluid motion in mind, let's examine what occurs within the coupling during engine idle, acceleration, vehicle cruise, and deceleration. Just

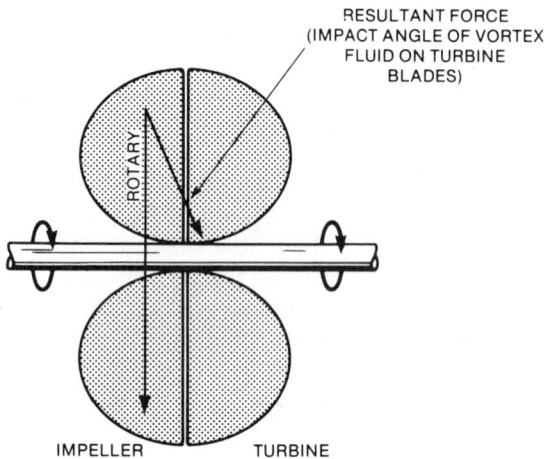

FIGURE 4-14 The impeller's vortex flow angle is very effective when turbine speed is nearly the same as that of the impeller.

remember that, although the text describes the fluid flow from one impeller blade to a turbine blade and back again, the entire process encompasses all the bladed sections of both units simultaneously.

Engine Idle

With the engine idling with the transmission in gear, the coupling does permit some torque transfer between the engine and transmission's input shaft, but no power. This 100 percent slippage in power between the two units is the result of several factors. First, the combined fluid motions produce a poor exit angle for impeller fluid entering the turbine. Second, since the impeller is rotating at engine idle, the fluid velocity it produces is insufficient to provide the impact energy necessary to overcome the load on the turbine, imposed by the vehicle's weight and its tire-to-road friction, on the transmission's input shaft. Although the input shaft does receive some torque transfer at idle, it does not move; so there is no introduction of power as such into the transmission.

However, there is a definite fluid circulation within the coupling at this time (Fig. 4-15). The vortex flow from the impeller leaves the tailing edge of a blade and enters the leading edge of an adjacent turbine blade. From here, the fluid passes along the turbine blade to its tailing edge, where it exits across to the leading edge of an impeller blade. In other words, the fluid from the impeller just circulates through the turbine and then returns to the center of the impeller for recycling.

Engine and Vehicle Acceleration

As the driver accelerates the engine with the transmission in gear, torque and power begin to pass between the impeller and turbine and the input shaft and vehicle start to move. When the impeller accelerates with the engine, its blades begin to pump larger and larger quantities of fluid into the turbine. As the point is reached when the combined impact energy of this fluid overcomes the load on the turbine and input shaft, they both begin to rotate, very slowly at first.

The reasons for the slow acceleration of the turbine is twofold. First, although there is a large volume of vortex flow entering the turbine, the angle it strikes the blades is not very efficient to impart much of the energy contained

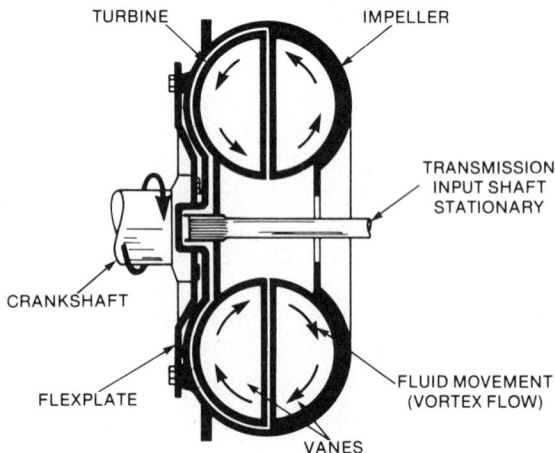

FIGURE 4-15 Fluid circulation within the coupling during engine idle.

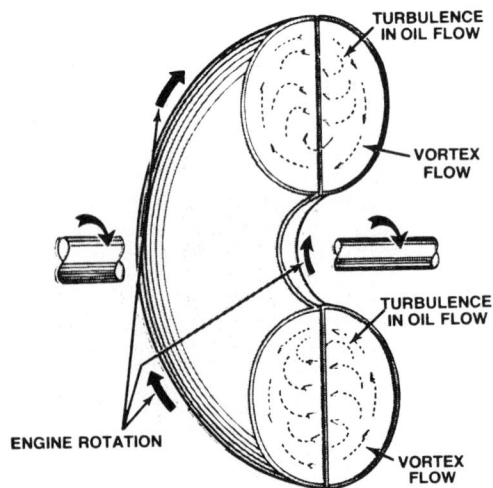

FIGURE 4-16 Without a guide ring, there will be fluid turbulence in the center of the unit. (Courtesy of Chrysler Corp.)

in the moving fluid to the vanes. Second, since the fluid's velocity during acceleration is very high, it strikes the leading edge of the flat turbine blade surface and the area between the blades and bounces back toward the impeller (Fig. 4-16).

Fluid bounce back within the coupling is a negative characteristic for two reasons. First, the bounce back of fluid causes it to swirl about in all directions—particularly in the center sections of both units. This swirling action or turbulence creates interference with the normal fluid circulation between the impeller and turbine and back again. Second, the fluid bouncing off the flat surface of a turbine blade not only represents a loss of impact energy but the deflecting fluid tends to return into the impeller, where it tries to slow its rotation. For these reasons, the fluid coupling is very ineffective in transferring torque and power whenever there is a large difference in speed between the impeller and turbine.

Vehicle Cruise

When a vehicle approaches low cruising speed, which may be between 25 and 40 miles per hour, the turbine approximates the rpm of the impeller. At this point, the coupling enters its most effective period, known as the coupling phase. At this time, the impeller's vortex flow is now low because both members pump fluid toward their outer diameters (Fig. 4-17). Therefore, each member produces a pressurized wall of fluid, which resists any shearing action. Only enough fluid leaves the tailing edge of an impeller blade and enters the leading edge of a turbine blade to maintain fluid circulation. This permits the fluid coupling to continue transmitting torque and power into the transmission from the engine with very little loss.

However, when the coupling is transmitting torque and power, one member (even during the coupling phase) must turn faster than the other. In other words, even during the coupling phase, about 10 percent slippage must exist between the impeller and turbine. This is necessary so the faster rotating member (usually the impeller) can maintain the circulation of fluid between the two members to keep all the bladed areas of both units full of fluid and provide some impact effect on the vanes. If both members turn at the same speed, circulation of fluid ceases and so will the transfer of torque and power within the coupling.

Vehicle Deceleration

The fluid coupling is also able to keep the engine connected to the transmission when a vehicle is decelerating with the engine operating at low speed or idle. Under these conditions, the vehicle begins to drive the engine through the drive train and fluid coupling. As a result, the engine, due to its compression, exerts a braking effect on the vehicle.

Within the coupling itself, the turbine now acts as the impeller, or pump, to create the fluid flow. Therefore, fluid passes from the turbine to the impeller, imposes a push on its blades, and returns to the turbine. This action maintains the vortex fluid flow necessary to maintain powerflow through the coupling. The coupling will continue this process until the engine accelerates the impeller until it overspeeds the turbine.

TORQUE CONVERTER DESIGN

Since its initial introduction, there have been a number of different torque converter designs in regard to the number and shape of internal members. **Internal members,** in this case, are simply the number of main independent parts within the converter, such as the impeller, turbine, and stator. Although some converters at one time had as many as five members, the most common type today has only three: an impeller, a turbine, and a stator (Fig. 4-18).

Although the impeller and turbine within the converter serve the same functions as those in a fluid coupling, their design is somewhat different. The design changes are mainly in the number of blades each of these members has and vane curvature. Both changes increase the overall efficiency of the converter when it acts as a fluid coupling and during its torque multiplication phase. The stator is only necessary to converter efficiency when it multiplies engine torque.

Impeller

A typical impeller of a torque converter is shown in Fig. 4-19. This unit is welded into the rear section of the converter housing and has a series of curved blades and a

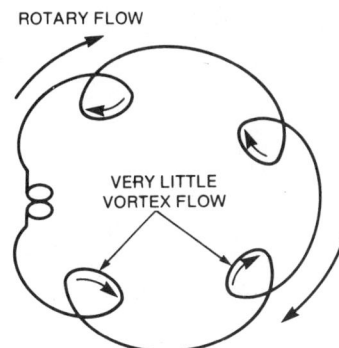

ROTARY FLOW

VERY LITTLE
VORTEX FLOW

FIGURE 4-17   *Vortex flow between the impeller and turbine is low during the coupling phase.*

split guide ring. All the blades curve backwards in respect to the normal clockwise rotation of the impeller. This design gives additional acceleration and energy to the fluid before it initially leaves the impeller's rim and enters the turbine. A flat blade also does this but not as effectively as the curved vane.

The guide ring is nothing more than what appears to be a second but smaller hollowed-out donut (Fig. 4-20). Half of the ring fits into the impeller blades near their center. The other half sits inside the turbine in about the same location.

The purpose of both halves of the guide ring is to

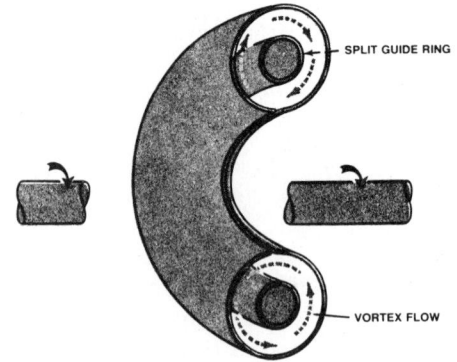

FIGURE 4-20   The guide ring is necessary to reduce fluid turbulence.

reduce oil turbulence at the center of the impeller and turbine. The ring accomplishes this task by channeling the flow of fluid in a circular pattern as it moves between the two members. Fluid turbulence, remember, was one of the factors that made the fluid coupling very ineffective during vehicle acceleration (see Fig. 4-16).

Turbine

The design of a converter turbine is similar to one found in a coupling. That is, the converter turbine turns in close proximity to the impeller, splines to the input shaft, absorbs the kinetic energy from the moving fluid, and converts this energy to rotary motion (Fig. 4-21). Also,

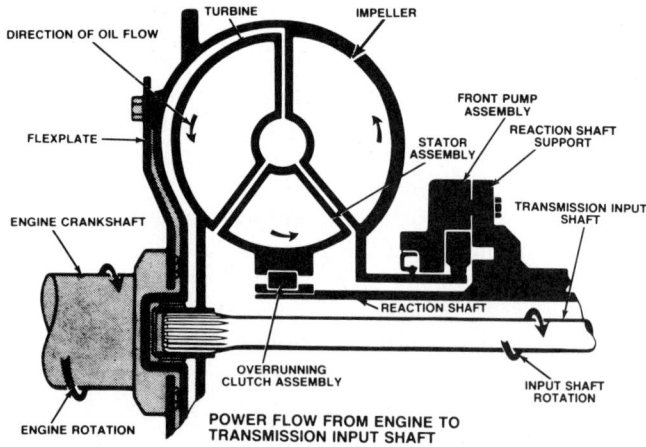

FIGURE 4-18   The three-member torque converter consists of an impeller, turbine, and stator. (Courtesy of Chrysler Corp.)

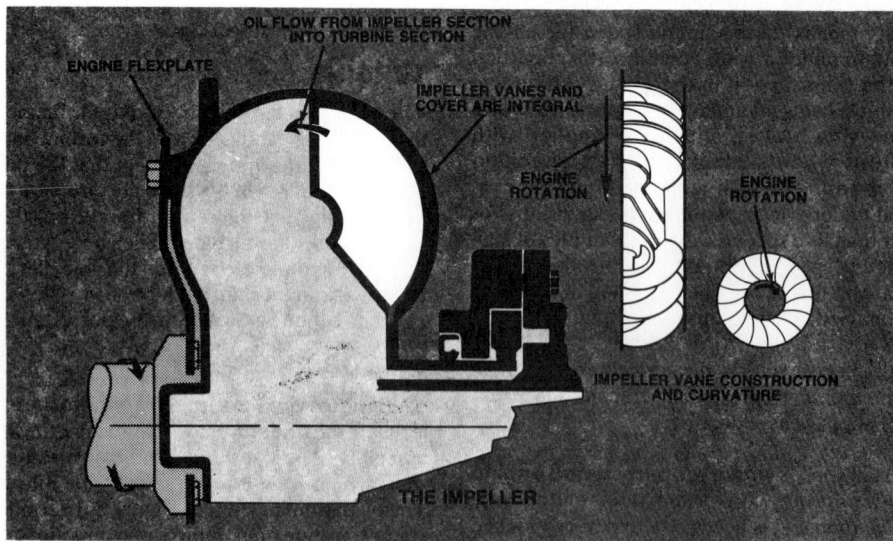

FIGURE 4-19   The impeller blades have a slight backward curvature to the normal clockwise rotation of the unit. (Courtesy of Chrysler Corp.)

the turbine has a given number of blades, or vanes, set radially around inside the housing.

However, to make the converter turbine more efficient, the bladed section has a number of design changes. For instance, instead of the turbine having the same number of blades as the impeller, as found in a fluid coupling, the converter has several more vanes in its turbine. These additional blades increase the turbine's efficiency.

Also, instead of being flat, the turbine blades have curvature that reduces shock losses due to sudden changes in fluid direction between the impeller and itself (Fig. 4-22). But at the same time, the curvature must allow

FIGURE 4-23   Moving fluid, if it strikes a flat blade surface, does not impart much of its energy to the vane.

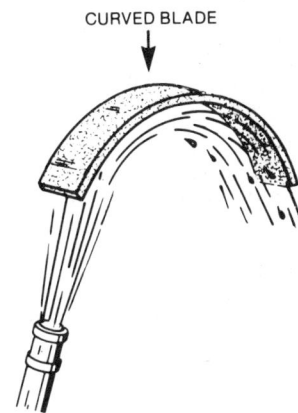

FIGURE 4-24   A curved blade absorbs a great deal of energy from the moving fluid.

FIGURE 4-21   The turbine and input shaft of a typical torque converter.

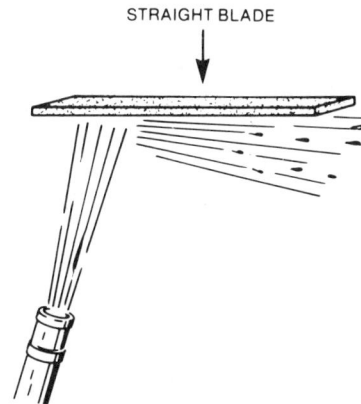

FIGURE 4-22   Turbine blade curvature is such that it reduces shock losses and absorbs energy from the moving fluid. (Courtesy of Chrysler Corp.)

the blades to absorb as much energy as possible from the fluid as it passes through the turbine.

The curvature of the leading edge of each turbine blade prevents shock losses as the fast moving fluid begins its travel through the unit. A flat blade, as used in the coupling, causes some of the incoming fluid to strike and bounce off the blade's flat surface (Fig. 4-23). This break-up in the otherwise smooth fluid flow results in a loss of fluid energy and interference with the normal rotation of the impeller. A curved blade has a leading edge designed to gradually change the direction of the fluid as it moves along its surface.

By curving the tailing half of each turbine blade in a direction opposite to impeller rotation (see Fig. 4-22), the blades absorb more fluid energy because this curvature changes the direction of the fluid flow. As previously mentioned, kinetic energy is a measurement of the amount of work required to slow or stop a moving object, the fluid. By completely changing the fluid's direction, the curved vane slows its velocity and, therefore, receives

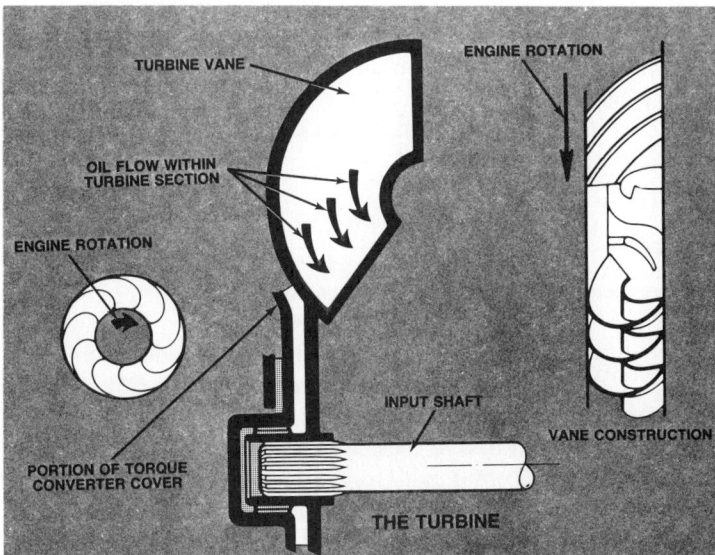

more impact force from the moving oil than it would if the blade were straight (Fig. 4-24).

To assist in keeping the fluid on track as it follows the blade, half of the guide ring is set into the midsection of all the turbine vanes (see Fig. 4-22). This ring, as mentioned, reduces shock losses and turbulence in the center section of the turbine by keeping the fluid traveling in a circular direction as it moves through the turbine.

Stator

The **stator** is the third member within the torque converter and fits between the turbine and impeller (Fig. 4-25). In this position within the converter, the vanes mounted on the stator can control the flow of returning fluid from the turbine's center to the middle of the impeller.

The control of the returning fluid is important for two reasons. First, by curving the turbine blades in such a manner as to reverse the fluid flow as it leaves this unit, it would reenter the center of the impeller in a direction that would slow its rotation (Fig. 4-26). The vanes of the stator overcome this problem by changing the fluid's direction back to that produced by the impeller. Second, by redirecting the fluid back into the impeller, the stator plays an important role during the converter's torque multiplication phase.

A stator assembly, complete with its vanes and overrunning clutch, is shown in Fig. 4-27. Note that the vanes or blades are straight and wide from the center of the

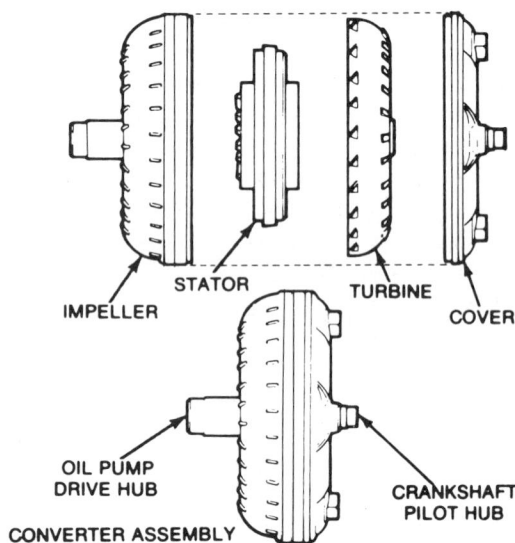

FIGURE 4-26 Due to blade curvature, the fluid leaving the center of the turbine enters the impeller in a direction that would normally hinder its rotation. (Courtesy of Ford Motor Co. of Canada Ltd.)

stator to its outside circumference. The vanes also curve quite sharply from the front of the stator to the back, where the blades are narrow. This vane design is responsible for redirecting the returning fluid from the turbine to the impeller into a helping direction.

The overrunning clutch is constructed into the center section of the stator. The **overrunning**, or **one-way clutch**, is built into the stator to permit the bladed outer portion of the unit to rotate (overrun) in a clockwise direction, but the same clutch stops any counterclockwise movement. The construction and operation of this form of clutch is covered in the next chapter.

TORQUE CONVERTER OPERATION

When functioning, the torque converter provides four separate operating phases, which include the maximum slippage, torque multiplication, and coupling. Just keep in mind that each one of these four phases occurs within the converter automatically, one after another, without the driver doing any more than stepping down releasing pressure on the accelerator pedal with the transmission in gear. The only obvious operational change that occurs while a converter passes through the four phases is the overspeeding of the engine when the converter slips during engine idle and acceleration.

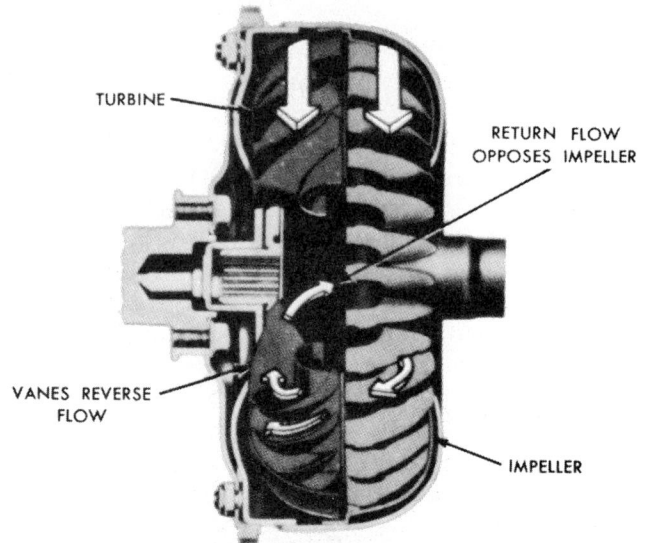

FIGURE 4-25 The stator fits between the turbine and impeller. (Courtesy of Ford Motor Co. of Canada Ltd.)

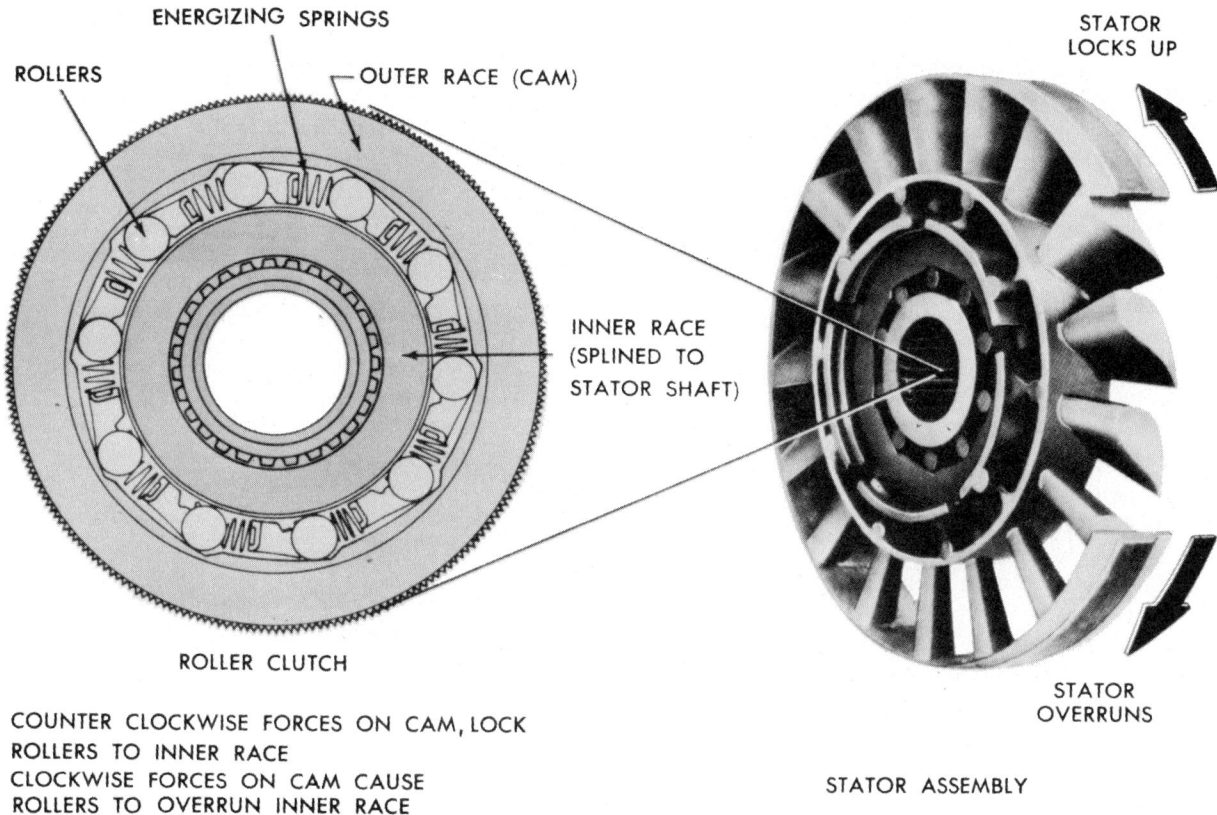

ENERGIZING SPRINGS

ROLLERS

OUTER RACE (CAM)

STATOR
LOCKS UP

INNER RACE
(SPLINED TO
STATOR SHAFT)

ROLLER CLUTCH

COUNTER CLOCKWISE FORCES ON CAM, LOCK
ROLLERS TO INNER RACE
CLOCKWISE FORCES ON CAM CAUSE
ROLLERS TO OVERRUN INNER RACE

STATOR
OVERRUNS

STATOR ASSEMBLY

**FIGURE 4-27   The design of a typical stator assembly.
(Courtesy of General Motors Corp.)**

Maximum Slippage

In a similar manner to the fluid coupling, a torque converter provides maximum slippage, or a 100 percent break in power, when the engine is idling with the transmission in gear. Although there is some torque transfer between the impeller and turbine, it is insufficient to overcome the vehicle's load imposed on the input shaft. As a result, for all practical purposes, powerflow is broken between the engine and transmission. At this time, the impeller will be turning at about 500 to 650 rpm, but the turbine will be stationary (0 rpm).

However, since the impeller is turning, it does create some vortex and rotary flow within the converter (Fig. 4-28). Due to slow impeller speed, the small amount of vortex flow passes from the impeller blades to those of the turbine, where it pushes on the vanes and then follows their curvature and exits from the center of the unit. From there, the fluid flows through the stator vanes and reenters the midsection of the impeller. Since there are only small amounts of vortex flow entering the turbine, the fluid's impact force is insufficient to cause turbine and input shaft rotation. Thus, the converter provides 100 percent slippage at this time.

Torque Multiplication

When the driver accelerates the engine to set the vehicle in motion, the torque multiplication phase of the converter begins. Since the impeller accelerates with the engine, it now starts to pump large quantities of fluid into the turbine. By virtue of its velocity and mass, the flowing fluid now possesses a great deal of kinetic energy.

After leaving the tailing edge of the impeller blades, the fluid enters the turbine vanes at their leading edges. Although the fluid's impact angle is not the most efficient at this point to impart the greatest amount of push on the blades, it is enough to set the turbine in motion and the vehicle begins to move.

As the fluid follows the curvature of the turbine blades, still imparting a push to the vanes, it exits from their tailing edges in a counterclockwise direction. Although this fluid still has some kinetic energy remaining, it represents a negative factor since the returning fluid is traveling in a direction that is opposite to the rotation of the impeller.

However, since the stator fits between the turbine and impeller, the returning fluid strikes the front face of the stator blades and attempts to turn them counterclockwise

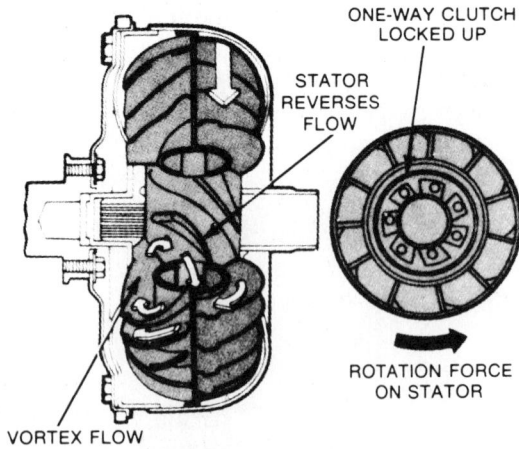

ONE-WAY CLUTCH
LOCKED UP

STATOR
REVERSES
FLOW

ROTATION FORCE
ON STATOR

VORTEX FLOW

FIGURE 4-28 Fluid circulation within the converters during maximum slippage at curb idle. (Courtesy of Ford Motor Co. of Canada Ltd.)

With the returning fluid now moving in the same direction as that being pumped by the impeller, the fluid applies its remaining energy to the impeller blades (Fig. 4-30). This action is the basis for the torque multiplication within the torque converter. Each time the total vortex flow passes from the impeller to the turbine, the moving fluid imparts a push on the turbine blades. The fluid will still have some kinetic energy remaining as it leaves the turbine on its way back to the impeller; and the stator reverses its direction before the fluid reenters the impeller.

Since the returning turbine fluid is flowing in the same

(Fig. 4-29). But the one-way clutch stops any stator movement in this direction. As a result, the curved stator vanes change the direction of the returning fluid before it reenters the impeller. The force required to change the fluid's direction is absorbed by the transmission case, thus there is very little loss of energy from the fluid as it exits the stator.

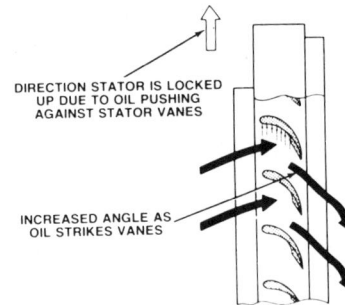

DIRECTION STATOR IS LOCKED
UP DUE TO OIL PUSHING
AGAINST STATOR VANES

INCREASED ANGLE AS
OIL STRIKES VANES

FIGURE 4-29 The stator's operation during the torque multiplication phase of the converter. (Courtesy of Chrysler Corp.)

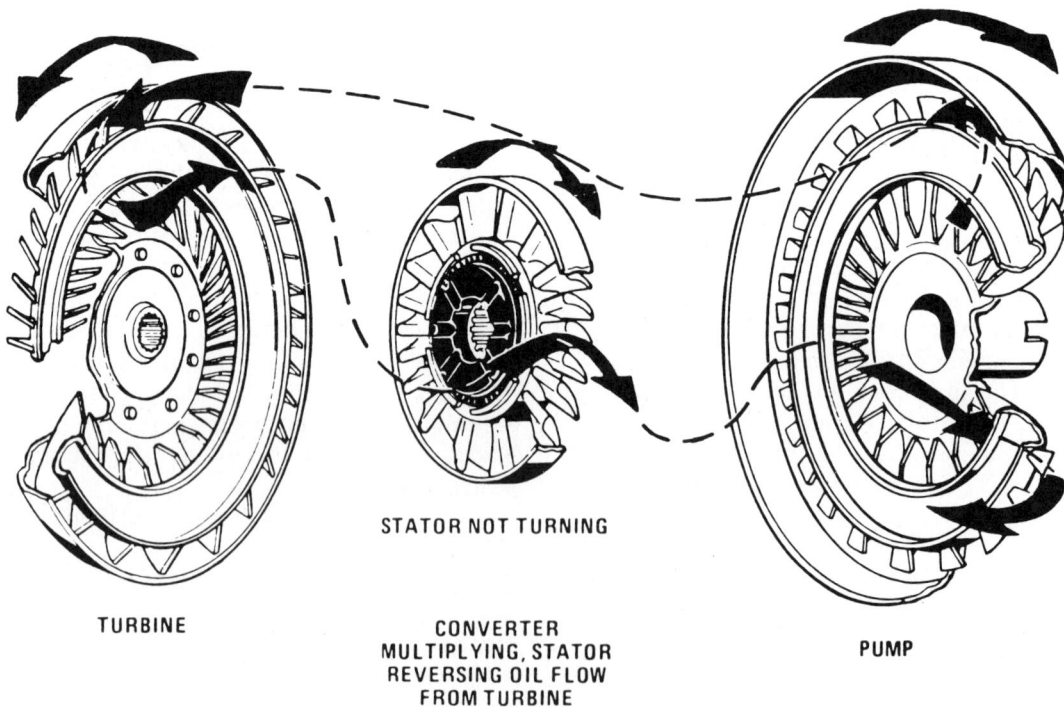

STATOR NOT TURNING

TURBINE

CONVERTER MULTIPLYING, STATOR REVERSING OIL FLOW FROM TURBINE

PUMP

FIGURE 4-30 Vortex flow circulation during a converter's torque multiplication phase. (Courtesy of General Motors Corp.)

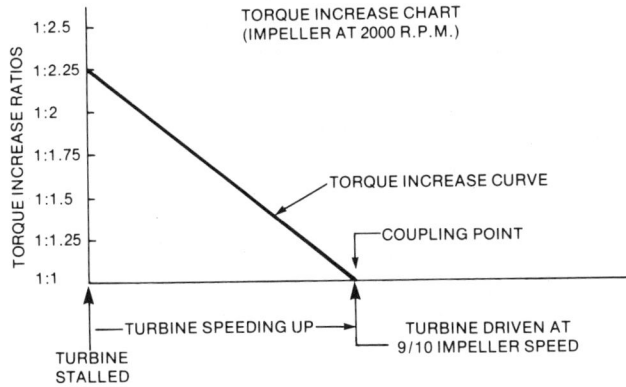

FIGURE 4-31 A chart showing a typical converter's torque mutliplication curve.

direction as impeller flow and the stator vanes have directed its movement against the impeller blades at an efficient force-producing angle, the recirculating oil no longer hinders impeller and engine rotation. Instead, the returning fluid assists the engine in driving the impeller. This action increases engine torque, which, in turn, makes it easier for the engine driven impeller to accelerate the fluid.

The end result is the turbine blades receive fluid flow possessing a greater velocity than it would without the effect of the stator. The total turbine torque produced by the impact energy of the accelerated fluid is therefore equal to energy remaining in the recycled fluid from the stator plus that produced by the impeller. In other words, the total torque producing force of the fluid flow into the turbine (**C**) is equal to impact force still remaining in the fluid returning from the turbine through the stator (**A**) plus the fluid accelerated by the impeller (**B**), or

$$C = A + B$$

Consequently, when the turbine is rotating at very slow speeds or even stopped with the impeller at high speeds, the torque on the input shaft attached to the turbine may be several times the torque of the engine. Most passenger car torque converters, for example, provide a maximum torque increase of between 1:2 to about 1:2.5 at stall speed (Fig. 4-31). **Converter stall speed** is a condition in which the impeller rotates as fast as the engine will drive it at wide-open throttle. However, with the transmission in gear and the vehicle's service brakes applied, the turbine is held stationary. Under these conditions, the converter absorbs a given amount of torque or reaches its capacity; the rpm of the engine stabilizes; and the converter just slips.

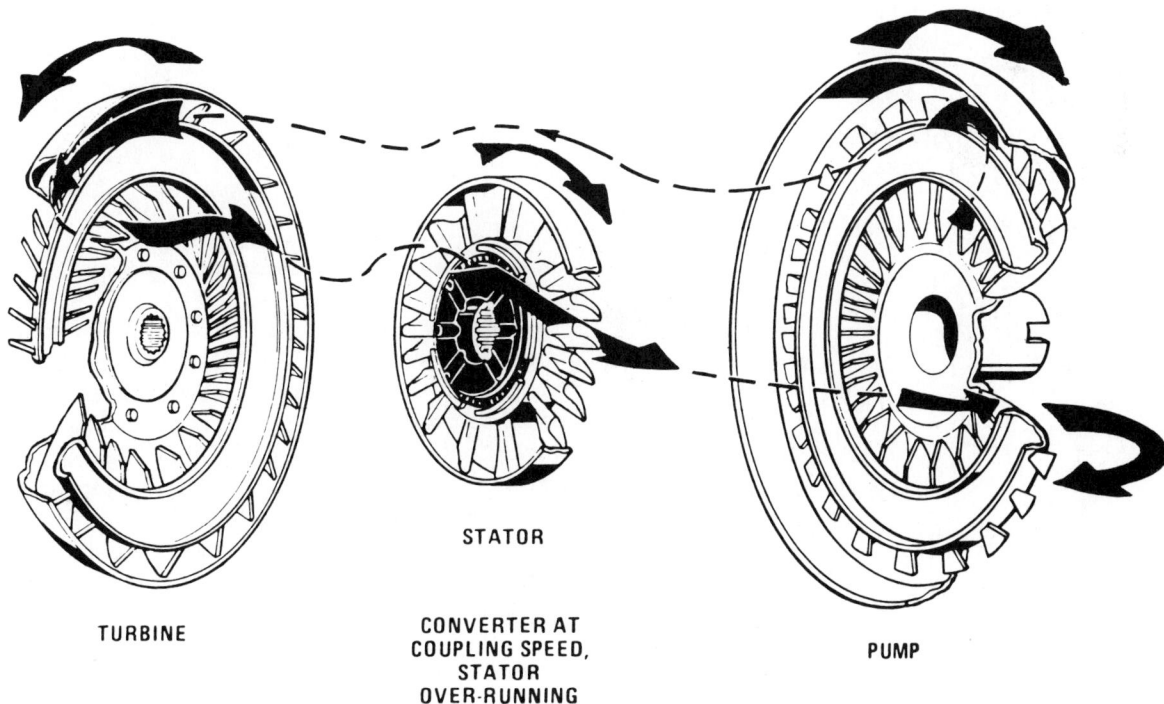

TURBINE   CONVERTER AT COUPLING SPEED, STATOR OVER-RUNNING   PUMP

FIGURE 4-32 Fluid flow within the converter during its coupling phase. (Courtesy of General Motors Corp.)

Coupling Phase

Under moderate acceleration, the torque converter enters its coupling phase as the vehicle approaches road speeds between 25 and 40 mph. At this time, two actions are occurring within the converter. First, the turbine has approached but not reached impeller speed (Fig. 4-31). Second, the stator begins to overrun because it is no longer needed due to the smooth but gradual decrease in torque multiplication by the converter.

The turbine during the coupling phase operates at about 9/10 of impeller speed, or 90 percent. This speed difference is necessary so that the impeller can continue to pump fluid into the turbine in order to keep all the bladed sections full of fluid so there can be torque and power transfer between the two units. If the turbine and impeller operate at the same speed, powerflow is broken between the engine and transmission.

During the coupling phase, rotary flow is high, but vortex flow is low. The high rotary flow is the result of high impeller and turbine speed. The reduction in vortex flow is caused by the counterpumping action by the turbine, which creates a second vortex flow in an opposite direction to that from the impeller. This creates a resistance to the normal vortex flow from the impeller. However, there is still some vortex flow, as shown in Fig. 4-32.

The stator must overrun (turn in the same direction as the turbine and impeller) during the coupling phase. If it does not, its blades or vanes interfere with the normal fluid flow between the turbine and impeller. This creates a drag within the converter, which slows everything down.

The stator rotates freely clockwise, at this time, due to the action of its one-way clutch. The clutch begins to free-wheel because fluid now impacts against the back faces of the stator blades (Fig. 4-33). Since during this phase of

converter operation there is very little vortex flow but a great deal of rotary flow, the small amount of returning turbine-to-impeller fluid assumes a nearly straight-through flow between the stator blades. In other words, the returning fluid tries to turn the stator blades clockwise instead of counterclockwise as it did during the torque multiplication phase. This action along with the strong rotary flow from both units creates a turning fluid mass that carries the stator clockwise and resists any shearing action.

Deceleration

In a similar way as the fluid coupling, the converter transfers torque during vehicle deceleration so the engine can act as a braking medium. During this situation, the stator also overruns and the turbine acts as the pump to force fluid flow into the impeller.

TORQUE CONVERTER COOLING

During its torque multiplication phase, the converter produces a great deal of heat. This is the result of the shearing action between the fluid molecules as the converter slips. In other words, considerable heat is generated in the fluid as a direct result of friction.

If the temperature of the **fluid** remains high enough, the oil begins to oxidize rapidly. Consequently, some form of cooling system is necessary to lower **and** maintain the temperature of the fluid below its oxidation point. Manufacturers use two basic types of cooling systems for this purpose—air and coolant. Both of these were discussed in detail in Chapter 3.

CONVERTER SIZE AND CAPACITY

The overall size of the converter and its stator blade angles are design features that determine the unit's capacity (its rate of torque absorption), stall speed, and coupling point. Small converters with steep stator blade angles reduce the volume of fluid flow between the impeller and turbine; consequently, the impeller operates at a faster rpm before its stall speed occurs. Furthermore, small, low capacity converters have a coupling point that occurs at a higher rpm and produces higher torque increase ratios than larger units (Fig. 4-34).

Converters of larger size are low capacity units with shallow stator blade angles. This design provides the converter with a low stall speed and coupling point. But the unit produces sufficient torque multiplication for average vehicle acceleration under moderate loads.

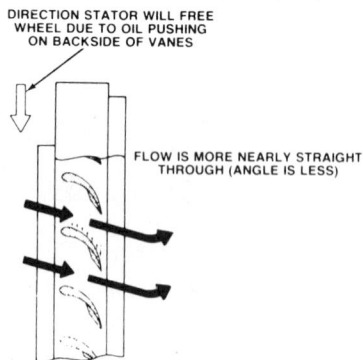

DIRECTION STATOR WILL FREE WHEEL DUE TO OIL PUSHING ON BACKSIDE OF VANES

FLOW IS MORE NEARLY STRAIGHT THROUGH (ANGLE IS LESS)

FIGURE 4-33 The stator's operation during the coupling phase of the converter. (Courtesy of Chrysler Corp.)

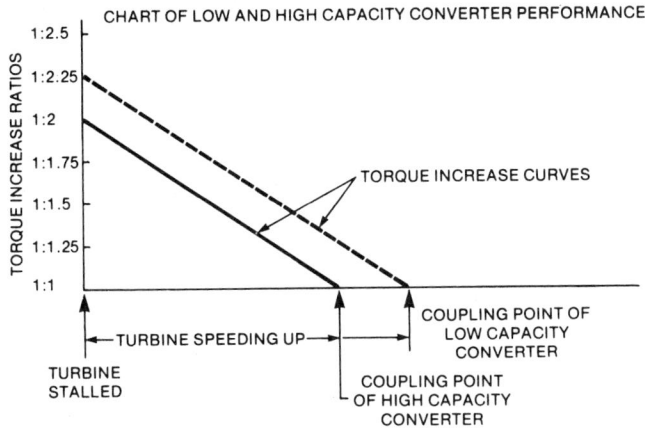

CHART OF LOW AND HIGH CAPACITY CONVERTER PERFORMANCE

FIGURE 4-34   A chart illustrating low and high capacity converter performance.

Converter manufacturers use either size or stator blade angle or a combination of the two to develop units that allow minimum creep at idle and provide a stall speed about 30 percent lower than the engine rpm producing maximum torque. If a converter reaches its stall speed at the same point where the engine reaches its maximum torque, the unit multiplies high torque values, which overheats the unit.

In addition, excessively high stall speeds, which are an inherent characteristic of low capacity converters, if not needed for vehicle performance, make the engine burn excess fuel and produce more noise. The high stall speed causes an elevated coupling point, which forces the engine to operate at increased rpm for longer periods; this reduces engine life. Consequently, converter designs are a compromise between the performance characteristics of both the low and high capacity units.

Vehicles with small engines utilize low capacity converters, while those with large engines use large, high capacity units. This mates the unit's stall and coupling speeds with the size of the engine to fully take advantage of this torque and speed range. For example, smaller four- and six-cylinder engines usually use relatively low capacity converters. This unit permits these engines to operate at near their maximum torque output rpm, which is usually at a much higher speed than large V-8 engines, before the converter stalls. Also, this converter design permits a relatively high coupling point. Finally, a low capacity converter like this one absorbs engine torque with greater slippage, while it provides more fluid energy for torque multiplication.

A large engine requires the high capacity converter. This unit provides a smaller overall torque multiplication curve and low speed coupling point. Even under a heavy acceleration, the torque multiplication curve does not stretch itself as far as a low capacity unit and coupling may occur at 45 mph.

In many cases, the vehicle manufacturer installs the same transmission in several identical automobiles with different engine sizes. To mate the converter with the engine, the manufacturer changes either its size, stator blade angle, or a combination of the two. Therefore, it is important that the mechanic does not alter the converter because it has design characteristics required by the engine. If the converter malfunctions, replace it with a unit that meets all the factory specifications.

CONVERTERS WITH VARIABLE-PITCH STATORS

Design

A method used for a number of years by manufacturers to provide a vehicle with the benefits of both low and high capacity converters incorporates a variable-pitch stator (Fig. 4-35). The converter uses a hydraulic reaction piston and control system that moves the stator (reaction) blades from the high to the low angle position. An electrical switch, usually connected to the throttle linkage, signals the hydraulic system when it is time to activate the piston, which, in turn, changes the blade angle.

Operation

When there is a large difference in speed between the impeller and turbine, the stator blades move to the high angle position (Fig. 4-36). This action produces a low

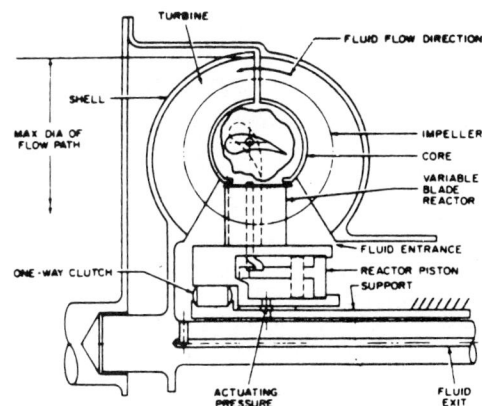

FIGURE 4-35   A converter with a variable-pitch stator. The reaction piston changes the angle of the stator (reaction) blades. (Reprinted with permission from the SAE Handbook, © 1978 Society of Automotive Engineers, Inc.)

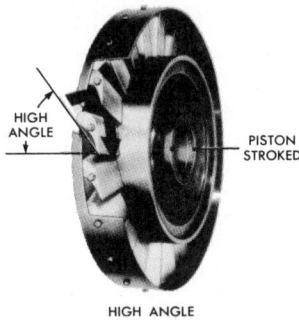

FIGURE 4-36   The variable stator blades in the high angle position. (Courtesy of Hydro-matic Division of General Motors Corp.)

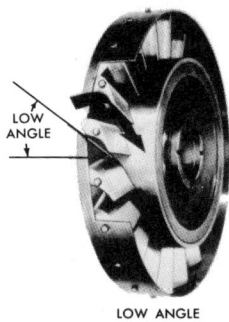

FIGURE 4-37   Variable stator blades in the low angle position. (Courtesy of Hydro-matic Division of General Motors Corp.)

capacity converter with considerable fluid flow restriction. As a result, the converter has a high stall speed and torque multiplication.

When the converter approaches the coupling point under normal acceleration, the stator blades move to the low angle position (Fig. 4-37). This action produces a high capacity converter with less fluid flow restriction. The converter remains in the low angle position even when the vehicle decelerates below cruising speed and stays there as long as the throttle valves in the carburetor are not opened past the 40 degree point. This minimizes creep if the vehicle slows to curb idle.

Advantages of Using Variable-Pitch Converters

The following advantages are prominent in converters with variable-pitch stators:

1. With the stator blades in the low angle position, high capacity condition, the converter allows little vehicle creep at curb idle and provides a very efficient coupling at cruising speeds.

2. With the stator blades in the high angle position, low capacity condition, engine speed increases for higher stall speed and torque multiplication.

3. The driver can extend the high angle position by keeping the accelerator pedal depressed. This action extends multiplication range of the torque converter to higher speeds and raises the converter's coupling point to as high as 90 mph (150km/h). The increased coupling point is very useful when a vehicle is pulling heavy loads.

4. The control system easily and smoothly changes the converter from low to high capacity by moving the piston within the unit itself.

Although this converter type provided the above-mentioned advantages, it has not been used for many years. General Motors employed this concept extensively for many years but discontinued the converter after its 1967 production year.

CONVERTER INSTALLATION

As previously stated, the engine turns the converter by means of the flex drive plate, which is bolted to the crankshaft; manufacturers employ several methods to support, center, and drive the unit. For instance, a bushing, a special type of round bearing pressed into the pump housing, supports and centers the hub attached to the rear side of the converter (Fig. 4-38).

The short hub on the front side of the converter (or special studs) center and support the front portion of the converter. One converter type uses a hub that fits snugly into a specially machined counterbore in the back of the crankshaft (Fig. 4-39).

Another converter style uses drive studs that have specially machined shoulders that mate with precision

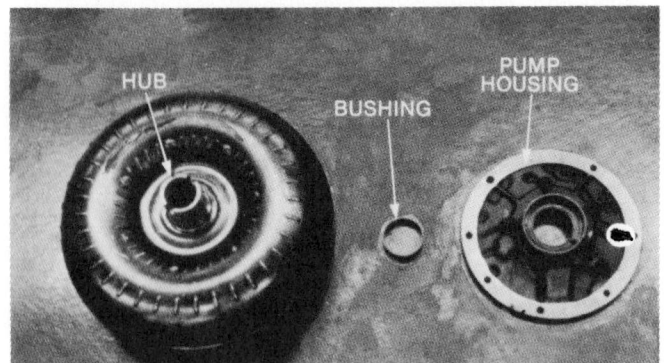

FIGURE 4-38   The converter's hub, its bushing, and the pump housing.

FIGURE 4-39 The front hub of many converters is supported in a counterbore located in the end of the crankshaft.

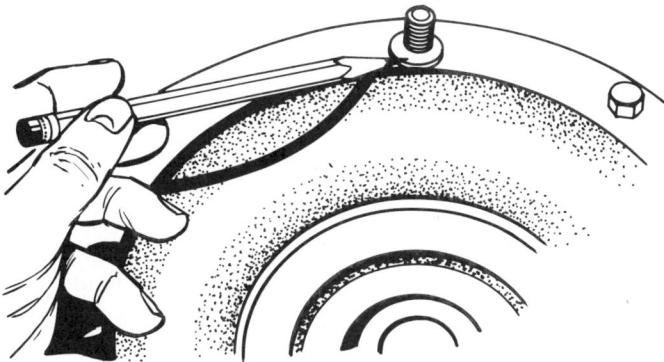

FIGURE 4-40 Converter drive studs support the front end of some converters.

drilled holes in the flex plate. The studs not only drive the converter through the flex plate but center and support it as well (Fig. 4-40).

Converters that center on the crankshaft attach to the flex plate by bolts or capcrews. These pass through the flex plate and thread into nuts or nutplates, fastened to the converter.

LOCK-UP CONVERTERS

Function

The lock-up torque converter is not new to the field of transportation. This device has been used for a number of years in truck and other industrial applications. But due to the fact that this converter costs more to produce, it has not until recently been used in passenger cars. However, with the rise in the cost of gasoline and the increased problems in obtaining good fuel economy, the lock-up converter pays for itself in a short period of time.

The **lock-up converter** (Fig. 4-41) has a turbine that operates at impeller or housing speed when the unit is in its coupling phase; due to this feature, this converter provides three benefits over the conventional type. First, with lock-up converter, the vehicle achieves better fuel economy because during the coupling phase the turbine **no longer** rotates at 9/10 impeller speed (see Fig. 4-31). Remember, with the conventional converter, about 10 percent slippage must exist between the turbine and impeller during the coupling phase in order for torque and power to transfer between the two units. This loss or

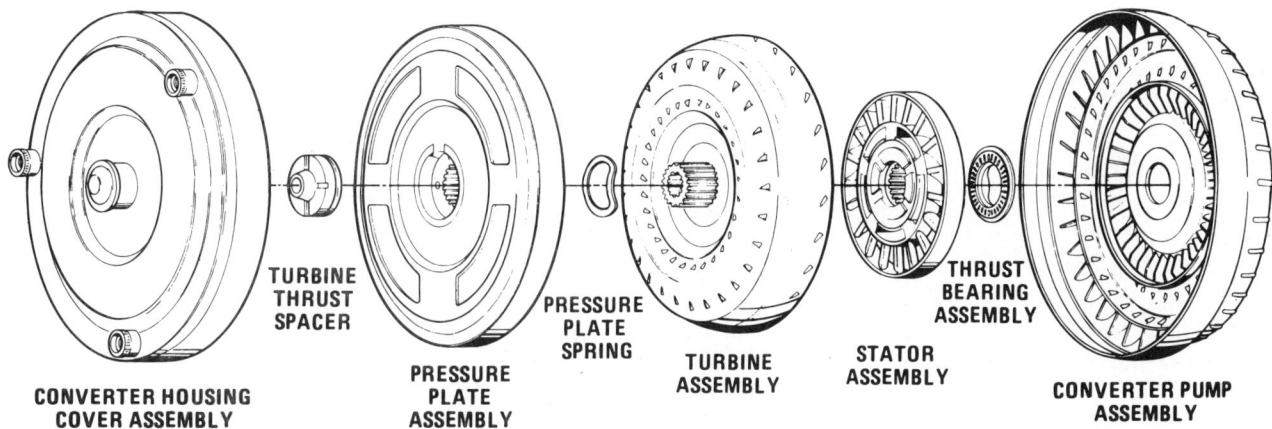

TURBINE THRUST SPACER

PRESSURE PLATE SPRING

THRUST BEARING ASSEMBLY

STATOR ASSEMBLY

CONVERTER HOUSING COVER ASSEMBLY

PRESSURE PLATE ASSEMBLY

TURBINE ASSEMBLY

CONVERTER PUMP ASSEMBLY

FIGURE 4-41 A typical lock-up torque converter. (Courtesy of General Motors Corp.)

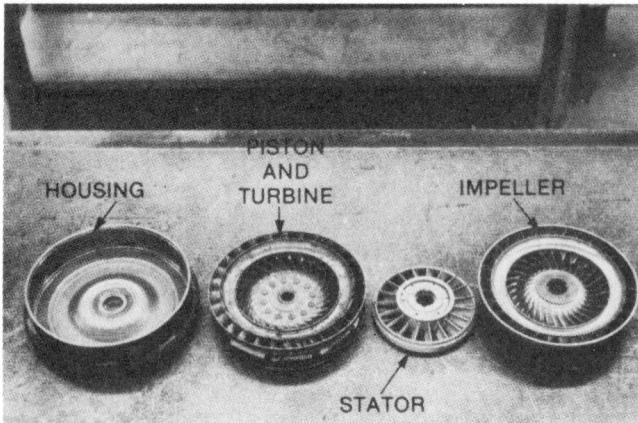

FIGURE 4-42   A Chrysler or American Motors lock-up converter consists of a housing, piston and turbine, stator, along with the impeller.

FIGURE 4-43   The converter housing has a friction lining bonded inside it.

FIGURE 4-44   The movable piston and turbine of a lock-up converter.

slippage reduces the fuel economy of the vehicle by about 4 percent in city driving and 6 percent on the highway.

Second, the lock-up converter lowers the transmission fluid's operating temperature somewhat during the coupling phase. In the conventional converter, fluid continues to circulate during this time in order to maintain torque and power transfer and any oil movement always increases its temperature.

With the lock-up converter, the turbine is driven by the housing by means of a clutch. Therefore, fluid circulation is no longer necessary during the coupling phase. As a result, the temperature of the fluid drops down.

Third, the lock-up converter reduces required engine rpm at vehicle cruising speeds. Not only does the lower engine rpm save fuel, but it allows the engine to operate with less noise, while at the same time, extends its life. Although these benefits seem small, in the long run they account for a significant savings for the owner of the vehicle.

Types

Basically there are two types of lock-up converters, based on the methods used to provide the unit with this feature. Several manufacturers, for example, use a hydraulically operated clutch to lock the turbine to the housing, while another company uses a centrifugal clutch for the same purposes.

CHRYSLER AND AMERICAN MOTORS HYDRAULICALLY CONTROLLED LOCK-UP CONVERTERS

Design

Figure 4-42 shows a Chrysler or American Motors style lock-up converter. This type of converter consists of a lined housing, piston and turbine combination, stator, and the impeller. On the inside of the metal housing just behind where the turbine operates, the manufacturer bonds a ring of friction material (Fig. 4-43). When the machined surface of the piston contacts this lining, it and the attached turbine will operate at housing speed, which is the same as that of the engine.

The piston attaches to the turbine and is responsible for locking this unit to the housing during the converter's coupling phase. To perform this task, the piston (Fig. 4-44) has a smooth machined surface on one side, which bears against the friction lining during the coupling phase. On its opposite side, the piston is in contact with the turbine through a series of tabs, machined into the piston's outer circumference. These tabs ride between cushion springs mounted on the turbine.

These cushion, or damper springs (Fig. 4-45), fasten to the outer circumference of the turbine between the piston tabs. The springs are necessary in the lock-up converter to absorb any engine torsional vibrations transmitted to the turbine through the housing. In the conventional converter, the fluid itself absorbed these vibrations, thus preventing them from reaching the transmission.

There are three remaining components to this lock-up converter: the vaned turbine, stator, and impeller. All of these parts have a similar design and operate in much the same manner as their counterparts found in a conventional three-element converter.

Operation

When the Chrysler or AMC lock-up converter is operating in the torque multiplication phase, the piston keeps the turbine away from the friction lining (Fig. 4-46). During this period, the hydraulic control system sends fluid to the impeller and to the left side of the piston. The resulting pressure forces the piston and turbine to the right and away from the housing. The turbine and impeller, at this time, operate in the same manner as those in a conventional converter.

But as this converter enters the coupling phase, the turbine begins to rotate at housing speed. At this point, the hydraulic control system cuts off fluid to the left side of the piston. Next, fluid directed between the impeller and turbine forces it and the piston against the friction lining. This action locks the turbine to the housing so that it operates at the same speed (Fig. 4-47). For additional information as to how the hydraulic system controls the operation of the lock-up function of this converter, refer to Chapter 11.

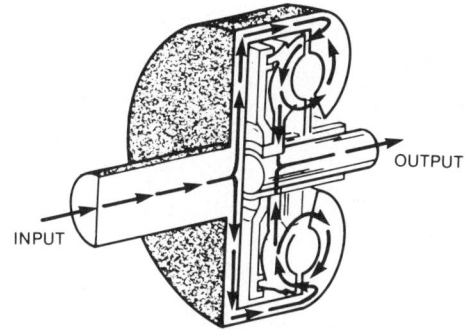

FIGURE 4-46 The Chrysler or AMC converter operating in its torque multiplication phase.

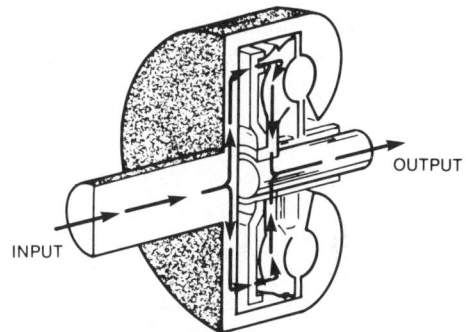

FIGURE 4-47 The Chrysler or AMC converter operating in its coupling phase.

GENERAL MOTORS HYDRAULICALLY CONTROLLED LOCK-UP CONVERTER

Design

Figure 4-48 illustrates a typical General Motors hydraulically controlled lock-up converter. This unit has about the same design as the Chrysler and AMC converter just described. However, some of the components have different names. For example, the housing is called the housing cover assembly; the piston is referred to as the pressure plate assembly; and the impeller is called the converter pump assembly. Also, the pressure plate, like the piston in the Chrysler unit, has damper springs, which this drawing does not show.

Operation

Figure 4-49 shows the operation of the General Motors lock-up converter during its torque multiplication phase. Note that the control system has directed fluid to the left

FIGURE 4-45 The damper springs are located around the outside of the turbine.

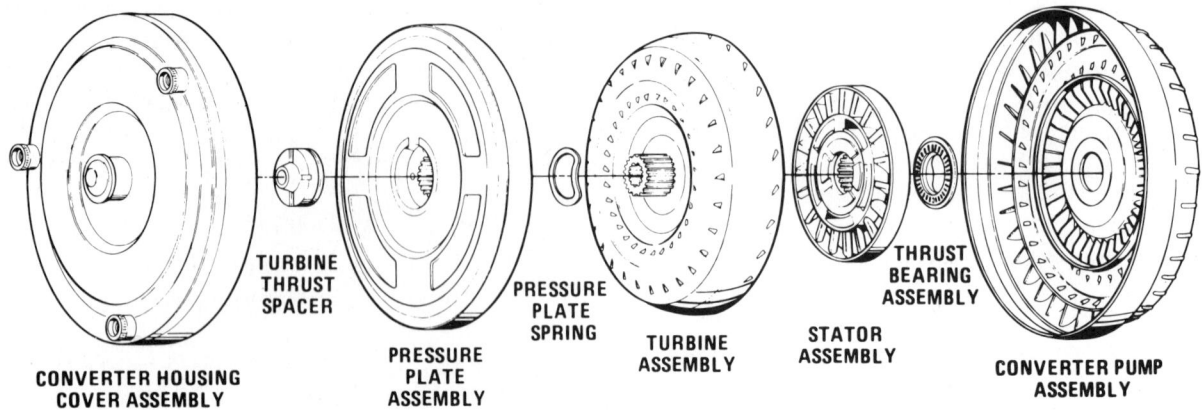

FIGURE 4-48 Exploded view of a GM lock-up converter. (Courtesy of General Motors Corp.)

side of the pressure plate assembly. The resulting force keeps the pressure plate and turbine toward the right, away from the cover assembly. The remaining components perform the same functions as in any conventional three-element converter.

When the converter reaches the coupling phase, fluid is cut off to the left side of the pressure plate assembly. Moreover, the control system directs fluid to the right side of the pressure plate, which forces the pressure plate against the lining inside the cover. Now, the turbine is locked to the cover and operates at engine speed (Fig. 4-50).

FORD'S CENTRIFUGAL CLUTCH LOCK-UP CONVERTER

Design

Ford's lock-up converter is not hydraulically controlled as were the last two units just discussed. Instead, Ford uses a centrifugal-type and roller-type clutch to provide the lock-up feature. All the remaining converter

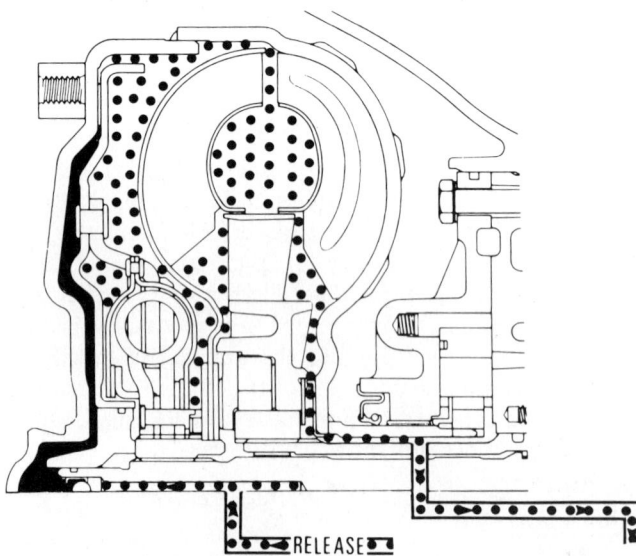

FIGURE 4-49 Lock-up converter operation during the torque multiplication phase. (Courtesy of General Motors Corp.)

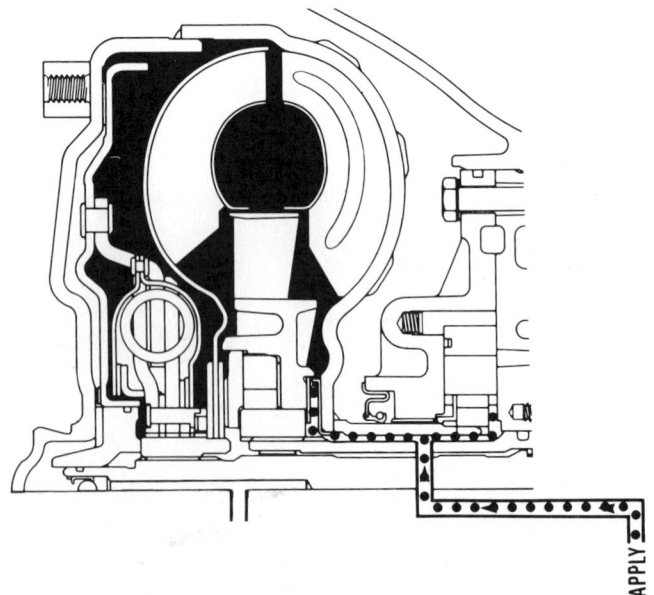

FIGURE 4-50 Operation of the converter during its lock-up phase. (Courtesy of General Motors Corp.)

components have a similar design and operation to those found in a conventional unit.

Figure 4-51 shows the entire converter including the centrifugal clutch, damper assembly, and coasting one-way clutch. The centrifugal clutch consists of a series of sliding friction shoes arranged around the circumference of the damper assembly. These shoes move outward due to the action of centrifugal force. When the shoes are out far enough, they contact the inside of the converter cover.

There are a number of springs in the damper assembly. These absorb any torsional vibrations transmitted through the cover from the engine, thus preventing them from entering the turbine and transmission.

The damper assembly also incorporates a mechanical overrunning clutch. This unit, known as the coasting one-way clutch, mounts on the damper between it and the hub on the turbine. With this arrangement, the clutch mechanically connects the turbine to the damper so they

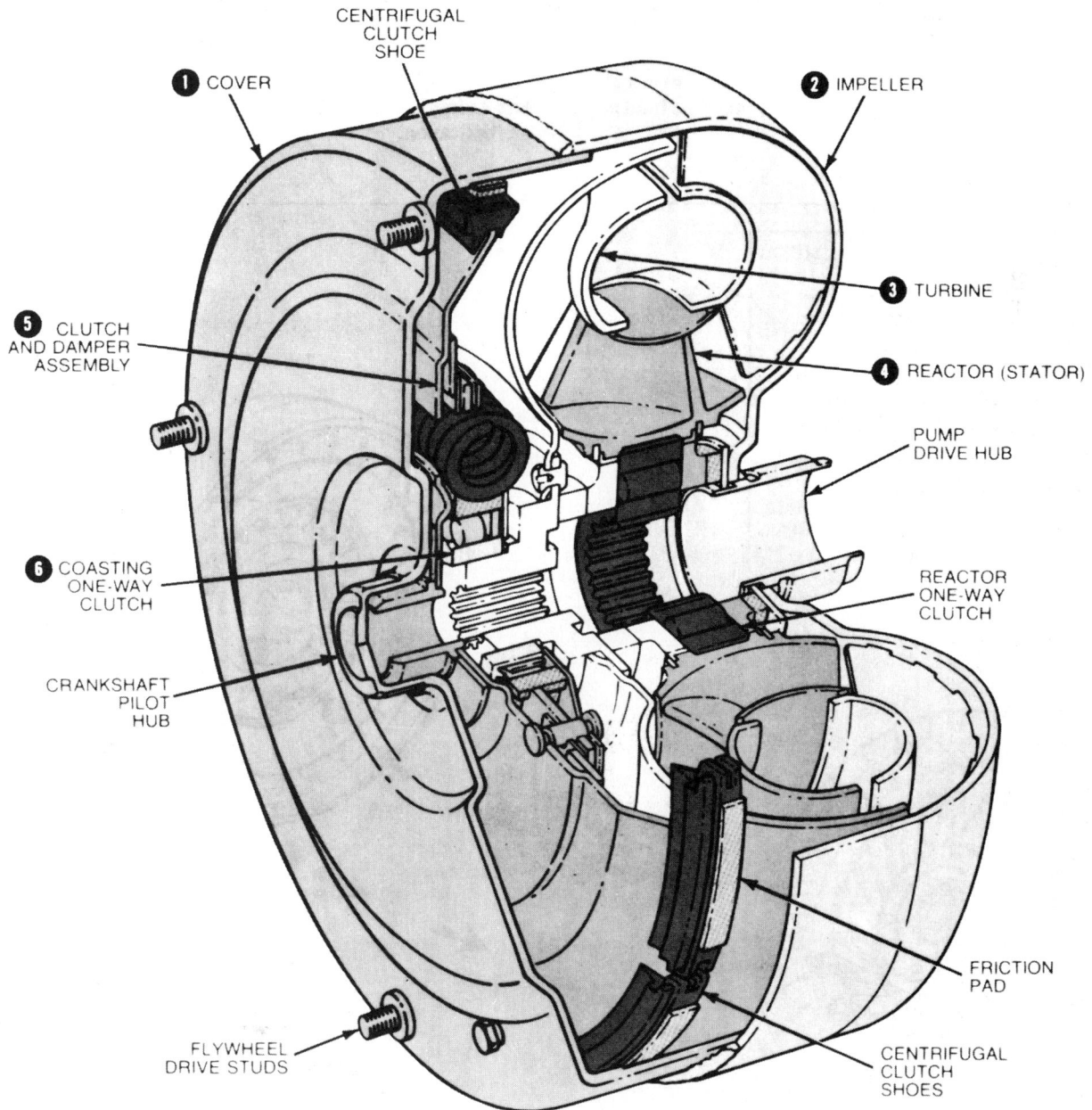

FIGURE 4-51   **The design of a centrifugal clutch lock-up converter. (Courtesy of Ford Motor Company)**

operate together during turbine acceleration to its lock-up point. But when the vehicle is coasting, the one-way clutch permits the turbine to turn freely from the damper assembly.

Operation

When the Ford centrifugal lock-up converter is operating at engine idle or very low speed, it functions in much the same manner as any conventional unit. However, as the speed of the turbine increases to a given amount during vehicle acceleration, the friction shoes move outward due to the action of centrifugal force. When the shoes are fully applied against the inside of the converter housing, torque and power flow from the housing through the damper clutch, torsional springs, one-way clutch, and to the turbine and its shaft. This action locks the turbine shaft to the housing, so it operates at engine speed.

However, the design of the centrifugal clutch is such that it provides some slippage if the torque demands on the engine increase with load. In other words, at times when the vehicle is under a load and accelerating, the centrifugal clutch permits a split between mechanical drive through itself and the normal hydraulic drive within the converter. Therefore, the determining factors as to when the centrifugal clutch applies and to what degree are turbine speed and vehicle load.

As the turbine slows down as the driver reduces foot pressure on the accelerator pedal, the shoes begin to move inward until they are free of the housing. When this occurs, the converter action returns to that of a conventional unit. But during this process, if the vehicle begins to drive the turbine via the drive train, the coasting one-way clutch freewheels, thus breaking the mechanical drive through the damper.

FORD'S OVERDRIVE CONVERTER

Design

As mentioned in Chapter 2, Ford uses a special converter in its automatic overdrive transmission (Fig. 4-52). This unit resembles a conventional three-element converter

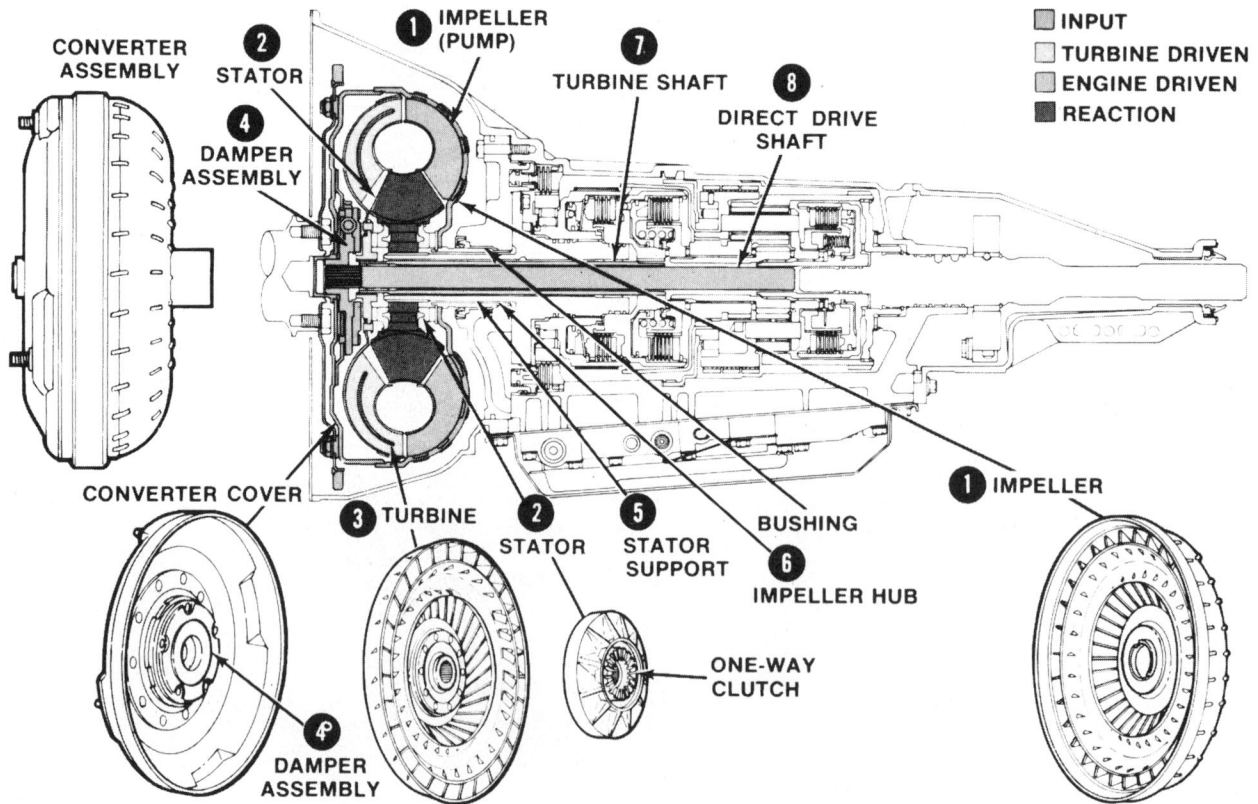

FIGURE 4-52   The converter and direct drive shaft used in Ford's automatic overdrive (AOD) transmission. (Courtesy of Ford Motor Company)

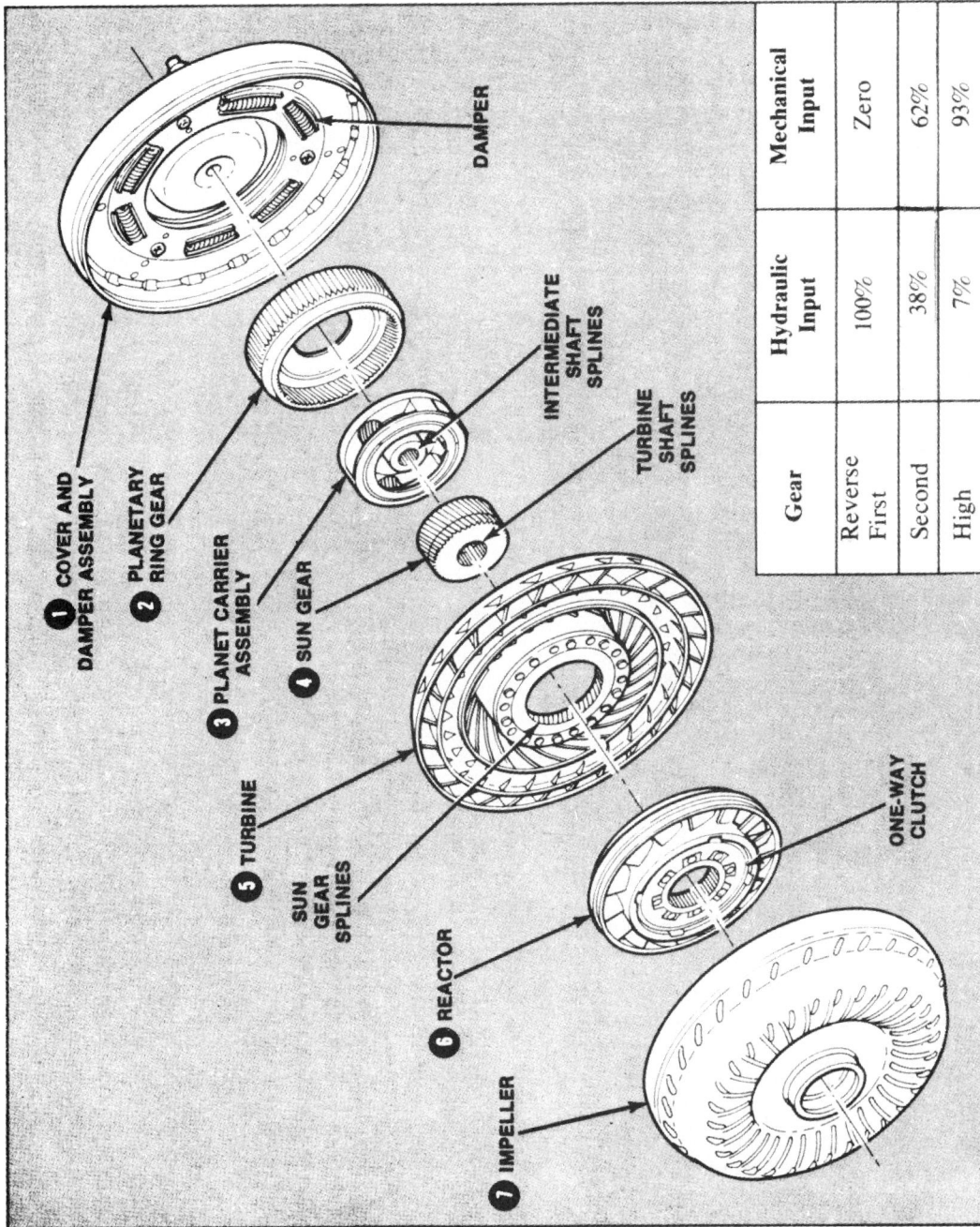

Gear	Hydraulic Input	Mechanical Input
Reverse First	100%	Zero
Second	38%	62%
High	7%	93%

FIGURE 4-53   The design of a splitter gear type converter used in Ford's transaxle. (Courtesy of Ford Motor Company)

except for the installation of the damper assembly installed onto the converter cover.

The damper assembly fits on the inside of the cover between it and the turbine. The purpose of this unit is to absorb through its springs any engine torsional vibrations passed to it from the cover assembly, thus preventing them from entering the transmission via the direct drive shaft.

The direct drive shaft has splines on one end that mate with ones in the damper. The other end of the shaft also has splines that fit into the direct clutch drum. With this arrangement, the converter cover and attached damper assembly turn the direct drive shaft and clutch drum at engine speed.

Operation

In operation, the transmission uses the input (turbine) shaft to hydraulically transmit torque and power from the turbine within the converter during reverse, first, second, and a portion of third. However, in third, the incoming torque and power are split between the turbine and direct drive shafts. At this time, the turbine shaft provides 40 percent of the input, while the direct-drive shaft supplies 60 percent.

In fourth gear (overdrive), the direct-drive shaft is mechanically providing 100 percent of the input torque and power from the engine to the transmission. The turbine shaft is also turning, but since the multiple-disc clutch drum it attaches to is now not applied, torque and power input from this source stops. For further information as to how the converter interacts with the planetary to produce the three forward speeds and an overdrive, review the section in Chapter 2 on Ford's AOD transmission.

FORD'S SPLITTER GEAR CONVERTER

Ford uses another type of torque converter that also mechanically transmits torque and power into the transmission. However, in the converter shown in Fig. 4-53, this mechanical input is through a simple planetary gearset inside the unit and not a separate shaft as in the AOD transmission.

Basically, this converter transmits torque and power through hydraulic impact energy in reverse, first, and part of second. In second, part of the torque and power input is transmitted mechanically into the transmission via the simple planetary gearset within the converter. Because of this division or split of the torque and power input, the converter is sometimes referred to as a splitter gear type.

Design

The splitter type converter shown in Fig. 4-53 consists of impeller, reactor (stator), turbine, simple planetary gearset, and damper assembly. The impeller, reactor, and turbine have a design similar to those found in any conventional converter with one exception. The turbine has splines in the center that mate with ones located on the sun gear.

The simple planetary gearset has a ring gear, planet carrier assembly, and a sun gear. The ring gear has splines on its outer circumference that match those machined into the damper assembly. With this arrangement, the ring gear turns with the housing and damper at engine speed.

The planet carrier splines to the transmission's intermediate shaft. Also, the teeth on its pinions mesh with those inside the ring gear.

The sun gear also has external splines, which mesh with those inside the turbine. Moreover, on the inside of the sun gear is a second set of splines that mate with those on the turbine shaft.

The damper assembly rivets to the inside of the cover. This unit has a series of springs that absorb torsional vibrations from the engine, thus stopping them from passing to the ring gear. And as previously mentioned, the ring gear splines into the center section of the damper assembly.

Operation

In first ratio, the power flow is as follows:

1. The engine drives the impeller in a clockwise direction.

2. The impeller through hydraulic action turns the turbine in a clockwise direction.

3. The turbine turns the turbine shaft clockwise.

4. The turbine shaft provides 100 percent of the input of torque and power into the planetary gear train.

In second ratio, the flow of power is as follows:

1. The cover of the converter drives the impeller and ring gear clockwise mechanically at engine speed.

2. The impeller through hydraulic action rotates the turbine also in a clockwise direction.

3. The turbine transmits 38 percent of torque and power through the sun gear, carrier, and to the intermediate shaft.

4. The cover and attached damper transmit 62 percent of the torque and power through the ring gear, carrier, and to the intermediate shaft.

5. The intermediate shaft transmits all the torque and power into the planetary gear train within the transmission.

6. The turbine also drives the turbine shaft clockwise. But it does not transmit any torque or power into the planetary gear train at this time due to the action of a one-way clutch.

In third ratio, the power flow is as follows:

1. The conveter's cover mechanically drives both the impeller and ring gear in a clockwise direction.

2. The impeller through hydraulic action rotates the turbine also in a clockwise direction.

3. The turbine drives the intermediate shaft clockwise via the sun gear and carrier.

4. The damper mechanically drives the intermediate shaft clockwise by means of the ring gear and carrier.

5. The intermediate shaft turns one member of the planetary.

6. If the converter is in its coupling phase, the turbine drives the turbine shaft at nearly the same speed as that of the intermediate shaft. The turbine shaft, in turn, drives another member of the planetary.

7. The turbine shaft provides 7 percent of the input of torque and power into the planetary, while the intermediate shaft supplies 93 percent.

SUMMARY

1. Every motor vehicle requires some form of coupling between its engine and transmission.

2. The fluid coupling or torque converter transmits engine torque to the transmission.

3. A torque converter can multiply engine torque and control engine speed.

4. A fluid coupling or converter acts as a clutch to smoothly connect or disconnect the flow of power from the transmission.

5. The torque converter and coupling use fluid to dampen out engine torsional vibrations.

6. The torque converter or fluid coupling along with its flex plate acts as the engine flywheel.

7. The torque converter mechanically drives the hydraulic pump.

8. The coupling is the simplest form of fluid clutch.

9. A fluid coupling consists of a housing, impeller, and turbine.

10. The impeller is the coupling's pump and attaches to the housing.

11. The coupling's turbine is free to rotate within the housing but connects by splines to the input shaft.

12. The turbine and impeller blades of a fluid coupling are flat.

13. Rotary flow is the result of the impeller blades pushing on the trapped fluid within the metal housing.

14. Vortex flow is produced by centrifugal force acting upon the fluid trapped between the blades of the impeller.

15. The relative strengths of the rotary and vortex fluid motions determine the actual exit angle of the oil as it leaves the impeller's rim and enters the turbine.

16. With the engine idling, the coupling does permit some torque transfer between the engine and the transmission's input shaft, but no power.

17. As the engine accelerates, torque and power begin to pass between the coupling's impeller and turbine.

18. The coupling enters its most effective period when the turbine approaches impeller speed.

19. During vehicle deceleration, the turbine acts as the impeller or pump to create fluid flow within the coupling and converter.

20. The torque converter consists of an impeller, stator, and turbine.

21. The impeller of a torque converter has curved vanes and a split guide ring.

22. The turbine of a converter has one additional curved blade than the impeller.

23. The stator fits between the converter's impeller and turbine.

24. The stator plays an important role during the converter's torque multiplication phase.

25. The stator is responsible for redirecting the returning fluid from the turbine into a helping direction.

26. A torque converter provides maximum slippage or a 100 percent break in power when the engine is running with the transmission in gear.

27. As the driver accelerates the engine to set the vehicle in motion, the torque multiplication phase of the converter begins.

28. Converter stall speed is a condition in which the impeller rotates as fast as the engine will drive it at wide-open throttle.

29. During the converter's coupling phase, the turbine is operating at 9/10 impeller speed and the stator is overrunning.

30. The torque converter produces a great deal of heat in the fluid during its torque multiplication phase.

31. The capacity of a converter is determined by its size and stator blade angle.

32. Vehicles with smaller engines use low capacity converters.

33. Vehicles with large V-8 engines have high capacity converters.

34. A converter with a variable-pitch stator provides a vehicle with the benefits of both low and high capacity units.

35. To operate properly, the converter must be centered and supported on the back of the engine crankshaft.

36. The lock-up converter has a turbine that operates at impeller, or housing, speed when the unit is in its coupling phase.

37. The lock-up converter increases fuel economy, reduces fluid temperature, and extends engine life.

38. A hydraulically controlled lock-up converter consists of lined housing, piston and turbine combination, and an impeller.

39. When a hydraulically controlled lock-up converter is operating in the torque multiplication phase, the piston keeps the turbine away from the friction lining.

40. As the hydraulically controlled lock-up converter enters its coupling phase, the piston moves the turbine into contact with the housing.

41. Ford's centrifugal lock-up converter does not use hydraulic action to drive the turbine at housing speed.

42. Ford's lock-up converter consists of a centrifugal-type and a roller-type clutch along with the other three members of any conventional unit.

43. During the coupling phase of the Ford converter, the clutch shoes move against the inside of the housing, thus locking the turbine to the housing.

44. The converter used with Ford's AOD transmission has a damper mounted on the inside of the cover; this damper splines to the direct drive shaft.

45. During fourth gear operation in an AOD transmission, the direct drive shaft delivers 100 percent of engine torque and power into the transmission.

46. Ford's splitter gear converter consists of a simple planetary gearset along with the other three members of a conventional unit to mechanically and hydraulically transmit engine torque and power into the transmission.

REVIEW

This section will assist you in determining how well you remember the material contained in this chapter. Read each item carefully. If you cannot complete the statement, review the section in the chapter that covers the material.

1. When the automatic transmission replaces the manual-shift transmission, the _____ replaces the foot-operated clutch.

    a. clutch

    b. band

    c. turbine

    d. torque converter

2. The fluid coupling provides a break in torque necessary for _____ and _____ .

    a. engine starting, stationary operation

    b. acceleration, coasting

    c. stationary operation, acceleration

    d. engine starting, acceleration

3. The member of the fluid coupling mechanism that causes fluid flow is the _____ .

    a. turbine

    b. impeller

    c. stator

    d. flex plate

4. The portion of the fluid coupling which absorbs the energy from the moving fluid is the _____ .

    a. housing

    b. impeller

    c. turbine

    d. flex plate

5. The fluid flow caused by the blades pushing on the trapped fluid inside the converter housing is _____ .

    a. vortex   c. centrifugal

    b. rotary   d. both a and c

6. The angle of the _____ flow changes with turbine speed.

   a. vortex

   b. rotary

   c. vertical

   d. both b and c

7. During stationary operation of a fluid coupling or converter, the _____ is rotating and the _____ is stationary.

   a. turbine, impeller

   b. impeller, turbine

   c. housing, impeller

   d. impeller, housing

8. When the torque converter reaches its coupling point, the turbine and impeller speed ratio is about _____.

   a. 6:10

   b. 7:10

   c. 8:10

   d. 9:10

9. For the fluid coupling mechanism to transfer torque, both members must be _____.

   a. rotating at the same speed

   b. rotating

   c. stationary

   d. rotating at different speeds

10. If the driver reduces speed so that the vehicle begins to drive the engine through the fluid coupling or converter, the _____ causes fluid flow within the unit.

    a. impeller

    b. turbine

    c. stator

    d. housing

11. The component which changes the fluid coupling into a torque multiplying device is the _____.

    a. turbine

    b. impeller

    c. stator

    d. housing

12. The stator's _____ allows it to turn when the stator is not necessary for torque multiplication.

    a. overrunning clutch

    b. blades

    c. support

    d. assembly

13. When the torque converter is increasing torque, the stator is _____.

    a. moving clockwise

    b. moving counterclockwise

    c. stationary

    d. locked to impeller

14. Torque converters that are low capacity have _____ stall speeds.

    a. low

    b. high

    c. zero

    d. moderate

15. Converter capacity is a result of the unit's _____ and _____ angles.

    a. size, turbine blade

    b. size, impeller blade

    c. size, stator blade

    d. both a and b

16. Transmissions that utilize torque converters, as mentioned in this chapter, do not _____.

    a. require foot-operated clutches

    b. have impellers

    c. have stators

    d. have housings

17. When a torque converter has the lock-up feature, the _____ locks to the housing during the coupling phase.

    a. impeller

    b. stator

    c. turbine

    d. overrunning clutch

18. Vehicles with larger engines use _____ capacity converters.

    a. high

    b. low

    c. moderate

    d. none of the above

19. Special drive studs can drive, _____, and _____ the converter.

    a. balance, support   c. balance, center

    b. center, support   d. center, balance

20. The _____ drives the transmission's hydraulic pump.

   a. turbine

   b. stator

   c. hub on the converter housing

   d. impeller

For the answers, turn to the Appendix.

When the average person encounters the term "clutch," he usually thinks of the foot-operated device required with a manual-shift transmission. However, an automatic transmission also has clutches that, in some applications, perform similar functions to the foot-operated unit.

FUNCTIONS OF AUTOMATIC TRANSMISSION CLUTCHES

Depending on the design, clutches can perform several functions, even within the same automatic transmission. For example, a clutch assembly directly or indirectly connects and disconnects members of the planetary gear train to the transmission's input shaft. Remember, the input shaft splines to the turbine within the torque converter, and if the turbine rotates with the clutch applied (activated), torque transfers from the turbine to the planetary gear train member.

In addition, a transmission manufacturer can build a gear box that uses a clutch to hold a member of the planetary or the stator within the converter stationary, either in one or both operating directions of the unit. In the case of a planetary member, the clutch replaces the use of a band to hold the member because the clutch does not require any adjustment after prolonged use and has the ability to withstand large torque loads for its size with a minimum of slippage.

CLUTCH TYPES

By their design, automatic transmission clutches are known as the multiple-disc wet clutch or the mechanical overrunning type. The disc type is hydraulically operated and consists of a number of drive and driven clutch plates housed in a drum (Fig. 5-1). This device can either be used to connect a planetary member to the input shaft or hold a a component stationary in both directions.

The overrunning clutch is strictly a mechanical unit that operates automatically to hold a transmission component stationary in one direction only. For the stator in the torque converter, an overrunning clutch prevents the unit from turning counterclockwise, while allowing it to freewheel in the clockwise direction (Fig. 5-2).

PRIMARY DISC CLUTCH DESIGN

For the sake of simplicity in discussing the various kinds of multiple-disc clutches used in automatic transmissions, this text refers to them as either primary, secondary, or

CHAPTER 5

Clutches Used in Automatic Transmissions

stationary. A **primary clutch** is one that has a drum which attaches directly to the input shaft, whereas the drum of a **secondary clutch** does not. Finally, a **stationary clutch** has no drum at all because the transmission case itself houses all the clutch parts.

Drum

A primary clutch assembly consists of a drum, hydraulic piston, return spring(s), residual check ball, drive and driven clutch discs, and a clutch hub (Fig. 5-1). Drum manufacturers usually use a good grade of cast iron and several machine processes to form the **drum,** or **cylinder,** which forms the housing for the remaining clutch parts. Also, in some cases, manufacturers produce a stamped steel drum as the clutch housing to reduce its weight and cost. In any case, the machined area inside the drum is large enough to retain the hydraulic piston, clutch discs, and the clutch hub. Lastly, the clutch drum, or cylinder, as mentioned, attaches to the input shaft and rotates at input shaft speed.

Hydraulic Piston

Inside the drum at the very bottom is a machined area (Fig. 5-3) that forms the cylinder or guide for the **hydraulic piston,** which is responsible for applying a force

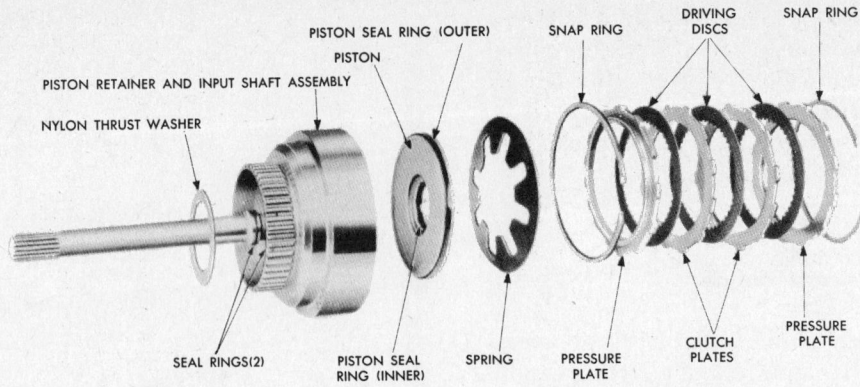

FIGURE 5-1 A primary multiple-disc clutch assembly. (Courtesy of Chrysler Corp.)

to the clutch discs. The piston (Fig. 5-4) moves within the machined cylinder area due to the action of hydraulic pressure acting on its base. The moving piston, in turn, applies a force to all the clutch discs, thus compressing them all together.

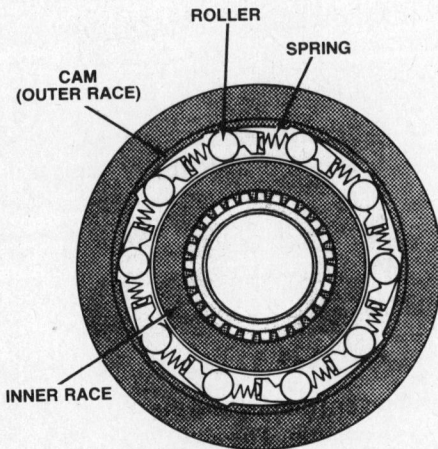

VIEW FROM ENGINE SIDE

¾ VIEW FROM ENGINE SIDE OF STATOR SHOWING VANE CURVATURE

FIGURE 5-2 A stator-mounted, overrunning clutch. (Courtesy of Chrysler Corp.)

Return Spring

A type of **return spring** is necessary to push the piston to the bottom of the drum when there is no applied hydraulic pressure. Moreover, since the spring resists any piston movement, as it attempts to react to the application of hydraulic pressure, the spring cushions clutch application.

Some automatic transmissions use only one large compression spring for this purpose; others use a series of small compression springs or a diaphragm spring like the one shown in Fig. 5-1. In either case, a **snap ring** that fits into a narrow groove cut into the drum secures the spring retainer and spring(s) to the drum. As a result, the force of the spring(s) when compressed exerts against the movable piston and also the retainer, snap ring, and drum.

Residual Check Ball

On the hydraulic piston (Fig. 5-4) or within the cavity at the base of the drum is a residual check ball and seat. The **residual check ball** allows any fluid trapped below the

FIGURE 5-3 The cylinder section of the drum forms the guide for the piston.

FIGURE 5-4   A residual check valve located in the clutch piston.

FIGURE 5-5   Operation of a residual check valve. (Courtesy of General Motors Corp.)

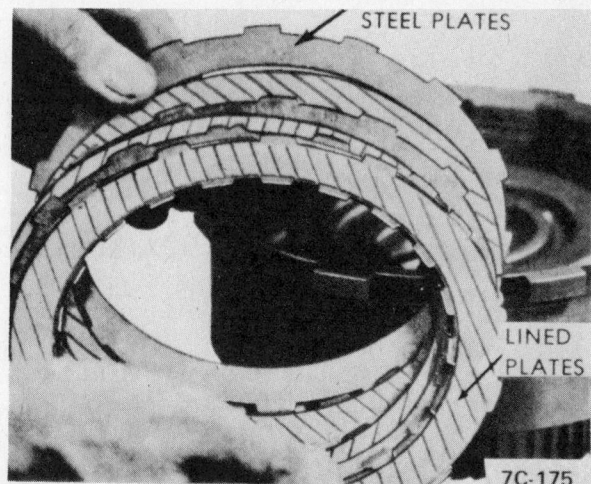

clutch piston to escape whenever there is no applied hydraulic pressure. This action is necessary if the clutch assembly is to have a long life. If the clutch did not have the check ball, high drum rotation would create sufficient centrifugal force in the residual fluid under the piston to cause it to partially force the discs together. The partially compressed clutch discs will then overheat due to increased friction and wear out prematurely.

However, during clutch operation, hydraulic pressure under the clutch piston forces the residual (exhaust) check ball against its seat (Fig. 5-5). This action seals off the area against leakage past the check ball and seat. When hydraulic pressure below the piston drops to zero with the drum turning, the residual check ball opens due to centrifugal force created by the rotating drum. With the ball off its seat, any residual fluid escapes from under the piston.

Steel Drive Plates

A primary clutch assembly has two types of clutch discs or plates: the drive steel and the driven lined, friction discs (Fig. 5-6). Manufacturers make the drive plates from rolled carbon-steel stock, which is run through straightening rolls before being stamped into flat circular discs.

The stamped discs also have teeth or lugs around their outer circumference. These projections fit into slots or grooves machined into the inside wall of the clutch drum. With this arrangement, any rotation of the drum forces the installed plates to turn, but they can slide back and forth in the drum whenever the clutch is not in use.

The steel plates undergo a tumbling process before being placed into service to provide them with a dull or matted surface on both sides. This type of finish is desirable as an aid to breaking in the new friction plates. After this period, the bearing surfaces of the steel plates become polished.

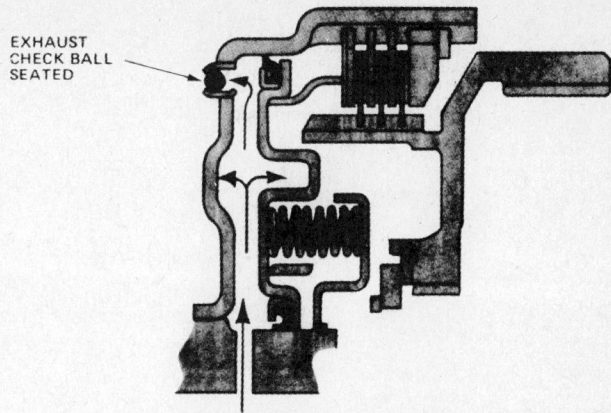

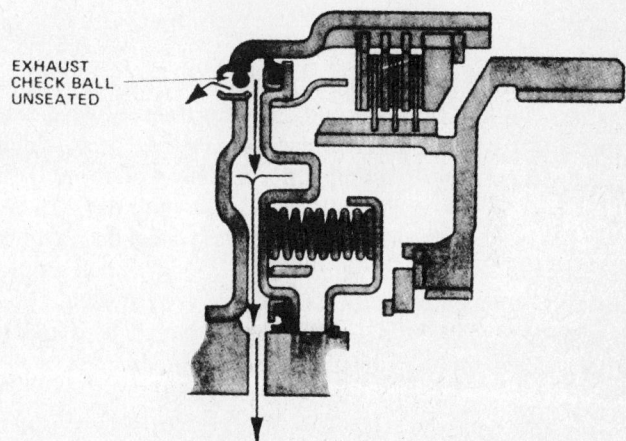

FIGURE 5-6   Typical steel and lined friction plates found in a clutch assembly. (Courtesy of General Motors Corp.)

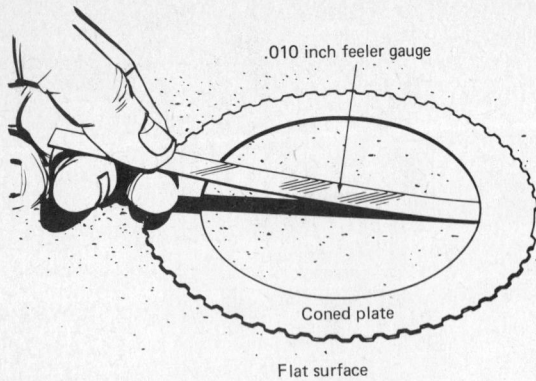

FIGURE 5-7   A mechanic checking the coning of a steel plate with feeler gauge.

Some clutch assemblies have one or more steel plates that are not flat. For example, early Ford three-speed cast iron automatics used in the rear clutch steel plates, which had on their inner circumference a coning of about .010 inch (Fig. 5-7). However, these gave way in Ford's FMX and CW transmissions to a single wavy steel disc, which served the same function of cushioning clutch application. Other manufacturers also use a wavy steel plate that may or may not have external teeth or lugs to lock it to the drum for the same purpose (Fig. 5-8).

Driven Friction Plates

The driven friction plates are flat steel discs, which have a type of lining bonded to both their front and back bearing surfaces along with teeth or lugs stamped into their inner circumference (see Fig. 5-6). The plate manufacturer molds a friction material under high temperature and

FIGURE 5-8   A wavy steel cushion plate.

pressure into the shape and size of the lining. Then the finished lining is bonded to both sides of the friction disc.

The teeth, or lugs, fit into slots, or grooves, cut into the outer circumference of the clutch hub (Fig. 5-9). With this design, the friction plates can slide back and forth on the hub; and the hub and the discs are free to rotate inside the drum when the clutch is not in service.

Manufacturers use either a full-metallic, semi-metallic, or organic material for the lining on friction plates. For instance, one transmission model that has two multiple-disc clutches has friction plates with a metallic lining in one unit and either a semi-metallic or organic lining in the other. Other transmission manufacturers use an organic lining on all friction plates within their units.

Metallic and semi-metallic linings are necessary where the clutch plate may encounter large torque loads and high temperatures. The most common metal used for this purpose is sintered or powdered copper mixed with graphite. A semi-metallic lining consists of powdered copper mixed with asbestos, lead powder, and resin binders. (NOTE: It is a common practice to refer to the lining bonded to friction plates as being composition because it is usually made up of more than one material.)

An organic lining is primarily paper pulp or cellulose.

FIGURE 5-9   A clutch hub with its integral ring (annulus) gear.

However, a composition lining can be made from a mixture of different materials such as paper, an asbestos fiber or powder, and the proper bonding agents. Also, for some heavy-duty installations, the manufacturer adds metal flakes to the base materials. These metal flakes are slightly abrasive in nature and, when used in friction plates, prevent the steel discs from becoming glazed or highly polished.

For many years, the trend has been to use a specially compounded composition plate material that, when used with a certain fluid type, provides a given friction characteristic to the clutch. This friction characteristic determines how harsh or soft the application of the clutch will be. If the amount of friction between the components is too high during application of the clutch, the change in gear ratios will be harsh. On the other hand, if friction is too low between the parts, the clutch will slip excessively, which softens the gear change but leads to premature clutch plate failure.

It should be obvious then that friction plates having a given lining must function with a certain type of fluid. This assures a certain shift quality and long clutch plate life. An incorrect fluid type causes too harsh or soft a clutch application resulting in poor shift quality and rapid clutch component failure.

Friction clutch plate lining may also have a number of circular or flat grooves formed into its surface. This pattern of grooving affects the friction characteristic of the plate by controlling the normal flow of fluid and vapors away from the discs as they all come together. In addition, grooving helps to cool the plates during clutch application. The type of grooving (either circular, vertical, or checkerboard) is determined to a great extent by the type of lining material and fluid used, expected steel plate surface condition, and anticipated clutch engagement speeds.

Clutch Hub

The final component of a primary clutch asembly is the **clutch hub** (Fig. 5-9). The hub, along with the slots or grooves that index with projections on the friction plates, may have a series of drilled holes between these indentations for the purpose of lubricating the discs. In addition, the hub attaches directly to a planetary gear train member; therefore, when the clutch assembly activates, torque passes through all the plates, to the hub, and finally into the planetary gear train member.

PRIMARY CLUTCH OPERATION

Applied

Whenever the transmission requires the primary clutch to connect a planetary member to the input shaft, hydraulic pressure enters the clutch drum behind the clutch piston through a special passage (Fig. 5-10). The hydraulic pressure acting on the piston causes it to move within its bore in the drum. This piston movement does two things: First, movement of the piston compresses the return spring(s). The spring(s) reach full compression as the piston reaches the end of its travel in the drum. Second, as the piston moves, it forces the drive and driven plates together which, in turn, causes the driven plates to operate at input shaft speed. And since the hub and planetary member together connect to the driven plates, the unit also rotates at input shaft speed.

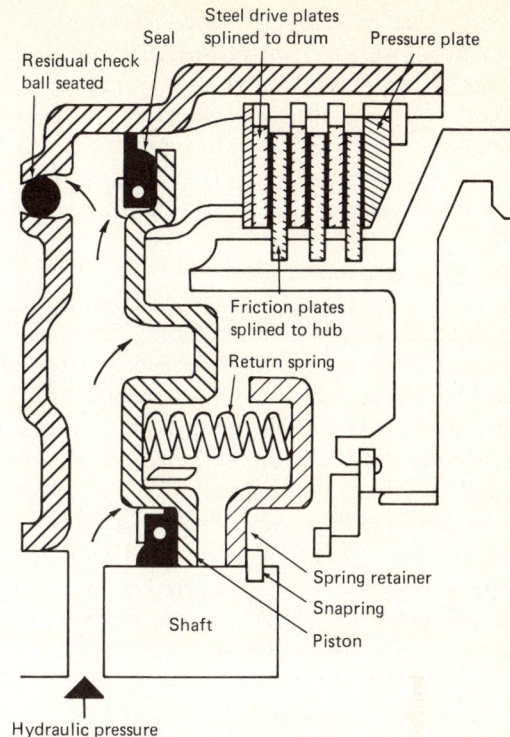

(a) Clutch applied

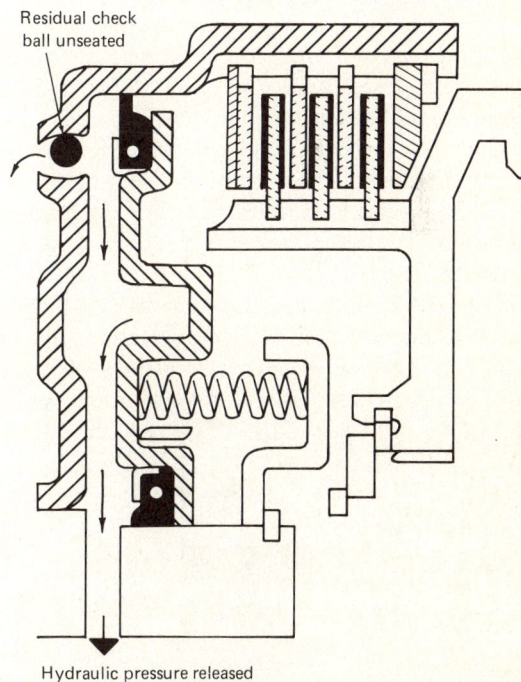

(b) Clutch released

FIGURE 5-10 **Hydraulic pressure causes the clutch apply piston to move in its bore to lock the drive and driven plates together at (A). Part (B) shows the clutch in the released position.**

While the hydraulic pressure causes the piston movement, the pressure plate and snap ring on the opposite side of the clutch assembly provide the reaction necessary to lock the discs together. Also, synthetic rubber seals between the piston and the drum prevent fluid leakage and the loss of hydraulic pressure. The design and function of seals will be covered in another chapter.

The fluid within the assembly cools, lubricates, and determines the harshness of clutch application. As the plate clearance begins to lessen during clutch application, the force on the fluid layers trapped between the discs begins to increase. As a result, the temperature of the fluid begins to increase rapidly due to the resulting friction between the individual fluid layers and between the oil film and the plates themselves.

Most of the heat generated during clutch application leaves the unit in the fluid as the force from the moving piston "squeezes" the excess fluid from between the plates. The heated fluid drains into the pan, where it is cooled by the surrounding oil. The temperature of this fluid has been lowered by the action of the transmission's cooling system and air flow around the pan.

The fluid between the clutch plates also reduces lining wear caused by friction. The fluid layers prevent the actual direct contact between the steel drive and composition driven plates as the clutch piston force "squeezes" the majority of the fluid from between the plates. Even as the plates come together, the composition lining absorbs and retains some of the fluid. Therefore, a thin layer of fluid remains on the surfaces of the discs. As the plates lock together, this residual fluid allows only spot contact between the discs. Thus, from the beginning of clutch application until the discs lock together, the fluid controls the amount of actual plate contact area and therefore the quantity of friction.

The type of fluid in the clutch determines the harshness of its application. At the beginning of clutch application, the drive and driven plates are rotating with a great difference in speed between the two. In fact, one may be

stationary while the other is operating at 500–750 rpm or higher. Also, the discs may be turning in opposite directions. In any case, as the clutch applies, the difference in speed reduces as the assembly picks up the load, imposed on the driven plates through the planetary, until all the discs rotate at the same speed.

As the speeds of the drive and driven plates approach the same rpm, the amount of friction between them increases or decreases, depending on the type of fluid used. With Type F fluid, friction increases to provide a firm, aggressive clutch application. With Dexron fluid, the amount of friction decreases, which results in a smooth, soft clutch application.

Released

To disengage the clutch assembly, the hydraulic system cuts off fluid pressure to the piston. With no hydraulic pressure on the back of the piston, the return spring(s) move the piston to the bottom of its bore in the drum. Then the residual check ball opens to allow any trapped fluid to escape. The drive and driven plates separate, and the clutch hub and planetary member cease rotation. As the plates are separating, fluid under pressure from the transmission's lubricating system again enters the clearance area between all the discs.

SECONDARY CLUTCH DESIGN

Except for a few differences, the secondary clutch assembly resembles a primary (Fig. 5-11). For instance, the drum of a primary clutch attaches directly to the input shaft; the drum (piston retainer assembly) of a secondary assembly connects via the lugs on its outer circumference to a planetary member. The drive discs of the primary clutch are steel, but the drive discs of a secondary unit are lined with a friction material.

In addition, the primary clutch hub connects to a

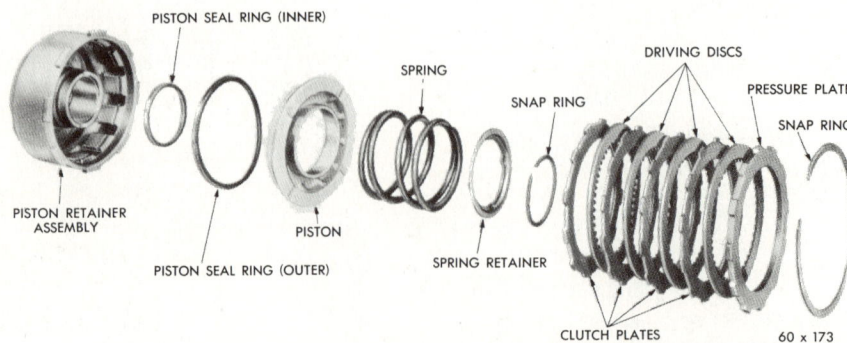

FIGURE 5-11   A secondary clutch assembly. (Courtesy of Chrysler Corp.)

FIGURE 5-12   The hub of the secondary clutch
assembly is part of the primary drum.

planetary member whereas the secondary hub is part of
the primary drum (Fig. 5-12). With this arrangement, the
secondary clutch hub turns with the primary drum at
input shaft speed. It is through this design that the
secondary clutch receives its driving torque and power
from the input shaft.

SECONDARY CLUTCH OPERATION

When the hydraulic system applies the secondary clutch,
torque and power also pass through this unit to a
planetary member from the input shaft. However, in this
case, torque moves from the secondary clutch hub driven
by the primary drum to the friction drive plates, to the
steel driven discs, to the secondary drum (piston retainer
in Fig. 5-11), and finally to the planetary member.

STATIONARY CLUTCH DESIGN

As its name implies, the design of a stationary clutch is
such that instead of locking a planetary member to the
input shaft, it holds a member stationary. In some installations, the clutch locks the member directly to the case.
While in other transmissions, the assembly works in
conjunction with an overrunning clutch to prevent
rotation of a planetary member in one direction only.
This latter design provides a very smooth change of gear
ratios.

Case

The stationary clutch assembly does not have a rotating
drum to house its operating components. Instead, the rear
portion of the transmission case (Fig. 5-13) forms the

housing for all the clutch parts. Also, since the housing
does not turn, a residual check ball is not necessary in this
installation to offset the effects of centrifugal force.

But some manufacturers do install a ball-check valve in
a stationary clutch piston for another reason. For
example, the Ford C-6 transmission has a check valve in
its stationary, reverse-clutch piston to provide a complete
apply-fluid leakdown. By permitting all the fluid to leak
out of the assembly, the next clutch application requires a
complete refill; this softens engagement of the unit.

Clutch Piston, Return Springs, and Retainer

Figure 5-13 shows the location of a stationary clutch in
the rear section of the transmission case. At the very back
of the case is a recessed bore, which accommodates the
clutch piston. The piston (Fig. 5-14) moves back and forth
in this bore but does not turn. Holding the piston at the
furthest end of its travel are a series of compression
springs. One end of each spring bears directly against the
clutch piston; the other end fits into a retainer, which a
snap ring secures to the transmission case.

Reaction and Friction Plates

The steel plates of a stationary clutch are known as
reaction plates because they do not rotate. However, in
much the same manner as their counterparts in rotating
clutch assemblies, these reaction steel plates have a series
of lugs located on their outer circumference (Fig. 5-15).
These lugs index with slots cut into the transmission case.
With this arrangement, the installed plate forms the
reaction (stationary) element of the assembly.

The friction or lined plates have teeth, or lugs,

FIGURE 5-13   The location of the stationary clutch
assembly.

FIGURE 5-14 A stationary clutch piston, return springs, and retainer.

machined into their inside circumference. These projections fit into the clutch hub (Fig. 5-16). The types of material used for stationary friction plates are about the same as those used to line the discs of rotating clutch assemblies.

Clutch Hub

The stationary clutch hub attaches directly to a planetary member. In Fig. 5-16, the hub shown is actually made as part of the ring gear. Therefore, when this clutch assembly applies, the ring gear stops turning, and the unit locks the gear firmly to the case.

STATIONARY CLUTCH OPERATION

Applied

When the stationary clutch mentioned above is necessary to provide the transmission with a gear ratio, hydraulic pressure enters the area between the back of the piston

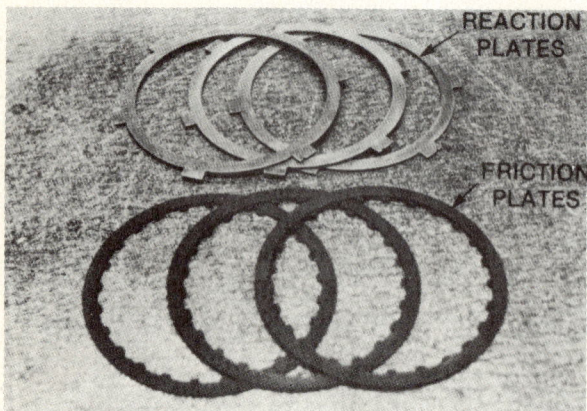

FIGURE 5-15 The reaction and friction plates of a stationary clutch assembly.

and the case. The force of the fluid moves the piston within its bore, compressing the return springs. The piston movement, as it compresses the springs, applies a force on the reaction plates which, in turn, move against the friction plates. If the friction plates are in motion, they begin to slow down as the reaction discs and case begin to pick up any load imposed on the ring gear. This process continues until the plate movement "squeezes" almost all the fluid out from between the discs and all the components become stationary, including the ring gear.

Released

When the stationary clutch is no longer necessary to hold the planetary member to the case, hydraulic pressure to the piston ceases. With no pressure acting on the piston, its return springs move it to the rear of the case. This action permits all the plates to separate as lubricating fluid returns between them. The friction plates, hub, and attached ring gear are now free to rotate inside the rear of the case.

FIGURE 5-16 Stationary clutch hub and ring gear.

TORQUE HOLDING CAPACITY OF MULTIPLE-DISC CLUTCHES

The maximum torque holding or transferring ability of a multiple-disc clutch is determined by a number of factors. These include amount of applied hydraulic pressure, surface area acted on by the pressure, frictional surface area, clutch diameter, and, in some cases, the type of return spring.

Increasing or decreasing hydraulic pressure does affect the torque-holding capacity of a clutch assembly. Increasing the pressure raises the torque capacity of the unit but can result in a very harsh shift quality. Too low an applied pressure creates excessive slippage within the clutch, which improves the shift quality but can result in premature clutch plate failure. In actual practice, the hydraulic system of the transmission has a design that applies the correct amount of pressure necessary for a desired shift feel, while at the same time promoting long friction plate life.

The actual overall size of the piston determines the surface area upon which hydraulic pressure can act. This, of course, has design limitations due to the available space inside the drum or, in the case of a stationary clutch, the case.

Manufacturers can increase the frictional surface area by the addition of more friction plates to a clutch assembly. This, of course, requires the redesigning of the clutch drum or case or changing the thickness of other clutch components such as the pressure plate or piston. In any case, this is a common practice among transmission manufacturers when a given transmission type is installed into various vehicles having different configurations or engine sizes. In other words, the type of vehicle and engine size determine just how many friction plates a clutch requires in order to have a given torque capacity and long service life.

Clutch drum diameter determines how large the piston can be and therefore its surface area and also the overall size of all the plates. A larger drum will have an increased torque capacity, while a smaller drum has a reduced capacity. However, drum diameter has design limitations due to the overall size of the transmission case, which

determines just how much internal operating room the drum will have.

The use of a dished-type spring also increases clutch capacity. This type of return spring is commonly known as a "Belleville spring" (refer to the clutch assembly shown in Fig. 5-1). The Belleville spring adds mechanical advantage to the piston through leverage as it forces the clutch discs together, thus increasing the capacity of the assembly.

MECHANICAL OVERRUNNING CLUTCHES

Function

Automatic transmissions also use one or more overrunning (one-way) clutches to control torque converter and planetary gear train operation. In practice, this device can either hold a given component against rotation in one direction while allowing the same part to freewheel in the other, or the device can permit the turbine to drive a planetary member clockwise but still permit it to spin freely counterclockwise.

As mentioned earlier, the stator within the torque converter mounts on an overrunning clutch (Fig. 5-17). This unit prevents the rotation of the stator blade assembly in a counterclockwise direction during the torque multiplication phase. However, the one-way

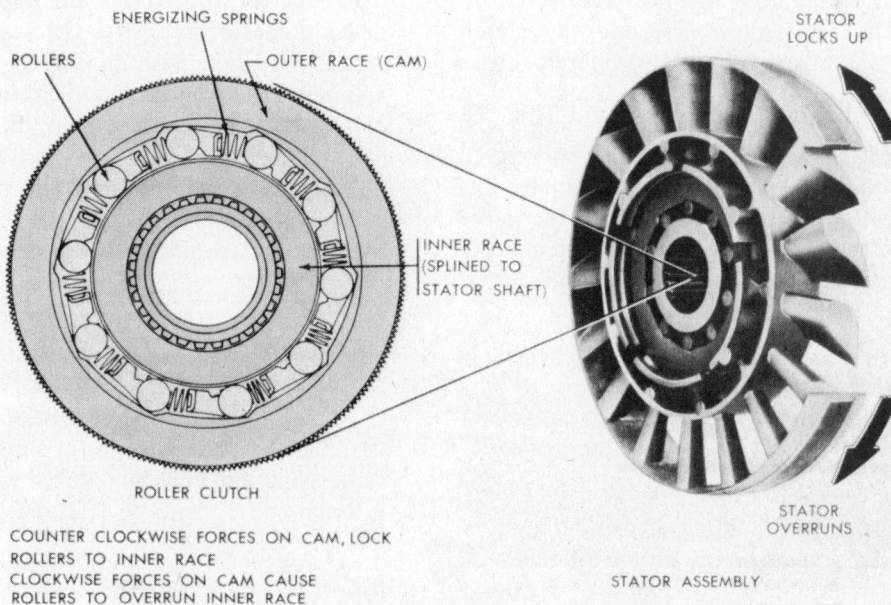

FIGURE 5-17   A stator-mounted, overrunning clutch. (Courtesy of Hydro-matic Division of General Motors Corp.)

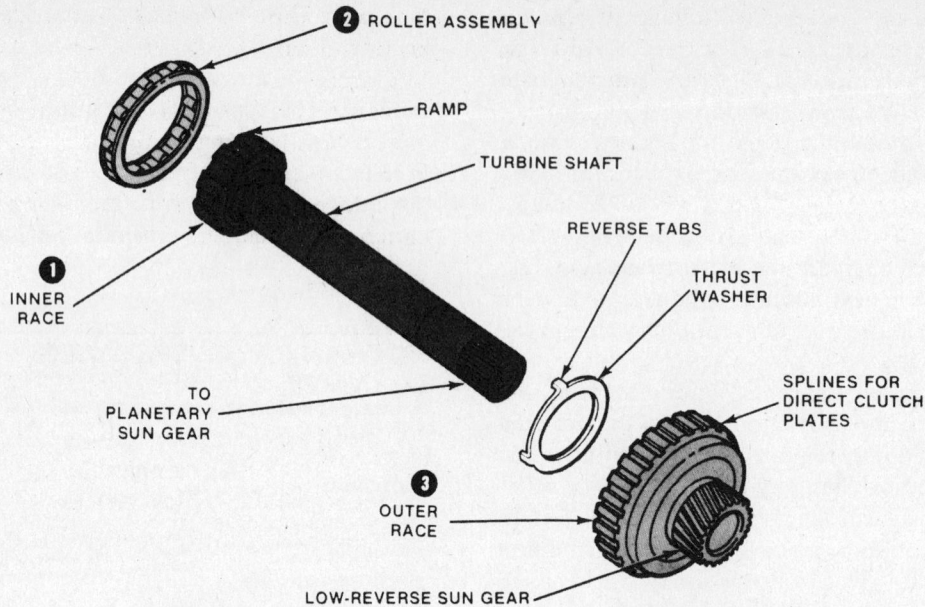

FIGURE 5-18   A one-way clutch is used to connect a planetary member to the turbine within a Ford ATX transaxle. (Courtesy of Ford Motor Company)

clutch allows the same assembly to turn freely clockwise during the converter's coupling phase.

Fig. 5-18 illustrates the use of a one-way clutch as a coupling device in a Ford ATX transaxle. In this installation, the clutch permits the converter, through the turbine shaft, to drive the low and reverse sun gear in a clockwise direction during drive low ratio. However, on deceleration in drive low ratio, the one-way clutch permits the sun gear to freewheel in a counterclockwise direction.

Another holding application for an overrunning clutch is shown in Fig. 5-19. In this situation, the one-way clutch allows the free rotation of a planetary member, the carrier, in a clockwise direction. However, if due to the application of a reverse torque the carrier attempts to twist counterclockwise, the overrunning clutch will lock it to the case.

ADVANTAGES OF USING ONE-WAY CLUTCHES

For the purposes of either holding or driving a transmission component, the overrunning clutch offers several advantages over the multiple-disc assembly or the brake band. For instance, the device is completely automatic in its operation; therefore, the clutch requires no mechanical linkage or hydraulic devices. This makes the clutch normally trouble free for the life of the transmission.

Since the one-way clutch is automatic and does not need any controls for its operation, the precise timing necessary for the application of a multiple-disc assembly or a band ceases to be a problem. In other words, before the use of a one-way clutch to control a planetary member, transmissions required a large number of hydraulic controls to time the application and release of disc clutches or bands for the same purpose. This is no longer necessary with an overrunning clutch because it acts and reacts to the application of torque in either a clockwise or counterclockwise direction.

In addition, a one-way clutch has a large torque-holding capacity for its size. This makes this unit ideal for use in transmissions that must operate under large torque loads but still provide smooth, trouble-free performance.

ROLLER CLUTCH DESIGN

Overrunning (one-way) clutches may be of two different designs, the roller or sprag. Both designs perform the same function and rely on a wedging action between internal components to lock the clutch up.

Inner and Outer Races

A roller clutch consists of an inner and outer race, a set of rollers, and a number of energizer springs (Fig. 5-17). The inner race, in this installation, has splines machined into its inner circumference. These splines mate with those cut into the stator support of the transmission. With this arrangement, the inner race is held against any rotation by the fixed stator support.

The outside circumference of the inner race is machined with precision accuracy as perfectly round as possible. The rollers will turn upon this area of the race when the clutch overruns.

The outer race fastens into the stator assembly. Located on this race is a series of high spots or cams evenly spaced around its outer circumference. The cams and the ramps leading to them house the rollers and energizing springs and provide the area into which the rollers will wedge themselves when the clutch locks up.

Rollers and Energizing Springs

A number of hardened steel rollers, one for each cam, lie in the areas between the beginning of the ramps to the enlarged portion of the cams. The inner race holds these

rollers in place. In addition, a roller retainer or cage maintains the rollers and springs in their proper relative positions.

Each roller has a wavy, energizing spring that maintains a force on it in the direction of clutch engagement. In other words, the springs force the rollers in a direction, which results in the rollers wedging in the low areas of the ramp between the inner and outer races.

ROLLER CLUTCH OPERATION

When discussing the operation of any one-way clutch, there are two modes or phases of operation: locked up or engaged, and overrun or freewheel. In the lock-up phase, the clutch prevents the rotation of a component in one direction. This discussion will focus on the roller clutch preventing the counterclockwise rotation of a stator within the torque converter.

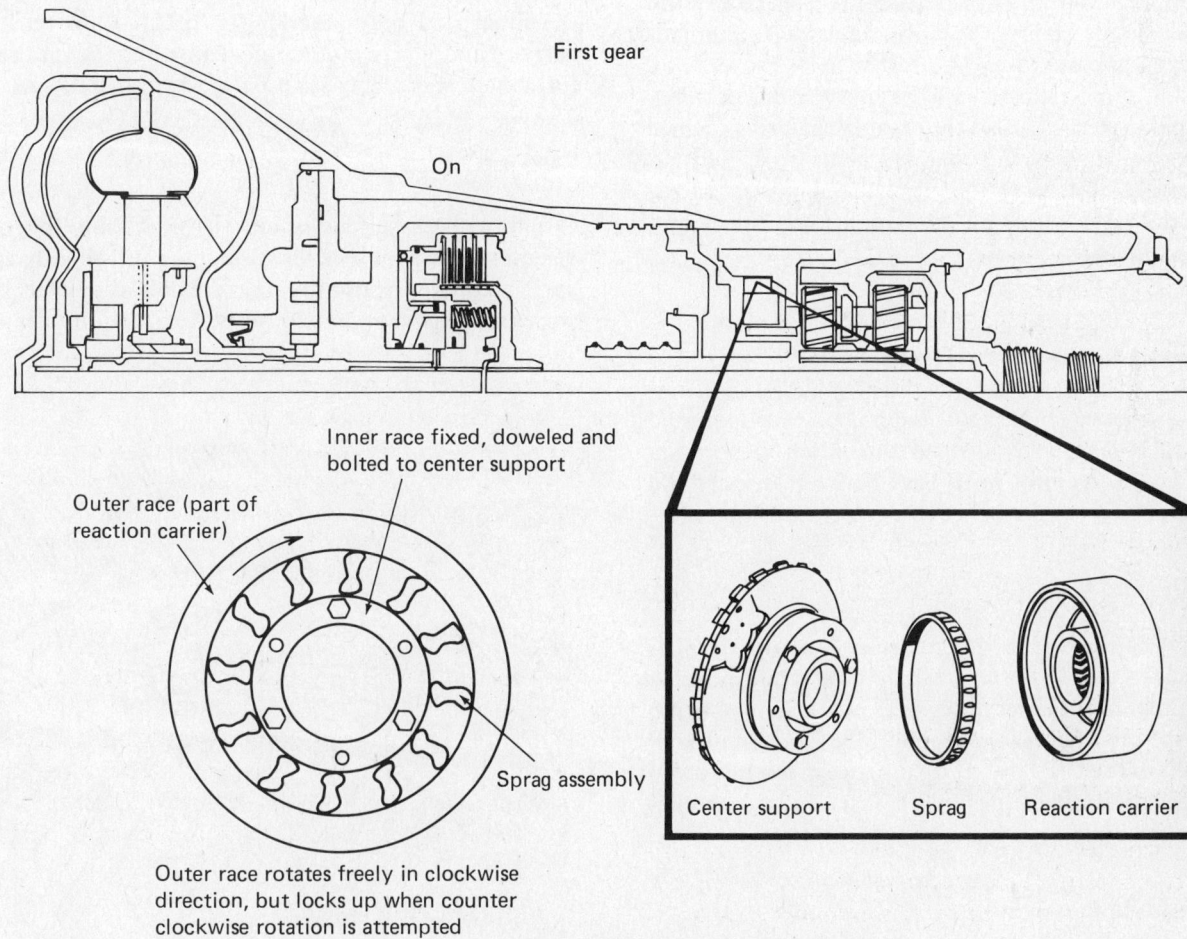

First gear

On

Outer race (part of reaction carrier)

Inner race fixed, doweled and bolted to center support

Sprag assembly

Outer race rotates freely in clockwise direction, but locks up when counter clockwise rotation is attempted

Center support　　Sprag　　Reaction carrier

FIGURE 5-19　A carrier-mounted, sprag-type one-way clutch. (Courtesy of Hydro-matic Division of General Motors Corp.)

During the overrun phase, the clutch allows a part to freewheel in a given direction. Again, using the stator as an example, the clutch will permit this unit to turn freely in a clockwise direction during the coupling phase of the converter.

Lock-Up Phase

When the returning turbine fluid strikes the front face of the stator blades (Fig. 5-17), the stator assembly, outer race, and cam attempt to rotate counterclockwise. However, the rollers, assisted by the energizing springs, move toward the low areas of the ramps, away from the high points of the cam. This action causes the rollers to wedge between the inner and outer races. As a result, the stator assembly locks to the transmission case through the fixed stator support.

Overrun Phase

As the returning turbine fluid strikes the back face of the stator blades, the stator assembly, outer race, and cam begin to move clockwise. The rollers now move away from the low ramp areas toward the enlarged end of cams. This action also compresses the energizing springs. Since the rollers are now in the enlarged cam areas, they no longer wedge between the inner and outer races. Consequently, the stator assembly can rotate freely in a clockwise direction.

SPRAG CLUTCH DESIGN

The sprag type overrunning clutch has no race-mounted cam or rollers (Fig. 5-19). Instead, this clutch consists of a sprag and cage assembly positioned between an inner and outer race.

Sprag and Cage

A sprag is a cam-shaped segment, which looks like a roller that has been overly flattened on two opposite sides. A given number of these sprag segments are held in position by a cage. Furthermore, each of the sprags is kept positioned in the cage and upright between the inner and outer races by strips of spring steel inserted into notches cut into the sprags.

Inner and Outer Races

In the transmission illustrated in Fig. 5-19, the inner race bolts into the center support. The outer circumference of the race is machined smooth and round so that the sprag

segments can ride freely over the surface when the clutch overruns. Also, when the clutch locks up, the inner race and center support absorb the load required to stop the carrier's counterclockwise rotation.

The outer race is built into the reaction carrier assembly. The inner circumference of this race is also machined smooth and round to permit the sprags to move freely over the surface when the clutch overruns.

SPRAG CLUTCH OPERATION

Lock-Up Phase

In a similar manner as the roller clutch, the sprag unit responds to the direction of applied torque and spring action. In the carrier mounted assembly shown in Fig. 5-19, the lock-up phase begins when the planetary gears attempt to roll the carrier counterclockwise. When this occurs, the larger diameter of each sprag wedges between the inner and outer races (Fig. 5-20). As a result, the carrier cannot turn counterclockwise and the sprag clutch locks it to the center support of the transmission.

Overrun Phase

When planetary gear action is such that the carrier attempts to move clockwise, the sprags tilt slightly against spring force positioning the smaller diameter of the segments between the two races. This action releases the

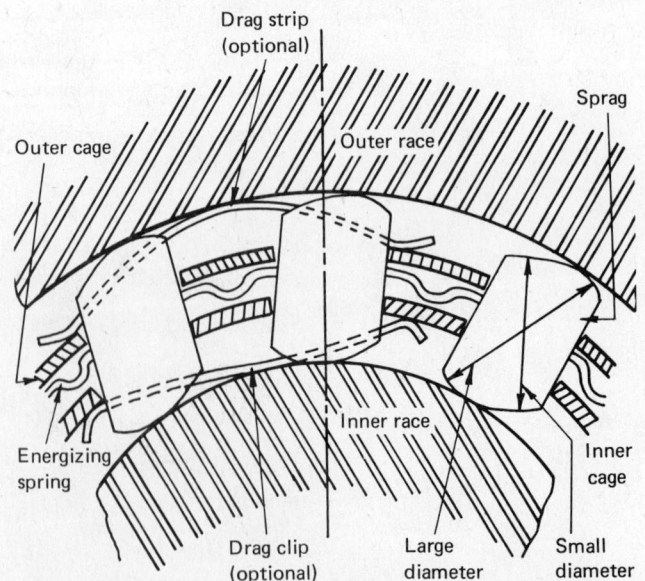

FIGURE 5-20   Sprag action during the lock-up phase of the clutch.

outer race so that it and the carrier can turn freely in a clockwise direction while the inner race remains stationary with the center support.

ONE-WAY CLUTCH ACTION DURING DECELERATION

The only real drawback to using a one-way clutch to hold a planetary member is that the sprag or roller unit can only stop rotation of the gear or component in one direction. As a consequence, neither of these clutches can hold a planetary member during deceleration or under vehicle coast conditions. The reason for this is that during a coast condition, the action of the planetary gearing causes the involved member to move in a direction that releases the clutch—permits it to overrun.

To offset this problem in order for the engine to provide braking during deceleration, manufacturers in many transmissions incorporate a multiple-disc clutch or a brake band to hold the same planetary member normally held by the one-way clutch. However, the disc clutch or band will hold the unit stationary in both directions. In other words, during normal driving conditions, the one-way clutch controls the planetary member, but during deceleration, the disc clutch or band applies to prevent the planetary member from freewheeling.

In the majority of transmissions, the driver has control over the application of this disc clutch or band when and if the braking effect of the engine is desired. In the transmission illustrated in Fig. 5-19, a band fits around the reaction carrier for this purpose. With the band applied, the reaction carrier will be held stationary to the case when gear action attempts to cause it and the sprag clutch to overrun in a clockwise direction. If not used, this band allows the sprag clutch to overrun the carrier, and torque transfer along with the braking effect will cease between the drive wheels and the engine.

SUMMARY

1. Wet-type automatic transmission clutches can either hold a planetary member stationary or connect one to the transmission's input shaft.

2. A primary-type clutch has a drum that directly attaches to the transmission's input shaft.

3. A secondary clutch drum attaches to a planetary member.

4. A stationary clutch assembly has no drum because the transmission case itself houses the clutch components.

5. A primary clutch assembly consists of a drum, hydraulic piston, drive steel plates, driven friction plates, and a hub.

6. The primary and secondary drum forms the cylinder or guide for the hydraulic piston; in the stationary clutch, this area is part of the transmission case.

7. The hydraulic piston changes hydraulic pressure to mechanical force to apply the clutch, and return springs move the piston to the bottom of its bore when the clutch is no longer necessary to transmission operation.

8. The residual check valve allows fluid to escape from under the piston when the clutch is off.

9. Drive or driven steel plates are usually flat and have external teeth or lugs.

10. The drive or driven friction plates are flat and have internal teeth or lugs.

11. Friction plates can have a bronze or composition lining.

12. The type of lining and the fluid used with it determine the friction characteristics of the clutch.

13. The internal teeth of the friction plates mate between indentations in the clutch hub.

14. When a clutch actuates, hydraulic pressure moves the piston within its bore; piston movement forces the steel and friction plates together.

15. The fluid within the clutch assembly cools, lubricates, and determines the harshness of clutch application.

16. A primary clutch, when activated, directly connects a planetary member to the input shaft.

17. When a secondary clutch activates, it connects a planetary member to the primary clutch drum.

18. A stationary clutch assembly, when activated, holds a planetary member stationary.

19. An overrunning clutch stops the rotation of or drives a component in one direction only.

20. The two types of overrunning clutches are the roller and sprag.

21. The roller type consists of an inner and outer race, a cam, a set of rollers, a roller retainer, and a set of springs.

22. The sprag type consists of a smooth inner and outer race, a series of sprags, and a cage.

23. When the roller clutch holds a component in one direction, its rollers are wedged in the low area between the inner and outer races.

24. If a sprag clutch holds a member, the sprags jam between the inner and outer races.

REVIEW

This section will assist you in determining how well you remember the material contained in this chapter. Read each item carefully. If you can't complete the statement, review the section in the chapter that covers the material.

1. The two types of clutches found in the automatic transmission are the multi-disc, wet type and the _____.

   a. friction

   b. liquid

   c. overrunning

   d. antifriction

2. Within a primary clutch assembly, the drive plates are _____.

   a. the steel type

   b. the friction type

   c. half the steel and half the friction type

   d. none of the above

3. Which clutch type does not require a drum?

   a. secondary

   b. stationary

   c. primary

   d. both b and c

4. The valve that releases the trapped fluid from under a clutch piston is known as the _____.

   a. pressure regulating valve

    b. residual check valve

    c. by-pass valve

    d. compensating valve

5. Excessive temperatures generated during clutch application leave the clutch plates via the _____.

    a. drum

    b. hub

    c. piston

    d. escaping fluid

6. Friction plates have teeth cut into their _____.

    a. inner circumference

    b. outer circumference

    c. linings

    d. both d and c

7. Composition lining can be made up of asbestos, metal, and _____.

    a. steel

    b. copper

    c. lead

    d. paper

8. The clutch type that, when activated, connects a planetary member to the input shaft is a _____ assembly.

    a. stationary

    b. secondary

    c. primary

    d. overrunning

9. The device that applies the force to "squeeze" the clutch plates together is the _____.

    a. hub

    b. piston

    c. retainer

    d. springs

10. The clutch type that holds a planetary member in one direction only is the _____.

    a. primary

    b. secondary

    c. stationary

    d. overrunning

For the answers, turn to the Appendix.

CHAPTER 6

Bands and Servos

As mentioned, in order for an automatic transmission to provide a vehicle with a number of gear reductions such as low and second along with reverse, one member of its planetary system must be held stationary. This provides the reaction necessary so that the remaining gear train components can develop the desired reduction and output motion.

The last chapter discussed the use of either a multiple-disc or a one-way clutch to hold a particular planetary member. In this regard, the disc clutch was found capable of holding a gear train member in either a clockwise or counterclockwise direction, while the one-way unit permitted a lock-up of a given component in one direction only.

BAND FUNCTION

Manufacturers have, since the introduction of the automatic transmission, used one or more bands in their respective units. The function of the **band** is to hold a planetary member in both directions during a gear reduction. In operation, the band stops or causes to stop the movement of a planetary member through the use of a constricting or squeezing action.

In an automatic transmission the band fits around a **drum** which, in turn, attaches to the planetary member

(Fig. 6-1). When the band applies around the drum, it and the planetary member cease to rotate. However, as the band releases, the drum and gear-train member are free to turn in either direction.

The reader may wonder why manufacturers continue to use a band for this purpose instead of a clutch; the reason is twofold. First, the band can hold a component in both directions; a one-way clutch cannot. Second, a band and its operating unit, the servo, take up very little space within the transmission case. In fact, many bands encircle a drum that houses a multiple-disc clutch. With this design, the clutch can connect the planetary member to the input shaft during one ratio, and the band can hold the same member to the case during another driving range.

DISADVANTAGES TO THE USE OF A BAND

There are three main disadvantages to the use of a band as a holding device.

1. In order for the band to possess a holding capacity equal to a clutch, the assembly requires a rather large drum diameter for the band to ride on. This results in an increase in the size of the transmission case.

2. Due to the effect of wear on the lining, a band requires periodic adjustment—in most cases, during its service life. If the adjustment is not made, the band may slip causing the planetary member to turn when it shouldn't.

3. A band used in a forward upshift ratio, like second gear, requires a transmission to have a somewhat more complicated hydraulic system than one using a clutch for this purpose. The reason for the more complex system is that in order to lock the band around the drum, its activating piston has to move much further than the one that locks the plates together in a disc clutch. This creates a timing problem on both upshift and downshift operations of the band, especially if the band application must coincide with the operation of a disc clutch.

BAND DESIGN

Backing

There are two parts that make up a band: the backing and the lining (Fig. 6-2). Manufacturers form the backing from a good grade of cast iron or spring steel in such a

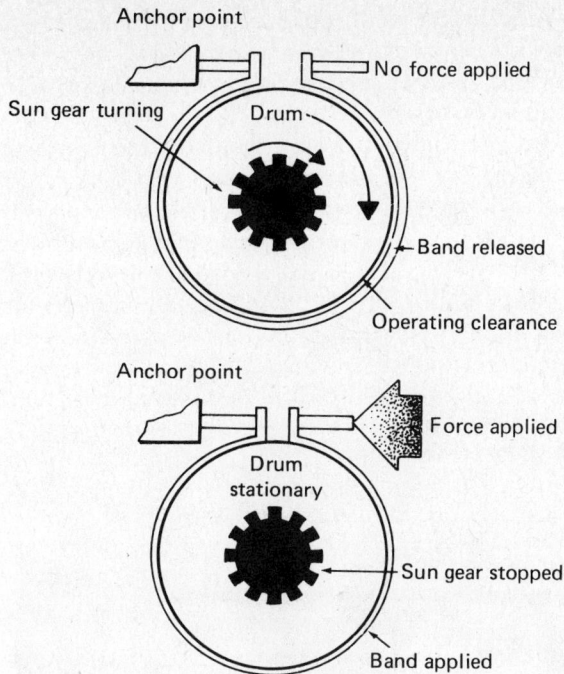

FIGURE 6-1 A band contracts around a drum to keep it from turning.

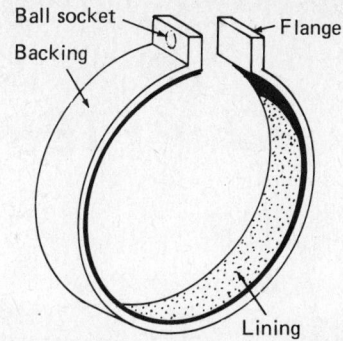

FIGURE 6-2 The construction of a typical band.

manner that the band is elastic. In other words, the ends of the band will spring apart after being forced together (see Fig. 6-1). This action maintains the band in the released, open position when not in use.

In order to mechanically close the opening in the band, the manufacturer casts or welds a reinforced flange to each end of the band. Each flange contains a ball socket or slot with a retaining pin. In either case, the socket or slot accommodates a piece of linkage from either the band's operating piston or from the case.

Lining

The inner circumference of a band is large enough to accommodate a friction lining (Fig. 6-2) but, at the same time, allows sufficient free-opening clearance for the rotating drum. The lining material is very similar to what is found on the friction clutch plates. In other words, the lining is composed of semi-metallic and organic materials.

A composition semi-metallic lining is formed from a mix of copper and lead powders, asbestos, resin binders, and in some cases, metal flakes. This type of lining is used in areas where the band is expected to hold rather large torque loads and operate under high temperatures.

However, a semi-metallic lining does create a problem. That is, the lining will tend to scrape or wear away the drum surface during repeated applications. For this

reason, a semi-metallic lined band is only used in areas of a transmission in which the application is against a stopped or slowly turning drum, such as low and reverse.

An organic lining is usually made from paper pulp or cellulose-based compound. The resulting lining is therefore relatively soft and conforms well with the drum surface. More importantly, the softer material will not normally cause wear problems on drum surfaces. For this reason, a paper-lined band is used in applications to stop a rotating drum, even one turning at several thousand revolutions.

Both semi-metallic and organic lining may have a pattern of grooves cut into its surface. This grooving provides for the controlled escape of fluid and vapors from between the band and drum during application. Moreover, the grooves along with the type of drum material and surface finish, fluid, and lining material determine the band's friction characteristics.

BAND TYPES

A typical automatic transmission may use one or more of the three different types of bands on the market. The different types of bands are the single-wrap, heavy duty; the single-wrap, light duty; and the split (Fig. 6-3). The type and size of the band used for a given transmission configuration depends on the torque loads the unit is expected to handle.

Single-Wrap, Heavy Duty

The **single-wrap, heavy duty band** is utilized in an area in which it must handle a rather high torque load. By single wrap, this means that the backing is a single piece of metal, either cast iron or steel. In addition, the cross-sectional area of the band is rather thick, with reinforced linkage receptacles. Moreover, the band width is wider than the light-duty counterpart, and the thick lining is usually semi-metallic with grooves.

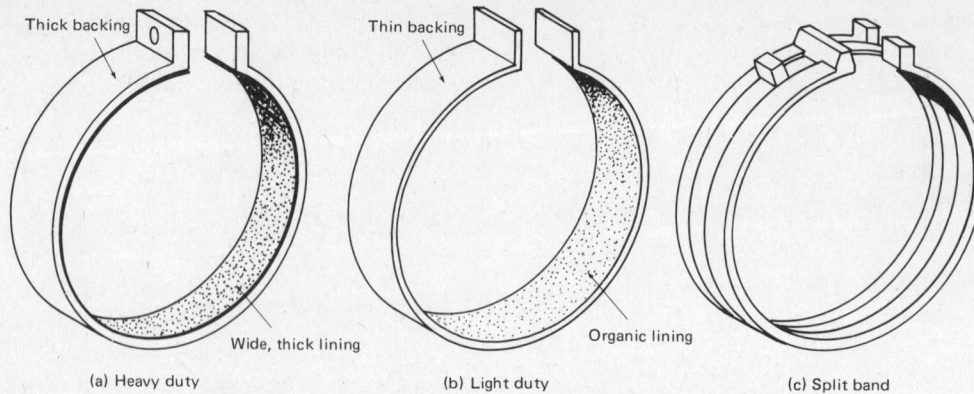

Thick backing

Thin backing

Wide, thick lining

Organic lining

(a) Heavy duty

(b) Light duty

(c) Split band

FIGURE 6-3   Bands may have (a) the single-wrap, heavy duty, (b) the single-wrap, light duty, or (c) the split, double-wrap design.

Single-Wrap, Light Duty

A **single-wrap, light duty,** as its name implies, is found only in areas where torque loads are low. This band is made from steel and usually has a thin organic lining bonded to its inner circumference. Furthermore, because this band's backing is rather thin, it is commonly known as a flexible type. A **flexible band** is one which, when removed from a transmission, does not retain its circular shape.

Split

For heavy-duty installations, most manufacturers install a **split band** instead of a solid type for three reasons. First, the split band matches the circular shape of the drum better than a single wrap. Therefore, it provides a greater holding ability with equal application force. Second, a double-wrap band, as it is sometimes called, engages more smoothly with the drum surface than the single wrap does. Consequently, the band provides a smoother shift quality. Third, a split band responds more effectively to the action of self-energization than either the heavy- or light-duty single-wrap units. This chapter will discuss self-energization later on.

In place of one solid piece of curved metal, the split band is a unit formed of three narrow individual segments (Fig. 6-4). During the forming process of this type of band, the manufacturer connects the two servo ends of the two outer segments, along with one end of the center segment, together with a metal strip. The free end of the center segment fastens directly to the band's apply linkage through a reinforced flange. The free ends of the outer segments also have flanges that accommodate an anchor, which pivots in the transmission case. Finally, the split band usually has a semi-metallic type lining bonded to the inner circumference of each band segment.

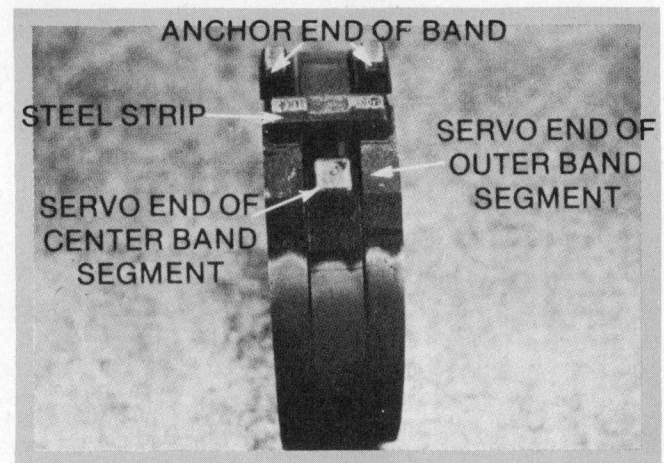

ANCHOR END OF BAND

STEEL STRIP

SERVO END OF OUTER BAND SEGMENT

SERVO END OF CENTER BAND SEGMENT

FIGURE 6-4   The construction of a split band.

BAND LINKAGE TYPES

As mentioned earlier, the backing portion of the band has two flanges, which accommodate several pieces of linkage. One piece of linkage acts as an anchor that attaches to the case (Fig. 6-5). The other link connects a flange to the servo, which contains the piston responsible for forcing the ends of bands together. The links or rod connected to the servo and the band are known as operating linkage.

Simple Rod with Adjuster

There are four types of operating linkages used in the various types of automatic transmissions. The first type is nothing more than a rod that connects between the servo

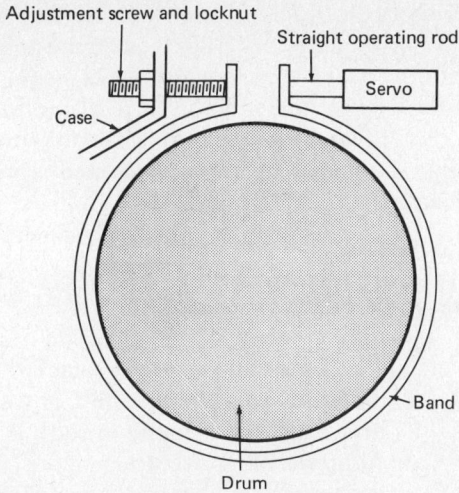

FIGURE 6-5 A straight operating rod delivers the motion of the piston to the band.

piston and band flange. This rod transfers the motion of the servo piston to the free operating end of the band (Fig. 6-5).

The first type of operating linkage (rod) has a given length. Therefore, there must be another method used to adjust band clearance. In this situation, the adjuster is incorporated into the anchor end of the band linkage.

An automatic transmission that uses the simple rod is one in which the servo placement is such that its piston can act directly on the band. Also, manufacturers use the

rod linkage in installations where the servo has a piston that can apply sufficient force to prevent band slippage, regardless of the applied torque on the band.

Lever Type with Adjuster

The second type of operating linkage uses a lever arrangement (Fig. 6-6). This kind of linkage is necessary when the servo location is such that its piston cannot act directly in a straight line on the band to apply it. The lever and its attached linkage change the direction of the piston's motion so that it can apply the band. At the same time, the lever can multiply piston force on the band if the longer side of the lever is on the servo side of the fulcrum. The fulcrum, in this case, is the pivot pin placed through the apply lever.

As in the case of the first type of operating linkage, the second also requires some form of adjuster to provide the proper amount of band clearance. The adjuster for the lever arrangement may be part of the anchor linkage, which is easily manipulated from outside the transmission case. Or, on some transmissions, there is an adjustment screw on the lever. In this situation, pan removal is necessary to perform a band adjustment.

Cantilever Type with Adjuster

The third kind of operating linkage is known as a **cantilever**. With this form of linkage, the band does not have an anchor. Instead, the cantilever itself attaches to

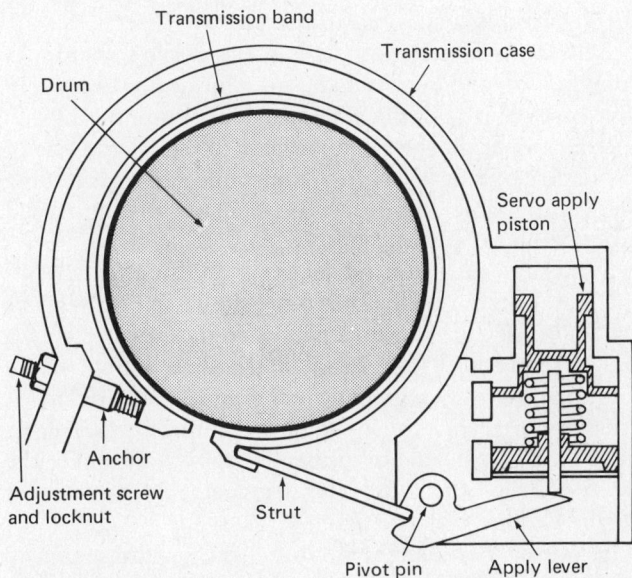

FIGURE 6-6 Lever-type operating linkage is necessary when a servo piston cannot act in a straight line on the band.

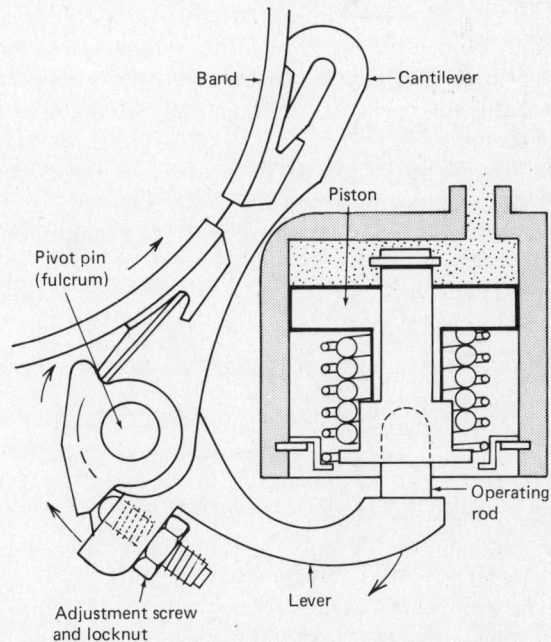

FIGURE 6-7 The cantilever linkage acts on both flanges at once to tighten a band around a drum.

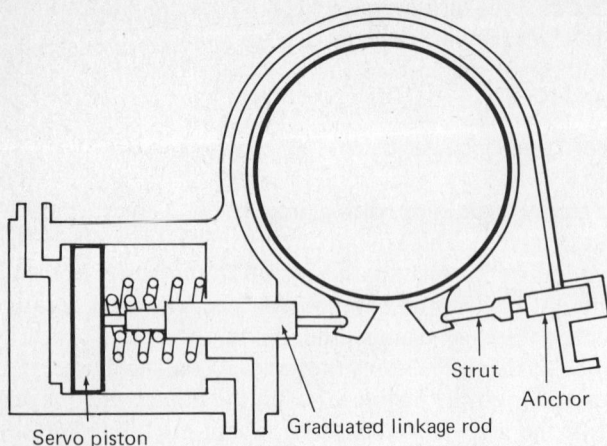

FIGURE 6-8 In some transmissions, a rod of graduated length is used to adjust the band's operating clearance.

both band flanges (Fig. 6-7). Therefore, the cantilever, its linkage, and the servo bear the torque load from the band, necessary to stop and hold the rotating drum.

Note in Fig. 6-7 that the lever arrangement is such that it also provides a directional change in applied motion and multiplies the force of the servo piston. Notice also that the band adjustment screw threads into the lever, so band clearance is set from inside the transmission pan.

Graduated Rod Type

The final type of operating linkage is quite similar to the first (Fig. 6-8). The main difference is that there is no adjuster built into the operating linkage or anchor. Instead, the servo has a graduated rod that fits between its piston and the band flange. This graduated rod is of a given length.

To provide the band with the proper clearance, adjustment is made by selecting and then installing one of several different rods. In some cases, the manufacturer

will identify each of the different length (graduated) rods through the use of a number of rings located around its band end. With the proper rod in place, the servo piston moves a specific distance to apply the correct amount of force to the band. When the piston moves back, the band opens, providing the correct operating clearance between itself and the drum.

SERVO AND ANCHOR STRUTS

Often, the name given to the linkage that attaches the servo piston to the band or from the case anchor to the band is "strut." If the linkage is not round (a rod), it is a strut. The strut can fit between the lever end and the band flange (Fig. 6-6) or between the anchor and flange (Fig. 6-8).

SELF-ENERGIZING ACTION OF BANDS

Whenever possible, engineers design the location of the servo and its operating linkage along with the case anchor point to take advantage of the self-energizing effect of drum rotation on the band when it is applied. **Self-energizing action** is the tendency of the band, after contacting a revolving drum, to roll with the drum (Fig. 6-9). This action occurs during initial band application and only if the servo piston tightens the band in the direction of drum rotation. If the drum turns in the opposite direction to servo piston travel, the effect of self-energization is lost.

Self-energizing action assists the servo's piston in applying force to the band and therefore helps to prevent excessive band-to-drum slippage. This action works well enough to permit the manufacturer to reduce the size of the band's operating servo piston while still maintaining the same tightening force as a larger piston would produce on a band without the energizing effect.

With these facts in mind, let's examine the effect of self-energizing action on the split band shown in Fig. 6-9. The action begins as the servo piston applies a force to the center segment of the band in the direction of drum rotation. As the segment first contacts the drum, it attempts to revolve in the same direction. Furthermore, since the opposite end of the center segment fastens to the servo ends of the two outer segments, the entire unit begins to turn with the drum.

However, because the opposite, outer band segments anchor to the case, the entire unit begins to wind or cinch up tightly around the drum. This action, along with the servo's applied piston force, tends only to permit a minimum amount of slippage between the band and drum

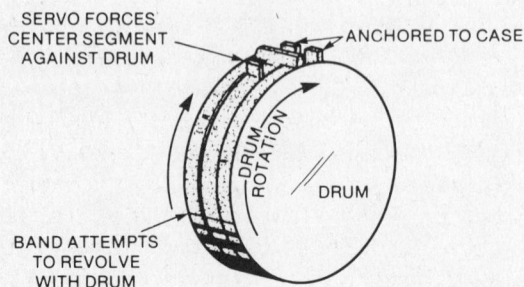

FIGURE 6-9 "Self-energizing action" begins as the servo piston forces the center band segment against the rotating drum.

because any slippage causes the band to cinch up even more around the drum.

Although our example used a split band, other band designs will react nearly the same but not as effectively. As a result, in installations in which the manufacturer expects very high torque loads, or which, for a design reason, needs to reduce servo piston size, the split band will replace a solid type for increased holding power.

BAND SERVOS

Function

For every band in the transmission, there will be a hydraulic servo. A **servo,** simply speaking, is a device that converts hydraulic pressure to mechanical force. This force is necessary to apply the band around the drum tightly enough to stop its rotation.

SERVO DESIGN

Housing

A servo assembly consists of a housing, piston, and return spring. The manufacturer can design the servo housing into the transmission case (Fig. 6-10), or it may be a detachable unit (Fig. 6-11). In either situation, the servo housing encloses all the remaining components and has a bore or cylinder that acts as the guide for the piston. In addition, the housing has apply- and sometimes release-pressure ports machined into it.

FIGURE 6-10   A servo housing cast into the transmission case.

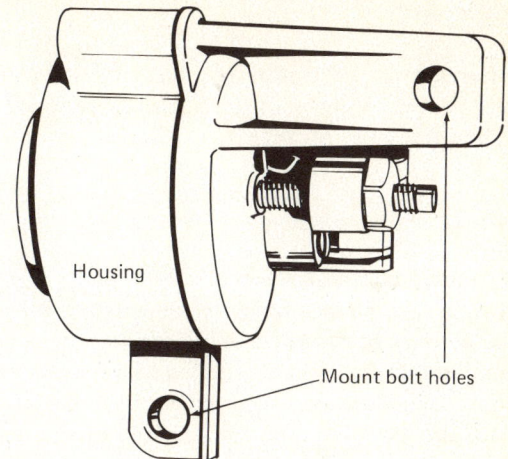

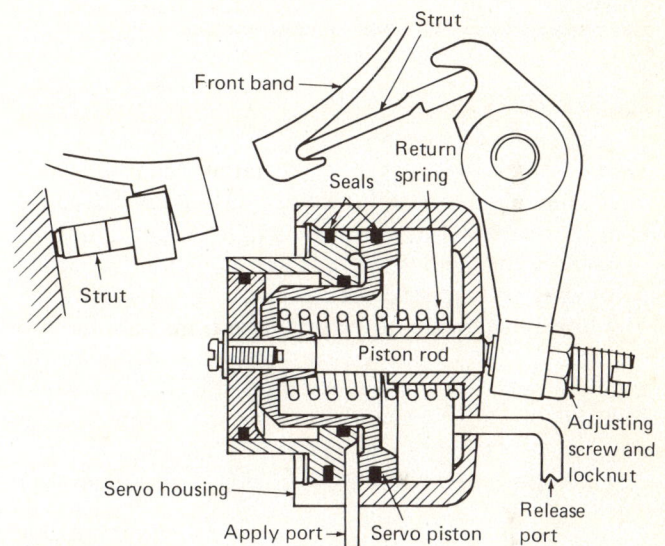

FIGURE 6-11   A servo that bolts to the transmission case.

Piston

The **piston** is the device that actually converts the hydraulic pressure into mechanical force to apply the band around the drum. The piston is round to match the opening in the servo housing and has one or more seals to prevent excessive leakage as it moves within the bore.

The area of the piston determines how much output force it will produce. For example, a piston with an area of 2 square inches operated by a pressure of 50 pounds per square inch (psi) will produce 100 pounds of force on the band (output force is equal to piston area times pressure). On the other hand, a piston with an area of 4 square inches acting under 50 psi produces 200 pounds of force, which is twice as much as the 2-inch piston. It should be quite obvious that the larger the piston area at a given pressure, the greater is the output force on the band.

Piston Rod

Attached to the piston is some form of rod. This rod transfers the motion of the piston directly to the band or to it via a lever arrangement or strut.

Spring

The majority of all servo assemblies contain at least one spring. The **return spring** is responsible for moving the piston back to its off (unapplied) position whenever there is no hydraulic pressure applied to the servo. However, in some installations, hydraulic pressure is also used along with the spring's force to move the piston back.

SERVO TYPES

Single-Acting

When considering types of servo designs, there are three kinds: the single acting, double acting, and the controlled load. A **single-acting servo** is one in which hydraulic pressure activates. During the release phase, the spring alone moves the piston back in its bore in the housing. The hydraulic system of the transmission directs fluid to the apply pressure port of the servo (Fig. 6-12). In the example shown in the illustration, the fluid pressure passes through the hollow piston rod to the apply side of the piston—the area between its head and the cover.

The fluid pressure acting on the area of the piston's head causes it to move in the housing bore, thus compressing the spring. At the same time, the movement of the piston begins to tighten the band around the drum. This process continues until the band locks around the drum with a given amount of force.

When the band application is no longer necessary for transmission operation, the hydraulic system stops fluid pressure to the servo. With no pressure applied, the return spring moves the piston back in its bore to starting position while, at the same time, the band releases its grip on the drum.

Double-Acting

A **double-acting servo** is one in which hydraulic pressure not only activates the piston to apply the band but can be used to release it as well. In the transmission shown in Fig. 6-13, the servo has a piston with two diameters. The larger diameter of the piston operates in the bore within the transmission case whereas the smaller diameter moves in the bore in the cover. Note also in the drawing the two seals of different diameters used to seal the two pistons.

During the application of the band, hydraulic pressure enters the servo, passes through the piston rod, and acts on the smaller piston area. With hydraulic pressure acting on the piston, it moves and applies the band in much the same way as a single-acting servo.

However, the release of the band by the servo can be accomplished in several ways (Fig. 6-14). For example, by cutting off hydraulic pressure to the apply side of the small piston, the return spring would push it back in its

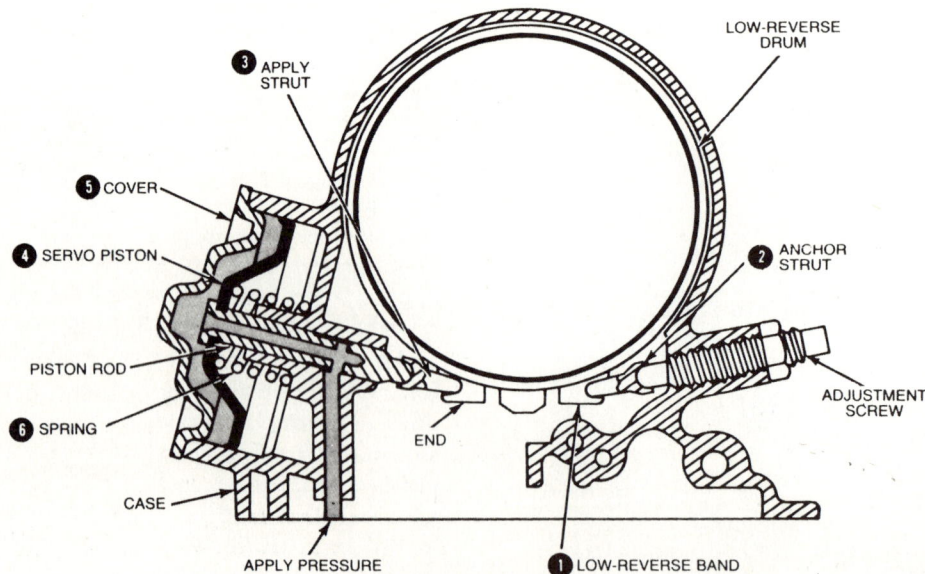

FIGURE 6-12 The design and operation of a single-acting servo. (Courtesy of Ford Motor Company)

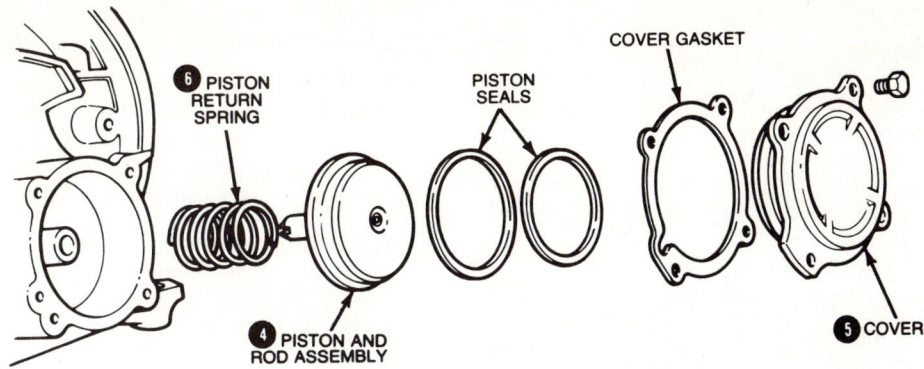

FIGURE 6-13 The design of a double-acting servo. (Courtesy of Ford Motor Company)

bore, thus opening the band. Or the hydraulic system can maintain the pressure on the apply side and direct the same amount to the release side. The release pressure acts on the large diameter piston. This action, along with the spring force, easily overcomes the effect of pressure on the apply piston, and the entire unit moves back, releasing the band.

Controlled-Load

A **controlled-load servo** is one designed to cushion the application of the band around the drum. There are basically two types of these units, one that cushions mechanically and the other hydraulically.

In the mechanical type of controlled-load servo, the piston has a spring-loaded plug located at its center; this plug is responsible for cushioning band application (Fig. 6-15). When the band is open, the smaller diameter piston spring extends the plug in its bore in the main piston. If the hydraulic system admits pressure to the area above the main servo piston, it and the plug move down, taking up all the band-to-drum operating clearance.

As the main piston continues to move, it begins to compress the plug spring because the plug and the band are resisting any further movement. However, fluid

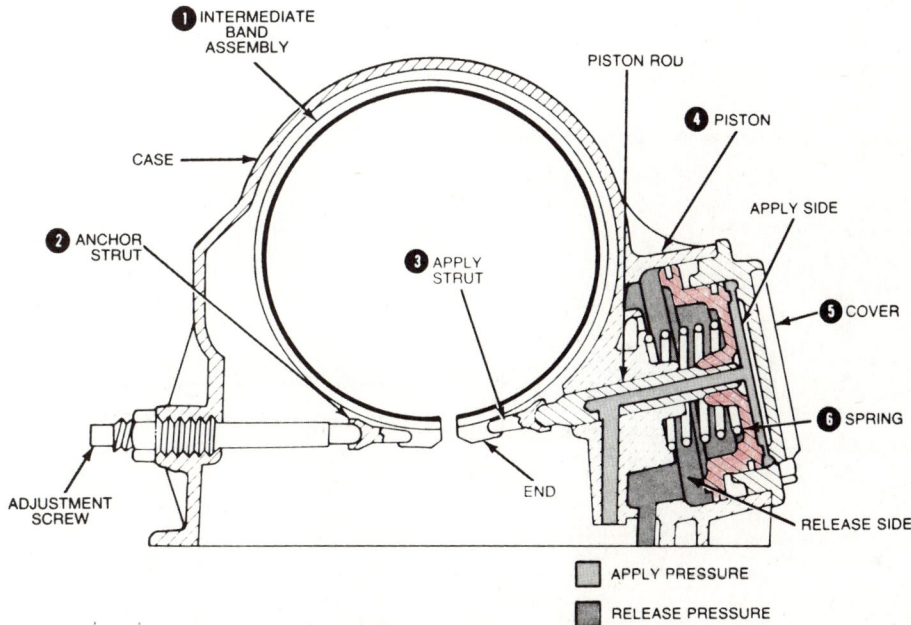

FIGURE 6-14 The operation of a double-acting servo. (Courtesy of Ford Motor Company)

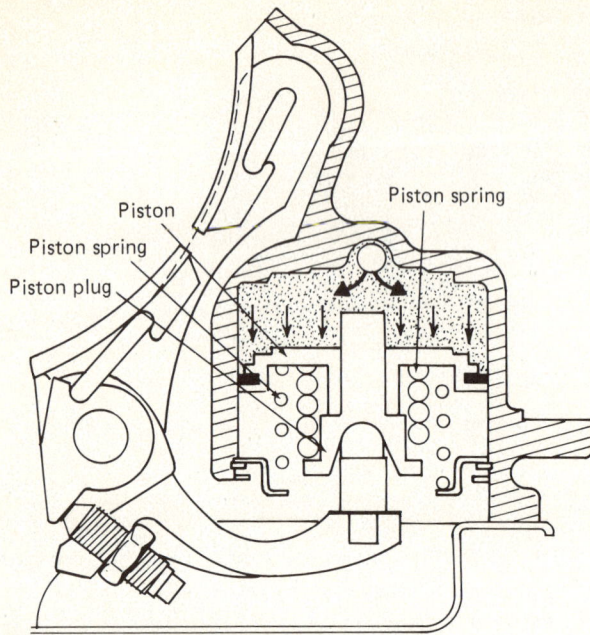

FIGURE 6-15 The design and operation of a mechanical controlled-load servo. (Courtesy of Chrysler Corp.)

pressure continues to move the piston down over the plug until a shoulder on it stops the main piston's travel. The period of time from the initial piston movement until it is down against the plug's shoulder represents a delay in the band tightening period. This action cushions the total band application.

A hydraulic type of controlled-load servo is shown in Fig. 6-16. This unit consists of a two-land main piston, which operates in a stepped housing bore. Each of the piston lands has a different diameter and has a groove cut into it, which accommodates a seal. The center section of the piston forms a guide or bore for a smaller piston and

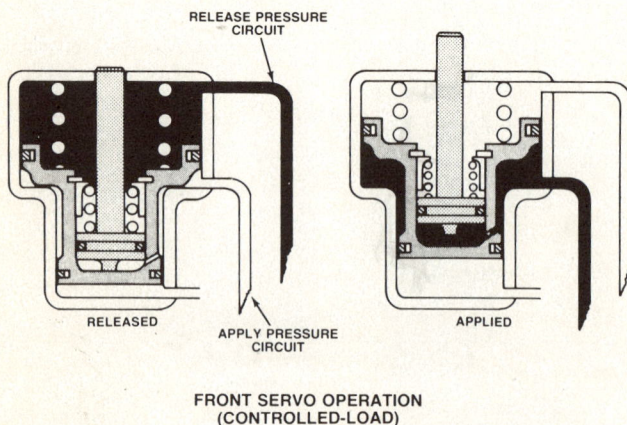

FIGURE 6-16 The design and operation of a hydraulic controlled-load servo. (Courtesy of Chrysler Corp.)

stem. This second, smaller piston as well as the main piston are controlled by hydraulic pressure.

This piston arrangement requires two return springs. The larger diameter spring maintains the main piston at the bottom of its travel when the band is open whereas the smaller spring attempts to keep the inner piston at the base of its bore in the larger unit during the same period of time.

During band application, the hydraulic system directs fluid pressure to the apply side of the servo. The fluid fills the cavity between the two lands and pushes against the area of the piston below the large land. This action causes the main servo piston to move upwards in its bore within the housing. Once the band applies, the inner bore of the main piston will try and fill with fluid admitted through the orifice, the calibrated opening in the piston wall. However, band load prevents the small piston from moving up.

To release the band, the hydraulic system applies pressure to the return spring side of the piston while still maintaining fluid flow to the apply port. As a result, pressure and the spring move the piston down in its bore. During the downward travel, the inner piston cylinder fills with a small amount of pressurized fluid that was trapped in the cavity between the two lands and the main piston bore. The trapped fluid pushes the small piston upward, compressing the small return spring.

If and when the servo applies the band again, the fluid, trapped under the small piston and the cavity, will act to cushion the band application. In order for the small piston to move back to the bottom of its bore in the larger one so the main piston can fully apply and load the band, the trapped fluid has to move back out from beneath the inner piston through the orifice, which takes a period of time to accomplish. Consequently, this action delays the loading, full tightening period of the band and therefore cushions band application.

BAND OPERATION

Applied

With all the facts on band and servo construction in mind, let's turn our focus of attention to just what occurs when a band applies and releases its hold on a drum. When a band is necessary to operate the transmission in a particular ratio, the hydraulic system sends fluid under pressure to the apply port of the servo. The servo piston begins to move and force the ends of the bands together. If the drum is in motion, the band also responds to self-energizing action (Fig. 6-17).

As the band's squeezing action takes up the operating clearance between the band's lining and the drum, its

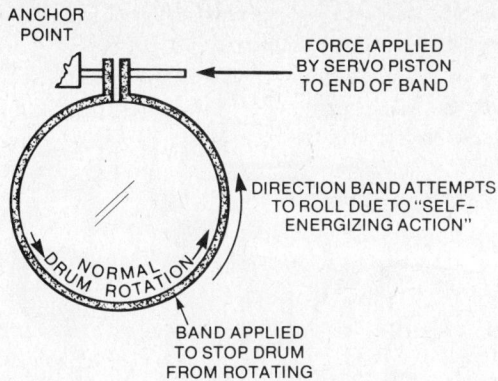

FIGURE 6-17 The force of the servo piston applies the band around the drum and stops its rotation.

speed begins to decrease; that is, if the drum was in motion prior to band application.

At this point, the band and drum operating temperature begins to rise rapidly due to the increase in friction between the fluid layers, band, and drum. As in the disc clutch, the fluid removes the excess heat and controls the amount of friction between the band's lining and the drum. The layers of fluid between the lining and drum absorb a good deal of the heat and carry it away as the oil squeezes out from under the band. Then the heated fluid returns to the pan, where its temperature is reduced.

Because the fluid layers between the lining and drum act as a wedge, band-to-drum friction is low. As the band tightens around the drum, the fluid layers lubricate, preventing actual contact between the two units, until the drum stops rotating. By this time, most of the excess fluid is gone from between the two parts and some actual contact occurs. But the contact is usually only minimal or spotty due to the type of lining material, lining surface roughness, and residual fluid absorbed into the composition material.

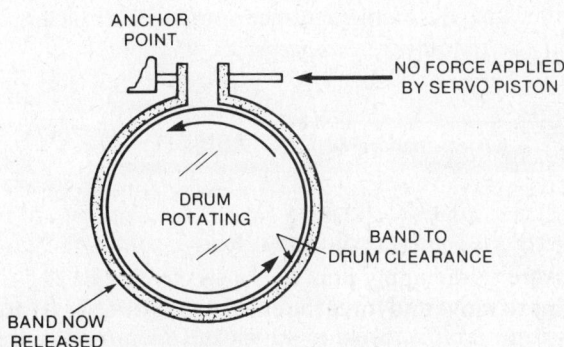

FIGURE 6-18 As the servo releases its force on the band, it springs open.

Released

When the hydraulic system stops sending pressurized fluid to the servo apply port or directs it to the release port, the band begins to release (Fig. 6-18). The servo's operating piston moves back out of the way, and the band's natural elasticity causes it to open. Next, the drum begins to rotate within the band as soon as operating clearance appears between the two components. Once the clearance returns, fluid fills this area and prevents any band-to-drum contact while at the same time cooling the components.

TORQUE CAPACITY OF A BAND

Transmission engineers design a band and drum arrangement so that it provides a given holding capacity on the affected planetary member, under all operating conditions. The torque capacity of any band-to-drum configuration is determined by the friction characteristics produced when in operation along with a number of design considerations.

The friction characteristics involved include:

1. **Friction material.**

2. **Drum material.**

3. **Type of fluid used.**

The design considerations that alter torque capacity are:

1. **Drum diameter.**

2. **Band and lining width.** Fig. 6-19 shows two bands that the same type of transmission can use. The manufacturer installs the larger band on the right if the

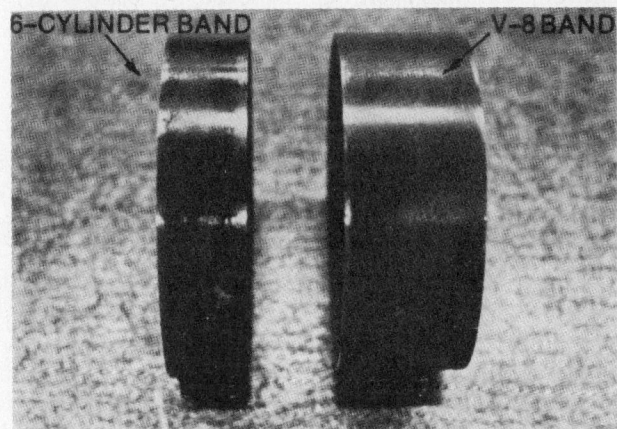

FIGURE 6-19 Two bands used in the same transmission design but with different widths.

transmission bolts up to a V-8 engine. But if the vehicle has a 6-cylinder engine, the transmission comes from the factory with the narrow band on the left. The wider band, therefore, carries the increased torque output produced by the V-8 engine.

3. **Amount of hydraulic pressure applied to the servo.** The higher the applied pressure, the greater will be the output force acting on the band.

4. **Diameter of servo piston.** The larger the piston is within design limitations, the greater will be the piston's output force on the band.

5. **Operating linkage type.** If the operating linkage contains a lever, the force on the band by the servo is higher.

6. **Servo position.** If the position of the servo is such that the band is influenced by self-energization, its holding capacity is higher.

7. **Normal direction of drum rotation.** If the direction of the drum rotation is counterclockwise, the band's capacity is increased due to self-energizing action. However, if the same drum turns clockwise, this action is lost. When a transmission uses the same band as it shifts from low to reverse, the hydraulic system increases the fluid pressure to the servo, which increases the piston's output force on the band. This action makes up for the loss of self-energizing effect on the band.

BAND ADJUSTMENTS

Function

When an automatic transmission consistently uses a band during its normal operating sequence, the band usually requires periodic adjustment or it may no longer be able to hold the drum and planetary member from rotating. The reason for this adjustment is twofold: (1) the servo piston can only move so far before it runs out of travel, and (2) the band-to-drum clearance increases as the lining wears. Therefore, band adjustment is periodically necessary to maintain the proper operating clearance. Servo piston travel can easily take up this clearance and still provide the force on the band to lock the drum stationary.

It would appear that a band should operate with the least amount of clearance possible; however, this is not the case. A given amount of operating clearance is necessary to allow sufficient fluid to circulate between the band and drum for adequate cooling and lubrication. For

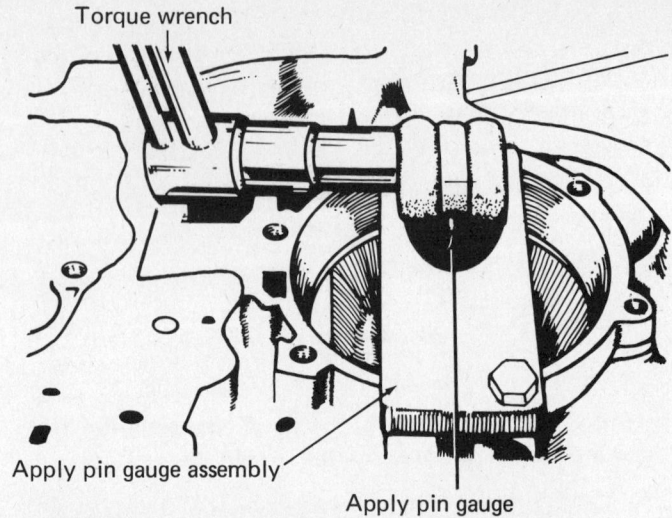

FIGURE 6-20 Checking the rear-servo apply pin length of a T-400 transmission using a special gauge.

FIGURE 6-21 An internal band adjuster located on the servo's operating linkage.

this reason, the band clearance should never be less than what the manufacturer specifies.

TYPES OF BAND ADJUSTMENT

Selective (Graduated) Piston Rod

There are three methods commonly employed by transmission manufacturers to adjust the operating clearance of bands: through the use of a (1) graduated rod, and (2) internal or (3) external adjuster. When a transmission uses a selective graduated rod, the band-to-drum clearance becomes a factor determined by the diameter of the drum, thickness of the band and lining, the stroke of the piston, and the length of its operating rod. With this arrangement, when the band-to-drum clearance becomes excessive due to lining wear, the mechanic has no alternative but to replace the band, piston rod, or both.

However, during transmission overhaul of a unit that uses this type of adjustment, the mechanic must check the band-to-drum clearance to see if it is to specifications. Fig. 6-20 shows the use of a torque wrench and apply pin gauge assembly on a T-400 transmission to check the operating clearance of its rear band. If the clearance is excessive but the band lining is in good condition, the technician can decrease the operating space by installing a longer piston rod (apply pin). In other words, the manufacturer supplies selective servo piston rods (pins) of various graduated lengths in order to adjust band clearance during the overhaul process.

Internal Adjuster

Many transmissions have a band adjuster located inside the fluid pan. In order to gain access to it, the pan requires removal. Fig. 6-21 shows such an adjustment screw located on the operating linkage of the servo.

To properly adjust this type of arrangement, a special wrench and gauge are necessary (Fig. 6-22). The gauge fits between the piston rod and adjusting screw (refer back to Fig. 6-11), while the wrench is necessary to apply a given amount of torque to the adjuster. Then the mechanic backs off the adjuster a number of turns as specified to provide the proper band-to-drum clearance.

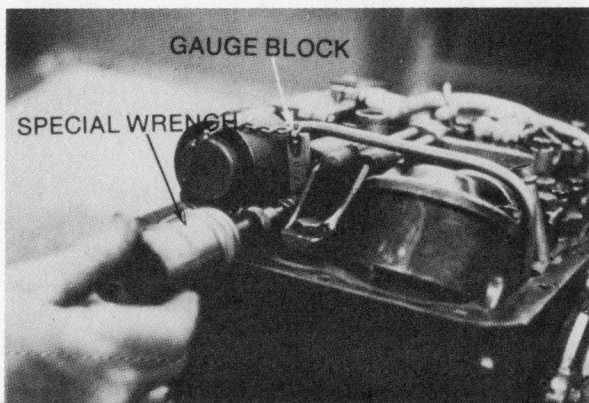

FIGURE 6-23   An external band adjuster located on the outside of the transmission case.

External Adjuster

Figure 6-23 shows an anchor type of band adjuster located on the outside of the transmission case. The adjusting screw is actually part of the anchor that prevents the band from turning in the transmission case. By turning the adjusting screw one way or the other, the band clearance increases or decreases because the anchor end moves toward or away from the servo end of the band.

To adjust the band-to-drum clearance with this arrangement, the mechanic tightens the band adjuster a specified amount with a torque wrench (Fig. 6-24). This action tightens the band around the drum. Next, the adjuster is backed off (loosened) a specific number of turns to set the band clearance.

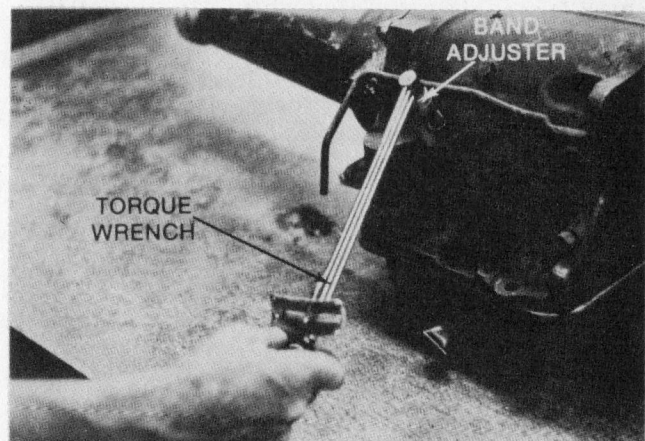

FIGURE 6-22   An internal band adjustment using a special wrench and gauge block.

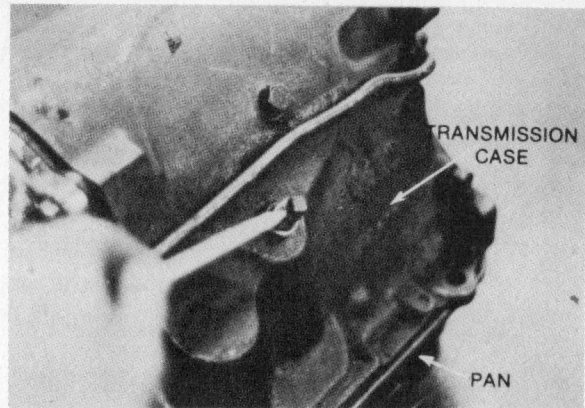

FIGURE 6-24   An external band adjustment using a torque wrench.

SUMMARY

1. The function of a band is to hold a planetary member in both directions during a gear reduction.

2. There are two reasons for the continued use of bands in automatic transmissions.

3. There are three disadvantages to the use of a band as a holding device.

4. There are two parts that make up a band, the backing and the lining.

5. Bands are made out of cast iron or steel in such a manner as to make them elastic.

6. Band lining is composed of either a semi-metallic or organic material.

7. A semi-metallic lining tends to wear away the drum surface.

8. An organic lining is used in applications where the drum rotates at high speeds.

9. Grooving of the lining improves the friction characteristic of the band.

10. There are three types of bands: the single-wrap, heavy duty; the single-wrap, light duty; and split.

11. A flex band does not retain its shape when removed from the transmission.

12. A split band is used in heavy-duty applications instead of the single-wrap type for three reasons.

13. The split band has three individual segments, each with its own lining.

14. The links, lever, or rod connected to the servo and the band is known as operating linkage.

15. There are four types of operating linkage: the simple rod with adjuster, lever type with adjuster, cantilever type with adjuster, and the graduated rod.

16. Strut is another name for band linkage for other than round rods.

17. Self-energizing action of the band is its tendency to roll in the direction the drum is turning.

18. A servo converts hydraulic pressure into mechanical force to apply the band.

19. A servo assembly consists of a housing, piston, and return spring.

20. The area of the piston determines how much output force it will produce.

21. There are three kinds of servos: the single-acting, double-acting, and controlled-load.

22. A single-acting servo piston is applied by hydraulic pressure and released by spring force.

23. A double-acting servo has a piston activated by hydraulic pressure and released by pressure and spring force.

24. A controlled-load servo cushions the application of a band.

25. The small piston of a controlled-load servo can operate mechanically or hydraulically.

26. There are two phases of band operation, applied and released.

27. Due to a number of factors, a band will provide a given amount of holding capacity.

28. There are two reasons why a band requires a periodic adjustment.

29. There are three methods used to adjust bands: the graduated rod, and internal or external adjusters.

REVIEW

This section will assist you in determining how well you remember the material contained in this chapter. Read each statement carefully. If you cannot complete the statement, review the section in the chapter that covers the material.

1. A band holds a planetary member stationary in _____.

   a. one direction only

   b. two directions

   c. a clockwise direction

   d. a counterclockwise direction

2. The band wraps around a _____.

   a. drum

   b. servo

   c. piston

   d. planetary member

3. The device that operates a band is a _____.

   a. actuator

   b. adjuster

   c. spring

   d. servo

4. The band type that is most effective in holding a planetary member is the _____.

   a. single-wrap, light duty

   b. single-wrap, heavy duty

   c. split

   d. both a and b

5. Bond lining is usually a _____ material.

   a. lead

   b. bronze

   c. steel

   d. composition

6. The piece of linkage that connects the band to the case is the _____.

   a. anchor

   b. strut

   c. link

   d. b or c

7. Fluid flows between the band and drum to reduce their operating temperature and _____.

    a. friction

    b. wedging effect

    c. efficiency

    d. none of the above

8. The self-energizing effect on the band occurs during band _____.

    a. release

    b. application and release

    c. application

    d. loading

9. The type of servo that cushions band application is the _____.

    a. single-acting

    b. dual-acting

    c. standard duty

    d. controlled-load

10. The single-acting servo has _____.

    a. one piston

    b. two pistons

    c. two springs

    d. always an external housing

11. Band adjustment is necessary in many automatic transmissions because of _____.

    a. band wear

    b. servo wear

    c. weak servo springs

    d. insufficient servo travel

12. If the band-to-drum clearance is less than what the factory specifications call for, the band may burn out due to a lack of _____.

    a. wear

    b. lubrication

    c. servo travel

    d. elasticity

For the answers, turn to the Appendix.

A modern automatic transmission contains an average of about 25 sealing devices. The purpose of these devices, gaskets, seals, and steel or Teflon rings is to control the amount of external and internal leakage of transmission fluid.

External leakage, fluid that leaks to the outside of the transmission itself does not cause a loss of system pressure. But it is very obvious to the vehicle owner because of the messy oily spots left on the driveway or garage floor. Not only is this type of leakage an annoyance in that it creates a tedious, difficult clean-up job, but it also forces the vehicle owner to check and add fluid to the transmission on a regular basis. If this is not done, the fluid level will drop to a point where the transmission will malfunction and most likely sustain some damage.

An **internal leak,** on the other hand, is the excessive loss of fluid from a sealed component inside the transmission itself. This form of leakage does result in a loss of pressure and may develop over a period of time. Consequently, the driver may not become aware of the problem until the transmission really begins to malfunction. In other words, the driver may feel the transmission slipping a little bit, on initial engagement or during an upshift, due to the loss of pressure caused by the internal leak; but he will usually tolerate these conditions as long as the vehicle continues to operate reasonably well so that he can avoid the high cost of repair.

LEAKAGE CONTROL— A LARGE TASK

The control of external and internal fluid leaks is a large undertaking for the automatic transmission manufacturer for three reasons:

1. Confining the fluid over a long distance

2. Confining the fluid when components are in motion

3. Confining the fluid under high temperature conditions.

Confining The Fluid Over a Long Distance

First, in order for the fluid to lubricate or activate the various transmission components, it must, in many cases, travel some distance and through various components before reaching its final destination. For example, in one model of automatic transmission, the fluid moves quite far from its starting point before reaching the rear lubrication system (Fig. 7-1). As in all hydraulic systems, initial fluid movement is from the reservoir to the pump via the oil screen or filter. The pump applies a force to the

CHAPTER 7

Hydraulic System Sealing Devices

fluid, which causes flow to the regulator valve and a buildup of pressure. From the regulator valve, the fluid moves into and out of the torque converter.

As the fluid leaves the converter, it passes through the converter check valve and enters the cooler circuit. The fluid now flows through an external line to the transmission cooler inside the radiator. After moving through this unit, the fluid finally enters a return line to the rear lubrication system of the transmission.

In order to prevent external or internal leakage of the fluid moving toward the rear lubrication system, the manufacturer uses a number of sealing devices. For this particular circuit, a total of about six devices are necessary to control the leakage.

Confining the Fluid When Components Are in Motion

The second sealing difficulty arises due to the fact that while some parts do not move, others have a rotary or reciprocating (back and forth) motion. In other words, the two sealed components may have relative motion between them. If, for instance, the two parts are stationary with a seal or gasket between them, the sealing job is less difficult. For this type of installation, manufacturers use some form of static seal, which prevents any leakage between the two parts.

But on the other hand, rotary or reciprocating motions between sealed components do present a real problem. The sealing devices in this case must control leakage when the parts are moving and even when they are stationary. The manufacturer uses a form of dynamic seal to check fluid leakage in these types of installations. These dynamic seals can stop leakage completely, but in some cases, the seal allows a controlled leak for lubrication purposes.

Confining the Fluid Under High Temperature Conditions

The final factor that makes sealing a problem is the high temperatures under which certain components operate. These high temperatures prevent the use of certain

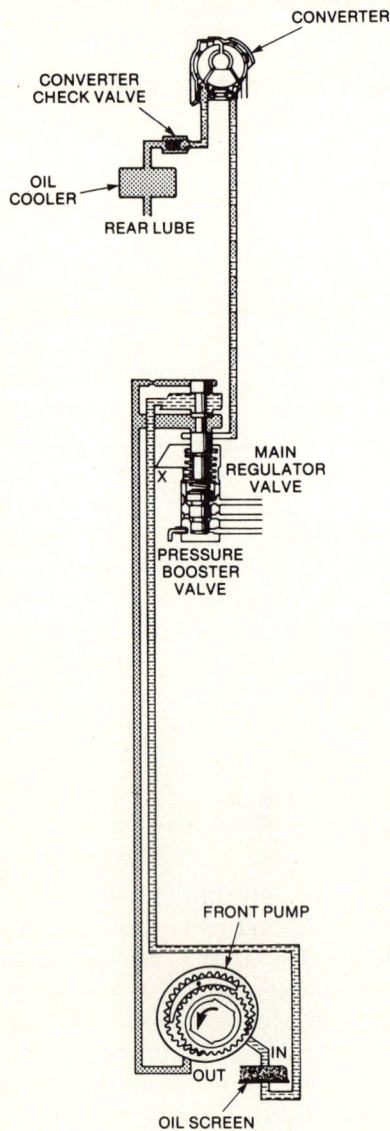

FIGURE 7-2   The synthetic rubber seal controls leakage far better than a metal sealing ring, but it deteriorates when subjected to high temperatures.

excellent seal designs. For instance, synthetic rubber seals (Fig. 7-2) deteriorate faster if subjected to high temperatures. This is the reason why manufacturers use metal sealing rings instead of synthetic seals in various locations within the transmission. The metal ring does not check fluid leakage as well as the synthetic type in most cases, but it can function well in areas of high temperature without failing.

STATIC SEALING DEVICES

A **static sealing device** is one that prevents leakage between two parts that bolt or fasten together by some means and therefore do not move in relation to one another (Fig. 7-3). In other words, the device provides what is known as a **positive seal.** The most common sealing devices used in automatic transmission in static applications are the gasket, lathe-cut, and O-ring seals.

Gasket

A gasket is a device that fills in the space between two machine surfaces to provide a tight seal between them. The gasket, in this situation, prevents leakage by filling in

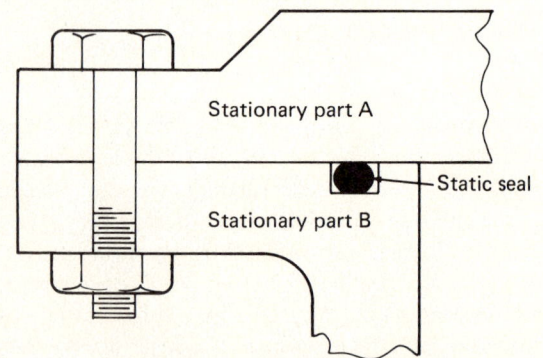

FIGURE 7-1   The fluid has to travel some distance from the pump to the rear lubrication system.

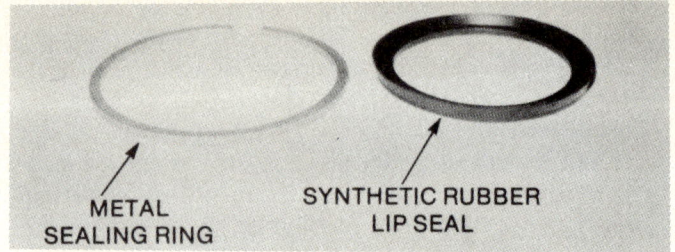

FIGURE 7-3   A static seal prevents leakage between two stationary objects which fasten together.

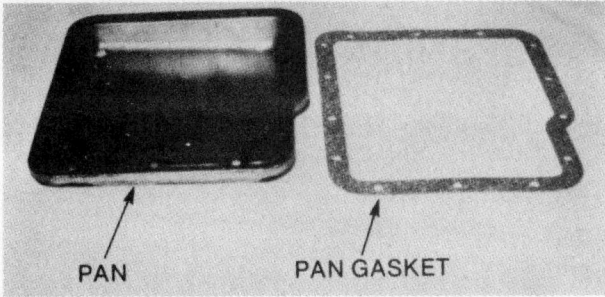

FIGURE 7-4   The gasket prevents leakage between the oil pan and the transmission case.

the surface irregularities found on the mating surfaces of each part. These irregularities would otherwise form passageways, through which the fluid could flow to create a leak between the two components.

The manufacturer uses several types of material to produce gaskets. The most common materials are: paper, cork, or a composition material of compressed particles of paper, cork, plastic, or other synthetic material. Lastly, three common areas of the automatic transmission where the manufacturer utilizes gaskets are the oil pan, front pump, and valve body.

Figure 7-4 shows the gasket that fits between the bottom of the transmission case and the oil pan to prevent leakage between the two parts. The gasket material used in this type of installation is cork or a composition material. Furthermore, the gasket is generally about 1/16 to 1/8 inch thick, and it easily compresses as the pan bolts are tightened.

Because of the flexibility of the material used and its thickness, pan gaskets do a rather good job of sealing even under less than ideal mating-surface conditions. For example, poor mating-surface conditions occur when the pan itself becomes somewhat distorted. The pan, which is usually formed of sheet metal, will easily distort if it contacts some object in the road or if someone over-

tighens its attaching bolts. Obviously, if pan distortion is excessive, the gasket will not seal and replacement of the pan is necessary.

A paper or composition gasket also fits between the reaction shaft support or stator support of the hydraulic pump and the transmission case (Fig. 7-5). This gasket performs two functions. First, like the pan gasket, it takes up the space between the two mating surfaces to prevent fluid leakage. But in this type of installation, the leakage would be internal because the pump has a second seal that controls external fluid loss. Second, open holes in the gasket form channels for the fluid as it moves from the transmission case to the pump and back.

Manufacturers also use the paper-type gaskets between the separator plate and the valve body halves (Fig. 7-6). These gaskets prevent leakage caused by surface irregularities, but at the same time, the large number of openings in them direct the flow of fluid from one valve-body half through the separator plate and to the different circuits found in the other remaining section of the valve body. Therefore, in many installations, a gasket is necessary on each side of the separator plate. However, on some valve bodies, where leakage is not a problem, the manufacturer will install only one gasket, while in still other designs, no gasket is needed at all.

The manufacturer, as mentioned earlier, may use the same transmission with several engine types. To mate the transmission with the engine, one of the things the manufacturer does, in many cases, is to install different design gaskets and/or separator plates between the valve body halves. One design, for instance, blocks off specific circuits between the valve-body halves, while another type may open these circuits and obstruct still others (Fig. 7-7). Consequently, the mating of the proper gasket to the valve body is of utmost importance. If a mechanic should install the wrong gaskets or place them in the wrong location, the transmission will malfunction or not operate at all.

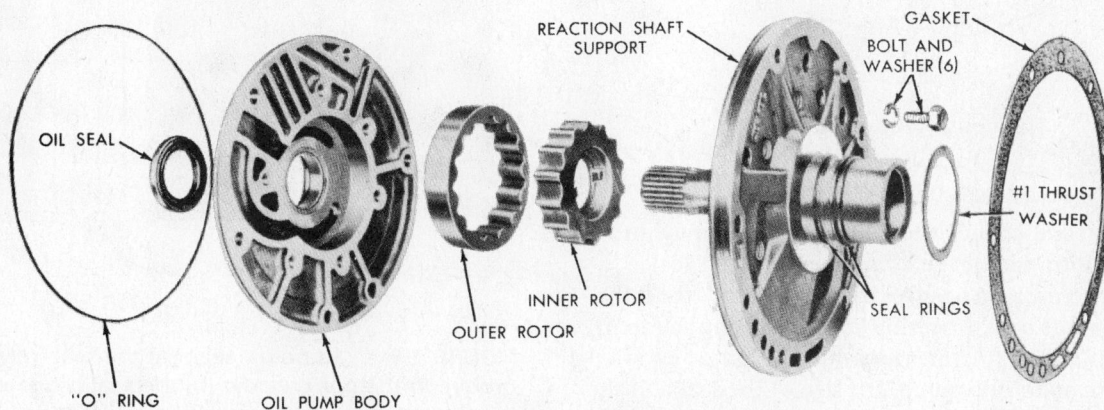

FIGURE 7-5   A typical hydraulic pump and its case gasket. (Courtesy of Chrysler Corp.)

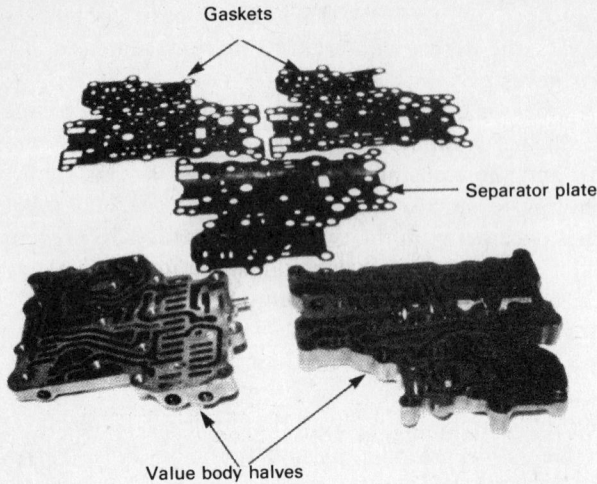

FIGURE 7-6 The gaskets prevent leakage between the separator plate and the two halves of the valve body.

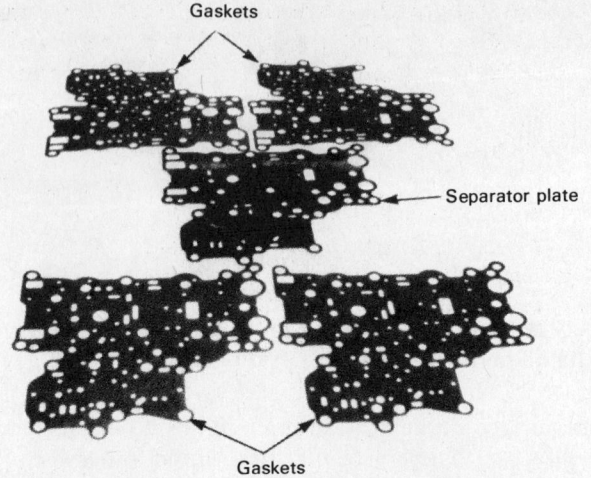

FIGURE 7-7 The use of different types of gaskets for the same valve body is one method used to mate a transmission with a specific engine style.

Lathe-Cut Seals

The second type of static sealing device is the square- or lathe-cut seal formed from a synthetic rubber, Neoprene. This material, used extensively by manufacturers, is not as brittle as pure rubber. Furthermore, the Neoprene can withstand more heat and contact with various chemicals much better than a pure rubber seal.

The lathe-cut seal is circular in shape with a square cross section (Fig. 7-8). To produce this design, the manufacturer generally molds this seal type into a tube shape and then cuts it into individual rings on a machine, the lathe, from which the seal derives its name. The unit now

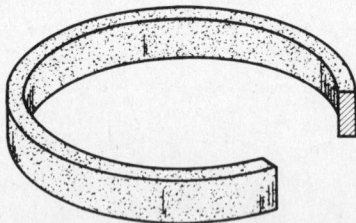

FIGURE 7-8 The design of a lathe-cut seal.

has the correct diameter and width to fit properly into a groove cut into one of the components it will seal.

In order to provide positive leakage control, the lathe-cut seal depends on a squeezing or compressing action. If, for example, a manufacturer uses this seal to stop external leakage of a hydraulic pump, the size of the seal and the transmission case bore diameter determine the extent of seal compression. In this installation, the seal's outer

diameter is larger than the bore in the transmission case and it fits rather snugly into the bottom of a groove cut into the outer circumference of the pump body [Fig. 7-9(a)].

As the installer pushes the pump assembly into its bore in the transmission case, the seal has to compress down somewhat into the groove in order for the pump to enter the cavity [Fig. 7-9(b)]. With the pump in place, the natural tendency of the compressed seal is to expand and try to return to its original shape [Fig. 7-9(c)]. As a result of this seal elasticity, the inner and outer circumferences of the unit provide a positive control of fluid leakage at its contact areas. In other words, the seal prevents fluid leak-

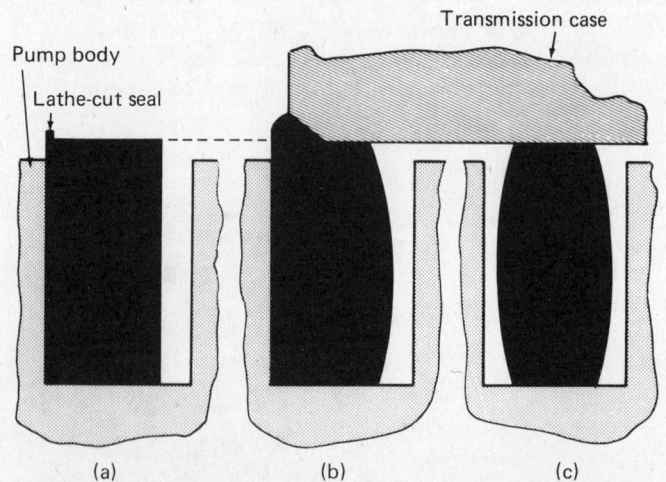

FIGURE 7-9 Lathe-cut seal: (a) the seal installed in its groove in the pump body; (b) the seal has to compress as the pump enters the case bore; and (c) the seal attempts to expand back to its original shape.

age from between its outer circumference and the transmission case bore and also at its inner surface and the groove in the pump body.

O-Ring Seals

The final type of static seal is the O-ring, which by its design also provides positive leakage control. The manufacturer molds this seal from Neoprene rubber into a tube shape, but the finished unit has a round cross section (Fig. 7-10). Moreover, by being larger in diameter than the bore it has to fit into, the O-ring functions by a compressing action to stop leaks in a manner similar to the lathe-cut seal.

FIGURE 7-10   The design of an O-ring seal.

Figure 7-11 shows an O-ring used on a front-pump assembly to prevent external leakage. The seal in this installation fits into a groove cut in the pump body. Also, the groove itself is not as deep as the thickness of the seal. As a result, the seal protrudes above the surface of the groove and pump body.

As the technician inserts the pump into its bore in the transmission case, the O-ring has to compress slightly to allow the pump to enter the cavity. The compression squeezes or packs the seal into the groove in the pump body. As it attempts to return to its original shape, the O-

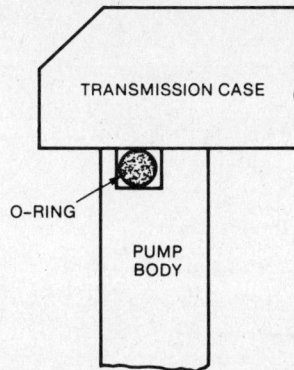

FIGURE 7-11   An O-ring seal used on a hydraulic pump assembly. (Courtesy of Chrysler Corp.)

FIGURE 7-12   An O-ring seal compressed into its groove by the installation of the pump.

ring forms a positive seal between the pump body and its bore in the transmission case (Fig. 7-12).

PROPER SELECTION OF SEALING DEVICES

Automatic transmission overhaul gasket and seal kits usually contain more of these items than are necessary to do the repair job. The reason for this is that the supplier loads the kit with all seals and gaskets necessary to rebuild all the models of the same transmission. In other words, there may appear to be duplicate gaskets and seals, but upon close examination, there will be noticeable, slight differences between them.

As mentioned earlier, one of the duplications may be valve-body gaskets. The kit may contain two to four gaskets, yet the valve body uses only two. The extra ones are for use on another model of the same transmission. The best and surest way to make sure you are using the correct gasket(s) is to save the old ones and match them up with the new units prior to installation.

The same rule applies to lathe and O-ring seals. Although the old seals may be somewhat distorted, you can compare them with the new ones to check their cross section as well as diameter.

If you install a new seal with the wrong dimensions, it will not perform its function. For example, a seal with too large a cross-sectional dimension will be cut or damaged during its installation (Fig. 7-13). On the other hand, a seal having too small a cross-sectional dimension cannot be compressed sufficiently during installation. A common cause of this is the use of a ring with too small a diameter for the installation so the seal has to be stretched in order to be inserted into its groove. This reduces its cross-sectional dimension. In either case, the seal usually will not control fluid leakage as it should.

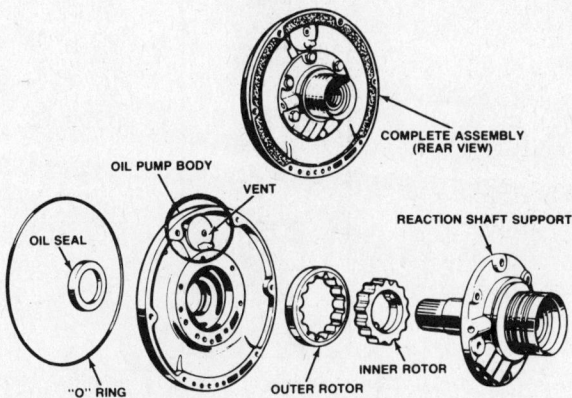

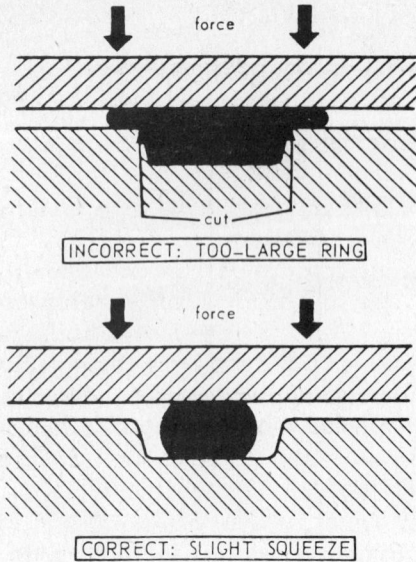

FIGURE 7-13   The effects of using an O-ring seal having the wrong cross-sectional dimension.

DYNAMIC SEALING DEVICES

Purpose

As mentioned, **dynamic sealing devices** must control fluid leakage at installations in which one or both components are moving (Fig. 7-14). In the example, an O-ring seal fits between Part A, which is stationary and Part B, which moves back and forth. In other words, the seal controls fluid leakage as Part B slides over Part A.

In applications as those mentioned above, the sealing device may provide a positive seal, while in others, a nonpositive type leakage control is all that is necessary. A **nonpositive seal** will permit a certain amount of controlled fluid leakage for lubrication purposes. The seal designs that are both static as well as dynamic are the

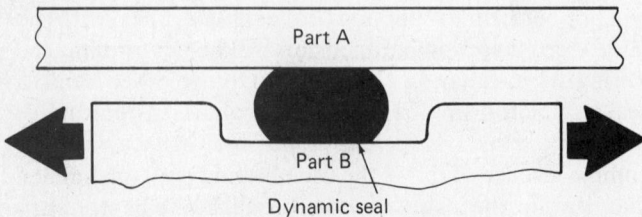

FIGURE 7-14   A dynamic seal controls fluid leakage at installations in which one or both components are moving.

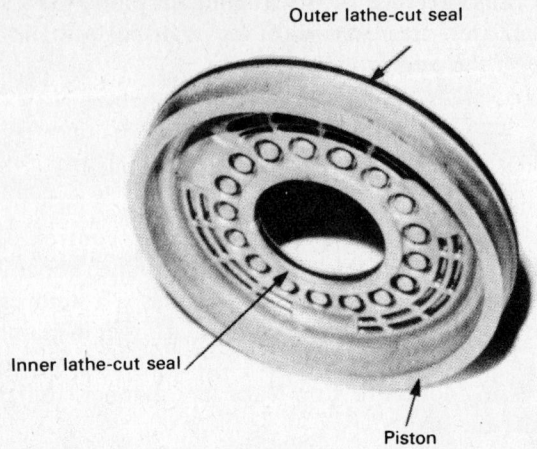

FIGURE 7-15   A dynamic lathe-cut seal installed on a clutch piston.

lathe-cut and O-ring; the devices that are primarily dynamic are the lip and metal-clad seals along with metal and Teflon rings.

LATHE-CUT SEALS IN DYNAMIC APPLICATIONS

Design and Installation

Figure 7-15 displays two lathe-cut seals installed on a clutch piston. In this installation, the seals will provide positive control of fluid leakage at both the inner and

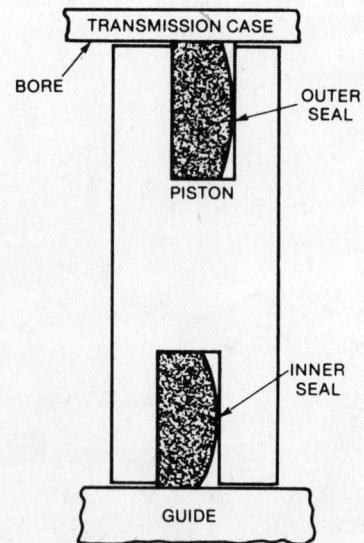

FIGURE 7-16   Two lathe-cut seals compressed into their grooves in a clutch piston.

outer circumference of the piston. Moreover, the seals perform this function with the piston stationary or moving back and forth in its bore.

As in a static application, a slight compression of these clutch seals provides the dynamic, positive control of fluid leakage. To accomplish this compressing action, the manufacturer makes the large, outer seal's diameter larger than the size of piston bore. On the other hand, the inner diameter of the inside seal is smaller than the size of the piston guide.

Thus, as the mechanic installs the piston into its bore and over its guide, both seals must compress slightly into their respective grooves (Fig. 7-16). But the elasticity of the seals attempts to restore them both to their original shape. As a result, the seals stop fluid leakage at their respective contact surfaces both statically and dynamically.

Operation

Figure 7-17 illustrates the action of one lathe-cut seal as a piston moves in its bore. As hydraulic pressure applies force to the piston, it moves in its bore fast enough to cause a dragging or bending of the portion of the seal pressed against the wall of the bore. In other words, the surface of the seal pressed against the wall does not move as far or as fast as the piston.

As hydraulic pressure is released, the bent portion of the lathe-cut seal moves back to its original position. This helps to draw the piston back to the bottom of its bore in the housing.

Since the lathe-cut seal bends and does not roll or move in its groove, it is well suited in installations where there is a good deal of reciprocating, or back and forth, motion. However, this design is never used where there is rotary motion because this would tend to pull the lathe-cut seal away from its seat in the groove.

O-RING SEALS IN DYNAMIC APPLICATIONS

Design and Installation

The O-ring seal, when installed in a reciprocating, dynamic application, also provides positive fluid leakage control due to its compression. However, manufacturers only use this type of seal for clutch pistons, servos, or other parts that have a short stroke or back and forth motion. The reason for this is that the O-ring tends to roll or twist if subjected to considerable movement. Consequently, this tendency to roll limits its dynamic application to parts with decreased motion in order to prevent damage to the seal.

Figure 7-18 shows an O-ring installed in a dynamic application. The O-ring in this installation fits into a groove in a movable component. In this case, the difference in bore diameter and seal size forces the seal to compress somewhat when the part is placed into its bore. Thus, the seal prevents leakage at two locations—its inner and outer circumference.

Operation

As the component moves in its bore, the motion forces the O-ring against one side of its groove (Fig. 7-19). This action packs the seal against the corner of the groove. As a result, the seal now controls fluid leakage at three locations: at its top, bottom, and one side; and it is capable of withstanding very high pressure without leaking.

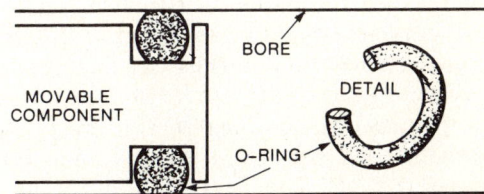

FIGURE 7-18 An O-ring seal installed in a dynamic application.

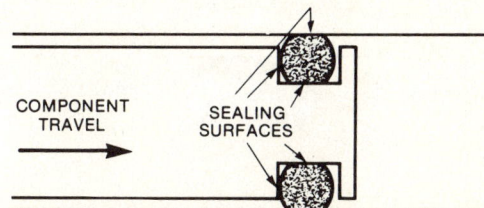

FIGURE 7-19 An O-ring forced against one side of its groove by component movement in one direction.

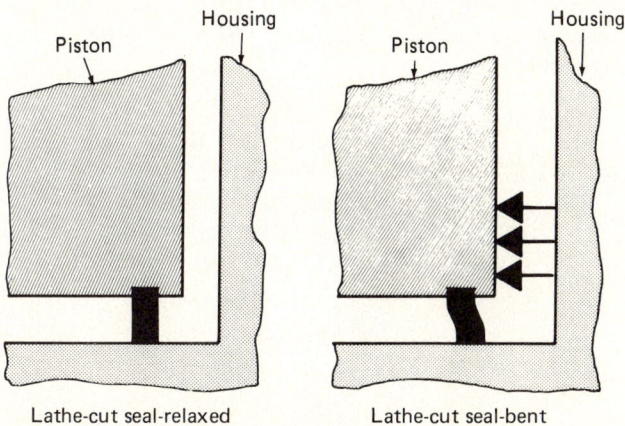

FIGURE 7-17 A lathe-cut seal bends as the piston moves in its housing bore.

FIGURE 7-20   The O-ring forced against the other side of its groove by component movement in the opposite direction.

If the part moves in the other direction, the resulting motion moves the seal in the opposite direction (Fig. 7-20). This action now packs the seal against the other side of the groove to prevent leakage at this point. In other words, the O-ring has the ability to provide leakage control on three sides no matter which direction the part moves.

As previously mentioned, the O-ring cannot tolerate excessive reciprocating motion. Also, the O-ring, like the lathe-cut seal, cannot perform its function where rotational forces exist. When subjected to rotary motion, the O-ring does not seat properly in its groove; consequently, it cannot control leakage at its contact points.

DYNAMIC LIP SEALS

Function and Design

A **dynamic lip seal** controls fluid leakage in applications in which a component can be subject to either reciprocating or rotary motion. But this seal can only prevent leakage in one direction.

The lip seal, like the O-ring and lathe-cut, is made from Neoprene and has a circular shape (Fig. 7-21). But this seal design also has an extended lip that is larger in diameter than the bore it fits into. Therefore, in order for the lip of this seal to enter the bore, it must deflect or compress somewhat downward. And as the lip attempts

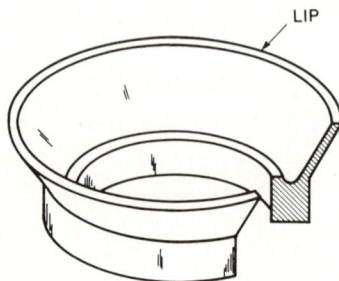

FIGURE 7-22   A lip seal installed in its groove in a clutch piston.

to return to its original shape, it applies tension against the wall of the bore, thus controlling fluid leakage at the contact points.

Installation

Figure 7-22 shows a lip seal seated in its groove in the outer circumference of a clutch piston, which is subject to reciprocating motion. In this installation, the seal controls leakage in two ways. First, as previously stated, the lip itself imposes tension on the bore wall due to its elasticity. Second, as hydraulic pressure activates the piston, it also applies a force on the lip that presses it even harder against the wall of the bore (Fig. 7-23).

Therefore, in order for the lip seal to function in a dynamic application, the lip must always face toward the pressure source. If it does not, hydraulic pressure will not be able to force the lip against the wall of its bore. In fact, pressure will deflect the lip away from the wall, and a large leak will result. This makes it imperative that the mechanic install this seal correctly with its lip facing the proper direction.

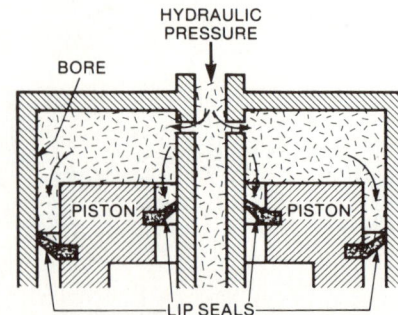

FIGURE 7-21   The design of a typical lip seal.

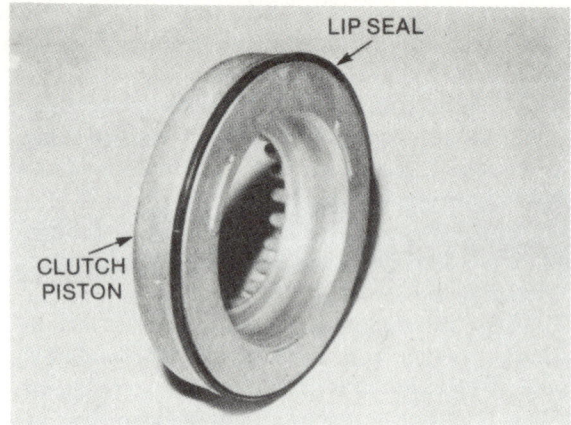

FIGURE 7-23   Hydraulic pressure acting on the lip forces it tightly against the wall of the bore.

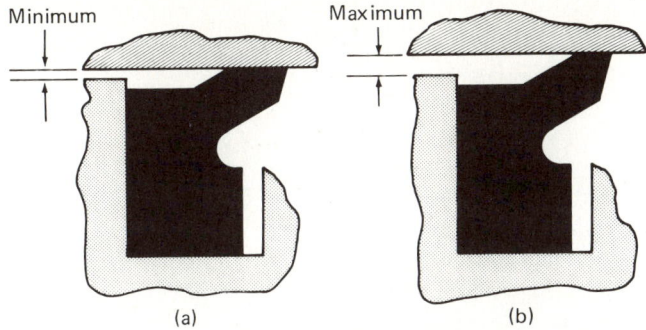

FIGURE 7-24   A lip seal can control leakage even when the piston is not completely centered in its bore.

Advantages to Its Use

Other than using two methods of providing leakage control, the lip seal offers another important advantage. That is, the seal provides flexibility that can be very helpful when the clutch piston's or other moving part's circumference is not an equal distance from the wall of the bore, all the way around. Figure 7-24(a) shows a piston in its minimum clearance position, and Fig. 7-24(b) shows it in its maximum clearance position. If the lip of the seal is in good condition, it provides adequate leakage control between the piston and its bore in either position.

DYNAMIC METAL-CLAD LIP SEALS

Purpose

The **metal-clad lip seal** also controls fluid leakage in areas where reciprocating motion occurs between two parts. However, its predominate use in automatic transmissions

is to seal rotating shafts. Unlike the plain lip seal described earlier, the metal-clad type is not designed to function under high pressure. Its main function is to retain the fluid that leaks past the shaft bushings found in such components as the front pump and extension housing and to keep foreign external contaminants from entering the transmission.

Design and Installation

The design of the metal-clad seal is such that it can control both static and dynamic forms of leakage. The simple metal-clad seal is nothing more than a Neoprene lip seal bonded to a circular metal shell (Fig. 7-25). The shell is larger in diameter than the bore it fits into, and in some cases, the manufacturer coats the shell's outer circumference with a sealing material such as rubber or resin. And because of the differences in diameter between the shell and the bore, the mechanic must press or drive the seal into position. This interference fit, in addition to the sealing compound when used, prevents a static leak between the bore wall and the seal's shell.

The Neoprene lip section of the seal must control both static and dynamic forms of leakage in addition to the entrance of foreign materials. Therefore, its design has to vary somewhat with different applications. For example, a shifter-shaft seal will usually have a single lip and no garter spring (Fig. 7-26). With this seal installed, the lip faces toward the fluid source, and the lip's elasticity alone prevents both static and dynamic leakage between the shaft and the outside of the transmission.

Because the front pump or extension housing seal has a harder task to perform, its lip has a garter spring underneath it (Fig. 7-27). This spring helps to keep the lip pressed against the rotating component. In other words, the spring assists the elasticity of the lip in controlling both static and dynamic fluid leakage.

To stop foreign materials such as dust, dirt, or moisture from entering the seal area, some metal-clad units have a double-lip and a garter spring [Diagram (a) of Fig. 7-28]. The transmission manufacturer installs

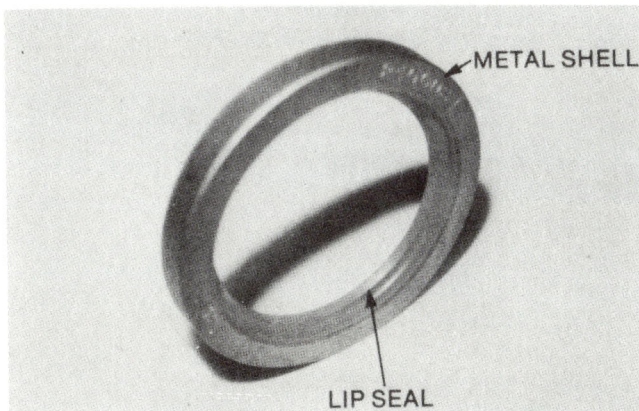

FIGURE 7-25   The design of a typical metal-clad seal.

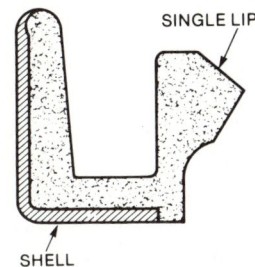

FIGURE 7-26   A metal-clad seal with a single lip and no garter spring.

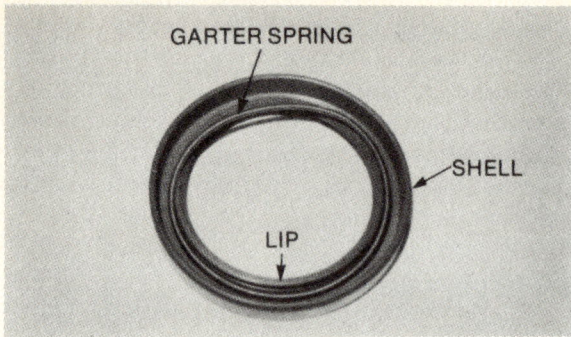

FIGURE 7-27　A metal-clad seal with a single lip and a garter spring.

In a number of metal-clad seal applications, the manufacturer must use supplemental mechanisms to control fluid leakage. As previously mentioned, with the lip facing the direction of the fluid source, the elasticity of the lip and, in many cases, the garter spring prevent fluid leakage. But in some installations, these two methods are not sufficient to control leakage within acceptable limits.

To provide the seal with additional capacity, the manufacturer will mold ribs or depressions on the outside lip surface (Fig. 7-30). The function of these is to direct the fluid away from the lip's surface or back toward its source. In other words, these alterations direct some of the fluid away from the lip-contact surface; therefore, normal lip elasticity along with the tension of the garter

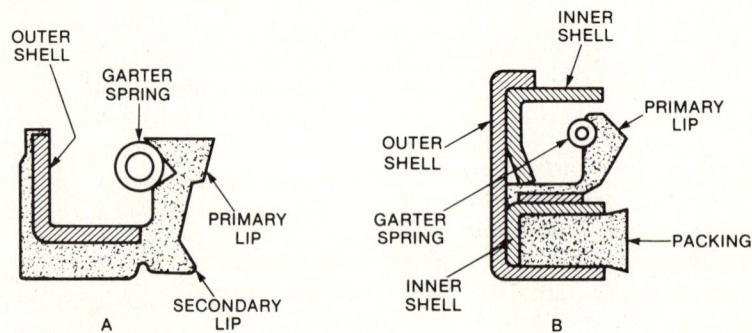

FIGURE 7-28　Metal-clad seals with a second lip or packing plus a garter spring.

these double-lip seals at locations such as the front pump but more commonly at the extension housing. With this design, the spring-loaded, primary lip controls fluid leakage, while the non-spring-loaded, secondary lip prevents the entrance of foreign material.

In some extension housing applications, the metal-clad seal has a spring-loaded, primary lip; but it uses a packing material instead of a secondary lip [Diagram (b) of Fig. 7-28]. In this design, the packing serves the same function as the secondary lip. The packing used in most cases is a felt material that fits into a groove made in the metal shell.

Figure 7-29 shows another type of extension housing seal that has a dust boot. This seal also has a spring-loaded, primary lip to control fluid leakage along with a secondary lip to stop the entrance of foreign material into the seal area. In addition, this seal has a dust boot attached to the shell. This synthetic boot has a ring that resembles a lathe-cut seal molded into its end. The sealing ring also encircles the slip yoke end of the drive shaft to provide extra protection against the entrance of foreign matter into the seal area.

spring will then be able to control fluid leakage, even under less than ideal conditions.

Operation

Figure 7-31 shows a metal-clad lip seal pressed into the rear portion of the extension housing. Under static conditions or when the transmission is operating, this seal

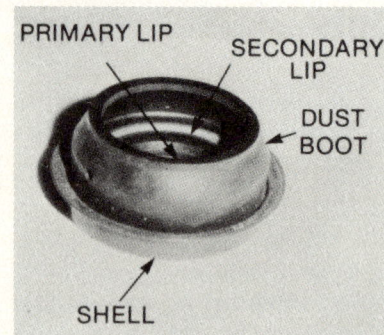

FIGURE 7-29　A metal-clad seal with a dust boot.

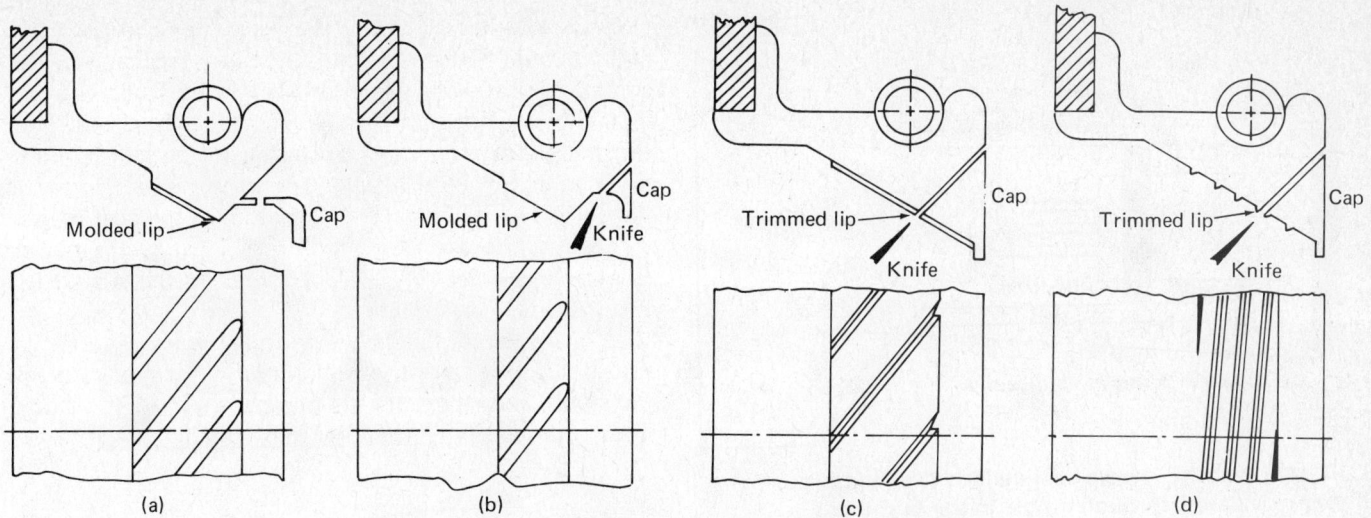

| (a) | (b) | (c) | (d) |

FIGURE 7-30   Supplemental sealing mechanisms. The seals shown in (a) and (b) have a molded lip. The excess material is removed by tearing the cap from the molded part in (a) and by a knife in (b). The helical ribs in both designs terminate at the contact point of the static lip. Seals shown in (c) and (d) are trimmed up seals; that is, a knife trimming operation forms the contact lip as the excess material is removed. The helical ribs protrude at the contact point and must be compressed to prevent the seal from leaking when the shaft is not rotating prior to initial operation. (Reprinted with permission from the SAE handbook, ©1978 Society of Automotive Engineers, Inc.)

provides a positive seal between the housing and the revolving slip-yoke section of the drive shaft.

With the output shaft revolving, the slip yoke that splines to it turns in a bushing pressed into the extension housing (Fig. 7-32). This bushing supports the slip yoke and the rear section of the output shaft.

Since the bushing receives fluid for lubrication from the hydraulic system, there is a given amount of fluid leakage past the bushing, where it supports the slip yoke. The fluid that leaks into the seal side of the housing is deflected back by the lip and passes back through a hole in the bushing to the output shaft side of the housing. From here, the excess fluid drains back into the reservoir.

METAL SEALING RINGS

Purpose

Metal sealing rings used in automatic transmissions provide nonpositive control of fluid leakage. In other words, these rings do permit some small amounts of fluid to pass by them (Fig. 7-33).

The metal sealing rings have two purposes. First, as previously mentioned, manufacturers use this ring type to control fluid leakage at locations where high temperatures are present. Second, the rings can provide adequate sealing at stationary or rotating shafts, which carry fluid to various components, but they do permit

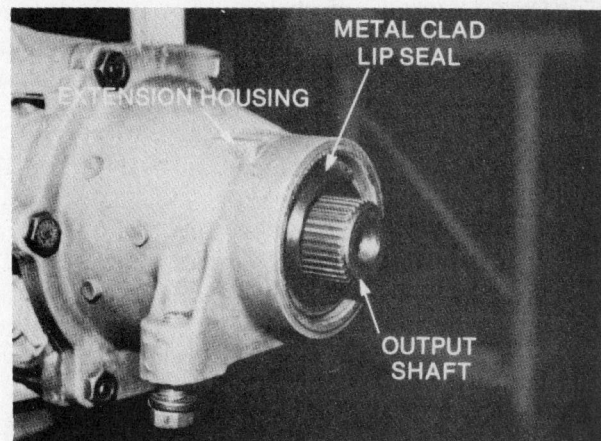

FIGURE 7-31   A metal-clad seal installed in an extension housing.

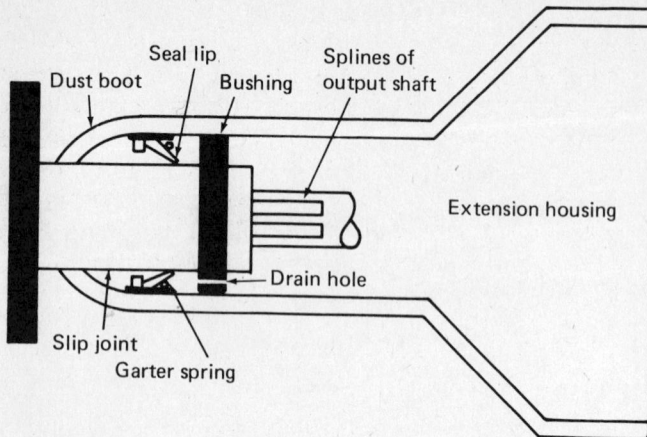

FIGURE 7-32 The lip seal deflects the fluid back, where it passes through a hole in the bushing.

some leakage of fluid that lubricates the shaft journals and bushings as needed. Stated another way, these rings act as dams to direct fluid from oil passages in a shaft to, for example, a clutch drum but allow some seepage for lubrication.

Figure 7-34 shows two typical steel sealing ring applications. The upper picture shows the rings installed over a clutch support shaft. The five pictured rings stop excessive loss of fluid to the apply circuits of the forward and direct clutches. The lower picture illustrates three seal rings that fit into grooves in the governor distributors; these units prevent excessive losses of fluid as it moves to and from the governor assembly.

Design

The metal sealing ring has about the same design and construction as the common piston ring. The manufacturers form these rings primarily from grey cast iron, but for some transmission component application, they also use aluminum. Furthermore, the iron sealing rings have a surface coating of phosphate or oxide; occa-

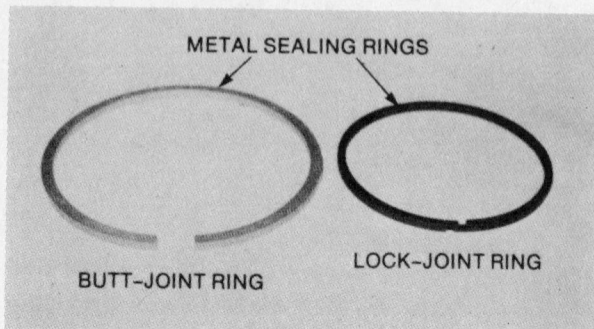

FIGURE 7-33 Typical steel sealing rings.

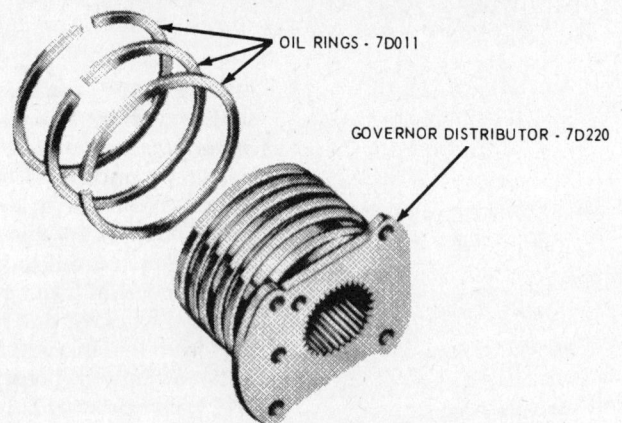

FIGURE 7-34 Typical steel ring applications. (Courtesy of Ford Motor Company)

sionally, the coating will be a metallic plating such as tin or chrome.

The metal rings are produced to a given cross-sectional size and diameter so they fit properly into their grooves and the bore area. Lastly, the process used to manufacture the ring is such that it gives it elasticity. In other words, the metal ring can, after being compressed, return to its original shape. As a result, the ring presses against the bore wall with a given amount of tension that is necessary to control leakage at this point.

Types

Manufacturers produce metal sealing rings with two types of joint ends—the butt and lock (Fig. 7-35). The butt-joint design is such that when the ring is in its bore a small clearance exists between the ends. This clearance compensates for heat expansion of the ring and permits some fluid leakage past the joint for lubrication. Consequently, the manufacturer installs this ring type in high temperature locations where leakage is not a problem or where it is necessary for component lubrication.

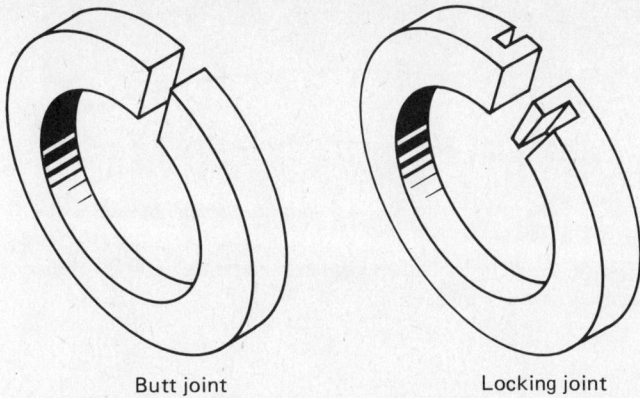

Butt joint   Locking joint

FIGURE 7-35   The two types of steel rings, the butt and locking joint.

The lock-joint type, on the other hand, has small tangs, which hold the ends of the ring together or in partial compression. But the design of the tangs is such that they do permit some ring tension on the bore wall and at the same time allow for heat expansion of the ring. The transmission manufacturer utilizes this ring design for two reasons: (1) to reduce the leakage past the ends of the rings after their installation, and (2) the lock joint holds the ring in compression enough so that the mechanic can install the component it fits on into a blind hole, without breaking the ring.

Operation

Figure 7-36 illustrates the use of two steel rings that seal against the loss of fluid between the two shafts. The shafts rotate at different speeds turning the various forward ratios and in opposite directions in reverse.

Figure 7-37 illustrates the action of the two rings installed on the primary sun gear shaft shown in Fig. 7-36. The rings fit into grooves on either side of a fluid passage and must control leakage between the sun gear shaft and its bore wall in the output shaft. The elasticity of the ring is great enough to initially control fluid leakage at point A long enough for the force of fluid pressure to move it against the B side of the groove. This ring action allows fluid to flow down area C, between the side of the ring and groove, and to the lower area at D.

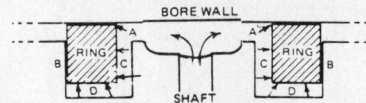

FIGURE 7-37   The action of two steel rings around a port opening in a shaft. (©Research, El Monte, California)

The fluid pressure, now acting on area D, forces the ring tightly against the bore wall. This action causes the metal ring to seal leakage at point A. In addition, since the combined forces of fluid pressure and ring elasticity keep it in contact with the bore wall, the ring rotates with the bore wall and output shaft as required.

TEFLON SEALING RINGS

A number of automatic transmission manufacturers use Teflon or a similar plastic material instead of metal rings in a number of dynamic applications such as accumulator pistons and some rotating and nonrotating shafts. These nonmetallic rings serve the same function as the metal ones, operate in much the same manner, and do an excellent job of sealing when working properly. Moreover, this type of ring provides a lower cost for the

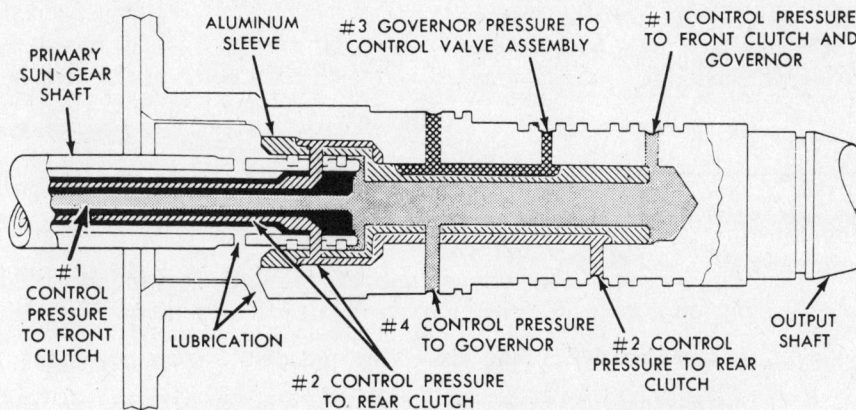

FIGURE 7-36   Steel rings used to seal in the fluid between two rotating shafts. (Courtesy of Ford Motor Company)

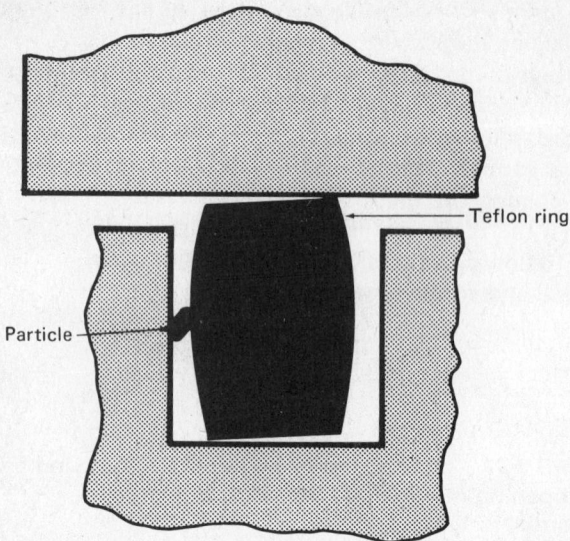

FIGURE 7-38   If a particle embeds itself in the side of a Teflon seal, the seal cannot properly function in its groove.

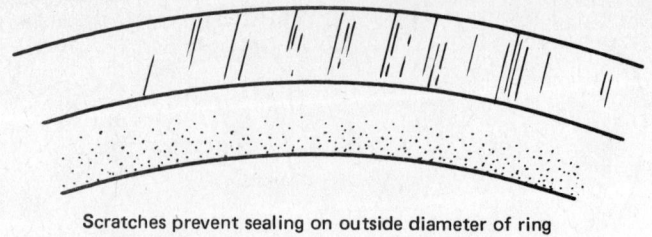

Scratches prevent sealing on outside diameter of ring

FIGURE 7-39   A Teflon seal can become scratched very easily.

original assembly-line installation, and many mechanics prefer them over the metal ones because they do not break as easily.

However, many experts in the field recommend the metal rings as a replacement for the Teflon for the following reasons:

1. Metal particles can embed themselves in the Teflon ring and therefore prevent it from sealing in its groove (Fig. 7-38). Actually, the problem causes the ring not to move against area B shown in Fig. 7-37 and often locks up the ring in its groove.

2. The Teflon seal is soft and can scratch very easily (Fig. 7-39). This prevents the ring from sealing at point A of Fig. 7-37.

3. It is very difficult to check a Teflon ring for wear, as you can a metal type, by placing it in the bore without damaging or scratching the material.

Finally, Teflon rings come in two styles, the one piece and angle joint. The solid, one-piece style is stretched and placed into its groove. This design eliminates the usual amount of leakage from the joint gap.

The angle joint or scarf-cut ring (Fig. 7-40) allows the mechanic to open it slightly for installation and removal. Therefore, this ring style allows minimum fluid leakage at the cut. Also, the angle-joint ring conforms well to irregularities in its bore wall.

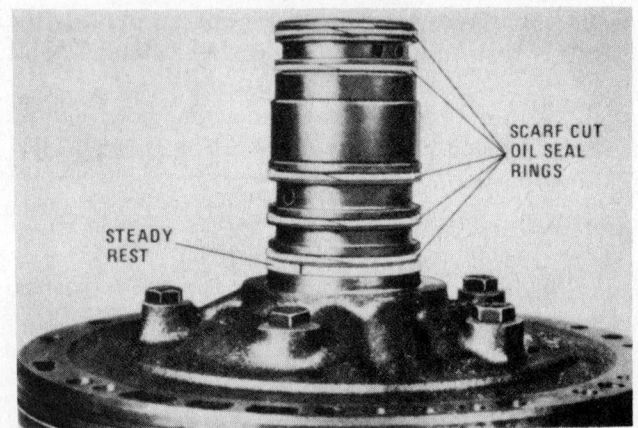

FIGURE 7-40   A typical Teflon, scarf-cut ring installation. (Courtesy of General Motors Corp.)

SUMMARY

1. The purpose of the various kinds of sealing devices found within the automatic transmission is to control the amount of external and internal fluid leakage.

2. External leakage is fluid loss to the outside of the transmission.

3. Internal leakage is the excessive loss of fluid under pressure from a sealed component inside the transmission.

4. The control of leakage is a large undertaking for the automatic transmission manufacturer.

5. The distance the fluid travels and the number of parts it moves through make sealing a problem.

6. High temperatures also makes sealing against leakage a problem.

7. The control of leakage around moving parts is more difficult than around those that are stationary.

8. A gasket is a static device that fills in the space between two machined surfaces to provide a tight seal.

9. Manufacturers make gaskets out of paper, cork, or a composition material.

10. Gaskets may have holes that permit fluid to move from one part to another.

11. Manufacturers use gaskets at such locations as the pan, pump, and valve body.

12. Lathe-cut, O-ring, and lip seals are made of Neoprene, a synthetic rubber.

13. Manufacturers use the lathe-cut seal in both static and dynamic applications. This device depends on a compressing action in order to provide positive leakage control.

14. The O-ring functions in both static and dynamic applications, in much the same manner as the lathe-cut, and provides positive control of fluid leakage.

15. Lip seals can function in applications where both reciprocating and rotary motions are present.

16. Manufacturers use the metal-clad seal in several locations on the transmission to stop external leakage.

17. Metal-clad seals prevent both static and dynamic leakage.

18. Some metal-clad seals have a garter spring that assists the lip in checking fluid leakage.

19. Some metal-clad seals have a second lip or packing that prevents the entrance of foreign material into the seal area.

20. Metal rings function well in high temperature areas but do not provide positive control of fluid leakage.

21. The metal sealing ring has about the same design and construction as the common piston ring used in an engine.

22. The metal sealing ring may have a butt or lock joint.

23. Manufacturers install Teflon rings instead of metal ones in some dynamic applications.

REVIEW

This section will assist you in determining how well you remember the material contained in this chapter. Read each item carefully. If you cannot complete the statement, review the section in the chapter that covers the material.

1. The average automatic transmission will have about _____ sealing devices.

   a. 6

   b. 12

   c. 18

   d. 25

2. High temperatures will deteriorate _____ seals.

    a. cast-iron

    b. aluminum

    c. synthetic rubber

    d. both a and c

3. The gasket provides a positive seal in _____ applications.

    a. static

    b. dynamic

    c. both a and b

    d. no

4. Lathe-cut seals provide positive sealing in _____ applications.

    a. no

    b. dynamic

    c. static

    d. both b and c

5. The seal that is circular in shape with a round cross section is the _____.

    a. lathe-cut

    b. O-ring

    c. lip

    d. metal

6. The devices that control fluid leakage due to a squeezing action of the entire seal are the _____.

    a. square-cut and metal

    b. metal and Teflon

    c. O-ring and square-cut

    d. lip and O-ring

7. Which seal type does not operate well in all dynamic applications?

    a. lip

    b. metal

    c. O-ring

    d. lathe-cut

8. The seal that utilizes hydraulic pressure to help it in providing a better control of leakage is the _____.

    a. lip

    b. lathe-cut

    c. gasket

    d. O-ring

9. The seal design that uses the garter spring is the _____.

    a. lip

    b. metal-clad

    c. gasket

    d. O-ring

10. The secondary lip, found on many metal-clad seals, controls the entrance of _____ to the seal area.

    a. fluid

    b. foreign material

    c. air

    d. both a and b

11. Metal sealing rings provide _____ sealing.

    a. nonpositive

    b. positive

    c. no

    d. extra

12. In some applications, the manufacturer will install a(n) _____ seal instead of a metal one.

    a. O-ring

    b. square-cut

    c. Teflon

    d. lip

For the answers, turn to the Appendix.

CHAPTER 8

Hydraulic Fundamentals

A basic automatic transmission consists of a planetary gear train assembly along with a very complex hydraulic control system. As mentioned in an earlier chapter, the **planetary** is a mechanical means of providing a vehicle with a given number of forward gear ratios plus a reverse. The **hydraulic control system** senses engine load and vehicle-speed combinations and then activates a certain gear ratio within the planetary through the application of a multi-disc clutch or band assembly. Moreover, the hydraulic system is responsible for directing fluid not only to the torque converter so that it can function but also to the lubricating circuits of the transmission.

It should be obvious that without the effects of hydraulics the automatic transmission could not possibly function. Also, a vehicle would not have the effective braking systems, as we know them today, or power steering that reduces driver fatigue in operating a vehicle.

An automatic transmission technician or mechanic working on the other above-mentioned systems must possess a working knowledge of hydraulics in order to understand the operation of these devices for trouble-shooting purposes. For instance, the transmission technician must master the theory of hydraulics to be able to understand and interpret the hydraulic system diagrams supplied by the gearbox manufacturer. With this knowledge, along with the diagrams and a good set of hydraulic gauges, the technician can diagnose about 90 percent of all automatic transmission malfunctions. But these trouble-shooting tools will be of no value to the mechanic unless he has a working knowledge of hydraulics.

HYDRAULICS

The word hydraulics comes from a Greek word, **hydros,** that means water or liquid. In modern technology, **hydraulics** is an exacting branch of physics that deals with practical applications (such as the transmission of energy or the effects of flow) of a fluid in motion. A **fluid** can either be a gas or a liquid. But because our subject matter is hydraulics, the use of the term fluid designates the liquid used in the system. In the case of automatic transmissions, the liquid is ATF (automatic transmission fluid).

CATEGORIES OF HYDRAULICS

There are two subdivisions or categories of hydraulics — hydrodynamics and hydrostatics. The automatic transmission uses both of these divisions in its torque converter and hydraulic control system.

Hydrodynamics

Hydrodynamics is the study that deals with the motion of fluids and the forces acting on solid bodies immersed in the liquid. Chapter 4 of this text covered the principles of hydrodynamics as it applies to torque converter operation. This section explains, for example, that a fluid in motion possesses kinetic energy by virtue of its mass, or weight, and the kinetic (impact) energy of the moving fluid on the turbine blades is responsible for setting the turbine and the input shaft of the transmission into motion. In this manner, torque and power transfers between the engine and transmission.

Hydrostatics

Hydrostatics deals with the characteristics of a fluid at rest and especially with the pressure in a liquid or exerted by it on some immersed body. More commonly, this phase of hydraulics is known as "pressure hydraulics" in which the motion of the fluid is incidental; what is important is the fact that pressure develops in a confined fluid by the application of a force on the liquid. And the fluid, in turn, can transmit the resulting pressure along with force to another object immersed in the confined liquid. The hydraulic control system of the transmission uses pressure hydraulics not only to determine when a shift must occur but also to apply a multi-disc clutch or band that causes the planetary to produce a gear ratio.

CLASSIFICATION OF SUBSTANCES

The basic principle of hydrostatics is that a fluid resists the compressing action of an applied force; therefore, the liquid can transmit pressure and force. To comprehend fully why a liquid like ATF is not compressible, the reader must be aware of a few fundamental facts relating to the physical makeup of all objects. Science classifies all objects (substances) as being either solid, liquid, or gas due to the substance's arrangement of molecules (Fig. 8-1).

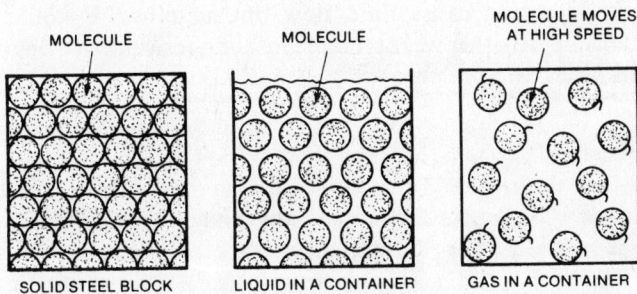

FIGURE 8-1   The molecular structure of solids, liquids, and gases. In each example, the round marbles represent the molecules within the object.

A **molecule** is the smallest particle, or building block, into which any substance can be divided and still retain its original physical properties. For instance, if a scientist divided a grain of salt in two and divided each subsequent grain again until he finished the division as finely as possible, the smallest particle having all the original properties of salt would be the molecule. This molecule would be almost one millionth of an inch in diameter and would need to be enlarged about 100 times before it could be seen in a microscope.

Solid

As mentioned, the arrangement of the molecules determines if a substance is either solid, liquid, or gaseous. A solid object, such as a piece of steel, has a rigid molecular structure. The molecules are all right next to one another and have a strong attraction to each other. This molecular arrangement resists any attempt to physically change the steel object's shape. For this reason, solid objects, for practical purposes, are not compressible.

Liquid

In a liquid such as ATF, the molecular structure is not as rigid as it is in a solid material. The molecules, in this case, do move a very slight amount in relation to one another and have less attraction for each other. This property allows a liquid to conform to its container's shape. However, the distance between the molecules in a liquid substance remains relatively close; consequently, it would take many tons of force to compress the liquid even a small amount. The combination of the flexibility of the liquid, the ability of it to conform to a container, and the relative incompressibility of a solid are the three characteristics of ATF that permit it to transmit force and motion in the hydraulic system.

Gaseous

In a gas such as steam, the molecules are far apart and move about at high rates of speed. Furthermore, the molecules can move freely in relation to one another and tend to repel each other. These factors give a gas its unlimited expansion quality, yet the gas substance is compressible by normal means because of the distance between the molecules.

PASCAL'S LAW

The basic foundation for pressure hydraulics is Pascal's law. After extensively experimenting with liquids, a French scientist, Blaise Pascal, in 1653 discovered the hydraulic lever. Through controlled laboratory experiments, Pascal proved, after applying a force to a confined liquid, that the liquid has pressure that can transmit force and motion. By experimenting with weights and pistons of various sizes, Pascal also found that a hydraulic system could produce a mechanical advantage or force multiplication and that the relationships between force and distance were exactly the same using hydraulics as with a mechanical lever.

From the data Pascal collected in his laboratory, he formulated a law that states: "Pressure on a confined fluid is transmitted equally in all directions and acts with equal force on equal areas." This law is too complex to completely understand without some explanation. In order to simplify Pascal's law, the following sections deal with each concept separately and thoroughly.

Force

Force is a pushing or pulling effort that attempts to cause motion, but it does not always produce this effect. In the

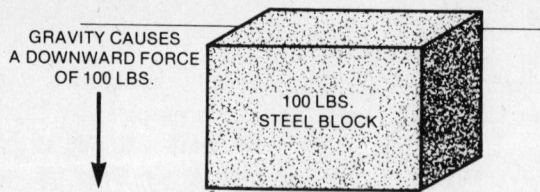

FIGURE 8-2 A steel block exerting 100 pounds (444.8 N) on the floor.

U.S. Standard System, the unit of measurement is the pound or ounce.

In the metric system, the unit of measure for force is the newton (N). One pound of force in the U.S. Standard System is equal to 4.448 newtons in the metric system. To convert pounds to newtons, just multiply the number of pounds by 4.448. Therefore, 100 pounds would be 444.8 newtons (100 × 4.448 = 444.8 N).

Gravity and friction are two good examples of force; other types are presented later in this chapter. The force of **gravity** is nothing more than the mass, or weight, an object has. For instance, if a steel block weighing 100 pounds is sitting on the floor, it exerts a downward force of 100 pounds (444.8 N) on the floor (Fig. 8-2).

The force of **friction** is present when two objects attempt to move against each other. In the example above, if a person attempted to push the same 100-pound block of steel across the floor, he would encounter some resistance to its movement. This resistance to motion is the force of friction present between the block and the floor.

Pressure

Pressure is the force exerted by a liquid on a given unit of surface area. In other words, pressure is force per unit area, or force divided by area.

The unit of measurement for pressure in the U.S. Standard System is pounds per square inch (psi). The metric equivalent of psi is kilograms per square centimeter (kg/cm^2), or kilopascals (kpa). One pascal is the equivalent of 1 newton per square meter (N/m^2). One kilopascal is 1,000 pascals. Pressure in pounds per square inch (psi or lb/in^2) can be converted to kilopascals by multiplying the amount by a factor of 6.895; 100 psi equals 689.5 kilopascals (kpa).

Figure 8-3 illustrates an example of fluid under pressure. In this situation, the round container, or cylinder, is full of fluid. Above the level of the fluid is a tight-fitting piston. If any force attempts to move the piston down trying to compress the fluid, pressure results.

Of course, the applied force will not create pressure in the fluid if it is not confined, for example, if either the

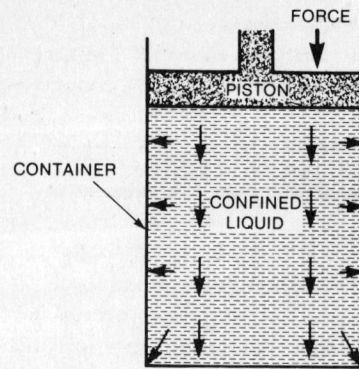

FIGURE 8-3 Pressure results when a force is applied to a confined liquid.

piston or container leaks. The fluid, in this case, would simply leak past the opening. Therefore, the force of the piston would cause fluid flow but no pressure would develop. In other words, there must be a resistance to flow in order to have pressure.

Area

Area is the measurement of any plane surface having given boundaries. The unit of measure for area in the U.S. Standard System is square inch (in^2). In the metric system, the unit is square centimeters (cm^2).

To further understand the concept of area, study Fig. 8-4. If the 4-inch square box shown in the illustration is divided into 1-inch squares, there are 16 squares in all, or an area of 16 square inches (16 in^2).

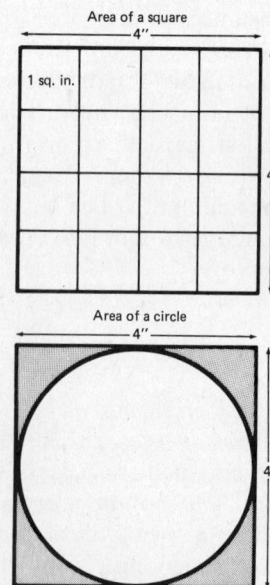

FIGURE 8-4 The areas of a square and circle.

If a circle were fitted into the 4-inch square, it would have an area of about 12.6 in². Notice in Fig. 8-4 that the circle has less area because a portion of the square is not within the circle. However, the area of a circle is still measured in square inches. This is true of any plane (two dimensional) surface such as a diamond, triangle, or rectangle.

Pressure on a Given Area

As mentioned, pressure is the result of a force applied onto a given area. In Fig. 8-3, force was applied to a fluid in a container having a given area. If the amount of force and area of the container were known, the amount of pressure in the fluid could be calculated using the formula: pressure (P) equals force (F) divided by area (A), or P = F/A.

In Fig. 8-5, a force of 160 pounds exerted by the weight of a fluid is distributed on the bottom surface area of a container having an area of 16 square inches. The pressure on this area is found by using the formula P = F/A. Therefore, 160 pounds divided by an area of 16 square inches (160 lb/16 in²) equals 10 pounds per square inch (68.96 kpa). This means that every square inch of the bottom of this container has a force applied to it of 10 pounds (44.48 N).

Equal Force on Equal Areas

Figures 8-3 and 8-5 point out another important part of Pascal's law. That is, "Pressure acts with equal force on equal areas." In another simple example of a hydraulic

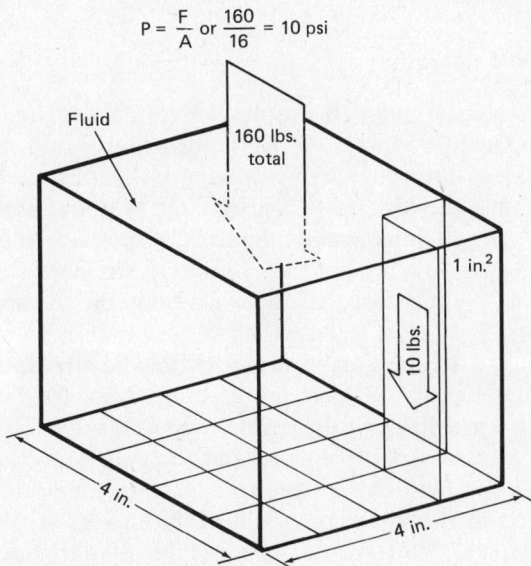

$$P = \frac{F}{A} \text{ or } \frac{160}{16} = 10 \text{ psi}$$

FIGURE 8-5   Every square inch of the surface area has 10 pounds (44.48 N) of force applied to it.

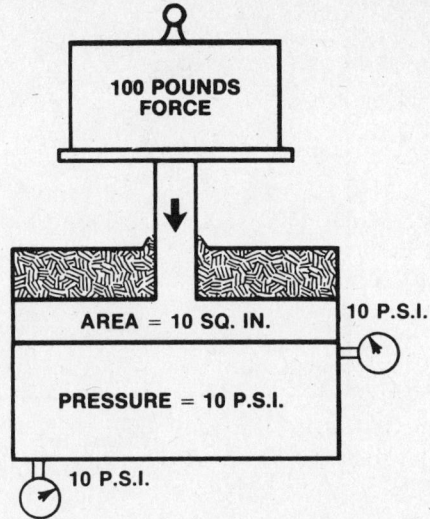

FIGURE 8-6   Pressure acts with equal force on equal areas. (Courtesy of Chrysler Corp.)

system (Fig. 8-6), an applied force of 100 pounds on a piston area of 10 square inches creates a pressure of 10 psi (68.9 kpa). More importantly, the 10 psi reading on each of the attached gauges indicates the equal pressure throughout the fluid—at the top, sides, and bottom of the cylinder.

TYPES OF APPLIED FORCE

Before attempting to calculate the pressure and force in different configurations of hydraulic systems, it is a good time to discuss the type of forces that can be applied to a piston. These forces include atmospheric, vacuum, spring, and mechanical.

Atmospheric

Atmospheric force, more commonly referred to as atmospheric pressure, is another example of an applied force due to gravity. However, in this case, it results from the weight of a blanket of air under which we live (Fig. 8-7). This weight, or force, creates a definite pressure on our bodies and everything else that exists around us.

The pressure is the result of the weight of the air itself. At sea level, for example, a cubic foot of air weighs about 0.08 pound, or 1.25 ounce. This does not seem to be very much; however, the blanket of air, our atmosphere, is over 50 miles thick. There are, in fact, many thousands of cubic feet of air stacked on top of one another, all adding their weight to the total.

The total weight, or downward push (force), of the atmosphere results in a pressure of 14.7 psi at sea level, or

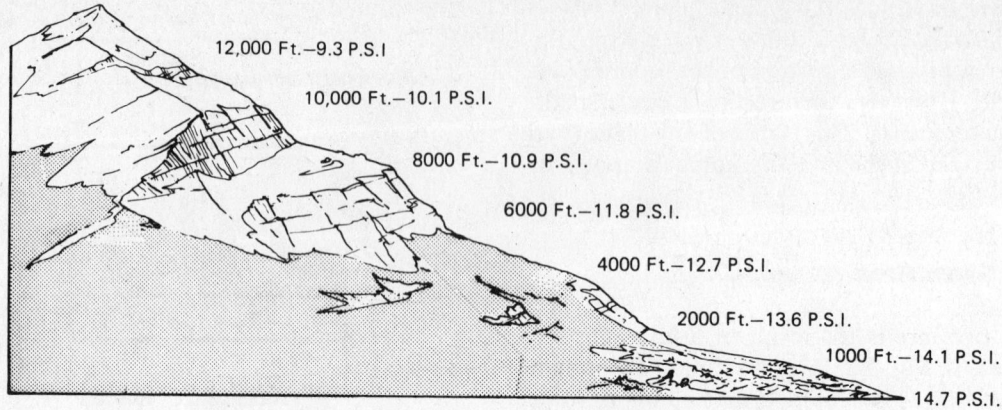

FIGURE 8-7   The weight of the air produces atmospheric pressure.

about 2,160 pounds on every square foot. This pressure exerts itself at the same point equally in all directions—downward, sideward, and upward.

From sea level upward to plateaus and mountains, atmospheric pressure decreases. This pressure reduction is due to a smaller amount of air at higher altitudes. Consequently, the total weight of the air decreases and so does atmospheric pressure. Figure 8-7 illustrates the approximate atmospheric pressure at various elevations above sea level.

Vacuum

A **vacuum** is any pressure lower than atmospheric. For instance, when a person sucks on a straw immersed in water (Fig. 8-8), this action removes the air from the straw. As a result, a low-pressure area or vacuum of less than 1 atmospheric (which is equal to atmospheric pressure) exists on the end of the straw.

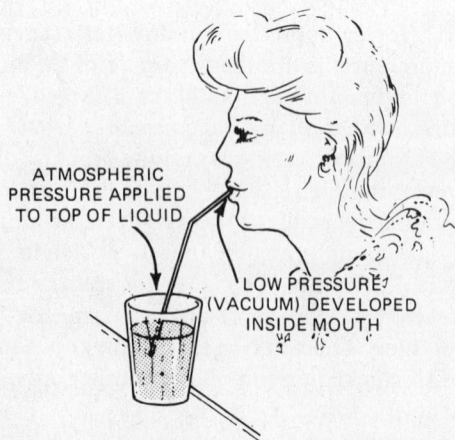

FIGURE 8-8   The vacuum principle.

Atmospheric pressure above the liquid in the container is greater than what is on the discharge end of the straw. The weight, or pressure, of the outside air on the surface of the liquid in the container forces the liquid up the straw in order to fill the void, or vacuum. The most important thing to understand is that the vacuum does not pull the fluid up the straw. Instead, atmospheric pressure pushes the liquid into the vacuum in an attempt to equalize the pressure.

Atmospheric pressure and vacuum are important factors dealing with the operation of hydraulic pumps and one of the signaling devices many of the transmissions utilize. This chapter later explains how this device uses both a vacuum and atmospheric pressure to alter spring force on a piston-type valve.

Spring

When dealing with hydraulic valves, a person will encounter spring force (Fig. 8-9). **Spring force** is the push or pull a spring produces when compressed or stretched. In Fig. 8-9, the compressed spring exerts force at both its ends—one pushing against the back of the valve and the other against the base of the cylinder. A stretched spring exerts a pulling force on both its ends on the spring anchors.

Figure 8-10 illustrates how a vacuum and atmospheric pressure interact to alter a spring's force on a valve. In this situation, the force of the spring is against the back of a diaphragm (a flexible piston) and the rear portion of a sealed chamber. This portion has a vacuum port that connects to the intake manifold of the engine. With this arrangement, the spring side of the diaphragm has vacuum applied to it when the engine is running.

On the other side of the diaphragm is another chamber, but it has a port that is open to the atmosphere. This port

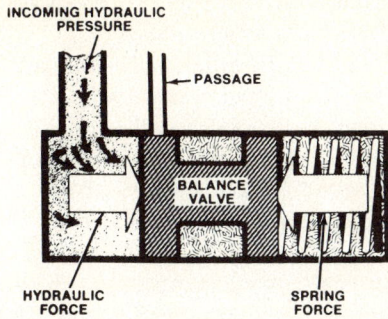

FIGURE 8-9 Spring tension (force) exerted on a valve. (Courtesy of Chrysler Corp.)

CALCULATING THE PRESSURE IN A HYDRAULIC SYSTEM

It is fairly easy to figure the pressure created in a sealed or closed hydraulic system that uses a piston to apply the force to the fluid. As mentioned earlier, a closed or sealed system is necessary in order for pressure to build up. In other words, the system must have an ending point, regardless of the size of the cylinder(s) or length of any connecting pipes.

If the system didn't have a terminating point or had a leak, pressure could not build up. Instead, the piston pump, when it moved, would just cause fluid flow. Later,

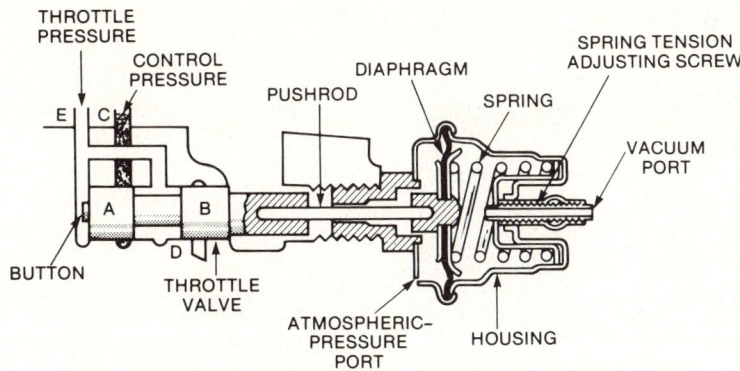

FIGURE 8-10 Atmospheric pressure and a vacuum interacting to alter a spring's force.

allows atmospheric pressure to act on the diaphragm and force it to the right whenever there is a vacuum on the spring side. This action increases the tension of the spring.

Whenever vacuum drops, the compressed spring expands and forces the diaphragm to the left. This action through the pushrod moves the valve to the left against the hydraulic pressure acting on its button end. In other words, when there is a vacuum on the diaphragm, spring tension increases; and hydraulic pressure moves the valve to the right. As vacuum drops, spring tension decreases as it forces the valve to the left.

Mechanical

The last type of force is mechanical or physical (Fig. 8-11). In this case, a valve receives a force by some type of linkage, activated by the driver. A typical example of this is the kickdown linkage, connected from the carburetor to the valve body inside the transmission. This linkage applies a mechanical force to the kickdown valve in order to cause a force downshift of the transmission into a lower gear ratio.

when we examine the automatic transmission's hydraulic system, you will become aware of the many controlled leaks a typical system will have. However, when using an engine-driven hydraulic pump, these leaks are insignificant (unless excessive) and are necessary to lubricate certain areas of the transmission. But in a piston-type

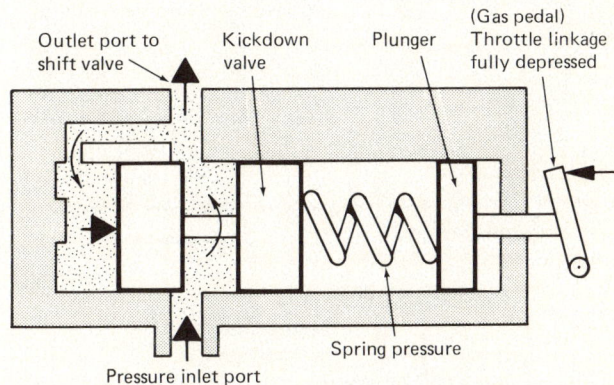

FIGURE 8-11 A mechanical or physical force applied to a piston. Movement of the linkage applies varying amounts of force to the valve by altering its spring tension.

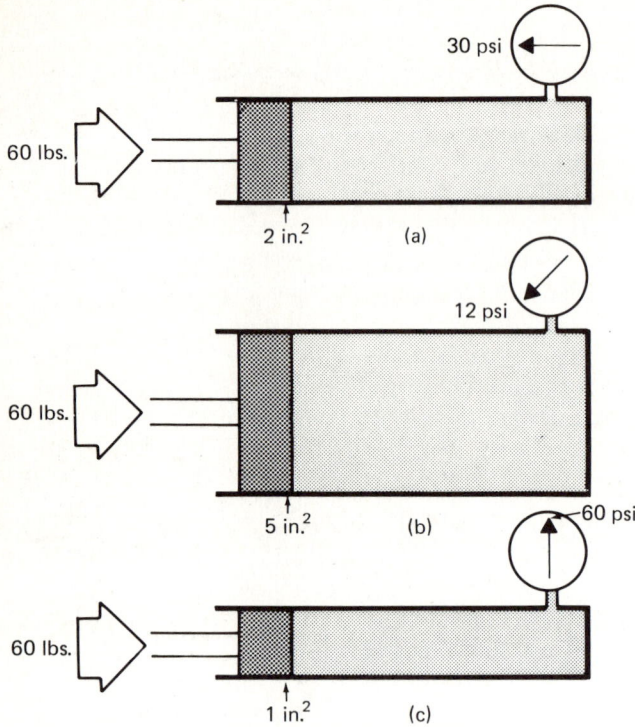

FIGURE 8-12 Calculating the pressure in hydraulic systems having pistons of different areas.

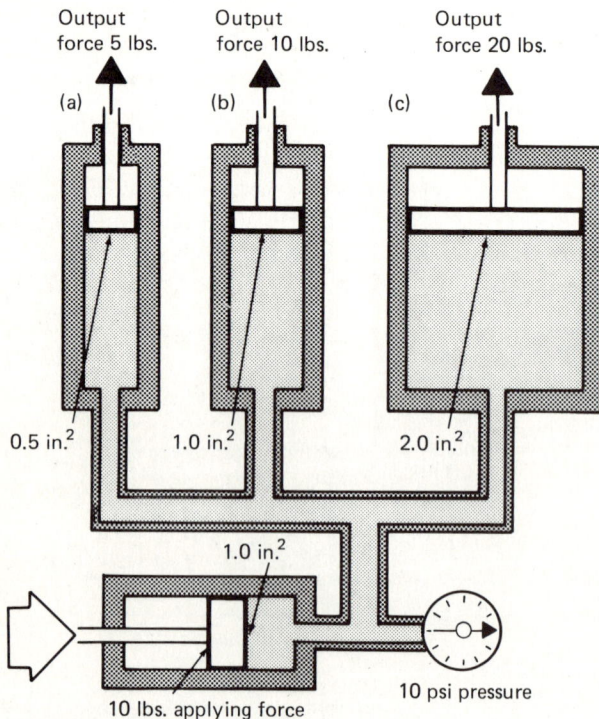

FIGURE 8-13 Output force equals system pressure multiplied by piston area.

pump system in which the pump operates very slowly, any leak causes a severe loss in pressure.

With these facts in mind, let's calculate the pressure created in several simple systems where a piston applies the input force. Figure 8-12(a) illustrates a simple system in which the piston has an applied force of 60 lb and has an area of 2 square inches. Using the formula for pressure of P = F/A, the results are P = 60/2 or P = 30 psi (206.85 kpa).

However, note what happens if the piston area is different but the force remains the same. In Fig. 8-12(b), the pressure P = 60/5, or 12 psi (82.7 kpa). Obviously, the pressure is not as great when a larger piston is used with the same amount of applied force. This is due to the fact that the applied force is spread over a larger area.

Figure 8-12(c) shows what occurs if the piston is smaller with the same amount of applied force. In this case, P = 60/1, or 60 psi (413.7 kpa). The pressure increases because the force is spread over a smaller area.

Another consideration to altering pressure is changing the amount of applied force. If more or less force is applied, the pressure will increase or decrease. In other words, for a given piston area, a larger force results in greater pressure, while less force causes a lower pressure.

CALCULATING FORCE IN A HYDRAULIC SYSTEM

It has been pointed out that if the force and area of a system's piston are known, the pressure can be calculated. Output force can also be figured if system pressure and piston size are known. To do this, the formula for pressure is utilized in a different form: F = P × A. The letters have the exact same meaning as before. However, the revised formula states that output force is equal to system pressure multiplied by piston area.

If, for example, we apply a system pressure of 10 psi to a 1-square-inch piston [Fig. 8-13(b)], the output force is 10 pounds (10 × 1 = 10). Now apply the same 10 psi to a 2-square-inch output piston [Fig. 8-13(c)], and the force equals 20 pounds (10 × 2 = 20).

If we reduce the size of an output piston, the force is less [Fig. 8-13(a)]. For example, a pressure of 10 psi applied to a 1/2 square-inch piston produces an output force of 5 pounds (10 × 1/2 = 5).

TRANSMITTING MOTION WITHIN THE HYDRAULIC SYSTEM

The discussion so far has been mainly about creating pressure and using it to transmit and change the amount of output force through the use of a confined fluid. But what about the factors of input and output motion? If input and output forces are the same, the amount of input and output motion should also be the same. However, if the hydraulic system increases output force, it will decrease output motion and vice versa.

Using our previous example that 10 pounds of force on a 1-square-inch piston develops a pressure of 10 psi, this applied to a 1-square-inch output piston produces 10 pounds of output force. This input-to-output force relationship remains the same no matter how far the pistons move. If the input piston moves 10 inches, the output piston also travels 10 inches (Fig. 8-14).

But if the output piston size is doubled to 2 square inches, the output force is also doubled. However, the output motion is half that of the input piston. Therefore, if the input piston travels 10 inches, the output moves only 5 inches (Fig. 8-15).

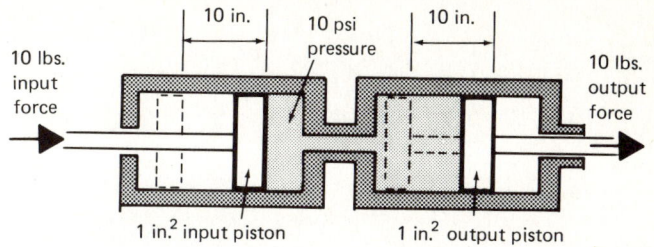

MECHANICAL ADVANTAGE THROUGH THE USE OF HYDRAULICS

A fluid in a hydraulic system can provide a mechanical advantage as a common lever does. As shown in Fig. 8-15, changing the area of either the input piston or the output piston changes output force. In the example, the output piston is double the size of the input. Thus the force at the output piston is twice as much as that at the input, which is a mechanical advantage. In this case, the mechanical advantage is 1:2.

However, as you recall from Chapter 2, mechanical advantage results in a loss of motion (Fig. 8-16). In the illustration, 100 pounds of force is used to lift a 200-pound force through the use of a lever. In other words, the lever provides another 1:2 mechanical advantage. However, the input force has to travel 2 inches to move the output force (the box) 1 inch. This gives us a ratio of distance traveled of 2:1. The main point here is that when there is a mechanical advantage, watch for an accompanying loss in distance traveled by the output force.

In a hydraulic system, the same rule applies. When the size of the output piston increases, the distance it travels must decrease. This loss of motion is in direct proportion to the difference in size between the input and output pistons. In Fig. 8-15, the input piston moves 10 inches to only 5 of the output, a 2:1 distance ratio.

BASIC HYDRAULIC SYSTEM

To show that pressure hydraulics is not difficult to understand, let's build a simple hydraulic jack system and see how it operates. The basic system consists of a reservoir,

FIGURE 8-14   If the input and output pistons are the same size, both their motions are equal.

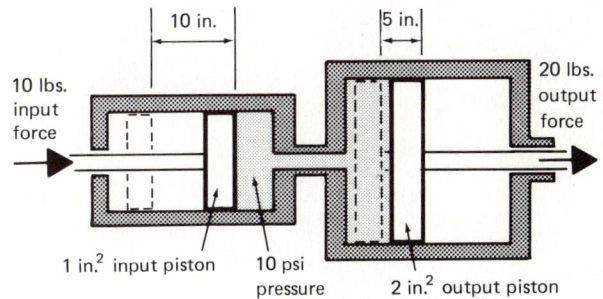

FIGURE 8-15   If the area of the output piston is doubled, so is the output force, but output motion is cut in half.

pump, valving, and an actuating mechanism. These components will change a small input force into hydraulic pressure, then convert the pressure to a larger output force to raise the load.

Reservoir

The first thing the system requires is a storehouse for the fluid until the system needs it. This storehouse is the **sump,** or **reservoir** (Fig. 8-17). In other words, the **reservoir** acts as the fluid source for the hydraulic system.

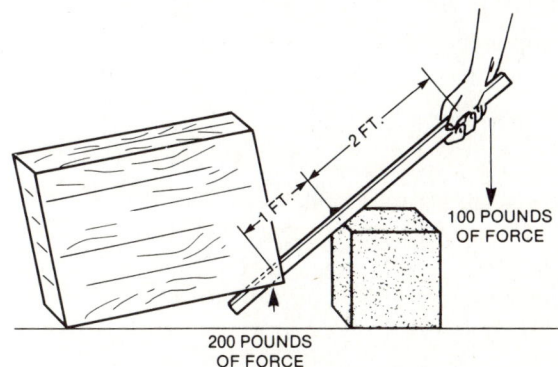

FIGURE 8-16   A lever producing a mechanical advantage.

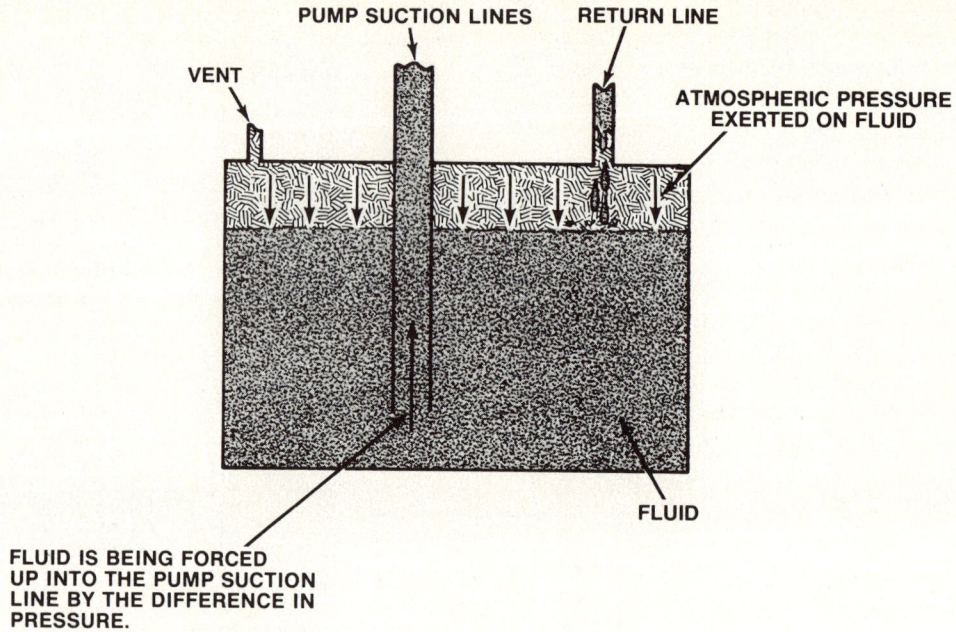

FIGURE 8-17 A fluid reservoir. (Courtesy of Chrysler Corp.)

The reservoir, in order to perform its function in the system, needs a vent line, suction line, and return line. In order for the fluid pump to operate properly, the fluid must move freely from the reservoir to this unit. The **vent line** allows atmospheric pressure to enter the reservoir. As the pump operates, it creates a vacuum or low pressure area from the pump down to the reservoir via the suction line. At the same time, atmospheric pressure pushes fluid up to the pump due to the difference in pressure between its suction port and the outside air. The **return line** is necessary because, with a system that is constantly operating, the fluid eventually has to return to the reservoir for recirculation through the system.

Pump

The **pump** applies force to the fluid and creates flow (Fig. 8-18). Remember, force is necessary to create pressure in the system, but the pump's force can also cause fluid flow within the system as needed. If the pump flow does not meet any resistance, there is no pressure buildup in the

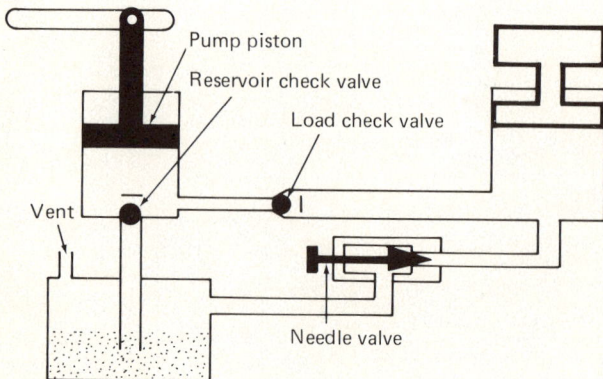

FIGURE 8-18 A basic hydraulic system.

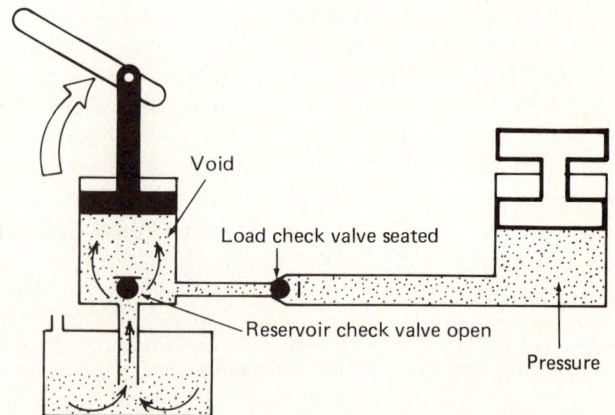

FIGURE 8-19 The operation of a hydraulic jack during its intake phase.

system, just a free flow of fluid. In other words, there must be resistance to flow in order to create pressure. Finally, the pump used in this basic system is a piston type, which a person will operate through a jack handle.

Valving

When the pump begins to displace fluid, the system will require some sort of **valving** to control the direction of its flow. In this system, just two valves are necessary, the check and needle. The **check valve** is nothing more than a one-way valve, which allows fluid flow in one direction only, and is necessary to permit repeated stroking of the pump. The **load check valve** prevents the load from coming down on the intake stroke. The **reservoir check valve** stops pressure loss to the sump during the pump's power stroke.

To lower the load, a **needle valve** is a necessary part of the system. The location of this valve is in the return line between the large output piston and the reservoir. Opening this valve allows the load on the output piston to act as the force on it that pushes the fluid back to the sump.

Activating Mechanism

Once the fluid has passed through the pump, valves, and lines, it will end up under the activating mechanism. This is the place where the hydraulic pressure moves a piston, causing it to perform some sort of mechanical work. The **activating mechanism,** the output piston in Fig. 8-18, is actually the dead end that the fluid flow encounters in a system, and this termination of flow causes the pressure buildup. The pressure then works against the surface area of the activating mechanism, causing a force application on the load.

Force lifts a vehicle, the load, upward. Types of activating mechanisms found in an automatic transmission include servo and multi-disc clutch pistons. In these situations, the output force applies a band around a drum or compresses the clutch plates tightly together.

SIMPLE HYDRAULIC SYSTEM OPERATION

Intake Phase

In order for a simple hydraulic jack to function to lift and lower a load, it requires three operating phases: intake, pump, and lowering. As the operator moves the jack handle up on the intake stroke, the small piston moves up (Fig. 8-19). This action creates a low pressure area (vacuum) in the area below the piston. Atmospheric pressure over the fluid in the reservoir pushes fluid through the reservoir check valve and into the area below the piston. At the same time, the load on the output piston tries to lower it, but this action creates a force on the fluid in a reverse direction, which seats the load check valve and traps the fluid in this portion of the system.

Pump Phase

When the operator moves the jack handle downward during the pump stroke, the piston moves down and applies a force to the fluid beneath it. This action creates a pressure in the fluid (Fig. 8-20). The pressure immediately closes the reservoir check valve and opens the load check. Then the pressurized fluid flows under the activating mechanism, the output piston. As a result, the piston raises the load, a vehicle, another notch.

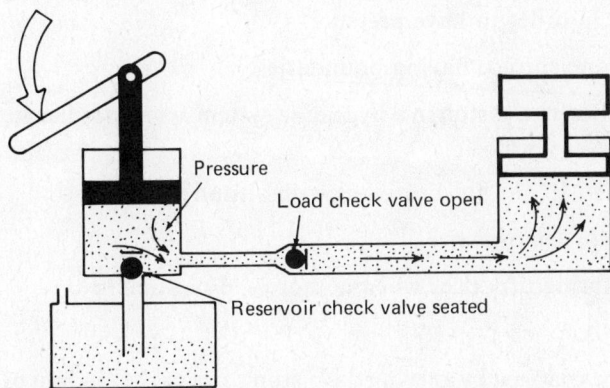

FIGURE 8-20 *The operation of a hydraulic jack during its pump phase.*

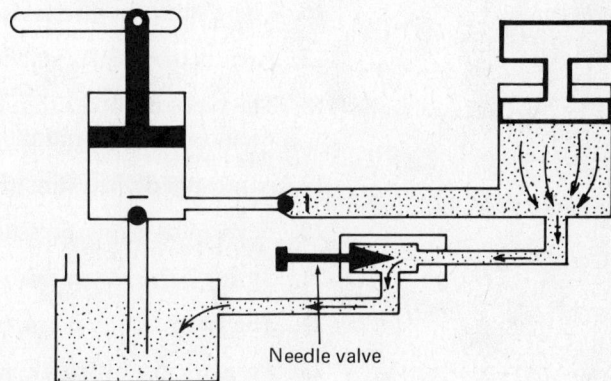

FIGURE 8-21 *The operation of a hydraulic jack during its lowering phase.*

Lowering Phase

To lower the vehicle or load, the operator unscrews the needle valve slowly (Fig. 8-21). Opening this valve allows the load to act as a force that pushes the unneeded fluid back to the reservoir. The vehicle will lower in relation to the amount of fluid returned to the sump; the operator has control of this flow by just how much he opens the needle valve.

The simple system explained above has an output of both force and motion. Therefore, the system uses pressure and continuous fluid flow. The pressure applies the force, and the flow creates the motion.

SUMMARY

1. An automatic transmission consists of a planetary gear train and a hydraulic control system.

2. Without the effects of hydraulics, the automatic transmission won't function.

3. A mechanic must have a working knowledge of hydraulics to understand transmission operation and how to troubleshoot a malfunctioning unit.

4. Hydraulics is a branch of physics that deals with practical applications of a fluid in motion.

5. Hydrodynamics is a study dealing with the motion of fluid and the forces acting on solid bodies immersed in the liquid.

6. Hydrostatics is a study that deals with the characteristics of a fluid at rest and under pressure.

7. Liquids resist compression due to their molecular structure.

8. A molecule is the smallest particle of any substance.

9. The arrangement of the molecules determines if a substance is solid, liquid, or gaseous.

10. Because ATF is not compressible, it can transmit force and motion.

11. The basic foundation of pressure hydraulics is Pascal's law.

12. Pascal's law states: "Pressure on a confined liquid is transmitted equally in all directions and acts with equal force on equal areas."

13. Force is a push or pull on an object measured in pounds or newtons.

14. Gravity and friction are two examples of force.

15. Pressure is force per unit area measured in psi or kilopascals.

16. There must be a resistance to flow in order to have pressure.

17. Area is the measurement of any plane surface having boundaries.

18. The types of forces that can be applied to a piston in a hydraulic system are atmospheric, vacuum, spring, and mechanical.

19. Atmospheric pressure at sea level is 14.7 psi but decreases with altitude.

20. A vacuum is any pressure lower than atmospheric.

21. Spring force is the push or pull a spring produces when compressed or stretched.

22. The formula for pressure is $P = F/A$.

23. Pressure in a hydraulic system can be changed by altering the piston's size or the amount of applied force.

24. Output force in a hydraulic system is equal to pressure multiplied by the piston's area.

25. If input and output forces are the same in a hydraulic system, so should be the combined motions of the two pistons involved.

26. A fluid in a hydraulic system can provide a mechanical advantage, as does a common lever.

27. Mechanical advantage results in a loss of motion.

28. A simple hydraulic system consists of a reservoir, pump, valving, and an activating mechanism.

29. The reservoir is the storehouse of fluid for the system.

30. The pump applies force to the fluid and creates flow.

31. The activating mechanism changes hydraulic pressure to output force to do some form of useful work.

32. A simple hydraulic jack has three operating phases: intake, pump, and lowering.

REVIEW

This section will assist you in determining how well you remember the material contained in this chapter. Read each item carefully. If you cannot complete the statement, review the section in the chapter that covers the material.

1. The science that deals with the effects of moving liquids is _____.

   a. physics

   b. chemistry

   c. hydraulics

   d. pneumatics

2. The smallest particle or building block of a substance is the _____.

   a. molecule

   b. atom

   c. electron

   d. neutron

3. The property of a liquid that makes it not compressible is its _____.

   a. viscosity

   b. molecular structure

   c. fluidity

   d. flow rate

4. Pascal discovered _____.

   a. the hydraulic lever

   b. hydraulics

   c. molecules

   d. force

5. If a small input force raises a larger output force, _____ occurs.

   a. pressure

   b. force

   c. force multiplication

   d. viscosity

6. Force per unit area is _____.

   a. pressure

   b. torque

   c. force

   d. force multiplication

7. In hydraulics, one common type of force is that of a _____.

   a. spring

   b. solid

   c. ram

   d. inertia

8. If a column of fluid has a force of 100 pounds applied to it by a two-square-inch piston, system pressure will be _____.

   a. 50 psi

   b. 100 psi

   c. 200 psi

   d. 250 psi

9. The output force of a hydraulic system is equal to system pressure times _____.

   a. input piston area

   b. output piston area

   c. pressure

   d. input force

10. If a hydraulic system that uses an input-output piston arrangement produces a force multiplication, the two pistons _____.

    a. move in opposite directions

    b. move the same distances

    c. move different distances

    d. do not move at all

11. The storehouse for the hydraulic system is the _____.

    a. sump or reservoir

    b. pump

    c. actuating mechanism

    d. valving

12. The device in a hydraulic system that produces force on the fluid is the _____.

    a. valving

    b. sump

    c. pump

    d. actuating mechanism

13. To control the direction of flow, a hydraulic system needs _____.

    a. valving

    b. a pump

    c. an actuating mechanism

    d. a sump

For the answers, turn to the Appendix.

CHAPTER 9

Automatic Transmission Hydraulic Pumps

The pump is the heart of the pressure hydraulic system; without it the system could not operate. The **hydraulic pump** converts an external source of power into a force that creates pressure when applied to the fluid within a closed system. The pressure, in turn, transmits motion and force to the activating mechanisms such as a multi-disc clutch or servo piston.

SYSTEM DESIGN AND DEMANDS DETERMINE PUMP TYPE

In most cases, the hydraulic system's design, pressure, or flow requirements determine the type of pump needed. For example, the simple hydraulic system, presented in the previous chapter, used a piston-type pump. This pump required a strong arm as the power source to operate it. In other words, the pump mechanism did use an external source of power or energy, in this case human strength, to apply a force to the fluid. The pump developed no power or force of its own; the pump simply converted an external source of energy or power into a force necessary to create pressure in the fluid.

The pressure and flow requirements of the simple jack system were not too great. The jack's piston pump only had to produce enough pressure to cause the activating mechanism to raise the vehicle. Moreover, the pump had to displace (expel) enough fluid to raise the output piston,

the activating mechanism, a small amount with each operating stroke. Therefore, a person's strength could raise a motor vehicle, by means of the hydraulic system in the jack.

The design of the automatic transmission's hydraulic system is such that it requires a rotary-type pump (Fig. 9-1). The **rotary transmission hydraulic pump** uses the engine as its external source of power. In other words, this pump converts engine power to a force that creates system pressure and causes fluid flow within the system.

In relation to the task of the pump in the simple hydraulic jack, the automatic transmission's rotary counterpart has a much larger job to do. In regard to pressure, the transmission pump must produce enough to properly operate all the clutch and band servos, many of which can be activated at the same time. Enough pressure is also necessary to maintain fluid flow through the various system restrictions, the torque converter, plus the cooling and lubrication systems of the transmission.

The pump must provide sufficient flow to produce the motion within the clutch and band servos and supply fluid to the converter, cooling, and lubricating systems. However, as mentioned, there are a number of controlled leaks in the lubrication system where the fluid escapes out and around bushings and at certain parts that use cast iron sealing rings. As long as the pump is operating efficiently and the leaks are within tolerances, they will have no effect on system pressure or other flow requirements.

ROTARY PUMPS

To meet these special pressure and flow requirements within its hydraulic system, the automatic transmission utilizes one or more durable rotary pumps. A **rotary pump** is one in which its internal components spin in a circle while producing system pressure and meeting the flow requirements of the system. This is just opposite to the piston pump that uses a reciprocating (back and forth) motion.

Design

The rotary pumps have two turning members of different sizes (Fig. 9-2); however, the vane-type rotary pump only has one. When there are two, the small inner member driven by the converter inside the outer member rotates at engine speed. The small member then turns the outer one within the pump housing. Furthermore, these two members rotate on different centers so that at one point there is very little clearance between them, just enough for lubrication. Then, the clearance varies through a half revolution to a point of maximum distance between the two.

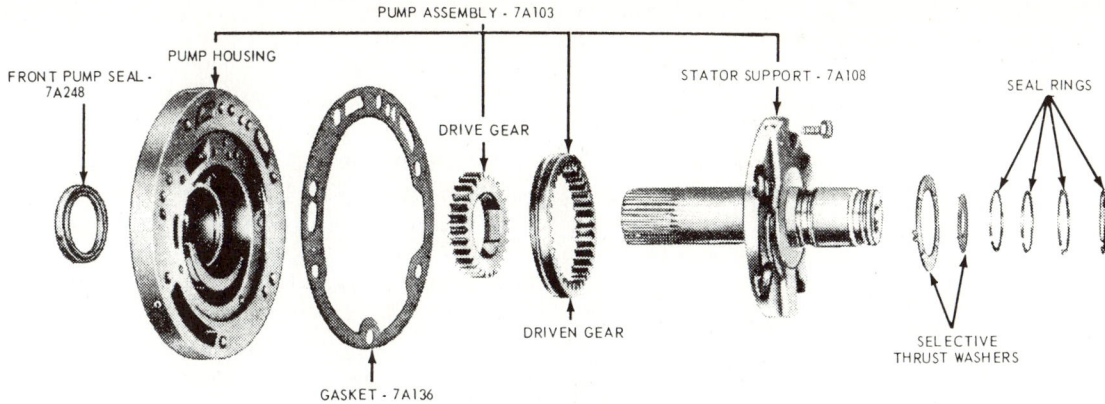

**FIGURE 9-1   A typical rotary gear-type pump.
(Courtesy of Ford Motor Company)**

The actual pumping mechanism, which can be either gear teeth or rotor lobes, forms sealed chambers between the two members. These chambers move around inside the pump housing carried along by the rotation of both members. Finally, these chambers that trap the fluid are constantly expanding and contracting as the two members turn.

Operation

As any one pump chamber passes the pump's inlet, its size increases or expands; this creates a vacuum between the pump inlet and the reservoir. The atmospheric pressure, present over the fluid in the reservoir, pushes the oil into the pumping chamber. Once the fluid fills the pumping chamber, the inlet portion of the pump's cycle for this particular chamber is complete.

As the filled chamber moves past the pump's inlet port, it traps the fluid. For the next few degrees of pump rotation past the inlet port, the chamber's size does not change. But further rotation causes the size of the chamber to decrease, and this reduction in size applies a force to the trapped fluid. Consequently, the chamber, by decreasing in size, squeezes the fluid out into the hydraulic system as the pumping chamber passes over the outlet port. Once this chamber travels past the outlet port, the delivery or displacement portion of the pump cycle is complete.

By having successive chambers follow one another through this process, the rotary pump produces a smooth, continuous output of fluid. In addition, since there is a minimum clearance at one point between the two

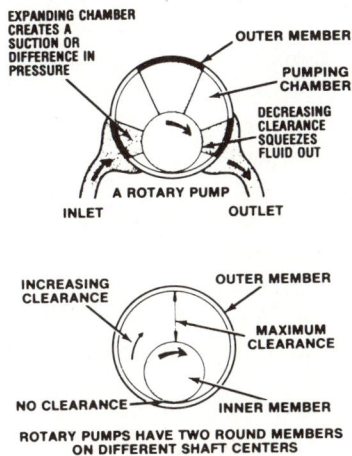

**FIGURE 9-2   Rotary pump design and operation.
(Courtesy of Chrysler Corp.)**

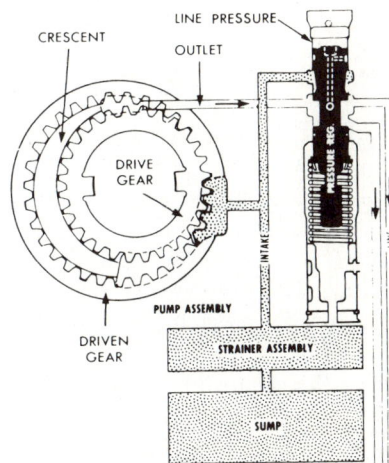

**FIGURE 9-3   The outer and inner members of an IX gear pump. The drive gear is the small inner member, while the large driven gear is the outer member.
(Courtesy of Hydra-matic Division of General Motors Corp.)**

members, the design prevents any chamber leakage between the outlet and inlet ports. As a result, the pump's delivery is positive even against high system pressure.

Types

The most common types of rotary pumps used today in automatic transmissions are the IX gear and rotor pumps, in addition to the variable capacity vane. The term "IX" is an abbreviation for internal-external, referring to the location of the pumping chambers in the rotating parts. For example, an IX gear pump (Fig. 9-3) has a large, outer member with internal teeth that form pumping chambers. These internal teeth mesh at one point with teeth located on the small, inner member; these external teeth also are pumping chambers.

GEAR PUMP DESIGN

Housing

An IX gear pump consists of a pump housing, drive and driven gears, and a stator support. The pump housing (Fig. 9-4), as its name implies, houses or encloses the pump members (gears) and bolts to the front or converter end of the transmission case. The pump manufacturer forms this component from a good grade of cast iron, and the completed housing contains a counterbore or cavity, inlet and outlet ports, plus several additional fluid passages.

The counterbore forms a guide for the driven large gear

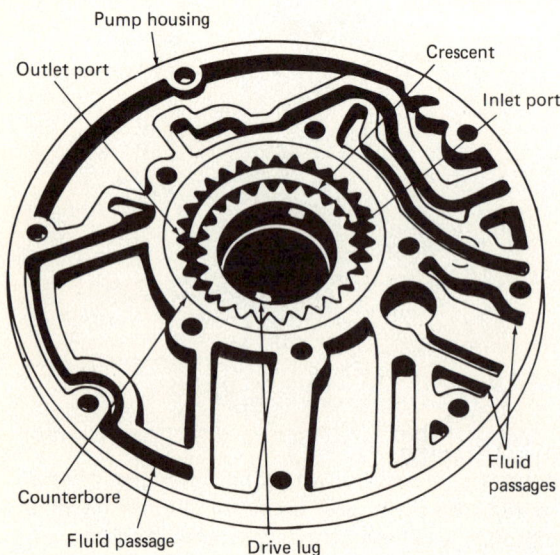

FIGURE 9-4   The design of a typical pump housing.

as it turns within the housing. The clearance between this gear and the wall of the counterbore is only enough to allow for lubrication.

Also cast into the counterbore, in the area where the two gears separate, is a crescent-shaped piece. This crescent forms part of the guide for the small gear as it rotates within the larger gear, thus preventing the meshing of the two gears on one side. Moreover, at the point of greatest separation, both gears fit closely to the crescent. This creates a space between the teeth on both gears and the crescent, which will become the pumping chambers. With this design, the crescent prevents the fluid moving from the outlet to the inlet port after being carried to the former in the pumping chambers. In other words, the crescent helps to separate the pump's inlet from the outlet even when the pump operates under high pressure.

Also found in the counterbore, in the area where the drive and driven gears first separate from mesh, is the inlet port. With the pump bolted to the transmission case, this port aligns with a passageway leading to the reservoir or sump. And as previously mentioned, during the inlet cycle of the pump, fluid enters this port and flows between the separating gear teeth.

Located on the opposite side of the counterbore where the rotating drive and driven gears mesh once again is the outlet port. Because the crescent prevents the fluid from returning to the inlet port, the meshing gears force the trapped fluid into the outlet port. And since the outlet port aligns with a passageway leading into the pressure side of the hydraulic system, pressure in it increases and fluid, as needed, flows to the various components of the transmission.

Depending on the design of the transmission, the fluid passages machined into the pump housing may feed several transmission components. For instance, most pump housings have inlet and outlet passages that feed the torque converter. In addition, many transmission models use the pump housing passages as part of the hydraulic circuit that carries fluid to a clutch or band, the cooling system, and parts of the lubricating system.

Drive and Driven Gears

The small drive gear has external spur (straight) type gear teeth. This gear also has flats or lugs cut into its inside circumference, which mate with similar machined areas on the rear torque converter hub. Consequently, whenever the converter rotates at engine speed, the small drive gear spins.

The teeth of the drive gear mesh with the internal teeth of the driven gear at one point in the counterbore, opposite to the crescent. Since the two gears are in mesh at this point, any rotation of the small drive gear will force the large driven one to turn.

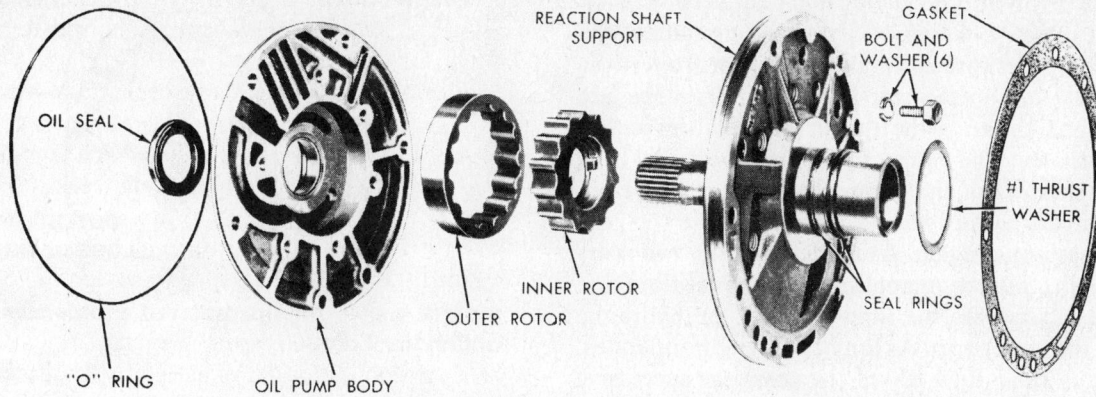

FIGURE 9-5   The stator (reaction shaft) support of a typical rotor-type pump. (Courtesy of Chrysler Corp.)

Stator Support

The stator support, also known as the cover or reaction shaft support, performs three functions (Fig. 9-5). First, this one-piece assembly covers and therefore seals the pump counterbore, where the two gears operate. Second, the long splined, round extension supports the inner race of the stator's overrunning clutch. Consequently, when the one-way clutch locks up, this section of the assembly has to withstand the counter torque applied to the stator blades. Third, on the opposite side from the splined stationary stator support is a shorter, round extension, which fits into the clutch assemblies. This section not only supports the clutches, but contains the fluid passages that feed fluid to these units.

GEAR PUMP OPERATION

When the torque converter drives the small, inner gear, it causes the large, driven one to revolve. Fluid enters the pump at the inlet port, where the two spinning gears separate from mesh (Fig. 9-6). The enlarging space between these separating teeth creates a low pressure area (vacuum) that fluid from the reservoir fills due to the pushing action of atmospheric pressure.

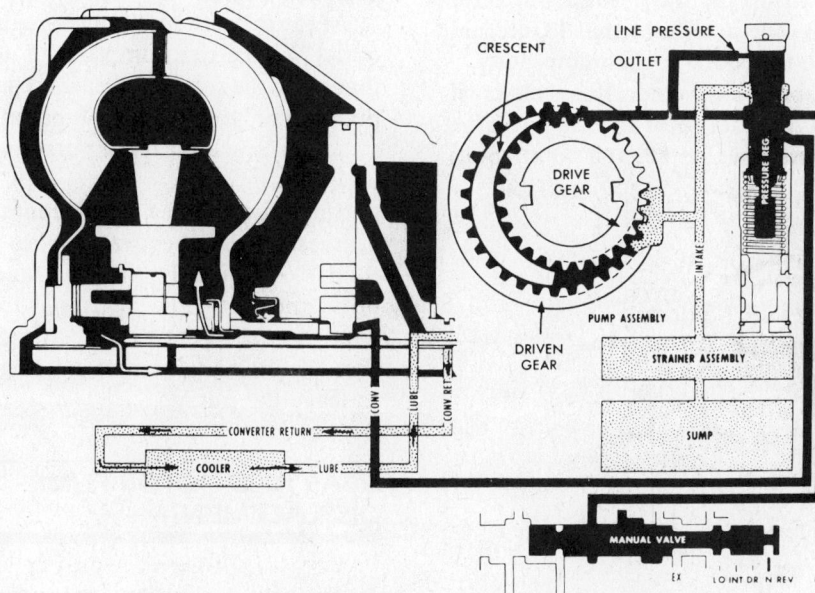

FIGURE 9-6   The operation of a typical gear-type hydraulic pump. (Courtesy of Hydra-matic Division of General Motors Corp.)

Next, the revolving gears carry the fluid, between their teeth, to the crescent. The crescent divides or separates the pumping chambers formed on the inner gear from those of the outer. And since the gear-to-crescent clearances are relatively small, this traps the fluid in the chambers so that it cannot leak out and return to the inlet port.

Further rotation of the pump gears brings the sealed pump chambers toward the outlet port, where the gear teeth will mesh once again. As the teeth of the two gears begin to mesh, this action applies a force to the fluid; the applied force squeezes the fluid out into the hydraulic system via the outlet port. As long as the engine operates, a continuous fluid flow leaves the pump as successive chambers come together at the outlet port.

ROTOR PUMP

Design

The main difference between the IX gear and rotor pump lies in the design of the pumping chambers. In the gear pump, the area between the teeth and the crescent forms the pumping chambers; in the rotor pump, the area between several rotor lobes makes up the pumping chambers.

Figure 9-5 shows a typical rotor pump that consists of an inner and outer rotor, pump body, and reaction shaft support. Except for the rotors, the remaining components have about the same design as their counterparts found in the gear pump. The inner rotor has 12 external lobes, and the converter drives it within the outer rotor but on a different center (Fig. 9-7). The outer rotor has 13 internal lobes and rotates in a counterbore in the pump body.

The two different operating center lines make it possible for a lobe tip on the inner rotor and the tip on the outer rotor to almost contact each other at a point about

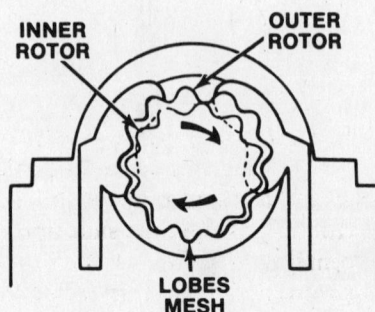

FIGURE 9-7   The design of the pumping chambers in a rotor pump. (Courtesy of Chrysler Corp.)

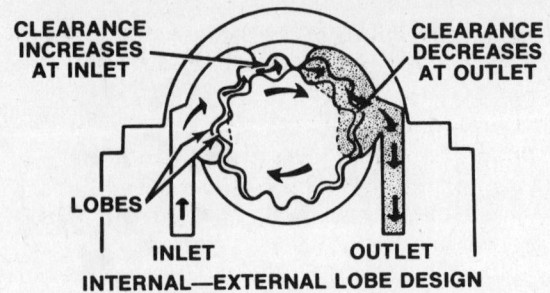

FIGURE 9-8   The operation of a rotor pump. (Courtesy of Chrysler Corp.)

180 degrees from where the lobes mesh. This close clearance between the two lobe tips actually separates the inlet from the outlet chambers of the pump. And when the outlet chamber delivers fluid to the outlet port, the meshing lobes on one end and the close lobe clearance on the other prevents the outlet fluid from leaking back to the inlet side.

Operation

As the converter turns the inner rotor at engine speed, the small rotor, in turn, drives the outer rotor (Fig. 9-8). Because the two rotors revolve on different centers, the lobes begin to separate at one point. This lobe separation creates a vacuum over the inlet port. As a result, fluid from the reservoir, due to the effect of atmospheric pressure, fills the area between about four lobes.

At the same time, the distance between the lobes on the outlet side of the pump begins to decrease. Remember that since the rotor tip clearance at the upper center of the pump is only enough for lubrication purposes, this construction seals the two pumping chambers from each other.

Since the trapped fluid cannot return to the inlet, the decreasing chamber size applies a force to the trapped fluid. Consequently, the fluid squeezes through the outlet port and into the hydraulic system. As the pump continues to revolve, the various lobes form chambers that expand and contract, causing a continuous fluid flow to leave the pump.

GEAR AND ROTOR PUMP DISPLACEMENT

Because the gear and rotor pumps have their outlet ports sealed from the inlet, they are both known as positive displacement. Positive displacement or delivery means that as long as the pump is revolving and the fluid enters its inlet port, it will deliver the fluid at a rate that will remain constant.

This process can continue even after the needs of the hydraulic system have been met. In other words, since the pump is engine driven, it can deliver much more fluid than the system requires, regardless of how high the pressure reaches.

The buildup of system pressure by the continuous pump flow could have no end, and the resulting high pressure could cause a number of severe problems within the transmission. For example, high pressure can overheat the fluid, break the transmission case along with servo and clutch pistons, and blow out seals and gaskets. In order to prevent these problems, manufacturers use a variable-capacity pump, relief valve, pressure regulating valve, or a combination of these items to control pump output. This chapter will discuss the construction and operation of the variable capacity pump in another section. The next chapter covers both the relief and pressure regulating valves.

PUMP CAPACITY

As mentioned, a positive displacement pump delivers the same volume of fluid with each rotation of its components, regardless of their speed. However, as pump speed increases, the total volume of fluid delivered per minute rises because the pump turns at more revolutions per minute. In other words, pump delivery rate increases with higher rpm, even though the pump displaces the same amount of fluid each time it turns through one revolution.

Pump capacity is the total volume of flow produced by a pump at a given pressure and rpm. This capacity must be large enough to provide the transmission with pressure and flow at all times. Correct capacity is especially critical at lower vehicle speeds when the transmission changes ratios frequently because, during these conditions, the hydraulic system requirements are high.

However, if pump capacity is too high, the unit requires excessive engine power to operate it. In addition, the relief or regulating valve will not be able to handle the excess pressure, and it may damage the transmission and hydraulic system.

Transmission engineers determine the needs of the hydraulic system and then determine what the capacity of the pump should be. The capacity of a typical transmission pump, operating at vehicle cruising speeds, is about 30 gallons of fluid per minute at a pressure of up to around 250 psi (1723.75 kpa).

VARIABLE CAPACITY VANE PUMPS

Function

A **variable-capacity vane pump,** as its name implies, alters its output as the needs of the transmission's hydraulic system change. By doing so, this pump conserves engine power and thereby reduces fuel consumption, especially at higher speeds.

The pump accomplishes its function by offering the advantages of both a high- and low-capacity pump. At low engine and pump speeds, for instance, the unit acts as a high-capacity pump that can produce the high pressure and fluid flow when the system needs are the highest. At high vehicle speeds where the hydraulic system requirements are reduced, the same unit becomes a low-capacity pump.

Design

The variable-capacity pump shown in Fig. 9-9 consists of a rotor, set of vanes, and a slide, which all operate inside a pump body and cover. The rotor is converter driven and operates inside the slide. The outer circumference of the rotor has a number of slots, which accommodate the vanes.

The vanes form the pumping chambers for the pump. These units move in and out of their respective rotor slots as they follow the curvature of the inner circumference of the slide. When clearance or space exists between adjacent vanes and the slide, a pumping chamber is formed.

The slide changes the capacity of the pump. This unit fits inside the enlarged counterbore of the pump body and can move within this cavity due to the action of the priming spring or hydraulic pressure. As the slide moves, it changes the rotor's centering within its inner circumference. By doing so, the capacity of the pump either increases or decreases.

Operation

Before the engine starts or when the hydraulic system requirements are very high, the priming spring moves the slide to the maximum output position (left-hand drawing of Fig. 9-10). In this position, the slide and rotor are not on the same centers.

Now as the rotor and vanes turn inside the slide, very efficient fluid carrying pumping chambers form in the expanding areas. As a result, fluid trapped between the vanes, at the inlet port, moves around inside the slide to

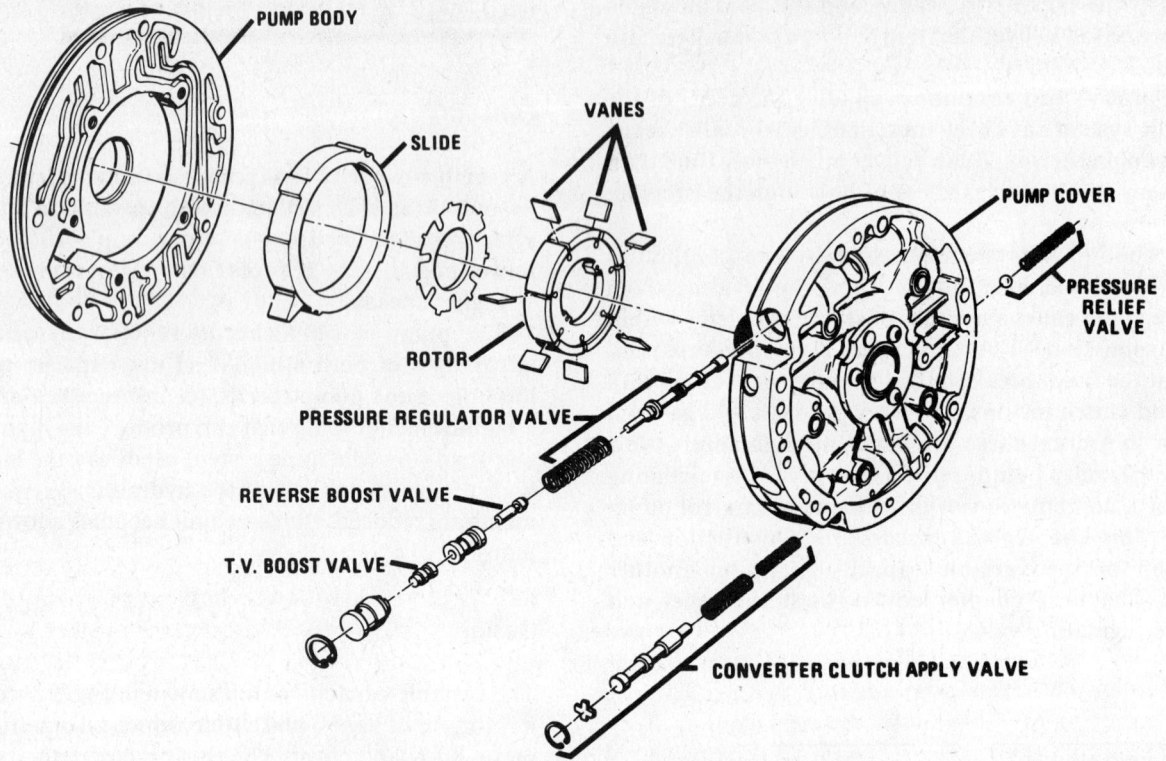

FIGURE 9-9   The design of a variable-capacity vane pump. (Courtesy of General Motors Corp.)

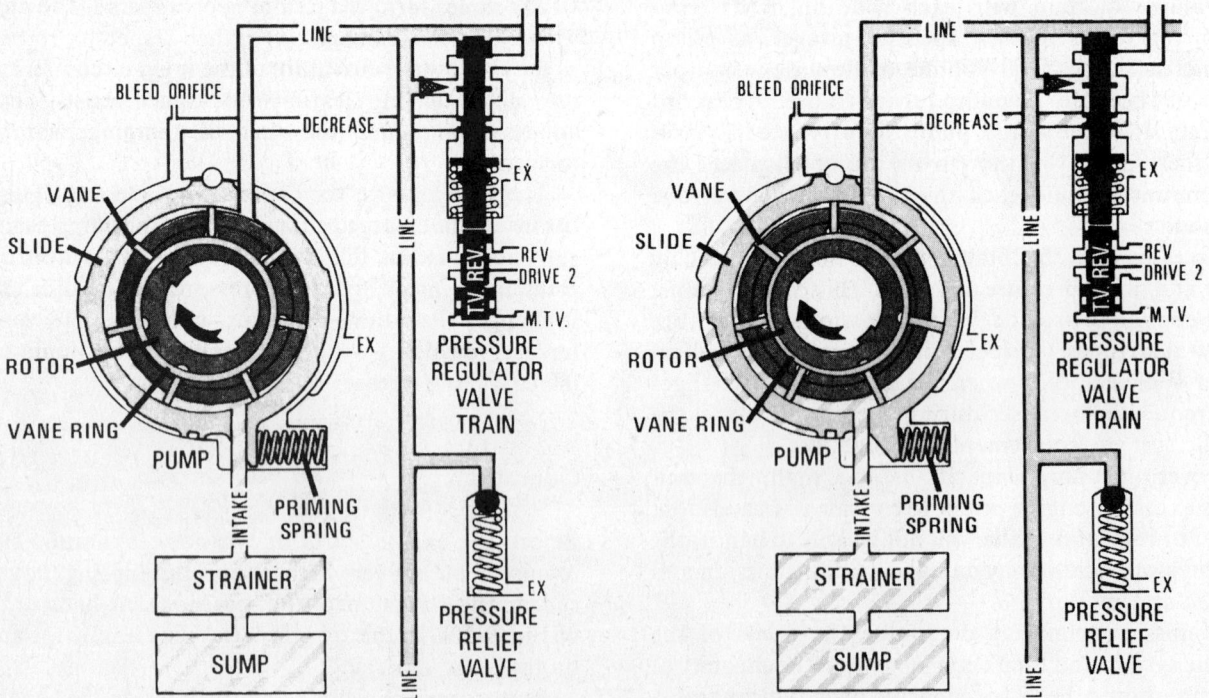

FIGURE 9-10   Operation of a variable-capacity vane pump. The unit's maximum output is shown on the left, minimum on the right. (Courtesy of General Motors Corp.)

the outlet port, where the decreasing rotor-to-slide clearance forces the liquid out into the system. Finally, in this slide position, the pumping chambers formed to the left of the inlet port are of maximum capacity, while the close clearance between the slide, vanes, and rotor seal the outlet port from the inlet. As a consequence, the pump produces maximum flow and can provide the system with high pressure.

However, when the system pressure reaches a pre-set amount, the slide moves back to the minimum output position (right-hand drawing of Fig. 9-10). As the slide moves back toward the center of the pump body, several different things occur within the assembly at the same time. First, the effective pumping chamber area decreases to the left of the inlet port. This reduces the amount of fluid moved between the inlet and outlet ports. Second, the formerly sealed area between the outlet and inlet ports is now opened up to make several fluid chambers. These chambers carry a given amount of fluid from the outlet back to the inlet ports. Consequently, pump capacity is reduced. Finally, if the slide moves into the center position, both the rotor and slide are on the same center, and the pump no longer produces any flow.

As mentioned, the priming spring keeps the slide in the maximum output position until system pressure reaches a pre-set value. When this occurs, the pressure regulator valve opens a passageway to an area between the slide and the pump housing. The entering fluid, under pressure, acts on the surface of the slide to move it against the tension of the priming spring. This moves the slide toward the center position, thus reducing system pressure to a specified value.

When system demands increase, the pressure regulator reduces the amount of fluid pressure against the slide. The priming spring then moves the slide to a new position that increases pump output and system pressure.

MULTIPLE PUMP SYSTEMS

Function

Up until the late 1960s, many automatic transmissions had two hydraulic pumps, a front and a rear (Fig. 9-11). The front pump, found in a multiple pump system, has a larger capacity than the rear unit. And the engine drives this pump through the hub on the converter. On the other hand, the transmission's output shaft turns the rear pump that develops flow only when the vehicle is in forward motion. With this arrangement, one or both pumps supply the fluid flow to the system.

Operation

With the engine running and the vehicle stationary or moving at very low speeds, the front pump supplies all the fluid flow to the system in the same manner as any gear, rotor, or vane pump. In other words, the front pump provides the increased fluid flow necessary to fill the converter, cooling, and lubrication systems. Furthermore, it produces the high pressure needed to activate the bands or clutches used in the high torque gear ratios such as low and reverse, and to prevent them from slipping. The front pump has more than enough capacity to meet all these requirements and does so until the vehicle reaches a predetermined speed.

At this road speed, the rear pump then produces sufficient fluid flow to relieve the front pump of its system pressure responsibilities, which are now relatively low— only enough to hold the clutches engaged. At this point, the front pump just "idles" or recirculates its fluid flow back to the reservoir or the pump's inlet port. But if the

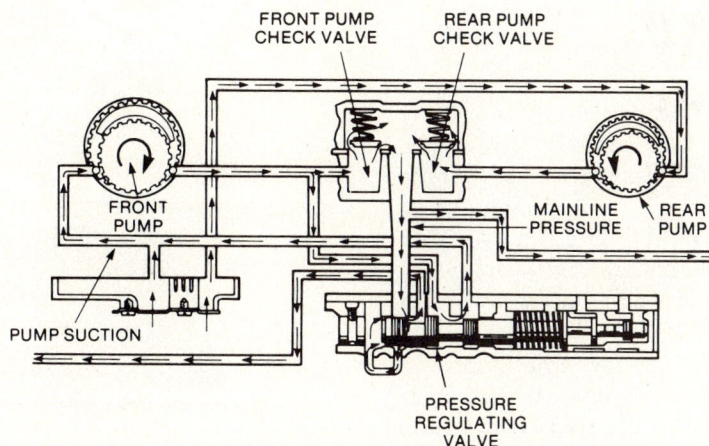

FIGURE 9-11   A multiple pump hydraulic system.

vehicle slows down, the front pump once again takes over the main flow and pressure requirements of the system. To coordinate the changeover period of the two pumps so the system will not lose any flow or sustain a pressure reduction, two check valves are necessary (see Fig. 9-11).

BENEFITS OF MULTIPLE PUMP SYSTEMS

A two-pump hydraulic system offers several advantages over the single-pump design. First, by using a rear pump that the output shaft drives, the vehicle can be push started. Second, since the rear pump must rotate whenever the vehicle is in motion, this reduces somewhat the amount of engine power needed to drive the front pump through the converter.

Vehicle can be Push Started

In a single-pump system, the one-engine driven unit supplies all the fluid flow necessary to operate the transmission at all times. If the engine, for whatever reason, cannot be started by normal means, the hydraulic system of the transmission will not have the fluid flow or pressure to apply the clutches or bands; the powerflow through the planetary system would be impossible. Therefore, the vehicle cannot be push started like one with a manual-shift transmission.

But the multiple pump system has a rear pump that operates whenever the vehicle is in motion. By pushing a vehicle with a transmission rear pump, the hydraulic system will have sufficient fluid flow and pressure to apply whatever clutch or band that is necessary to activate the planetary. Then by pushing the vehicle, drive shaft torque transmitted through the transmission will turn the engine over.

Reduces Engine Power to Drive Pump

With the two-pump system, the engine only has to use some of its power to drive the pump until the vehicle reaches a certain speed. Then the rear pump takes over the requirements of the hydraulic system, thereby reducing the load on the engine. In other words, the vehicle's momentum assists the engine not only in keeping the vehicle in motion but in driving the rear pump by means of the drive and output shafts.

Since hydraulic pump design in general has improved over the years, transmission manufacturers have discontinued the installation of rear pumps. The reason for this deletion was twofold: (1) two-pump systems produced more noise, (2) two-pump systems are more expensive to produce.

DISADVANTAGES TO THE SINGLE-PUMP SYSTEM

But there are two disadvantages of not having a rear pump in the automatic transmission. First, as previously stated, the vehicle cannot be push started. And second, the vehicle should not be towed more than a few city blocks with its drive wheels on the ground. Without a rear pump to circulate the fluid necessary for lubrication, towing a vehicle with the drive wheels revolving will cause severe frictional damage to the unit. Therefore, if a vehicle must be towed with the drive wheels on the ground, the engine must be operational or the drive shaft removed whenever possible.

PUMP RESERVOIR

The **pump reservoir,** or **sump,** is the storehouse for the fluid (Fig. 9-12). The reservoir for the automatic transmission pump is the oil pan attached to the bottom of the transmission case. This sheet-metal pan, depending on the type of transmission, holds about four quarts of ATF. In addition, a filler tube and dipstick that extends from the pan to the engine compartment provides a convenient method of filling and checking the fluid level in the transmission.

Other than acting as the pump and system reservoir, the oil pan helps to reduce the operating temperature of the fluid. Because the fluid is recirculating back to the pan, some of its heat transfers to the air that passes over the bottom of the reservoir.

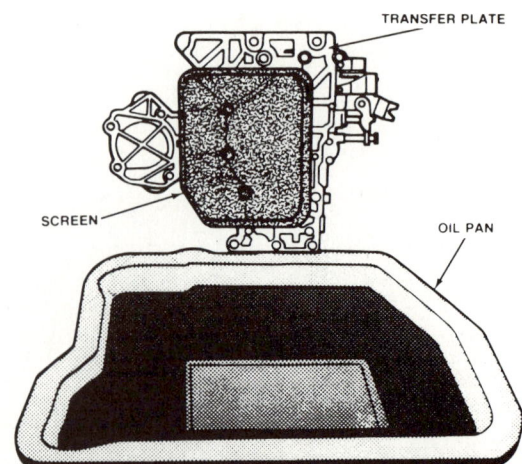

FIGURE 9-12 The transmission oil pan and a Dacron filter secured to the transfer plate of the valve body. (Courtesy of Chrysler Corp.)

PUMP AND SYSTEM FILTER

To protect all the lubricated components of the transmission from foreign particle damage, the pump and hydraulic system must have some form of filter (Fig. 9-13). Manufacturers now utilize three types of material for these filters: fine wire mesh, treated paper, or a synthetic fiber called Dacron.

In any case, the location of the main filter is between the reservoir, the pan, and the pump's inlet port. In this location, the filter traps any metal particles or clutch plate or band material circulating in the fluid before they enter the pump. (NOTE: Most of this type of foreign material, if it has any weight at all, will sink to the bottom of the fluid in the pan; so the filter is only necessary to trap the smaller particles that remain suspended in the oil.)

Many transmission types also have small screen-type filters installed in key locations in the hydraulic system to provide extra protection to a given component or valve. These types of filters are rarely ever serviced except during a transmission overhaul.

However, this is not the case for the main filter; it requires periodic replacement. The time interval between filter changes, of course, depends on where and how the operator drives the vehicle. But in any case, a dirty filter restricts the fluid flow to the pump, which can cause pump failure and a reduction of flow to hydraulic system components. Both problems definitely create erratic transmission operation—or the unit may not function at all.

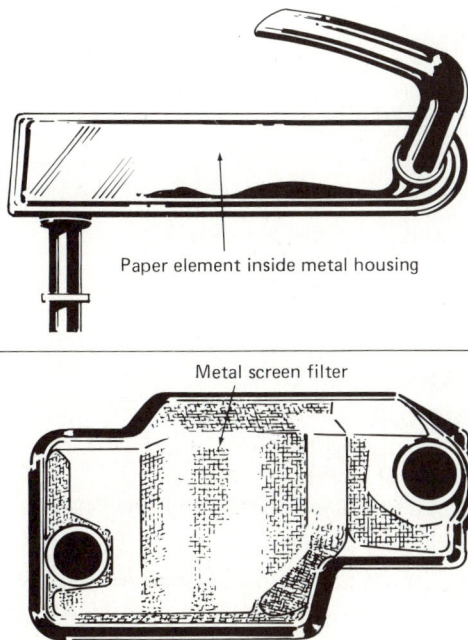

Paper element inside metal housing

Metal screen filter

FIGURE 9-13   Typical paper and metal screen fluid filters.

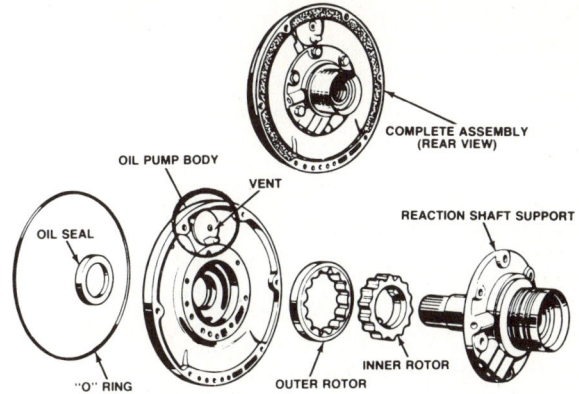

FIGURE 9-14   A transmission vent located in the pump body. (Courtesy of Chrysler Corp.)

PUMP OR RESERVOIR VENT

The reservoir must have a **vent** (Fig. 9-14). Manufacturers install the vent in the main transmission case above the fluid level or in the pump body. In either case, this vent allows atmospheric pressure to enter the transmission and reservoir area. The pressure is necessary to push the fluid up through the filter and into the pump's inlet port. If the vent clogs, the pump's action will slow down or stop completely.

PUMP EFFICIENCY

If, at any time, parts of the pump become worn or damaged, fluid can leak between the outlet and inlet ports. This leakage reduces pump capacity or its efficiency. **Efficiency,** in this case, is the actual pump output with respect to displacement or flow rate. In other words, if a given pump should produce 30 gallons per minute at vehicle cruising speeds but only delivers 15 gallons per minute, its efficiency is only 50 percent.

A loss of efficiency does not always mean a reduction in normal system pressure, but this depends a great deal on how much pump effectiveness is lost. Pressure can still build up as long as the pump continues to deliver fluid to the system and the oil is not leaking off somewhere else in the system. However, a loss of pump efficiency does, in most cases, reduce not only the application of pressure on but also the movement of the activating mechanisms. In other words, the pressure in the system with low pump efficiency may be normal until a band or clutch applies; then the pressure drops so low that the resulting force is unable to firmly apply either a band or multi-disc clutch. This results in slippage within the affected unit—and usually premature failure.

As mentioned, pump wear directly affects its efficiency. Also, a typical pump has more than enough capacity to compensate for the small amount of normal wear that takes place during its service life. Therefore, pump efficiency should remain stable throughout the life of the transmission.

However, a transmission pump can develop excessive wear for several reasons. The first and most common cause for pump wear is contaminated fluid. If the fluid becomes overheated enough, it will become oxidized and lose its ability to lubricate all transmission components properly. Since the pump is always operating with the engine, even when the vehicle is not moving, oxidized fluid wears it out very quickly.

The other common cause of pump wear is low fluid flow to the pump. This is usually the result of insufficient fluid in the transmission or a plugged filter. In either case, the end result is the same — the pump draws in a quantity of air instead of fluid. This causes poor pump lubrication and affects the operation of the hydraulic system.

SUMMARY

1. The hydraulic pump is the heart of the pressure hydraulic system.

2. The hydraulic system's design, pressure, or flow requirements determine the type of pump needed.

3. A hydraulic pump converts an external source of power into force necessary to create pressure in the fluid.

4. The transmission's hydraulic pump must produce a great deal of fluid flow and operate under high pressure.

5. A rotary pump is one in which its internal components spin in a circle while producing system pressure and providing the fluid flow necessary to operate the transmission.

6. The automatic transmission uses one or more rotary pumps.

7. Rotary pumps usually have two round members of different sizes, which operate on the same centers.

8. The pumping mechanisms of a rotary pump can either be gear teeth, rotor lobes, or vanes that form sealed chambers between the two members.

9. The most common types of rotary pumps used are the IX gear and rotor, in addition to the variable capacity vane.

10. An IX gear pump consists of a pump housing, drive and driven gears, and the stator support.

11. Located in the counterbore of the pump housing are the crescent, inlet, and outlet ports.

12. The small pump gear has external teeth, and the converter drives it.

13. The large gear has internal teeth, and it drives the small gear.

14. The stator support serves three functions.

15. Separating gear teeth or rotor lobes create a vacuum at the inlet port.

16. The gear crescent or rotor lobes prevent the fluid from leaking from the pump's outlet to its inlet port.

17. The meshing of gear teeth or rotor lobes applies a force to the fluid; this force causes fluid flow and system pressure.

18. Positive displacement means that a pump, when revolving, will always deliver fluid to the system at a constant rate.

19. Pump capacity is the total volume of flow produced by the pump at a given pressure and speed.

20. A variable capacity vane pump alters its output as the needs of the hydraulic system change.

21. The variable capacity vane pump consists of a rotor, set of vanes, and a slide, which all operate inside a pump body and cover.

22. Some transmissions have two hydraulic pumps. The second or rear pump supplies the transmission with sufficient flow and pressure so that the vehicle can be push started and relieves the front pump of its load at vehicle cruising speeds.

23. The fluid pan, located at the bottom of the transmission case, is the reservoir for the hydraulic system.

24. The main fluid filter prevents foreign particles from entering the pump's inlet.

25. An open transmission vent allows atmospheric pressure to enter the case above the level of the fluid.

26. Pump efficiency is the actual pump output with respect to displacement or flow rate.

27. Pump wear reduces its efficiency.

REVIEW

This section will assist you in determining how well you remember the material contained in this chapter. Read each item carefully. If you cannot complete the statement, review the section in the chapter that covers the material.

1. A hydraulic pump applies _____ to the fluid.

   a. pressure

   b. flow

   c. force

   d. displacement

2. The external source of power for the transmission's hydraulic pump is the _____ .

   a. engine

   b. converter

   c. input shaft

   d. driven gear or rotor

3. Pump members are _____ in a gear or rotor pump.

   a. the same size

   b. not the same size

   c. on the same operating center

   d. both b and c

4. Atmospheric pressure forces fluid into the pump during _____.

   a. the inlet cycle   c. the outlet cycle

   b. the delivery cycle   d. every third revolution

5. The most common types of IX pumps are the gear and the _____.

   a. tip

   b. vane

   c. lobe

   d. rotor

6. The inner pump member receives driving power directly from the _____.

   a. drive shaft

   b. torque converter

   c. engine

   d. driven member

7. The gear teeth or lobes apply force to the fluid as they _____.

   a. come together

   b. expand

   c. repel each other

   d. turn in opposite directions

8. The crescent, inlet, and outlet ports are in the pump's _____.

   a. stator support

   b. counterbore

   c. sump area

   d. case

9. The portion of the gear pump that prevents outlet fluid from returning to the inlet is the _____.

   a. teeth

   b. lobes

   c. tips

   d. crescent

10. Transmission hydraulic pumps are the _____ type.

    a. positive-displacement

    b. efficient

    c. full-delivery

    d. semi-displacement

11. If a pump is 80 percent efficient, its _____ may be cut back.

    a. pressure

    b. flow

    c. force

    d. rpm

12. In order to successfully push start a vehicle, the transmission must have a _____.

    a. rear pump

    b. front pump

    c. drive rotor

    d. filter

13. The filter prevents foreign particles from entering the _____.

    a. pump's outlet port

    b. pump's inlet port

    c. fluid pan

    d. vent

14. The vane pump described in this chapter is _____.

    a. efficient

    b. full delivery

    c. positive displacement

    d. variable capacity

For the answers, turn to the Appendix.

CHAPTER 10

Pressure Regulating Devices

The last chapter discussed the construction and operation of variable- and positive-displacement pumps. All of these pumps produce sufficient fluid flow normally to operate the automatic transmission at low engine speeds. However, at high engine rpm, the positive-displacement pump develops so much flow that the resulting pressure can damage the transmission and hydraulic system.

FUNCTION

To offset this inherent characteristic of gear or rotor pumps, manufacturers install one or more pressure regulating devices in the transmission's hydraulic system. A **pressure regulating device** bleeds off excess pump flow thereby reducing system pressure. These safety valves route the excess fluid flow either back to the intake side of the pump or directly to the reservoir. In either case, by bleeding off a given amount of fluid flow, the regulating device maintains hydraulic pressure to a safe operating limit.

TYPES

Within the hydraulic system of a typical automatic transmission can be two types of pressure regulating devices, the relief and the regulating valve. The **relief valve** is a very simple device that bleeds off excess pump flow; this, in turn, maintains system pressure to a **fixed amount**. In other words, with this device, the pressure may be lower, but never higher, than a predetermined amount. Finally, since the relief valve controls pressure to a set amount, it usually acts in a hydraulic system as a high-pressure safety valve that protects the transmission against the buildup of extremely high pressures.

The **pressure regulating valve** is a somewhat more complex device than the relief valve because it has more functions to perform. A regulator valve is used for pressure control and in this respect performs the same function as the relief valve. In other words, both valves bleed off excess fluid flow to maintain system pressure. The difference is that a regulator valve **can adjust** (change) the amount of system pressure to meet the demands made on the transmission by engine torque loads and vehicle driving conditions, while the relief valve controls pressure to a set value depending on its spring tension.

If the regulator valve were to keep the pressure at a fixed amount, the multiple-disc clutches and bands could slip when the vehicle is accelerating. The reason for this is that, during this time, the engine is producing higher torque output, which the clutch or band must handle. However, because the regulator valve raises hydraulic pressure on the clutch or band servo piston, its force is increased on the plates or band. This ensures that either or both units will be able to perform their function, even under the heavy torque loads.

The pressure regulator also acts as the valving for the torque converter. After engine startup, the regulator opens a port so that fluid can flow to the converter. When the engine is shut off, the same valving seals the port to the converter, thus trapping the oil within the unit.

RELIEF VALVES DESIGN AND OPERATION

There are three basic types of relief valves: the piston, ball, and needle.

Piston Type

Figure 10-1 illustrates a simple piston type. This unit consists of a piston, spring, and housing. The piston moves back and forth in a special bore in the housing due to the effects of hydraulic pressure on its left (face) side and spring force on its right. As it moves within its bore, the piston acts as a valve to open or close the sump port.

The **spring** provides a specified (calibrated) amount of tension against the back of the piston. This tension (force)

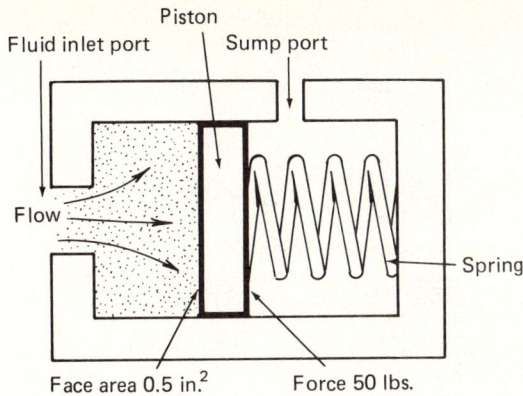

Piston
Fluid inlet port   Sump port
Flow
Spring
Face area 0.5 in.²   Force 50 lbs.

FIGURE 10-1   The design and operation of a piston-type relief valve.

attempts to keep the piston from opening the sump port against the action of hydraulic pressure. In other words, the strength of the spring determines just how much pressure is necessary before its fluid pushes the piston over far enough to open the sump port.

As mentioned, the housing contains the bore for the piston. The space between the piston and the bore surface is just enough to permit the piston to move on a film of fluid. The housing also has two port openings—a sump and a fluid inlet. The sump port is an opening in the housing to the reservoir. The size of this port is important; it must be large enough to carry away all the fluid pushed through it when the piston uncovers the opening. This feature is necessary if the relief valve is to protect a hydraulic system from the effects of very high pressure. The inlet port is at a terminating point in a passageway within the system and is open at all times to fluid flow.

With the pump operating, fluid enters the inlet port. When the fluid backs up against the piston, pressure increases and acts as a force on its face (Fig. 10-1). This causes the piston to move toward the right against the spring force. As the piston moves to the right, it uncovers the sump port. As a result, fluid returns through this port to the sump, thus bypassing the hydraulic system.

Since the open sump port creates a leak in the system, pressure drops. The amount of system pressure remaining before the spring moves the piston back to seal the pump port can be calculated by using the formula, pressure equals force/area (P = F/A). The force of the spring, 50 pounds (222 N) divided by the valve face area of 0.5 square inches is equal to 100 psi (690 kpa).

If pressure drops below 100 psi (690 kpa), the spring force pushes the piston to the left, closing off the sump port. This permits the pressure to build up in the system. In actual use, the piston moves just enough to maintain a pressure of 100 psi (690 kpa) in the system at all times.

Finally, in some hydraulic systems, a relief valve can control the fluid flow within a specific circuit. A **circuit**, as

used here, refers to a particular sealed passageway or tube that carries the fluid flow from one area to another such as from the pump's outlet port to a relief or regulating valve. The valve, in this case, prevents fluid flow until system pressure reaches a given amount. Then, the piston moves over and uncovers the sump port. But the sump port no longer carries the excess fluid to the reservoir. Instead, the port acts as the inlet for the special circuit the relief valve controls.

Ball Type

The ball-type relief valve is quite similar in design and operation to the piston type (Fig. 10-2). The main difference is that instead of the piston, the unit uses a round ball. Half of the round portion of the ball acts as the face, against which hydraulic pressure acts. The spring applies its tension to an area of the ball, opposite to the face.

In the bore that the ball moves in, there is a machined seat. When the ball is held against this seat by spring force, fluid flow from the pump or system cannot reach the sump port. In other words, the face of the ball and the seat act as a valve to control the fluid flow and pressure buildup. However, when the pressure against the face of the ball is high enough, the resulting force pushes the ball off the seat, thus uncovering the sump port. This action reduces system pressure to an amount determined by spring force divided by the exposed area of the ball face.

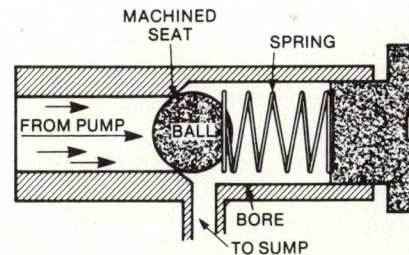

MACHINED SEAT   SPRING
FROM PUMP   BALL
BORE
TO SUMP

FIGURE 10-2   The design of a ball-type relief valve.

Needle Type

The needle-type relief valve has a pointed instead of a round valve face (Fig. 10-3). Hydraulic pressure acts on this area just as it did the flat and curved faces of the piston and ball valves. Note also in the illustration the machined seat that the spring holds the pointed face against. This combination seals off the sump port until system pressure forces the needle valve off the seat, against the tension of the spring.

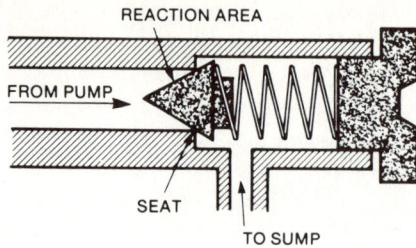

FIGURE 10-3   The design of a needle-type relief valve.

SPOOL VALVES

Before discussing the design and operation of the more complex pressure regulating device, this is an ideal time to discuss the design of a spool valve. And more importantly, how this unit responds to the application of a number of differing forces to control fluid flow to one or more circuits of the hydraulic system.

Design

A **spool valve,** Fig. 10-4, is a cylindrical-shaped device, which looks much like a common sewing thread spool. This valve consists of two lands, four faces, and one annular groove or valley. The purpose of the **lands** is to support the valve in its bore and cover and uncover ports located in the bore in which the entire assembly rides.

The **faces** of the valve form areas upon which hydraulic pressure or spring force can act. The valve shown in Fig. 10-4 has four faces; therefore, an applied force can push on the front and rear portions of each land. The direction in which hydraulic pressure will move the valve depends upon the area of the face, which we will see later is not always the same. Moreover, the face portion of the valve upon which hydraulic pressure acts to move it against the force of a spring is known as the **reaction area.**

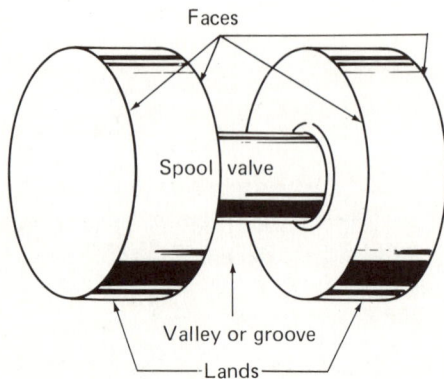

FIGURE 10-4   The design of a simple spool valve.

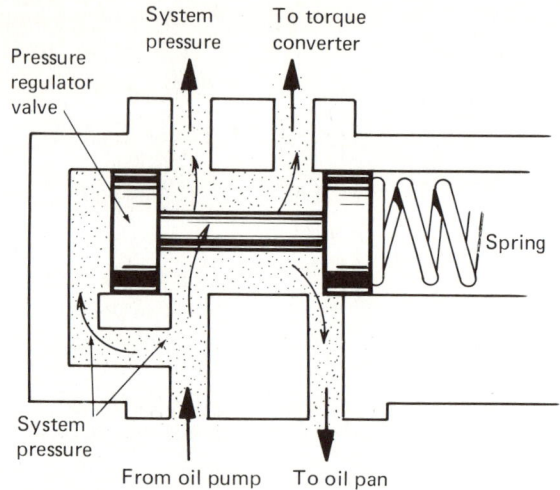

FIGURE 10-5   A spool valve acting as a pressure regulator. Note that all four ports in the bore are now partly to fully open.

The annular groove or valley is nothing more than a channel or passageway for the flow of fluid. As the spool valve moves in its bore, the lands uncover port openings that permit fluid to flow into the valley. This fluid can then pass through the groove and exit out of the valve through another uncovered port.

As mentioned, the spool valve operates in a bore that has a number of port openings (Fig. 10-5). These ports connect to various circuits that transport fluid to and from the valve. For instance, with the spool valve positioned as shown in the illustration, both lands have uncovered four ports. The left-hand lower port permits fluid from the pump to enter the valley between the lands. The right-hand lower port is only barely cracked open, which permits a portion of the pump's flow to return to the oil pan. Both upper ports are wide open; one is delivering fluid into the hydraulic system while the other is directing fluid to the torque converter.

SPOOL VALVE OPERATION

Simple Balanced Type

A balanced valve, whether it be a spool or any of the types of relief valves discussed earlier, can control both the direction of flow and the amount of pressure (Fig. 10-6). The valves can be balanced by spring force and hydraulic pressure or by pressure alone. Balanced, as used here, means that the valve will cease its movement in the bore when the applied forces on both ends are equal.

The balanced valve has the capability of moving either left or right. For instance, when the spring tension (force)

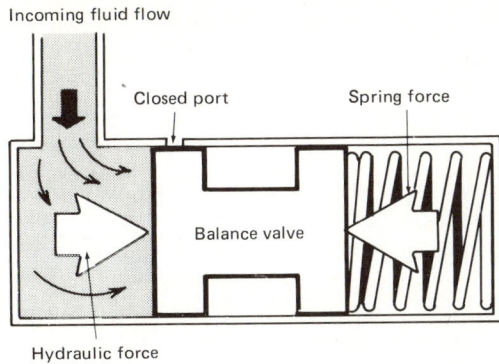

FIGURE 10-6 Hydraulic pressure and spring force both act on this balanced-type spool valve.

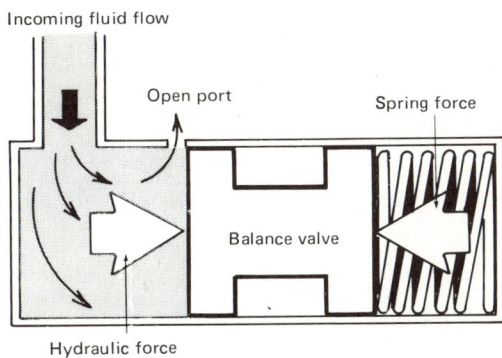

FIGURE 10-7 Spring force is overcome by hydraulic pressure; the valve moves, and its land uncovers the port.

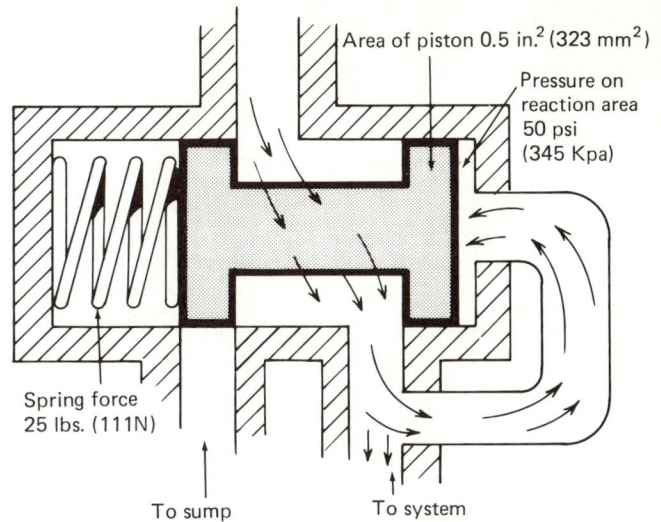

FIGURE 10-8 The spool valve is in balance when the force against the reaction area is the same as spring tension.

Differential Type

A **differential spool valve** has one land face larger than the other. In Fig. 10-9, the right-hand face has twice the area as the left-hand face. Now, if the system pressure acting on both faces is the same, the force against the larger one is double that on the smaller. For instance, if the area of the larger face is 1 square inch with a system pressure of 100 psi (690 kpa), the force produced is 1 × 100, or 100 pounds (445 N). The force on the smaller face is 0.05 square inch × 100, or 50 pounds (222 N). As a result, the spool valve moves toward the lesser force.

The same rule applies if the pressure enters the valley between the two lands of a differential spool (Fig. 10-10). The spool valve moves toward the right because the pressure creates a greater force on the larger face.

is greater than the hydraulic pressure, the valve moves left and closes the port (Fig. 10-6). However, when the fluid pressure on the reaction area is greater than spring force, the spool valve moves to the right and opens the port (Fig. 10-7).

If the applied pressure on the valve's reaction area drops, the same amount of spring tension on the spool will move it back to the left. In other words, the valve moves away from the end having the greatest applied force and stops its travel when both are equal or balanced out.

The amount of force against a land's face can be determined by multiplying the applied system pressure by the area. For example, if the applied pressure on the spool valve's reaction area is 50 psi (345 kpa) and its area is 0.5 square inches, the force against the land face is 0.5 × 50, or 25 pounds of force (111 N).

If the spring force in Fig. 10-8 is 25 pounds (111 N), the valve will be balanced. But if hydraulic pressure drops, spring force will move it to the right. On the other hand, as pressure increases, it overcomes spring tension, and the valve moves to the left.

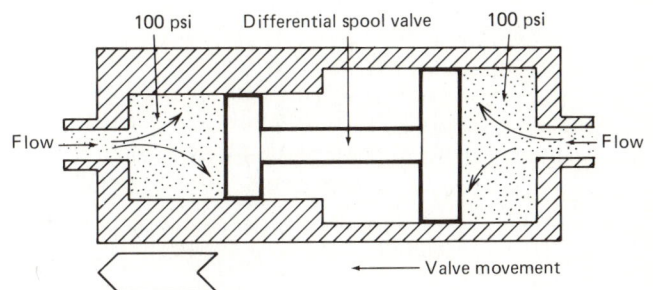

FIGURE 10-9 If equal pressure acts on two faces of a spool valve with different areas, the larger one has the greater applied force.

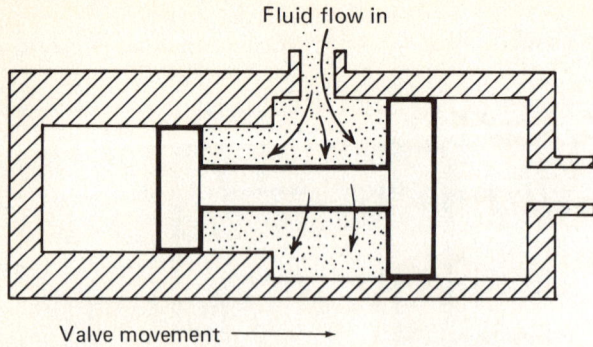

FIGURE 10-10   If equal pressure acts in the valley of a differential spool valve, it moves in the direction of the highest applied force.

Balanced Type with Auxiliary Force

In some installations, it is necessary to alter the spring's tension in order to change the balancing force on the reaction area. This is usually done by the addition of another piston or spool valve between the end of the spring and its housing bore (Fig. 10-11). This second device is known as either a boost piston or valve.

Notice also in Fig. 10-11 that the boost valve spool is the differential type and the housing has a port opening for auxiliary fluid to enter the valley between the lands. Since the left land face is larger than the right, the resulting force will push the spool valve to the left even though the pressure against the two land faces is the same.

To determine how much force application is on the larger face, we subtract the amount on the smaller face. For instance, if the pressure in the valley is 100 psi (690 kpa) and the larger face has an area of 2 square inches, the applied force is 2 × 100, or 200 pounds (890 N). The

applied force on the smaller face is 100 psi × 1 square inch, or 100 pounds (445 N). The force on the larger face is 200 pounds (890 N) minus 100 pounds (455 N), or 100 pounds (445 N). This simply means that the boost valve moves to the left with an applied force of 100 pounds (445 N).

force on larger face   =   200 pounds (890 N)
force on smaller face = −100 pounds (445 N)

applied force   =   100 pounds (445 N)

When the boost valve moves, it increases spring tension (force) on the main balanced spool valve. Consequently, in order to balance the main spool valve, a greater hydraulic pressure must act on its reaction area. This principle is used within the hydraulic system of the automatic transmission in pressure regulating and shift valves along with some servo pistons.

Manual Type

A **manual type** of spool valve is not a balanced unit. Instead, the valve moves within its bore due to the application of mechanical force (Fig. 10-12). In this example, the manual valve moves within its bore mechanically due

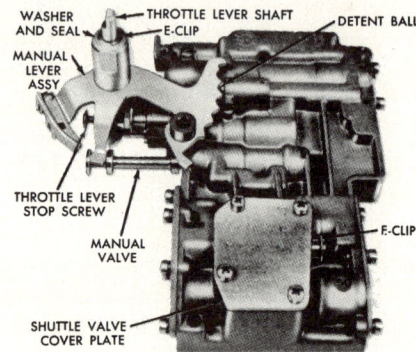

FIGURE 10-12   A manual valve and lever assembly. (Courtesy of Chrysler Corp.)

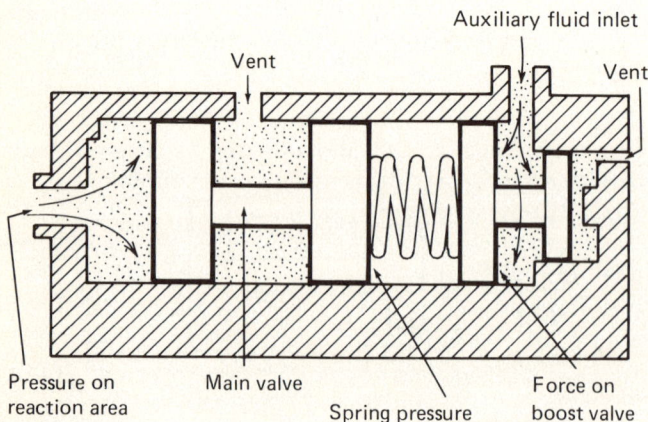

FIGURE 10-11   A balanced valve with a spring that has its tension changed by an auxiliary force applied by fluid pressure against a boost valve.

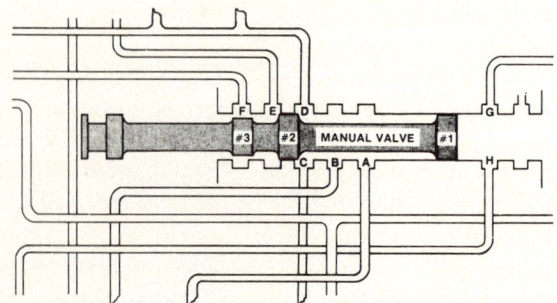

FIGURE 10-13   The design of a typical manual multiple-land spool valve. (Courtesy of Chrysler Corp.)

to the movement of the manual lever assembly. This assembly connects through linkage to the gearshift-selector lever in the driving compartment of the vehicle. Consequently, as the driver moves the gearshift selector, the attached linkage causes the manual lever to force the valve to move within its bore (Fig. 10-13). When moved to a specific point, the multiple lands cover and uncover given hydraulic circuits, thereby controlling the fluid flow within the transmission during its various operating phases.

MAINLINE PRESSURE

As stated earlier, there must be some resistance to the flow of fluid from the pump in order to develop pressure. If no resistance exists, the pump would produce a measurable amount of flow but no pressure.

However, there are many restrictions within the transmission's hydraulic system. These include the pump's outlet port along with the various lines and passages leading from the pump to the many control valves in the system.

The most important restriction in the system, and the one used to develop or control system pressure, is the regulator valve itself. The pressure regulator valve, as mentioned, is a variable restriction, an opening in the hydraulic system whose size can be changed in order to alter pressure.

As a result of all the restrictions and the action of the regulator valve, continuous pump flow produces what is known as **mainline pressure, control pressure,** or **drive oil.** In either case, mainline pressure in the closed transmission hydraulic system is the same everywhere, even in the regulator's input and output passages. This means that the fluid flow needed to operate the transmission itself really does not have to pass through the regulator valve, although it acts on its reaction areas. Finally, mainline or control pressure in a typical transmission will range from between 55 to 150 psi (380 to 1034 kpa) in its forward driving ranges and about 300 psi (2069 kpa) in reverse.

PRESSURE REGULATOR VALVE DESIGN

As stated earlier, an actual pressure regulating valve, as found in the hydraulic system of the automatic transmission, is much more complex in design than any of the examples of spool valves just presented. The main reason for this is that this valve has to vary the mainline pressure within the transmission to meet the varied operating conditions under which the unit must function.

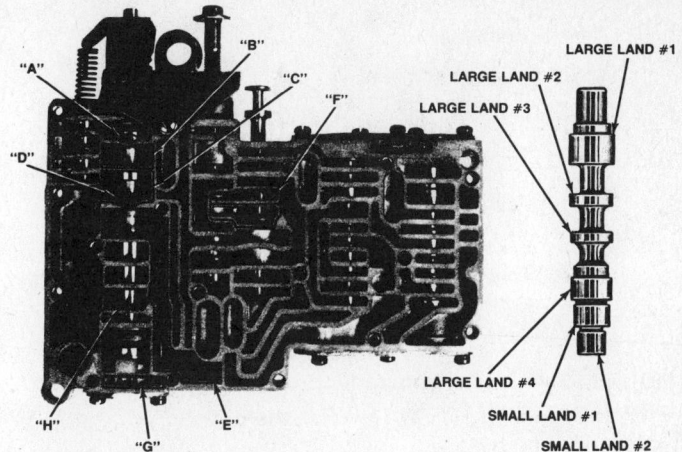

FIGURE 10-14   *A typical pressure regulating spool valve. (Courtesy of Chrysler Corp.)*

Valve

Figure 10-14, for example, shows a typical pressure regulator valve removed from its bore in the valve body. This valve has four large diameter lands and two smaller diameter lands. In addition, between each large land is a wide annular groove; two narrow valleys follow land #4, one between land #4 and #1, and the other between the smaller lands #1 and #2.

To keep all the references straight and understandable, this text numbers the lands of this pressure regulator sequentially from the spring, or uppermost, end of the valve. For instance, the large #2 land is the second one from the spring end, land #3 is the third from the spring, etc. And with the lands numbered as such, the reader can easily identify the locations of all the annular grooves or valleys.

Spring

With the numbering sequence in mind, the next step in understanding how this valve operates is to master the functions of the remaining components that are part of the regulating valve train (Fig. 10-15). These consist of a compression spring and bracket, line-pressure plug, and a throttle-pressure plug. The **compression spring** and **bracket** that has an adjustment screw will apply a rather heavy force to the face on the end of large land #1. This large spring force will attempt to maintain the valve in the closed position to regulate mainline pressure to a rather high amount, about 275 psi (1896 kpa). On the opposite end of the valve from the spring is the line-pressure plug, sleeve, and throttle-pressure plug.

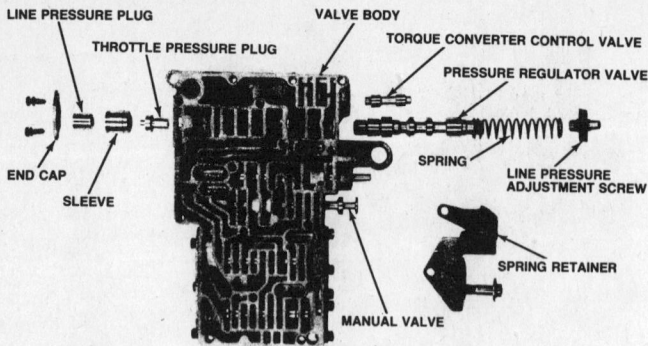

FIGURE 10-15 The components of the regulating valve train. (Courtesy of Chrysler Corp.)

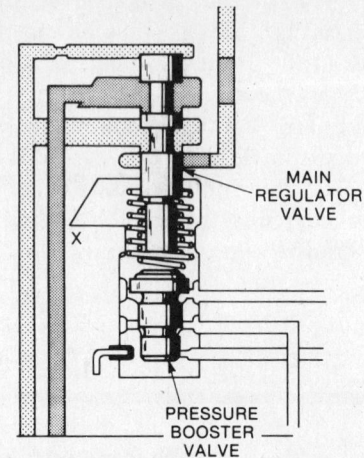

FIGURE 10-16 A pressure regulating valve with the throttle plug (booster valve) located on the spring side of the assembly.

Line-Pressure Plug

The **line-pressure plug** is a unique form of boost-type piston, which assists the regulator valve's reaction area in moving the main assembly against the heavy spring tension. Hydraulic pressure will apply its force to the left end face of this small plug, and this action will force it to move in its sleeve, which forms its bore in the valve body. The movement of the line-pressure plug will assist mainline pressure, on the main valve's reaction area, to move it against the tension of the spring. As a result of this combined action, mainline pressure is much lower than it would be without the plug's assistance.

Throttle-Pressure Plug

The throttle-pressure plug is another boost-type piston, which fits between the line-pressure plug and the regulator valve itself. This plug, when activated by the force of pressure, will resist any movement of the line-pressure plug. In this way, mainline pressure will be increased to meet the demands of the hydraulic system. The resulting increased mainline pressure is necessary to provide the additional force required during an upshift or for locking up a clutch or band to prevent slippage under heavy engine torque loads. In other words, the throttle plug, working with the rest of the regulating valve train, compensates or adjusts mainline pressure to meet the various needs of the transmission's hydraulic system.

Some manufacturers install the throttle plug or boost valve and the line-pressure plug, if used, in a different location. For example, Fig. 10-16 shows a regulator valve with the throttle plug (pressure booster valve) bearing against the spring. The spring in this installation has less tension than the one used in the previous example. And the booster valve, when moved by the force of the hydraulic pressure, increases spring tension. As a result of the increased tension, mainline pressure in the system increases accordingly.

PRESSURE REGULATOR VALVE OPERATION

Non-Boosted

With the engine operating, the hydraulic pump rotates and creates a vacuum at its inlet port (Fig. 10-17). This difference in pressure, between the sump and the pump's inlet, causes fluid to flow up from the reservoir, through the filter, and into the pump. After the pump applies a force to the fluid, it leaves the pump and enters the hydraulic system circuits.

One of these circuits routes fluid through the manual valve to the line-pressure plug and to the end of the small #2 land, which is the regulator valve's reaction area. The force of the hydraulic pressure on both of these areas attempts to move the regulator valve against the tension of the spring. But the reaction areas on both the valve and plug are relatively small, so on initial engine start-up, only a small combined force acts on the pressure regulator valve; this is enough though to move it slightly against the tension of the spring.

This slight valve movement causes the large #1 land to open a port and circuit, leading to the torque converter control valve. Now, fluid flows through the control valve to the torque converter and the cooling system. In other words, pump flow and the resulting pressure does not have to be very great to cause the regulating valve to move and open a feed circuit for the torque converter.

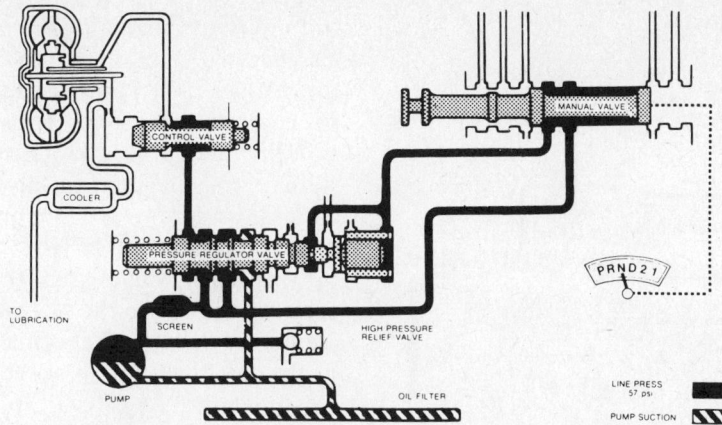

FIGURE 10-17   Operation of a pressure regulating valve train. (Courtesy of Chrysler Corp.)

As the engine accelerates, pump flow increases along with pump rotation. The resulting mainline pressure acting on the line-pressure plug and the regulator's reaction area causes it to move further toward the spring. This movement permits the #3 large land to open a port and passage leading to the inlet side of the pump. With the port open, fluid in the annular groove between large lands #2 and #3 begins to return to the pump's inlet.

Now, as the fluid is venting to the pump, mainline pressure drops due to the "controlled leak." With a reduction in the mainline pressure, the regulator valve stops its movement toward the spring because its force and that resulting from the hydraulic pressure applied to the reaction area and line-pressure plug are balanced. Therefore, hydraulic pressure regulation is just a repetition of the opening and closing of a port by the large #3 land, and the acompanying "controlled leak" bleeds off excess pump flow from the system to its inlet port.

Boosted

Our previous discussion of the pressure regulator valve operation pointed out how the spring's tension controlled the amount of system pressure. In other words, once the pump produced the flow necessary to create sufficient pressure to balance spring force to move the regulator valve to the vent position, mainline pressure ceased to increase beyond this point. But as also stated, mainline pressure in the system has to increase during certain times. Consequently, the pressure regulator requires a boost or compensating device. In our example in Figs. 10-15 and 10-17, the boost valve is the throttle plug.

As the driver slowly accelerates the vehicle, a special hydraulic pressure signal from the throttle valve begins to flow through a passage to and acts on the regulator's throttle plug (Fig. 10-17). Chapter 12 covers the operation of the throttle valve system; so for now, just remember that this device develops and sends a signal to the throttle plug. And the intensity of the signal is in proportion to the rpm of the engine or its load. The signal enters behind the face of the throttle plug through a passage and port between the two circuits feeding the regulator's reaction area and the line-pressure plug.

The presence of the throttle pressure, acting on the rear face of the throttle plug, forces it against the line-pressure plug, resulting in the movement of both units to the right. This action allows the spring to force the regulator valve to the right, which causes the "controlled leak" at the port to decrease or stop altogether. As this occurs, the mainline pressure increases in order to push the regulator valve over to the left so that it will be in a regulating cycle again with its port to the pump's inlet slightly open.

The actual amount of pressure increase will be in proportion to the value of the signal sent from the throttle valve to the throttle plug. At idle or very low engine speeds, this signal is low; therefore, the mainline pressure increase will be very small, if any. But both the signal and mainline pressure go up rapidly as the driver increases engine speed.

However, at a given engine speed, the increase in mainline pressure stabilizes; this action prevents excessively high and unnecessary pressures from developing in the system. In other words, the pressure signal on the throttle plug begins to raise mainline pressure as the driver accelerates the engine, and it continues to raise the pressure until the engine reaches a pre-set speed. Then the mainline pressure stabilizes and goes no higher. Finally, the amount of pressure increase and the engine rpm where it reaches its maximum depend on the type of transmission, hydraulic system requirements, and the type of engine used in the vehicle.

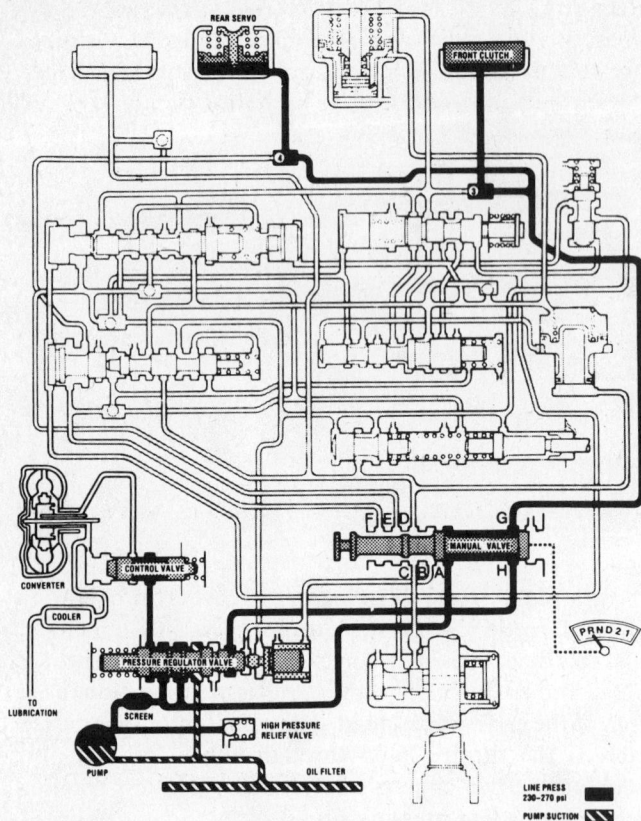

FIGURE 10-18 Mainline pressure regulation in reverse. (Courtesy of Chrysler Corp.)

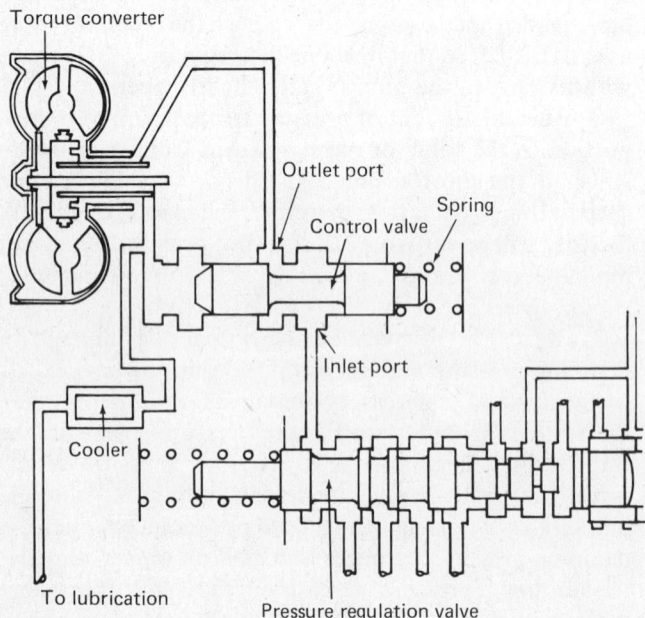

FIGURE 10-19 A typical torque converter control or regulating valve.

In most automatic transmissions, mainline pressure is also boosted up in reverse and sometimes in low. Observe the mainline (line) pressure in Fig. 10-18; it is 230 to 280 psi (1586 to 1931 kpa). The reason for this increase in mainline pressure is the position of the manual valve.

In reverse, the manual valve directs fluid to a port that applies pressure to the face on the small #2 land and cuts off fluid to the small #1 land (the normal reaction area of the regulator valve) and the line-pressure plug. The pressure applied to the #2 land face causes a much lesser amount of force on the regulator valve than that achieved by the combined forces of the reaction area and the line-pressure plug. As a result, mainline pressure has to be much higher in order to move the regulating valve against the tension of the spring.

TORQUE CONVERTER CONTROL OR REGULATING VALVE

Function

A number of automatic transmissions use a torque converter control or regulating valve (Fig. 10-19). This valve is in the system to limit the buildup of pressure in the converter to prevent it from ballooning. Typical converter pressure ranges from 7 to 57 psi (48 to 393 kpa).

Design and Operation

The control valve shown in Fig. 10-19 is a simple balanced spool valve having two lands and one groove. On the rear face of the right-hand land, a calibrated spring exerts its force in an attempt to move the valve to the left. On the end of the left-hand land is the reaction area, which hydraulic pressure will act upon; the resulting force will try to move the valve against spring force.

The bore for the control valve has two port openings. The lower port opening brings in fluid from the pressure regulator valve, and it is open at all times.

The upper port allows fluid in the valley between the lands to pass to the converter. This port is important in that the left land will restrict its opening as the control valve regulates converter pressure.

When the engine starts, the pump delivers fluid to the regulator. Once the regulator valve moves initially, fluid passes to the control valve. This fluid passes through the groove and out the upper port to the converter.

Once the converter, cooler, and lubrication circuits are full of fluid, pressure begins to build up in the system and on the control valve's reaction area. When the pressure reaches about 7 psi (48 kpa) on the reaction area, the resulting force moves the control valve to the right, compressing the spring. This action restricts the outlet port opening to the converter.

As the pressure continues to increase in the converter to 57 psi (393 kpa), the control valve moves further to the right until the left land blocks off the outlet port. At this time, no fluid can flow to the converter and its internal pressure decreases. When the pressure drops sufficiently, the spring will force the valve to the left and the land will partially uncover the outlet port.

SUMMARY

1. A pressure regulating device bleeds off excess pump flow, thereby reducing system pressure.

2. The relief and the regulating valves are two common types of regulating devices.

3. A relief valve controls system pressure to a fixed amount.

4. A pressure regulating valve alters system pressure to meet the demands made on the transmission.

5. The pressure regulator acts as the valving for the torque converter.

6. There are three types of relief valves: the piston, ball, and needle.

7. The piston relief valve consists of a piston, spring, and housing.

8. A relief valve can control the fluid flow within a specific circuit.

9. A circuit is a sealed passageway or tube that carries fluid from one component to another.

10. The ball-type relief valve consists of a ball valve, spring, and housing.

11. The pressure needed to open a relief valve is determined by dividing spring force by the area of the reaction force.

12. The needle-type relief valve has a pointed reaction face.

13. A simple spool valve consists of two lands, four faces, and one annular groove or valley.

14. A land of a spool supports the units and acts as a valve.

15. The spool valve face forms the reaction area for hydraulic pressure and the place upon which spring force applies itself.

16. The groove in the spool acts as a passageway for fluid through the valve.

17. A balanced valve can control the direction of fluid flow and the amount of system pressure.

18. A balanced valve will cease movement in its bore when the applied forces on both ends are equal.

19. The amount of force against a land's face can be determined by multiplying the applied pressure by the area.

20. A differential spool valve has one face larger than the other.

21. A boost valve, operated by an auxiliary pressure, changes the balancing force on a spool valve.

22. A manual-type spool valve moves in its bore due to the application of a mechanical force.

23. Mainline or system pressure is the result of all the circuit restrictions and the regulating valve on the pump's flow.

24. A typical pressure regulating valve consists of a multiple-land spool, spring, line-pressure (boost) plug, and a throttle (boost) plug.

25. The line-pressure plug assists in moving the pressure regulating valve against heavy spring tension.

26. The throttle-pressure plug blocks the movement of the line-pressure plug whenever pressure has to increase to meet system demands.

27. As a regulator valve land opens the port slightly, fluid enters a passage to the torque converter and cooling system.

28. Hydraulic system regulation is just a repetition of opening and closing of the pump inlet port; and the accompanying "controlled leak" bleeds off excess fluid.

29. The throttle-valve's pressure signal via the throttle plug will begin to raise system pressure as the driver accelerates the engine, and it will continue to raise system (mainline) pressure until it reaches a given value.

30. A torque converter control or regulating valve limits the buildup of pressure within the unit to prevent it from ballooning.

REVIEW

½his section will assist you in determining how well you remember the material in this chapter. Read each item carefully. If you cannot complete the statement, review the section in the chapter that covers the material.

1. A device that bleeds off excess system pressure and flow is the _____ .

   a. ball valve

   b. poppet valve

   c. relief or regulating valve

   d. transition valve.

2. The device that usually protects a transmission against the buildup of extremely high pressure is the _____ .

   a. relief valve

   b. spool valve

   c. transition valve

   d. throttle valve

3. The surface of any valve that the force of hydraulic pressure pushes on and attempts to move the valve against spring tension is the _____ .

   a. land

   b. groove

   c. port

   d. reaction area

4. If a valve can move in either direction in its bore as a result of forces being applied to both its ends, it is a _____ .

   a. balanced valve

   b. transition valve

   c. spool valve

   d. manual valve

5. A valve spring that has a certain tension is a _____ spring.

   a. compression

   b. calibrated

   c. estimated

   d. adjustable

6. System pressure of a transmission operating in forward driving ranges is between _____.

   a. 150–300 psi

   b. 100–250 psi

   c. 75–200 psi

   d. 55–150 psi

7. The portion of a spool valve that opens and closes port openings is the _____.

   a. lands

   b. grooves

   c. reaction area

   d. spring

8. If the hydraulic pressure, acting on the primary reaction area, is greater than the spring force on the pressure regulating valve, it will _____.

   a. close

   b. open

   c. balance itself

   d. both a and c

9. The application of pressure to a boost piston of any spool produces a change in the forces necessary to balance the _____.

   a. valve

   b. differential area

   c. pressure

   d. movement

10. The component of the regulating valve train that assists in moving the valve against spring tension is the _____.

    a. line-pressure plug

    b. throttle plug

    c. adjustment screw

    d. sleeve

11. The device that works with the rest of the regulating valve train to adjust hydraulic pressure is the _____.

    a. adjustment screw

    b. line-pressure plug

    c. throttle plug

    d. sleeve

12. Initial pressure regulator movement opens a port to the _____.

    a. sump

    b. torque converter

    c. reservoir

    d. clutches

13. When the regulator valve moves open to a point where the valve enters the regulatory cycle, a land opens a port to the _____.

    a. clutch

    b. converter

    c. pump

    d. valve circuits

14. The device necessary in some transmissions to limit the pressure in the converter is the _____.

    a. compensating valve

    b. limit valve

    c. torque converter control valve

    d. manual valve

For the answers, turn to the Appendix.

As the fluid leaves the rotating pump, it enters a number of different hydraulic circuits. A circuit, as mentioned, refers to the complete path of the fluid flow as it moves from one place to another. For example, the last chapter explained how the fluid flows through a circuit from the pump to the regulator valve, which controls mainline pressure to a safe operating limit and supplies fluid to another circuit that feeds the torque converter and lubrication system.

The remaining pump flow enters and fills the various other circuits within the transmission's hydraulic system. These circuits are controlled by a number of relay or switching valves along with devices that control the rate of fluid flow to an activating mechanism. Most of the time, these valves are located into one assembly known as the **valve body** (Fig. 11-1). However, in a few cases, some valving may also be found in the case circuitry, adjacent to the valve body.

CHAPTER 11

Hydraulic Circuit Relay Valves and Flow Control Devices

FUNCTION OF RELAY OR SWITCHING VALVES

A **relay valve** directs the fluid flow along a given path or circuit. The valve simply opens or closes a circuit connecting mainline pressure to a multi-disc clutch, band servo, or other hydraulic device. In other words, the relay valve is nothing more than an on-off switch that controls fluid flow, but it does so without restricting its movement or changing the pressure.

TYPES OF RELAY VALVES

Basically speaking, there are two types of relay valves in regard to how the unit activates: manual or hydraulic. The manual-type relay valve operates mechanically through some form of linkage controlled by the driver. The hydraulic type is either a balanced spool valve or operates by fluid flow.

The mechanically operated relay unit is the manual valve, which this chapter discusses first. The hydraulic units, that is, the shift and check valves, will also be covered.

MANUAL VALVE

Function

The **manual valve** (Fig. 11-2) is a mechanically operated relay valve. Its function, when activated by the driver, is to direct fluid that applies the appropriate multi-disc or band necessary to activate the planetary gear train. In other words, the valve sets up the operating conditions within the transmission once the driver moves the gearshift selector to a given driving range.

Since the driver mechanically operates this valve, some form of linkage must connect it to the gearshift-selector lever inside the vehicle. Figure 11-2 shows the manual lever assembly, located on the side of the valve body, that is responsible for actually moving the valve in its bore within the housing. In addition, external linkage of various kinds connects this manual lever to the gearshift selector. Consequently, as the driver moves the gearshift selector, the linkage causes the manual lever to move the valve in its bore.

Design

Since the manual valve has a spool design, it has several annular grooves and lands that control port openings. Figure 11-3 shows a typical manual valve that has two lands of uniform diameter and one of a lesser size with annular grooves between them. The large groove, between lands #1 and #2, connects the pump and regulator-valve circuits with whatever other driving-range circuit selected by the driver. The smaller groove, between lands #2 and #3, vents to the sump those forward-drive circuits that are not in use.

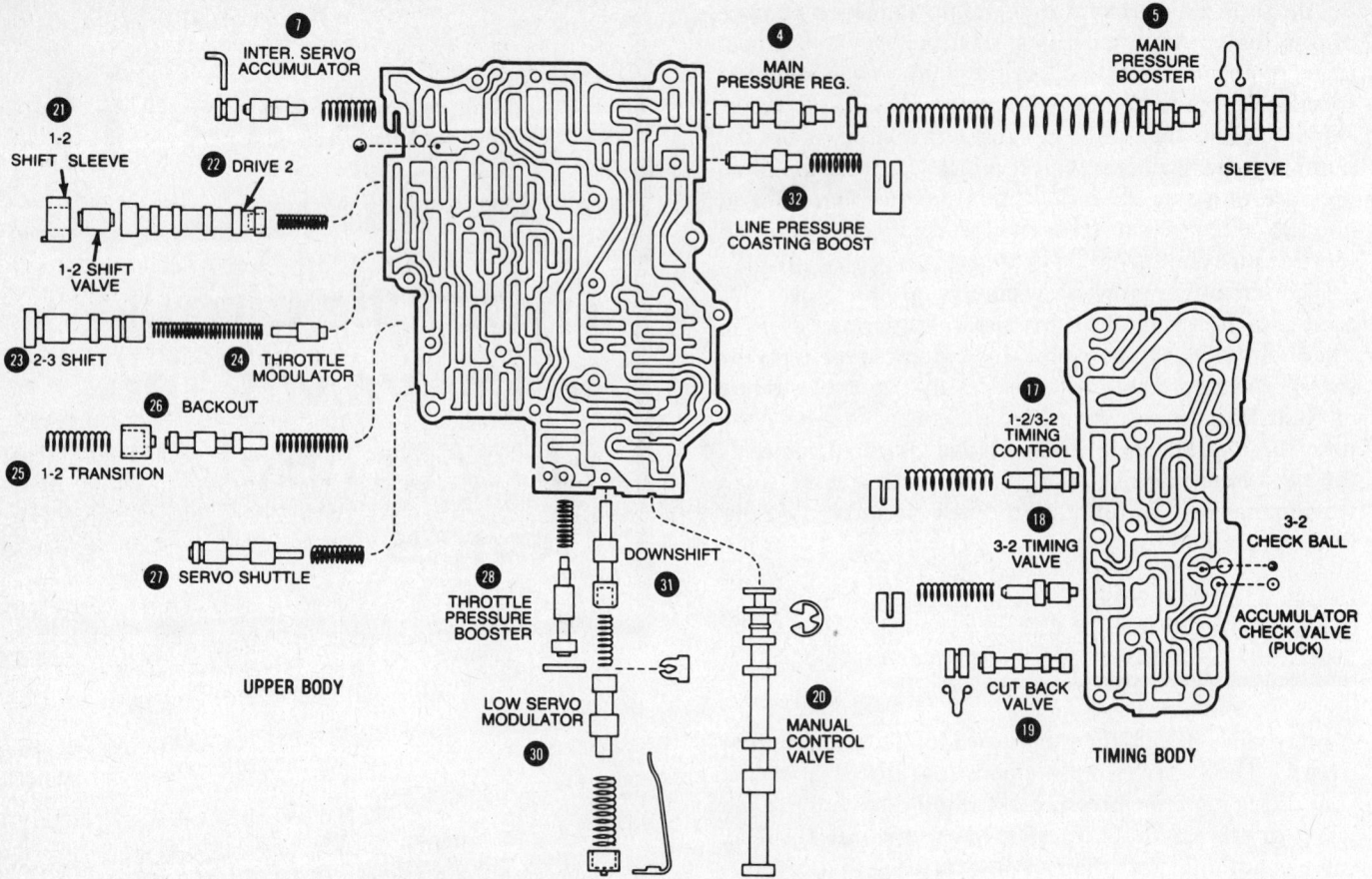

FIGURE 11-1 **The exploded view of a valve body.**
(Courtesy of Ford Motor Company)

Port D, which is open during all forward-driving ratios, supplies fluid to a circuit leading to a clutch, an accumulator, and related circuits. The force of this fluid causes the clutch to apply and the accumulator to move within its bore against spring tension. The **accumulator,** when incorporated into an apply circuit of a multi-disc or band, absorbs a portion of the apply pressure, which in turn controls shift quality. This device is covered in detail later on in this chapter. Finally, port D also supplies fluid to the throttle and 1-2 shift valves, which are necessary to automatically upshift the transmission from first to second ratio under various operating conditions.

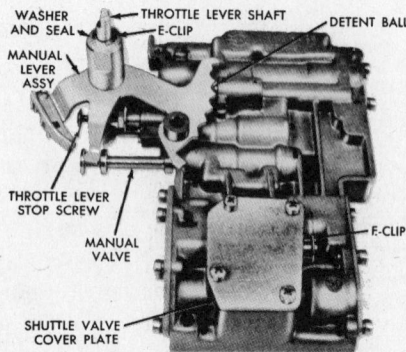

FIGURE 11-2 **The manual valve and lever assembly.**
(Courtesy of Chrysler Corp.)

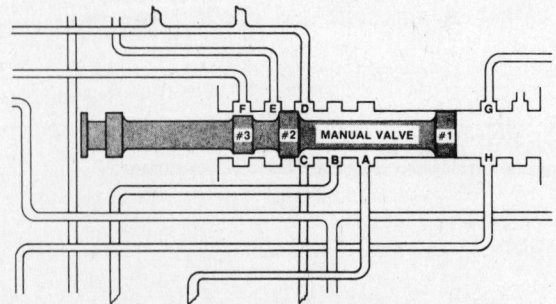

FIGURE 11-3 **The design of a typical manual valve.**
(Courtesy of Chrysler Corp.)

As the manual valve moves, its lands open and close ports located in the bore within the valve body; each of these ports connects to a circuit that carries fluid to a specific activating mechanism or another hydraulic valve. For instance, in Fig. 11-4, circuit A is the inlet port to the manual valve. This port connects to a circuit leading to the pressure regulator valve and hydraulic pump. Therefore, this port supplies fluid to the manual valve and its circuits and it has to be open no matter what position the gearshift-selector lever is in.

If port B is open, it feeds a circuit from the manual valve to the line-pressure plug and the reaction area of the pressure regulator valve. This fluid flow, when acting against these areas, provides the force to move the regulator valve against spring tension to control mainline pressure to within safe operating limits. Lastly, to regulate mainline pressure in all forward driving ranges, the lands of this manual valve keep port B open.

Port C, if open, supplies fluid to a circuit leading to the governor valve (Fig. 11-4). The governor is a type of balanced spool valve that senses vehicle speed from the transmission's output shaft. The valve then sends a hydraulic pressure signal to the shift valves that controls automatic shifting in relation to vehicle speed. Chapter 12 covers the construction and operation of this valve and its circuitry in detail.

If ports E and F are open, they supply fluid to circuits leading to the 2-3 and 1-2 shift valve governor plugs; this action prevents the shift valves from opening. For instance, if port E is open, the 2-3 shift valve cannot open because port E directs fluid to a circuit terminating on the back face of the governor plug. This action prevents the plug from opening the 2-3 shift valve. As a result, the transmission cannot upshift out of second ratio. In other words, port E provides the transmission with a second gear hold capability.

Also, if port F is open, the 1-2 shift valve cannot open. This action locks the transmission in low gear because the 1-2 shift valve governor plug has the force of fluid pressure acting on it. Consequently, the transmission cannot upshift out of low gear because the shift valve will not move to the open position.

Ports G and H are open during reverse gear ratio. Port G supplies fluid to a circuit leading to the reverse activating mechanisms; and port H directs fluid to a circuit back to the pressure regulator. Now, fluid from port G

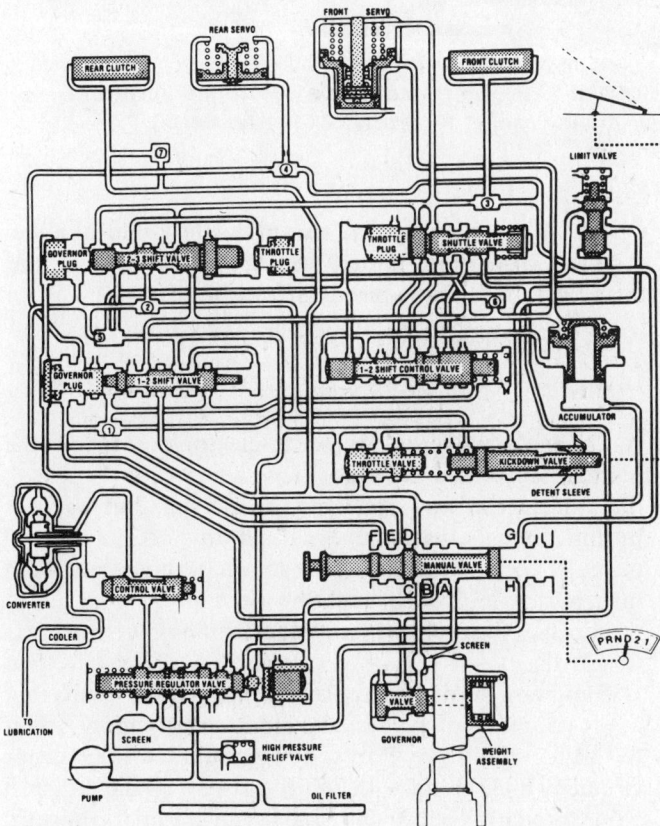

FIGURE 11-4   A diagram of the manual valve, 1-2 and 2-3 shift valves, throttle valve, and governor circuits. (Courtesy of Chrysler Corp.)

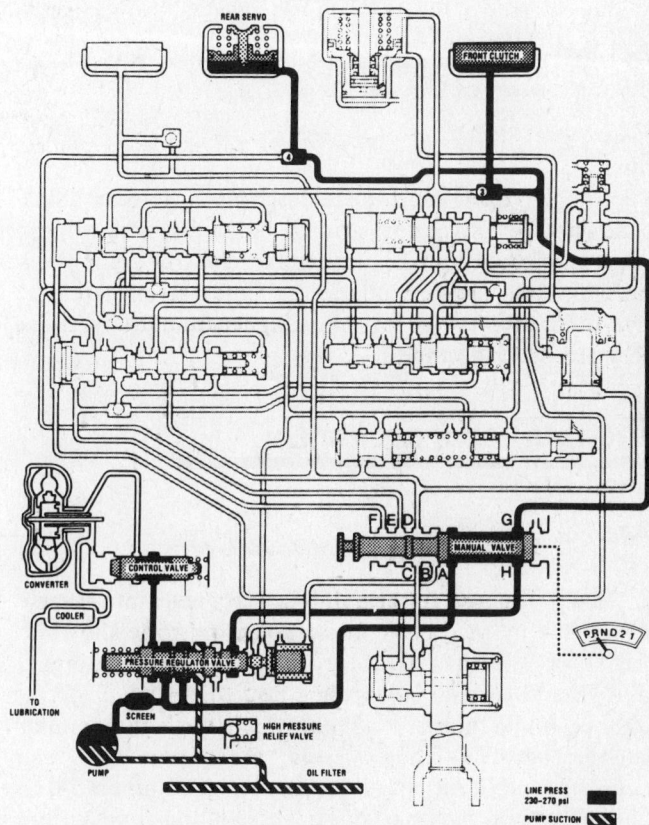

FIGURE 11-5   The manual valve positioned in reverse-gear ratio. (Courtesy of Chrysler Corp.)

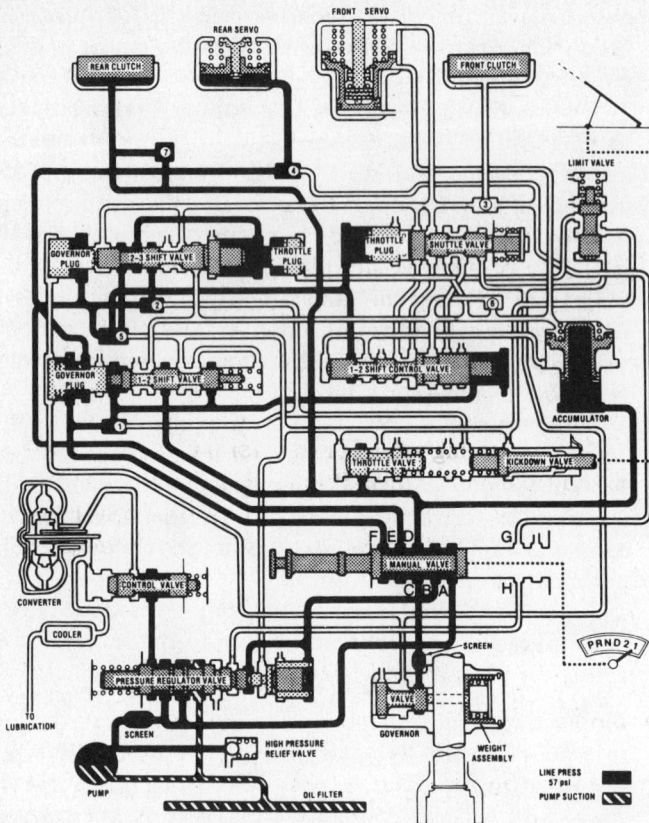

FIGURE 11-6   The manual valve positioned in low-gear ratio. (Courtesy of Chrysler Corp.)

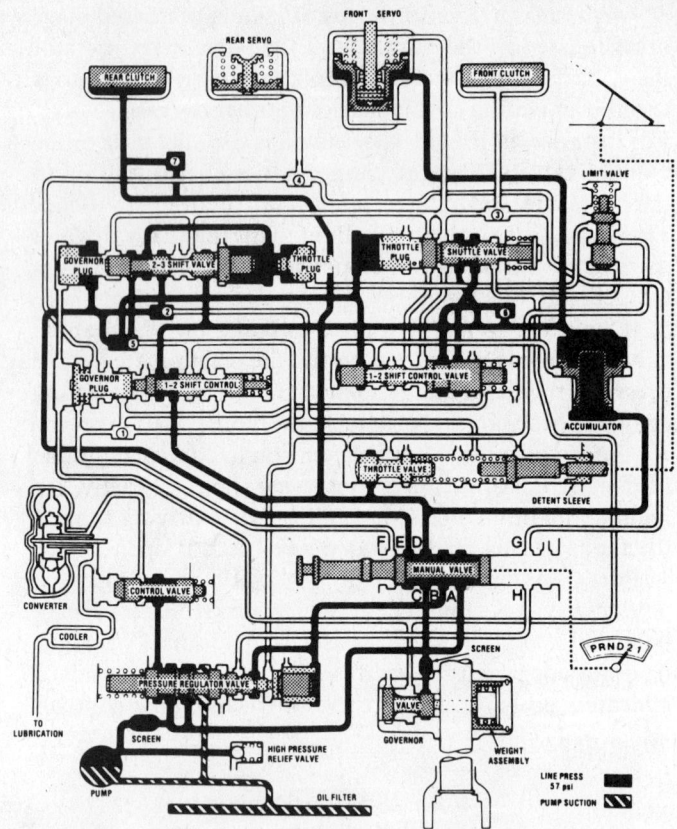

FIGURE 11-7   The manual valve positioned in second-gear driving range. (Courtesy of Chrysler Corp.)

applies the front multi-disc clutch and the rear band. Since port B, of the forward-drive circuits, is not open, port H passes fluid to the regulator valve train. This pressure acts only on one area of the valve; therefore, hydraulic pressure in reverse is higher than it was in the forward-driving ranges.

MANUAL VALVE OPERATION

Reverse

If the driver places the gearshift-selector lever into reverse range, the manual valve moves to the position shown in Fig. 11-5. The manual valve's land #1 now opens ports G and H, and land #2 blocks fluid flow to ports F, E, D, C, and B. Fluid moves from port A through the annular groove, between lands #1 and #2, to port G, which supplies fluid to the circuit leading to the front multi-disc clutch and rear band servo. The force of this fluid applies this clutch and band. At the same time, fluid also passes out of port H to the pressure regulator circuit to control mainline pressure in reverse. Finally, since ports F, E, D,

C, and B are open to the sump via the other annular grooves, any fluid under pressure remaining in the forward-drive circuits bleeds off to the reservoir.

Low

As the driver moves the gearshift selector lever to low, the manual valve assumes a position as shown in Fig. 11-6; the vehicle can now start off in low gear but will not upshift. In this situation, the manual valve's #2 land opens ports B, C, D, E, and F. This action permits fluid from inlet port A to flow through the large groove and out of these ports. Also, the #1 land blocks fluid flow from ports G and H.

Fluid now flows from ports D and E to the circuits that apply the rear clutch and band along with a supply circuit to the 1-2 shift valve. Port C directs fluid to the governor circuit, which along with the fluid sent to the 1-2 shift valve, prepares the transmission for an automatic upshift. However, since fluid travels through ports E and F to a circuit leading to the rear (differential) faces of both shift valve governor plugs, the transmission cannot upshift.

Furthermore, with the manual valve positioned in low,

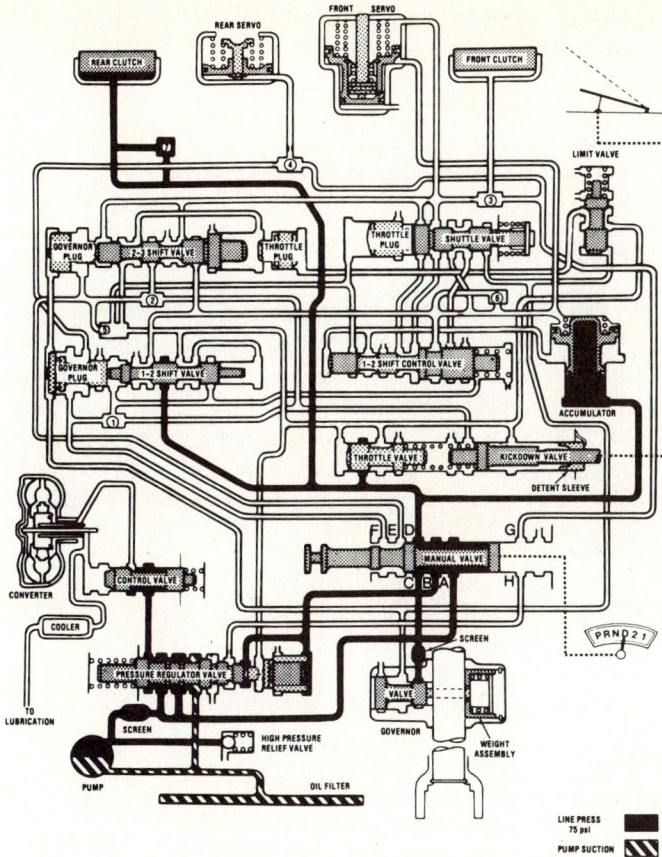

FIGURE 11-8   The manual valve positioned in the normal, drive range. (Courtesy of Chrysler Corp.)

ports G and H, used for reverse, are open to the reservoir. This action permits any remaining fluid under pressure from the connecting circuits to return to the sump. With no fluid flow to these circuits, there is therefore zero pressure, so the front clutch releases due to the action of its return springs. Moreover, system pressure lowers because port H is open to the sump; and port B once again supplies fluid flow into the circuits leading to the regulator's reaction area and the line-pressure plug.

Second

If the driver decides to use the second-gear selector position, the manual valve assumes a position that permits this transmission design to start the vehicle off in low; but the unit will upshift to second gear only (Fig. 11-7). In this valve location, land #2 of the manual valve blocks off port F but leaves ports D, E, B, and C open to mainline pressure. Furthermore, land #1 continues to block fluid passage to ports G and H, which therefore remain open to the reservoir.

Port D continues to send fluid to a circuit which applies the rear clutch necessary for low-gear operation

and also the 1-2 shift valve circuit. Port C supplies fluid to the governor circuit, which along with the fluid sent to 1-2 shift valve, prepares the transmission for a 1-2 automatic upshift. Consequently, when the vehicle reaches a given road speed, the 1-2 shift valve opens, and the transmission upshifts to second gear, as seen in Fig. 11-7, with the application of the front band.

But, since port E is open, fluid goes to the 2-3 shift valve governor plug to oppose the opening of the 2-3 shift valve. The force of this fluid acts on the back (differential) face of the governor plug, which balances any governor pressure on its front face. As a result, the 2-3 shift valve stays in the off or downshift position and the transmission remains in second gear.

The remaining valve ports perform the same functions as they did in low range. Ports G and H stay open to bleed off the reverse circuit fluid to the reservoir while port B supplies fluid to the regulator valve train circuits.

Drive

For normal, fully automatic operation, the driver places the selector lever into drive range; this action moves the manual valve into the location as shown in Fig. 11-8. In this valve position, land #2 only opens ports D, C, and B. Also, land #1 still blocks off fluid flow to the reverse circuit ports, G and H.

Mainline pressure from port A flows through the large annular groove to ports D and C. Fluid from port D passes through a circuit and applies the rear clutch that activates the planetary gear train for low gear ratio; this same port also directs fluid to the 1-2 shift valve circuit. Port C provides fluid to the governor circuit that will produce a pressure signal responsible for opening both the 1-2 and 2-3 shift valves. The pressure regulator's reaction area and its line-pressure plug circuits receive fluid flow from port B, and ports G and H stay open to the sump. Now the transmission is set up to automatically upshift from 1-2 and then from 2-3 when the vehicle reaches a given road speed.

SHIFT VALVES

Function

Every fully automatic transmission has one or more **shift valves** (Fig. 11-9). These balanced spool-type valves and their associated parts are in the hydraulic system for the purpose of making the transmission fully automatic in its operation. In other words, the shift valve automatically controls the upshift and downshift from one gear ratio to another without the driver manually altering the driving ranges himself.

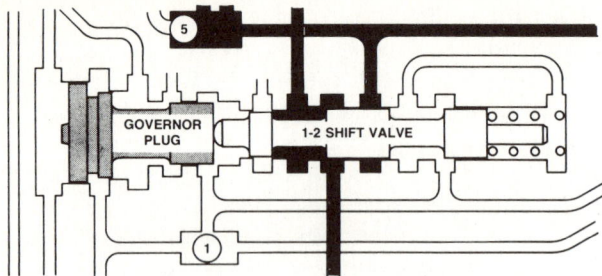

FIGURE 11-9   A typical shift valve and associated parts. (Courtesy of Chrysler Corp.)

The shift valves, in order to control the gear ratio changes, just act as on-off hydraulic switches; the number of these units found in the valve body depends on the number of forward gear ratios the transmission produces. For example, if the automatic transmission provides the vehicle with two forward speeds, the valve body only contains one shift valve. In this case, the shift valve controls the automatic upshift from low to direct and the downshift from direct to low. In other words, when this valve opens, mainline pressure from the manual valve moves through the shift valve to upshift the transmission. But if the same valve closes, it blocks fluid to the required clutch or band; consequently, the transmission downshifts from direct to low.

A three-speed fully automatic transmission has two shift valves (Fig. 11-10). In this case, the 1-2 shift valve controls the automatic 1-2 upshift and the 2-1 downshift; the 2-3 valve is responsible for the 2-3 upshift and the 3-2 downshift.

Figure 11-11 illustrates the hydraulic system of a four-speed automatic transmission with three shift valves. The 1-2 shift valve controls the automatic 1-2 upshift and the 2-1 downshift. The 2-3 shift valve is responsible for the 2-3 upshift and the 3-2 downshift; the 3-4 unit governs the 3-4 automatic upshift and downshift. Finally, since most all the shift valves have about the same design and operate in

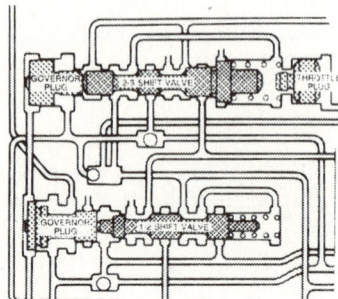

FIGURE 11-10   The 1-2 and 2-3 shift valves of a three-speed transmission. (Courtesy of Chrysler Corp.)

much the same manner, this section only covers one, a 1-2 shift valve and its circuitry.

SHIFT VALVE DESIGN

Spring

In general, a shift valve fits very closely in its bore in the valve body and moves back and forth on a pressurized film of fluid. A typical 1-2 shift valve train is shown in Fig. 11-12; it consists of a calibrated spring, a governor plug, and the shift valve itself. The calibrated spring that bears against the rear face of #6 land of the shift valve attempts to keep the valve train in the downshifted (closed) position.

Governor Plug

Located on the opposite end of the shift valve from the spring is the governor plug. This unit will have the force of governor pressure acting on the left face of its #1 land. This force tries to move the governor plug to the right. And since the governor plug bears against the extended end of the shift valve, any plug travel causes the shift valve to move to the right against spring tension.

The governor plug has three lands that have specific functions to perform during the several operating phases of the shift valve train. To make it easy for the reader to locate the lands and see what they do, this text numbers them sequentially from left to right. For example, the largest land to the left is #1; and the smallest to the right is #3.

Furthermore, the ports that these lands open and close or from which fluid flows to act on other land faces have a letter designation. For instance, beginning at the left side of the plug, the governor-pressure, inlet-port designation is I. Port F supplies fluid flow from the manual valve that acts on the right face of lands #1 and #2; this keeps the governor plug from moving in low range. Lastly, port J is the sump port or is open to the reservoir.

The left end of the large, #1 land governor plug is the fluid reaction area for the assembly. This area has a raised button, which prevents the plug from completely bottoming in its bore. Therefore, this construction creates a small space between the end of the bore and the reaction face area. Into this space, fluid from the governor circuit enters from port I and acts on the reaction area. And, as mentioned, the resulting force attempts to move the plug and shift valve against the tension of the spring.

The back, inside face of the #1 land forms a differential area. If the manual valve is in the low position, the force of fluid entering through port F acts on this area and the rear

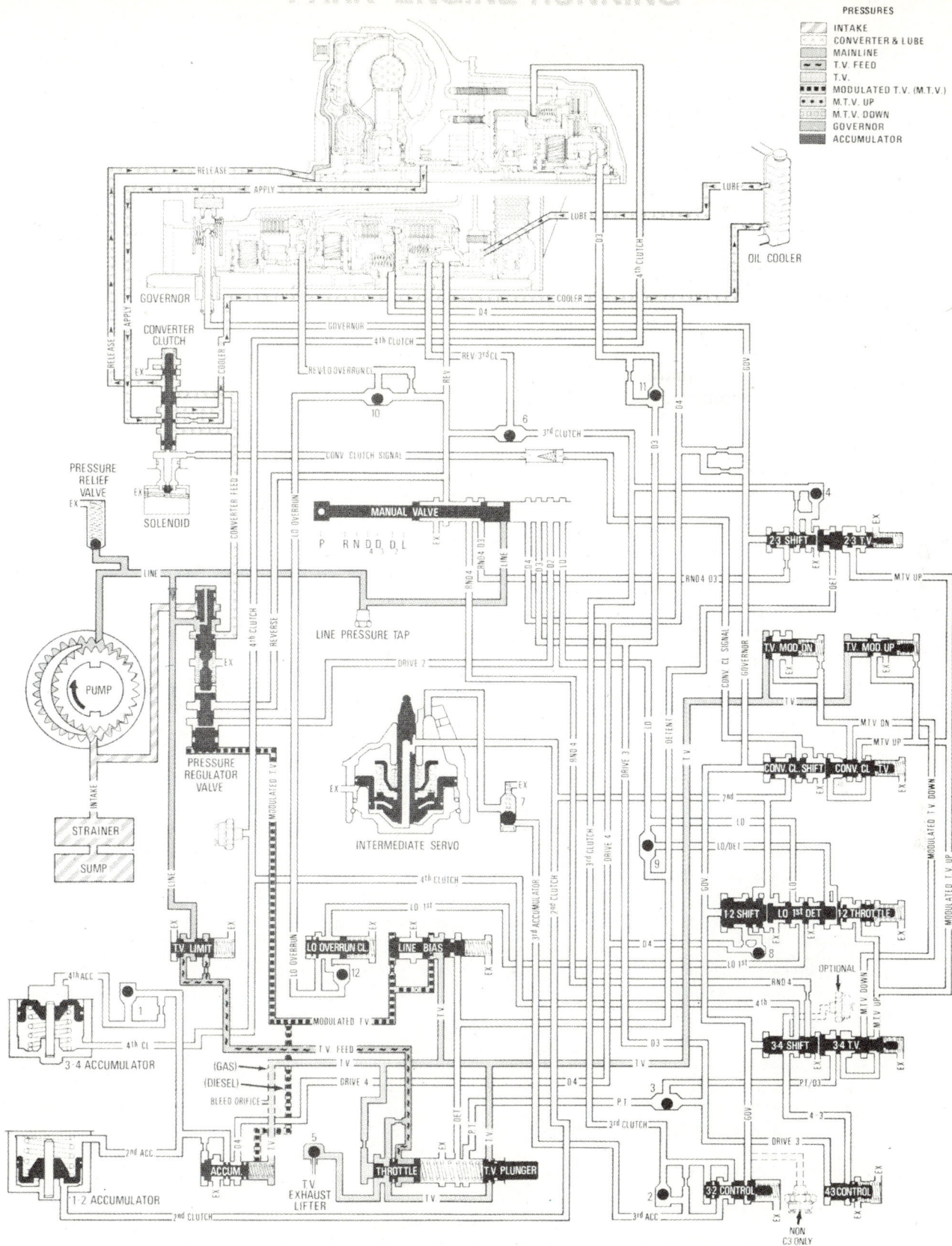

FIGURE 11-11   The hydraulic system of a four-speed fully automatic transmission. (Courtesy of General Motors Corp.)

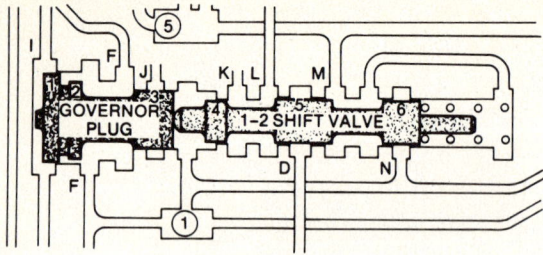

FIGURE 11-12   A typical 1-2 shift valve train, shown here in the closed position. (Courtesy of Chrysler Corp.)

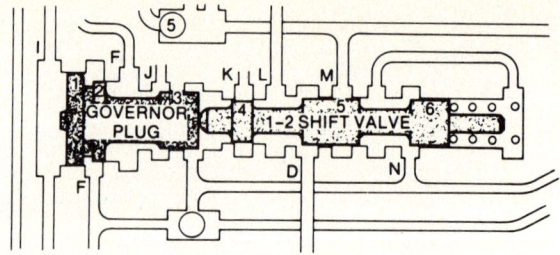

FIGURE 11-13   A 1-2 shift valve in the open position. (Courtesy of Chrysler Corp.)

face of the #2 land. The combined forces on these two faces balances off any governor pressure acting on the reaction area. As a result, the governor plug and shift valve cannot move against spring force. Finally, if the 1-2 shift valve train is in the open (upshift) position and the driver places the selector level in low, port F supplies fluid to the differential area on the rear face of the #1 land, which overcomes governor pressure on the reaction area. As a result, the shift valve closes and the transmission has a forced 2-1 downshift.

In the second and third gear ratios, land #2 of the plug prevents fluid from port F from passing through the large groove whenever the shift valve train is open. This action temporarily blocks mainline pressure from the apply circuit of the rear band servo used in low gear.

But if the driver moves the gearshift selector from drive to low, fluid from the lower F port enters the small groove between lands #1 and #2. This fluid acts on the rear face of the #1 land and assists the spring force in closing the 1-2 shift valve train against governor pressure. As a result, land #2 opens and permits mainline pressure from the lower F port to pass through the large groove, out the upper F port, and to the apply circuit of the rear band servo.

Land #3 of the governor plug controls the operation of the J sump port. For example, with the manual valve positioned in any other forward-driving range but low and with the shift valve open, land #3 opens port J. This action permits any remaining pressurized fluid from the rear servo apply circuit to vent through the J port to the reservoir. In other words, fluid from the band apply circuit bleeds off to the sump via the upper F port, large annular groove, and J port. Finally, if the shift valve train is in the closed position, the #3 land completely blocks the J sump port.

Shift Valve

The 1-2 shift valve, shown in Figs. 11-12 and 11-13, has three lands, which have the number designations of 4, 5, and 6; these lands cover and uncover the hydraulic ports

required for a 1-2 upshift or a 2-1 downshift. For instance, land #4 opens the sump port K whenever the 1-2 shift valve is in the downshift (closed) position (Fig. 11-12). The opening of port K permits the fluid, used in the second-gear apply circuit, to move back through port K and return to the reservoir.

Land #5 has two distinct functions. First, it controls the fluid flow through port D, which is in the inlet opening from the manual valve. If the 1-2 shift valve is in the closed position, this land blocks fluid flow from port D from reaching the second-gear apply circuits. When the valve opens (Fig. 11-13), land #5 uncovers the D port. This action permits fluid to pass through the annular groove, between lands #4 and #5, and out the L port to the second-gear apply circuit.

Second, land #5 controls the movement of fluid through the M port. The land opens the M port whenever the 1-2 shift valve is closed (Fig. 11-12). By opening this port, land #5 allows throttle pressure to pass through port M, around the annular groove, through a crossover passage, and to the chamber on the spring side of the 1-2 shift valve. Along with the force of the spring, the throttle pressure opposes the opening of the 1-2 shift valve train. But as the 1-2 valve opens, due to the pressure on the governor plug (Fig. 11-13), land #5 blocks port M; this action cuts off throttle pressure to the spring end of the shift valve.

Land #6 also has several functions. For example, it covers and uncovers port N. Port N, when opened by land #6 (Fig. 11-13), acts as a sump port for the trapped throttle pressure behind the shift valve as land #5 closes the M port. In addition, as long as the 1-2 shift valve is open, downshift pressure from the kickdown valve can enter through port N and assist the spring's tension in closing the shift valve. This action, of course, would downshift the transmission from 2-1.

Lastly, land #6 has the force of the spring imposed on its end face. The tension of the calibrated spring divided by the size of the reaction area determines how much governor pressure is necessary to balance this force to open the valve and upshift the transmission. Moreover, to limit the travel of the 1-2 shift valve toward the spring, the end face of land #6 has a machined shank. It prevents the 1-2 shift valve from bottoming in its bore, thereby positioning all the valve lands in their respective locations whenever the 1-2 shift valve is open.

SHIFT VALVE OPERATION

First Ratio

When the driver positions the gearshift-selector lever into drive range, the manual valve directs fluid flow to the rear clutch, governor, and the 1-2 shift valve. Looking at the hydraulic schematic shown in Fig. 11-14, note that fluid from the manual valve's D port applies the rear clutch, forces the accumulator valve up in its bore, but deadends at the 1-2 shift and throttle valves because they are both closed. In addition, fluid from the manual valve's port C flows to the governor valve. Now the transmission's planetary is in first ratio, the vehicle is ready to move, and

the circuits that are necessary for the 1-2 automatic upshift have fluid under pressure supplied to them.

Second Ratio

As the driver presses down on the accelerator pedal and the vehicle begins to move forward, the upshift circuits begin to function. First, fluid under pressure from the throttle valve enters port M, passes around the annular groove, and flows through the crossover passage to the spring end of the 1-2 shift valve. Now, the 1-2 shift valve has both spring force and throttle pressure acting on its right side, holding it over in the left end of its bore.

The actual amount of throttle pressure varies with how much the driver depresses the accelerator. With a light accelerator depression, the throttle pressure is low, but a heavy depression on the pedal results in high throttle pressure. The manner in which the throttle valve develops this pressure is described in Chapter 14.

As the vehicle's speed increases over about 5 mph, governor pressure, from the governor valve, flows

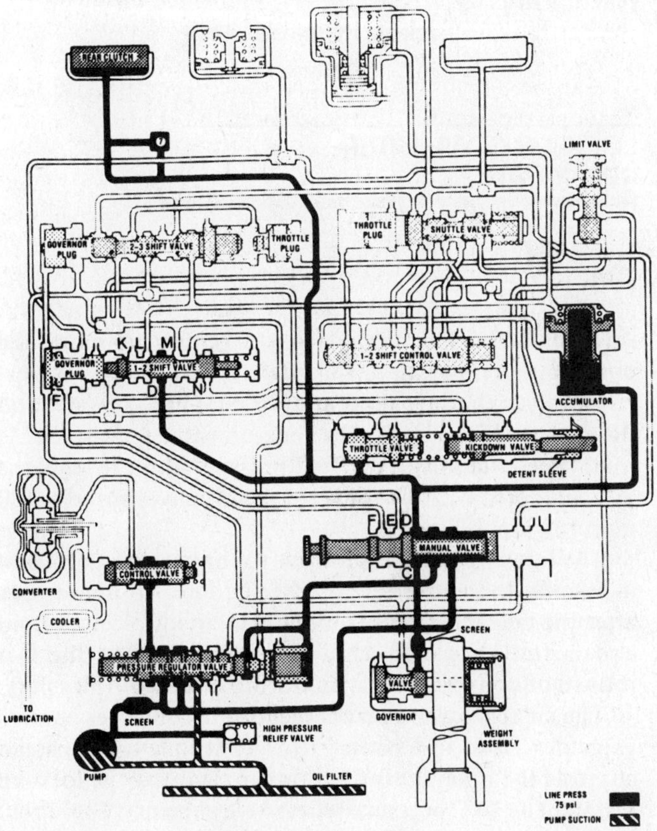

FIGURE 11-14   A diagram of a drive, first-gear ratio circuit. (Courtesy of Chrysler Corp.)

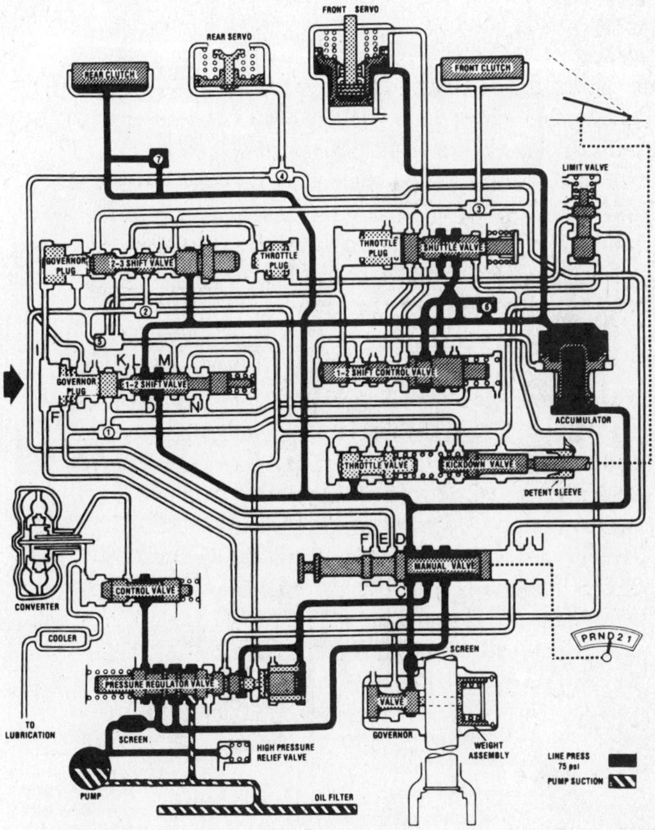

FIGURE 11-15   A diagram of a drive, second-gear ratio circuit. (Courtesy of Chrysler Corp.)

through port I to the 1-2 shift valve governor plug and attempts to open the valve train. When governor pressure builds up to the point where its force is greater than the combined efforts of throttle pressure and spring tension, the 1-2 shift valve train begins to move over to the right (Fig. 11-15).

As the valve begins to move right, the #5 land of the shift valve closes one port and opens another, while land #6 opens a sump port. Land #5 closes off the M port; this closure stops throttle pressure from acting on the spring end of the shift valve. The trapped throttle pressure now vents to the reservoir through the N port, which land #6 has opened. With zero throttle pressure acting on the spring end of the valve, the 1-2 shift valve now "snaps" full open rapidly, rather than just slowly moving to the right — its upshifted position.

As the valve snaps open, the #5 land uncovers the D port. This action permits fluid flow from the D port to pass through the annular groove between lands #4 and #5, out of the L port, and to the front band servo circuit. This action applies the band and forces the accumulator piston down in its bore.

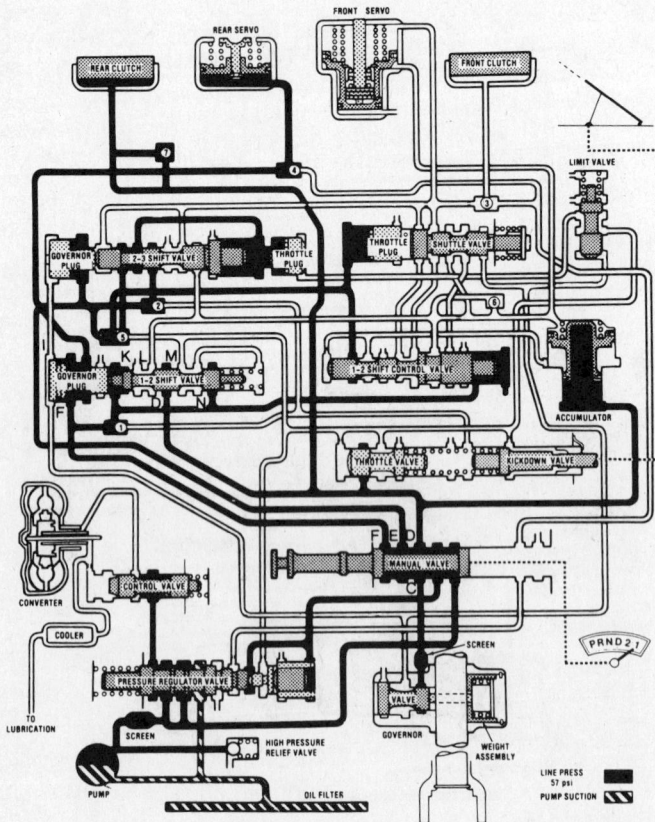

FIGURE 11-16   A diagram of a manual, low-gear ratio circuit. (Courtesy of Chrysler Corp.)

Second to First Ratio

As the vehicle slows to a stop, the 1-2 shift valve train moves back to the first gear position (Fig. 11-14). The 1-2 shift valve moves back in its bore due to a reduction in governor pressure acting on the governor plug and the force of the calibrated spring pushing on the right end of the valve train. Now, with the valve train back in the first-gear position, the 1-2 shift hydraulic circuits will cycle back to those shown in the illustration.

The #5 land blocks off port D fluid from reaching the L port, and the #4 land opens the K sump port. This action permits the fluid, utilized to apply the band and move the accumulator piston in second ratio, to bleed off to the reservoir. As the pressure drops off in this circuit, the band releases and the transmission downshifts to first gear ratio. The transmission is now ready to start the vehicle off in first ratio.

As mentioned previously, the driver can also force the 1-2 shift valve back by moving the selector lever to the manual low position (Fig. 11-16). This action causes the manual valve to open a port, which directs fluid flow to the 1-2 shift valve lower F port. This fluid acts on the rear face of the #1 land of the governor plug. This offsets the effect of moderate governor pressure on the plug's reaction area. As a result, the 1-2 shift valve spring pushes the entire train back to the closed position.

LOCK-UP CONVERTER RELAY VALVES

Function

As explained in Chapter 4, a lock-up converter requires a number of relay valves in order to activate it. Typical relay valves that are used are the lock-up, fail-safe, and the switch (Fig. 11-17). The lock-up valve senses vehicle road speed via governor fluid pressure and prevents the lock-up portion of the converter from activating below 40 mph (65 km/h).

The fail-safe valve is sensitive to throttle pressure, and it permits the converter to enter its lock-up phase only after the transmission's front clutch circuit receives fluid and is fully applied. Also, this valve permits the fast release of converter lock-up during a forced downshift.

The switch valve replaces the torque converter control valve described in Chapter 10. The switch valve instead controls the on and off fluid flow to activate the lock-up piston. The T/C pressure relief valve shown in Fig. 11-17 prevents excessively high pressure from developing in the converter, thus it takes over the original function of the torque converter control valve.

CONVERTER RELAY VALVE DESIGN

Lock-Up Valve

The lock-up valve is a balanced spool with three lands and a calibrated spring. The right-hand face of #1 land forms the reaction area upon which governor pressure acts. Land #2 opens and closes the vent port B.

Land #3 has two functions. First, the land covers and uncovers port D. Second, its left-hand face forms the area where the calibrated spring exerts its force. Notice also in Fig. 11-17 the extended shank on the end of land #3. This shank limits the travel of the lock-up valve in its bore as it opens so that the lands and ports are in proper alignment.

The calibrated spring maintains the lock-up valve in the closed position until the vehicle reaches a given road speed. As mentioned, the vehicle must be operating over 40 mph (65 km/h) before governor pressure is sufficient to overcome the spring's tension.

Fail-Safe Valve

The fail-safe valve is a balanced spool having two lands, one large groove, and a calibrated spring. Land #1 has several purposes. For instance, its right-hand face forms not only a reaction area for throttle pressure but for spring force to act as well. In addition, land #1 covers and uncovers port C as the valve moves in its bore.

Land #2 also has two purposes. First, it controls the opening and closing of the vent port G. Second, the left-hand face of land #1 forms another reaction area. In this case, the area has front clutch apply pressure acting on it. This pressure attempts to move the fail-safe valve against spring tension and the force of the throttle pressure.

The large groove permits two pathways for the fluid to flow. The first is from port C to port H; the second from port H to port G. The pathway the fluid takes depends upon the position of the fail-safe valve.

The calibrated spring attempts to keep the fail-safe valve closed. Its tension has the assistance of throttle

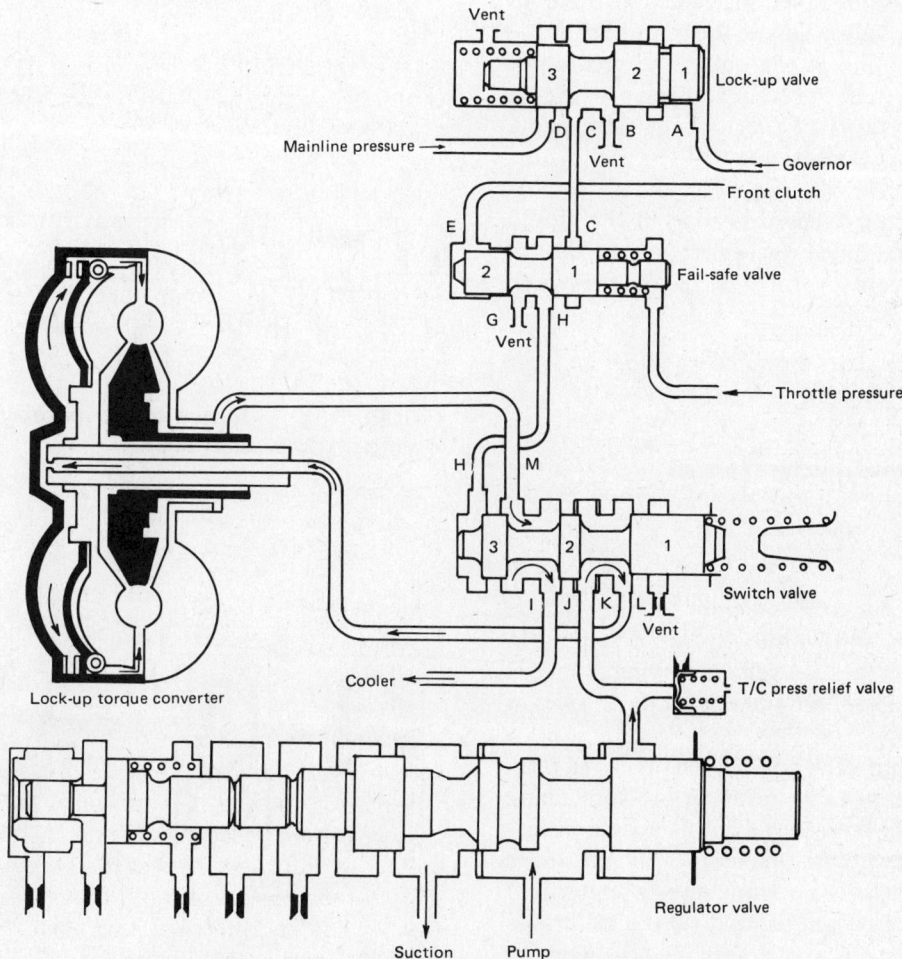

FIGURE 11-17   Torque converter relay valve positions during the unlock phase.

pressure for this purpose, both of which act on the rear face of land #1.

Switch Valve

The switch valve is a balanced unit with three lands, two grooves, and a calibrated spring. The #1 land controls the opening and closing of the L vent port. Also, spring tension exerts its force on the rear face of the land.

Land #2 controls a number of port openings. For instance, during the unlock phase of converter operation (Fig. 11-17), the land permits fluid to flow between the J and K ports via the groove between lands #1 and #2. During the converter's lock-up phase, Fig. 11-18, land #2 now permits the passage of fluid from ports I to J via the groove between lands #2 and #3.

Land #3 has two functions. First, the land opens and closes the M port. Second, the left-hand face of land #3 forms the reaction area for mainline pressure, the force of which acts against spring tension to open the valve. Note in Figs. 11-17 and 11-18, the raised button on the end of the land. This has the same function as the one found on land #2 of the fail-safe valve. This is, the button stops the land face from bottoming in the bore, thus creating a small pocket in which fluid can enter and begin acting on the reaction area. Land #1 of the lock-up valve has a chamfered edge on the reaction area that serves the same purpose.

The calibrated spring keeps the valve in the closed position. But when mainline pressure acts on the face of land #3, its force overcomes spring tension and the switch valve opens.

CONVERTER RELAY VALVE OPERATION

Unlock Phase

As mentioned, the lock-up portion of the torque converter will not function with the transmission not in third ratio or high gear and the vehicle moving at speeds in excess of 40 mph (65km/h). The reason for this lies in the fail-safe valve.

Although the lock-up valve could open in second ratio at higher vehicle speeds due to governor pressure acting on its reaction area, the fail-safe valve will be held closed. The reason for this is twofold: (1) since the transmission is not in third ratio, there is no front clutch apply fluid acting on the left face of land #2 and (2) the calibrated spring's force and that of throttle pressure acting on the right face of land #1 keep the valve in the position as shown in Fig. 11-17.

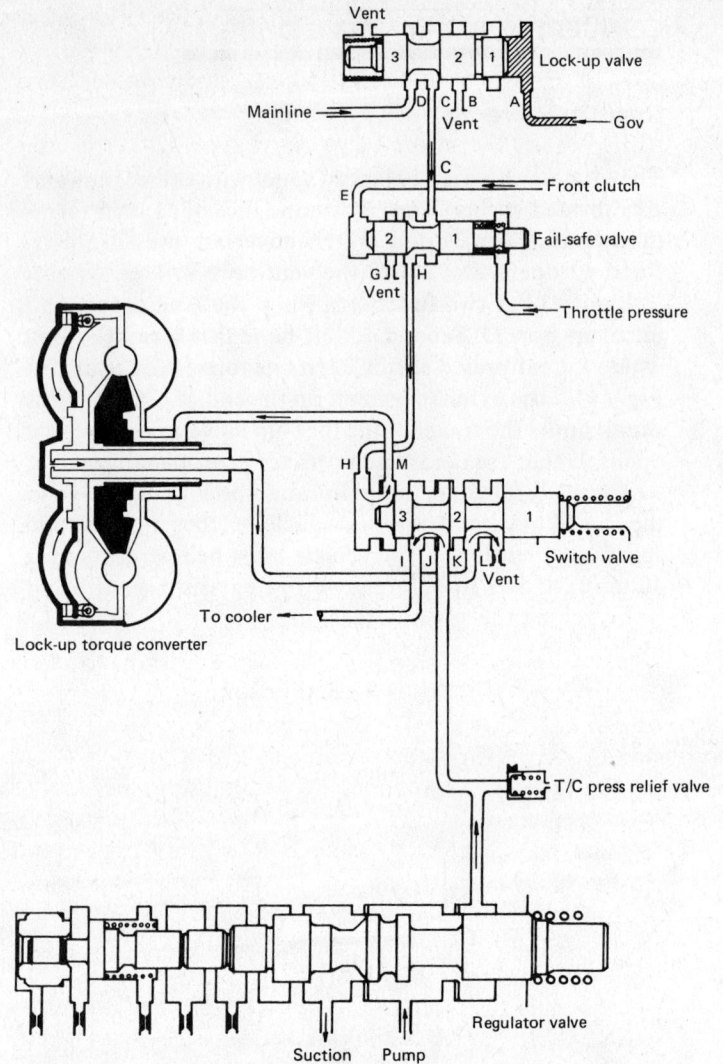

FIGURE 11-18 **Torque converter relay valve positions during the lock-up phase.**

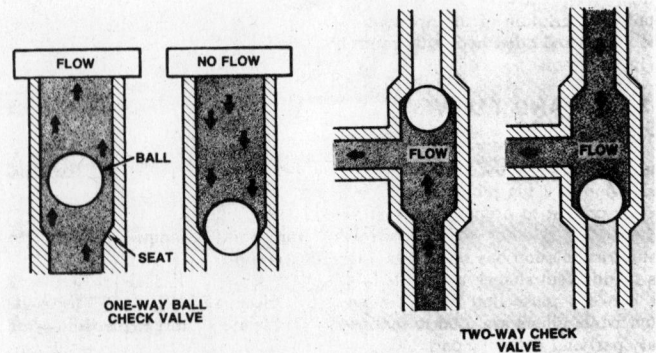

FIGURE 11-19 **One-way and two-way ball check valves. (Courtesy of Chrysler Corp.)**

Since the fail-safe valve remains closed, mainline pressure cannot pass through it and act on the reaction area of the switch valve. Therefore, it remains closed. As a result, the converter feed fluid leaving the regulator enters the J port of the switch valve, moves through the groove between lands #1 and #2, and out the K port. From here, the fluid enters a circuit in the transmission's input shaft. The fluid flows around the front of the lock-up piston, through the converter, and out again where it enters the switch valve's M port. The fluid passes through the annular groove between lands #2 and #3 and out the I port into a circuit leading to the cooling and lubricating system. Consequently, the lock-up piston remains disengaged.

Lock-Up Phase

During the lock-up phase, all three relay valves are open. The lock-up valve opens due to governor pressure acting on its reaction area, which moves the valve to the left against spring tension. As the valve opens, land #3 uncovers the D port. As a result, mainline pressure enters the D port, moves through the annular groove, and out the C port to the fail-safe valve (Fig. 11-18).

The fail-safe valve moves to the right due to the force of front clutch apply fluid acting on the left face of land #2, against the combined forces of throttle pressure and spring tension. With the valve open, mainline pressure from the lock-up valve passes through the groove between lands #1 and #2 and out the H port into a circuit leading to the switch valve.

This mainline pressure acts on the reaction area of the switch valve and moves it open against the force of spring tension. With the switch valve open, two things occur. First, the converter feed fluid from the front side of the lock-up piston now vents to the reservoir through the K port, annular groove, and L port. Second, mainline pressure reacts on the M port, causing fluid to flow into the converter out circuit. Third, normal converter input fluid flows from the regulator to the J port, through the annular groove, and out the I port into the cooling and lubrication system.

With the switch valve open, mainline pressure forces the lock-up piston to the left. When the piston makes firm contact with the friction surface, bonded to the converter cover, the turbine is now locked to the engine crankshaft.

CHECK VALVES

Function

The hydraulic system of most automatic transmissions contains one or more check valves. A **check valve** is the simplest form of relay valve that permits the flow of fluid in one direction only. The check valve is not a balanced type or operated by mechanical linkage. Instead, the direction of apply fluid flow, in the circuit the check valve controls, activates it. When opened by a given fluid flow, the check valve permits oil to move from one area of a circuit to another or it allows two hydraulic circuits to use one passage. By doing so, the check valve plays an important role in directing the various flow patterns used during the different operating phases of the transmission.

Types

There are three types of check valves commonly used in automatic transmissions: the one-way ball check, two-way ball check, and the poppet. Fig. 11-19 shows the one-way and two-way ball check valves. The one-way check valve is shown permitting the fluid to flow in a circuit in one direction only. The two-way valve allows two circuits, upper and lower, to use the common horizontal passage. The poppet-type check valve is illustrated in Fig. 11-20. This check valve also permits fluid flow in a circuit in one direction.

CHECK VALVE DESIGN AND OPERATION

One-Way Ball Type

The **one-way ballcheck valve** consists of a ball, seat, and stop pin. The ball (Fig. 11-21) is made of steel or some synthetic material and operates inside a hydraulic circuit. In the circuit bore is the valve seat. The location of the seat in the example is at port A, where the fluid will enter; port

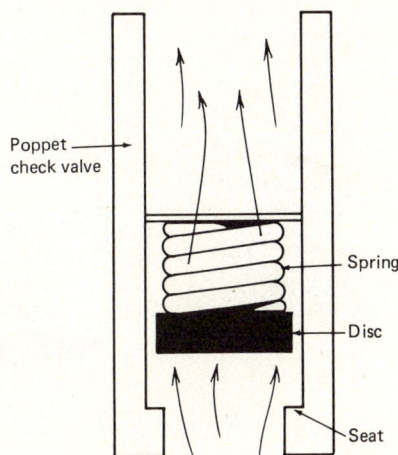

FIGURE 11-20   A typical poppet-type check valve.

Poppet check valve
Spring
Disc
Seat

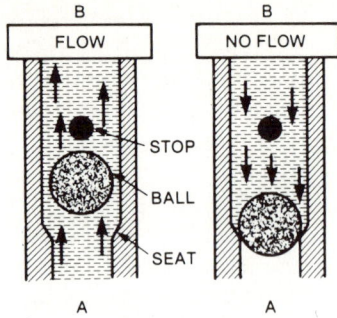

FIGURE 11-21   The design and operation of the ball-type one-way check valve.

B has no valve seat. Moreover, the stop pin, located between the two ports, prevents the check ball from reaching the outlet port B.

With this arrangement, fluid can move from port A through port B but not in the other direction. When fluid enters port A, for example, it pushes the ball off its seat and against the stop. The fluid then passes around the ball and out of port B. However, if fluid attempts to flow in the other direction, from port B through port A, no flow occurs because the oil movement forces the ball against its seat.

Two-Way Ball Type

The **two-way ball check valve** has a single ball with two seats and no stop pin. The seat locations are at the two different circuit flow sources. And flow from either source will move the ball against the opposite seat (Fig. 11-22). With this design, fluid flow can move through either of the inlet ports A or B and to the outlet port C, but fluid can only move from one of these inlet ports to the outlet at any given time.

During the operation of this valve, higher pressure acting on one inlet port causes fluid to seat the ball. This same action blocks fluid flow under a lower pressure from entering and discharging from the outlet port C. For instance, if fluid enters port A, it pushes the ball valve against the seat of port B. This action seals off the entrance of fluid under a lower pressure from port B and only allows fluid from port B to pass out of port C.

But on the other hand, if fluid under a higher pressure passes through port B, it will seat the ball, thus blocking port A. As a result, fluid flows through the valve and out of port C.

Poppet Type

The **poppet-type check valve** consists of a flat disc and a spring (Fig. 11-23). The disc fits over and seals an opening in the circuit slightly smaller than its diameter. In the illustration, the disc is located over port A of the valve. The light spring is responsible for holding the disc in alignment against the seat when there is no applied pressure.

When fluid flows through the A port of the valve, it forces the disc away from its seat. The fluid then flows around the disc and out of port B. However, if the fluid enters the B port, it cannot exit from the A port because its force has seated the disc over the opening.

CIRCUIT FLOW CONTROL DEVICES

In the last chapter, the text covered the use of pressure-relief and pressure-regulating valves. Both of these units primarily are in a hydraulic system to control its mainline

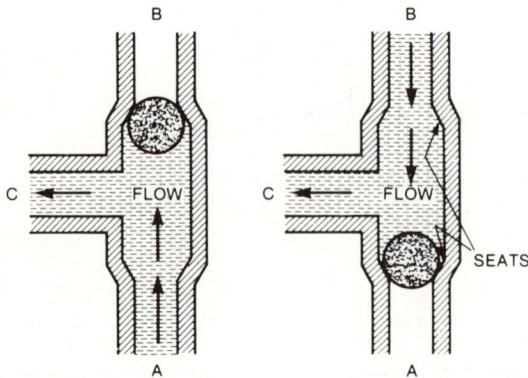

FIGURE 11-22   The design and operation of the ball-type two-way check valve.

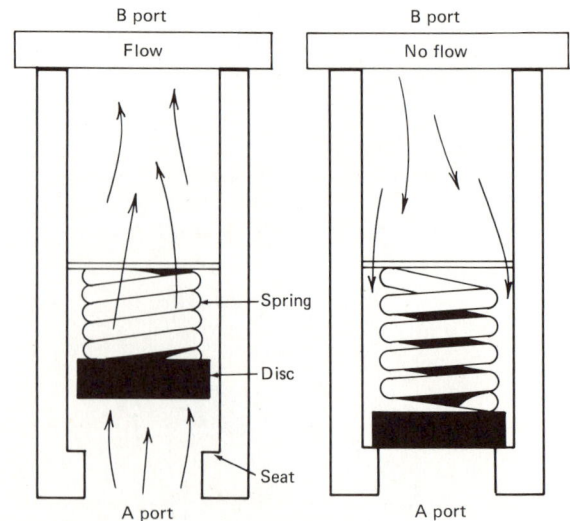

FIGURE 11-23   The design and operation of a poppet-type check valve.

pressure. However, manufacturers install several other types of flow control devices in a **given circuit** to control the rate of fluid movement. This action is taken to reduce for a period of time the buildup of pressure in a certain portion of the circuit for the purpose of either cushioning or delaying the full application of a band or multiple-disc clutch. These devices are the orifice and the accumulator.

ORIFICE

Function and Design

An **orifice** is simply a calibrated restriction placed in a given hydraulic circuit; this impedes the flow of fluid and therefore slows the buildup of pressure downstream from the orifice. Figure 11-24 shows an orifice placed into a hydraulic circuit. To show the effect of the orifice on this circuit, it has two hydraulic gauges installed, one on each side of the restriction.

Operation

As the fluid flows in the circuit, the orifice slows down the flow of fluid; this delays the buildup of pressure on the right side of the orifice (Fig. 11-24). The pressure reduction occurs because, as the fluid attempts to move through the orifice, not all of the oil can pass through the restriction at one time. Consequently, pressure builds up in the circuit in front of the orifice or on the pump side.

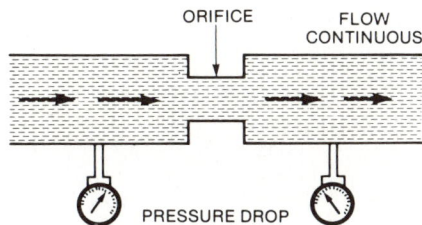

FIGURE 11-24   An orifice creates a flow restriction and slows the buildup of pressure in a circuit.

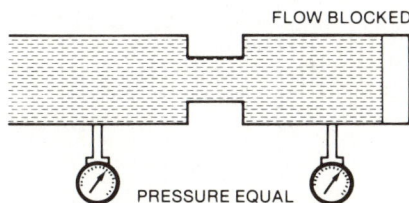

FIGURE 11-25   When flow stops, the pressure equalizes on both sides of the orifice.

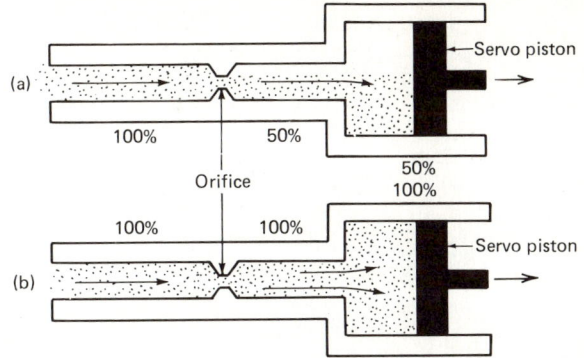

FIGURE 11-26   An orifice controlling the percentage of pressure buildup on a servo piston.

This difference in pressure between both sides of the orifice remains as long as fluid continues to flow through the restriction. But when flow stops through the orifice, pressure equalizes on both of its sides (Fig. 11-25). This situation occurs whenever the hydraulic pressure on the downstream side of the orifice has fully applied the activating mechanism located there.

Figure 11-26 shows the approximate percentage of pressure drop on a servo piston due to the action of an orifice of the proper size. Diagram A shows the percentage of pressure on the servo piston as it just reaches the end of its travel, about 50 percent. However, as the fluid completely fills the circuit to the right of the orifice, pressure rises to 100 percent or mainline (diagram B of Fig. 11-26).

By slowing down the flow rate to the servo piston, the servo causes a lower percentage of mainline pressure acting on it. This results in a cushioning effect on the band's application around the drum, so the driver does not experience a harsh gear change. But once the piston has stopped its travel and the circuit completely fills with

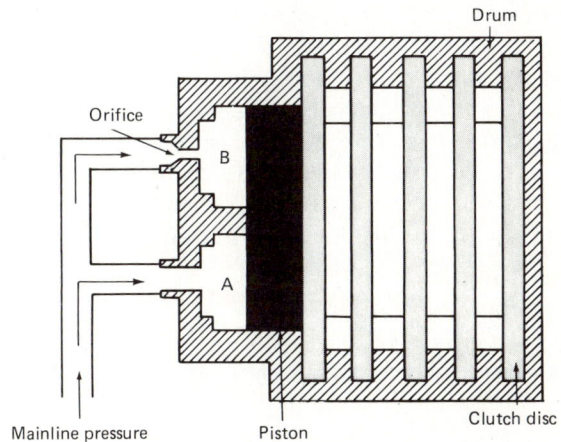

FIGURE 11-27   An orifice providing a cushioning or delaying action on the application of a multiple-disc clutch.

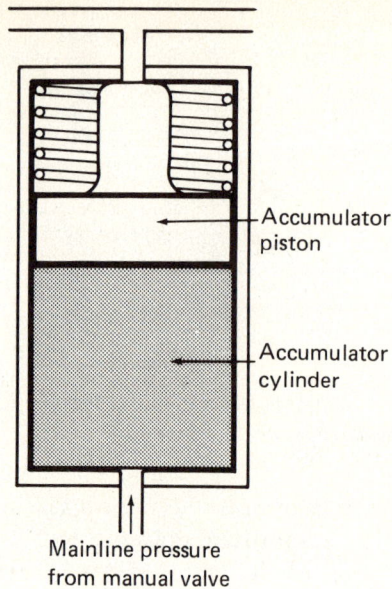

FIGURE 11-28   The design of a simple accumulator.

fluid, full mainline pressure locks the band securely around the drum.

An orifice has the same effect on the application of a multiple-disc clutch. Figure 11-27 shows a clutch piston with two apply circuits, A and B. The inner circuit A permits rapid fluid flow to one area of the piston for initial movement of the piston and clutch application. Circuit B, on the other hand, has an orifice that delays fluid movement and the buildup of pressure on a second piston area. With the arrangement, the full application of the clutch is delayed by the orifice until the B circuit is full of fluid. This cushions or delays the full application of the clutch.

It should be obvious that the orifice plays an important role also in the proper timing of the automatic gear changes. In the automatic transmission, the changing from one gear ratio to another requires a delicate control of the release and application of the multiple-disc clutches and bands. Therefore, a number of engineered small orifices are used throughout the hydraulic circuitry to obtain the correct timing. This produces a quality shift pattern so the engine does not flare up excessively or the gear ratios fight one another.

ACCUMULATOR

Function

The accumulator, like the orifice, is another flow control device. Both serve functions in the hydraulic system that are quite similar. The orifice and the accumulator, when

in operation, can function to cushion or delay the application of a multiple-disc clutch or band. The orifice does this by restricting fluid flow; the accumulator accomplishes the same task by absorbing clutch or band apply fluid.

Design

The simple accumulator consists of a cylinder, piston, and return spring. The transmission manufacturer may cast or machine the accumulator cylinder into a detachable housing, the transmission case, or the valve body. In all cases, the cylinder serves as the guide for the movable piston.

The piston may be the solid type, single-diameter; spool type, single-diameter; or spool type, multiple-diameter. Figure 11-28, for example, shows a piston type with a single diameter. All of these piston types move within the cylinder due to the force of hydraulic pressure acting on one or more of the surface areas.

The accumulator also has a calibrated return spring. This spring, the length of the piston's travel, along with whichever reaction area the hydraulic pressure acts on are the factors that determine how much apply fluid the accumulator will absorb.

Operation

When the driver moves the gearshift-selector lever to a forward-drive range, mainline pressure causes fluid to flow into the lower port of the accumulator from the

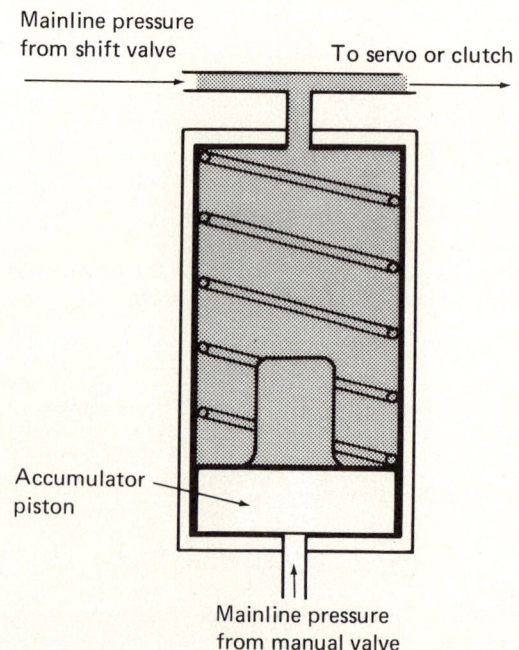

FIGURE 11-29   The operation of a simple accumulator.

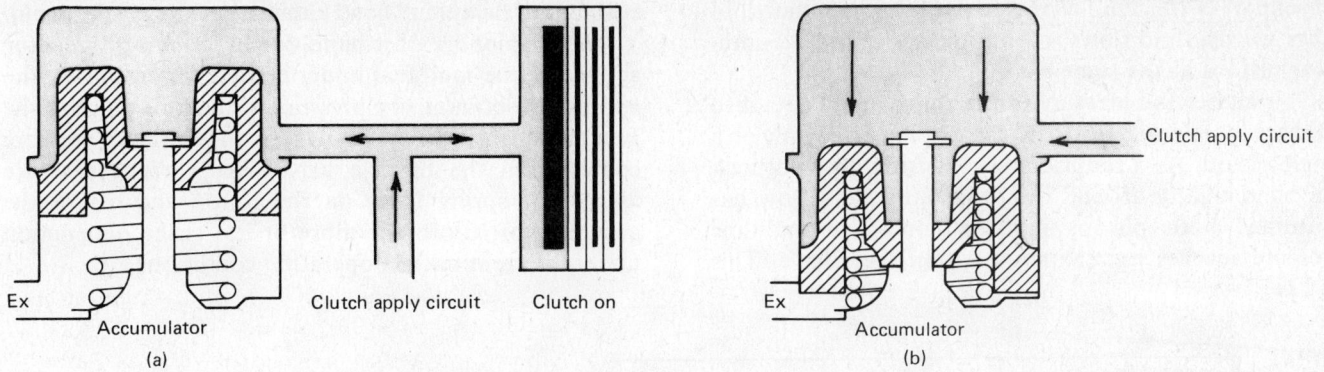

(a)   (b)

FIGURE 11-30 **The operation of an accumulator in a multiple-disc clutch apply circuit.**

manual valve. The fluid, under pressure, strokes or moves the piston up in the cylinder against the force of the spring.

When a shift valve moves to the open position, fluid enters a circuit to the clutch or servo and to the top port of the accumulator (Fig. 11-29). Once the fluid fills the spring chamber, pressure on both sides of the piston equalizes. As a result, the calibrated spring strokes the

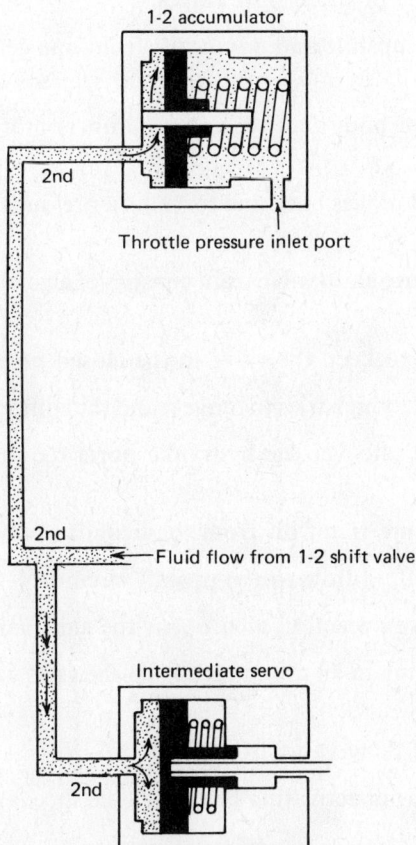

FIGURE 11-31 **The operation of an accumulator in a band servo apply circuit.**

piston to the bottom of the accumulator cylinder. As the piston moves down, it must take on, absorb, or accumulate fluid. Until the accumulator completely fills with fluid, full multiple-disc clutch or band application cannot take place. In this way, the application of the clutch or band is cushioned.

ACCUMULATOR OPERATION WITHIN A MULTIPLE-DISC CLUTCH CIRCUIT

Figure 11-30 illustrates an accumulator's operation in a multiple-disc apply circuit. When the shift valve opens, fluid flows into the clutch and accumulator circuit (A of Fig. 11-30). The initial flow of fluid builds up sufficient pressure against the clutch piston to move it far enough to overcome some of its spring force; this takes up the slack or clearance between the clutch plates. As the discs continue to compress together, pressure in the circuit builds up rapidly.

At a point just before maximum mainline pressure is reached, the accumulator piston strokes against its spring tension (B of Fig. 11-30). Because the accumulator piston absorbs apply fluid as it strokes, the final buildup of pressure in the clutch circuit is more gradual. This cushions or delays the full application of the clutch and smooths its engagement.

ACCUMULATOR OPERATION WITHIN A BAND SERVO APPLY CIRCUIT

Figure 11-31 shows an accumulator's operation in an intermediate band servo apply circuit. When the vehicle reaches given speed, the shift valve opens and supplies mainline pressure to the intermediate (second-gear) servo; part of this fluid diverts to the 1-2 accumulator. In other words, fluid flows to both the servo and accumulator pistons at the same time.

This causes the pressure in the entire apply circuit to increase more gradually than it would if it acted only on a single piston. As a result, the servo piston does not lock the band tightly around the drum until fluid flow has bottomed both pistons in their bores and mainline pressure reaches its maximum within the circuit. This cushions or delays full band application around the drum.

The cushioning of a band or clutch by accumulator action can be modified under certain conditions by the application of auxiliary pressure to the spring side of the accumulator piston. In Fig. 11-31, the auxiliary pressure comes from the throttle valve. The throttle pressure assists the spring force on the piston and causes the accumulator to provide more or less cushioning action under different vehicle operating conditions.

SUMMARY

1. Most relay valves are in the valve body.

2. A relay valve directs the fluid flow along a given path or circuit.

3. There are two types of relay valves: mechanically activated and hydraulically operated.

4. The manual valve is a mechanically operated relay valve.

5. A manual valve has a number of lands and grooves that control fluid flow through a number of port openings leading to multiple-disc clutches, bands, servos, or other hydraulic devices.

6. As a manual valve moves, its lands open and close various ports.

7. Every fully automatic transmission has one or more shift valves.

8. The shift valve automatically controls the upshift and downshift from one gear ratio to another.

9. The number of shift valves found in the valve body depends on the number of forward gear ratios the transmission produces.

10. A shift valve fits very closely in its bore and moves back and forth on a pressurized film of fluid.

11. A shift valve is a balanced type, and the entire shift-valve train consists of a governor plug, calibrated spring, and the shift valve itself.

12. The calibrated shift valve spring attempts to keep the valve in the closed position.

13. The governor plug has pressure acting on it, which tries to move it and the shift valve open.

14. The shift valve has lands that cover and uncover the hydraulic ports required for an automatic upshift and downshift.

15. When a shift valve closes, hydraulic pressure is cut off from the upshift apply circuit.

16. As the shift valve opens, its lands permit fluid flow to the upshift circuits.

17. Governor pressure acting on the shift valve's reaction area opens the shift valve.

18. Throttle pressure assists the calibrated spring in an attempt to keep the shift valve in the closed position.

19. A lock-up converter requires a number of relay valves to activate it.

20. The lock-up valve prevents the converter from activating below vehicle speeds of 40 mph (65km/h).

21. The fail-safe valve only permits the converter to lock up if the front clutch is fully applied.

22. The switch valve controls the fluid flow that activates the lock-up piston.

23. During the lock-up phase of the converter, all three relay valves are open.

24. The hydraulic system of an automatic transmission contains one or more check valves.

25. A check valve permits the flow of fluid in one direction only.

26. There are three types of check valves used in an automatic transmission: the one-way ball check, two-way ball check, and the poppet.

27. The one-way ball check and the poppet valves permit fluid to flow in a circuit in one direction only.

28. The two-way ball check valve allows two circuits to use one passage.

29. The one-way ball check valve consists of a ball, seat, and stop pin.

30. The two-way ball check valve has a single ball, two seats, and no stop pin.

31. The poppet check valve consists of a flat disc and spring.

32. The orifice is a calibrated restriction that slows the buildup of pressure in one portion of a circuit.

33. An orifice reduces fluid flow in a circuit and creates a back pressure.

34. The accumulator cushions the application of a clutch or band by absorbing apply fluid.

35. A simple accumulator consists of a cylinder, piston, and return spring.

36. The accumulator piston may be the solid type, single diameter; spool type, single-diameter; or spool type, multiple-diameter.

37. The cushioning of a band or clutch by accumulator action can be modified by the application of auxiliary pressure to the spring side of the accumulator piston.

REVIEW

This section will assist you in determining how well you remember the material contained in this chapter. Read each item carefully. If you cannot complete the statement, review the section in the chapter that covers the material.

1. An on-off valve type that directs flow along a given path is the _____.

   a. orifice

   b. accumulator

   c. relay

   d. both a and c

2. The valve the driver activates to set up the operating conditions within the transmission is the _____.

   a. manual valve

   b. accumulator

   c. one-way valve

   d. 1-2 shift valve

3. The portion of a spool valve that opens and closes the port openings is the _____.

    a. grooves

    b. lands

    c. shank

    d. reaction area

4. The valve that is responsible for automatically upshifting the transmission is the _____.

    a. manual valve

    b. orifice

    c. detect valve

    d. shift valve

5. A three-speed fully automatic transmission will have _____.

    a. three shift valves

    b. two shift valves

    c. one shift valve

    d. one shift and one manual valve

6. The component of the shift valve train, mentioned in this chapter, that moves the shift valve against spring tension is the _____.

    a. governor plug

    b. throttle plug

    c. shift valve

    d. calibrated spring

7. The device that produces a pressure signal in proportion to vehicle road speed is the _____.

    a. throttle valve

    b. accumulator valve

    c. shift valve

    d. governor valve

8. An upshift occurs when a shift valve _____.

    a. opens

    b. closes

9. A pressure that is in proportion to engine speed or load is _____.

    a. line

    b. throttle

    c. downshift

    d. governor

10. The device that controls fluid flow in one direction only is the _____ valve.

   a. manual

   b. check

   c. throttle

   d. governor

11. The device that controls pressure sources from two different directions is the _____ valve.

   a. two-way check

   b. one-way check

   c. orifice

   d. manual

12. A calibrated restriction placed in a hydraulic circuit is called a(n) _____.

   a. valve

   b. ball-check

   c. orifice

   d. obstruction

13. The device that absorbs apply fluid to cushion a band or clutch application is a(n)_____.

   a. orifice

   b. servo

   c. modulator

   d. accumulator

14. The relay valve that prevents the lock-up converter from activating below 40 mph (65 km/h) is the _____.

   a. lock-up

   b. fail-safe

   c. switch

   d. both b and c

For the answers, turn to the Appendix.

CHAPTER 12

Auxiliary Hydraulic Signals and Their Control Devices

As stated in Chapters 10 and 11, a fully automatic transmission requires a number of auxiliary hydraulic pressures (signals). These signals are responsible for adjusting mainline pressure and varying the upshift and downshift sequences of the transmission to meet different conditions. In other words, the signals provide the vehicle with smooth shift patterns, which efficiently reflect the load on the engine, speed of the vehicle, and the will of the driver.

TYPES OF HYDRAULIC SIGNALS

The hydraulic system of the automatic transmission uses three types of signals. These are directed to appropriate areas of the shift valves for the purpose of altering shift patterns to match the various conditions mentioned above. Also, one or more of the signals acts on the regulating valve train to adjust mainline pressure so that it will always be high enough to prevent slippage within a multiple-disc clutch or between a band and drum. The three basic types of signals are throttle pressure, kickdown pressure, and governor pressure.

Throttle Pressure

Throttle pressure is a hydraulic signal, derived from mainline pressure; its intensity is always proportional to engine load. This signal is also known as **TV pressure** or **modulator pressure.** The use of either of the terms depends on the language employed by the vehicle manufacturer.

In any case, the signal is utilized for three purposes within the hydraulic system of the transmission. They are as follows:

1. It supplies an auxiliary force to the regulator valve train. The signal acts on a boost valve within the regulator train to adjust the mainline pressure to match driving conditions.

2. Throttle pressure acts on the spring end of a shift valve to control upshift and downshift points in relation to engine load or torque output. For instance, at cruising speed or during acceleration under light engine load and moderate throttle pedal depression, the engine's torque output is low. In this situation, the transmission can shift to a higher gear ratio sooner than it would under heavy load.

    However, during hard acceleration or hill climbing, engine torque output is high due to the more fully depressed position of the accelerator pedal. As a result, upshifts must occur at higher speed.

3. A throttle pressure signal can act on an accumulator to control the cushioning of a clutch or band application.

Kickdown Pressure

Kickdown, or detent pressure, as it is sometimes called, is also derived from mainline pressure. This signal operates by itself or in conjunction with throttle pressure and spring force to move a shift valve to the closed position, against the force of governor pressure. This action forcibly downshifts the transmission into a lower gear ratio.

Governor Pressure

The **governor signal** is obtained from mainline pressure and is always proportional to vehicle speed. The signal opposes both throttle pressure and spring force on the shift valve to assist in timing upshifts and downshifts, in relation to vehicle speed. Moreover, governor pressure is used in many cases as an auxiliary force on a number of supporting control valves within the transmission's hydraulic system.

THROTTLE VALVES

Function

If the governor pressure mentioned above were the only signal sent to the shift valve(s), it would open these units at the same vehicle road speed each time. The upshifting of the transmission at the same vehicle speed would not be a problem as long as the engine rpm and load were always the same at this moment, but they are not. For example, if the engine is pulling the vehicle up a grade, it has a greater load on it than on a level road. Consequently, the engine must operate at a faster rpm in order to carry the additional load. Now, if the transmission upshifts too soon, the engine will not have the lower gear ratios to multiply its torque and thus be able to operate at the increased rpm necessary for high torque output. As a result, the engine is not able to carry the increased load successfully.

To overcome this problem, the transmission must have some means to sense engine load. To perform this function, the transmission has a throttle valve incorporated into its hydraulic system. This **throttle valve** (Fig. 12-1) produces a signal that is proportional to engine load, and the resulting pressure acts on the spring side of a shift valve to assist the spring's tension in opposing governor pressure.

Since the throttle valve responds directly to engine load, a reduced load results in low throttle pressure; increased load equals high throttle pressure. Presented below are some important relationships to remember concerning the throttle valve and throttle pressure.

1. Low throttle pressure is a result of

    a. low engine load

    b. high manifold vacuum

    c. a reduced opening of the carburetor throttle valves

2. High throttle pressure is a result of

    a. high engine load

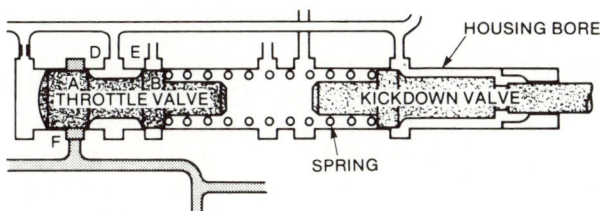

FIGURE 12-1   A typical throttle valve train.

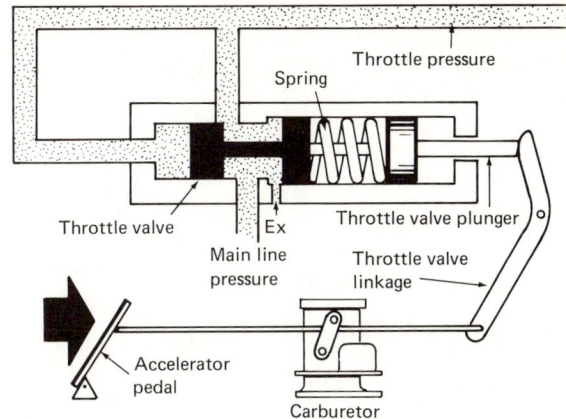

FIGURE 12-2   A throttle valve controlled by mechanical linkage.

b. low manifold vacuum

c. an increased opening of the carburetor throttle valves

Types

There are two types of throttle valves—those that are mechanically operated and those that are activated by engine vacuum. Figure 12-2 illustrates a throttle valve controlled by mechanical linkage; Figure 12-3 shows a similar unit controlled by a cable. In both installations, movement of the accelerator pedal is transmitted to the carburetor and then to the throttle valve. The resulting motion at the throttle valve alters the spring's tension. This, in turn, varies the amount of throttle pressure needed to balance the valve against spring force.

The main drawback to this system is that the linkage or cable requires periodic adjustment, otherwise a malfunction can occur. If, for instance, the linkage or cable is sufficiently out of adjustment to cause low mainline pressure, bands and multiple-disc clutches will begin to slip. If not corrected immediately, this slippage leads to premature transmission failure.

Figure 12-4 illustrates a throttle valve controlled by engine vacuum. In this unit, the flexing of the diaphragm alters spring force on the valve. The diaphragm moves in one direction or the other according to the influence of intake manifold vacuum and atmospheric pressure. The changing spring forces on the valve produce the differing amounts of throttle pressure.

Although the vacuum-operated throttle valve does not require any adjustments after the transmission leaves the factory, there are three common problems that result if the diaphragm malfunctions or it loses all or part of its vacuum source. If the diaphragm ruptures or completely

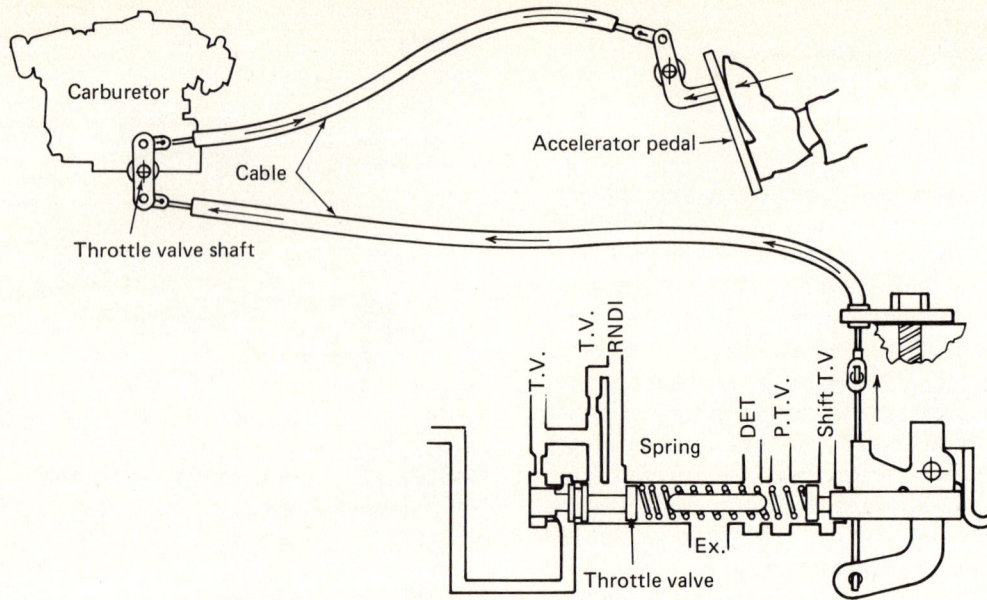

FIGURE 12-3   A throttle valve governed by a cable arrangement.

loses its vacuum signal, high mainline pressure results. The high pressure causes very harsh shifting and in many cases, the transmission will not automatically upshift.

A leaky diaphragm allows intake manifold vacuum to pull fluid from the transmission. This results in a very smelly exhaust that contains a lot of smoke. Also, the transmission loses a quantity of fluid; this loss may not become apparent to the driver until the transmission begins to slip, which causes damage to the multiple-disc clutches and/or bands.

A partial loss of vacuum to the diaphragm unit, due to a pinched hose or line, causes a temporary delay in mainline pressure buildup during rapid acceleration. The restricted hose or line prevents the vacuum unit from sensing any immediate change in manifold vacuum; consequently, the throttle pressure signal does not reflect the actual amount of vacuum the engine is producing. So the throttle pressure signal to the boost valve within the

regulator valve train does not have the effect of raising mainline pressure during hard vehicle acceleration. This can create slippage in bands and multiple-disc clutches on upshifts and especially during a forced downshift.

MECHANICALLY OPERATED THROTTLE VALVES

Design

Figure 12-5 shows a typical mechanically operated throttle valve train. This assembly consists of a kickdown valve, calibrated spring, and throttle valve, all of which operate in the same housing bore within the valve body. The kickdown valve moves in the bore by a mechanical force provided by the driver via the accelerator pedal. In

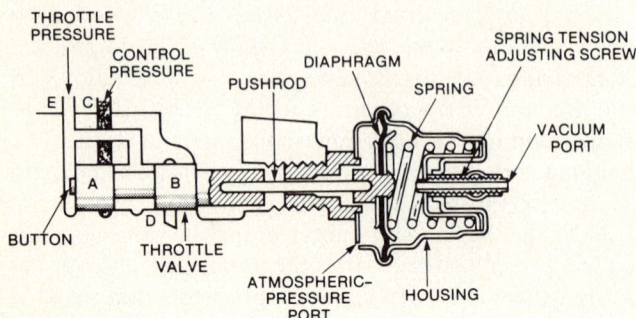

FIGURE 12-4   A throttle valve activated by engine vacuum.

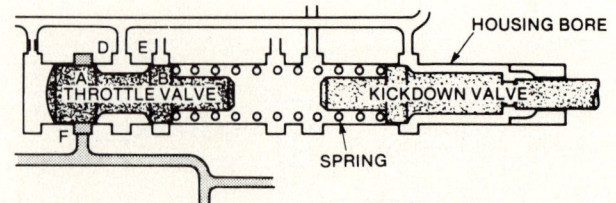

FIGURE 12-5   The design of a mechanically operated throttle valve assembly.

other words, the kickdown valve, by means of external linkage attached to the carburetor, moves in its bore as the driver depresses the gas pedal.

The kickdown valve itself performs two functions as it moves within its bore. First, as the driver depresses the accelerator pedal, the valve moves toward the left. This valve motion increases the tension of the spring, which fits between the kickdown and throttle valves. The increased spring tension causes the throttle valve to move against throttle pressure. Second, the kickdown valve has a hydraulic function, which will be covered later in this chapter.

The spring in this installation also serves two purposes. First, in the at-rest-accelerator-pedal position, the initial tension of the spring maintains the kickdown and throttle valves at the far ends of the bore. Second, since the throttle valve is a balanced type, the varying spring tension opposes throttle pressure to determine valve position in the bore.

The throttle valve itself has two lands of different lengths, but they both have the same diameter (Figs. 12–5 and 12–6). The back face of land A forms the reaction area upon which throttle pressure acts to balance the valve against spring force. Furthermore, land A controls the activity of port F. For instance, when the throttle valve cycles to the position where it produces no throttle pressure, land A blocks port F, which is the mainline pressure inlet (see Fig. 12–5).

Land B also performs several functions. For example, spring tension bears against the right-hand face of this land. Moreover, land B controls the opening and closing of port E, a sump port. As the driver releases foot pressure on the accelerator pedal, land B partially or fully opens the sump port to bleed off pressure in the throttle-valve circuit.

Operation

If the vehicle is stationary with its engine running at idle rpm, the throttle valve assumes the position as shown in Fig. 12-5. Under this condition, land A blocks mainline pressure into the valve at port F. Land B blocks the sump port E and only a slight balancing pressure exists in the throttle-valve circuit.

As the driver depresses the accelerator pedal, the linkage moves the kickdown valve to the left. Because the kickdown valve moves, the spring, located between the two valves, compresses somewhat and forces the throttle valve to the left. This action permits land A to crack open the mainline pressure port F.

Mainline pressure at port F causes fluid to pass through the annular groove and out port D to create throttle pressure (Fig. 12-6). Some of this fluid passes through the orifice and acts on the reaction area of the throttle valve. The resulting force, working on the reaction area, attempts to stabilize the valve's movement caused by the spring's tension. In other words, the amount of throttle pressure in the circuit and on the reaction area will be whatever it takes to balance the valve against spring force. Obviously, as the operator continues to depress the accelerator pedal, throttle pressure increases even more in order to balance the valve against the heavier spring force.

If at any time the driver depresses the accelerator pedal to the wide-open position, the throttle valve moves as far over to the left as it can (Fig. 12-6). This action permits land A of the throttle valve to fully open port F. Now, mainline pressure causes as much fluid flow as possible into the throttle pressure circuit. In this situation, throttle pressure will reach the same intensity as mainline.

As the driver releases the accelerator pedal so the engine returns to idle, the throttle valve cycles back to the position as shown in Fig. 12-5. With no outside force on the kickdown valve, the spring pushes it back to the left, which also reduces spring force on the throttle valve. Because of the reduced spring tension, throttle pressure, working on the valve's reaction area, immediately pushes it to the right. This action causes land A to block the entrance of fluid through port F.

In addition, land B opens the sump port and throttle pressure begins to bleed off to the reservoir. Pressure rapidly decreases in the circuit until it balances the movement of the valve against the remaining spring force. At this point, the throttle valve moves slightly over to the right and closes the sump port E and the throttle valve train has finished its return cycle to the closed (idle) position.

There are two concepts to remember about a throttle valve's operation. First, the amount of throttle pressure will always be in proportion to the tension of the spring. If spring tension is weak, throttle pressure is low; increased spring force raises the signal. Second, throttle pressure can increase until it reaches the same amount as mainline, but it can go no higher.

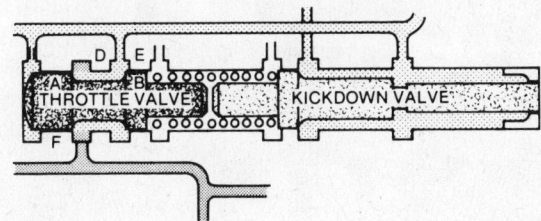

FIGURE 12-6   The mechanically operated throttle valve in the full-open position.

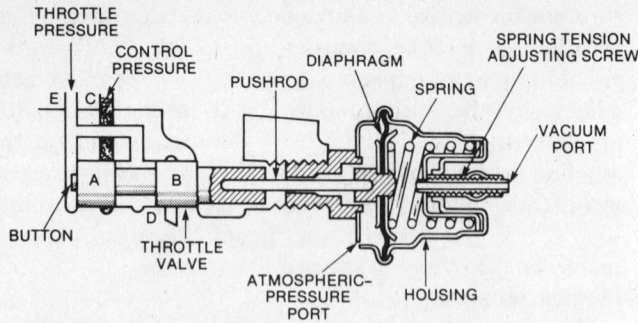

FIGURE 12-7 A typical vacuum-controlled, throttle-valve train.

VACUUM-CONTROLLED THROTTLE VALVE DESIGN

Throttle Valve

Figure 12-7 illustrates a typical vacuum-operated, throttle-valve train. This assembly consists of a throttle valve, pushrod, and vacuum-diaphragm unit. The throttle valve is a balanced spool type, which operates, in this example, in a special bore located in the transmission case.

This throttle valve has two lands and one annular groove. Land A has two functions. First, it controls the opening and closing of port C, which is the mainline (control) pressure inlet. Second, the left end of land A forms the reaction area for the valve. In other words, throttle pressure, acting on this face, moves the valve against the force of the spring. The button on the end of land A keeps the valve from bottoming in its bore and creates a space into which throttle pressure can enter when the valve is in the full left-hand position in its bore.

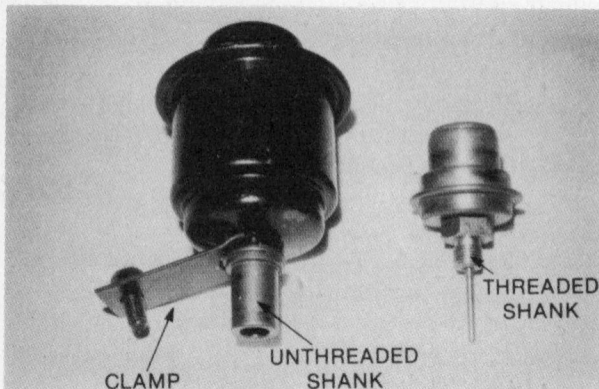

FIGURE 12-8 The threaded and clamp-on designs of vacuum-diaphragm assemblies.

Land B controls fluid movement through port D, a sump port. For instance, if land B opens port D, throttle pressure bleeds off to the reservoir. When port B blocks the sump port, throttle pressure increases in its circuits.

The groove, between lands A and B, channels the fluid from one port to another. For example, when the spring tension forces the throttle valve to the left, fluid passes from port C to port E via the annular groove. But as the valve moves toward the right, land B opens the sump port and land A closes the control pressure port. As a result, throttle pressure in the circuit bleeds off to the reservoir via port E, the annular groove, and sump port D.

Push Rod

The right-hand end of the throttle valve has a recess. This recess accommodates the push rod, which connects the throttle valve to the vacuum diaphragm. With this arrangement, any diaphragm or valve movement is transmitted immediately to the other component.

Vacuum-Diaphragm Assembly

The vacuum-diaphragm assembly consists of a housing, diaphragm, and spring. The manufacturer forms the housing in two sections using stamped sheet metal. Then after inserting the spring and diaphragm between the two halves, they are crimped together.

Each section of the housing also has a port designed into it. The left side (Fig. 12-7) has an atmospheric-pressure port that just vents this section to outside air. The other section has a vacuum port, which a line and hose connect to the engine's intake manifold.

Designed into the housing is a means of securing the entire assembly to the transmission case. One method, commonly used by manufacturers, is to secure a threaded, machined shank to one end of the sheet metal housing (Fig. 12-8). The threads on the shank mate with similar ones machined into the transmission case. The other method incorporates the use of an unthreaded shank, which has slots machined into it. A U-shaped clamp fits into the slots, and a bolt secures the clamp, shank, and housing onto the transmission case.

The diaphragm is a sealed, flexible piece of material made from chemically treated fabric and Neoprene rubber (see Fig. 12-7). The manufacturer installs and secures the diaphragm inside the housing so that it seals one side of the housing from the other, but it has sufficient flexibility to move slightly back and forth. Also, fastened to both sides of the diaphragm at its center is a metal piece, which is flat on one side but has an extended shank on the other.

The metal piece reinforces the center of the diaphragm

so that this section can withstand the two forces applied to it. The two forces, in this case, will be spring tension and throttle pressure.

The spring exerts its force on the right side of the diaphragm. To do this, the spring fits between the center of the diaphragm and one end of the housing; it exerts its spring force against both of these areas.

The second force, applied to the center of the left side, is that of throttle pressure. Throttle pressure, in this situation, first acts on the throttle valve itself which, in turn, applies a force to the push rod. And since the push rod also bears against the center section of the diaphragm, it receives the force of the pressure.

As previously mentioned, the diaphragm assembly has a spring installed between the diaphragm and housing. The spring's force works against throttle pressure to balance the throttle valve. In addition, the spring functions on the vacuum side of the diaphragm, and the intake manifold vacuum, acting on it, will reduce the spring's force on the throttle valve. And, finally, in some installations, the manufacturer installs an adjustment screw in the housing, which can increase or decrease initial spring tension on the diaphragm and throttle valve. Consequently, by moving this adjustment screw, a mechanic can increase or decrease throttle pressure.

OPERATION OF A VACUUM-CONTROLLED THROTTLE VALVE

Zero Vacuum Phase

Without the diaphragm in operation, the throttle valve is a simple balanced valve with throttle pressure balancing spring force (see Fig. 12-7). The spring forces the throttle valve to the left, which permits land A to open port C. Now, control or mainline pressure causes fluid to flow through the annular groove and out of port E to create throttle pressure. This pressure, after filling the circuit

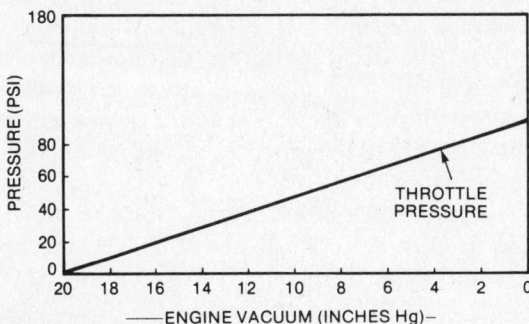

FIGURE 12-9 A chart of the engine's vacuum effect on throttle-valve pressure.

with fluid, immediately begins to push against the reaction area of the throttle valve and attempts to balance the valve against spring force.

If throttle pressure rises too high, it pushes the valve to the right, which causes land A to close port C and land B to crack open sump port D. As the TV pressure begins to bleed off to the reservoir, its force on the reaction area decreases. The spring pushes the valve to the left, which again opens port C and closes port D. Thus, throttle pressure balances the valve against spring force, and the amount of TV pressure, which in this case is necessary to balance spring force, will be as high as mainline pressure. Consequently, with no vacuum action of the diaphragm, mainline and TV pressure have the same intensity.

High Vacuum Phase

But as previously mentioned, vacuum on the diaphragm reduces spring force. For example, when the engine is running without a load, intake manifold is high. This vacuum is effective against the spring side of the diaphragm, and atmospheric pressure is active against the opposite side. Therefore, with high engine vacuum, atmospheric pressure pushes the diaphragm over to the right compressing the spring.

In effect, the high vacuum reduces the force of the spring on the throttle valve and therefore reduces TV pressure. This occurs because with a reduction of spring force on the throttle valve, TV pressure on the valve's reaction area pushes it to the right. This action permits land A to block port C and land B to open port D. As a result, TV pressure drops to zero or to whatever amount it takes to balance the reduced spring tension.

Low Vacuum Phase

As the engine is put under a load, its vacuum drops in proportion to the applied load and the spring offsets more of the atmospheric pressure's effect on the diaphragm. As a result, the diaphragm and throttle valve have a higher effective spring force that moves them both back to the left. This new position produces a higher throttle pressure in order to balance the valve against the increased spring force.

In effect, the increased spring force pushes the valve far enough toward the left to permit land A to open port C and land B to block port D. Now, TV pressure will increase until its force on the valve's reaction area balances the increased spring tension. In other words, the TV pressure is inversely proportional to the engine's vacuum. When engine vacuum is high, TV pressure is low; if engine vacuum is low, TV pressure is high (Fig. 12-9). Consequently, throttle pressure varies from 0 psi at 20 inches of vacuum to whatever mainline pressure is at zero vacuum.

NONCOMPENSATED VACUUM-DIAPHRAGM ASSEMBLIES

The vacuum diaphragm assembly just described is a noncompensated unit. Noncompensated means that the device has no method of adjusting for increases in altitude, and altitude changes do affect engine and transmission performance. For example, with increases in altitude, atmospheric pressure drops while engine vacuum remains relatively unchanged. Not only does this affect engine performance by reducing the engine's ability to pull in an air/fuel mixture but causes a noncompensated diaphragm assembly to produce a higher than normal throttle pressure.

Higher than normal TV pressure at this time causes several changes in transmission operation. First, the increased pressure, along with the loss in engine power, raises the shift points too high; they should occur at or near an rpm where the engine produces maximum torque output. Instead, the upshifts occur after the engine reaches maximum torque and it begins to drop off. Second, higher TV pressure raises the mainline pressure in the whole system, which makes clutch or band operation very harsh. In other words, the higher overall pressure increases the harshness of all upshifts and downshifts of the transmission at high altitudes.

ALTITUDE-COMPENSATED, VACUUM-DIAPHRAGM ASSEMBLIES

Design and Operation

To offset the effects of high altitudes on transmission performance, many manufacturers install an altitude-compensated, vacuum-diaphragm unit on their units (Fig. 12-10). Along with other components of a noncompensated unit, the compensated device has a bellows attached between the diaphragm and the housing. The bellows is made of pleated metal and resembles the pleats of an accordian. During the forming process, air is removed from the bellows, and it is then sealed against the entrance of atmospheric pressure. In other words, the inside of the bellows is under a vacuum.

Because the pressure inside the bellow is less than atmospheric, it tends to collapse or squeeze the bellows. Since the bellows is stretched and then inserted into the assembly, the effect of atmospheric pressure on the bellows makes it pull on the diaphragm with what is known as a collapsing force.

At sea level, a throttle valve train, using the compensated-diaphragm assembly, produces the same amount of TV pressure at any given engine vacuum as the noncompensated unit. The reason for this is that, at sea level, the spring force and the bellows' collapsing force on the diaphragm are equal to the spring's tension on the diaphragm of a noncompensated assembly.

At higher altitudes, the compensated valve train produces less throttle pressure. This reduction is a direct result of lower atmospheric pressure acting on the bellows, thus it does not have as much collapsing force. Consequently, the bellows expands and assists the reduced atmospheric pressure on the diaphragm in decreasing the spring force on the throttle valve.

As a result, throttle pressure decreases in relation to the

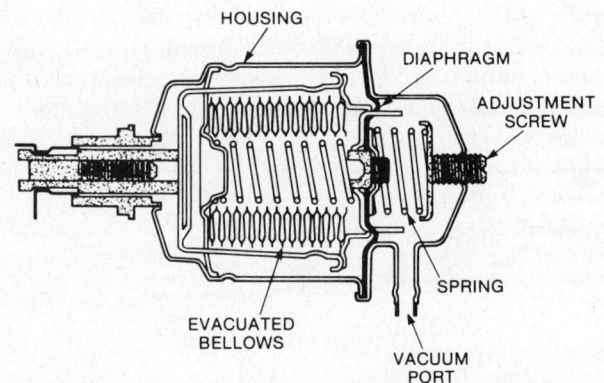

FIGURE 12-10   An altitude-compensated, vacuum-diaphragm assembly.

increase in altitude above sea level. For instance, at sea level, with an engine producing 18 inches Hg of manifold vacuum, mainline pressure of a typical noncompensated and a compensated unit will be the same, about 60 psi (414 kpa). At 5,000 feet, however, the noncompensated assembly will adjust mainline pressure to 70 psi (482 kpa). At the same altitude, the compensated device will adjust mainline pressure to 60 psi (414 kpa), or to about the same amount as at sea level.

As a result, throttle pressure decreases in relation to the increase in altitude to make the shift feel and upshift points comparable with sea level conditions. In other words, the engine's loss of power at high altitude is offset by a lower TV pressure, which causes the transmission shifts to remain as smooth as they were at sea level.

DUAL DIAPHRAGM ASSEMBLIES

Function

With the introduction of exhaust gas recirculation (EGR systems) in motor vehicles, those with vacuum-controlled, throttle-valve systems had to undergo some design changes. This became necessary to maintain not only the correct mainline pressure but also the shift points as well. Engines with an EGR system have a somewhat lower inherent manifold vacuum, which, as previously pointed out, causes a vacuum-controlled throttle valve to produce higher TV pressure. In turn, the high TV pressure creates late and harsh shifting.

Design and Operation

To overcome this problem, some transmissions have a vacuum diaphragm having a larger diameter, while others use the dual diaphragm unit shown in Fig. 12-11. Although the two diaphragms illustrated in the diagram have the same diameter, they operate on different sized areas. For instance, the left diaphragm is effective against the smaller area A, while the right diaphragm is effective against the larger area B. However, both diaphragms are part of one assembly, which bears against the spring on one side and the push rod on the other.

Note also in the illustration the two port openings, A and B. Port A connects to an intake manifold vacuum

source. Therefore, with the engine running, the spring side of the left diaphragm has engine vacuum applied to it. Port B connects via a line to the EGR circuit that uses ported vacuum, which is not nearly as high as engine vacuum. In any case, the ported EGR vacuum applies itself between the two diaphragms. This, in fact, creates a differential area that assists the engine vacuum on the left diaphragm and spring chamber.

During engine warm-up idle, or any closed throttle position, there is no EGR action and therefore zero vacuum acting on port B and the differential area. Under these conditions, engine vacuum acts on area A of the left diaphragm. As a result, atmospheric pressure entering the assembly on the right side of area B pushes the diaphragm assembly to the left, compressing the spring. The distance the diaphragm moves and compresses the spring depends on the strength of the engine vacuum signal.

During acceleration, the EGR system begins to function and a ported vacuum signal acts on the space between the two diaphragms. This vacuum signal acts on differential area B. As a result, atmospheric pressure provides an additional push against the spring. This action reduces mainline pressure to what it should be and corrects the late upshift characteristics of vehicles with EGR systems.

MECHANICALLY OPERATED KICKDOWN VALVES

Purpose

The manufacturer designs into every fully automatic transmission a hydraulic valve circuit, which permits the driver to downshift the transmission without moving the gearshift-selector lever into a lower driving range. In other words, by activating this valve circuit, the driver can force the transmission into a lower gear ratio for such things as passing another vehicle or climbing a steep grade. In some transmissions this valve circuit is the throttle valve, while in others a separate valve, known as kickdown or detent valve, accomplishes the task. These valves can operate mechanically, through linkage or cable, or electrically by means of a solenoid.

A mechanically operated kickdown valve can operate alone as it does when used in a hydraulic system that has a vacuum-operated throttle valve. Or, the valve can be part of the linkage-activated, throttle-valve train. Because both of these valve designs function in much the same way, this text only covers the latter type.

Figure 12-12 shows the kickdown and throttle-valve train presented earlier in this chapter. As mentioned, in this valve train, the kickdown valve did two things. First,

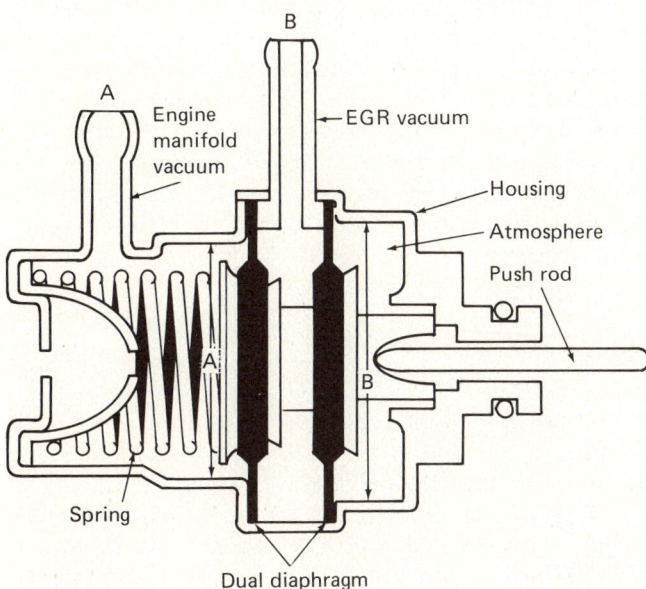

FIGURE 12-11 A dual-diaphragm modulator assembly.

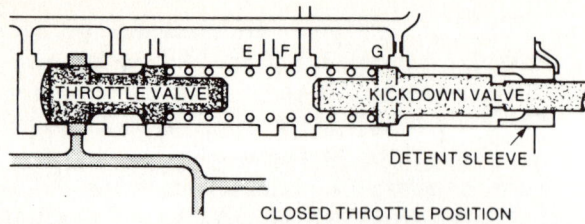

FIGURE 12-12 *The design of a kickdown and throttle-valve train.*

as throttle linkage moved the valve, it directly increased spring force that raised throttle pressure. Second, the valve performed a hydraulic function, which was to produce a kickdown pressure signal that, at certain road speeds, would downshift the transmission.

Design

To actually produce the signal, the kickdown valve utilizes only one land, which controls the fluid movement through three ports. For instance, at any given kickdown valve position, the land permits throttle pressure to push fluid through port G to fill the groove area on the right of the land itself. In addition, as the valve moves in as far as it can in its bore, the land opens port F, the kickdown pressure port (Fig. 12-13). This action connects ports F and G together. Lastly, as the kickdown valve moves back to the right, the driver has released pressure on the accelerator pedal, the land permits sump port E and kickdown pressure port F to combine; but it blocks port E from port G.

Operation

As the driver accelerates the vehicle, the kickdown valve moves in its bore and the throttle valve begins to produce TV pressure in its circuits (Fig. 12-12). One of these circuits directs fluid, via an orifice, to port G. Now, fluid enters this port and fills the groove area of the kickdown

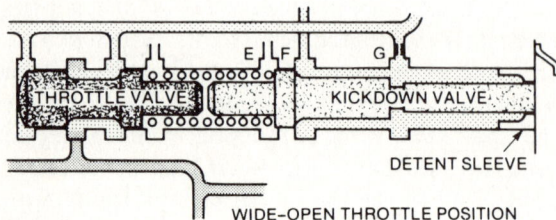

FIGURE 12-13 *The kickdown valve in the wide-open accelerator pedal position.*

valve on the right-hand side of its land. At the same time, port E is open to vent any residual kickdown pressure through port E to the reservoir. This valve and circuit activity continues until the driver fully depresses the accelerator pedal.

If the driver depresses the accelerator pedal to the full open or detent position, the kickdown valve land opens port F (Fig. 12-13). This action permits TV pressure to push additional fluid through port G, into the groove area, and out port F into the kickdown circuits. The fluid that enters the kickdown circuits is under kickdown pressure, and the circuits carry this pressurized fluid to the spring side of the shift valve, where its force assists the spring's tension in moving the valve to the downshift position.

Whenever the driver releases foot pressure on the accelerator pedal, the spring, between the two valves, forces the kickdown valve to the right (see Fig. 12-12). In this position, the valve land prevents fluid from port G from reaching port F. Furthermore, the land placement is such that port F fluid pressure begins to vent back to the reservoir via the open port E. As a result of the reduction in kickdown pressure on the spring end of the shift valve, governor pressure will once again open it and the transmission will upshift.

ELECTRICALLY OPERATED KICKDOWN VALVES

Purpose

The electrically operated kickdown valve performs the exact same function in the hydraulic system as the mechanical type just described. In other words, the valve train, when the driver pushes the accelerator pedal to the floor, also directs fluid under pressure to the shift valves to force them into the downshift position. But the electrically operated valve train does this task in a slightly different manner and requires more components in order to function.

Design

A typical electrically operated kickdown valve train consists of a detent switch, detent solenoid, hydraulic-detent valve, and detent regulator. The detent switch (Fig. 12-14) usually mounts directly on the firewall or the side of the carburetor, and throttle linkage activates the switch when the driver pushes the accelerator pedal all the way to the floor. As the linkage closes the detent switch, it completes an electrical circuit to the detent solenoid.

The detent solenoid is an electrically activated

hydraulic valve that controls the amount of mainline (control) pressure acting on the detent valve (Fig. 12-15). This solenoid consists of an electromagnet that has a coil energized by power from the detent switch. Inside the center of the coil is a spring-loaded, movable plunger; the manufacturer machines one end of it to form a valve that controls the activity of a sump port.

At any accelerator-pedal position except fully depressed, the detent switch is open, and the electrical circuit to the solenoid is broken. The tension of the plunger spring then forces the plunger down so that its valve contacts a seat, which is over the sump port. As a result, control fluid under pressure, directed to the detent valve, cannot leak off through the sump port to the reservoir.

If the driver fully depresses the accelerator pedal, the detent switch closes and the detent solenoid energizes (Fig. 12-16). Now, the electromagnet pulls the plunger up, compressing its spring, and the valve opens the sump port. As a result, the control pressure, acting on the detent valve, bleeds off to the reservoir.

The primary function of the hydraulic, detent valve is to move the detent-regulator valve against spring tension. As previously pointed out, if the detent switch deenergizes the solenoid, control pressure can build up and act against the left end of the detent valve, its reaction area (Fig. 12-15). This pressure forces both the detent and regulator valves to the right against spring tension. If, on the other hand, the solenoid energizes, control pressure on the detent valve bleeds off and the spring forces the regulator and detent valves to the left (Fig. 12-16).

The detent-regulator valve controls the amount of fluid flow that enters the detent (kickdown) pressure circuits; to do this, the valve has three lands that control

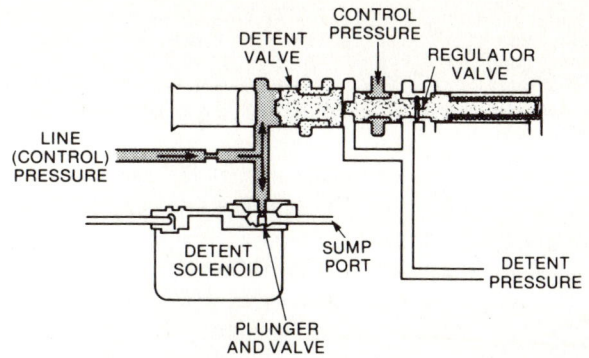

FIGURE 12-15   **The detent solenoid and its circuitry in the deenergized position.**

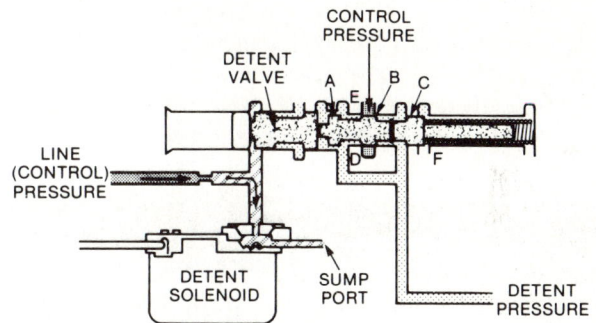

FIGURE 12-16   **The detent solenoid and its circuitry in the energized position.**

three port openings. For example, land A controls port D, which is the detent-pressure outlet port. If the regulator valve moves right (Fig. 12-15), land A blocks control pressure to this port. But as the valve moves left (Fig. 12-16), the land uncovers port D.

Land B has several purposes. For example, when the regulator valve moves left (Fig. 12-16), the land supports that end of the spool in the bore. In addition, the land has a drilled passage through it that carries fluid from the E port to the D port via the groove between lands A and B.

Finally, land C controls fluid flow through port F. For example, whenever the regulator valve moves right (Fig. 12-15), land C opens the sump port F, which permits detent fluid pressure from port D to pass through the annular groove, out port F, and to the reservoir. However, if the regulator valve moves left (Fig. 12-16), land C blocks port F and the detent pressure increases in its circuitry.

Operation

At any accelerator-pedal position except wide open, the electrical circuit to the solenoid is open and the detent-valve train produces no detent pressure to downshift the

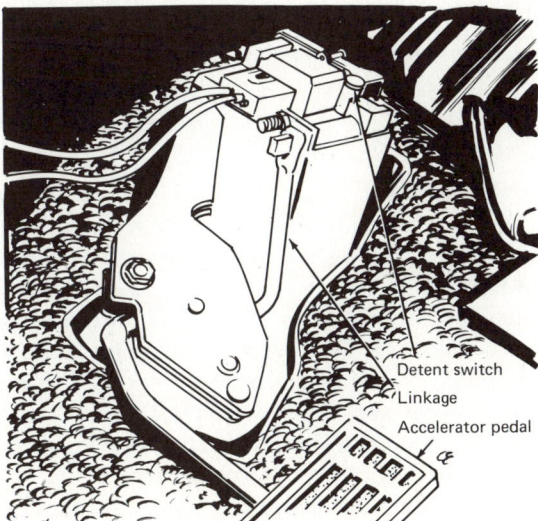

FIGURE 12-14   **A firewall-mounted, electrical detent switch.**

transmission (see Fig. 12-15). In this situation, the detent switch is open, which breaks the electrical circuit to the solenoid. The valve, located on the solenoid plunger, blocks off control pressure to the reservoir, which allows line pressure to force the detent valve to the right.

In turn, the detent valve pushes the regulator valve over against spring tension; as a result, the regulator valve produces zero detent pressure. Detent pressure is zero because land A now blocks fluid flow from port D. Also, land C opens the sump port F to port D, which reduces any residual detent pressure to zero.

As the driver moves the accelerator pedal to the fully depressed position, the detent switch energizes the detent solenoid and detent pressure increases (Fig. 12-16). Because the electrical circuit to the solenoid is complete, the plunger and its valve retract away from the seat covering the sump port. Now, control pressure on the end of the detent valve drops, which allows the detent-regulator valve spring to force both units to the left.

As a result of the regulator valve movement, land A opens port D, and land C blocks the sump port F. Now, fluid from port E enters the passage in land B, passes through the annular groove, moves out of port D, and travels through the circuits to the spring ends of the shift valves. At this time, detent pressure has the same intensity as control pressure; and at certain vehicle speeds, it is high enough to push the shift valves back against governor pressure.

Finally, in both types of kickdown systems, kickdown or detent pressure usually will not be able to downshift the transmission if the vehicle is moving at a very high rate of speed. In other words, the manufacturer provides the transmission with a method of preventing a high speed forced downshift either by regulating the intensity of kickdown or detent pressure or by the design of the shift valve itself. The manufacturer designs the system this way to protect the engine and transmission from damage, which can occur to them from an extremely high speed downshift.

GOVERNOR VALVES

Function

The governor assembly incorporates a valve that senses vehicle speed from the transmission's output shaft (Fig. 12-17). The valve, in turn, develops a hydraulic signal, which is in proportion to vehicle speed. This signal acts on the shift valve train to open it when the vehicle reaches a predetermined road speed.

In other words, the governor-valve signal changes as vehicle speed increases. When a vehicle is stationary or below about 10 mph, the governor pressure is zero or very

FIGURE 12-17 A single-stage governor assembly mounted on the output shaft.

low (Fig. 12-18). From about 10 mph on, the pressure signal increases in intensity until it reaches its maximum amount at higher vehicle speeds.

Types

The actual amount of governor-pressure signal increase, per mph of vehicle speed, depends on the type of governor used. For instance, a single-stage governor (Fig. 12-18) sends out a pressure signal that steadily increases with each mph of vehicle speed. The two-stage governor, in contrast, produces a pressure signal that rapidly grows at low vehicle speeds to a point and then slowly continues to increase with vehicle speed until it reaches its maximum intensity.

SINGLE-STAGE GOVERNOR

Design

Figure 12-19 shows a single-stage governor utilized on an early Ford two-speed transmission; this unit consists of a housing and plug-piston-type governor valve. The governor housing or body attaches to and rotates with the transmission's output shaft. Inside one of the extended sections is a bore for the governor valve. In addition, this bore has two ports machined into it—the inlet and exhaust. Lastly, the other, extended section forms a counterweight required to balance the rotating governor body.

The governor valve operates, within the bore in the governor body, by the balanced valve principle (Fig. 12-20). The two forces attempting to balance each other are mainline (control) pressure and centrifugal force. Control

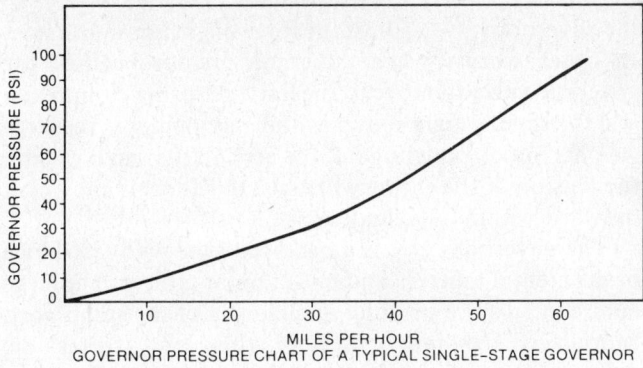

GOVERNOR PRESSURE CHART OF A TYPICAL SINGLE-STAGE GOVERNOR

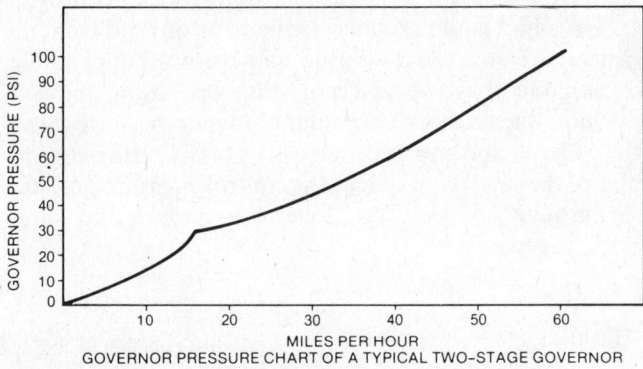

GOVERNOR PRESSURE CHART OF A TYPICAL TWO-STAGE GOVERNOR

FIGURE 12-18   A typical governor pressure curve for a single- and two-stage governor.

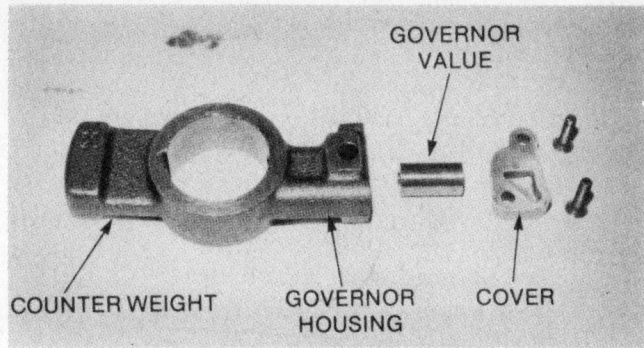

FIGURE 12-19   The components of a single-stage governor assembly.

pressure. Now, as the valve begins to move, it starts to close off the exhaust port.

With the exhaust port partially restricted, pressure develops in the governor circuit. The pressure increase is steady as long as the vehicle accelerates. When the governor pressure reaches a given amount, it opens the shift valve to upshift the transmission.

The governor's signal reaches its maximum as centrifugal force moves the plug valve past the exhaust port, thus sealing it off. At this point, the vehicle is moving at a high rate of speed and centrifugal force balances control pressure. Now, the governor valve produces a signal, which has the same intensity as control pressure.

pressure attempts to force the valve inward in its bore, while centrifugal force, which varies with the rotation of the output shaft, tries to move the valve outward in its bore against the force of control pressure. Finally, as these forces move the valve back and forth, it opens or closes the exhaust port.

Operation

When the vehicle is stationary, with the engine running and the transmission in forward-driving range, the governor valve assumes the position shown in Fig. 12-20. Control pressure, via an orifice, enters the governor body and tries to move the valve inward against centrifugal force. But since the output shaft is stationary, no centrifugal force acts on the valve. Consequently, control pressure forces the valve inward.

This valve movement opens the exhaust port wide open. Now, control pressure bleeds off to the reservoir. As a result, the governor valve produces a zero-pressure signal.

As the vehicle begins to pick up speed, the governor valve starts to move outward in its bore due to centrifugal force acting on it (Fig. 12-21). The centrifugal force attempts to balance the valve against the force of control

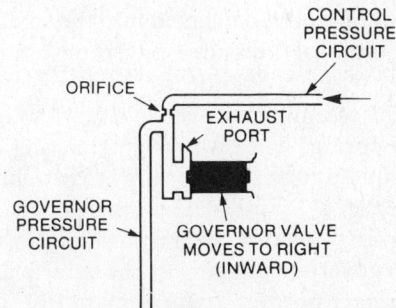

FIGURE 12-20   Single-stage governor valve operation with the vehicle stationary.

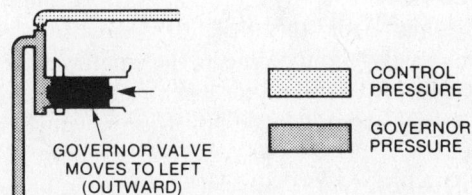

FIGURE 12-21   Single-stage governor valve operation with the vehicle moving.

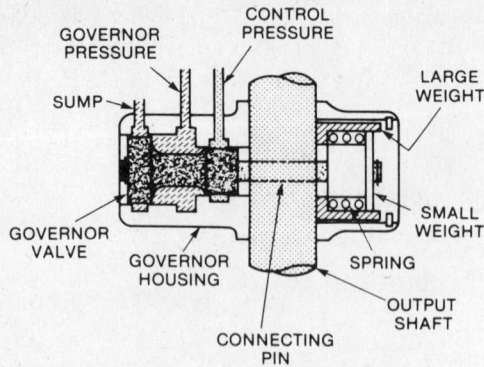

FIGURE 12-22 The design of a typical two-stage governor assembly.

TWO-STAGE GOVERNOR

Design

Figure 12-22 illustrates a two-stage governor for a three-speed automatic transmission. This unit consists of a housing, weight assembly, and spool-type governor valve. This type of governor housing also attaches to and rotates with the output shaft. In addition, the housing has two bores of different diameters. The large bore forms a guide for the weight assembly; the governor valve operates in the small bore. Finally, machined into the small bore are three ports: control pressure, governor pressure, and sump.

The weight assembly consists of a large weight, small weight, and spring. The two weights, acting as a unit, produce adequate governor pressure at low vehicle speeds to produce the 1-2 upshift. In other words, the two weights working together are responsible for the initial rapid rise in governor pressure as the vehicle accelerates from a stationary position (refer back to Fig. 12-18); the resulting pressure signal is high enough to open the 1-2 shift valve.

The small weight operates inside a cylinder within the large weight. At a vehicle speed of about 20 mph, the large weight cannot move any further in its bore and the small weight takes over the task of increasing governor pressure. Lastly, a pin connects the small weight directly to the governor valve so that when centrifugal force acts on the weight assembly, the governor valve will move in its bore.

The location of the governor spring is between the bottom of the small weight and a spring seat machined into the base of the cylinder within the large weight. The spring's tension will assist centrifugal force in balancing the governor valve against the force of governor pressure. In other words, when governor pressure causes the governor valve to move to the left, the spring compresses and the small weight moves within its cylinder in the large weight. But as centrifugal force acts on the small weight, the tension of the spring will assist this force in moving it outward within its cylinder.

The governor valve is a balanced-type with two lands of different diameters and one annular groove. The large land controls the opening and closing of the sump port. Moreover, the inner face of this land forms the differential reaction area upon which governor pressure applies its greatest force during the valve's operation.

The small land regulates the passage of fluid from the control pressure port to the annular groove. For example, during one phase of governor valve operation, the land position is such that the annular groove connects together the control- and governor-pressure ports. During another phase, the small land blocks the control-pressure port and the groove connects the governor-pressure and sump ports together.

Operation

Diagram A of Fig. 12-23 shows the position of a two-stage governor valve and weight assembly when the transmission is in a forward-driving range, but the vehicle is stationary. The small land blocks the entrance of fluid from the control-pressure port to the annular groove. The large land also cracks open the sump port that allows any governor pressure remaining in the circuits to bleed off to th reservoir via the annular groove.

As the output shaft begins to rotate and the vehicle starts to move forward, the governor weight assembly begins to move outward due to centrifugal force acting on both weights. As the weight assembly moves outward, it pulls the governor valve with it until the small, governor-valve land uncovers the control-pressure port (diagram B of Fig. 12-23). As the land uncovers the control-pressure

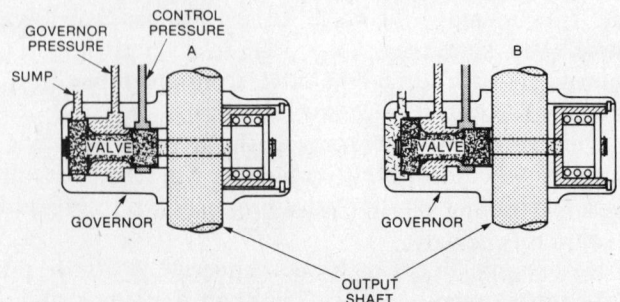

FIGURE 12-23 The operation of a two-stage governor assembly.

port, fluid begins to flow into the annular groove and out the governor-pressure port. When the governor circuit is full of fluid, pressure begins to increase rapidly.

At 15 to 20 mph, the large governor weight approaches the end of its travel in the governor housing, and the governor valve begins its balancing cycle. The increasing governor pressure begins to act on the differential reaction area of the left land, which is the larger of the two. The resulting force causes the governor valve to stabilize its movement momentarily. This action causes the small weight to move inward, compressing the spring, as the large weight comes to the end of its travel.

As the vehicle continues to increase in road speed, the governor valve moves more inward. The increased output shaft speed causes the small, inner weight, assisted by the spring, to move more outward in its cylinder within the large weight, bringing the governor valve with it. This valve movement causes the small land to open up the control-pressure port even more, which has the effect of increasing governor pressure.

At very high vehicle speed, the small weight pulls the governor valve to a position that opens the control-pressure port wide open. Now, at this point, control and governor pressure are equal in intensity because fluid from the control port has unrestricted flow to the governor port via the annular groove. Finally, the governor valve remains in this position as long as the small weight has sufficient centrifugal force acting on it.

At any point where the vehicle stops accelerating or reaches cruising speed, governor pressure stops increasing. This situation occurs because the governor valve becomes balanced between centrifugal and spring force on one side and governor pressure on the other. In other words, governor pressure, acting on the differential area of the valve, balances the effects of spring tension and centrifugal force on the small weight. As a result, the valve stops any further inward movement, which maintains the control-pressure port in a partially restricted condition. Consequently, governor pressure stabilizes to a fixed amount, which can be less than that of control pressure.

When the vehicle slows down, governor pressure decreases due to a reduction in centrifugal force. With less force on the weight assembly, governor pressure, acting on the differential area of the valve, immediately pushes it outward. This movement permits the large land to open the exhaust port. The amount of the exhaust port opening, of course, along with the resulting reduction in governor pressure, depends on how much centrifugal force and spring tension remain to balance governor pressure on the valve.

In summation, with a double-weight arrangement, the governor provides a rapid rise in governor pressure at low vehicle speeds. This rapid rise in pressure can open the 1-2 shift valve for a rather quick upshift from first to second

ratio. The smaller weight then takes over to gradually increase governor pressure, which will be necessary later for the 2-3 upshift at moderate to high speeds.

BALL-TYPE GOVERNOR

Design

Up to this point, this chapter has discussed both the single-stage and two-stage governor assemblies which use piston- or spool-type valves. However, there is another type of governor that uses two balls in place of the piston- or spool-type valve (Fig. 12-24).

This governor is a case-mounted assembly that consists of a governor shaft, secondary weight and spring, primary weight and spring, and primary and secondary check balls. The governor shaft has a gear mounted on one end that meshes with a drive gear on the output shaft; through this gear arrangement, the output shaft rotates the governor assembly within the transmission case.

The governor shaft also has a number of port openings. At its lower end are port openings for drive oil (mainline pressure) and governor oil pressure. Also, located under each of the check balls is an exhaust port.

Both the secondary and primary weights pivot on a pin that passes through the governor shaft. The **secondary weight** and **spring** control the operation of the primary check ball. The **primary weight** and **spring** govern the action of the secondary check ball. The secondary weight is lighter than its primary counterpart. Moreover the secondary spring has more tension than the primary.

The secondary check ball regulates governor oil (pressure) at low speeds; the primary check ball controls the pressure at moderate to high speeds.

Operation

When the vehicle is stationary with the engine running and the transmission in a forward-driving range, there is no governor pressure. Since the output shaft is not turning, there is no centrifugal force acting on the two weights. Therefore, their springs force them both outward. As a result, both ball check valves uncover the two exhaust ports.

In this situation, drive (mainline pressure) causes fluid flow into the governor shaft and governor pressure circuits. But the fluid bleeds off to the reservoir via the open exhaust ports. Thus, governor pressure is zero.

When the vehicle accelerates, the primary weight begins to move inward due to centrifugal force, but the secondary remains extended. As the primary weight

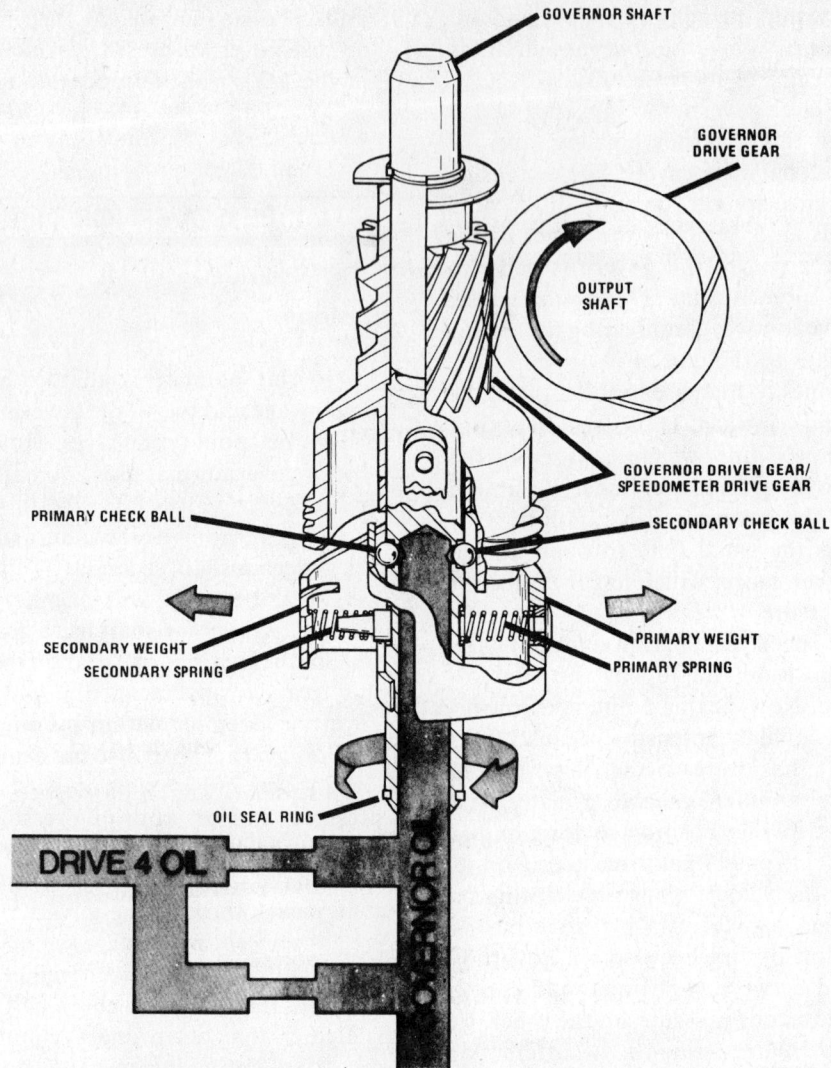

FIGURE 12-24 The design and operation of a ball-type, two-stage governor. (Courtesy of General Motors Corp.)

moves in, it begins to seat the secondary check valve, thus reducing the controlled hydraulic leak at this point. As a result, pressure begins to build up in the governor oil circuit.

At a given road speed, the primary weight is all the way in, and the secondary begins also to move inward. As it does so, the weight begins to seat the primary check valve. Governor pressure begins to increase even more as the primary ball seats because this action closes off the other sump port. When both ports are closed by the check balls, governor pressure has the same intensity as drive oil (mainline pressure).

The actual amount of governor pressure at any given time depends on the balancing effect of three forces on the check balls. For example, drive oil (mainline pressure), which, once it enters the governor shaft, becomes governor oil pressure, attempts always to unseat the check balls and bleed off via the vent ports to the reservoir. Resistance to the unseating of the check balls comes in the form of centrifugal force acting on both weights. However, this force is opposed by spring tension on both weights. So the position of the check balls is determined by the force of governor oil from below trying to open them, opposed by centrifugal force on the weights. The springs have the effect of offsetting centrifugal force on the weights, thus they actually assist governor pressure in opening the check balls by opposing the load placed on their upper surfaces.

TIMING VALVES FOUND IN THE HYDRAULIC SYSTEM

Manufacturers of automatic transmissions can use several types of timing valves to smooth out the operation of their units. These devices include such components as scheduling valves, shuttle valves, and timing valves. The main function of these valves, when used, is to help control the application and release of multiple-disc clutches and bands so that the transmission will shift smoothly under any engine operating load or vehicle speed condition. Because each manufacturer uses different types of timing devices, this text will not attempt to cover them in detail. When in doubt as to what a particular valve of this type does in a hydraulic system, refer to the technical manual for that automatic transmission.

SUMMARY

1. Auxiliary signals adjust mainline pressure and vary the upshifts and downshifts to meet different conditions.

2. The three types of auxiliary signals are throttle pressure, kickdown pressure, and governor pressure.

3. Throttle pressure is always proportional to engine load.

4. Kickdown pressure is used to forcibly downshift a transmission.

5. Governor pressure is always proportional to vehicle speed.

6. The throttle valve produces a pressure signal that is proportional to engine load.

7. Throttle valves can be mechanically or vacuum operated.

8. Linkage or cable operated throttle-valve systems require periodic adjustments, otherwise malfunctions can occur.

9. Vacuum-operated throttle valves require no external adjustments, but problems occur in the system if there is a loss of normal vacuum to the diaphragm or if it fails.

10. A typical mechanically operated, throttle-valve train consists of a kickdown valve, calibrated spring, and throttle valve.

11. Spring tension on the throttle valve increases or decreases as the driver moves the accelerator pedal.

12. Varying spring tension determines throttle-valve position and throttle-pressure intensity.

13. The typical throttle valve is a balanced unit with several lands, which control port activity.

14. The amount of throttle pressure will always be in proportion to the tension of the spring. If spring tension is weak, TV pressure is low. As spring force increases, TV pressure rises until it reaches the same intensity as control pressure, but it can go no higher.

15. A vacuum-controlled, throttle-valve train consists of a throttle valve, push rod, and diaphragm assembly.

16. The vacuum-operated throttle valve itself performs the same functions as its counterpart in the mechanically operated system.

17. The vacuum-diaphragm assembly consists of a housing, diaphragm, and spring.

18. The housing of the vacuum-diaphragm assembly has an atmosphere and a vacuum port.

19. The diaphragm is a sealed, flexible piece of material made of fabric and Neoprene rubber.

20. The function of the vacuum diaphragm is to reduce spring tension on the throttle valve.

21. With no vacuum acting on the diaphragm, throttle pressure will have the same intensity as mainline pressure.

22. High vacuum, acting on the diaphragm, reduces the force of the spring on the throttle valve and consequently lowers TV pressure.

23. High vacuum assemblies can be either noncompensated or compensated.

24. The compensated diaphragm assembly causes the throttle valve to produce lower throttle pressure when a vehicle is operating at high altitudes.

25. The compensating vacuum-diaphragm assembly uses a bellows to make up for changes in altitude on throttle pressure.

26. A dual-diaphragm vacuum assembly is used on EGR equipped vehicles to correct both a high mainline pressure and late upshifting condition.

27. The kickdown valve circuit permits the driver to downshift the transmission without moving the gearshift-selector lever.

28. The kickdown valve can operate by means of linkage, a cable, or electricity.

29. The mechanically operated kickdown valve opens by linkage or a cable as the driver pushes the accelerator pedal to the floor.

30. The electrically operated throttle-valve train consists of an electrical-detent switch, detent solenoid, hydraulic-detent switch, and detent regulator.

31. The detent switch that operates by mechanical linkage energizes the solenoid as the driver pushes the accelerator wide open.

32. The solenoid controls the amount of fluid flow going to the detent valve's reaction area.

33. The function of the detent valve is to move the regulator valve against spring tension.

34. The detent-regulator valve controls the amount of fluid that enters the detent-pressure circuits.

35. At any accelerator pedal position except wide open, the electrical circuit to the solenoid is open, and the detent-valve train produces no detent pressure to downshift the transmission.

36. As the driver moves the accelerator pedal to the fully depressed position, the detent switch energizes the solenoid and detent pressure increases.

37. The governor valve develops a pressure signal that is in proportion to vehicle road speed.

38. Governor pressure is zero or very low when a vehicle is stationary or below about 10 mph, but it then increases in intensity, with further vehicle acceleration, until reaching its maximum amount at high vehicle speeds.

39. A single-stage governor consists of a housing and governor valve.

40. A single-stage governor valve has two forces acting on it — control pressure and centrifugal force.

41. A two-stage governor consists of a housing, weight assembly, and governor valve.

42. The weight assembly consists of a large weight, small weight, and spring.

43. A pin connects the small weight to the governor valve. The small weight is responsible for opening the governor valve at speeds over 20 mph.

44. The governor spring assists centrifugal force in balancing the governor valve against the force of governor pressure.

45. The two-stage governor valve is a balanced unit with several lands and grooves.

46. The actual intensity of the governor-pressure signal depends on how much centrifugal force is acting on the weight assembly.

47. A two-stage governor produces a rapid rise in governor pressure at low vehicle speeds.

48. A ball-type governor consists of a governor shaft, secondary weight and spring, primary weight and spring, and primary and secondary check balls.

REVIEW

This section will assist you in determining how well you remember the material contained in this chapter. Read each item carefully. If you cannot complete the statement, review the section in the chapter that covers the material.

1. The valve that produces a hydraulic signal in proportion to vehicle speed is the _____.

   a. throttle valve

   b. kickdown valve

   c. detent valve

   d. governor

2. The output shaft of the transmission drives the _____ assembly.

   a. detent valve

   b. kickdown valve

   c. governor valve

   d. throttle valve

3. In a two-stage governor, centrifugal force acts on the _____.

   a. weight assembly

   b. governor valve

   c. spring

   d. connecting pin

4. The type of governor that produces a rapid rise in governor pressure at low vehicle speeds is the _____.

   a. single-stage

   b. two-stage

   c. three-stage

   d. both a and c

5. The valve that produces a pressure signal that is in proportion to engine load is the _____.

   a. governor valve   **c.** throttle valve

   b. kickdown valve   **d.** detent valve

6. In the throttle valve design discussed in this chapter, the component directly moved by throttle linkage was the _____.

   a. detent valve

   b. kickdown valve

   c. throttle valve

   d. spring

7. In throttle valve systems, the amount of throttle pressure always varies with _____.

   a. spring tension

   b. road speed

   c. manual valve position

   d. detent pressure

8. In a vacuum-operated throttle valve system, the diaphragm functions to _____ spring tension on the throttle valve.

   a. decrease

   b. increase

   c. remove

   d. stabilize

9. With zero vacuum on the diaphragm, TV pressure will be _____.

   a. low

   b. moderate

   c. zero

   d. the same as mainline pressure

10. At engine idle speed, the vacuum-operated throttle valve will produce _____ TV pressure.

    a. high

    b. low

    c. zero

    d. moderate

11. The vacuum diaphragm assembly, designed to operate the throttle valve effectively at both sea level and high altitudes, is the _____.

    a. modulated type

    b. noncompensated type

    c. compensated type

    d. adjusted type

12. The valve circuit that permits the driver to downshift the transmission without moving the selector lever is the _____.

    a. throttle   c. manual

    b. governor   d. kickdown

13. The kickdown valve produces its signal as the driver moves the accelerator pedal
_____ .

    a. partially down

    b. halfway down

    c. fully down

    d. to the idle position

14. The component of the electrically operated kickdown valve train that bleeds off control pressure to the detent valve is the _____ .

    a. regulator valve

    b. spring

    c. detent switch

    d. detent solenoid

15. As the driver fully depresses the accelerator pedal, the detent solenoid's plunger
_____ .

    a. remains stationary

    b. retracts away from the sump port

    c. extends toward the sump port

    d. vibrates

16. Dual-diaphragm vacuum throttle valves are used on some vehicles with a (an)
_____ system.

    a. EGR

    b. EEC

    c. air injection

    d. catalytic converter

For the answers, turn to the Appendix.

CHAPTER 13

Hydraulic System Diagrams

Up to this point, this text has covered a great deal of information concerning the design, construction, and theory of operation of the majority of all hydraulic system components. Let us now put this knowledge to use in a practical way and at the same time learn how to interpret hydraulic system diagrams.

This diagram is a map of an automatic transmission's hydraulic system. As such, it shows the various routes that fluid can take as it travels from one point in the system to another (Fig. 13-1). Because of this, these drawings are also commonly referred to as flow charts.

PURPOSE

The hydraulic system diagram is a very important and useful tool for the purpose of troubleshooting a malfunctioning automatic transmission because it provides the technician with a great deal of useful information. For instance, the chart shows the repairman all the different circuits of the transmission's hydraulic system. The diagram is, of course, only a graphic representation of all the circuits the manufacturer casts or machines into the various transmission components and, for practical reasons, it is not a true scale drawing.

If it were an accurate drawing, as to the length and direction of each circuit, the diagram would be quite large

and more difficult to interpret. Consequently, the charts show the mechanic that a transmission has a certain circuit along with the valving to control the flow of fluid within it, but the diagram will not inform him of the circuit's actual length or direction from one point to another. This makes the chart of little use to a mechanic in actually locating a given circuit physically in the transmission itself.

LAYOUT OF HYDRAULIC SYSTEM DIAGRAMS

Valving

As mentioned, the hydraulic system diagram provides a great deal of useful information; this is accomplished by means of its elaborate layout. For example, the chart shows all the valves used in the hydraulic system, and it illustrates each complete valve train in considerable detail. Figure 13-2 is a partial drawing of a chart showing a complete valve train with all its related components. This was taken from a diagram similar to the one shown in Fig. 13-1. The drawing clearly shows the number of lands, grooves, and reaction areas of the valve plus all the port openings and circuits, which the lands themselves control.

For the sake of simplicity, the charts usually show a valve train in either its closed or fully open position. The only exception to this general rule is in the case of regulating-type valves, which a chart usually pictures closed or in their regulating cycle position.

Pump

A complete diagram will also illustrate the hydraulic pump and its related circuitry (Fig. 13-3). This section of the schematic will show, in most cases, the type of pump employed by the hydraulic system, gear, rotor, or vane; the various port openings to the pump; and the number of circuits leading to and away from it. Furthermore, this portion of the chart will illustrate, but not in great detail, the sump and inlet screen, strainer, or filter of the system.

Hydraulically Operated Units

The last items, displayed on most hydraulic charts, are the hydraulic-operated units. In other words, a schematic usually shows all the band servos and clutch pistons necessary to apply the frictional components, which control the operation of the planetary gear train. And due to the complexity in design and operation of most servos over the clutch pistons, the drawings usually illustrate the

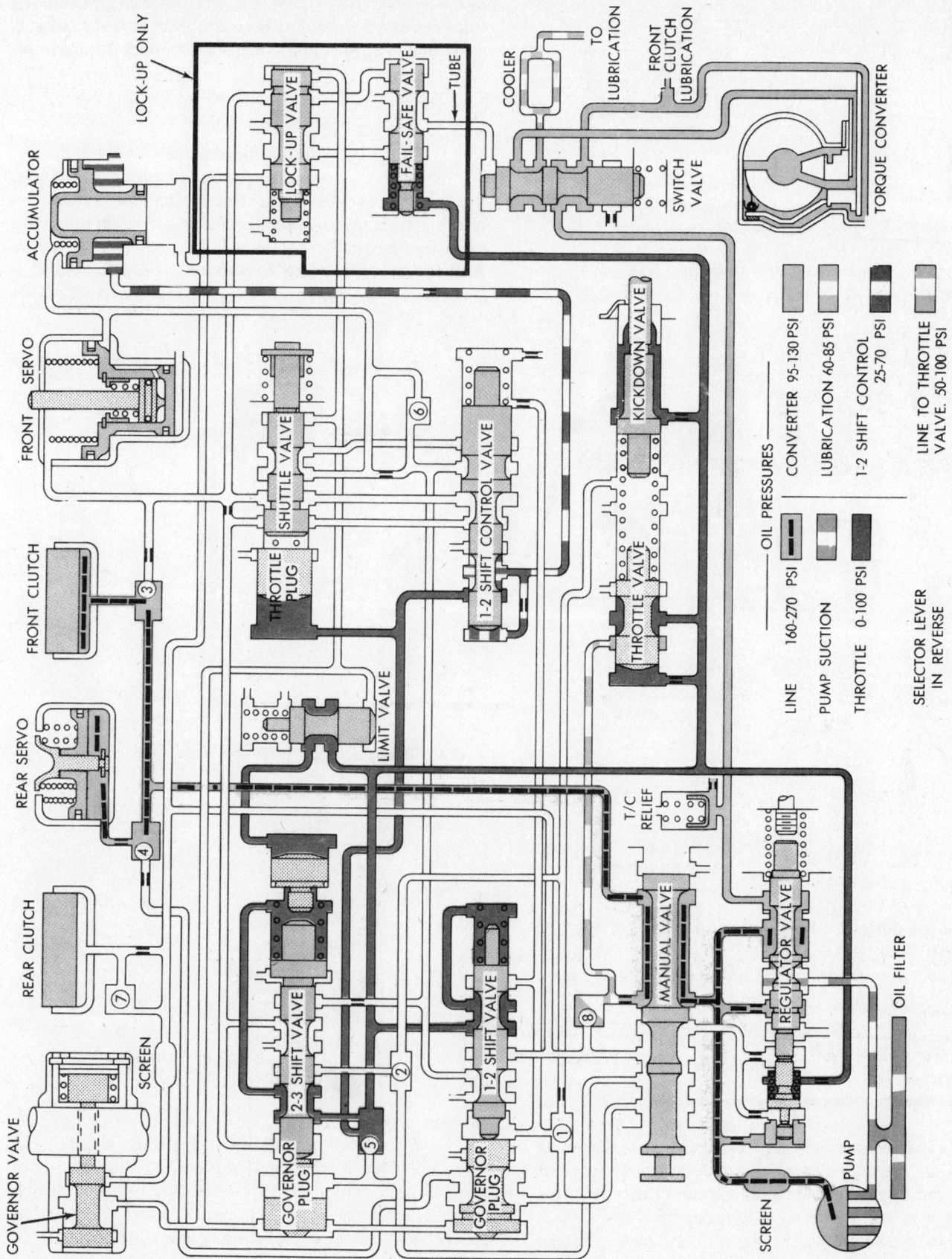

FIGURE 13-1 A hydraulic system diagram of a Chrysler 727 automatic transmission. (Courtesy of Chrysler Corp.)

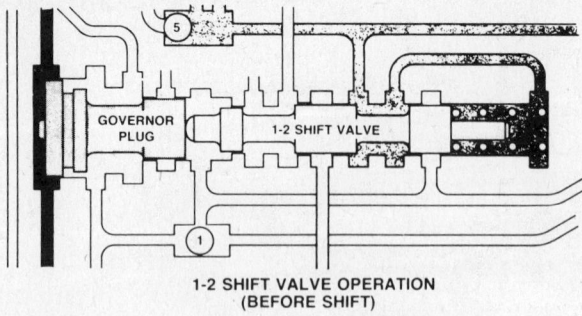

FIGURE 13-2 A drawing of a shift-valve train. (Courtesy of Chrysler Corp.)

transmission. In any case, the mechanic must know the standard symbols and at the same time be aware of and look for any special marks a given manufacturer may use.

Circuit

The standard symbols, used extensively by the industry, are those that represent the circuit, a circuit not in use, orifice, and exhaust vent. The symbol for a circuit is nothing more than two parallel unbroken lines, which run from one hydraulic component to another (Fig. 13-5). And as previously stated, these parallel lines do not necessarily represent the length or direction a circuit follows

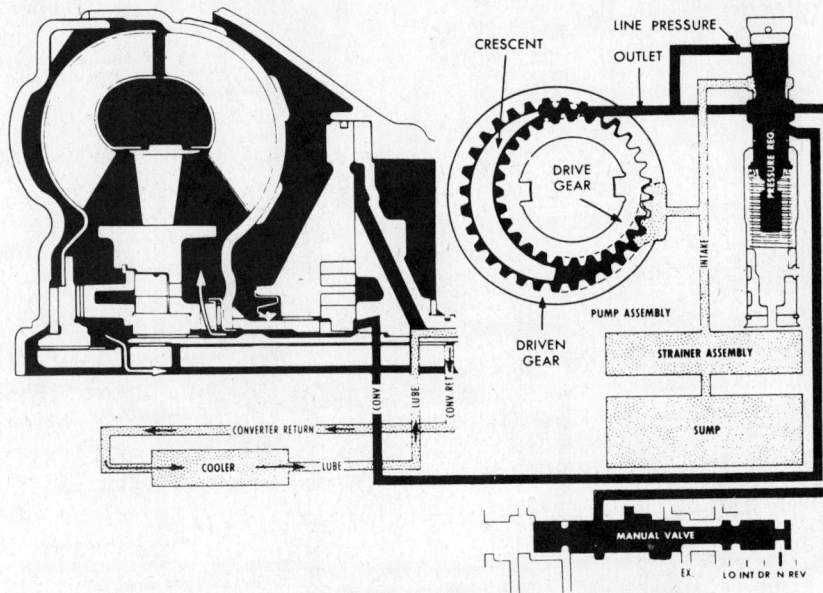

FIGURE 13-3 A partial diagram showing the pump and its circuitry. (Courtesy of Hydro-matic Division of General Motors Corp.)

servos in great detail — but not the clutch pistons (Fig. 13-4). Lastly, as in the case of the valve trains, the diagram shows the hydraulic-operated units in a single position — either off or in their fully applied position.

SYMBOLS USED ON HYDRAULIC DIAGRAMS

Before a technician can interpret a hydraulic diagram, he must understand the symbols used on that particular chart. Many transmission manufacturers use the same symbols or marks to represent certain devices and circuits on hydraulic diagrams. Others use their own symbols, which only relate to the chart for a particular type of

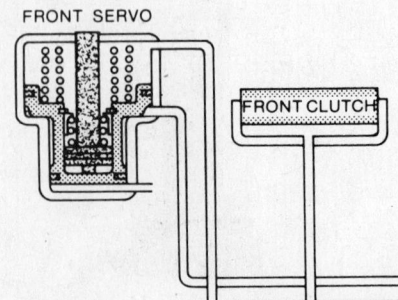

FIGURE 13-4 A partial hydraulic chart showing a servo and clutch piston.

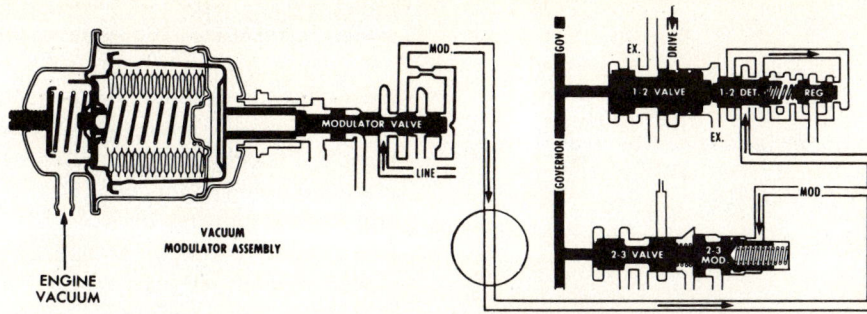

FIGURE 13-5   The two parallel lines within the circle represent the standard symbol for a circuit. (Courtesy of Hydro-matic Division of General Motors Corp.)

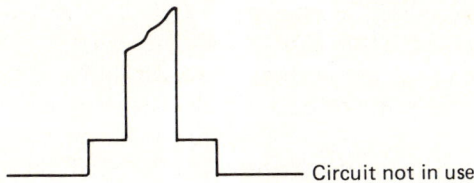

FIGURE 13-6   The symbol for a circuit that is not in use.

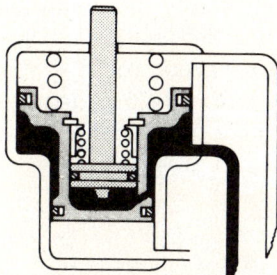

FIGURE 13-7   A partial diagram illustrating a servo assembly and a circuit that is not in use. (Courtesy of Chrysler Corp.)

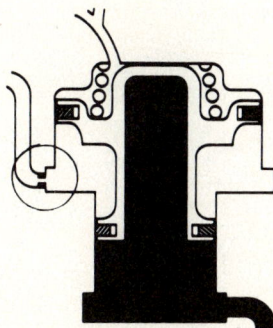

FIGURE 13-8   The mark, a circuit restriction within the circle, represents the standard symbol for an orifice. (Courtesy of Chrysler Corp.)

within the transmission itself. In other words, all the marks really show is that a circuit provides a path for the fluid from one unit to another.

Circuit Not in Use

Figure 13-6 illustrates a symbol for a circuit that is not in use. Manufacturers use this mark on a diagram for several reasons. For instance, the diagram may have this mark to show a circuit that an engineering modification has deleted, or a circuit machined into a transmission case for use with one particular valve body, where the manufacturer does utilize several different ones on the same unit. Also, the many manufacturers use the symbol on partial diagrams where they want to show only certain things such as in Fig. 13-7. This drawing shows the operation of a servo assembly; note that the circuit connected to the upper port opening is not in use.

Orifice

Figure 13-8 shows the symbol for the orifice. This mark, a restriction, appears in the circuit leading to the component affected by the action of the orifice. In the example, the orifice is in the circuit to slow down the operation of an accumulator.

Exhaust Vent

Figure 13-9 shows the marking for a sump or exhaust vent. In some instances, the manufacturer will place the word "vent" next to the symbol, or the letter "X" for exhaust. In any case, the symbol represents a port opening through which fluid will flow back to the sump or reservoir.

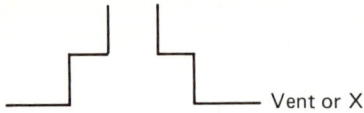

FIGURE 13-9   A symbol for an exhaust or sump vent.

Valving

There is not a standard symbol for any valve in the hydraulic system. Instead, each manufacturer uses a drawing of the valve itself. The reason for this action is that each different transmission design uses a certain number of special valves that perform certain tasks in that particular unit. Consequently, there can be no one symbol to represent any valve for all hydraulic diagrams.

A typical drawing of a 2-3 shift valve train, as it will appear on the hydraulic diagram, is shown in Fig. 13-10. Note that the drawing shows the complete valve train with its individual components. Also notice that this schematic shows all valve lands, grooves, and port openings very clearly. Lastly, to make it easier for the mechanic to locate the valve train on the completed diagram, the draftsman prints the name of each valve and, in some cases, its major components on or to one side of the unit itself.

Pump, Torque Converter, and Hydraulically Operated Units

There are also no standardized symbols for the hydraulic pump, torque converter, and the hydraulically operated units. All manufacturers use drawings to indicate these components on their respective diagrams, and the complexity of the part's design and operation determines how accurate is the artwork. For instance, Fig. 13-7 shows a line drawing of a servo. Note the detail in this illustration. This makes it easier for the technician to use the drawing to troubleshoot the unit if it happens to malfunction.

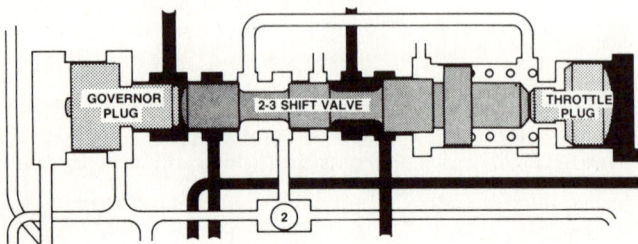

FIGURE 13-10   A drawing of a 2-3 shift-valve train. (Courtesy of Chrysler Corp.)

INDIVIDUAL CIRCUIT CODES

Because the hydraulic diagram incorporates all the transmission's many pressurized circuits, a mechanic would have a difficult time tracing a particular circuit if the manufacturer did not code them in some manner. Circuit codes are the means by which the draftsman designates one pressurized circuit from another. And the two most common methods used to code circuits are the use of (1) color or (2) lines, dots, and shade tones.

Color Method

The color-coded diagram is one in which the manufacturer uses a given color to represent the type of pressurized-fluid flow in a particular circuit. Figure 13-11 illustrates an example of typical color codes displayed in small boxes at the bottom of a hydraulic chart. In this particular example, blue is the color for mainline or control pressure; blue and white are the colors used for pump suction; converter pressure is yellow; and yellow-white are the colors used for the lubrication circuits.

Also note in Fig. 13-11 the pressure specifications next to the color-coded boxes. These specifications inform the technician of how much pressure should be in each circuit during a given phase of transmission operation represented by the schematic. For example, in this diagram of reverse ratio, the specifications are: line 230–270 psi, converter 10–75 psi, and lubrication 5–30 psi. However, in other cases, the specifications are not on

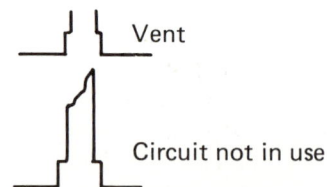

Oil pressures

Line 230-270 psi	Blue
Pump suction	Blue and white
Converter 10-75 psi	Yellow
Lubrication 5-30 psi	Yellow and white

Selector lever in reverse

FIGURE 13-11   The color-code chart used on a typical schematic.

the diagram itself but are found in tables located in a specific area of the vehicle's service manual. Finally, the mechanic should use these specifications as a guide when pressure testing the transmission for circuit malfunctions.

Pattern Method

Figure 13-12 shows the coding section of a hydraulic chart that has the pattern method of designating the different pressures found in the various circuits. Note that in this code different combinations of line or dot patterns represent control, lubrication, compensating, and converter pressures. Also, this diagram does not have the pressure specifications printed next to the coded boxes.

Finally, to make it easier for the mechanic to locate a hydraulic problem that occurs in a particular driving range, the manufacturer usually produces a single hydraulic schematic for each operating range of the transmission. In other words, the manufacturer makes a diagram for each forward-drive range, neutral, park, and reverse. Each of these charts only shows the mechanic what is occurring in the hydraulic system during that one particular phase of operation. By excluding from the diagram the codes of circuits that are not in use, the mechanic can more easily follow the chart and locate the malfunctioning component.

VALVE DEFINITIONS

In order to assist the mechanic in interpreting the hydraulic diagrams of an automatic transmission, the manufacturer will usually devote a section of the vehicle's service manual to define or describe the function of the various valves used in the hydraulic system. In some manuals, the diagrams will precede the hydraulic schematics—a much better practice.

The following valve definitions are typical of those found in an average service manual, and these describe the valves shown on the diagram in Fig. 13-13 along with the charts that follow it.

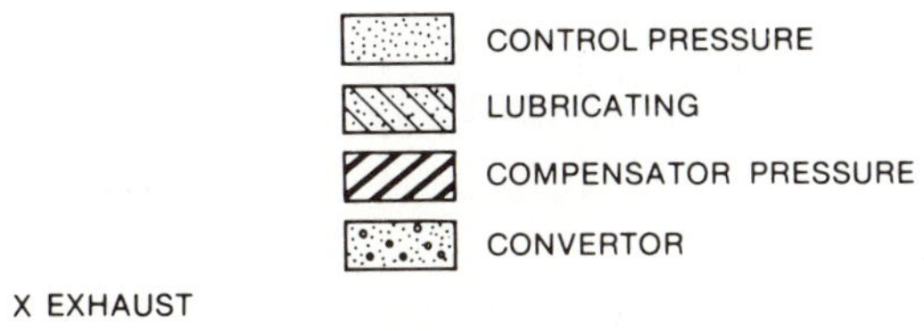

CONTROL PRESSURE
LUBRICATING
COMPENSATOR PRESSURE
CONVERTOR

X EXHAUST

FIGURE 13-12   *Pattern codes found on a hydraulic diagram.*

- **Torque converter relief valve:** A safety valve that will protect the torque converter from the buildup of excessively high internal pressure.

- **Pressure regulating valve:** This valve regulates the pressure of the pump's output and supplies the basic fluid pressure necessary in order to operate the transmission.

- **Manual valve:** The driver, by means of mechanical linkage, activates this valve; it, in turn, sets up the operating conditions within the transmission.

- **Governor valve:** It is a balanced valve that senses vehicle speed from the transmission's output shaft; the valve then sends a hydraulic speed signal to the shift valves, which control gear shifting in relation to vehicle speed.

- **Throttle valve:** This valve produces a pressure signal that is in proportion to engine speed or load; this signal acts on the shift valves to control gear shifting in relation to engine speed or load.

- **Kickdown valve:** A valve activated by full depression of the accelerator pedal; kickdown pressure from this valve will then cause the shift valves to close, which downshifts the transmission.

- **1-2 and 2-3 shift valve trains:** These valves trigger the automatic shifts, in response to governor and throttle signals, by directing fluid under pressure to the appropriate band or clutch that, in turn, causes the shift to occur.

- **1-2 shift control valve:** This valve aids in controlling the quality of the 1-2 upshift; it also aids in the timing and quality of the various 3-2 downshift ranges.

- **Accumulator:** The accumulator controls shift quality by absorbing the apply pressure of the rear clutch and front band.

- **Shuttle valve train:** This valve train helps to control the rate of front servo apply pressure during part-to-full throttle, 1-2 upshift and the downshift from 2-1.

- **Limit valve:** This valve prevents a part-throttle 3-2 downshift at high vehicle speeds.

- **Lock-up valve:** This valve senses vehicle speed via governor pressure and prevents the lock-up portion of the converter from activating below 40 mph (65 km/h).

- **Fail-safe valve:** The fail-safe valve is sensitive to throttle pressure, and it permits the converter to enter its lock-up phase only after the front clutch is fully applied.

- **Switch valve:** This valve controls the on and off fluid flow to the converter's lock-up piston.

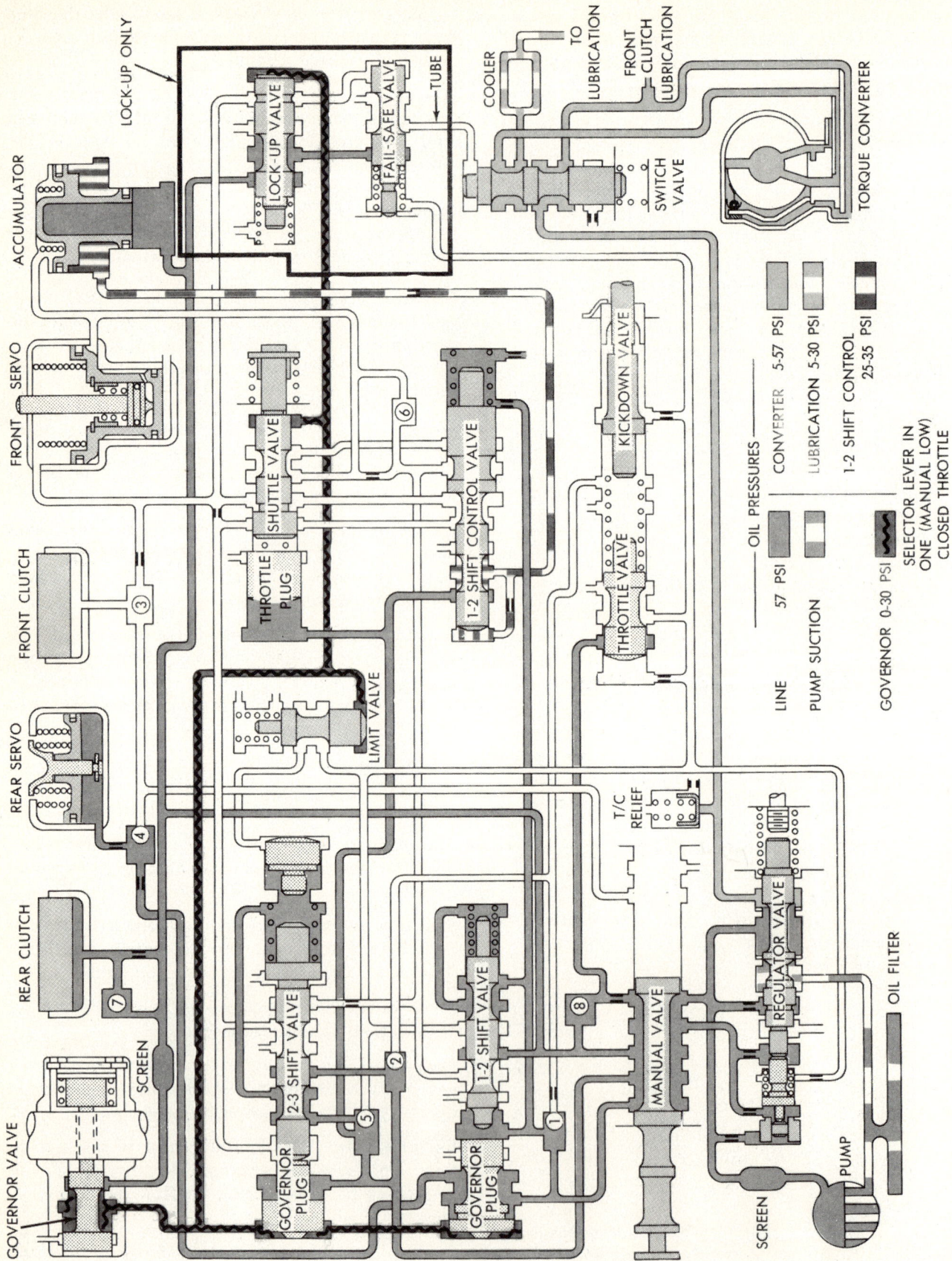

FIGURE 13-13 Knowing the function of each of these valves makes it much easier to interpret the total chart.

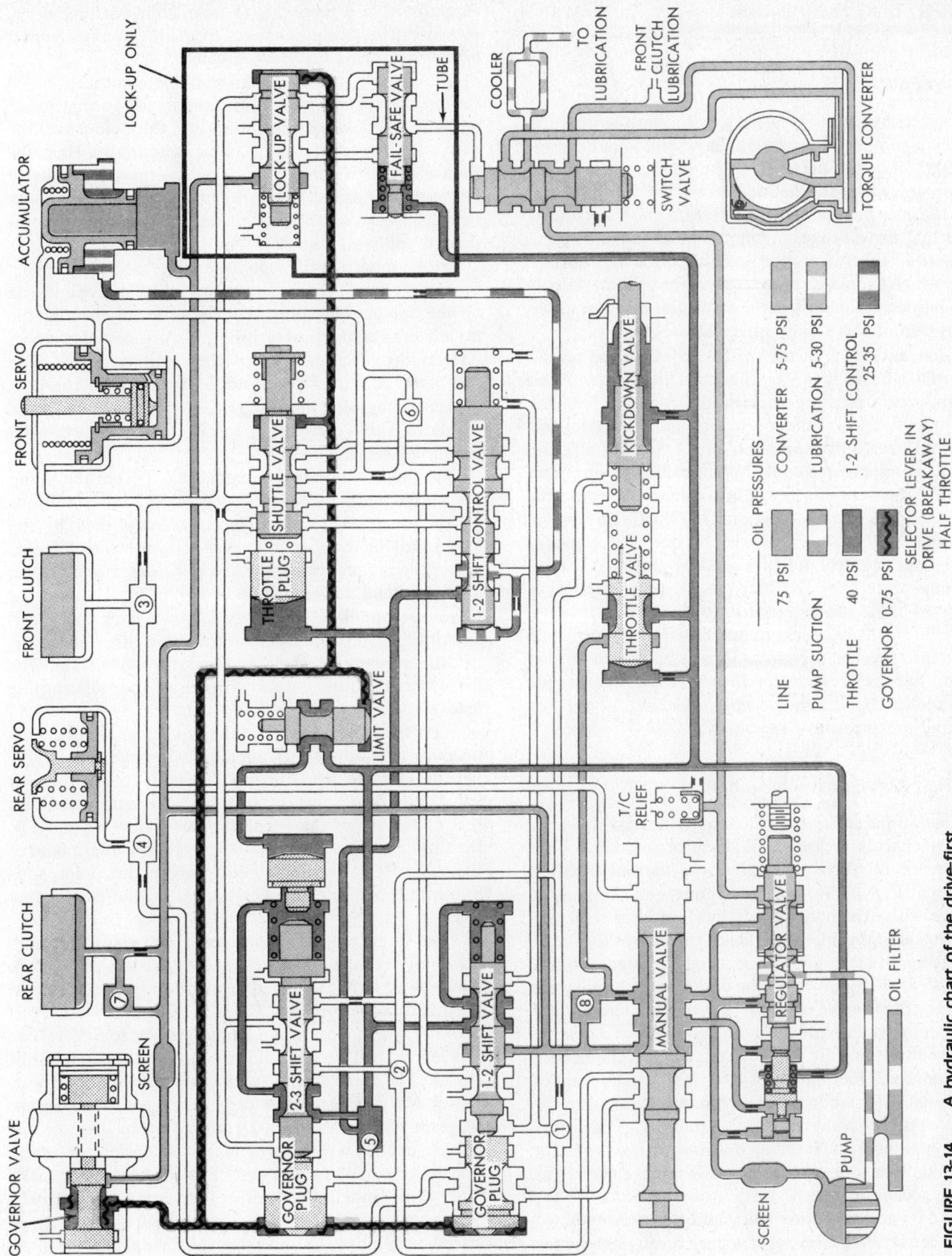

FIGURE 13-14 A hydraulic chart of the drive-first (breakaway) ratio for a Chrysler 727 automatic transmission. (Courtesy of Chrysler Corp.)

219

INTERPRETING A DIAGRAM

Steps to Follow

To be successful in interpreting a hydraulic diagram, the technician should follow several easy steps. First, locate the diagram(s) that presents the operating phase of the transmission where the malfunction occurs. For instance, if the transmission does not upshift from first to second in the normal drive range, the mechanic should study the **drive-second chart** to see the position of the appropriate valves, which should be open to cause the upshift. Also, it may be necessary to look at the **drive-first diagram** to see the position of these same valves before they open.

Second, take note of the special codes located on the diagram itself. In Fig. 13-14, the codes that represent the various circuit pressures are in the form of patterns consisting of various tones of shading and of vertical and horizontal lines. For example, a circuit that has all dark shading represents one containing throttle pressure; line or control pressure in a circuit is depicted by moderately solid shading. Pump suction is illustrated by a moderately shaded box followed by one that is white, while 1-2 shift control pressure is a dark-shaded box next to a solid-white one. The only circuit represented by a horizontal line is the governor. In this case, a dark wavy line against a shaded background illustrates the presence of this pressure in a circuit. Finally, notice that this diagram has the pressure specifications for each circuit, which the mechanic can use when checking the various operating pressures in the circuitry.

Reading a Drive-First Ratio Chart

Now that all the codes are understood, the next step is to study the chart to locate the fluid source and then follow the flow of oil through each of the various circuits. Looking at Fig. 13-14, note that pump suction begins at the filter where the pump draws fluid from the reservoir and moves it to the inlet port. This is shown by the code of alternating white and moderately shaded boxes. However, as the fluid leaves the outlet port (the pressure side of the pump), the code changes to a moderate, solid that represents mainline pressure.

As the fluid exits the pump, it first enters a circuit with four branches. One branch of the circuit carries fluid to the manual valve while the other three bring fluid to the various regulator valve ports for a number of purposes.

For example, upon initial opening of the regulator valve, fluid from one of these branches passes through the annular groove and exits from an outlet port into the torque converter circuit. This fluid passes the closed torque converter (T/C) relief valve, through the open switch valve, and into the front clutch lubrication system along with the torque converter. Note how the code

changed from a moderate to light, solid shading as the pressure in the circuits changed from line to converter as it passed through the regulator valve.

The fluid now passes through the converter, but a sufficient level remains always in the unit so that it can perform its function at all times. As the fluid leaves the converter, it enters a passage to the switch valve. Here, the fluid again passes through the open switch valve, where it becomes lubrication pressure. Note the change of codes from light, solid shading to one of alternating light-shaded and white boxes.

Looking back at the regulator valve, the fluid entering the far left-hand branch passes through an orifice and acts on the line-pressure plug. The fluid moving through the middle branch also passes through an orifice and acts on a differential reaction area of the regulator valve. The combined forces of the fluid acting on both the line-pressure plug and the differential reaction area of the regulator valve assist in governing line pressure in all forward-driving ranges.

The fluid, entering the manual valve from the pump output circuit, is used to set up the operating conditions within the transmission. This is accomplished by the driver moving the gearshift-selector lever, which, in turn, places the manual valve in given positions. For instance, in the manual valve position shown in Fig. 13-14, fluid leaves a manual valve port and moves back through a circuit to two restricted port openings on either side of the throttle-pressure plug of the regulator valve train. The force from this fluid is also used to assist in regulating line pressure in all forward-driving ranges.

With the manual valve in the drive position, its lands uncover two other port openings. The upper left port permits fluid to flow through a circuit to the 1-2 shift valve, governor valve, accumulator, lock-up valve, and rear clutch. Since the 1-2 shift valve is now off (closed), the fluid deadends at the valve. In other words, a land of the 1-2 shift valve blocks fluid flow at this point, thus preventing it from reaching the second-gear apply circuits.

The fluid entering the governor valve becomes governor pressure (indicated by the code of a wavy black line against a dark-shaded background) as the vehicle begins to move forward. From the governor valve, this pressure enters a branched circuit leading to four valves. The left branch carries fluid to both the 1-2 and 2-3 shift valves. The force from the pressure will open the shift valves when it reaches a predetermined value. In the diagram shown in Fig. 13-14 that governor pressure has not yet opened either valve against its spring tension.

The right-hand branch carries governor pressure to the limit valve and lock-up valve. The force of governor pressure acts on the face of the lower, limit-valve land and attempts to move the valve against spring tension. If the limit valve moves up sufficiently, it cuts off throttle pressure to the 2-3 shift-valve throttle plug.

Governor pressure acting on the right-hand face of the lock-up valve also attempts to move it against spring force. The spring will keep the lock-up valve closed until the vehicle reaches 40 mph (65 km/h), where governor pressure is strong enough to overcome its tension. But when the lock-up valve does open, fluid will pass through it to the fail-safe valve.

Line pressure from the manual valve causes fluid flow to the accumulator and rear clutch. The fluid acting on the accumulator forces its piston to move against spring tension. At this same time, fluid pressure acts on and attempts to apply the rear clutch piston. But because of accumulator action, the rear clutch will not fully apply until the accumulator piston bottoms in its bore. This action cushions clutch application.

The other open port at the manual valve directs fluid through a circuit to the throttle valve. Remember, the throttle valve produces a pressure signal from line pressure that is in proportion to engine speed or load. A circuit from the throttle valve carries this signal to the spring ends of both the 1-2 and 2-3 shift valves as shown by the dark, solid-shaded code. Throttle pressure acting on both valves assists their springs in keeping the units closed against the force of governor pressure.

Throttle pressure also acts on the shuttle and fail-safe valves. The force from this pressure permits or stops the movement of the valves in their bores. For example, throttle pressure on the shuttle valve attempts to move it against spring tension while this same pressure on the fail-safe valve assists spring force in attempting to keep it closed against the action of line pressure.

Throttle pressure passes through the limit valve to act on the plug, located on the spring end of the 2-3 shift valve. The resulting force on the plug assists the spring, along with throttle pressure acting on the 2-3 valve itself, in trying to keep it from opening due to the effects of governor pressure. In effect, this delays the opening of the 2-3 shift valve so that it does not open at the same time as the 1-2.

At this point, all the hydraulic circuits necessary to set the vehicle in motion and upshift the transmission are full of fluid and under pressure. Now, as the driver pushes down on the accelerator pedal, the vehicle moves forward in first ratio and the governor as well as the throttle valves begin to develop their respective signals. These will upshift the transmission at a given road speed and engine load.

Reading a Drive-Second Ratio Chart

Now let's look at the diagram of drive-second ratio (Fig. 13-15) and note what has occurred in the hydraulic system to cause the upshift from first to second. Governor pressure, acting on the 1-2 governor plug, has moved the valve train to the open position against the combined effects of throttle pressure and spring force. This valve movement blocks throttle pressure to the spring end of the 1-2 shift valve and opens a circuit through the shift valve to apply the front servo.

The servo-apply pressure, leaving the 1-2 shift valve, performs several functions. First, it charges a circuit leading to the 2-3 shift valve with fluid. The fluid in this circuit deadends at the 2-3 shift valve because it is now in the closed position, but when the valve opens, the fluid will upshift the transmission into third ratio.

Second, the open 1-2 shift valve supplies pressure and fluid flow to a circuit leading to the accumulator and front servo. The fluid under pressure first passes through an orifice before it reaches the accumulator; the orifice slows fluid flow and pressure buildup. The fluid then enters the upper port on the accumulator where it assists the spring in pushing the piston down against the line pressure below it.

At the same time, fluid begins to move the servo piston in its bore; however, the servo piston cannot tighten the band firmly around the drum until the accumulator piston bottoms in its bore and pressure builds up on its side of the circuit from the orifice. Action of the accumulator and orifice cushions or delays band application.

To assist in cushioning or delaying band application even more under certain conditions, the 1-2 shift control valve supplies 1-2 shift control pressure to the middle land of the accumulator piston. This pressure along with line pressure, beneath the lower land of the piston, controls the buildup rate of servo apply pressure necessary to activate both the servo and accumulator pistons during the 1-2 upshift. Therefore, the 1-2 control valve improves the quality of the 1-2 upshift under certain conditions.

TROUBLESHOOTING MALFUNCTIONS USING HYDRAULIC DIAGRAMS

Hydraulic charts can be of practical use to a mechanic in troubleshooting a malfunctioning transmission. For instance, if the unit will not automatically upshift from first to second, the first thing the mechanic should try to determine from the diagrams is what valve(s) the hydraulic system uses to upshift the transmission. In Fig. 13-15, the diagram indicates the valve train that actually upshifts the transmission is the 1-2. This valve train, when opened, directs fluid to the front servo that applies the band.

The next thing to consider is the forces the hydraulic system uses to control this valve's operation. In the case of the 1-2 shift valve, governor pressure opens the valve in opposition to its spring and throttle pressure. In other words, the governor pressure must be high enough to

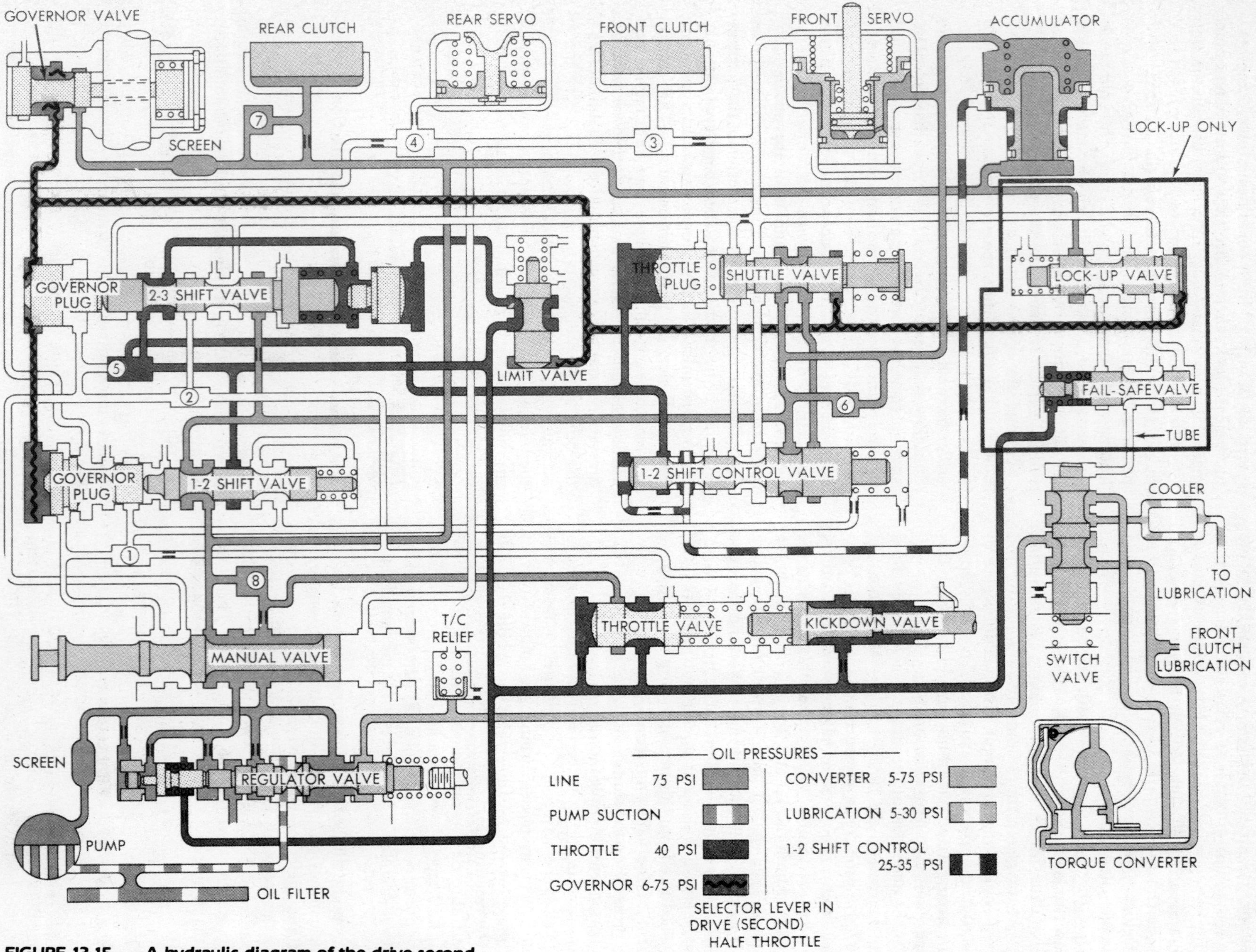

FIGURE 13-15 A hydraulic diagram of the drive-second ratio for the Chrysler 727 automatic transmission. (Courtesy of Chrysler Corp.)

overcome the tension of the spring and the force of throttle pressure or the 1-2 shift valve will not open.

Consequently, one of the possible problems that may cause the transmission to not automatically upshift from first to second is a malfunctioning governor. Referring back to the diagram in Fig. 13-15, note that the governor in this transmission has a screen in its inlet circuit. If this screen becomes clogged, the governor will not receive fluid from the manual valve and, as a result, it will not produce any governor pressure. Of course, the governor would also malfunction if its valve or weight assembly sticks in its bore in the housing. Finally, to determine if the governor is at fault, always refer to the procedure outlined in the service manual for testing governor operation. This procedure varies from one transmission type to another.

A second possible reason why the shift valve will not open is high throttle pressure on the 1-2 shift valve (see Fig. 13-14). In the case of the mechanically operated unit, excessively high throttle pressure may be the result of linkage or a cable that is out of adjustment or the throttle valve is stuck in its bore. If, on the other hand, the transmission has a vacuum-operated unit, the cause of the problem may be a ruptured diaphragm, vacuum leak, or a stuck valve. But in any case, the result is high throttle pressure that assists the shift-valve spring in preventing governor pressure from opening the shift valve. Lastly, to test and/or adjust the throttle valve, always refer to the service manual for the correct procedure to follow. These procedures also differ from one transmission to another.

If both the governor and throttle valves are functioning according to specifications but the transmission will not upshift, the shift valve may be stuck in its bore. In this situation, governor pressure will not be able to open it and the servo apply circuit will not receive fluid flow.

However, before tearing down the valve body to check for a jammed 1-2 shift valve, always check the service manual first to determine if it is possible to test the transmission's servo apply pressure. If a test is possible, always perform it because if the pressure is normal but the transmission still won't upshift, the problem is *not* in the shift valve at all.

For example, if the servo receives apply pressure but the transmission doesn't upshift, the band itself may be out of adjustment or worn out. In this case, the hydraulic system is doing its job, but the band is just not holding. And since the band does not hold the planetary, geartrain member necessary for second-ratio operation, the transmission remains in first gear.

In conclusion, hydraulic diagrams can save the technician a great deal of time and effort in trying to determine transmission malfunctions. If the mechanic takes a few moments to consider what is taking place in the hydraulic system during the time the malfunctions occur and then performs whatever tests are necessary to verify component operation, he will be able to pinpoint the problem in the transmission and not waste time guessing or using the trial and error method of troubleshooting.

SUMMARY

1. For the purpose of troubleshooting a malfunctioning automatic transmission, the hydraulic system diagram is a very useful tool because it provides the repairman with a lot of useful information.

2. If the schematic were to be a true scale drawing as to the length and direction of each circuit, it would be quite large and more difficult to interpret.

3. The hydraulic system diagram contains the pump circuit, all the valve circuits, the torque converter, and all the hydraulically operated units.

4. Before a technician can interpret a hydraulic chart, he must understand the symbols used on that particular schematic.

5. The standard symbols, used extensively by the industry, are those that represent the circuit, orifice, vent, and a circuit not in use.

6. There is not a standard symbol for any valve in the hydraulic system; instead, each manufacturer uses a drawing of the valve itself.

7. There are also no standard symbols for the hydraulic pump, torque converter, and the hydraulically operated units.

8. Because the hydraulic chart incorporates all the pressurized circuits of the transmission, a

mechanic would have a hard time tracing a particular one if the manufacturer did not code the diagrams in some manner.

9. The color-coded diagram is one in which the manufacturer uses a given color to represent the type of pressure in a particular circuit.

10. The pattern-coded diagram is one in which the manufacturer uses different combinations of lines, dots, or shade tones to represent the type of pressure in a given circuit.

11. In order to assist the mechanic in interpreting the hydraulic diagram of an automatic transmission, its manufacturer will usually devote a section of the service manual to define the various valves used in the hydraulic system.

12. The first thing a mechanic should do when using a diagram to troubleshoot a problem is locate the chart that covers the operating phase of the transmission where the malfunction occurs.

13. When interpreting a schematic, the mechanic should take note of the special symbols and codes located on the drawing.

14. When the technician understands all the codes, the next step in mastering the schematic is to locate the fluid source and then follow the fluid flow through each of the various circuits.

15. The first thing the mechanic should do when using the diagram to troubleshoot a transmission malfunction is to determine from the chart which valves are necessary to operate the unit during the phase of operation where the problem occurs.

16. The next step is to test, if at all possible, each valve and circuit, one at a time, to determine exactly where the problem originates.

REVIEW

This section will assist you in determining how well you remember the material contained in this section. Read each item carefully. If you cannot complete the statement, review the section in the chapter that covers the material.

1. The portion of the hydraulic schematic which is not drawn to scale is the _____.

   a. valves

   b. orifices

   c. circuits

   d. codes

2. The hydraulic charts usually show the _____ valve in its part-open position.

   a. regulator

   b. shift

   c. accumulator

   d. manual

3. The standard symbols on hydraulic charts are those representing the _____ and the _____.

   a. valve, circuit

   b. orifice, circuit

   c. orifice, mechanically operated unit

   d. mechanically operated unit, torque converter

4. Two parallel lines represent the _____.

   a. circuit

   b. orifice

   c. valve

   d. torque converter

5. To assist the mechanic in following a given circuit on the schematic, the circuits have special _____.

   a. symbols

   b. valves

   c. marks

   d. codes

6. Usually a section of the service manual will contain _____ that will help the mechanic interpret the hydraulic chart.

   a. valve definitions

   b. an index

   c. pressure specifications

   d. only one chart

7. When studying a schematic, first locate the fluid _____.

   a. valve

   b. source

   c. regulator

   d. type

8. If the transmission will not upshift from first to second, check the chart for the _____ that upshifts the unit.

   a. band

   b. clutch

   c. valve

   d. both a and b

9. Before disassembling a valve body for a stuck valve, check the service manual to see if a _____ should be performed to verify the valve's condition.

   a. special inspection

   b. test

   c. alignment

   d. both a and c

10. If the hydraulic system checks out according to factory specifications but still malfunctions, the problem is most likely _____.

   a. a defective clutch   **c.** a defective governor

   b. a defective band   **d.** either a or b

For the answers, turn to the Appendix.

CHAPTER 14

Transmission Operation

Up to this point, this text has covered the design and operation of the many individual components that make up an automatic transmission. The time has now come to put all this information together and see how all those parts operate simultaneously to produce the various transmission ratios. Because there are so many transmission types now in service and because of the space limits of the text, this chapter will concentrate on two examples, one using a Simpson gear train and the other the Ravigneau.

The example of a Simpson gear train will be the Chrysler TorqueFlite, which produces three forward gear ratios and uses a lock-up converter. The sample of a Ravigneau planetary will be Ford's automatic overdrive (ADO) transmission, which produces four forward ratios. To make it easier for the reader to follow the operation of these units, this chapter will first present a description of each transmission and then cover its operation, both mechanically and hydraulically.

TORQUEFLITE TRANSMISSION

Description

Although it has undergone a number of modifications, the Chrysler TorqueFlite transmission has been in constant use since 1960 (Fig. 14-1). This unit makes use of a one-piece aluminum case (the converter housing and main case are a single casting) and a bolt-on extension housing. The planetary is a Simpson design, which produces three forward speeds and reverse. Also, to operate the gear train, the transmission uses two multiple-disc clutches, an overrunning clutch, and two servo-operated bands. And finally, the typical gearshift-selector positions for this transmission are: park (P), drive (D), manual second (2), and manual low (l).

There are three basic models of the TorqueFlite transmission: the 904, 727, and the automatic transaxle. The A-904 was originally designed to be used with the slant-six engine in 1960 (Fig. 14-1). In 1962, Chrysler produced a heavy-duty version of the 904, the A-727, for V-8 applications (Fig. 14-2). Both units have had a long service life and are now adapted to handle a wide variety of Chrysler engine and vehicle configurations.

In 1978, Chrysler redesigned the basic TorqueFlite into an automatic transaxle. The designation for this unit is A-404. Since then, three other models have been produced: the A-413, A-415, and the A-470 (Fig. 14-3). Although all of the units operate in much the same manner, there are minor design differences.

Converter

The TorqueFlite converter attaches to the engine crankshaft through a flexible drive plate and serves as a hydraulic coupling that connects engine torque and power to the input shaft of the transmission. Furthermore, this three-member converter multiplies engine torque when vehicle operating conditions demand more torque than the powerplant can produce. However, it acts as an efficient fluid coupling at normal road load conditions and at higher vehicle speeds. Lastly, some of the TorqueFlite models now use the lock-up feature, where the turbine operates at converter housing speed during the unit's coupling phase.

Planetary Gear Train

The TorqueFlite planetary gear train consists of two simple planetary gearsets connected together by a common sun gear (Fig. 14-4). The front unit consists of an annulus (ring) gear and carrier. The rear clutch assembly, when activated, connects the front annulus gear to the transmission's input shaft. Therefore, the front annulus gear is the driving member of this gearset. The front carrier is the output member of the gearset because it splines directly to the output shaft.

The common sun gear can be a driving or stationary planetary member. This gear connects to the front clutch drum through a driving shell. If this clutch activates, the input shaft drives the sun gear clockwise.

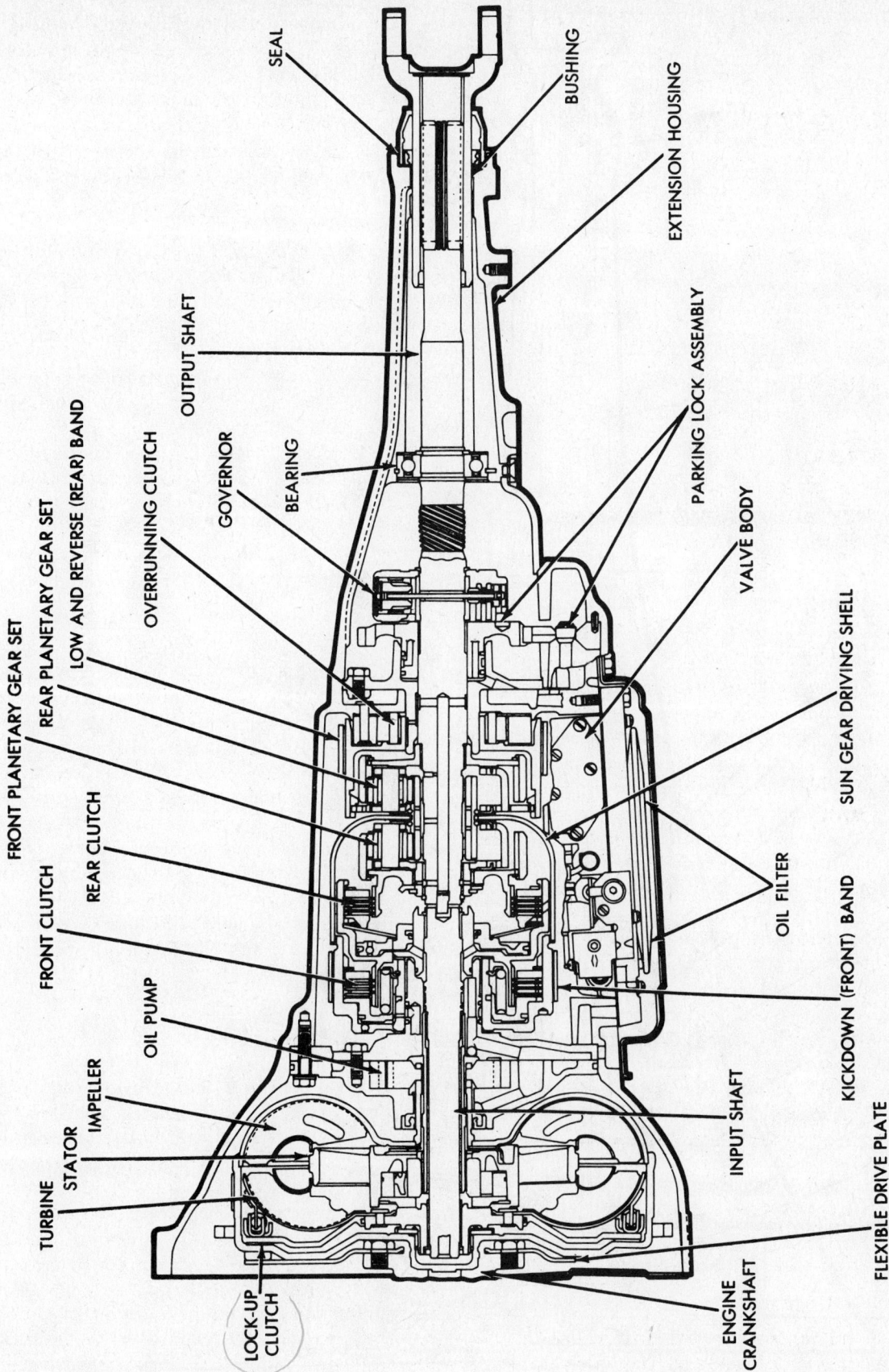

SEAL

BUSHING

EXTENSION HOUSING

OUTPUT SHAFT

FRONT PLANETARY GEAR SET

REAR PLANETARY GEAR SET

LOW AND REVERSE (REAR) BAND

OVERRUNNING CLUTCH

GOVERNOR

BEARING

PARKING LOCK ASSEMBLY

VALVE BODY

FRONT CLUTCH

REAR CLUTCH

SUN GEAR DRIVING SHELL

OIL PUMP

OIL FILTER

TURBINE

STATOR

IMPELLER

INPUT SHAFT

KICKDOWN (FRONT) BAND

FLEXIBLE DRIVE PLATE

ENGINE CRANKSHAFT

LOCK-UP CLUTCH

FIGURE 14-1   An A-904 TorqueFlite transmission.
(Courtesy of Chrysler Corp.)

227

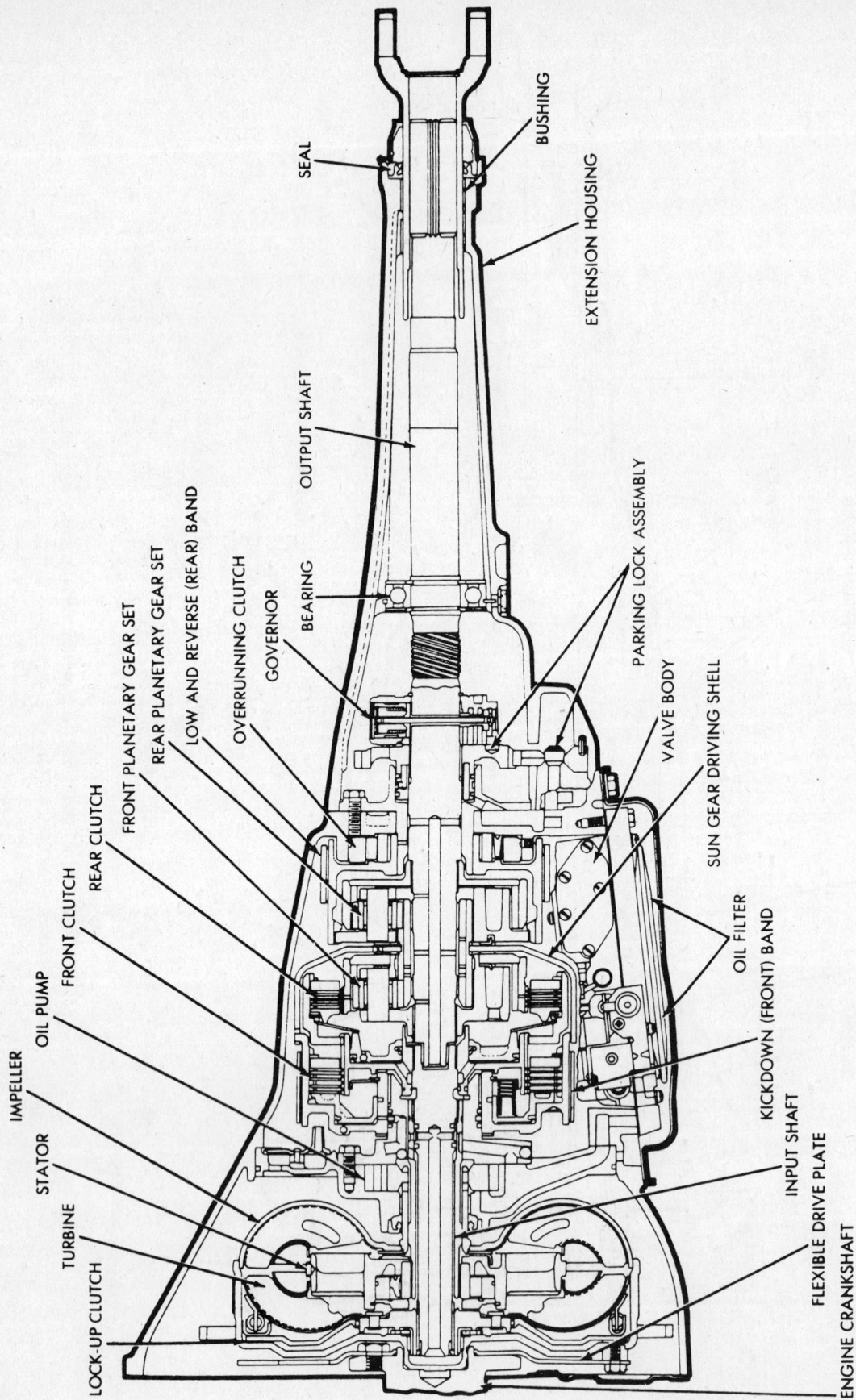

SEAL

BUSHING

EXTENSION HOUSING

OUTPUT SHAFT

PARKING LOCK ASSEMBLY

BEARING

GOVERNOR

OVERRUNNING CLUTCH

LOW AND REVERSE (REAR) BAND

REAR PLANETARY GEAR SET

FRONT PLANETARY GEAR SET

VALVE BODY

SUN GEAR DRIVING SHELL

REAR CLUTCH

FRONT CLUTCH

OIL FILTER

OIL PUMP

KICKDOWN (FRONT) BAND

IMPELLER

STATOR

INPUT SHAFT

TURBINE

FLEXIBLE DRIVE PLATE

LOCK-UP CLUTCH

ENGINE CRANKSHAFT

FIGURE 14-2   The A-727 TorqueFlite transmission.
(Courtesy of Chrysler Corp.)

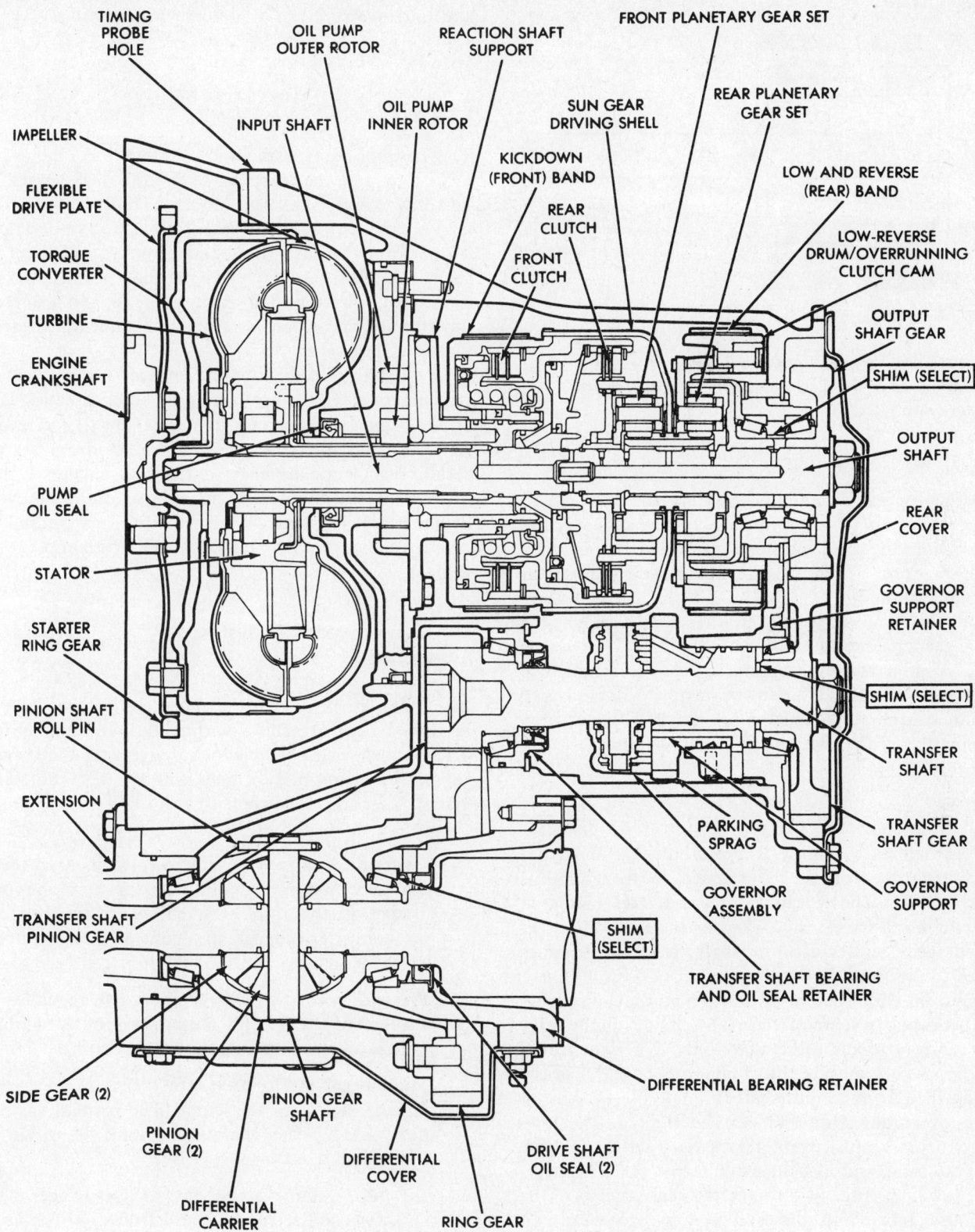

FIGURE 14-3   A typical TorqueFlite transaxle. (Courtesy of Chrysler Corp.)

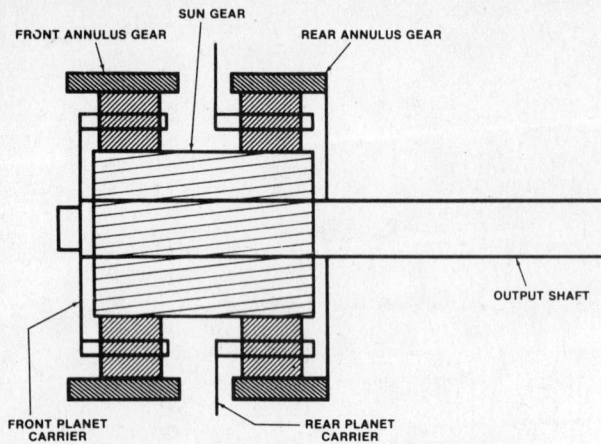

FIGURE 14-4   The complete TorqueFlite Simpson planetary gear train in schematic form. (Courtesy of Chrysler Corp.)

A servo-operated band also encircles the front clutch retainer or drum. If the hydraulic system activates the servo piston, the band will stop the drum's rotation. As a result, the drum, via the driving shell, holds the common sun gear stationary.

The rear planetary gearset consists of an annulus gear and a carrier. The annulus gear of this gearset is the output member because it splines to the output shaft. The carrier is the reaction (stationary) member. It attaches to a drum held by an overrunning clutch, a band, or a combination of the two, to the transmission case.

Clutches

The rear clutch assembly is a multiple-disc unit that transmits torque from the input shaft to the front annulus gear (Fig. 14-5). The hydraulic system activates this clutch in all forward-drive ratios. And as a result, the front annulus gear rotates at input shaft speed and supplies torque to the front planetary gearset.

The front clutch assembly is also a multiple-disc type. The hydraulic system of the transmission applies this clutch in reverse and direct drive only. When the clutch does apply, it connects the common sun gear, via the driving shell, to the input shaft (Fig. 14-6).

An overrunning mechanical clutch holds the low-reverse drum against counterclockwise rotation in drive (D) first ratio and also in manual low (1). The overrunning clutch (Fig. 14-7) fits in the rear portion of the transmission case and splines to the low-reverse drum; it attaches to the rear carrier. When the planetary gear train action attempts to force the rear planet carrier and low-reverse drum to turn counterclockwise, the overrunning clutch locks them to the transmission case. However, the same clutch permits free rotation of the rear planet carrier and low-reverse drum in a clockwise direction.

Bands

The TorqueFlite transmission uses two brake bands, the kickdown and the low-reverse. Figure 14-8 shows the kickdown band in this transmission. The task of this band is to hold the common sun gear stationary in second ratio. To perform this task, the band actually applies on the surface of the front clutch retainer that, in turn, locks to the sun gear driving shell. As a result, the sun gear stops any rotation and becomes a reaction member for the front planetary gearset.

The function of the low-reverse band is to prevent the rear planet carrier from turning in either direction (Fig. 14-9). To perform this task in manual low (l) and reverse (R), the band surrounds the low-reverse drum. As you recall, this drum attaches to the rear carrier. Consequently, if the band tightens around the low-reverse drum, it and the rear carrier will stop turning and the carrier becomes the reaction member of the rear planetary gearset.

Function of Hydraulic System Components

The final elements of the TorqueFlite that will be described next are those within the hydraulic system. These units are responsible for supplying the fluid flow to the various transmission operating devices, controlling system pressure, and up and downshifting the assembly at the proper time (Fig. 14-10).

1. **Hydraulic pump:** A positive-displacement, rotor- or gear-type pump, depending on transmission model, used to supply sufficient fluid flow and pressure to fill the converter, engage the front and rear clutches, apply the bands, and feed the lubrication system.

2. **Pressure regulating valve:** The valve regulates the amount of the pump's output to provide the basic fluid pressure to operate the transmission at all times.

3. **Manual valve:** The driver, by means of mechanical linkage, activates this valve. The manual valve, in turn, sets up the operating conditions within the transmission.

4. **1-2 and 2-3 shift valves:** These valves trigger the up and downshifts from 1-2 and from 2-3. The valves perform this function by responding to governor and throttle signals and then directing fluid under pressure to the appropriate band or clutch that causes the shift to occur.

FRONT CLUTCH
APPLIED

1.00

INPUT SHAFT

REAR CLUTCH
APPLIED

1.00

OUTPUT SHAFT

REAR CLUTCH APPLIED

FRONT CLUTCH APPLIED

INPUT SHAFT

OUTPUT SHAFT

POWER FLOW IN DIRECT DRIVE

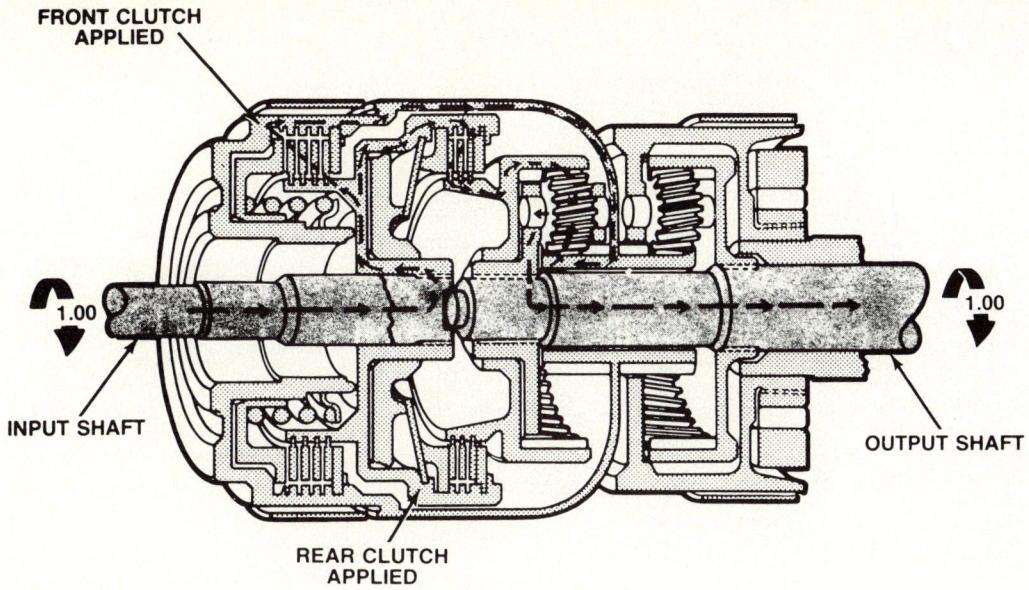

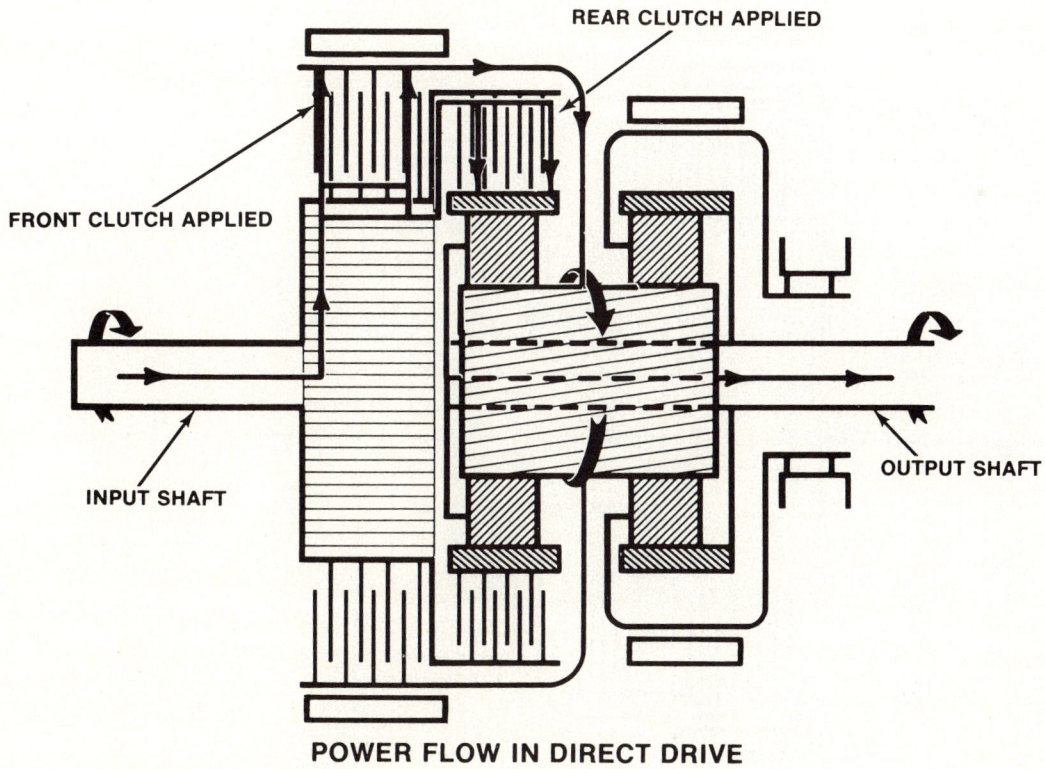

FIGURE 14-5   The rear clutch, when applied, connects
the front annulus gear to the input shaft. (Courtesy of
Chrysler Corp.)

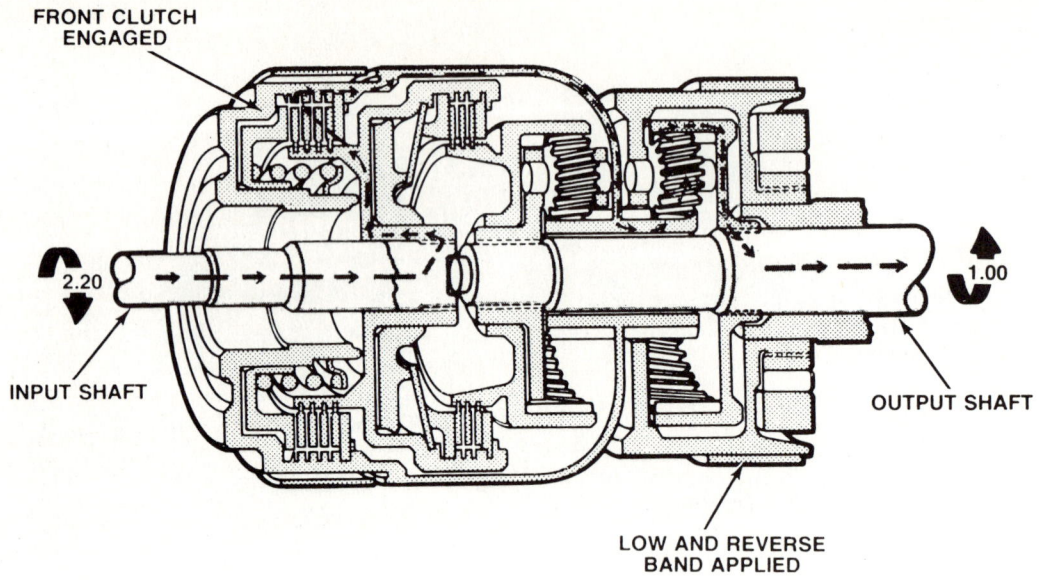

FRONT CLUTCH
ENGAGED

2.20

INPUT SHAFT

OUTPUT SHAFT

1.00

LOW AND REVERSE
BAND APPLIED

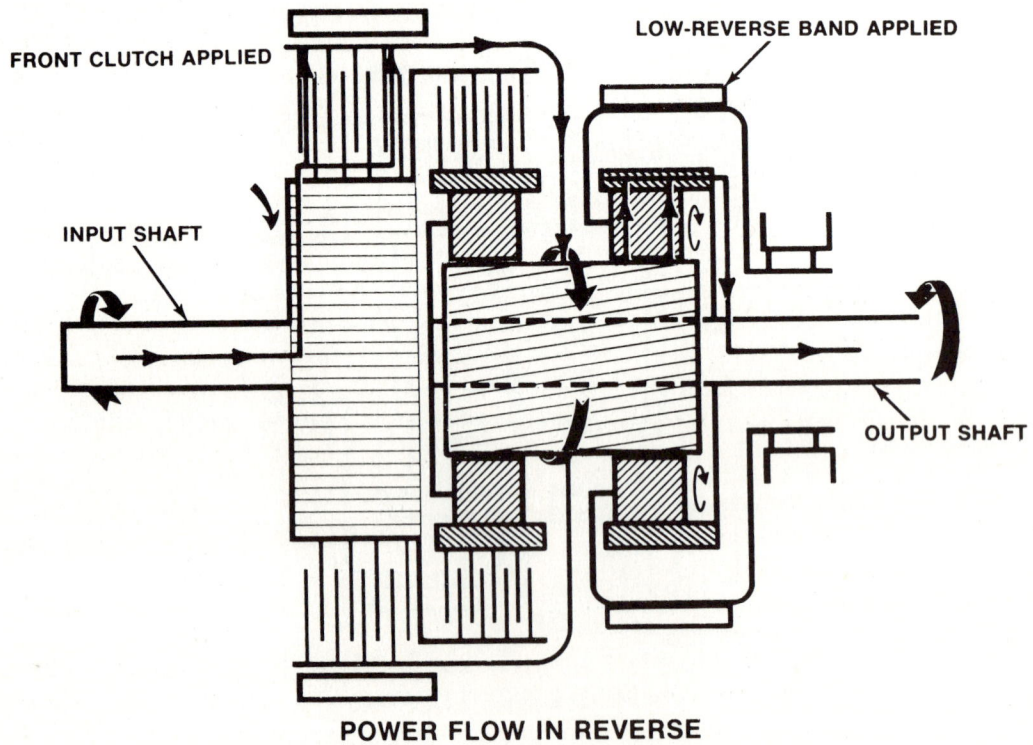

FRONT CLUTCH APPLIED

LOW-REVERSE BAND APPLIED

INPUT SHAFT

OUTPUT SHAFT

POWER FLOW IN REVERSE

FIGURE 14-6   When the front clutch applies, it
connects the common sun gear to the input shaft.

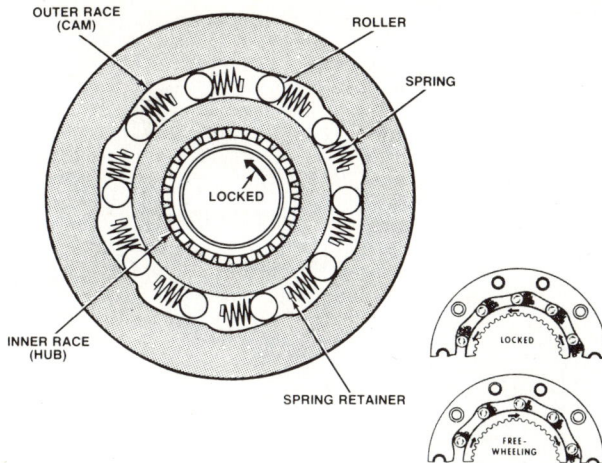

FIGURE 14-7 The overrunning clutch assembly of a TorqueFlite transmission. (Courtesy of Chrysler Corp.)

5. **Governor valve:** This is a balanced valve that senses vehicle speed from the transmission's output shaft. The governor valve then sends a hydraulic speed signal to the shift valves to control the changes in gear ratios in relation to vehicle road speed.

6. **Throttle valve:** This valve produces a hydraulic pressure signal that is in proportion to engine speed or load. The signal acts on the shift valves to regulate the gear changes in relation to engine load or speed.

7. **Kickdown valve:** The kickdown valve operates in conjunction with the throttle valve. When the kickdown valve is activated by full depression of the accelerator pedal, it produces a hydraulic signal that can cause a shift valve to close, which downshifts the transmission.

8. **Shuttle valve:** This valve helps to control the rate of front servo release during a part- to zero-throttle, 2-3 upshift.

9. **Accumulator:** The accumulator controls shift quality by absorbing some of the apply pressure of the rear clutch or front servo.

10. **Lock-up valve:** The lock-up valve reacts to governor pressure and controls fluid flow to the fail-safe valve. When the vehicle is operating below 40 mph (65 km/h), the lock-up valve remains closed, thus cutting off control pressure to the fail-safe valve. This action prevents converter lock-ups.

11. **Fail-safe valve:** The fail-safe valve has several functions. For example, the valve controls fluid flow from the lock-up valve to the switch valve. The fail-safe valve is sensitive to throttle pressure and only opens to permit lock-up in high gear after the transmission's front clutch circuit is completely pres-

surized. In addition, the fail-safe valve allows a rapid converter lock-up release when a forced 3-2 kickdown occurs.

12. **Switch valve:** The switch valve controls the on and off pressures, which are responsible for moving the lock-up piston within the converter. This valve replaces the control valve used on earlier models.

13. **Torque converter (T/C) pressure relief valve:** This valve is in the system to protect the torque converter from the effects of excessively high fluid pressure.

TORQUEFLITE OPERATION

Neutral and Park: Hydraulic Operation

Figure 14-11 illustrates what is taking place in a given TorqueFlite hydraulic system with the engine operating at idle rpm and the gearshift selector in neutral (N). The fluid first moves from the pan, through the filter, and to the suction side of the pump. After the pump pressurizes the fluid, the fluid enters a branched circuit; one branch leads to the high pressure relief valve. The other branch leads to the pressure regulating valve.

After passing through a screen, the fluid enters two ports of the regulating valve and then moves through a circuit to the manual valve. With the manual valve in the neutral position, the fluid passes through the large groove and enters a circuit leading to the line-pressure plug and primary reaction area of the pressure regulating valve. The resulting fluid pressure, acting on these areas, moves the regulating valve against spring tension.

When the regulating valve initially moves, its large, number-one land uncovers a port leading to the torque converter control valve (the switch valve in later designs). Now, fluid can pass through the annular groove, between lands one and two of the regulating valve, to the open control valve. From the control valve, fluid enters the torque converter, cooler, and lubrication circuits.

The pressurized fluid leaving the converter acts on an area of the control valve. The resulting pressure tends to move the control valve against spring tension. As a result, the control valve moves in its bore and regulates converter pressure. In the later designs, the T/C pressure relief valve takes over this function.

As system pressure continues to increase, the regulator valve moves further against the tension of its spring. This further movement permits the large, number-three land to open a circuit to the suction side of the pump. Consequently, pump pressure, in the groove between lands two and three, begins to leak off to the sump. The resulting controlled leak regulates the pressure in the system to about 57 psi.

KICKDOWN BAND
APPLIED

INPUT SHAFT

1.45

1.00

OUTPUT SHAFT

REAR CLUTCH
ENGAGED

KICKDOWN BAND APPLIED

REAR CLUTCH APPLIED

INPUT SHAFT

OUTPUT SHAFT

POWER FLOW IN DRIVE SECOND

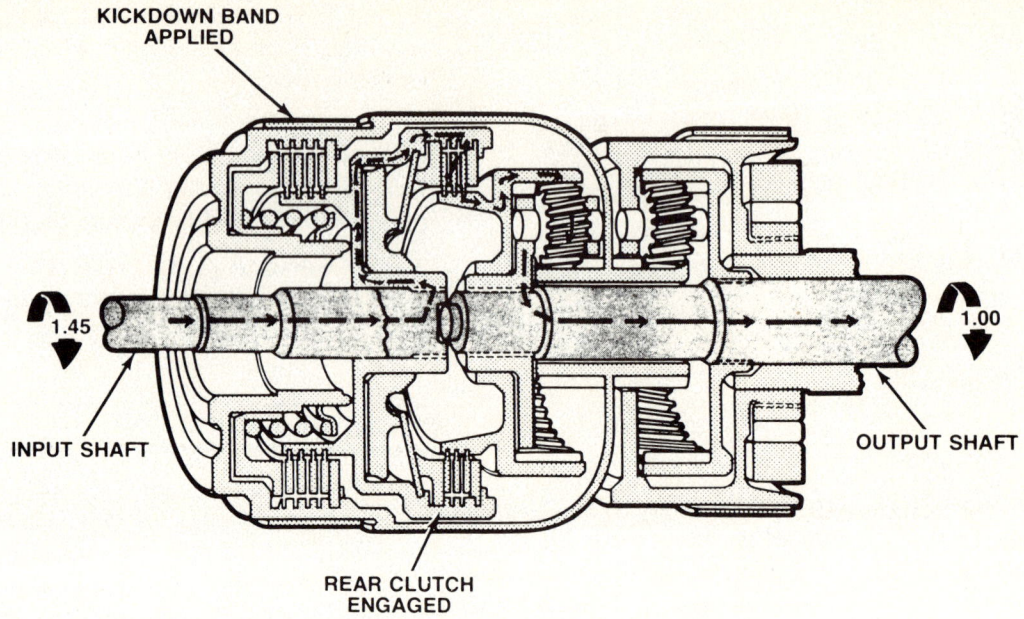

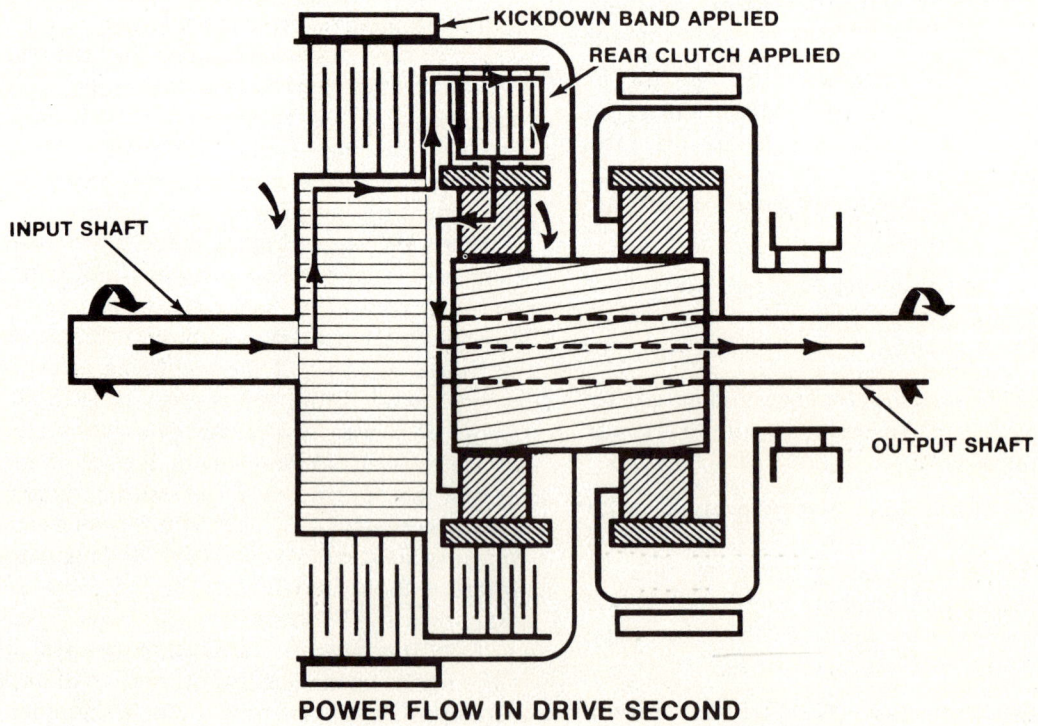

FIGURE 14-8   The kickdown band holds the common
sun gear stationary. (Courtesy of Chrysler Corp.)

**POWER FLOW IN DRIVE BREAKAWAY
OR MANUAL LOW**

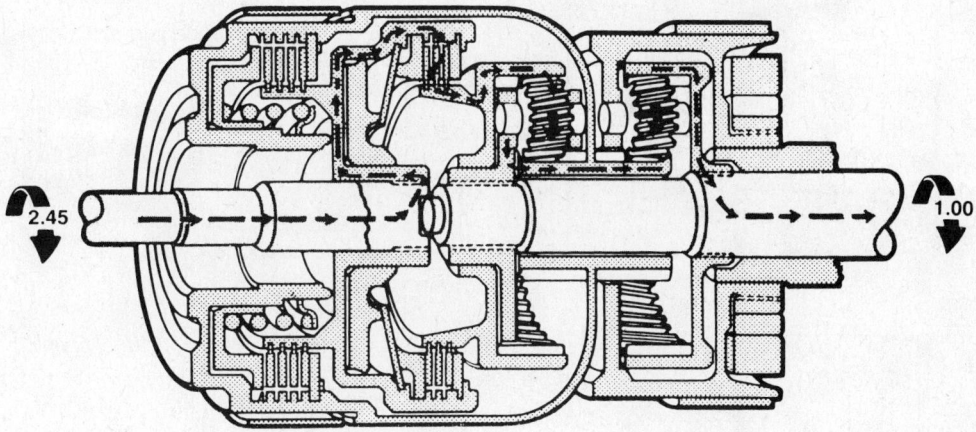

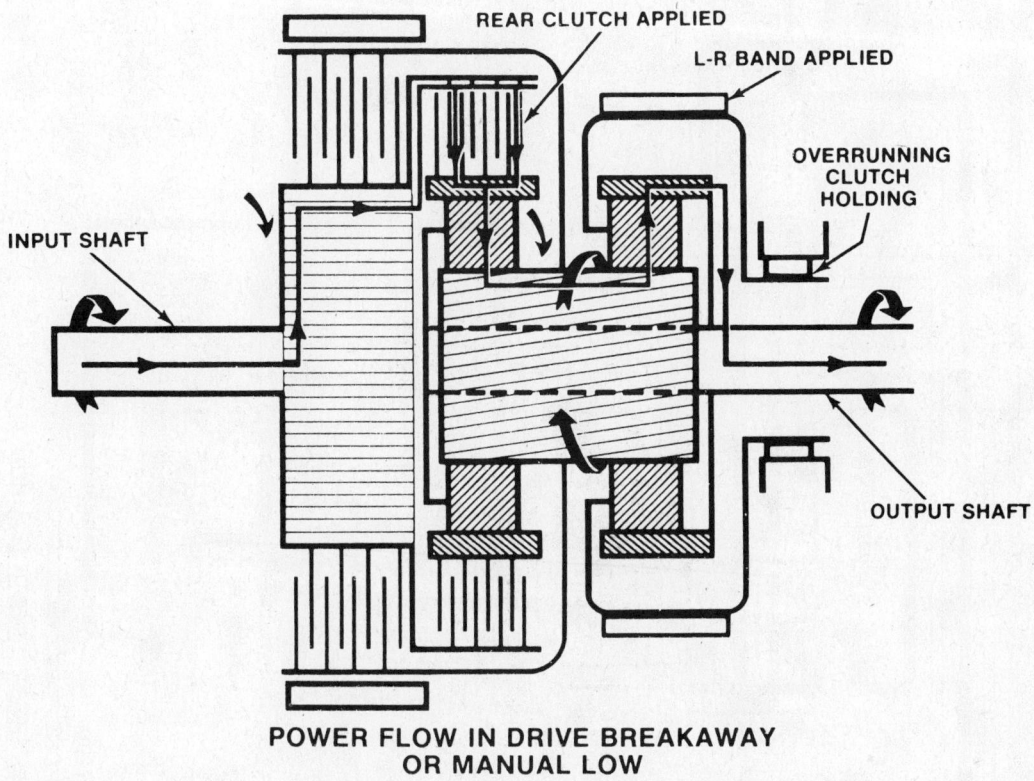

FIGURE 14-9   The low-reverse band holds the rear
carrier against rotation in either direction. (Courtesy of
Chrysler Corp.)

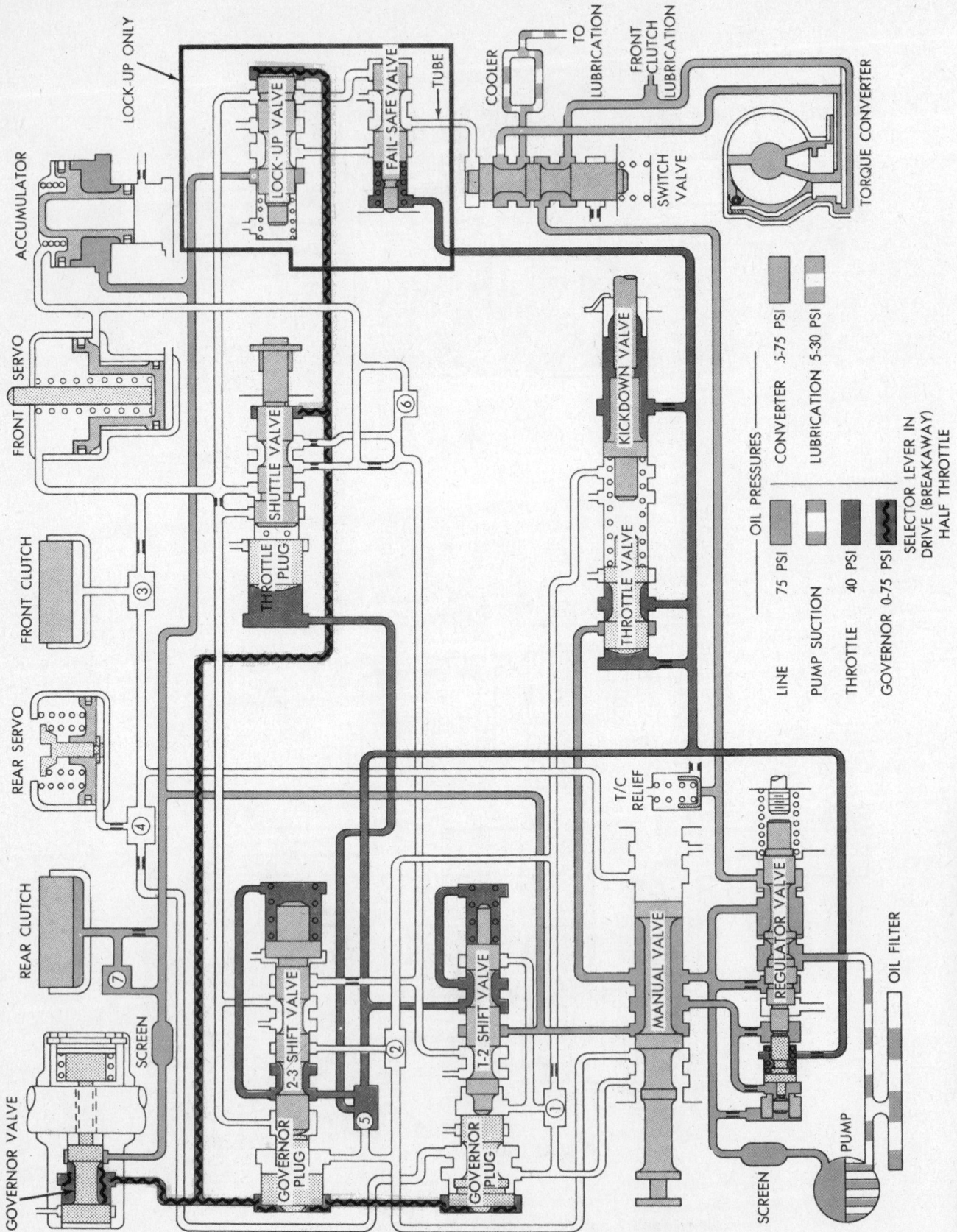

FIGURE 14-10 The hydraulic system components of a TorqueFlite transmission. (Courtesy of Chrysler Corp.)

236

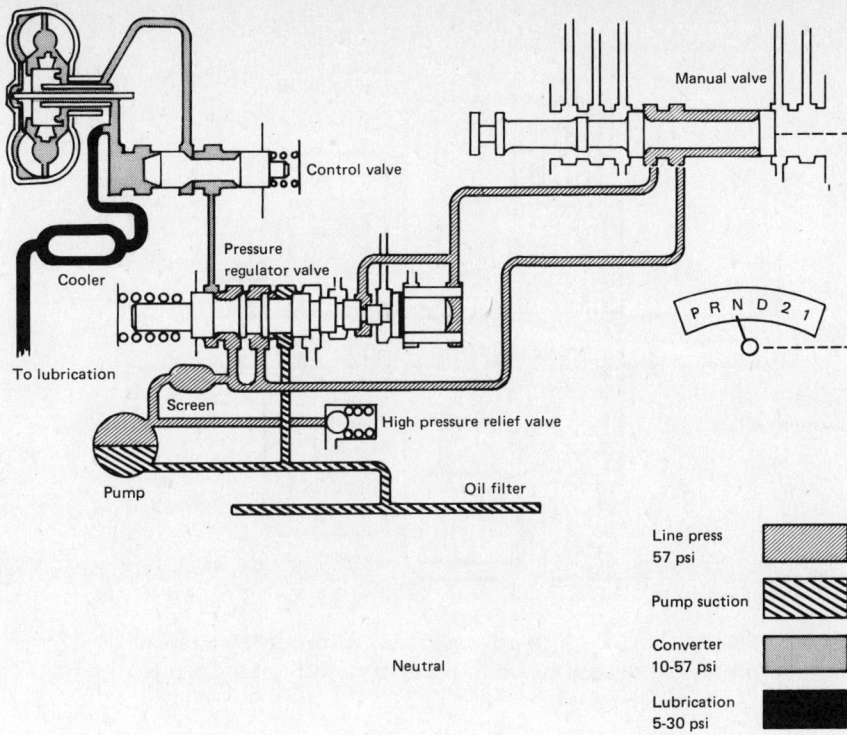

Manual valve

Control valve

Cooler

Pressure
regulator valve

To lubrication

Screen

High pressure relief valve

Pump

Oil filter

P R N D 2 1

Line press
57 psi

Pump suction

Converter
10-57 psi

Lubrication
5-30 psi

Neutral

**FIGURE 14-11   A TorqueFlite hydraulic schematic with
the manual valve in neutral. (Courtesy of Chrysler Corp.)**

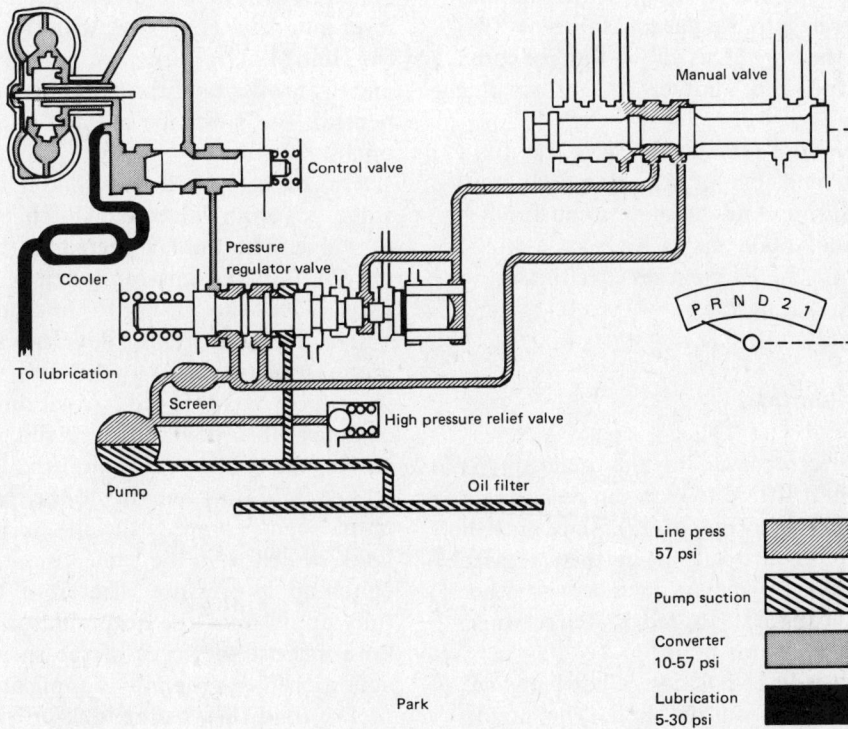

Manual valve

Control valve

Cooler

Pressure
regulator valve

To lubrication

Screen

High pressure relief valve

Pump

Oil filter

P R N D 2 1

Line press
57 psi

Pump suction

Converter
10-57 psi

Lubrication
5-30 psi

Park

**FIGURE 14-12   A TorqueFlite hydraulic schematic with
the manual valve in park. (Courtesy of Chrysler Corp.)**

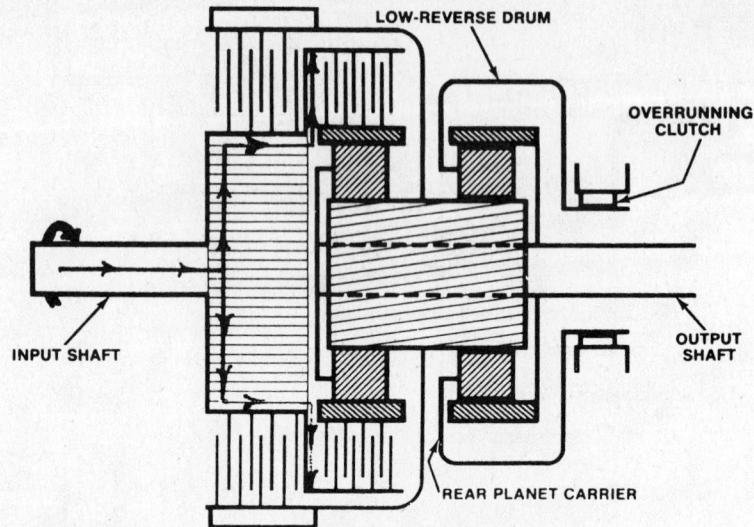

FIGURE 14-13   The powerflow of a TorqueFlite transmission in neutral or park. (Courtesy of Chrysler Corp.)

If, on the other hand, the gearshift-selector lever is in the park (P) position, line pressure will decrease (Fig. 14-12). The difference in system pressure between neutral and park is due mainly to the position of the manual valve. As the valve moves into the park position, its two lands shift position in the bore. This allows fluid to come between one large land and another one of smaller dimension.

Since the size of the bore remains the same, the fluid leaks past the smaller land, out of the valve body, and returns back to the sump. And because some fluid is flowing, pressure cannot build up as high as it did in neutral. In other words, the leak causes a reduction in pressure in the park operating range.

Neutral and Park: Powerflow

With the gearshift-selector lever in the neutral (N) position, the output shaft is free to turn but receives no driving torque from the engine (Fig. 14-13). Since both of the multiple-disc clutches are off or in their release position, there is no driving planetary member for either gearset. The input shaft can rotate the rear clutch retainer, but the powerflow ceases at this point.

In the part (P) selector-lever position, a pawl engages with the parking gear on the output shaft. This action locks the output shaft to the transmission case. This locking arrangement prevents the vehicle from rolling down an incline due to its weight.

Drive, Breakaway: Hydraulic Operation

Figure 14-14 shows what occurs within the hydraulic system whenever the driver places the gearshift-selector lever into drive (D). Now, with the engine running at idle, the control of line pressure, converter feed, fluid cooling, and transmission lubrication is the same as it was in neutral. But since the manual valve has moved, it now opens several new ports.

One of the ports directs fluid in a circuit to the 1-2 shift valve, governor valve, rear clutch, accumulator, and lock-up valve. The fluid, directed to the 1-2 shift valve, will upshift the transmission when it opens; then the shift valve blocks the oil flow to the second-gear apply circuit.

The fluid going to the governor valve will be converted into a pressure signal by this unit, which is in proportion to vehicle road speed. A circuit from the governor channels this governor pressure signal to both shift valves, the shuttle valve, and the lock-up valve.

The fluid flow, moving to the rear clutch and accumulator, begins to apply the clutch. However, the accumulator, which is in the same circuit, absorbs some of the clutch apply pressure. Therefore, the rear clutch cannot fully apply until the accumulator piston bottoms in its bore against the force of the spring. This accumulator action cushions the initial application of the rear clutch.

The fluid flow to the lock-up valve will eventually be used to activate the converter into its lock-up phase. However, this cannot occur until both the lock-up and fail-safe valves open.

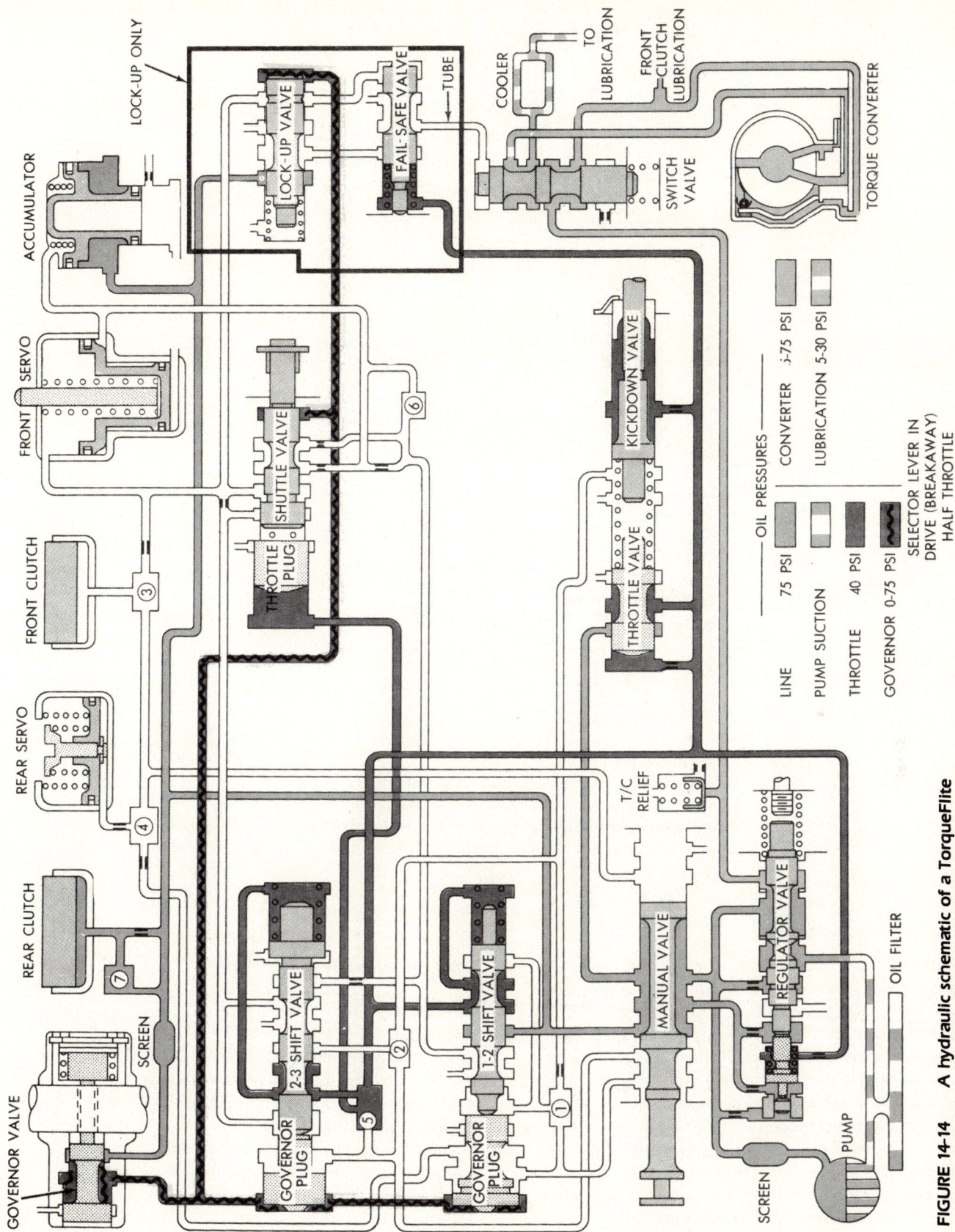

FIGURE 14-14 A hydraulic schematic of a TorqueFlite transmission in the drive-breakaway range. (Courtesy of Chrysler Corp.)

239

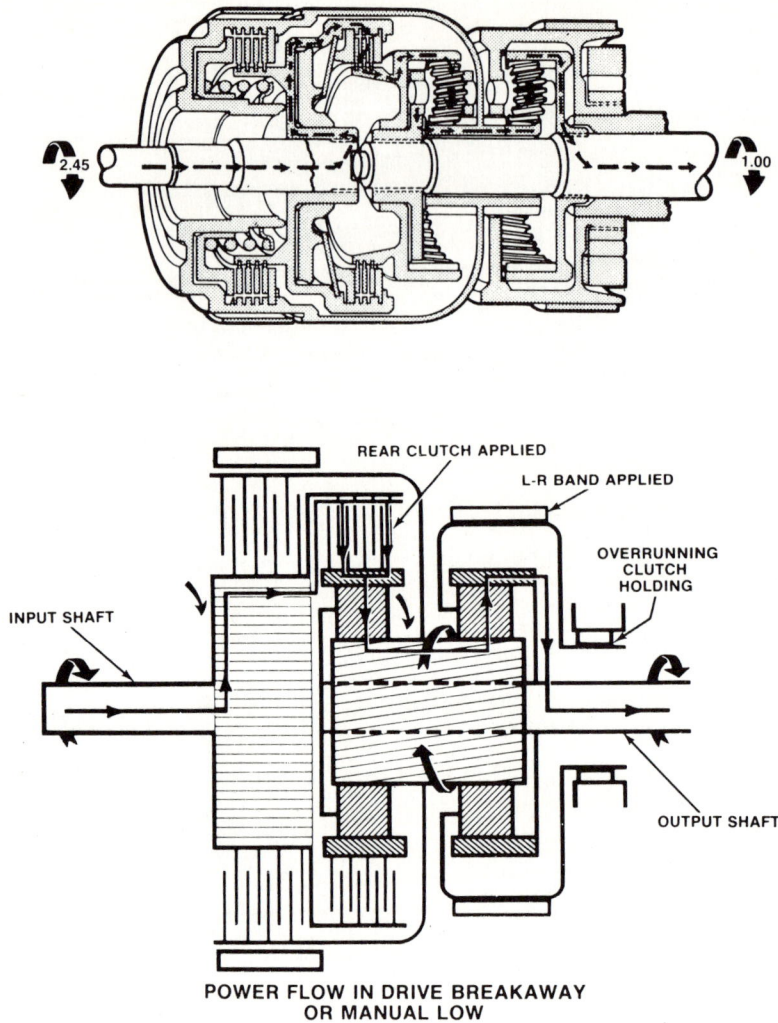

POWER FLOW IN DRIVE BREAKAWAY
OR MANUAL LOW

FIGURE 14-15 In drive-breakaway, the low-reverse band is not applied. However, in manual low (I), the band applies for extra holding power and to provide engine braking during deceleration of the vehicle. (Courtesy of Chrysler Corp.)

Another circuit, opened by the manual valve, is to the throttle valve. The throttle valve will develop a pressure signal from the incoming fluid that is in proportion to engine speed or load. This signal will increase as the driver depresses the accelerator pedal. Circuits carry the throttle pressure signal to the regulator valve, shift valves, shuttle valve, kickdown valve, and the fail-safe valve.

Throttle pressure acts on the regulator valve to increase line pressure during vehicle acceleration. This action prepares the hydraulic system for the upcoming shift to compensate for the normal reduction in pressure, as other hydraulically operated units activate.

Drive, Breakaway: Powerflow

With the gearshift-selector lever placed in drive (D), the transmission is in breakaway, or the first gear ratio (Fig. 14-15). As previously mentioned, the hydraulic system now activates the rear clutch that connects the input shaft to the front annulus gear. The input shaft drives this gear clockwise at turbine speed.

Since the weight of the vehicle is on the output shaft, it will not turn at this time. Consequently, the front planet carrier and rear annulus gear are also momentarily stationary. This action causes the front annulus gear to

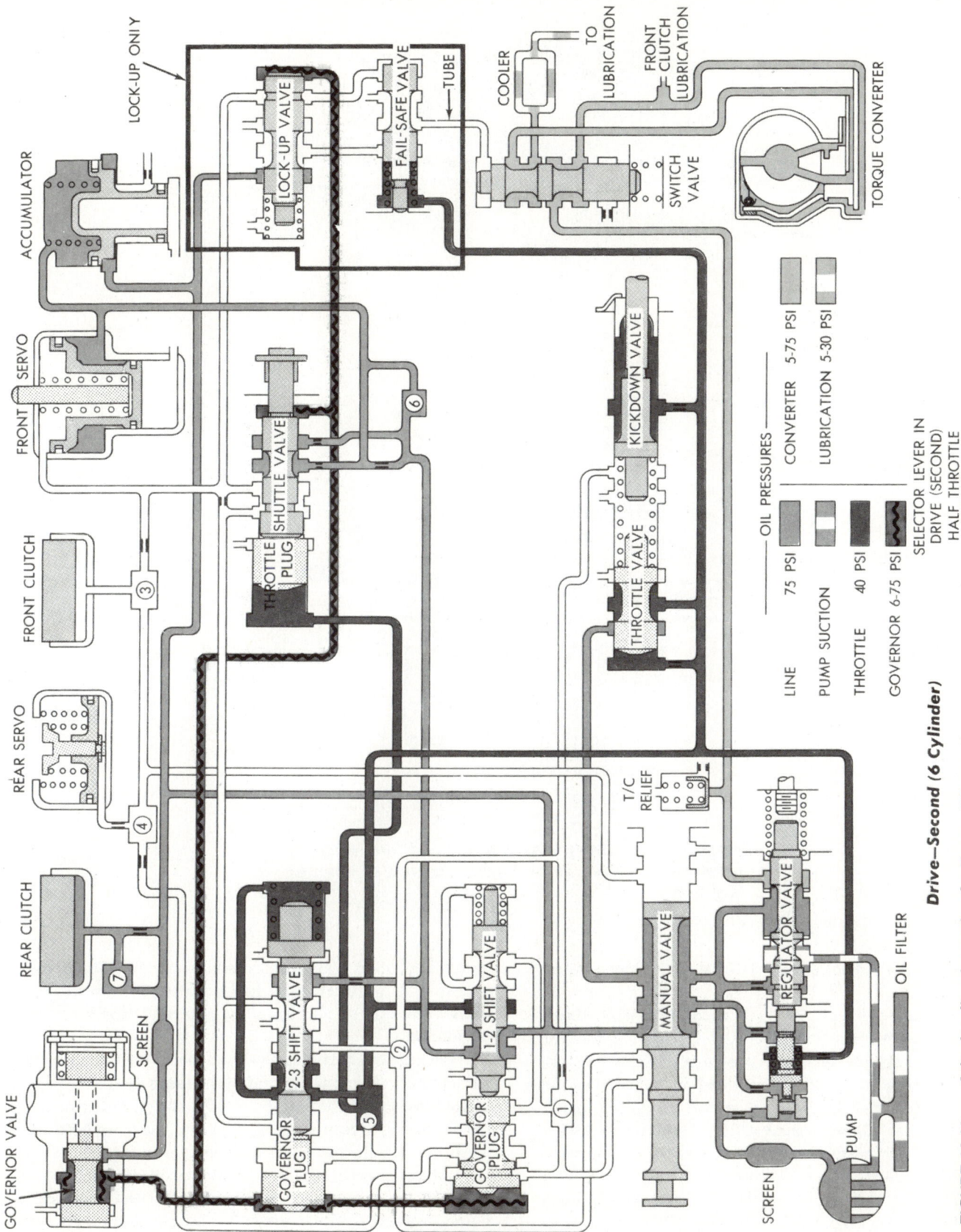

FIGURE 14-16 A hydraulic schematic of a TorqueFlite transmission in drive-second ratio. (Courtesy of Chrysler Corp.)

Drive—Second (6 Cylinder)

OIL PRESSURES

LINE	75 PSI	CONVERTER 5-75 PSI
PUMP SUCTION		LUBRICATION 5-30 PSI
THROTTLE	40 PSI	
GOVERNOR	6-75 PSI	SELECTOR LEVER IN DRIVE (SECOND) HALF THROTTLE

KICKDOWN BAND
APPLIED

1.45

INPUT SHAFT

OUTPUT SHAFT

1.00

REAR CLUTCH
ENGAGED

KICKDOWN BAND APPLIED

REAR CLUTCH APPLIED

INPUT SHAFT

OUTPUT SHAFT

POWER FLOW IN DRIVE SECOND

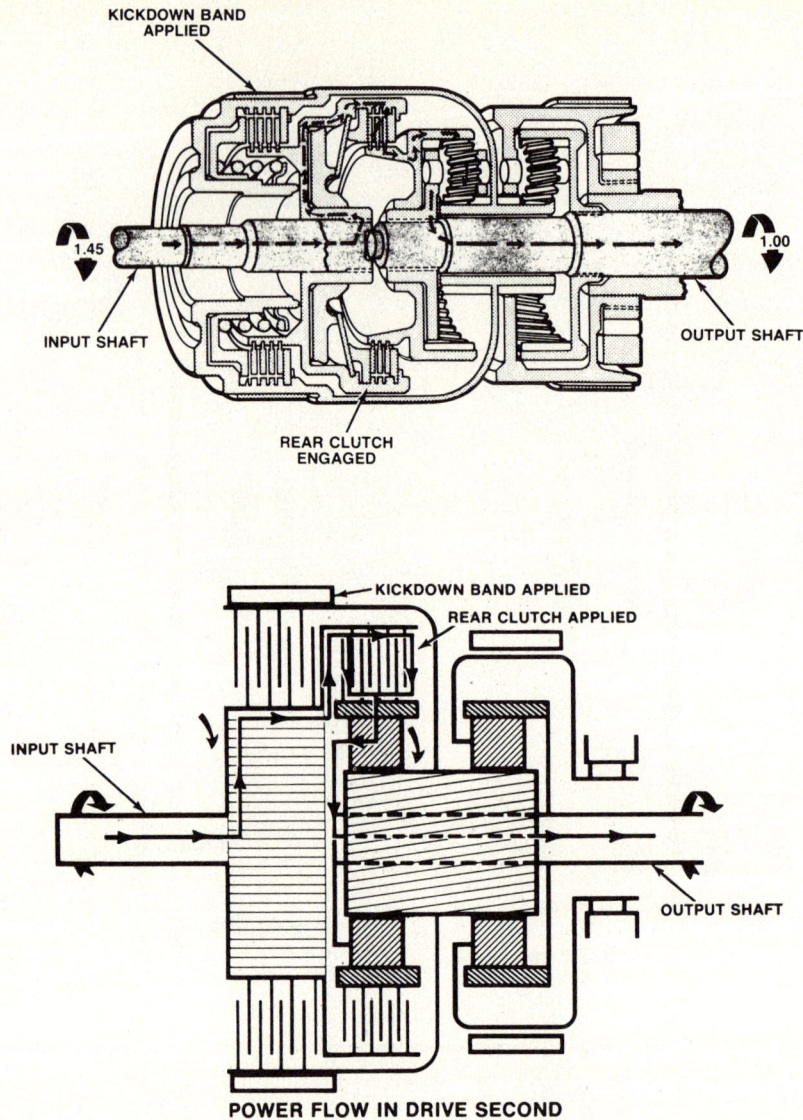

FIGURE 14-17   The powerflow within a TorqueFlite transmission in drive, second ratio. (Courtesy of Chrysler Corp.)

drive the front carrier pinions clockwise on their support pins.

The front pinions, which are momentarily idling, rotate the sun gear counterclockwise, and this action forces the rear planet pinions to turn clockwise. Now, with the weight of the vehicle imposed on the rear annulus gear, the rear pinion rotation on this gear attempts to force the rear planet carrier to turn counterclockwise.

But the rear planet carrier locks into the low-reverse drum that splines into the inner race of the overrunning clutch. With this arrangement, any counterclockwise rotation of the rear carrier and the low-reverse drum causes the overrunning clutch to lock up. And with this

clutch locked up, the rear carrier is stationary. This action results in a torque transfer and powerflow from the rear planet pinions to the annulus gear and finally to the output shaft.

This output shaft rotation also forces the front carrier to turn at the same speed and in the same direction as the output shaft. Consequently, the front annulus gear and carrier are both turning clockwise, but the front carrier is spinning at a slower speed than the annulus gear. In this situation, the actual gear ratio and torque multiplication between the input and output shafts is the result of a combination of ratios provided by the front and rear planetary gearsets.

Drive, Second: Hydraulic Operation

Figure 14-16 illustrates the condition of the hydraulic system when the transmission automatically upshifts from breakaway to second. During this phase of hydraulic operation, governor pressure has now overcome the effects of spring force and throttle pressure on the 1-2 shift valve, and it opens. Now, fluid passes through this valve and flows into the second-gear apply circuit.

A branch of this circuit terminates at the 2-3 shift valve. At this time, a land of the 2-3 shift valve blocks the fluid from the third-gear apply circuit. In other words, this portion of the second-gear apply circuit just supplies the fluid necessary to complete the 2-3 upshift when the 2-3 shift valve opens.

After supplying the 2-3 shift valve, fluid follows the circuit through an orifice, where it goes first to the shuttle valve and then to the accumulator and front servo. The fluid entering the accumulator will assist its spring in pushing the piston down against line pressure which is below the piston. At the same time, pressurized fluid begins to move the servo piston in its bore. But the servo piston will not tighten the band firmly around the drum until the accumulator piston bottoms in its bore. This action cushions band application.

Drive, Second: Powerflow

In drive, second ratio, the transmission makes the upshift by keeping the rear clutch applied and also by activating the front servo, which applies the kickdown band (Fig. 14-17). The kickdown band holds the front clutch retainer that locks to the sun gear driving shell. As a result, the common sun gear is stationary or locked to the transmission case.

Since the rear clutch is still on, the input shaft drives the front annulus gear clockwise at input shaft speed. Because the sun gear is now stationary, the annulus rotation causes the planet pinions of the front unit to turn clockwise. The pinions, in turn, "walk" the carrier around the held sun gear in a clockwise direction. This action transmits torque to the output shaft that splines directly to the front planet carrier. Note that in this gear ratio the front planetary gearset alone produces the reduction and torque multiplication.

Drive, Third: Hydraulic Operation with the Converter Unlocked

Figure 14-18 shows what takes place in the hydraulic system when the upshift occurs from second to direct ratio. During this phase of operation, governor pressure has overcome the effects of spring force and throttle pressure on the 2-3 shift valve, and it has opened. Now, fluid from the 1-2 shift valve circuit passes through a groove in the 2-3 shift valve and flows into the direct apply circuit.

Through this apply circuit, fluid flows to the area between the 2-3 governor plug and the shift valve, the shuttle valve, front servo, front clutch, and lock-up valve.

Depending on accelerator pedal position, the fluid to the shuttle valve may have little or great effect on the quality of the 2-3 upshift. For instance, under moderate acceleration conditions, the shuttle valve assumes a position as shown in Fig. 14-18. In this case, the front servo release and clutch apply fluid must pass through two orifices before acting on these units. The orifices, in this situation, control the timing of the band's release to the application of the clutch.

However, if the driver would lift his/her foot off the accelerator pedal during the 2-3 upshift, the band may release before the clutch applies. This would result in a rough upshift due to clutch-band fight. The shuttle valve takes care of this problem by moving to the left as throttle pressure drops off. This permits the direct apply fluid to bypass the first orifice, flow through the valley within the valve, and then flow to the front servo and clutch. This results in a fast band release and a smooth 2-3 upshift.

The fluid that enters the release side of the servo begins to move the servo back in its bore, assisted by spring force. At the same time, fluid passes through the second orifice and begins to apply the front clutch. With the front band released and the front clutch applied, the transmission is in direct drive.

The direct apply circuit also delivers fluid to the lock-up valve. This fluid passes through a groove in the valve to the fail-safe valve. Once the front clutch is applied and pressure builds up in the direct apply circuit, it is great enough to open the fail-safe valve against throttle pressure and spring force. However, since the lock-up valve is still closed, the converter remains in its unlocked phase of operation.

Drive, Third: Powerflow

As previously stated, when the 2-3 shift valve opens, the front band releases and the front clutch applies (Fig. 14-19). But the rear clutch, during this time, remains applied. With the application of the front clutch, the input shaft now drives the front clutch retainer, which locks to the driving shell. The driving shell, in turn, turns the sun gear at input shaft speed. And finally, since the rear clutch is still on, the input shaft also drives the front annulus gear clockwise at input shaft speed.

Remember from the discussion in Chapter 2 that, whenever the input shaft drives any two members of the

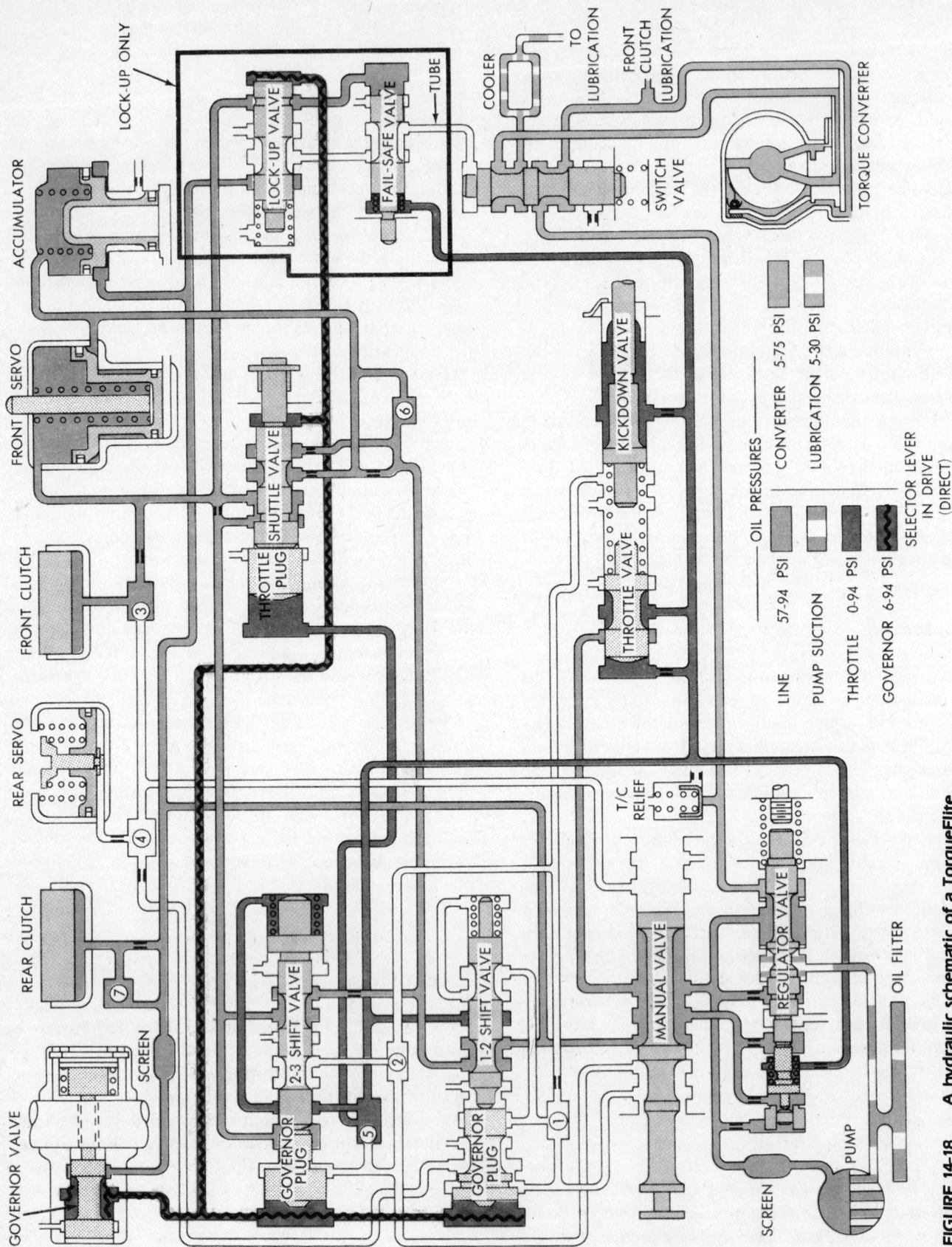

FIGURE 14-18 A hydraulic schematic of a TorqueFlite transmission in direct drive with the converter unlocked. (Courtesy of Chrysler Corp.)

FRONT CLUTCH
APPLIED

INPUT SHAFT

1.00

1.00

OUTPUT SHAFT

REAR CLUTCH
APPLIED

REAR CLUTCH APPLIED

FRONT CLUTCH APPLIED

INPUT SHAFT

OUTPUT SHAFT

POWER FLOW IN DIRECT DRIVE

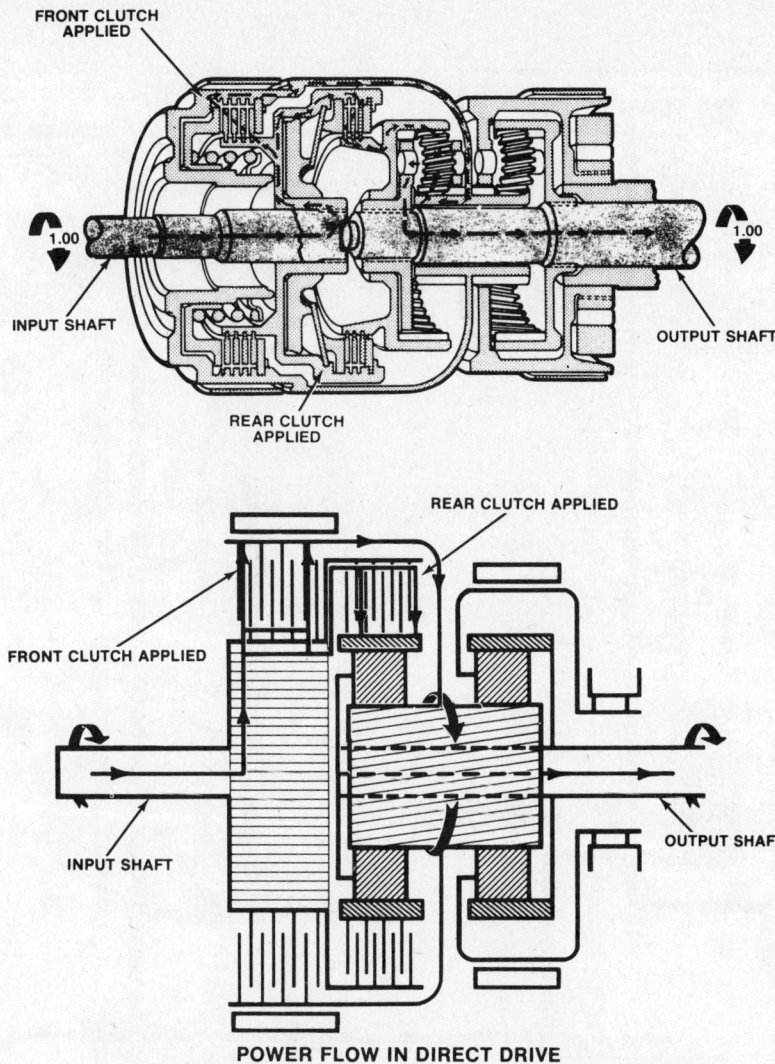

FIGURE 14-19   The powerflow within a TorqueFlite transmission in drive, direct ratio. (Courtesy of Chrysler Corp.)

planetary gear train in the same direction and at the same speed, direct drive results. In other words, when the two members rotate at the same speed and in the same direction, the gear train locks up and turns as a solid unit.

This condition results from the fact that the front and rear planet pinions cannot turn at all in direct drive. The only rotational units are the input and output shafts. The remaining parts just spin as a single unit.

Drive, Third: Hydraulic Operation in Direct Lock-Up

Figure 14-20 illustrates what occurs within the hydraulic system in drive, lock-up. During this phase of transmission operation, the vehicle is in direct drive and at a

vehicle speed above 40 mph (65 km/h). Under these conditions the lock-up, fail-safe, and switch valves are all open.

The lock-up valve opens due to governor pressure acting on its reaction area, which moves the valve to the left against spring tension. As the valve opens, fluid from the drive circuit passes through the large annular groove to the fail-safe valve.

The fail-safe valve has already opened since the transmission is in direct drive with the front clutch applied. With an open fail-safe valve, drive apply fluid passes through its annular groove to the switch valve.

The drive fluid acts on the reaction area of the switch valve and moves it open against the force of spring tension. With the switch valve open, two things occur.

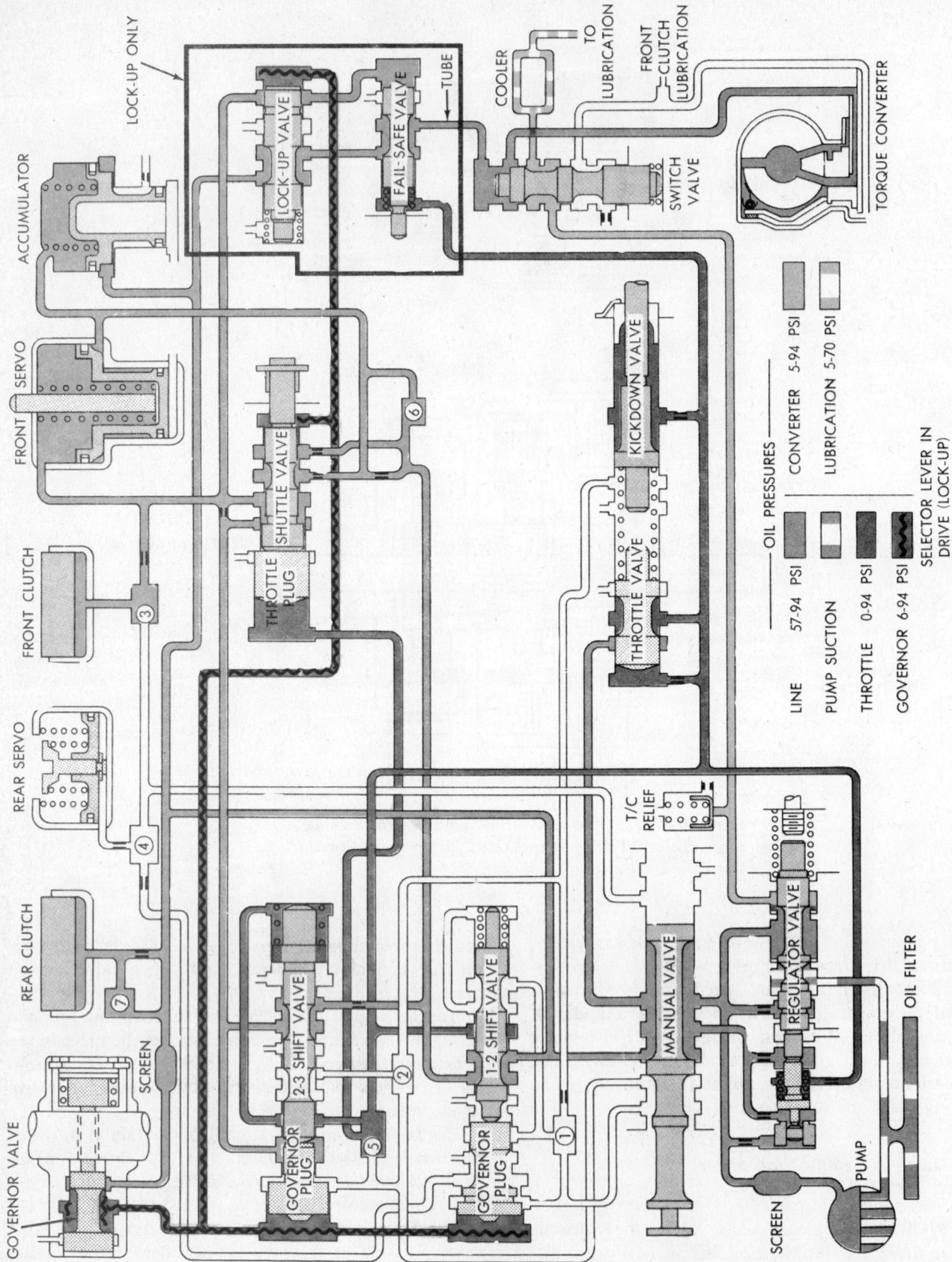

FIGURE 14-20   A hydraulic schematic of a TorqueFlite transmission in drive lock-up. (Courtesy of Chrysler Corp.)

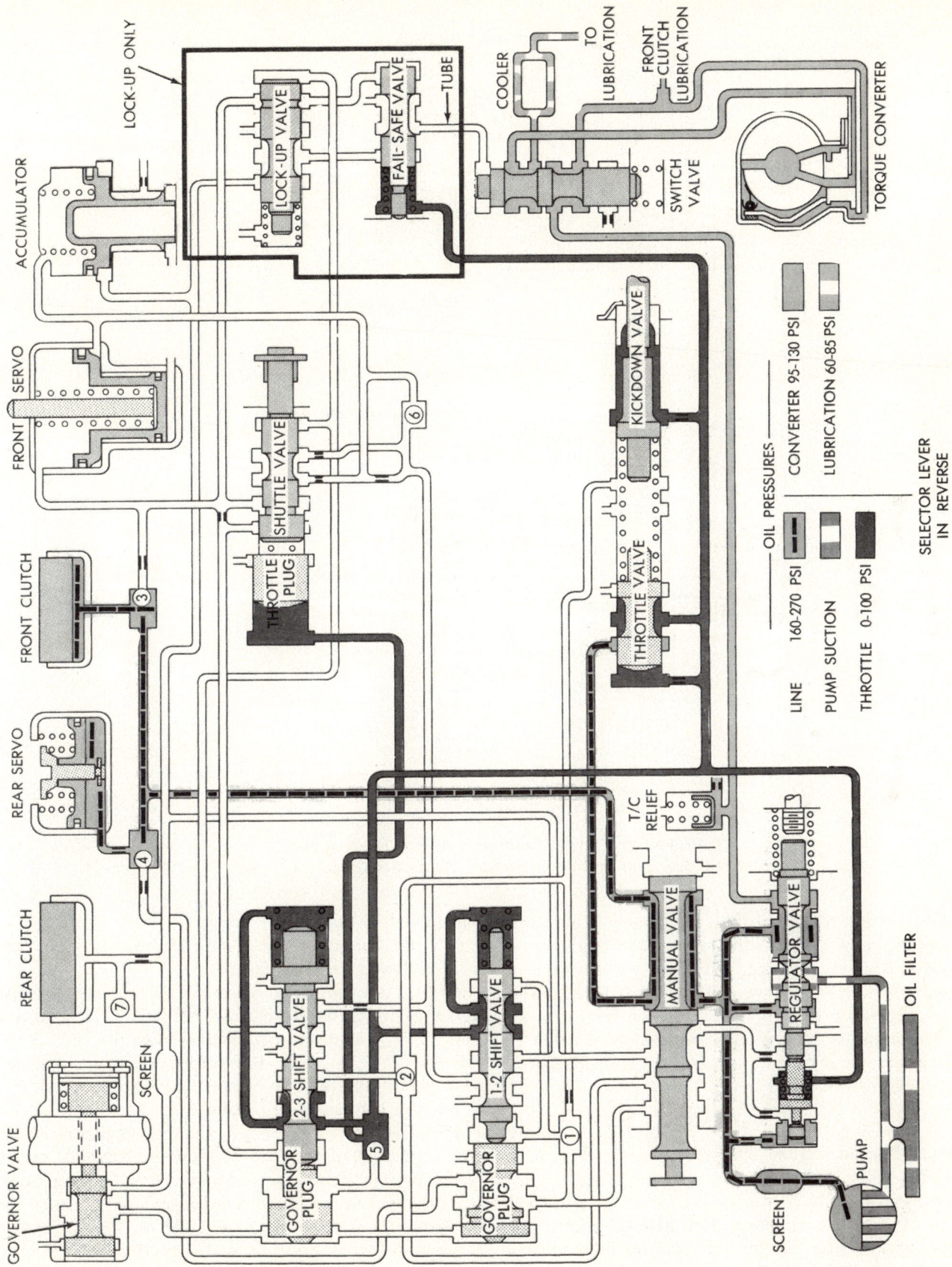

FIGURE 14-21 A hydraulic schematic of a TorqueFlite transmission in reverse. (Courtesy of Chrysler Corp.)

FRONT CLUTCH
ENGAGED

2.20

INPUT SHAFT

1.00

OUTPUT SHAFT

LOW AND REVERSE
BAND APPLIED

FRONT CLUTCH APPLIED

LOW-REVERSE BAND APPLIED

INPUT SHAFT

OUTPUT SHAFT

POWER FLOW IN REVERSE

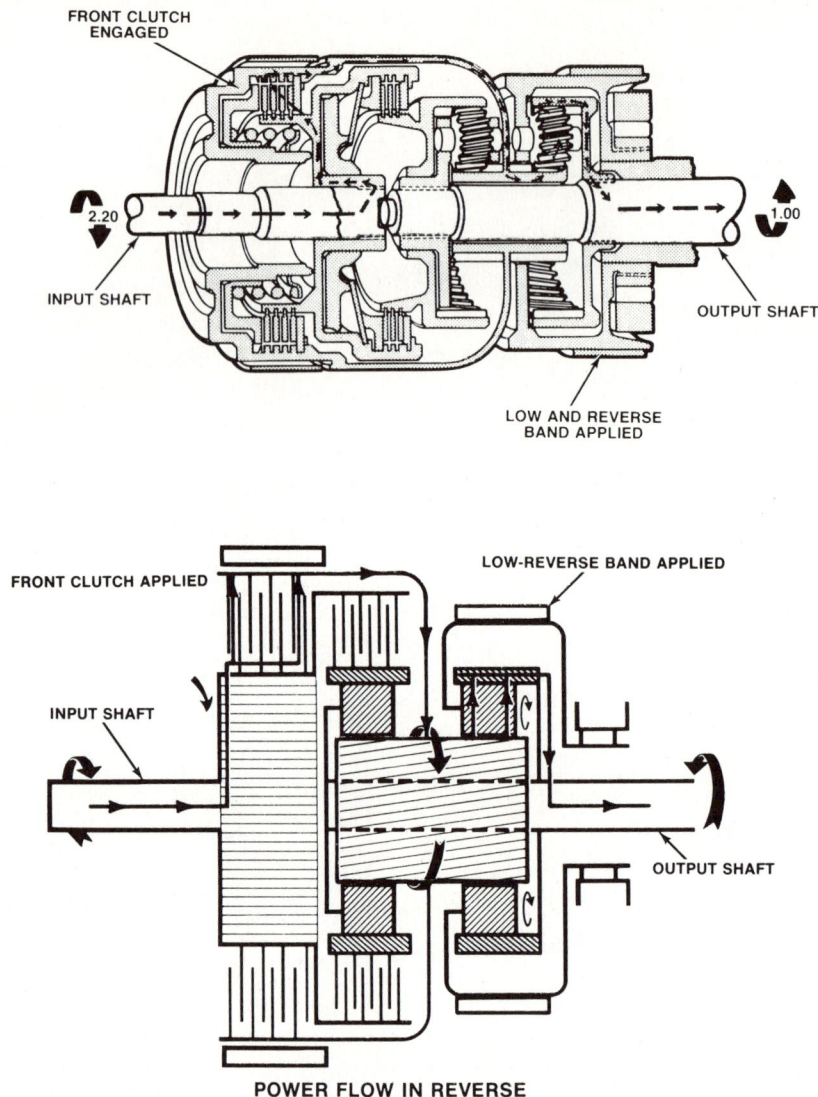

FIGURE 14-22   The powerflow within a TorqueFlite transmission in reverse. (Courtesy of Chrysler Corp.)

First, the converter feed fluid from the front side of the lock-up piston now vents to the reservoir through the switch valve. Second, direct apply fluid passes out of the switch valve and into the converter out circuit. This fluid forces the lock-up piston to the left. When the piston makes firm contact with the friction surface bonded to the converter cover, the turbine is locked to the engine crankshaft.

Reverse: Hydraulic Operation

Figure 14-21 shows the condition of the hydraulic system when the driver places the gearshift-selector lever into reverse (R). With the manual valve moved into this

position, it directs fluid through a circuit to the front clutch and rear servo. The resulting pressure applies the front clutch and rear band.

At the same time, fluid is now cut off to the left face of the throttle plug, which bears against the regulator valve. This action would normally increase line pressure to an extremely high amount. But the reverse manual valve position directs fluid to a small reverse plug piston, situated next to the throttle plug.

However, the reverse plug has a much smaller reaction area than that of the throttle plug. As a result, more system pressure is necessary to move the valve far enough to regulate line pressure. Therefore, line pressure in reverse range is between 160 and 270 psi, depending on accelerator pedal position with its effect on throttle pressure.

Reverse: Powerflow

In reverse gear, the hydraulic system activates the front clutch and rear band. As a result, the input shaft drives the sun gear clockwise at input shaft speed. Also, the rear band holds the rear planet carrier stationary (Fig. 14-22).

As the input shaft drives the sun gear clockwise, it turns the rear planet pinions counterclockwise on their support pins. Since the carrier is stationary, the pinions, in turn, rotate the rear annulus gear. Consequently, the annulus gear that splines to the output shaft now drives it counterclockwise.

In reverse, the rear planetary alone handles the transmission of torque. There is input to the front unit through the sun gear, but no other member is stationary. Therefore, torque through the front unit cannot occur. In other words, the front planetary gears just idle during the reverse phase of transmission operation.

FORD AUTOMATIC OVERDRIVE (AOD) TRANSMISSION

Description

To increase the fuel economy of certain automobile models, Ford, in 1980, was the first to produce an automatic transmission with a built-in overdrive. This unit, known as the automatic overdrive (AOD) transmission, has four speeds forward in which fourth gear is a 0.67:1 overdrive. This overdrive ratio has the effect of reducing engine rpm to sustain a given road speed by 33 percent, which results in better fuel economy.

In the AOD transmission, Ford uses the compound Ravigneau planetary gear train. This is nothing new for Ford because it has used this design in its FX, MX, and FMX three-speed units since their initial introduction in 1958 and some two-speed units prior to this. Chapter 2 explains the design and operation of the Ravigneau

planetary as used in the FMX transmission and AOD units.

Basically speaking, the AOD transmission is similar to the FMX, but there are design differences. For example, the AOD has a one-piece aluminum case with an integral converter housing. The FMX has a cast-iron transmission case with a bolt-on aluminum converter housing.

The AOD transmission has four multiple-disc clutches, two bands, and two one-way clutches to control planetary operation. The FMX uses two multiple-disc clutches, two bands, and a single one-way clutch for this purpose. Furthermore, the bands in the AOD are non-adjustable; in the FMX, they are.

Selector-Lever Positions

The AOD transmission has six selector-lever positions (Fig. 14-23). The functions of each of these positions are as follows:

- **P (Park):** The transmission is in neutral; however, the output shaft is locked to the case, via a linkage-operated parking pawl engaging with a gear. This parking gear is integral with the output shaft-mounted ring gear.

- **R (Reverse):** The transmission operates in reverse at a lower gear ratio. Powerflow is from the engine through the torque converter, thus permitting additional torque multiplication beyond what the planetary provides.

- **N (Neutral):** The transmission again functions in neutral, but the output shaft is *not* locked to the case. Therefore, the parking brake should be applied if the vehicle is left stationary in neutral, since the rear wheels are free to turn.

- **O/D (Overdrive):** This is the normal driving range of the transmission. It will automatically upshift and downshift through all the forward gears: first, second, third (direct), and overdrive. From a standing start, the transmission starts the vehicle off in low gear and successively upshifts to second, direct, and overdrive as road speed increases. Also, during coast down or braking, the transmission downshifts through all the gears direct, second, and first as the vehicle comes to a stop.

- **3 (Overdrive Lockout):** The transmission operates only in first, second, and direct. The overdrive is "locked-out" in this position.

- **1 (Manual Low):** The transmission operates only in low gear. Furthermore, the low-reverse band is applied so there is engine braking through the transmission. If the driver moves the selector lever from O/D, or 3, to 1,

```
┌───┐
│ P │   — PARK
│ R │   — REVERSE
│ N │   — NEUTRAL
│(D)│   — OVERDRIVE
│ 3 │   — OVERDRIVE LOCKOUT
│ 1 │   — MANUAL LOW
└───┘
```

FIGURE 14-23 *The selector-lever positions of an AOD transmission. (Courtesy of Ford Motor Company)*

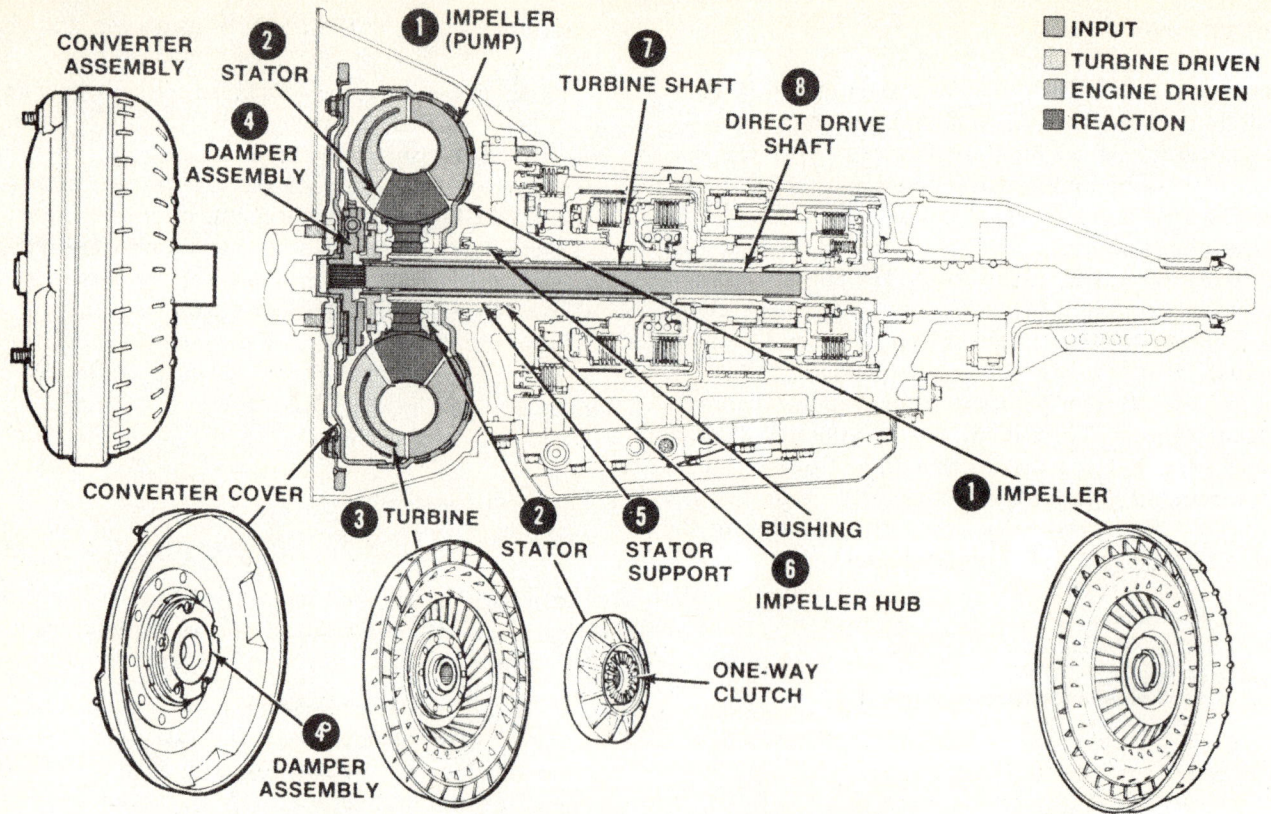

FIGURE 14-24   The location of the AOD torque converter components. (Courtesy of Ford Motor Company)

the transmission will downshift to second above about 25 mph and then to manual low as vehicle road speed drops below 25 mph.

Torque Converter

The AOD torque converter is essentially a three-element unit; however, it provides two power inputs to the gear train. The location by key number and function of the converter components are as follows (Fig. 14-24):

① **Impeller:** The impeller, also called the converter pump, is driven through the welded converter cover by the drive flex plate. Thus it is the input member of the converter and forms the rear housing for the entire welded assembly.

② **Stator:** This is the reaction or stationary member of the converter that is necessary for torque multiplication. It controls the flow of the returning fluid between the turbine and impeller so that it can be used to multiply torque.

③ **Turbine:** The turbine is the output vaned member of the torque converter. It is set in motion by fluid pumped against its vanes by the impeller.

④ **Damper Assembly:** This unit mounts in the converter cover and is necessary to smooth or cushion out engine vibrations from the transmission during direct and overdrive ratios.

⑤ **Stator Support:** This unit is part of the transmission oil pump and is responsible for not only supporting the stator but absorbing its reverse twist when the converter multiplies torque.

⑥ **Impeller Hub:** This component is built into the converter housing and supports it in a bushing within the oil pump. Also, the hub is responsible for driving the pump gears.

⑦ **Turbine Shaft:** It transmits torque from the turbine to the reverse or forward clutch in reverse, first, and second gears. This is a fluid drive since the impeller must drive the turbine using transmission fluid. Also,

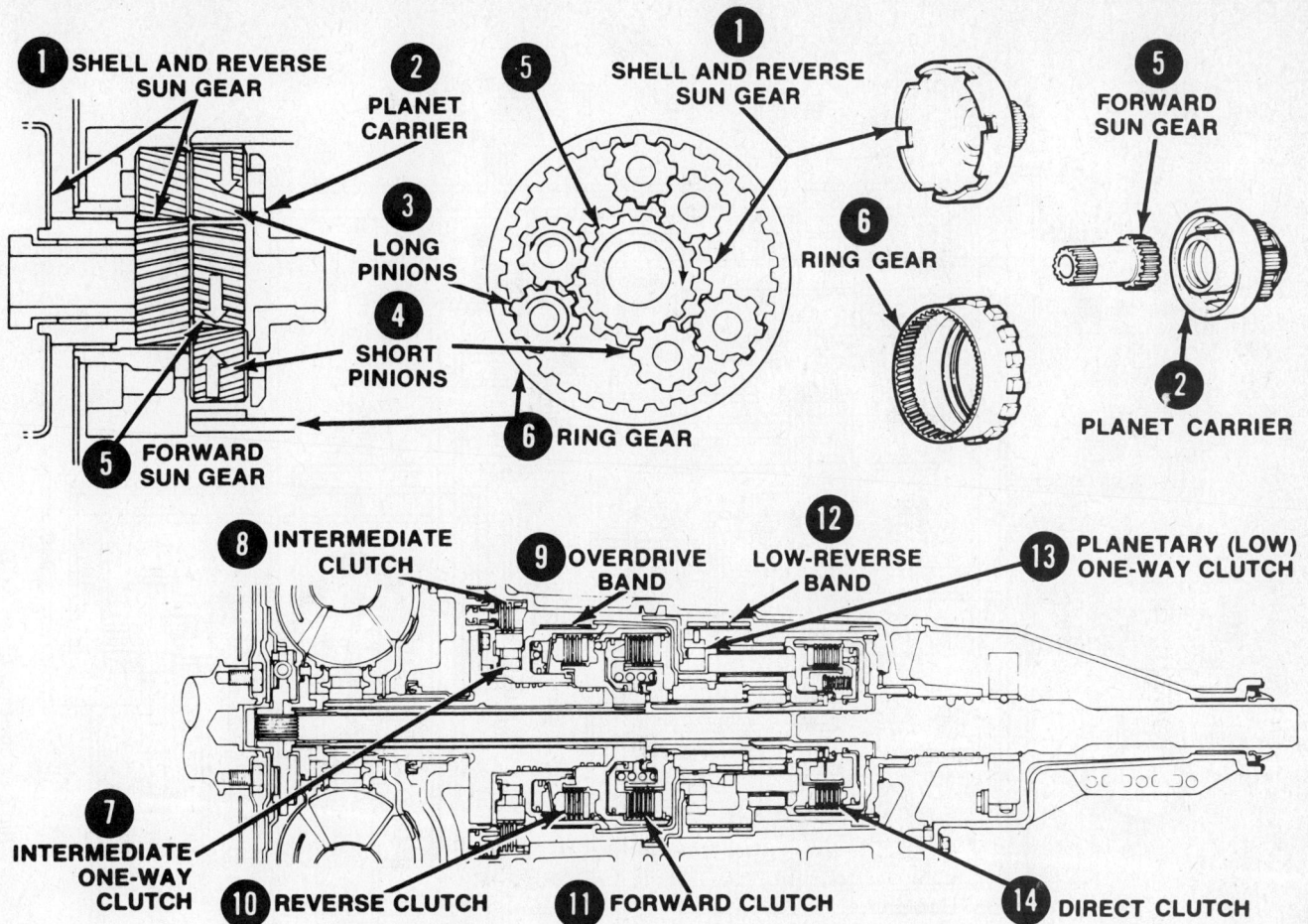

FIGURE 14-25 The planetary, clutches, and bands of the AOD transmission. (Courtesy of Ford Motor Company)

in direct (third) ratio, the turbine shaft provides 40 percent of the input torque to the planetary.

⑧ **Direct Drive Shaft:** This unit forms a mechanical connection between the damper assembly and the direct clutch assembly. The shaft, in third ratio, delivers 60 percent of the input torque to the planetary while in overdrive it provides a direct mechanical coupling from the converter cover to the planetary, via the direct clutch.

Planetary Gear Train, Clutches, and Bands

Figure 14-25 illustrates the planetary gear train clutches, and bands of the AOD transmission. To make it easier for the reader to locate these parts during the upcoming discussion, each has a circled key number.

① **Reverse Sun Gear:** This unit is integral with the input shell and turns separately from the forward sun gear. It is driven by the turbine shaft whenever the reverse clutch is applied. Also, the reverse sun gear can be held stationary by the application of the intermediate clutch or overdrive band.

② **Planet Carrier:** The AOD planetary has a single carrier with long and short pinions. The carrier can be held stationary by either the low-reverse band or low one-way clutch.

③ **Long Pinions:** These components of the carrier are in constant mesh with the ring gear, reverse sun gear, and the short pinions.

④ **Short Pinions:** The short pinions are in constant mesh with the forward sun gear and long pinions. They do not mesh with the ring gear, but the pinions can drive the ring gear through the long pinions.

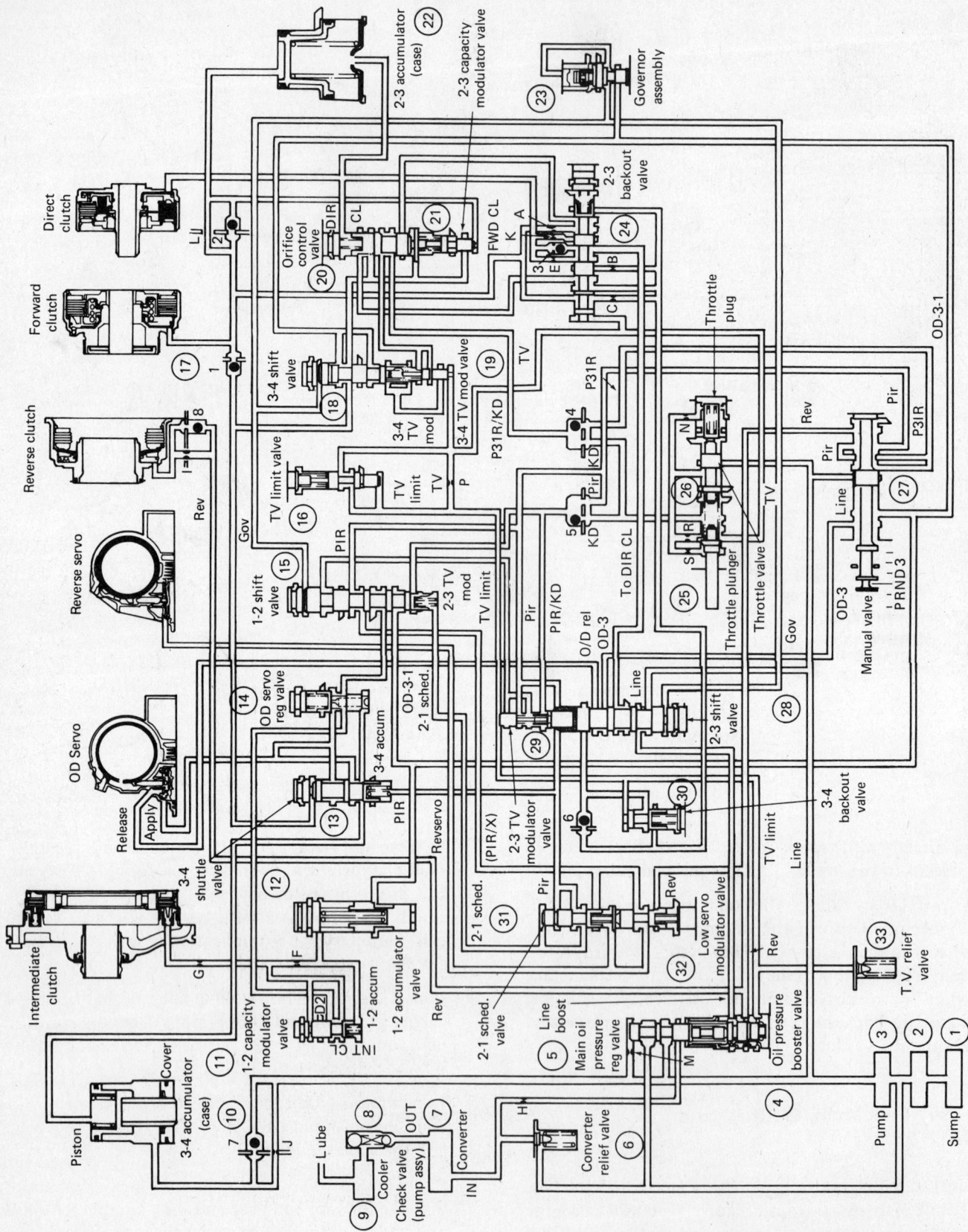

FIGURE 14-26 The hydraulic components of an AOD transmission. (Courtesy of Ford Motor Company)

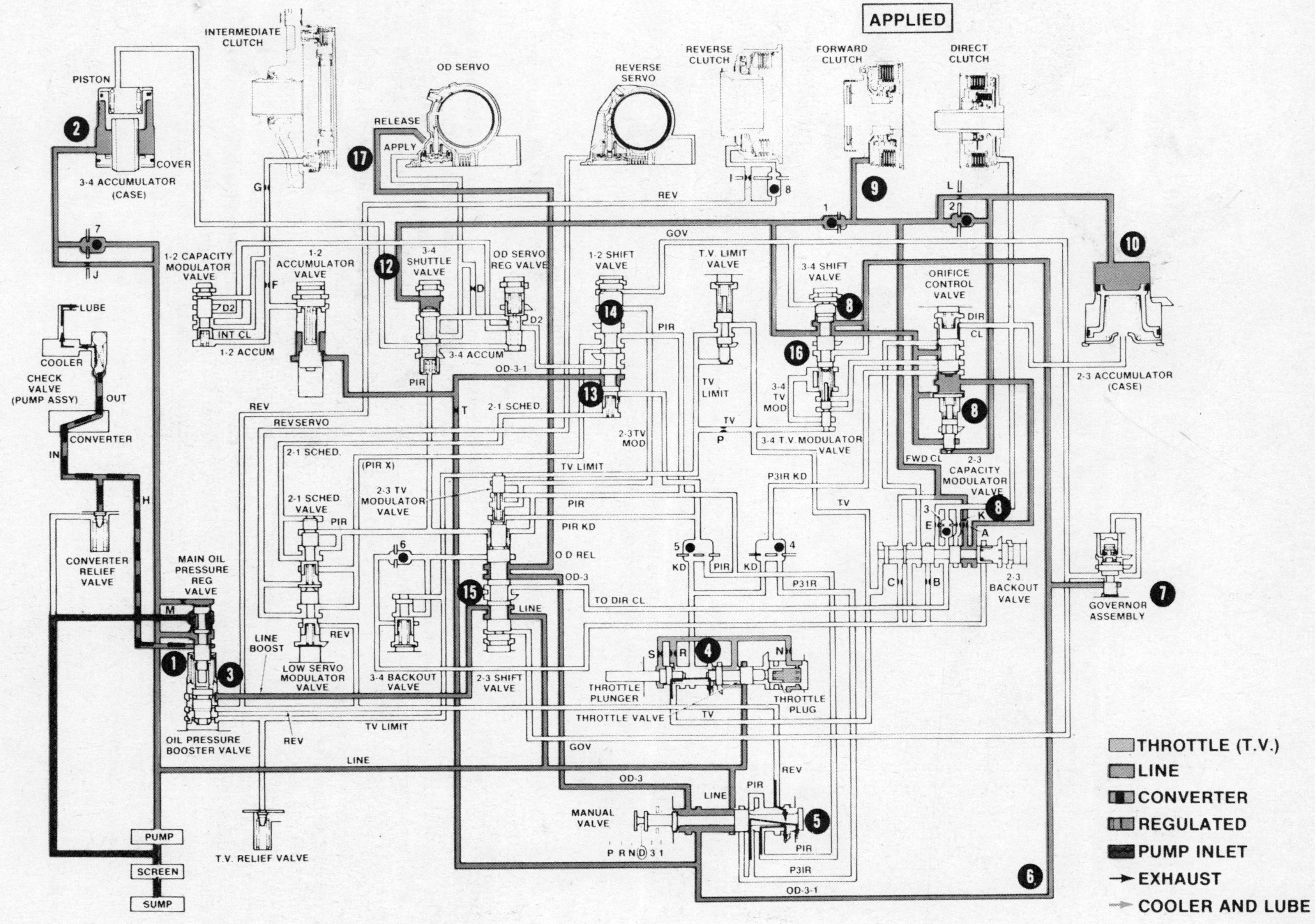

FIGURE 14-27   A hydraulic schematic of an AOD transmission in overdrive range, first gear. (Courtesy of Ford Motor Company)

253

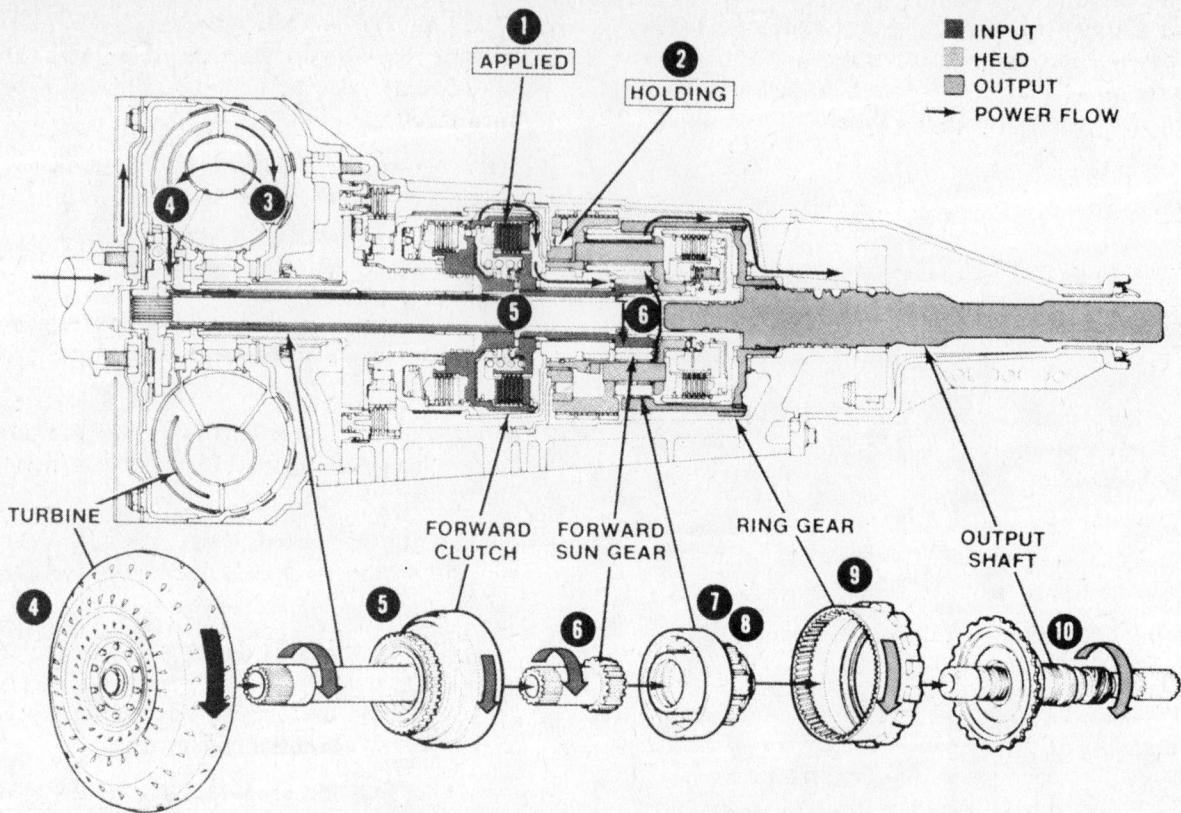

FIGURE 14-28 The first gear powerflow of an AOD transmission. (Courtesy of Ford Motor Company)

⑤ **Forward Sun Gear:** This component is driven whenever the forward clutch applies; that is, in first, second, and third ratios. Also, this sun gear meshes with the short planet pinions.

⑥ **Ring Gear:** The AOD transmission uses only one ring gear. It meshes with the long planet pinions and splines to a flange on the output shaft. Thus, the ring gear forms the only output member of the planetary.

⑦ **Intermediate One-Way Clutch:** Its inner race locks onto the reverse clutch drum and holds this unit against counterclockwise rotation in second gear, provided the intermediate friction clutch is on. The one-way clutch allows the reverse drum to freewheel clockwise in third and second with the selector lever in O/D range.

⑧ **Intermediate Clutch:** It is a multiple-disc clutch that when applied in second, third, and overdrive holds the intermediate one-way clutch outer race stationary (locks it to the case).

⑨ **Overdrive Band:** This unit wraps around the reverse clutch drum to hold it from rotation in either direc-

tion. Therefore, when applied in overdrive, it holds the drum and the input shell and reverse sun gear assembly stationary.

⑩ **Reverse Clutch:** It is a multiple-disc clutch that applies in reverse to connect the turbine shaft to the input shell and reverse sun gear assembly. As a result, this sun gear will be driven at turbine speed.

⑪ **Forward Clutch:** It is a multiple-disc clutch that connects the turbine shaft to the forward sun gear when applied in first, second, and third gear ratios. As a result, this sun gear will be operating at turbine speed during these ratios.

⑫ **Low-Reverse Band:** This unit wraps around the outer drum surface of the planet carrier. When applied in manual low and reverse, the band holds the carrier from turning in either direction.

⑬ **Low One-Way Clutch:** A roller-type overrunning clutch that prevents the planet carrier from turning counterclockwise during the first gear ratio. If the carrier turns clockwise, the clutch overruns or freewheels.

⑭ **Direct Clutch:** It is a multiple-disc clutch that when applied in direct and overdrive gear ratios connects the direct drive shaft to the planet carrier. As a result, the carrier operates at engine speed.

Hydraulic System Components

Figure 14-26 shows the main components of the hydraulic system of the AOD transmission. To make it easier for the reader to locate these parts during the upcoming discussion, each has a circled key number.

① **Sump:** The oil pan is the sump for the transmission and contains a supply of ATF needed to operate the hydraulic system.

② **Screen:** This unit protects the pump inlet from the entrance of dust and other foreign material that may cling to the fluid.

③ **Oil Pump:** It circulates fluid to the hydraulic system whenever the engine is running.

④ **Oil Pressure Booster Valve:** This unit increases or decreases line pressure in relation to the opening of the throttle. Also, this valve provides different line pressures depending on selector position or gear ratio.

⑤ **Main Regulator Valve:** It controls line pressure within the hydraulic system.

⑥ **Converter Relief Valve:** This unit prevents excessive buildup of pressure in the torque converter.

⑦ **Converter:** It acts as a fluid clutch and torque multiplier.

⑧ **Check Valve:** This valve prevents fluid in the converter from draining back when the engine is shut off.

⑨ **Cooler:** It removes the heat generated in the fluid within the torque converter and transmission.

⑩ **3-4 Accumulator:** This device cushions the 3-4 upshift by controlling the rate of overdrive servo application.

⑪ **1-2 Capacity Modulator Valve:** And

⑫ **1-2 Accumulator Valve:** Both units operate together to smooth out the 1-2 upshift.

⑬ **3-4 Shuttle Valve:** This valve controls the operation of the OD servo regulator valve and 3-4 accumulator.

⑭ **OD Servo Regulator Valve:** It determines the apply pressure for the overdrive servo on the 4-3 downshift.

⑮ **1-2 Shift Valve:** This device controls the automatic 1-2 upshift and 2-1 downshift.

⑯ **TV Limit Valve:** This valve regulates the throttle pressure that acts on the pressure booster valve and the 2-3 shift valve by limiting TV pressure at wide-open throttle.

⑰ **Check Valve:** There are eight ball-check valves in this system used as shuttle or one-way valves.

⑱ **3-4 Shift Valve:** It controls the automatic 3-4 upshift and 4-3 downshift.

⑲ **3-4 TV Modulator Valve:** This unit regulates the throttle valve (TV) pressure that acts on the 3-4 shift valve.

⑳ **Orifice Control Valve:** This valve prevents the transmission from shifting out of second ratio directly into overdrive.

㉑ **2-3 Capacity Modulator Valve:** It controls the back-pressure on the 2-3 accumulator in order to smooth the 2-3 upshift.

㉒ **2-3 Accumulator:** This device cushions the application of the direct clutch to provide a smooth 2-3 upshift. Also, it cushions forward clutch application to smooth out its initial engagement.

㉓ **Governor:** This unit provides a hydraulic signal to the system that is in proportion to vehicle road speed.

㉔ **2-3 Backout Valve:** It controls the fluid flow rate to the direct clutch. Also, the valve operates with the orifice control valve to govern the flow rate to the forward clutch.

㉕ **Throttle Plunger:** This component varies the spring force on the throttle valve with accelerator pedal travel. Also, it operates the kickdown system at wide-open throttle.

㉖ **Throttle Valve:** It provides a signal to the hydraulic system that is in proportion to engine load.

㉗ **Manual Valve:** This valve is activated by linkage from the gearshift-selector lever. As the valve moves, it directs fluid to apply clutches or servos and provide the automatic functions of the hydraulic system.

㉘ **2-3 Shift Valve:** This valve controls the automatic 2-3 upshift and 3-2 downshift.

㉙ **2-3 TV Modulator Valve:** It regulates TV pressure acting on the 1-2 and 2-3 shift valves.

㉚ **3-4 Backout Valve:** This valve smooths out the 3-4 upshift if it should occur at a closed throttle.

㉛ **2-1 Scheduling Valve:** It determines the 2-1 downshift speed when the gearshift-selector lever is moved to manual low from direct or overdrive range.

㉜ **Low Servo Modulator Valve:** This device regulates the low and reverse servo apply pressure to provide a smooth 2-1 downshift in manual low range.

㉝ **TV Relief Valve:** This valve prevents excessive TV limit pressure if the TV limit valve fails to regulate properly.

AOD OPERATION

First Gear: Hydraulic Operation

Figure 14-27 shows what is occurring within the hydraulic system of the AOD transmission in overdrive range, first gear at closed throttle. By key number, the sequence of events that take place within the system is as follows:

① The main regulating valve is controlling the pressure of the fluid coming from the pump. Fluid also flows from the regulator valve through the converter, cooler, and lube circuits.

② The 3-4 accumulator piston moves within its bore due to the force of line pressure.

③ Since the 2-3 shift valve is outward in its bore, line boost pressure causes line pressure to increase.

④ All throttle valve pressure is exhausted, and there will be no TV pressure in the system until the driver depresses the accelerator pedal.

⑤ At the manual valve, the REV, P-1-R, and P-3-1-R passages are all open to the sump.

⑥ The OD-3-1 passage from the manual valve fills the forward clutch and governor system with fluid.

⑦ Until the vehicle begins to move, there will be no governor pressure.

⑧ Fluid flows through the 3-4 shift valve, orifice control valve, the A orifice, and to the FWD CL passage.

⑨ The FWD CL passage becomes pressurized and begins to apply the forward clutch.

⑩ Forward clutch apply pressure also forces the 2-3 accumulator piston in its bore against spring tension; this action controls the amount of apply pressure, resulting in a smooth forward clutch engagement.

⑪ The fluid from the OD-3-1 circuit also acts on the 1-2 accumulator piston, forcing it to move in its bore.

⑫ Fluid from the OD-3-1 circuit, flowing through the 3-4 shift valve, holds the 3-4 shuttle valve inward.

⑬ Line pressure from the OD-3-1 circuit acts on a dif-ferential area of the 1-2 shift valve to oppose governor pressure.

⑭ As the driver depresses the accelerator pedal, TV limit pressure becomes high enough to move the 2-3 TV modulator valve into regulation; additional pressure will now be applied to a differential area between the outer lands of the 1-2 shift valve.

⑮ Fluid from the LINE passage acts on a differential area of the 2-3 shift valve to oppose governor pressure.

⑯ Line pressure is effective on the 3-4 shift valve, holding it outward.

⑰ The OD-3 passage directs fluid to the release side of the overdrive servo.

First Gear: Powerflow

With the selector lever in D, or 3 range, the transmission's hydraulic system applies the forward clutch, ① of Fig. 14-28, and the planetary (LOW) one-way clutch ② is holding. The forward clutch locks the turbine shaft to the forward sun gear, while the one-way clutch holds the carrier from rotating counterclockwise.

By key number, the powerflow through the planetary is as follows:

③ The engine turns the impeller within the converter clockwise.

④ The impeller hydraulically drives the turbine clockwise.

⑤ The turbine shaft turns the forward clutch clockwise.

⑥ The forward clutch drives the forward sun gear clock49í ⁵ pinions turn the long pinions clockwise.

⑨ The long pinions drive the ring gear clockwise at a slower speed than the turbine shaft.

⑩ The ring gear drives the output shaft clockwise at a gear ratio of 2.4:1.

Second Gear: Hydraulic Operation

Figure 14-29 illustrates what is occurring within the hydraulic system of the AOD transmission in second gear in overdrive range. As listed by key number, the sequence

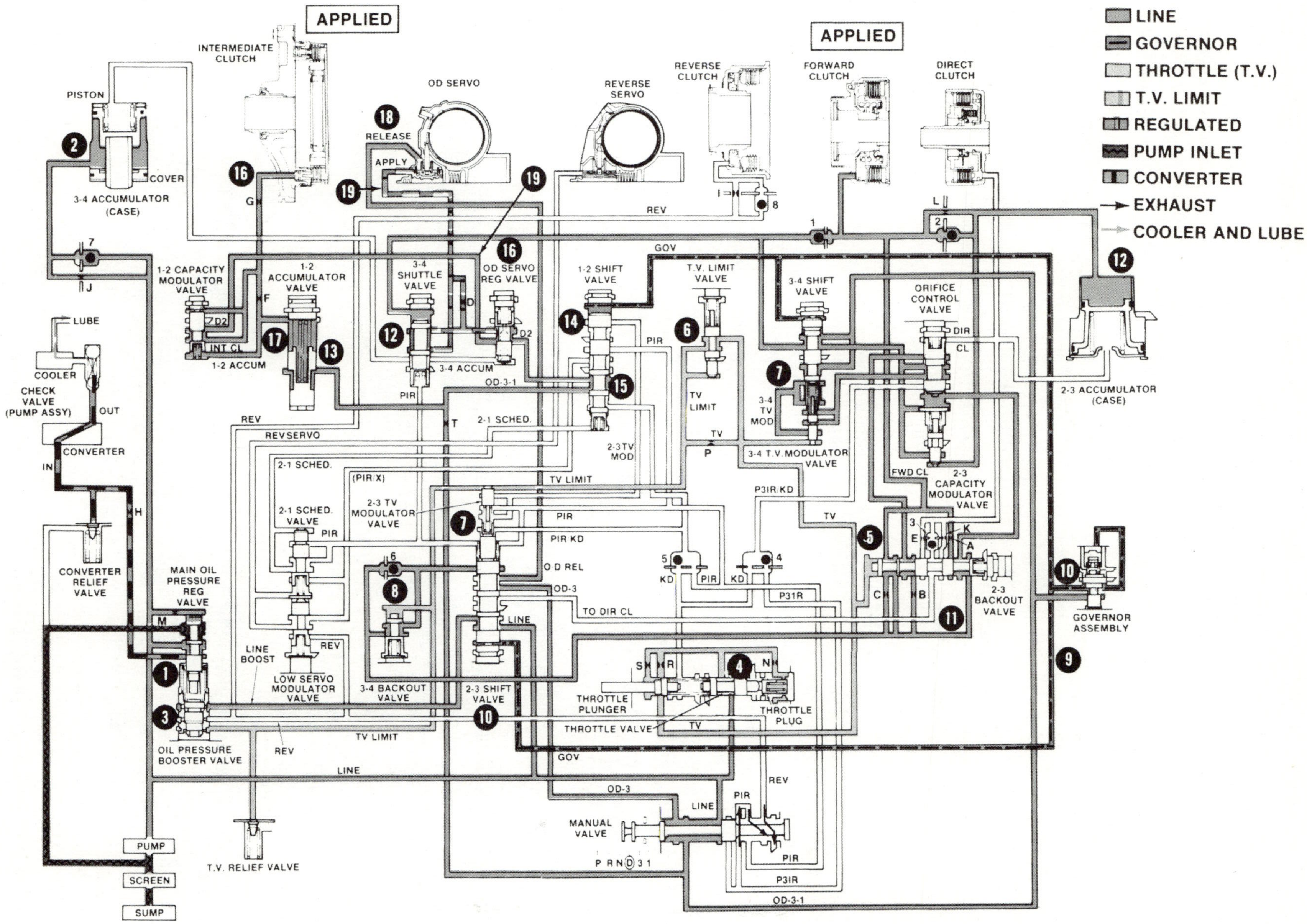

FIGURE 14-29 A hydraulic schematic of an AOD transmission in overdrive range, second gear. (Courtesy of Ford Motor Company)

of events which take place within the system occurs in the following manner:

① The main oil pressure regulating valve is controlling the pressure of the fluid from the pump. Also, fluid is flowing from the regulator valve to the converter, cooler, and lube circuits.

② The 3-4 accumulator piston remains in the same position as in low gear.

③ Since the 2-3 shift valve has not moved, LINE BOOST pressure is causing an increase in control pressure. Also, since the accelerator pedal is depressed, there is sufficient TV pressure to force the pressure booster valve against the regulator valve. This action increases line (control) pressure as TV pressure intensifies.

④ The throttle plunger has moved in; consequently, the throttle valve is producing a regulated TV pressure in the range of 30 psi.

⑤ TV pressure moves the 2-3 backout valve outward in its bore.

⑥ Throttle valve pressure causes fluid flow to and through the TV limit valve, until the pressure reaches about 85 psi.

⑦ Because there is relatively low throttle pressure, the TV modulator valves are not producing modulated pressure to delay the upshifts.

⑧ TV limit pressure forces the 3-4 backout valve outward in its bore.

⑨ The OD-3-1 circuit to the governor and forward clutch remain full of fluid at line pressure.

⑩ The vehicle is in motion, so the governor directs a speed signal to all the shift valves.

⑪ Due to the action of TV pressure on the 2-3 backout valve, forward clutch apply pressure is rerouted a different way through it, bypassing orifice A.

⑫ Clutch apply pressure continues to hold the 2-3 accumulator piston down in its bore and the 3-4 shuttle valve in.

⑬ OD-3-1 circuit pressure remains effective on the differential area of the 1-2 accumulator.

⑭ Governor pressure has moved the 1-2 shift valve in its bore.

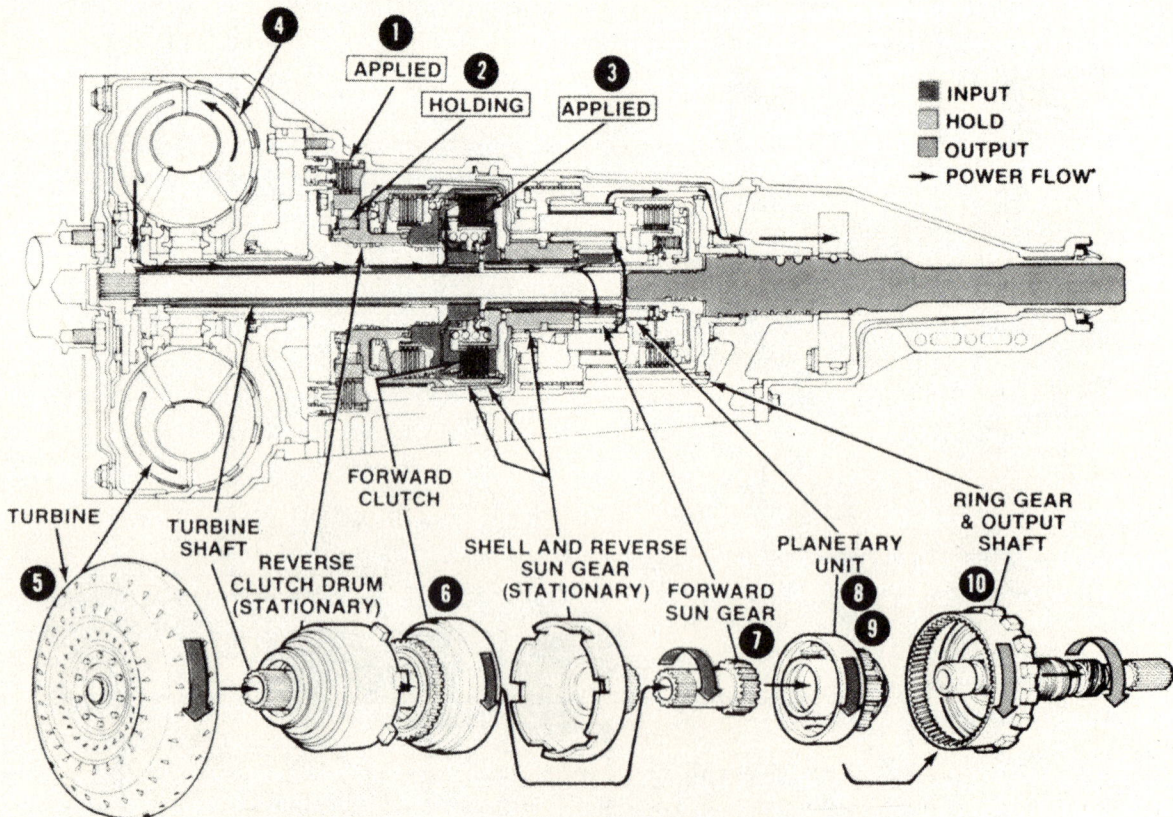

FIGURE 14-30 *The second gear powerflow of an AOD transmission. (Courtesy of Ford Motor Company)*

⑮ The OD-3-1 circuit now supplies fluid through the 1-2 shift valve to the OD servo regulator valve.

⑯ Fluid from the OD-3-1 circuit flows through the 1-2 capacity modulator to the INT CL passage. This applies the intermediate clutch.

⑰ The clutch apply pressure is also effective, through orifice 5, to the spring end of the 1-2 accumulator. The accumulator has moved down to cushion the intermediate clutch application.

⑱ The OD-3 and OD REL passages supply fluid under pressure to the release side of the overdrive servo.

⑲ The OD servo regulator valve provides a regulated amount of fluid pressure to the apply side of the overdrive servo to keep it released.

Second Gear: Powerflow

With the selector lever in D, or 3, range and the transmission in second gear, the hydraulic system applies the intermediate clutch (① of Fig. 14-30). This clutch locks the one-way clutch outer race stationary. Also, the one-way clutch ② prevents the reverse clutch drum, shell, and reverse sun gear from turning counterclockwise. The forward clutch ③ remains applied as in low gear.

By key number, the powerflow through the planetary is as follows:

④ The engine turns the converter's impeller clockwise.

⑤ The impeller hydraulically drives the turbine clockwise.

⑥ The turbine shaft rotates the forward clutch clockwise.

⑦ The forward clutch turns the forward sun gear clockwise.

⑧ The forward sun gear drives the short pinions within the planetary counterclockwise.

⑨ The long pinions are driven clockwise and "walk" the carrier around the stationary reverse sun gear.

⑩ The long pinions also spin the ring gear and output shaft clockwise at a reduced speed. The gear ratio is 1.47:1.

Third Gear (Direct): Hydraulic Operation

Figure 14-31 shows what takes place within the hydraulic system of the AOD transmission in third gear in overdrive range. By key number, the sequence of events which take place within the system is as follows:

① The regulator valve is controlling the high volume of fluid output from the pump. The converter, cooler, and lube system are receiving a constant supply of fluid from the regulator valve.

② The 3-4 accumulator remains up in its bore in preparation to cushion the 3-4 upshift.

③ TV limit pressure is increasing line pressure in response to engine load. LINE BOOST pressure, at this time, is exhausted.

④ The OD-3 and OD-3-1 passages at the manual valve remain charged with fluid at line pressure. (The P-1-R, P-3-1-R, and REV passages are exhausted.)

⑤ The throttle valve is producing high TV pressure, which holds the 2-3 backout valve outward in its bore.

⑥ The 3-4 TV modulator valve is regulating modulated TV pressure to the 3-4 shift valve.

⑦ The TV limit valve is regulating TV limit pressure to about 85 psi. This pressure moves the 3-4 backout valve out in its bore and causes the 2-3 TV modulator valve to begin its regulating cycle.

⑧ Line pressure remains available to the governor, which regulates a pressure signal in response to the road speed.

⑨ The forward clutch remains applied, and the fluid on top of the accumulator piston is still under pressure.

⑩ The OD servo regulator is controlling apply pressure to the servo, and line pressure continues to apply to intermediate clutch.

⑪ OD REL circuit now receives fluid from the forward clutch apply circuit instead of the OD-3 passage. This is due to the opening of the 2-3 shift valve by governor pressure.

⑫ Th OD-3 passage now supplies fluid to the DIR CL passage via the 2-3 shift valve.

⑬ Direct clutch apply fluid passes through orifice K.

⑭ The direct clutch applies due to the action of the fluid from the DIR CL passage.

⑮ The force of the fluid from the DIR CL passage moves the 2-3 accumulator piston up in its bore to cushion the shift.

⑯ Fluid above the accumulator piston seats the #2 check ball and flows past the 2-3 capacity modulator valve. It operates in regulation during the 2-3 upshift to control the accumulator back pressure. When the shift is complete, the valve returns to its normal in position.

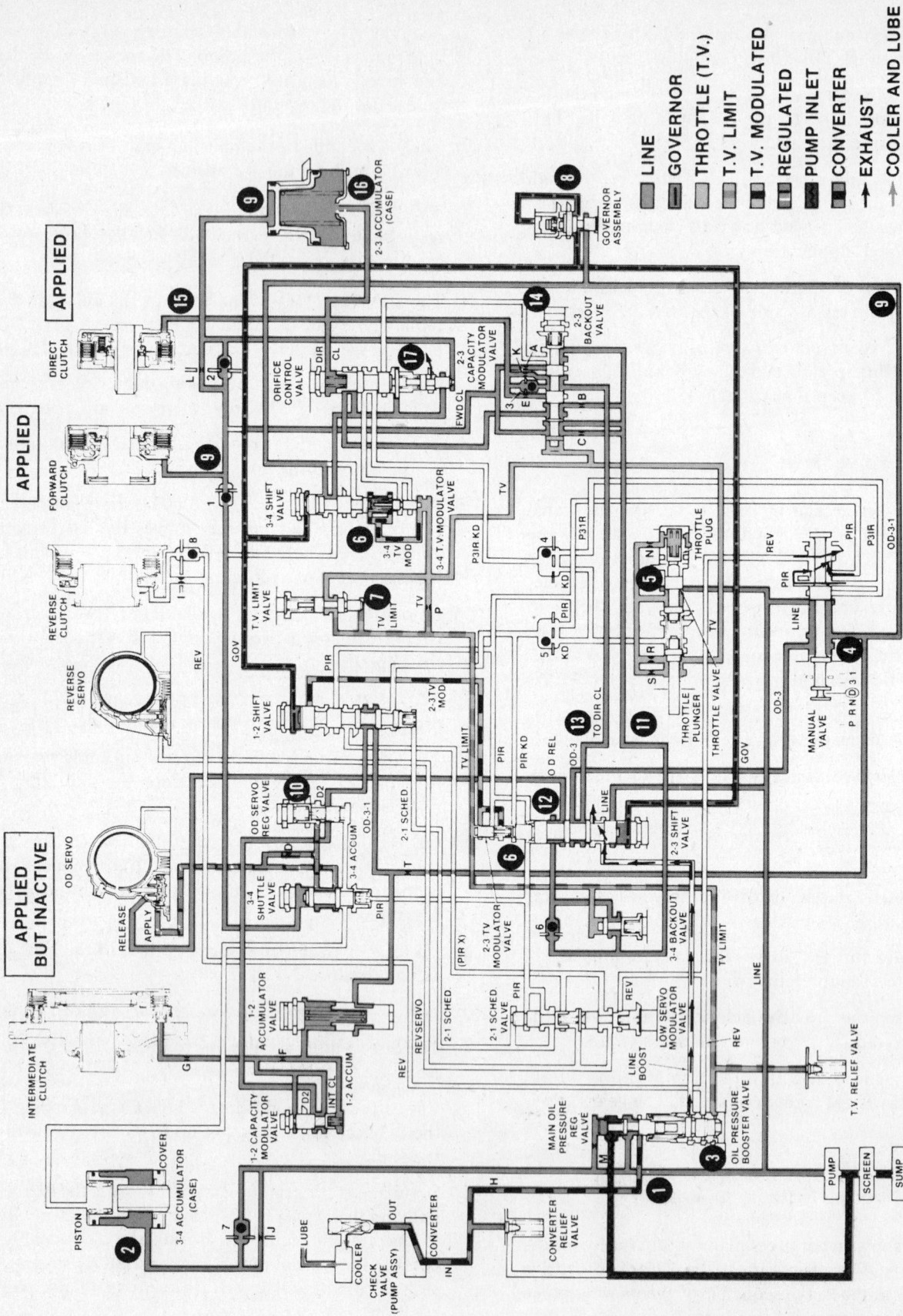

FIGURE 14-31   A hydraulic schematic of an AOD transmission in overdrive range, third gear. (Courtesy of Ford Motor Company)

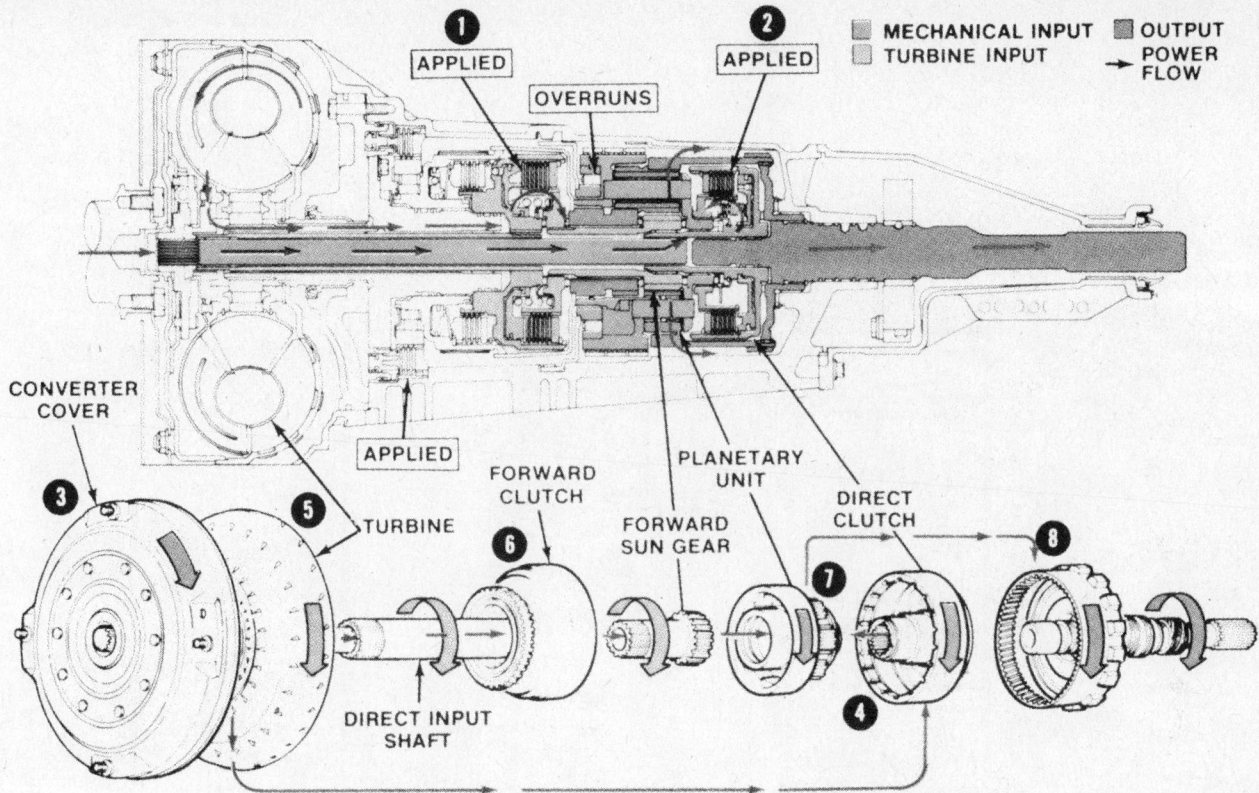

FIGURE 14-32 The third gear powerflow of an AOD transmission. (Courtesy of Ford Motor Company)

Third Gear (Direct): Powerflow

When an AOD transmission upshifts to third gear (direct drive), the hydraulic system applies the forward clutch (①) of Fig. 14-32) and the direct clutch ②. The forward clutch connects the turbine (tube) input shaft to the forward sun gear. The direct clutch locks the direct drive (inner) input shaft to the planet carrier.

By key number, the powerflow in third gear is as follows:

③ The cover of the converter turns the direct drive shaft and the direct clutch clockwise at engine speed.

④ The direct clutch drives the planet carrier clockwise at engine speed.

⑤ The impeller drives the turbine hydraulically at almost engine speed in a clockwise direction.

⑥ The turbine shaft rotates the forward clutch and the forward sun gear.

⑦ With two members of the planetary gear train being driven clockwise at the same time, they are, in effect, locked together. So the entire gear train locks up and rotates as a unit.

⑧ The long planet pinions drive the ring gear and output shaft clockwise in direct drive or third gear. For all practical purposes, the gear ratio is 1:1.

Fourth Gear (Overdrive): Hydraulic Operation

Figure 14-33 illustrates what takes place within the hydraulic system of and AOD transmission in fourth (overdrive) gear. By key number, the sequence of events which take place within the system is as follows:

① The main oil pressure regulating valve is controlling line pressure and directing fluid into the converter, cooling, and lube systems.

② Line pressure remains applied under the 3-4 accumulator piston.

③ The OD-3 and OD-3-1 passages also remain charged with fluid under pressure at the manual valve. Thus, there is fluid supplied to the governor and to the OD servo regulator valve.

④ The throttle valve continues to regulate a moderate amount of TV pressure, around 30 psi. TV and TV limit pressure force both the 2-3 and 3-4 backout valves outward in their bores.

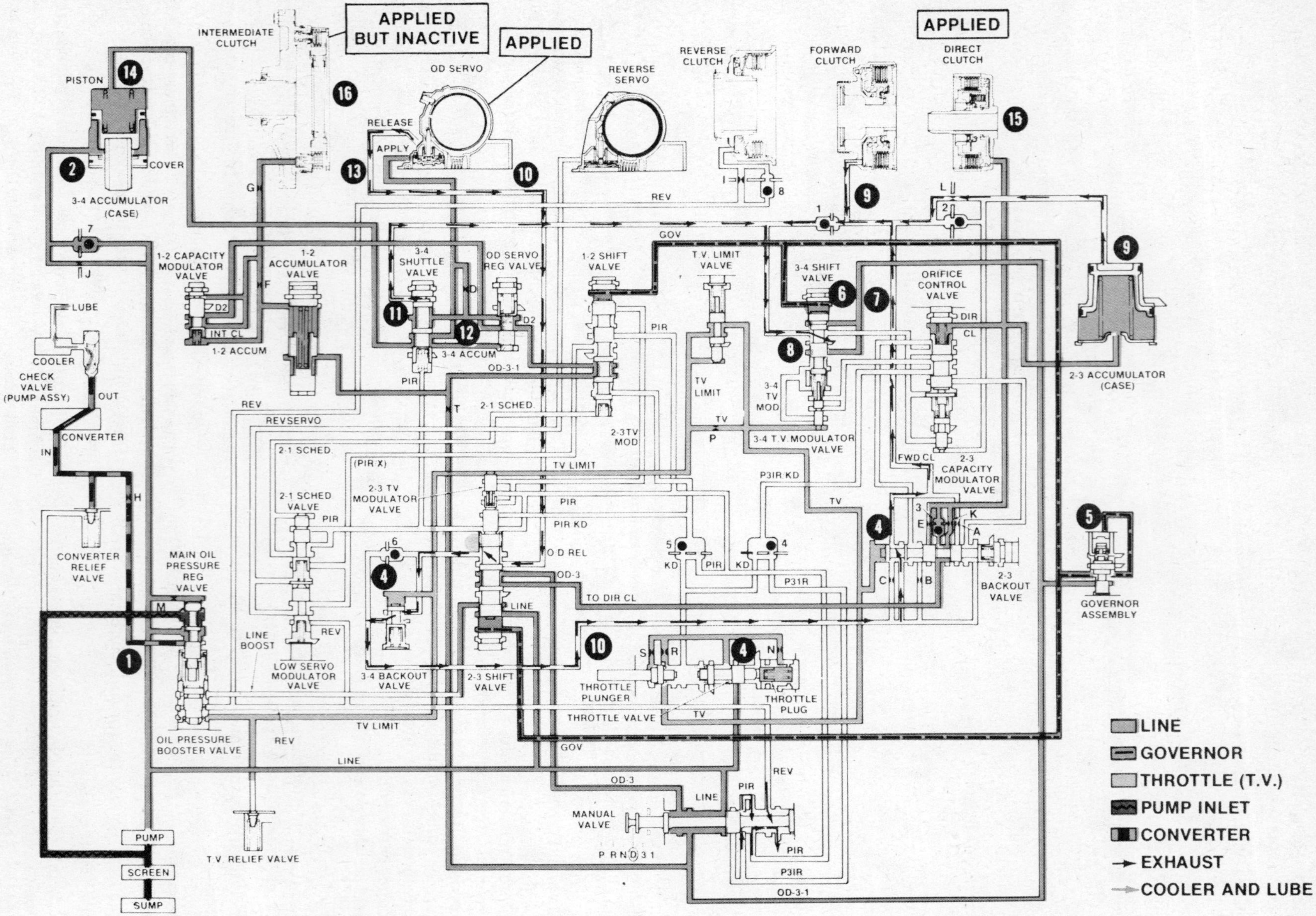

FIGURE 14-33 A hydraulic schematic of an AOD transmission in overdrive (fourth gear). (Courtesy of Ford Motor Company)

⑤ The governor is producing a relatively high pressure signal due to an increased road speed.

⑥ Governor pressure forces the 3-4 shift valve inward in its bore.

⑦ The 3-4 shift valve now blocks the forward clutch fluid apply passage.

⑧ The 3-4 shift valve opens an exhaust passage for the forward clutch apply circuit.

⑨ Fluid pressure from the forward clutch and from on top of the 2-3 accumulator exhausts to the sump.

⑩ The OD servo releases pressure exhaust through the 2-3 shift valve, 3-4 backout valve, and 2-3 backout valve.

⑪ Fluid pressure at the end of the 3-4 shuttle valve exhausts to the sump, and the valve moves outward in its bore.

⑫ OD-3-1 pressure is now acting on the inner end of the OD servo regulator valve. This pressure, through a drilled hole in the valve, also acts on the outer end of the valve and moves it inward in the bore. Therefore, the valve stops regulating and sends out line pressure.

⑬ Fluid under full line pressure is applied to the apply side of the overdrive servo through orifice D.

⑭ Servo apply pressure is also effective above the 3-4 accumulator piston, which moves it down to cushion the upshift.

⑮ The direct clutch remains applied via fluid supplied through the DIR CL passage from the 2-3 shift valve.

⑯ The intermediate clutch also remains applied, but it is inactive at this time.

Fourth Gear (Overdrive): Powerflow

When the AOD transmission upshifts to fourth (overdrive) gear, the hydraulic system applies the overdrive band (① of Fig. 14-34) and the direct clutch ②). The overdrive band holds the reverse sun gear stationary through the reverse clutch drum and driving shell. The direct clutch connects the planet carrier to the engine, via the converter cover.

By key number, the powerflow in fourth is as follows:

③ The converter cover turns the direct drive shaft clockwise at engine speed.

④ The direct drive shaft spins the direct clutch.

⑤ The direct clutch rotates the planet carrier clockwise at engine speed.

⑥ The long pinions "walk" the carrier around the stationary reverse sun gear in a clockwise direction.

⑦ The ring gear and output shaft are driven at a faster speed than the direct drive shaft by the rotation of the long planet pinions. The gear ratio is 0.67:1.

Reverse Gear: Hydraulic Operation

Figure 14-35 shows what occurs within the hydraulic system of an AOD transmission in reverse gear. By key number, the sequence of events that take place within the system is as follows:

① The main oil pressure regulating valve is controlling pump output. Fluid is flowing from the regulator valve to the converter, cooler, and lube circuits.

② Line pressure forces the 3-4 accumulator piston up in its bore.

③ LINE BOOST and REV boost pressures act on the oil pressure booster valve to raise the line pressure.

④ The manual valve supplies fluid under pressure to the REV, P-1-R, and P-3-1-R passages. The valve's position also exhausts the OD-3 and OD-3-1 passages.

⑤ With the throttle valve at idle, TV pressure is zero.

⑥ As the driver accelerates the engine to move the vehicle, the throttle valve charges the TV passage with fluid. TV limit pressure increases line pressure by acting on the oil pressure booster valve. Maximum line pressure occurs in reverse with three pressures acting on the booster valve.

⑦ Line pressure forces the low servo modulator valve over in its bore. This opens the REV passage to the REV SERVO passage.

⑧ Line pressure from the REV SERVO passage applies the reverse servo.

⑨ Full line pressure from the REV passage applies the reverse clutch. Orifice 1 delays the clutch application slightly, so the band always applies first. This smooths the initial clutch engagement.

⑩ Other passages in the valve body receive fluid under pressure from the P-1-R circuit such as the 2-1 SCHED and 2-3 TV MOD. But the pressure in these passages has no effect on reverse operation.

⑪ The P-3-1-R passage also receives fluid under pressure, which passes to the 3-4 shift valve through the orifice control valve. This action has no effect on reverse operation but does provide a fourth gear lockout for driving range 3.

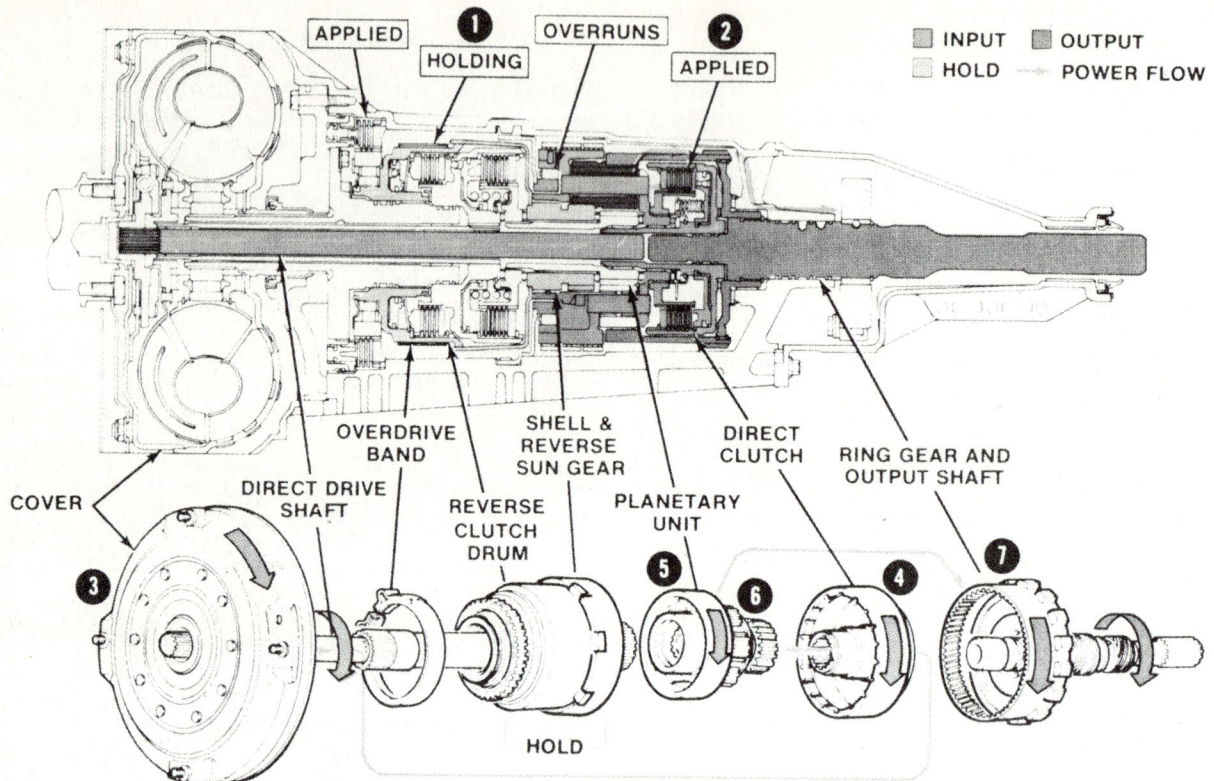

FIGURE 14-34   The fourth gear (overdrive) powerflow of an AOD transmission. (Courtesy of Ford Motor Company)

⑫ Since the manual valve exhausts the OD-3-1 passage, the governor is inactive in reverse.

Reverse Gear: Powerflow

When the driver shifts the AOD transmission into reverse, the hydraulic system applies the reverse clutch (① of Fig. 14-36) and the low-reverse band ②. The reverse clutch connects the turbine input shaft to the reverse sun gear. The low-reverse band holds the planet carrier stationary.

By key number, the powerflow is as follows:

③ The engine spins the impeller clockwise.

④ The impeller hydraulically drives the turbine clockwise.

⑤ The turbine shaft rotates the reverse clutch, shell, and the reverse sun gear clockwise.

⑥ The reverse sun gear turns the long planet pinions counterclockwise.

⑦ The long planet pinions drive the ring gear and output shaft counterclockwise at a 2:1 gear ratio.

ELECTRONICALLY CONTROLLED AUTOMATIC TRANSMISSIONS

Function

The use of electronics in the operation of the automobile has become extensive. For example, since the mid-1970s, manufacturers have been using a number of electronic components in automobile ignition and fuel systems in order to reduce harmful exhaust emissions and to increase vehicle fuel economy. Also, many vehicles now have partial-to-full electronic digital instrumentation on the dashboard in place of the mechanically operated speedometer and electrically activated needle-type gauges.

The shift to the use of electronic components is mainly due to the accuracy under which they operate. In other words, an electronic part can perform a given function much faster and with a great deal more accuracy than, let's say, linkage or a contact point switch, whether they be operated mechanically, electrically, or pneumatically. Also, electronic parts can be less expensive to produce

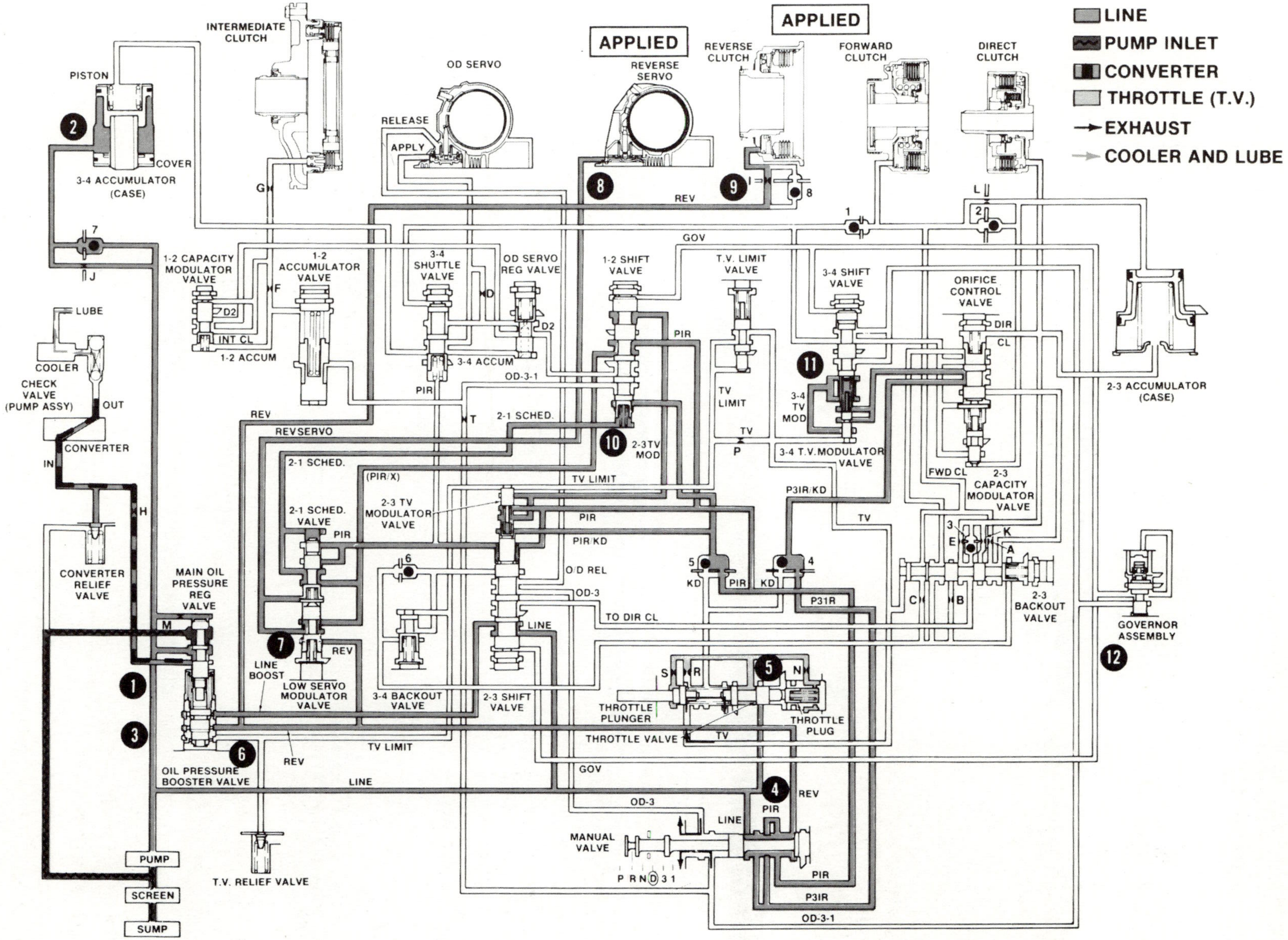

FIGURE 14-35 A hydraulic schematic of an AOD transmission in reverse gear. (Courtesy of Ford Motor Company)

265

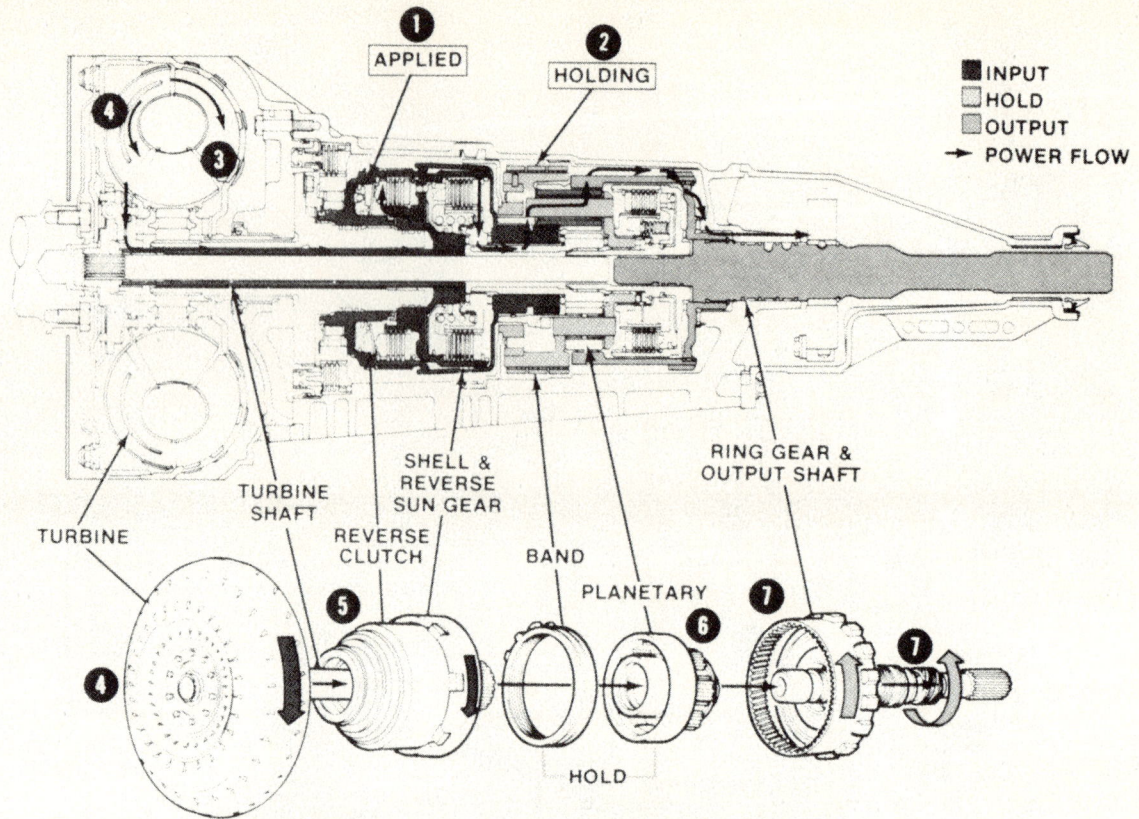

FIGURE 14-36   The reverse gear powerflow of an AOD transmission. (Courtesy of Ford Motor Company)

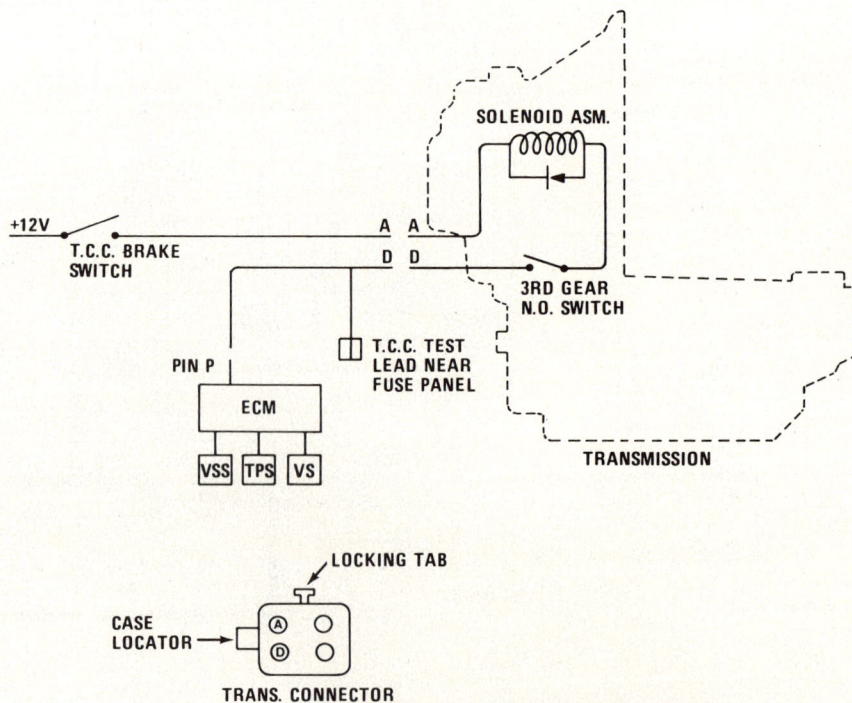

FIGURE 14-37   A schematic showing the components of a typical computer command control TCC system.

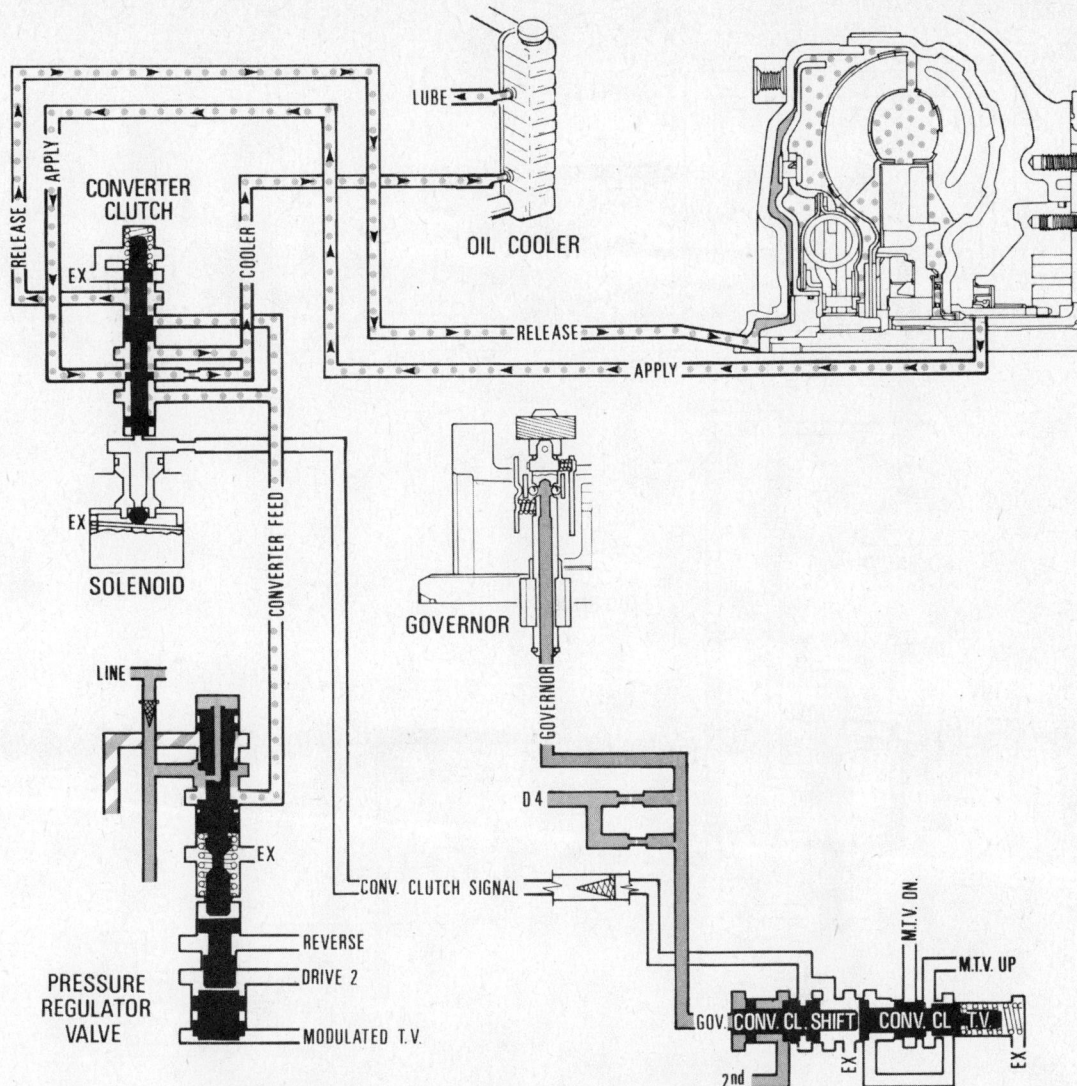

FIGURE 14-38   A schematic of the TCC hydraulic components in the release position. (Courtesy of General Motors Corp.)

and install on a vehicle to perform a given function than one that operates by other means.

In regard to the operation of the automatic transmission, the use of electronic control has not been too extensive. However, a number of vehicles have a transmission that utilizes some electronic control over one aspect of its operation. Automatic transmissions used with the General Motors Computer Command Control (CCC) system are a good example. In these units, the lock-up section of the converter is electronically controlled.

Design

Although the CCC system consists of a large number of components, the ones used to activate the lock-up phase of converter operation include the following (Fig. 14-37):

1. **TCC Brake Switch:** This device prevents an engine stall, anytime the brakes are applied, by releasing the torque converter clutch (TCC).

2. **3rd Gear Switch:** It breaks the solenoid circuit to the ECM until the transmission upshifts into third gear, thus preventing converter lock-up in the lower gear ratios.

3. **Vacuum Switch (VS):** This unit sends engine vacuum (load) information to the electronic control module.

4. **Throttle Position Sensor (TPS):** It signals the electronic control module information on changes in carburetor throttle valve position from its closed to wide-open position.

5. **Vehicle Speed Sensor (VSS):** This device provides a series of electrical pulses to the electronic control module, indicating vehicle speed.

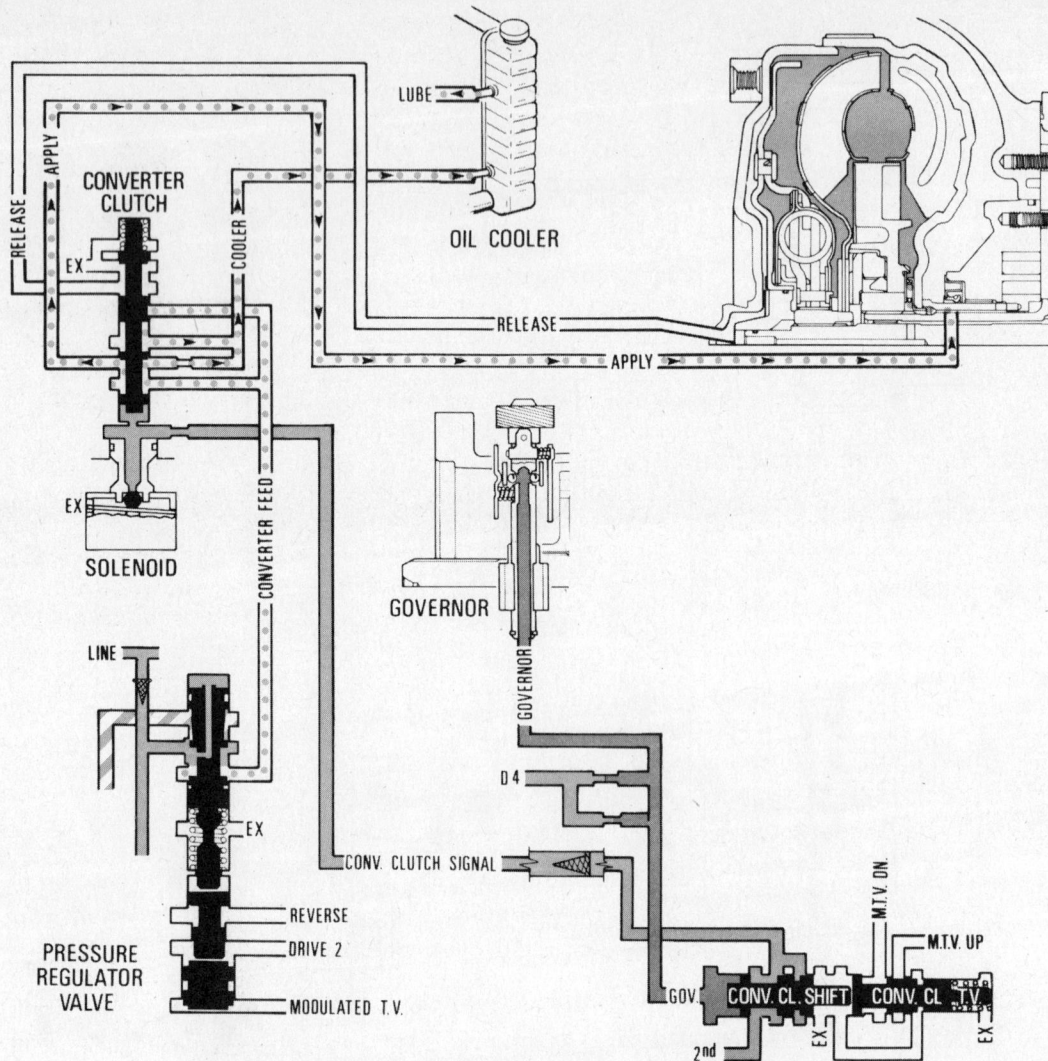

FIGURE 14-39 A schematic of the TCC hydraulic components in the applied position. (Courtesy of General Motors Corp.)

6. **Electronic Control Module (ECM):** This unit processes the data information from the VS, TPS, and VSS units. Depending on the data received, the ECM is programmed to open and close the ground circuit for the solenoid assembly.

7. **Solenoid Assembly:** A solenoid-operated hydraulic valve that controls fluid flow to the torque converter clutch. The electrical current needed to activate the unit has to pass through the closed brake switch, the solenoid windings, and to ground via the ECM.

Operation

In park, neutral, reverse, first, and second gear, the converter lock-up clutch is in the released position. This action is due not only to the de-energizing of the solenoid but also partially to the action of valving within the hydraulic system (Fig. 14-38). The solenoid remains not energized, regardless of ECM action, until the transmission upshifts to third, so lock-up cannot occur during other gear ranges.

In the hydraulic system, the clutch apply valve is held in the release position by its spring. In this position, the apply valve accepts converter feed fluid from the pressure regulator valve and directs it into the release passage. From there, it passes through the turbine shaft and into the cavity between the converter clutch pressure plate and the cover. This moves the clutch pressure plate from the converter cover, releasing the converter clutch.

After the fluid releases the converter clutch, it flows from behind the pressure plate and through the converter. From it, the oil moves backward through the apply passage to the apply valve.

This valve then directs the fluid into the transmission cooler inside the radiator. The oil, returning from the cooler, at that point passes into the transmission lubrication system.

The converter clutch shift valve controls the operation of the apply valve. This shift valve, when activated, directs converter clutch signal oil to the apply valve. If this oil is permitted to act on the apply valve, it will move against its spring tension.

However, at this point, a spring acting on the converter clutch shift valve train tends to keep it closed. This action blocks second (2nd) oil coming from the 1-2 shift valve.

The converter clutch shift valve opens due to the force of governor pressure. When governor pressure is sufficient to open the valve, second oil enters the converter clutch signal passage. The oil then passes through an orifice to the converter clutch apply valve. But since the solenoid is not energized, the converter clutch signal oil is exhausted, so the apply valve and converter clutch both remain in their release positions.

When the 3rd gear switch and ECM energize the solenoid, the converter clutch signal oil overcomes the converter clutch apply valve spring, shifting the valve and changing the direction of converter feed oil (Fig. 14-39). The converter apply valve now directs converter feed oil, from the regulator valve, into the apply passage.

The apply oil flows between the stator shaft and the converter hub to charge the converter with fluid. This forces the converter clutch pressure plate against the converter cover causing a mechanical link between the engine and turbine shaft. At the same time, the converter clutch apply valve sends some converter feed oil through an orifice to the transmission cooler in the radiator. Fluid returning from the cooler then moves to the transmission's lubricating system.

In conclusion, the reader should understand that there are, with the above-mentioned system, a number of factors that cause the ECM to de-energize the solenoid. These include engine operating conditions resulting in high vacuum (closed throttle) or low vacuum (wide-open throttle), as well as the application of the brakes or a reduction in vehicle speed.

SUMMARY

1. The TorqueFlite transmission uses a Simpson gear train.

2. The AOD transmission uses a Ravigneau planetary.

3. The aluminum TorqueFlite transmission has been in service since 1960.

4. There are three basic models of the TorqueFlite transmission: the 904, 727, and the automatic transaxle.

5. The TorqueFlite transmission consists of a converter, Simpson planetary gear train, two multiple-disc clutches, an overrunning clutch, and two bands.

6. There are basically 13 components within a typical TorqueFlite hydraulic system, which are responsible for controlling the operation of the entire unit.

7. The hydraulic pressure in a TorqueFlite transmission is lower in park than in neutral.

8. In neutral and park, there is no powerflow through the TorqueFlite transmission because none of the clutches are applied.

9. In drive, breakaway range of the TorqueFlite, the input shaft drives the front annulus gear, while the overrunning clutch holds the rear planet carrier from counterclockwise rotation.

10. In second ratio, the hydraulic system of the TorqueFlite applies the kickdown band; the rear clutch remains activated.

11. When the TorqueFlite is in second ratio, the front annulus gear is the driving member; the sun gear is stationary; and the carrier is the output member.

12. In third ratio, the hydraulic system of the TorqueFlite releases the kickdown band and applies the front clutch; and the rear clutch remains activated.

13. With the TorqueFlite in third ratio, its input shaft drives the front annulus and sun gear clockwise.

14. In third ratio with a vehicle speed above 40 mph (65 km/h), the TorqueFlite's lock-up, fail-safe, and switch valves open.

15. In reverse, the TorqueFlite hydraulic system applies the front clutch and rear servo.

16. When the TorqueFlite is in reverse, the input shaft drives the sun gear; the rear carrier is held; and the rear annulus gear is the output member of the gear train.

17. The Ford AOD transmission was first used in 1980 to increase a vehicle's fuel economy.

18. The AOD transmission has six selector-lever positions, P, R, N. O/D, 3, and 1.

19. The AOD torque converter has three elements and provides two power inputs into the gear train.

20. The AOD transmission has a Ravigneau planetary, controlled by two one-way clutches, two bands, and four multiple-disc clutches.

21. The hydraulic system of the AOD transmission, mentioned in this text, has 33 components.

22. With the AOD transmission in first gear of the overdrive range, the hydraulic system applies the forward clutch.

23. In overdrive range, first gear of the AOD transmission, the turbine shaft drives the forward sun gear, while the carrier is held.

24. In overdrive range, second gear of an AOD transmission, the hydraulic system applies the forward clutch and the intermediate clutch.

25. In second gear of the D, or 3, range of the AOD transmission, the turbine shaft drives the forward sun gear, while the reverse sun gear is held.

26. In overdrive range, third gear of the AOD transmission, the hydraulic system applies the forward clutch, the intermediate clutch, and the direct clutch.

27. In third gear (direct drive) of the AOD transmission, the turbine shaft drives the forward sun gear; the direct drive shaft turns the planet carrier.

28. When the AOD transmission is in overdrive, the hydraulic system applies the overdrive band, the direct clutch, and the intermediate clutch.

29. In overdrive gear, the direct drive shaft of the AOD transmission turns the carrier, while the reverse sun gear is held stationary.

30. In reverse gear, the hydraulic system of the AOD transmission applies the low-reverse band and the reverse clutch.

31. During reverse operation of the AOD transmission, the turbine shaft drives the reverse sun gear; the carrier is held stationary.

32. The use of electronics in the operation of the automobile has become extensive.

33. The shift to the use of electronic components is mainly due to the accuracy under which they operate.

34. Some vehicles have an automatic transmission with electronic control of its lock-up converter operation.

35. The CCC system components used to control the lock-up portion of converter operation include the brake switch, 3rd gear switch, vacuum switch, throttle position switch, vehicle speed sensor, ECM, and the solenoid assembly.

36. In the released position with the solenoid not energized, the TCC apply valve is held in place by spring tension.

37. The converter clutch shift valve controls the operation of the apply valve.

38. The converter shift valve opens due to the force of governor pressure.

39. When the 3rd gear switch and ECM energize the solenoid, the converter clutch signal oil overcomes the converter clutch apply valve spring, shifting the valve and changing the direction of converter feed oil.

REVIEW

This section will assist you in determining how well you remember the material contained in this chapter. Read each item carefully. If you cannot complete the statement, review the section in the chapter that covers the material.

1. The original version of the TorqueFlite was the _____ model.

   a. 727

   b. 904

   c. 404

   d. 413

2. In the TorqueFlite transmission, the output member of the rear planetary is the _____.

   a. annulus

   b. sun

   c. carrier

   d. pinions

3. The TorqueFlite transmission can have either a _____ type pump.

   a. plunger or vane

   b. gear or vane

   c. rotor or vane

   d. rotor or gear

4. The TorqueFlite transmission has lower line pressure in park than in neutral due to the position of the _____ valve.

   a. regulating

   b. throttle

   c. manual

   d. governor

5. In the drive, breakaway range of the TorqueFlite, the manual valve directs fluid to apply the _____ _____.

   a. rear clutch   c. kickdown band

   b. front clutch   d. low-reverse band

6. In the TorqueFlite transmission, the holding planetary member in second ratio is the
   _____ _____ .

   a. front carrier

   b. rear annulus

   c. sun gear

   d. rear carrier

7. The TorqueFlite converter cannot lock up unless the transmission is in _____ ratio.

   a. reverse

   b. first (breakaway)

   c. second

   d. third

8. The driving planetary member within the TorqueFlite transmission during reverse ratio is the _____ _____ .

   a. front annulus

   b. sun gear

   c. front carrier

   d. rear annulus

9. The AOD converter provides _____ power inputs to the gear train.

   a. 1

   b. 3

   c. 2

   d. 4

10. In the AOD transmission, the ring gear attaches to the _____ .

    a. output shaft

    b. turbine shaft

    c. direct drive shaft

    d. forward sun gear

11. The component that prevents the rotation of the carrier in both directions is the _____ .

    a. low one-way clutch

    b. direct clutch

    c. intermediate one-way clutch

    d. low-reverse band

12. The unit within the hydraulic system that cushions the 3-4 upshift is the _____ .

    a. 3-4 accumulator   c. check valve

    b. 3-4 shift valve   d. OD servo regulator valve.

13. In first gear of the overdrive range of the AOD transmission, the hydraulic system applies the _____ .

    a. reverse clutch

    b. intermediate clutch

    c. forward clutch

    d. direct clutch

14. The held planetary member in first gear within the AOD transmission is the _____ .

    a. forward sun gear

    b. ring gear

    c. reverse sun gear

    d. carrier

15. The driving planetary member in second gear within the AOD transmission is the _____ .

    a. carrier

    b. reverse sun gear

    c. forward sun gear

    d. ring gear

16. In third ratio, the hydraulic system applies _____ clutches.

    a. 4

    b. 3

    c. 2

    d. 1

17. In third ratio, the direct drive shaft of the AOD transmission turns the _____ .

    a. carrier

    b. ring gear

    c. forward sun gear

    d. reverse sun gear

18. The held member of the planetary of an AOD transmission in overdrive is the _____ .

    a. ring gear

    b. forward sun gear

    c. carrier

    d. reverse sun gear

19. The driving planetary member in reverse operation of the AOD transmission is the _____ .

    a. ring gear

    b. forward sun gear

    c. reverse sun gear

    d. carrier

20. The device that signals engine load information to the ECM of the TCC system to operate the torque converter clutch is the _____.

    a. brake switch

    b. vacuum switch

    c. throttle position switch

    d. vehicle speed sensor.

21. The ECM can only energize the solenoid of the TCC system in _____ gear.

    a. third

    b. first

    c. second

    d. both a and c

22. The fluid used to open the clutch apply valve in the TCC hydraulic system comes from the _____.

    a. governor

    b. throttle

    c. converter clutch shift valve

    d. 2-3 shift valve

For the answers, turn to the Appendix.

Section II

SERVICE

To properly maintain and repair an automatic transmission, a shop must have certain equipment and service tools. Although some of this equipment and tooling may have some use in other areas of automotive repair, the majority of the items discussed in this chapter will be found only in an automatic transmission repair shop. It is not the intent of this section to cover all the equipment and special tools necessary to rebuild every kind of automatic transmission. But it will present an overview of those tools commonly used to repair the majority of units and perform the many service techniques outlined within this section.

CHAPTER 15

Shop Equipment and Tools

TORQUE CONVERTER FLUSHER

Function

In any automatic transmission repair facility, a torque converter flushing machine (Fig. 15-1) is a valuable piece of equipment. This machine's function is to recycle used converters and eliminate costly comebacks. To accomplish these tasks, the machine provides a special flushing process necessary to thoroughly clean a sealed converter.

The flushing process does two things. First, it removes any accumulated wear particles of gears, bushings, bands, and clutch plates trapped within the converter. These particles would otherwise recirculate with the fluid through a newly rebuilt transmission and damage the unit. Second, the special solvent, used in the flushing process, tends to dissolve and carry away any built-up varnish from within the torque converter. Varnish, remember, is a byproduct of oxidized fluid, and if allowed to remain in the converter, it will contaminate the new fluid.

Design

The flushing machine shown in Fig. 15-1 has an electric motor and timer, universal drive assembly, solvent reservoir, two filters, and a solvent pump. The electric motor turns an adjustable drive shaft that engages into the universal drive assembly and also drives the solvent pump. The timer switch is not only the on-off control for the electric motor and its driven components, but it permits the operator to time the flushing operation. In other words, the technician can set the timer for up to 30 minutes of cleaning, and at the end of this time period, this unit will shut the machine off.

The universal drive assembly (Fig. 15-2) has two functions. First of all, its drive tip engages into the input shaft splines of the turbine. And with the drive shaft inserted into the sleeve of the machine, the electric motor

can spin the turbine inside the converter. This spinning action causes the turbine to produce a vortex flow in the cleaning solvent, contained inside the converter.

Second, the universal drive assembly has two quick-disconnect fittings that connect hoses to the assembly. A hose from the solvent pump attaches to the upper fitting, and when the machine is on, solvent moves into the fitting through the inlet chamber and drive shaft and finally into the area under the turbine. The lower discharge fitting connects a hose to the drive assembly from the machine's sump. When the machine is in operation, solvent passes into the lower chamber from the impeller side of the converter, moves out of the fitting, and returns via the hose to the solvent sump.

The sump is the solvent reservoir for the machine. It contains a quantity of cleaning agents such as mineral spirits, Varsol, or safety solvent. Also, a line attached to the sump carries the cleaning agent to the pump whenever the operator activates the timer switch.

The machine also has two filters (Figs. 15-1 and 15-3). A magnet holds the first filter to the bottom of the solvent tank, where it screens out any large particles from the cleaning agent. The location of the second filter is in the line between the tank and the pump. This cartridge-type filter removes and suspends any smaller particles so that the cleaning agent will not carry them into the pump and eventually back into the converter.

Finally, the solvent pump has the responsibility of

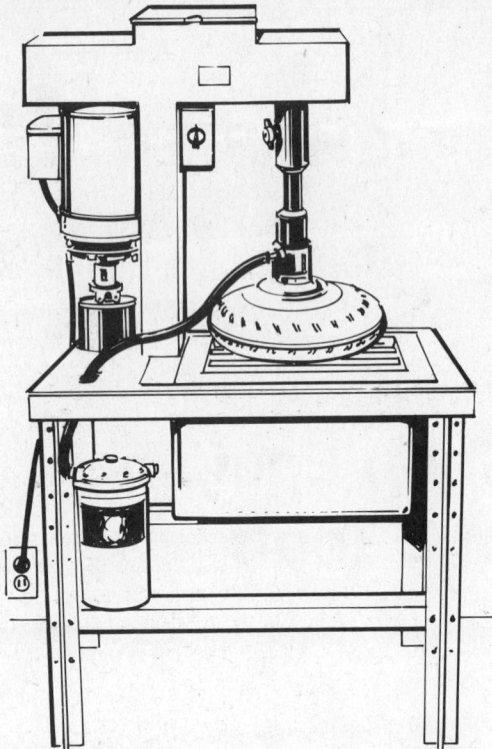

FIGURE 15-1   A typical torque converter flushing machine.

supplying the converter with a pressurized supply of cleaning agent. When the electric motor is on, it drives the pump. The pump, in turn, moves the fluid first from the solvent tank, then through the filters, and finally into the converter via the universal drive assembly.

Operating Characteristics

This type of flushing machine utilizes the vortex method to circulate the agent required to clean the inside of the converter. With the machine in operation, the pump supplies cleaning fluid to the inlet chamber, via the quick-disconnect fitting. The fluid moves through the hollowed drive shaft and under the turbine at a pressure of approximately 30 psi (Fig. 15-4). This action keeps the converter charged with solvent.

At the same time, the drive tip has engaged the turbine splines, and it is turning the turbine at about 250 rpm. The turbine, when spinning, is a large centrifugal pump due to its physical structure. As a result, the spinning turbine creates a vortex flow or flow due to centrifugal force.

This vortex flow begins at the inside center of the turbine and terminates at the curved edge where the cleaning agent moves outward from all the vaned passages at a tremendous velocity. This vortex flow then

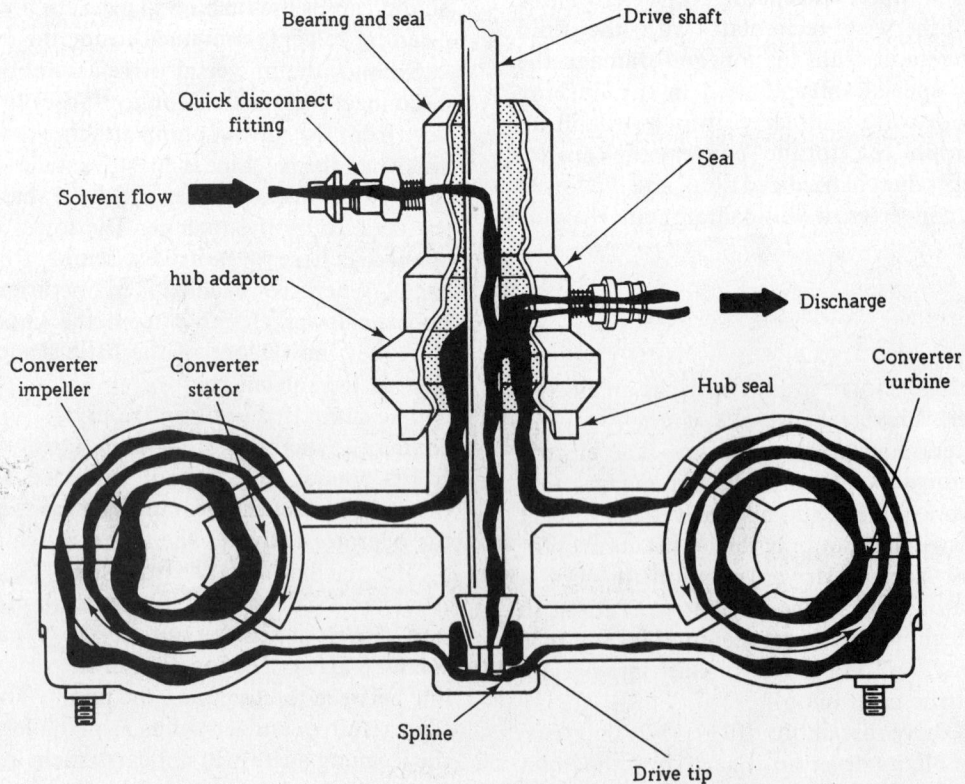

FIGURE 15-2   Universal drive assembly of the flushing machine.

moves through the impeller and stator, and returns to the turbine. This process continues over and over as long as the timer is on. As a result of this flushing action, the cleaning agent lifts up the particles and varnish and carries them out of the discharge port and into the sump and filters.

CONVERTER END-PLAY GAUGES

Because all modern automotive torque converters are sealed units, it is impossible for the mechanic to visually inspect their internal components for wear. Consequently, special equipment is necessary. Figure 15-5 shows two pieces of such equipment used to check for internal wear in a sealed converter. The gauge on the left is a factory device that can test only one converter type. The piece of equipment on the right, however, is a universal gauge that can check the majority of automotive converters.

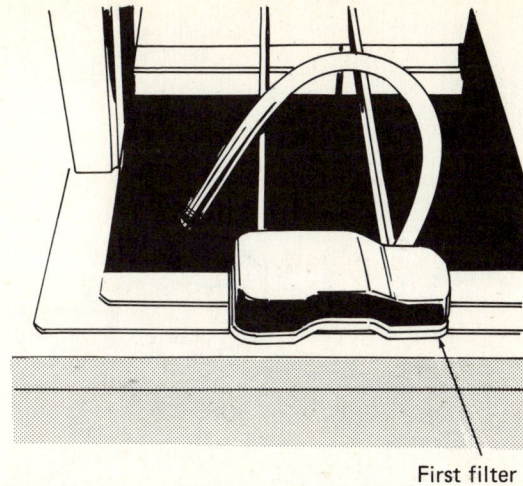

First filter

FIGURE 15-3   This converter flusher has two filters, one located at the bottom of the reservoir and the other situated between the reservoir and the sump. The first filter has been raised from its location at the bottom of the sump. The second filter's location can also be viewed in Fig. 15-1.

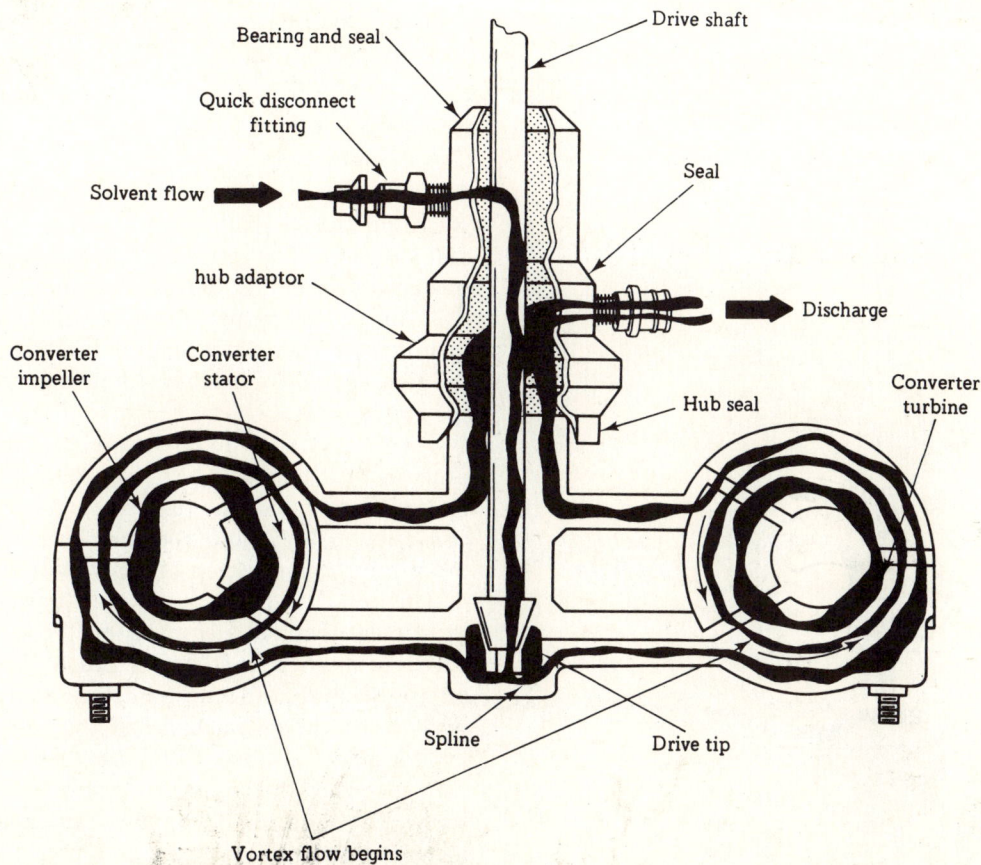

FIGURE 15-4   Flusher operation.

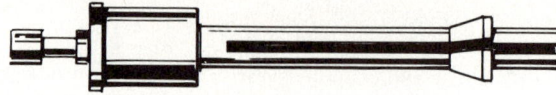

T-400 converter
end-play gauge fixture

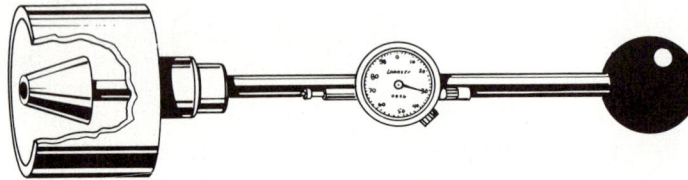

Universal end-play gauge

FIGURE 15-5   Several typical torque converter end-play gauges.

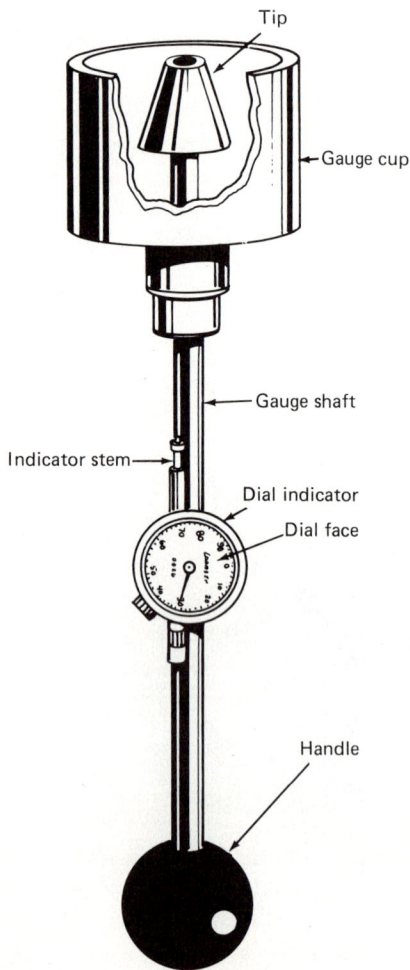

Tip

Gauge cup

Gauge shaft

Indicator stem

Dial indicator

Dial face

Handle

FIGURE 15-6   Design of the universal end-play gauge.

Both of these devices check the end-play of the turbine. Now, since bearings or thrust washers control this end-play, any excessive movement indicates abnormal wear of these parts. Consequently, if the end-play is beyond specifications, the mechanic should not reuse the converter without first having it rebuilt.

Universal Gauge Design

The universal end-play gauge (Fig. 15-6) consists of a gauge cup, gauge shaft, tip and handle, plus an adjustable dial indicator. The gauge cup slides into a slot machined in a converter mount bracket that mounts on the side of the torque converter flushing machine (Fig. 15-1) mentioned earlier in this section. In addition, the center of the gauge cup has a machined hole that acts as a guide for the gauge shaft.

On one end of the guide shaft is a tapered tip and on the other end is a handle. The design of this tip is such that it makes contact with the turbine splines inside the sealed converter. The handle permits the mechanic to easily move the guide shaft up or down within the gauge cup.

The dial indicator is an adjustable, precision instrument that actually measures the amount of turbine end-play in thousandths of an inch. The instrument is adjustable in that the mechanic can move it to any given position on the guide shaft and then lock it in place.

This device also has a dial face that the technician can rotate, to zero the gauge, by simply loosening a thumbscrew. The dial itself has a scale that reads from 0 to 0.100 inch. An indicator stem that senses turbine end-play contacts the bottom of the gauge cup and moves the needle in proportion to the amount of end-play. With this arrangement, the mechanic can easily determine the total end-play by the amount of needle deflection across the scale.

TORQUE CONVERTER DRAIN PLUG INSTALLING SETS

Many sealed torque converters come from the factory without a drain plug; this makes servicing of the converter much more difficult. For instance, when it is time to change the fluid in the transmission and converter, the mechanic cannot drain the dirty fluid from the converter. Instead, the technician will have to pump the fluid out, which is a time-consuming process. In addition, when a mechanic wants to clean the converter and check its end-play, he will have a hard time draining the old fluid out of a converter that has no plug. To offset these service problems, special equipment is available to drill and tap converters for installation of a drain plug.

Figure 15-7 shows a drill, guide, and tap set used for the purpose of installing drain plugs in General Motors torque converters. This set consists of a mount bracket and attaching hardware, drill and tap guides, in addition to a drill and tap. The mount bracket has two slotted openings, one at each end of the unit, which permit the mechanic to attach it to different-size converters using the hardware provided with the set. The unit also has a leg or extension welded at an angle to the main bracket.

This extension serves several functions. First, it houses a locking receptacle for the drill or tap guides. Second, it is the portion of the bracket that actually determines the general area of the converter where the mechanic will drill and tap to install the plug.

The drill and tap guides serve about the same function but for different tools. For example, the drill guide, when inserted into the receptacle, acts as an alignment tool for the drill so that it will maintain the proper angle and not wobble during the drilling process. As a result, the mechanic can machine a nearly perfectly round hole in the converter at the correct angle and to the proper size.

On the other hand, the tap guide, when installed into the receptacle, serves as a support for the tap during the threading process. In other words, this guide keeps the tap in alignment so that the threads are cut into the converter housing at the correct angle. The guide also prevents the tap from wobbling, which can cause the tap to enlarge the hole excessively.

The drill, of course, machines the hole in the converter housing. This drill is special in that it has to be the correct size to fit into the guide properly. At the same time, it must also cut the correct size hole for the tap.

The tap used with this set is ⅛-inch pipe tap. This

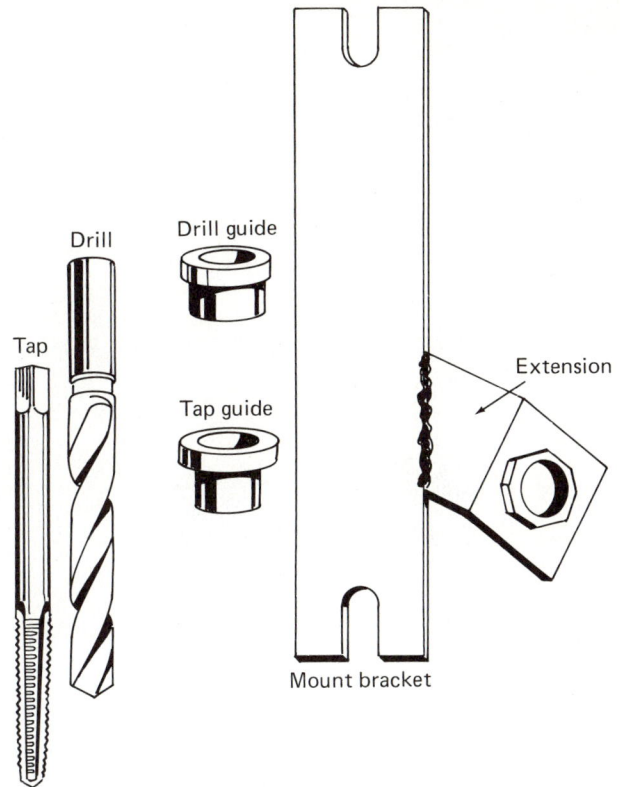

FIGURE 15-7   *Drill and guide set used to install drain plugs in torque converters.*

device cuts a tapered, internal pipe thread into the converter housing. This thread design is necessary to prevent leakage around the threads after the mechanic installs the plug.

Finally, it is very important that the technician clean the torque converter after this machining process is complete. Otherwise, metal chips trapped in the converter will circulate in the fluid and severely damage the transmission. Therefore, a mechanic should never drill and tap a converter that is still in the vehicle, because there is no proper way to clean out the unit.

AUTOMATIC TRANSMISSION DYNAMOMETERS

Many repair shops that specialize in rebuilding automatic transmissions use a special transmission dynamometer to test remanufactured and malfunctioning units. This machine (Fig. 15-8) simulates the operating conditions

FIGURE 15-8   A typical automatic transmission dynamometer.

the transmission encounters while in service in a motor vehicle without the need of the unit being in the vehicle itself. In other words, the machine bench checks the rebuilt or malfunctioning transmission for various problems with the unit operating under various load conditions, and the technician can spot and repair any defects in the transmission *before* installing it in the vehicle.

Various Problems Uncovered by the Dynamometer

The dynamometer can reveal many types of defects in an automatic transmission. The types of malfunctions detected during a test and the frequency of their occurrence will depend on two things: (1) the skill of the machine operator, and (2) the type of transmission tested at any given time. With these facts in mind, let's look at some of the problems most commonly discovered by this piece of equipment and some of their causes.

1. **Low or no hydraulic pressure:** This problem can result from such things as a defective hydraulic pump, missing or stuck valves in the valve body, loose-attaching bolts, and internal leakage due to defective gaskets or sealing rings.

2. **Improper or no upshift pattern:** A common defect brought on by such things as a malfunctioning governor, throttle valve, or stuck shift valves within the valve body.

3. **No downshift:** In this situation, the transmission remains in high or intermediate when it should automatically downshift to low. This condition occurs

when there is high governor pressure caused by a stuck governor valve or a shift valve (or valves) stuck in the open position.

4. **Bushing seizure or failure:** A problem usually caused by the lack of proper lubrication or by foreign materials trapped between the bushing and the component it supports.

5. **No drive forward or reverse:** A condition, if the transmission has normal hydraulic pressure, commonly caused by a defective band, defective clutch assembly, or by a manual valve that is out of adjustment or not properly connected to its actuating linkage.

6. **External leaks:** A problem that is easily seen and usually results from a defective gasket or seal.

7. **Unacceptable noise levels:** This situation can be the result of worn planetary gear train components, bushings, and thrust washers. In addition, the noise may be hydraulic pump whine brought on by high hydraulic pressure, worn pump components, or air in the system.

8. **Internal unbalance that sets up a vibration pattern:** This condition may be due to such problems as an unbalanced torque converter, worn bushings, or transmission shafts that are bent or out of alignment.

9. **Servo lugs (ends) broken off a band:** This particular condition will cause a transmission not to operate at all in at least one gear ratio. The lug itself snaps off the band as a result of metal fatigue, compounded

usually by high hydraulic pressure on the servo piston as it applies the band.

10. **Wrong valve body or a valve body that is contaminated with gum or varnish:** These particular conditions will cause a wide variety of transmission malfunctions including high or low hydraulic pressures, erratic or no upshifting or downshifting, or a possible no-drive condition in forward or reverse.

Finally, a bonus feature of pretesting a rebuilt transmission is that it thoroughly lubricates the transmission internally. This action is desirable under any condition, but it is most important when the unit may be in storage from one week to six months.

Design

The dynamometer shown in Fig. 15-8 has a drive head section, cradle mount assembly, instrument panel, utility panel, main-control panel, and a brake load section. The drive head section (Fig. 15-9) contains an electrically powered, main-drive motor; input-shaft assembly; a pair of vari-pitch, vee-belt pulleys; a hydraulic, speed control cylinder, and an input tachometer drive assembly. The main-drive motor supplies the driving torque necessary to rotate the input-shaft assembly. A start-stop switch on the control panel activates this motor, which operates at one predetermined speed.

The input-shaft assembly drives the torque converter of the transmission or its input shaft. The converter drive (Fig. 15-10) or direct drive keys into one end of the input-shaft assembly. Therefore, when the operator turns the start-stop switch to the "start" position, the main-drive motor rotates the torque converter or the input shaft of the transmission.

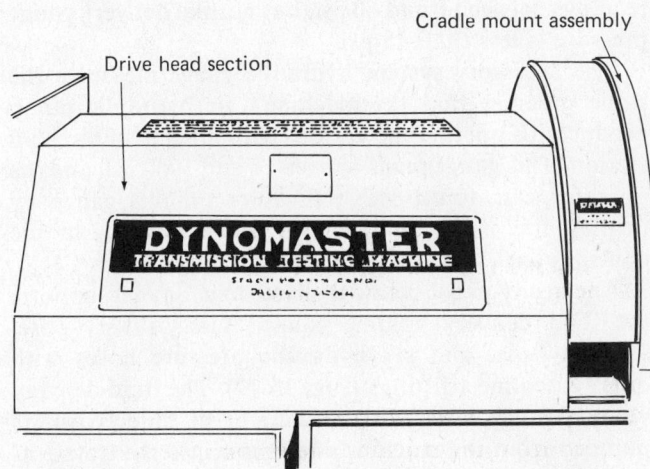

FIGURE 15-9   Drive head section of the dynamometer.

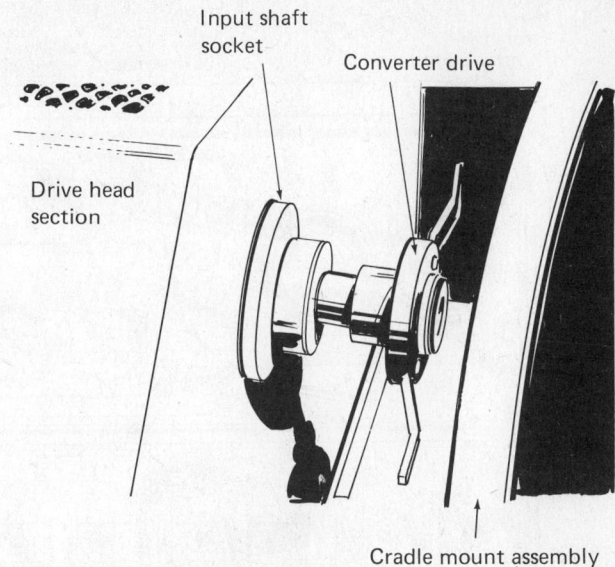

FIGURE 15-10   Input shaft, converter drive, and cradle mount assembly of the dynamometer.

The pair of vari-pitch vee-pulleys alters the speed (rpm) of the input-shaft assembly. One of these pulleys attaches to one end of the main-drive motor; the other fastens to the free end of the input-shaft assembly. A wide vee-belt rides between these two pulleys and transmits the driving torque from the motor to the input shaft assembly.

The speed control cylinder has the responsibility of actually changing the width of the pulleys. By altering the width of these pulleys, the belt moves either up or down in the vee-pulley slot. As a result, the working size of the pulleys changes and so does the speed ratio between the motor and the input shaft. In other words, the drive motor runs at a constant rpm, but when the operator moves the input-speed control handle, the hydraulic speed control cylinder alters the pulley width to raise or lower the input shaft speed.

The input tachometer assembly directs a signal to the input tachometer mounted on the instrument panel, indicating the rpm of the input shaft. A small vee-belt that operates between a pulley on the input shaft and one on the tachometer assembly drives a small generator within this unit. The generator creates an electric signal in proportion to input shaft speed; this signal, in turn, causes the tachometer gauge needle to move in proportion to the speed of the input shaft.

The cradle mount assembly (Figs. 15-9 and 15-10) supports and aligns the front section of the automatic transmission in the machine. The transmission itself bolts to a specially machined mount plate; the mount plate, in turn, fits over dowel pins on the cradle mount assembly. Finally, attaching bolts secure the mount plate to the

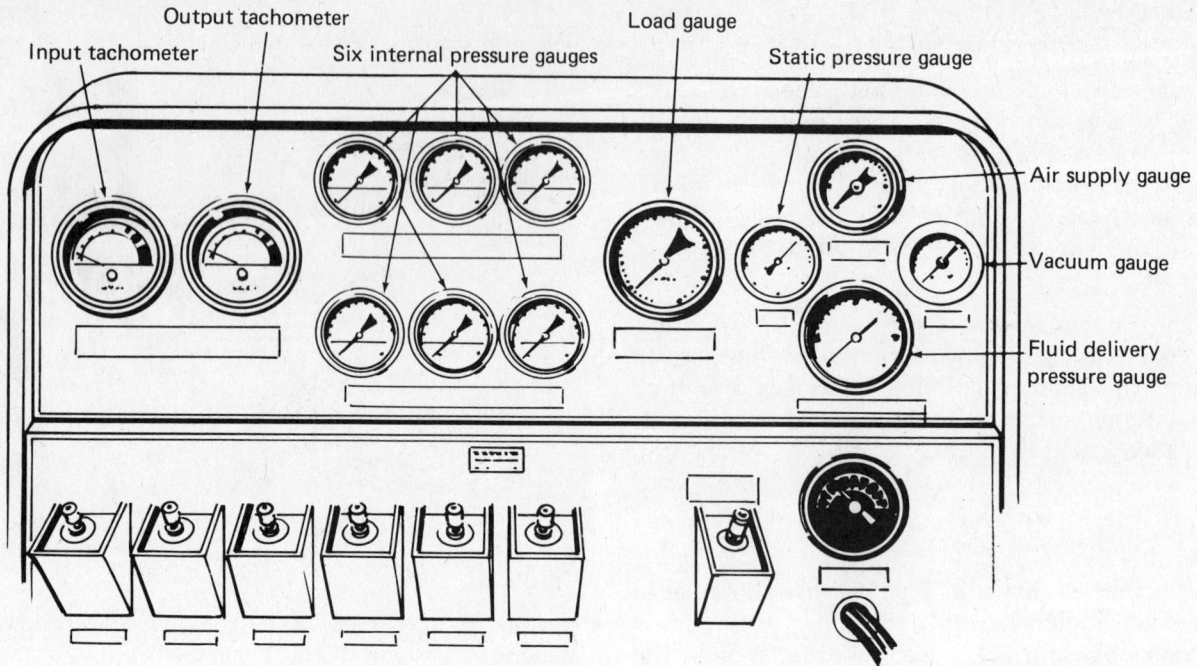

FIGURE 15-11 Instrument panel of the transmission tester.

cradle assembly, which also can move back and forth on a slide built onto the bed of the machine. This cradle movement, along with a variety of mount plates, permits the operator to adapt the machine so that it can test different types of transmissions.

The instrument panel (Fig. 15-11) houses two tachometer gauges, a set of transmission internal fluid gauges, a dynamometer load application gauge, main air supply gauge, vacuum gauge, transmission fluid delivery pressure gauge, and an accessory-system, static hydraulic pressure gauge. The first tachometer measures the speed (rpm) of the input shaft, and the second tachometer records the relative speed of the output shaft in rpms.

The six transmission fluid-pressure gauges, when connected to the transmission via pressure hoses and fittings, measure the various internal hydraulic pressures developed within the transmission. The actual number of the hydraulic circuits measured by these gauges depends on the transmission design. For example, the Ford C-6 automatic transmission has two test points: one is for control pressure and the other is for throttle pressure. On the other hand, a Chrysler TorqueFlite transmission has five pressure test points: line, front-servo release, lubrication, rear-servo apply, and governor.

The dynamometer load application gauge shows the amount of pressure applied to the machine's load section. The normal test load measured by this gauge is 1200–1600 psi. The maximum is 3000 psi.

The main air-supply gauge measures the amount of shop air supplied to the machine. The gauge itself has a dial that reads from 0 to 160 psi, and a reading of 75 psi is adequate to operate the machine's air/hydraulic pump.

The vacuum gauge measures the amount of negative pressure the machine's vacuum pump system produces. The gauge reads between 0 and 30 inches Hg (mercury), and it will indicate to the operator any sizable leak in the modulator system of a transmission being tested on the machine.

The transmission fluid delivery gauge indicates the pressure of the hydraulic fluid used to fill a test transmission from the delivery hose. This gauge has a range of readings between 0 and 60 psi, but normal delivery gauge pressure is less than 45 psi.

The accessory system hydraulic gauge measures the static pressure that the machine's air/hydraulic pump produces to operate the various components of the load section. This gauge reads between 0 and 1000 psi, and the operator can adjust static pressure on this gauge by turning the input air-regulator valve located on the control panel.

The utility panel below the instrument panel supports the fluid delivery volume gauge, fluid delivery hose, vacuum hose, and six hydraulic pressure hoses with quick-disconnect fittings (Fig. 15-12). The fluid delivery volume gauge measures the amount of fluid in quarts pumped from the machine's reservoir into the transmission being tested.

The fluid-delivery hose, to the left of this gauge,

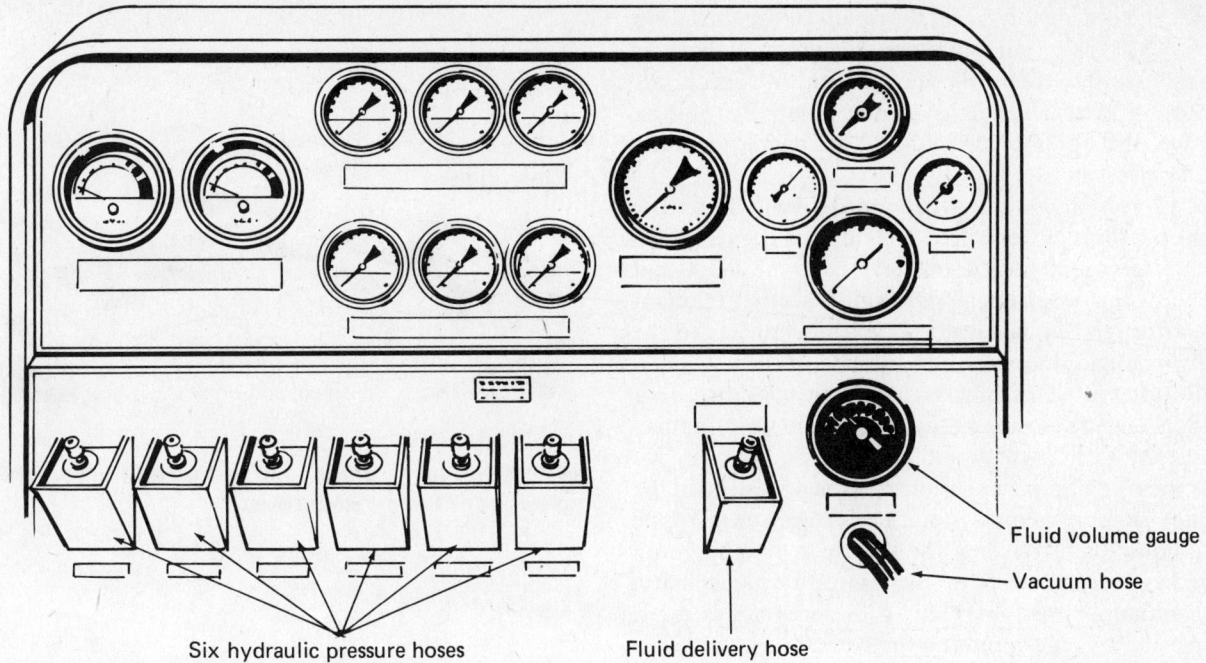

Six hydraulic pressure hoses

Fluid delivery hose

Fluid volume gauge

Vacuum hose

FIGURE 15-12   Dynamometer's utility panel.

connects to the machine's delivery pump and carries pressurized fluid to the transmission. The hose itself pulls out from the machine, and it is long enough in length to reach the dipstick tube near the front of most transmissions.

The operator connects the vacuum hose to the transmission before testing it on the machine. The open end of this hose fits over the vacuum modulator fitting. If the transmission design is such that it has no modulator, the operator will not connect this hose and will leave the vacuum pump switch in the off position.

The six hydraulic pressure hoses attach, via their quick-disconnect fittings, to the hydraulic test points of the transmission. As previously mentioned, these hoses also connect to the pressure gauges on the instrument panel; therefore, the hoses carry the various test pressures to the gauges.

Mounted on the main control panel (Fig. 15-13) are the input air-regulator valve, 12-volt dc switch, vacuum pump switch, transmission fluid supply switch, dynamometer load apply switch, disc-brake hydraulic valve,

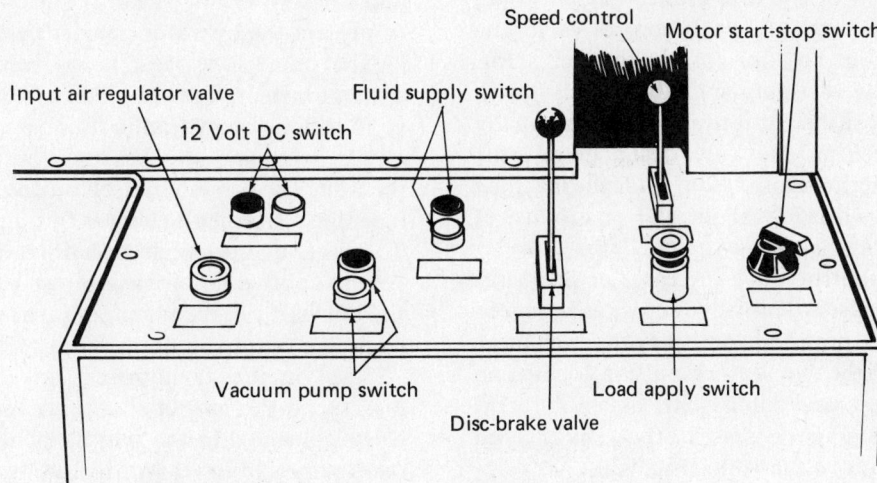

Speed control

Motor start-stop switch

Input air regulator valve

Fluid supply switch

12 Volt DC switch

Vacuum pump switch

Load apply switch

Disc-brake valve

FIGURE 15-13   Main control panel of the dynamometer.

input-speed control valve, and the input-motor start-stop switch. The input air pressure regulator valve, as previously stated, controls the amount of input air pressure delivered to the air/hydraulic pump. By rotating this valve, the operator can adjust the machine's static hydraulic pressure.

The 12-volt dc switch is an on-off switch for the machine's 12-volt system. This system tests the electrical kickdown circuit on such transmissions as the T-300 and 400. The switch, when on, allows a small electrical current to pass through two test leads. One lead connects to the kickdown solenoid terminal on the side of the transmission; the other attaches directly to the transmission case—the grounded side of the circuit. The current flow, in turn, activates the kickdown circuit.

The vacuum pump switch controls the operation of the machine's vacuum pump system. It is only an on-off-type switch; consequently, when the switch is on, the pump produces a given amount of vacuum up to its capacity. This vacuum, as previously stated, is necessary to test a transmission's vacuum modulator system.

The transmission fluid supply switch activates the pump system used to transfer the fluid from the machine's reservoir to the supply hose on the utility panel. This switch is also an on-off switch and does not control the pressure of the pump. Finally, except when the operator services the transmission with fluid, the switch is usually left off to prevent the pump circuit from overheating.

The dynamometer load apply switch is a regulatory-type switch. It controls the amount of accessory-system hydraulic pressure directed to the hydraulic dynamometer. And as previously mentioned, the operator, by adjusting this switch, applies a hydraulic pressure of between 1200 and 1600 psi to the dynamometer during an average test sequence on a transmission.

The disc brake control valve, as its name implies, controls the operation of the disc brake, located in the brake load section. By activating this control valve, the operator can stop the rotation of the output shaft. Furthermore, this valve design is such that it will stay in the "brake apply" position until the operator manually moves it to the off position. This valve design permits the technician to work on the transmission while the machine is running without worrying about the possibility of getting hurt by a rotating output shaft.

The input-speed control valve increases or decreases the rpms of the input shaft. This valve activates the speed control cylinder, and the speed control cylinder changes the relative widths of the two vee-belt pulleys, located on the main drive motor and input shaft assembly. This action either increases or decreases input shaft speed relative to the position of the control valve.

The input-motor start-stop switch activates the main drive motor. It is only an on-off switch. Therefore, the switch cannot regulate motor speed.

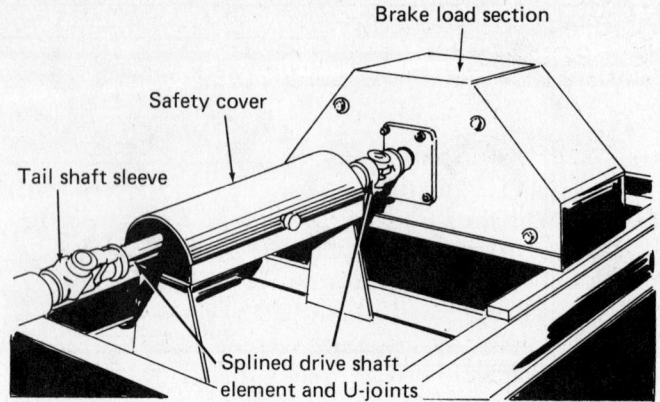

FIGURE 15-14 Brake load section and output shaft element of the dynamometer.

The brake load section of this machine (Fig. 15-14) houses the hydraulically operated dynamometer, the disc-brake assembly, the dynamometer oil reserve tank, the splined driveshaft element, and the output tachometer drive assembly. The hydraulically operated dynamometer, which connects to the driveshaft via pulleys and a notched belt, applies a load to the driveshaft. And as previously stated, the amount of this load depends on the position of the load switch on the control panel.

The disc-brake assembly is also a hydraulically operated unit. Its function is to actually stop or hold the driveshaft during the various test procedures. This action is necessary so that the operator can shift the transmission from a forward to reverse ratio without damaging the unit and can stall-test it. The brake also acts as a safety device to protect the operator against personal injury.

The oil reserve tank is the reservoir for the closed dynamometer hydraulic system. The capacity of this tank is 32 gallons of hydraulic fluid.

The driveshaft attaches to the output shaft of the transmission. The driveshaft itself on one end connects into the machine's driveshaft element through a U-joint assembly. The other end of the driveshaft has a splined section that keys to another U-joint assembly. This assembly keys into the tailshaft sleeve, which, in turn, splines into the transmission's output shaft.

The output tachometer drive assembly connects directly to the output shaft element. This unit sends an electrical signal to the output tachometer on the instrument panel. In other words, this assembly is also a small generator that develops an electrical signal in proportion to output shaft speed, and this signal operates the tachometer gauge.

PORTABLE DIAGNOSTIC TEST EQUIPMENT

Hydraulic Gauge Sets

Figure 15-15 shows a hydraulic gauge set to test the internal pressures of an automatic transmission that is still in a motor vehicle. This set consists of two gauges with hoses and a set of adapters. The small gauge tests such pressures as throttle and governor. It has a 2½-inch dial that measures pressure from 0–100 psi (0–690 kpa).

The larger gauge measures the amount of line (control) pressure within an operating transmission. This unit has a 3½-inch dial. The face of the dial has pressure divisions from 0–300 psi (0–2069 kpa).

Both gauges connect to hoses that are six feet long. With this hose length, the technician can temporarily install the gauge inside the vehicle. He can then observe its reading while performing operational tests on the vehicle and transmission.

This set also includes several adapters, small hoses, and reducing bushings. The mechanic will use these components, in conjunction with either or both gauges, for several reasons: (1) to connect the gauges to different types of automatic transmissions, (2) to connect the gauges to a transmission's test points, obstructed by the vehicle's frame, body, or exhaust pipes.

Vacuum Pump and Gauge Assembly

The hand-operated vacuum pump and gauge assembly (Fig. 15-16) serves several useful purposes. First, with this device a mechanic can test a modulator diaphragm assembly for vacuum leaks with the unit on or off the transmission. Second, with this tool and the large hydraulic gauge previously mentioned, the technician can test and adjust the control (line) pressure of a transmission, equipped with a vacuum modulator assembly, with the unit still in the vehicle.

The single piece of equipment shown in Fig. 15-16 has a vacuum gauge, vacuum pump, inlet and outlet ports, release trigger, and a hand grip and pump handle. The vacuum gauge connects externally via a hose to the component being tested and internally to the vacuum pump. The dial scale of the gauge has readings ranging from 0 to 30 inches Hg, and 0 to 76 mm Hg.

The vacuum pump itself is inside the main housing. The pump handle mechanically activates this pumping mechanism, and it, in turn, evacuates the air from any component being tested.

This pump requires two valve controlled ports—an

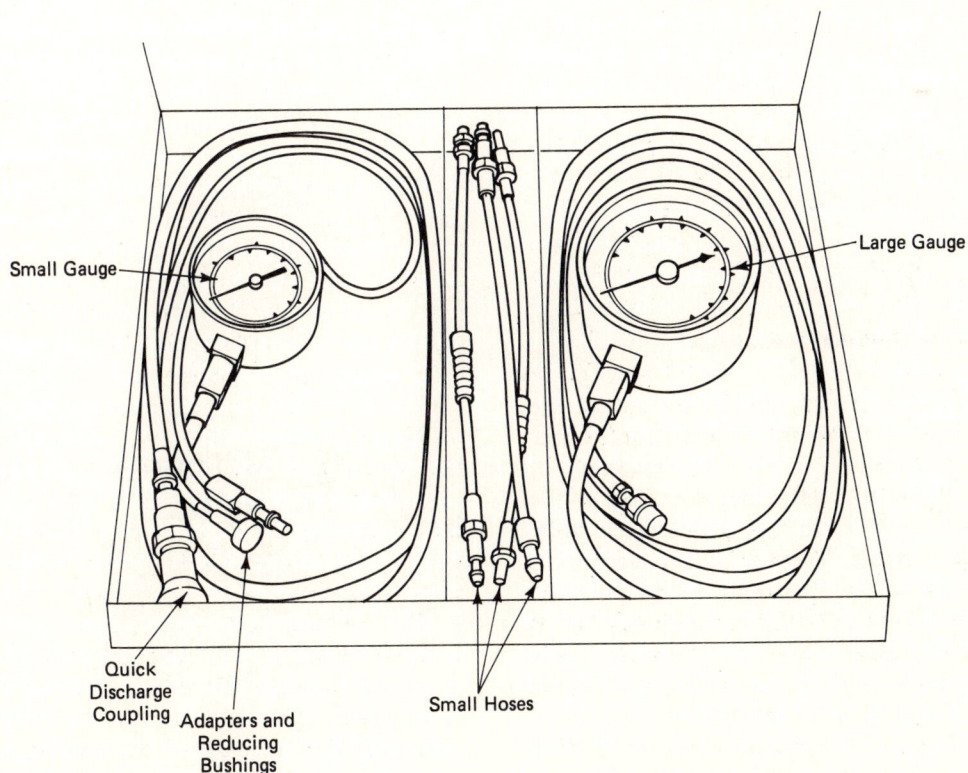

FIGURE 15-15   Typical hydraulic test gauge set.

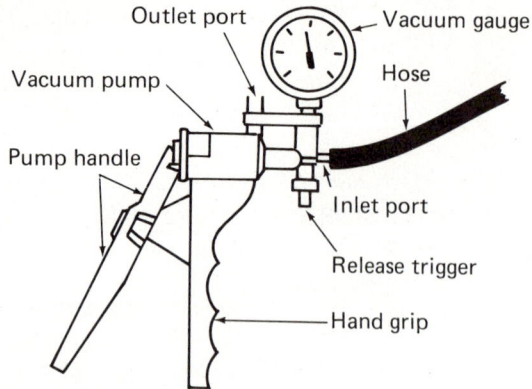

FIGURE 15-16   Hand-operated vacuum pump and gauge assembly.

inlet and outlet. The inlet port connects, via a hose, to the unit being tested. The outlet port, on the other hand, provides an opening for the pump directly into the atmosphere.

The release trigger has one important function: it vents all or part of any vacuum built up by the pump assembly to the atmosphere. In other words, by coordinating movements of the pump handle and trigger, the operator can easily pump up a *specific* amount of vacuum with this piece of equipment. As previously mentioned, the mechanic utilizes this vacuum, whether it be 5 inches or 20 inches Hg, to check or adjust modulated control pressure.

The hand grip and handle merely provide a means by which the mechanic can hold and activate the pump assembly. The palm of the mechanic's hand fits over the pump handle with the fingers encircling the hand grip. This construction permits ease of control with one hand, leaving the other free to perform other tasks.

TRANSMISSION CLEANING EQUIPMENT

Automatic transmission repair facilities primarily use four types of cleaning equipment to remove dirt, grease, oxidized fluid, and corrosion from units being rebuilt. These pieces include the steam cleaner, the jet cleaner, the safety-type parts washer, and the abrasive blaster. Most repair shops, for example, use the steam cleaner to remove dirt and grease from a transmission after the mechanic removes it from the vehicle, and just prior to tearing it down. In operation, this unit converts a mixture of water and soap into steam that quickly dislodges the dirt and grease from the outside surfaces of the transmission case.

The jet cleaner, on the other hand, will clean both the outside and inside surfaces of large transmission

components. When in service, this machine uses a preheated chemical solution and water under pump pressure to perform the cleaning task inside the machine itself. Simply speaking, the operation involves the spraying of the hot solution onto the parts, through a set of nozzles, as the parts revolve on a turntable assembly. This combined action removes most remaining dirt, grease, and oxidized fluid in six to ten minutes.

The one piece of cleaning equipment most often found in all repair shops is the safety-type, solvent filled, parts washer. This unit usually serves the function of cleaning smaller components that are too fragile to be cleaned in or that can get lost in a jet cleaner. The parts washer, depending on the solvent used and the type of machine, removes dirt, grease, and varnish by either washing the parts by hand under a stream of clean solvent or by soaking the parts in the solvent, agitated by air.

Many shops now also use an abrasive-type blasting machine to clean certain transmission components. The machine's blasting process serves two functions: (1) certain abrasive blasting will remove paint, rust, and corrosion from metal surfaces without contaminating the surface, changing its critical dimensions, or removing any of its base metal; (2) abrasive blasting removes the glaze built up on clutch drums and steel clutch plates. This action provides an excellent friction characteristic to these components and therefore reduces band or clutch slippage.

FIGURE 15-17   Design of a typical steam cleaner.

Steam Cleaners

Figure 15-17 shows a typical steam cleaner used in automotive repair shops. This unit consists basically of a water pump with control switch, heating coil and burner assembly, soap tank and valve, plus a gas valve and pressure hose. An electric motor drives the water pump by means of a vee-belt and pulleys. The pump itself pressurizes tap water sufficiently to push it through the heating coil. In addition, the pump switch—an on-off toggle switch—controls the operation of the pump by either connecting or disconnecting electrical power to its motor.

The heating coil and burner turn the tap water from the pump into steam. The water passes through the coil under pump pressure. The coil itself sits above a gas-fired burner assembly that heats the coil. In other words, cold water enters the coil's inlet, and steam under very high pressure exits from the coil's outlet.

The soap tank and valve supply soap or cleaning agent to the tap water before it enters the pump. The tank itself is just a reservoir for a mixture of water and a cleaning

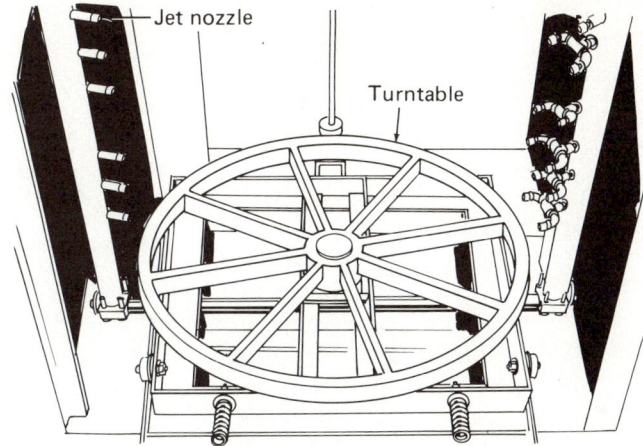

FIGURE 15-19   Motor driven turntable of the jet cleaner.

agent. The valve just regulates the amount of this mixture that leaves the tank on its way to the pump.

The gas valve controls the combustible gas flow to the burner underneath the coil. This valve is an on-off type that the operator turns on after tap water is flowing through the coil. A safety-type shut-off valve activated by the burner's pilot light prevents gas flow through this main gas valve if the pilot should go out.

Attached to the coil's outlet port is a pressure delivery hose. This hose carries the pressurized steam to a combination handle and nozzle assembly that connects to the opposite end of the hose. The operator uses this assembly to direct the flow of steam onto the component he is cleaning.

Jet Cleaners

The jet cleaner assembly shown in Fig. 15-18 consists of an insulated reservoir, a gas burner, a turntable assembly, a set of oscillating jets, and a filter assembly. The insulated reservoir of this machine has a fluid capacity of 300 gallons of cleaning agent. The cleaning agent, in this case, is water mixed in the correct proportion with a chemical that will clean either iron, steel, aluminum, or a combination of these metals.

This mixture must be very hot in order to thoroughly clean the parts; the gas burner assembly performs this function inside of the reservoir itself. This assembly consists of piping, pilot light, and thermostat. The burner piping fits into the lower section of the reservoir, and when the safety-type pilot light ignites the gas in the burner, the heat generated in its piping heats the cleaning agent contained in the reservoir. Finally, a thermostat controls the operation of the burner assembly so that the cleaning agent never drops below a predetermined temperature.

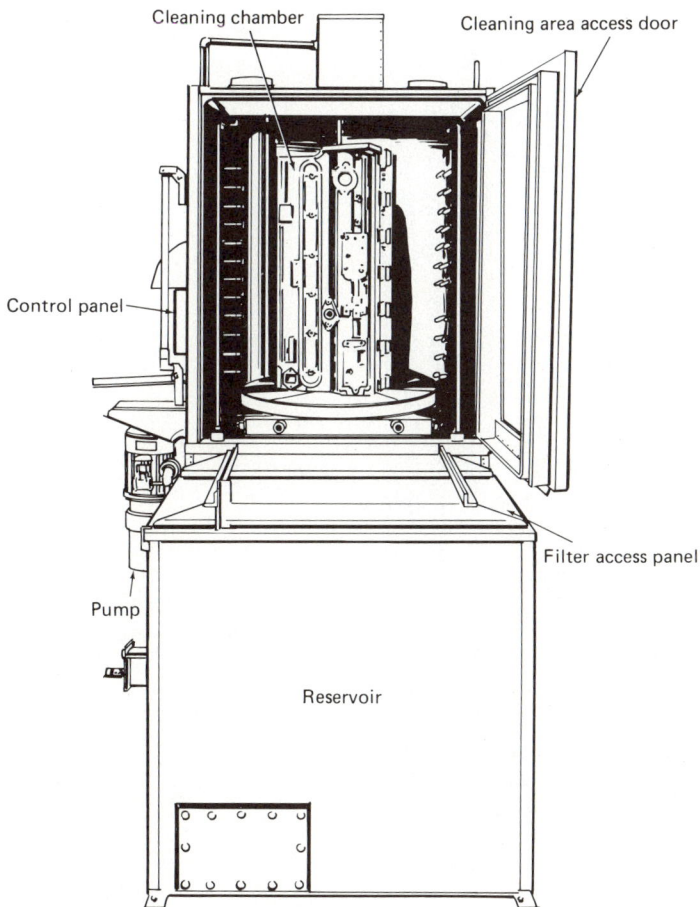

FIGURE 15-18   Typical jet-type cleaner.

The turntable (Fig. 15-19) fits inside the cleaning chamber. This device serves as the holding platform for parts that the machine will clean. Larger components sit on the turntable directly; smaller parts fit into a basket that the turntable supports.

So that the machine thoroughly cleans the parts contained on the turntable, an electric motor revolves the turntable at 1/8 rpm. This action permits the cleaning agent from the nozzles sufficient time to reach and clean all areas of the parts in a short period of time. Finally, a turntable switch located on the control panel permits the operator to stop the turntable's rotation in order to load or unload parts from the platform.

A set of oscillating jet assemblies (Fig. 15-20) are also inside the cleaning chamber. The jet nozzles themselves fit onto two hollow tubes that oscillate back and forth. This oscillation results from the action of an electric motor and mechanical linkage. Finally, another switch on the control panel activates a motor-driven pump that supplies pressurized hot cleaning solution to the nozzle assemblies as they oscillate.

The filter assembly (Fig. 15-21) sits below the level of the cleaning chamber but above the reservoir. The cleaning solution, draining from the cleaning chamber, passes through this stainless steel filtering system. The filter screens and waste container remove and store dirt and particles so that they will not recirculate back through the pump and damage it or contaminate the newly cleaned parts.

Parts Washers

Figure 15-22 illustrates one type of safety-type parts washer. This unit has a reservoir (soaking tank) and pump, twin filtering system, air agitation system, drying shelf, and a cover controlled by a fusible link. The soaking tank is part of this machine's cabinet. The tank itself holds a quantity of safety cleaning solvent, and the recommended type is one that has a high flash point and will not harm a mechanic's skin, hands, or clothing.

Located in the lower right side of the soaking tank is

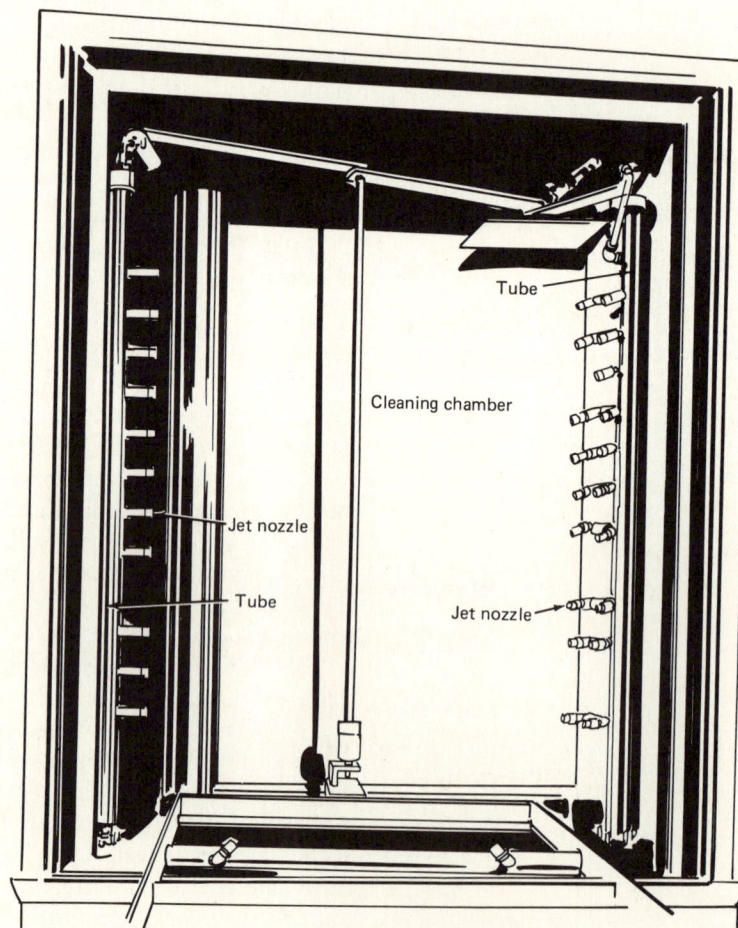

FIGURE 15-20 Set of oscillating jet assemblies.

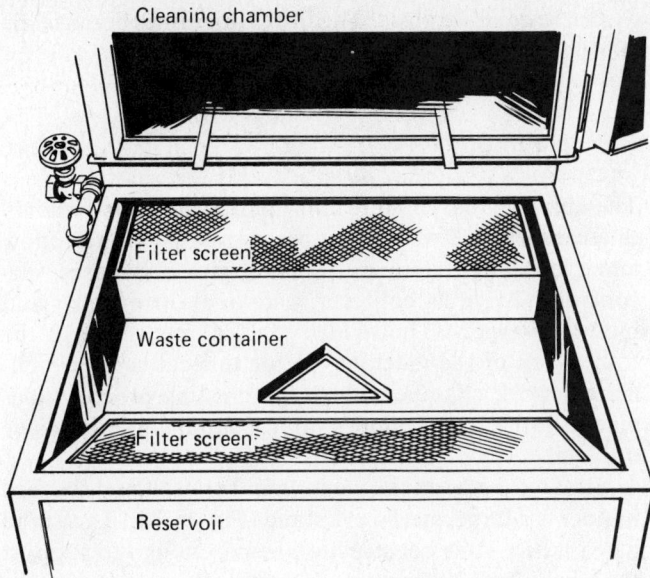

FIGURE 15-21   Filter assembly of the jet cleaner.

the solvent pump. The function of this pump is to transfer solvent from the tank, via a lower hose, to a braided metal upper hose situated above the level of the solvent in the main tank. The mechanic will use this flowing solvent to wash and rinse off parts he is cleaning.

An on-off electrical switch and light assembly, located in front of the pump but on the outside of the cabinet, activates the pump. When the operator turns the switch on, the pump begins to function and the indicator light glows. This indicator light helps to remind the operator to turn the pump switch off when pump operation is no longer necessary to the cleaning process.

The design of the twin filtering system is such that it keeps all contamination from entering the pump's reservoir so that the pump transfers only clean solvent; consequently, the system protects the pump from particle damage. The system consists of a barrier filter and cartridge filter. The barrier filter separates the main tank reservoir from the pump reservoir. This device stops all larger particles from entering the pump's reservoir; as a result, they sink to the bottom of the main tank.

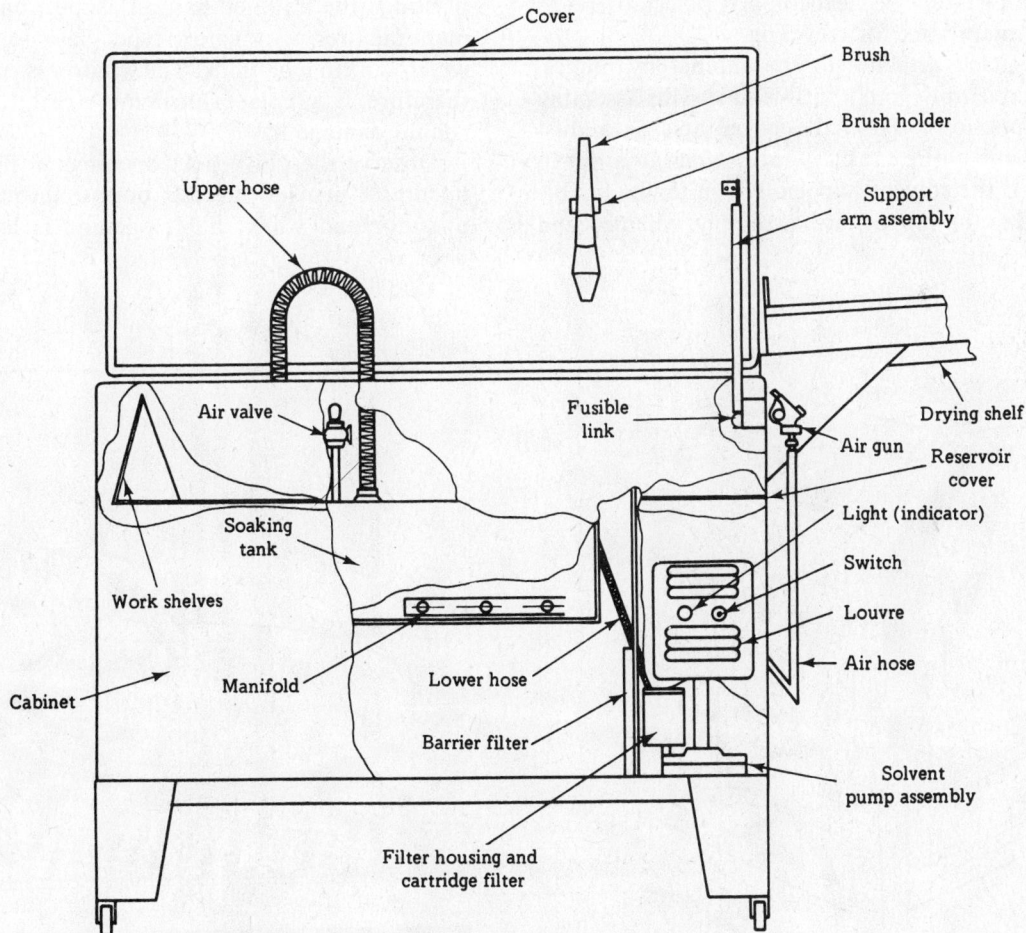

FIGURE 15-22   Design of a typical safety-type parts washer.

On top of the pump body is the cartridge-type filter. This filter design is such that it traps any smaller particles of contamination that might get by the barrier filter and flow into the pump's inlet port. Finally, because the cartridge filter traps these particles, the operator will have to replace it on a periodic basis, or the filter will clog and cause pump damage.

The air agitation system consists of an air valve and manifold. The air valve connects to the external air supply via a pressure hose. The valve itself regulates the flow of shop air going through a pipe to the air manifold.

The air manifold rests about at the midpoint in the main tank, beneath the level of the solvent. The function of this manifold is to evenly distribute the air, supplied to the manifold by the air valve, to the surrounding solvent. In other words, if the operator opens the air valve, air passes through the holes in the manifold, causing the solvent to agitate.

The sheet-metal drying shelf mounts to the side of the cabinet. Its construction is such that the drain area of the shelf tapers down slightly from its outer end to where it attaches to the side of the cabinet. This design permits the solvent, draining from the cleaned parts placed there, to return to the main tank for recycling.

The cover itself attaches to the cabinet by hinges, support arm assembly, and a fusible link. This assembly serves two functions. First, the cover acts as a dust protector for the entire tank assembly when it is not in use. Second, if the solvent happens to ignite, the fusible link, attached to the support arm assembly, will melt and cause the cover to close. The fire extinguishes because the closing cover cuts off the oxygen to the flames.

Abrasive Blasters

The abrasive blaster shown in Fig. 15-23 consists mainly of a cabinet-type work chamber with inspection window and glove port openings, blasting gun and hoses, air-control valve, dust collector, and an electric motor and blower assembly. The welded steel cabinet serves as the framework of the machine. Inside this cabinet is a well-lighted work chamber where the blasting process takes place and a feed hopper that contains the abrasive material.

Situated between the work chamber itself and the feed hopper is a large, steel work stand (Fig. 15-24). This stand has a lattice-style construction heavy enough to support the work. At the same time, however, this design permits the heavier abrasive particles to fall directly back into the hopper after being used to clean the parts.

So that the operator can observe the parts he is blasting, the cabinet has an inspection window. The manufacturer uses safety-type glass for this window, which is 3/16 inch thick. The window is rubber mounted; therefore, it is replaceable in minutes if the abrasive blast should damage it.

Observe the glove port openings in Fig. 15-23; these openings provide access points through which the operator can work. Each opening is large enough to

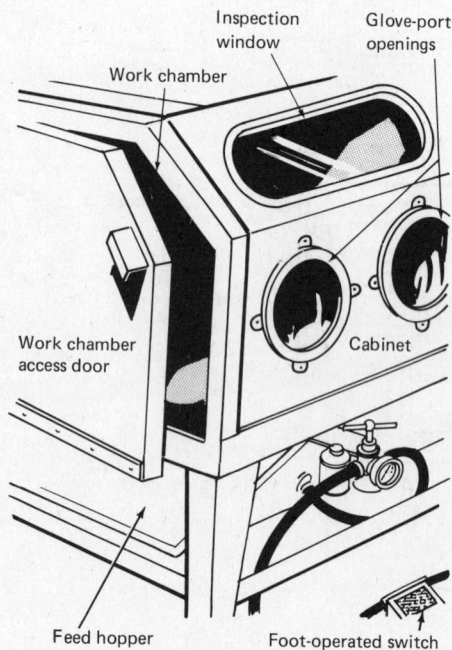

FIGURE 15-23   Typical abrasive blaster.

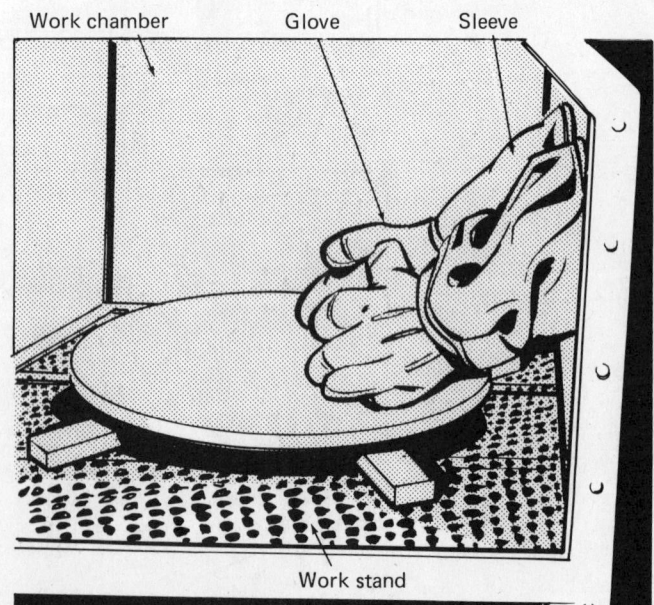

FIGURE 15-24   Work chamber and stand area of the abrasive blaster.

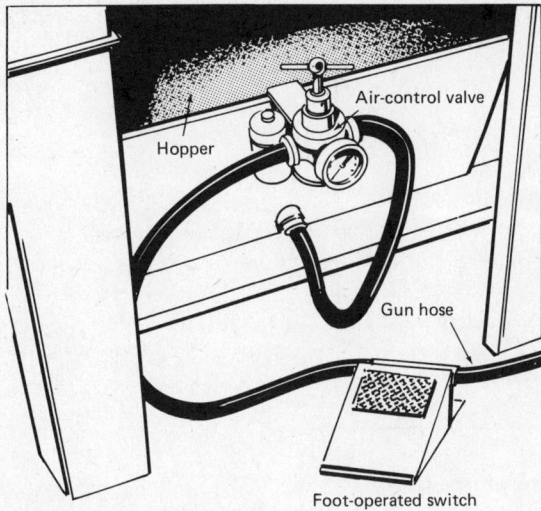

FIGURE 15-25 Air control valve and foot switch of the abrasive blaster.

provide maximum freedom of movement for one of the operator's arms. To protect the operator's hands and arms from abrasive blast, the manufacturer equips both port openings with gauntlet-type sleeves and gloves (Fig. 15-24).

The blasting gun and nozzle, held by the operator through one of these port openings, directs the abrasive blast onto the component being cleaned. This gun has two hoses attached to it. One of these hoses connects the gun to the shop's air supply via the air-control valve (Fig. 15-25). In addition, a foot-operated switch controls the flow of this compressed air from the control valve to the blasting gun. Finally, the other hose carries the abrasive material from the hopper to the blasting gun.

The access door (Fig. 15-23) has two functions. First, it provides a large enough opening for the operator to slide components into the work chamber and onto the work stand. Second, through this access door, the operator can pour the abrasive material into the feed hopper, which is below the work chamber.

The air-control valve (Fig. 15-25) regulates the pressure used by the gun during the blasting process. The valve, in most cases, reduces the amount of air pressure supplied by the shop's air compressor to a value that is suitable for the type of abrasive being used and for the type of work the operator is cleaning. For example, the maximum pressure for glass beads is 70 psi and for silica sand 45 psi, but the operator may have to lower these pressures when cleaning precision parts to avoid damaging them.

The dust collector, motor, and blower assembly (Fig. 15-26) act as a unit to remove and trap dust and dirt from

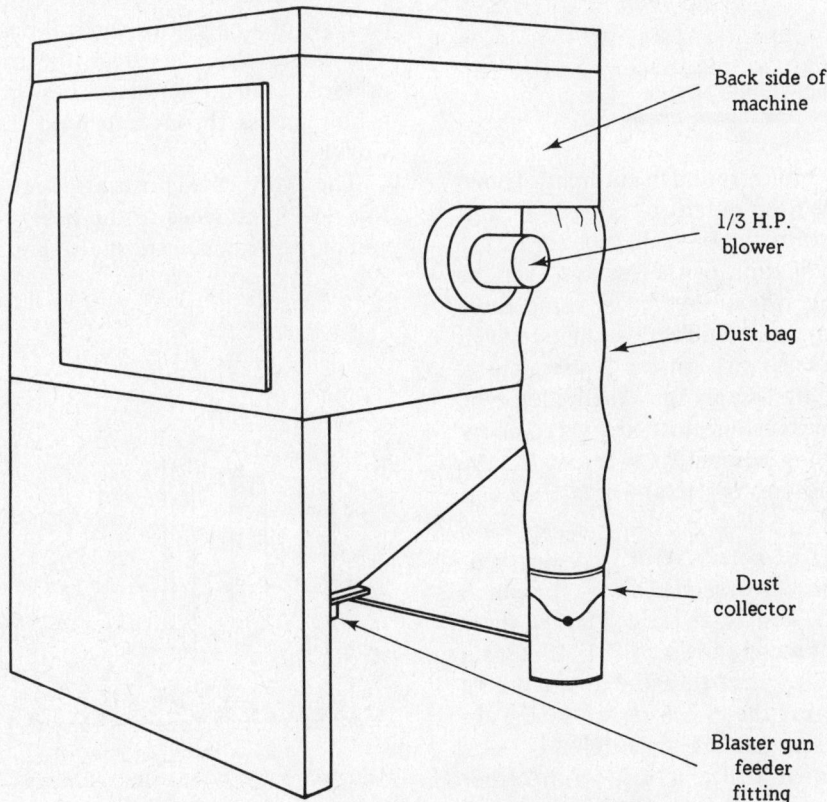

FIGURE 15-26 Dust collector, motor, and blower assembly installation on a typical abrasive blaster.

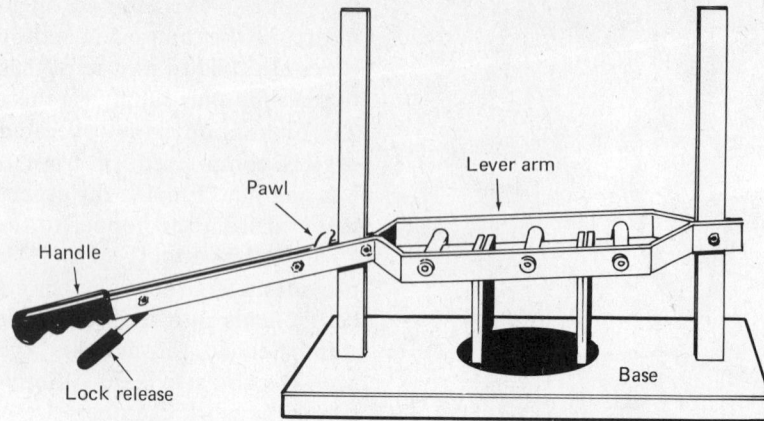

FIGURE 15-27   Universal-type clutch return-spring compressor.

the work chamber. The collector itself attaches to the lower end of the dust bag that fits onto the blower housing. With this construction, all fine particles of dust and dirt pulled from the work chamber by the motor and blower enter the dust bag before reaching their final destination—the collector. This action removes all smaller particles of dust and dirt and permits only heavier abrasive particles to return to the hopper for reuse by the blasting gun.

CLUTCH SPRING COMPRESSORS

Multiple-disc clutch assemblies found in automatic transmissions must have some form of return spring or springs to move the piston back to the base of its bore as the clutch releases. The installation of the springs over the piston is such that they will always exert some force (tension) on the retaining device and the piston itself even when the piston is in the released position. Consequently, with the exception of the diaphragm (bellville) type spring, this initial spring tension must be overcome by some form of compressing equipment in order for the mechanic to remove the spring's retaining device and finally the piston.

Many different types of clutch return spring compressors are now in current use to service open-type clutch assemblies. Figures 15-27 and 15-28 show a few of these units. The piece of equipment shown in Fig. 15-27 is a universal type. This means that this unit is adaptable enough to fit and compress the clutch return springs in most automatic transmission clutch assemblies.

The two devices shown in Fig. 15-28, on the other hand, are factory-type compressors. This means that these units fit and properly compress only the springs found in one type of transmission. Therefore, in order to

rebuild all the various types of automatic transmissions, the average repair shop must have a rather large quantity of these factory compressors or use the universal type.

Universal Type

Figure 15-29 shows a lever-type universal compressor in position over an open-type clutch drum; this device consists of a base, two vertical posts, and the lever itself. The metal base serves as the framework of the compressor. Each corner of this base has a machined hole that a bolt passes through to hold the base securely to the workbench.

The two vertical posts attach to the base. The left post, Fig. 15-29, attaches to the base via a bolt welded to the end of the post. This post also has a series of holes drilled

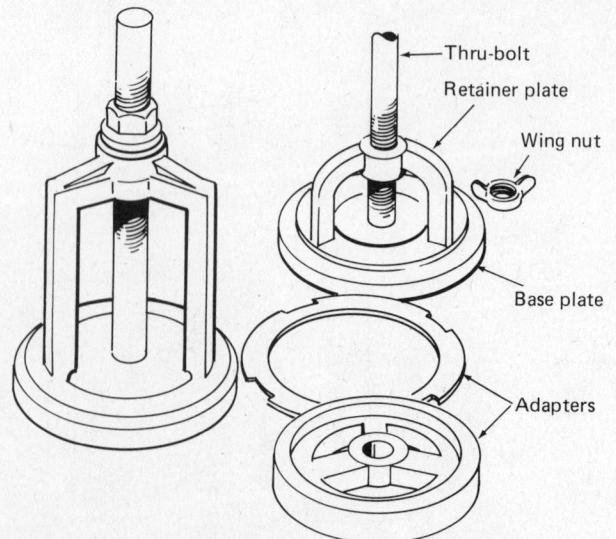

FIGURE 15-28   Factory-type clutch return-spring compressors.

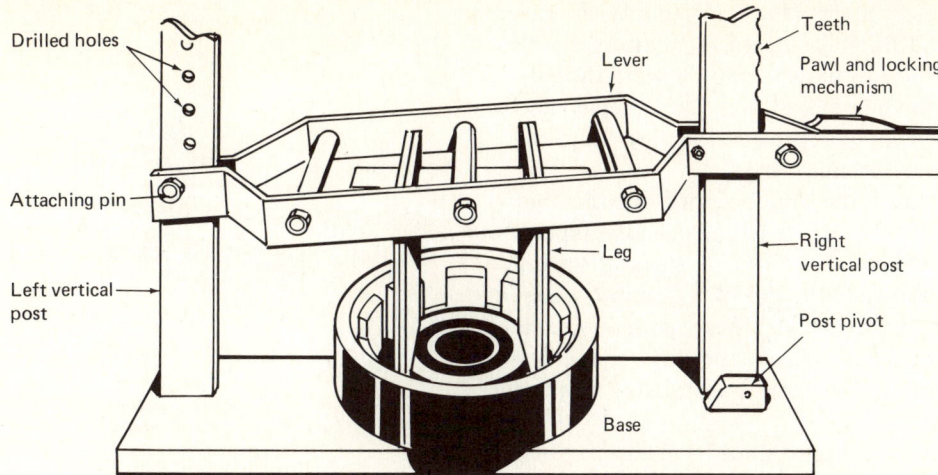

FIGURE 15-29   Lever-type universal compressor positioned over an open-type clutch drum.

into it; these holes accommodate an attaching pin that secures the fulcrum end of the lever to the post. Furthermore, this series of holes permits the mechanic to raise or lower the fulcrum point as necessary to compensate for different drum heights.

The right vertical post pivots back and forth on its mounting bracket, and it has a series of teeth machined into its back edge. The pivot point permits the mechanic to lower this post, making it easier for him to install and remove a clutch drum from the machine. The teeth index with a pawl, which is part of a locking mechanism built into the handle section of the horizontal lever. This mechanism locks the lever to this post, which frees the mechanic's hand so that he can remove the spring's retaining device during the compressing procedure.

The design of the horizontal lever is such that it provides the needed mechanical advantage to assist the mechanic in overcoming the tension of the clutch springs. As previously stated, one end of the lever—the fulcrum point—fastens via a pin to the left vertical post. The other end of the lever has a handle assembly where the mechanic applies his downward force that the lever will multiply (Fig. 15-27).

Between the fulcrum point and the handle of the lever are two legs that mount on a pivoting metal bar. These two legs are also adjustable in that the mechanic can slide them both back and forth on the metal bar. This design provides this universal machine now with two adjustments: the legs that move to compensate for different spring and retainer sizes, and a movable fulcrum point to take care of varying drum height differences (Fig. 15-29).

Factory Type

Figure 15-30 shows a factory-type spring compressor installed over a typical open-type clutch drum; this unit uses the principle of the inclined plane to multiply force and overcome spring tension. The machine itself has a base plate, thru-bolt with wing nut, adapters, and a retainer plate. In Fig. 15-28, the metal base plate has a machined hole in its center that acts as the guide for the thru-bolt; and the head of this bolt bears against the bottom of the base plate. When this tool is in service, the base plate rests under the clutch drum; or in the case of a clutch located in the rear of th mentioned, the head of this bolt bears against the lower side of the base plate, with its threaded shank passing upward through the clutch drum. The threaded section of the thru-bolt also passes through a hole in a retainer plate that fits over the spring retainer; the wing nut then threads down over the thru-bolt until it

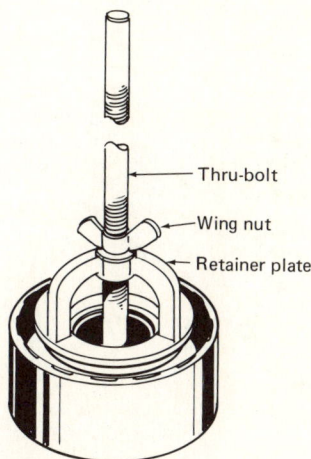

FIGURE 15-30   Factory-type compressor installed over an open-type clutch drum.

contacts the retainer plate (Fig. 15-30). With this arrangement, any additional clockwise rotation of the wing nut forces the retainer plate to move toward the base plate. And since the plate fits over the springs and retainer, this movement compresses them so that the mechanic can remove the snap ring.

The adapters fit over the clutch spring retainer but under the retainer plate of the machine. These adapters are necessary because of the different sizes of spring retainers found in the *same* model of automatic transmission. In other words, with these adapters, this compressing equipment will fit all the different clutch assembly styles found in the same transmission over its years of production.

Some types of clutch drums do not have an open hub like those just discussed. With this design, the transmission's input shaft attaches directly to the drum. As a result, the compressing equipment shown in Figs. 15-27 and 15-28 will not function on these assemblies.

To service the closed-type drums, most shops now use an arbor press and a specially designed retainer press plate (Fig. 15-31). The arbor press supplies the additional force needed to overcome spring force; the hydraulic or rack and gear-type presses are the most common devices used for this purpose.

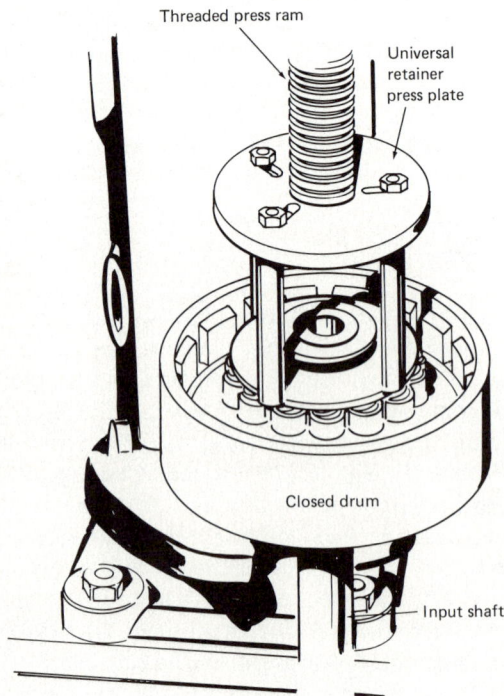

FIGURE 15-32   Typical threaded bushing puller assembly.

The retainer press plate can have either the universal or factory design. The universal unit has a slotted press plate with legs that adjust in or out within these slots. Consequently, the mechanic can adjust this assembly to fit several different sizes of clutch spring retainers. The factory plate, however, has fixed legs; therefore, they will conform only to one type of spring retainer. This means that the shop must have a number of these units in order to service the many types of closed-type drum assemblies, or use the universal design.

BUSHING EQUIPMENT

The average automatic transmission has many sizes and types of bushings installed at various locations within the unit. These **bushings** are specially designed bearings that

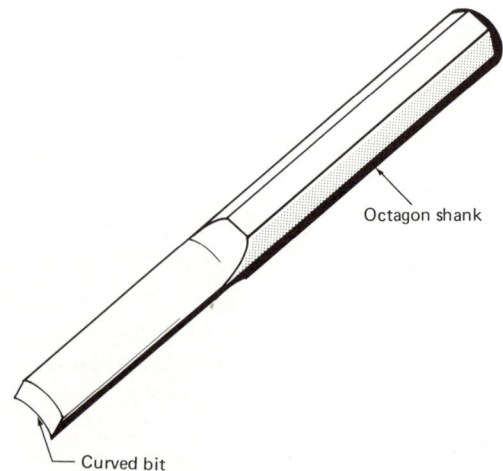

FIGURE 15-31   Arbor press and retainer plate in position over a closed-type clutch drum.

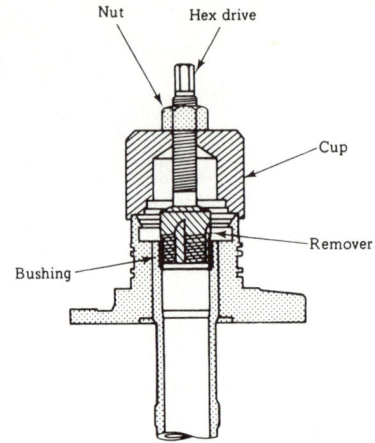

FIGURE 15-33   Bushing chisel.

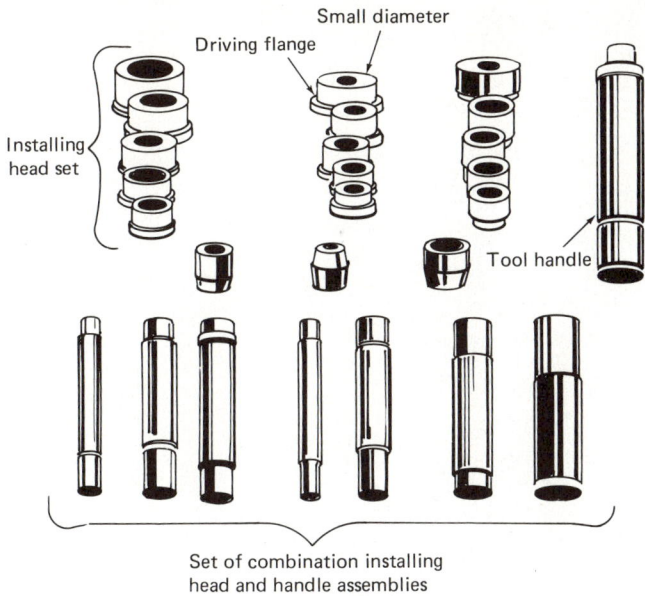

FIGURE 15-34   Universal bushing removing and installing set.

support the revolving shafts, planetary gears, and drums of the transmission. The bushing itself has a round, outer steel backing. Cast on the inner circumference of this backing is a softer bearing material such as a babbit metal composed of tin, copper, and antimony, or an alloy of copper and aluminum.

To prevent the bushing from shifting in its bore once the mechanic installs it, the backing is slightly larger in diameter than the bore it fits into. This provides an interference (press) fit between the backing and the bore. Consequently, in order to remove and replace these interference bushings during a transmission overhaul, special equipment is necessary.

The equipment required to remove an old bushing depends on whether its bore is open or closed. For example, if the bore is open at both ends and has the same diameter all the way through, the bushing will pass out of the unrestricted end during the removal procedure. If, on the other hand, the bushing's bore has a seat or bottoms out (a blind hole), the mechanic must remove the bushing from the open or installation end of the bore.

The pieces of equipment shown in Figs. 15-32 and 15-33 are typical of those used to remove a bushing from a blind bore. The tool illustrated in Fig. 15-32 performs this function by pulling the bushing from its bore. The tool shown in Fig. 15-33, however, splits the backing of the bushing in two pieces so that the mechanic can pry it out easily with a screwdriver or pull it out with needle-nose pliers.

Threaded Puller

The bushing puller shown in Fig. 15-32 consists of a remover, nut, and cup. On the one end of the hardened steel remover is a tap that cuts a series of threads into the inside circumference of the bushing. Cutting these threads into the bushing is necessary to firmly lock the remover to the bushing itself so that the remover will not pull out of the bushing during the extraction procedure. Finally, flutes or grooves run the length of the tap threads; they permit metal chips to escape so the chips will not jam and damage the threads of the tap.

On the other end of the remover is a hex drive and a threaded shank. A mechanic uses a wrench on the hex drive to thread the remover into the bushing. The threads of the shank along with the nut provide the pulling force needed to remove (pull) the bushing from its bore.

The cup section of this tool performs two design functions. First, it transmits the pulling force, provided by the nut moving downward on the threaded shank, directly to the surface of the component containing the bushing. Second, the legs of the cup straddle the bushing's bore, thus allowing sufficient clearance for the remover and bushing to clear the cup's inner surface during the pulling process.

Chisel

The chisel pictured in Fig. 15-33 has a design that permits the mechanic to cut the bushing without damaging the blind bore it fits into. The shank of this cutter has an octagon shape that provides the mechanic with better hand control of the chisel during the cutting operation. Also, the curved bit pushes the bushing material away from the work; therefore, the mechanic can see more easily where to guide the chisel. Lastly, both cutting edges of the tool have a slight chamfer where the bit terminates at the narrow end of the lower shank. This design tends to reduce bore damage as the chisel cuts through the bushing.

Bushing Driver Set

The bushing equipment shown in Fig. 15-34 can remove bushings from open bore installations and can install bushings in both open and closed bores. This universal set consists mainly of various sizes of installing heads, a tool handle, and a set of combination installing head and handle assemblies. The round, multiple-diameter installing head supports a given size bushing as the mechanic installs it in the bore. The smaller diameter of each head fits into and supports the inside surface of the bushing. The outer diameter, its driving flange, bears against the rather thin metal portion of the bushing between its inner

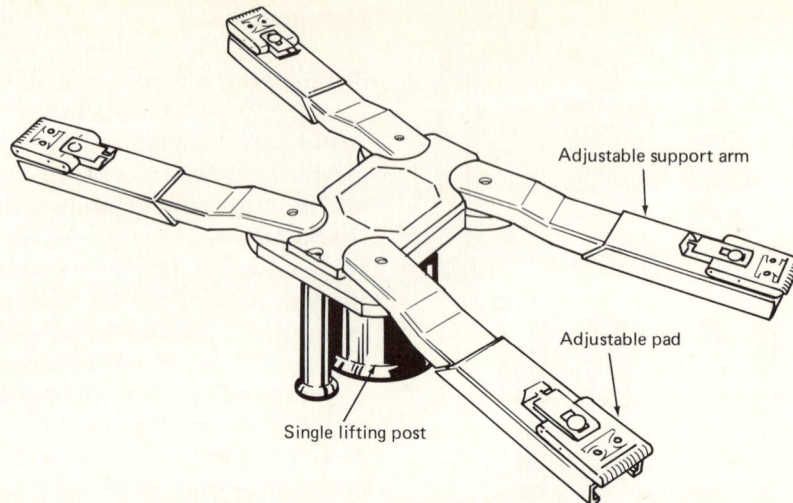

FIGURE 15-35   Typical single-post lift.

and outer circumferences. But the outside circumference of the driving flange is slightly smaller than that of the bushing. This design allows the installing head to pass unrestricted through the housing bore when the mechanic uses it to remove a bushing.

The tool handle of the set is the component that receives the force necessary to remove or install a bushing. This force may be in the form of hammer blows or from an arbor press. The handle itself has a small-diameter machined shank that slides into a hole in each installing head. Furthermore, to protect the head of the handle from the mushrooming effect of hammer blows, the manufacturer grinds a small chamfer around the outer circumference of the head. Finally, to provide a positive gripping point for the mechanic's hand, the manufacturer also knurls the handle section of the tool.

For the installation of small bushings, the combination installing head and handle assembly, as the name implies, combines both these components into one unit. In other words, on each end of this assembly is an installing head that fits into a given size bushing. With a bushing installed in one end, the opposite end would them form the driving head where the mechanic applies the pushing force. But in this case, the driving force has to come from a press. *A hammer blow would mushroom over the machined surfaces of this assembly; consequently, a bushing would no longer fit over the damaged end.*

LIFTING EQUIPMENT

In order to remove, repair, or install an automatic transmission in a motor vehicle, the mechanic must have some method of raising the vehicle before he can work under it.

The two most common pieces of equipment used for this purpose are the hydraulic lift or hoist and the floor jack. Of the two, the lift is the most efficient and safest because this unit permits the mechanic to raise the vehicle high enough to work under it standing up. The hydraulic floor jack does not.

Hydraulic Lifts

Two types of hydraulic lifts are in common use today, the single-post and the twin-post lifts both of which usually operate by hydraulic fluid under pressure from a compressed air source. The **single-post hoist** (Fig. 15-35) has one hydraulically operated lifting post centered under the vehicle. Attached to the lift post are four adjustable support arms and racks or pads, which fit underneath the vehicle's frame. With this arrangement, the wheels of the vehicle hang free as the hoist raises it off the shop floor.

Because of its design, the single post lift is not very efficient for transmission repair work. The center post and attaching arms make it difficult for the mechanic to reach and remove certain transmission components. Furthermore, with some transmission and vehicle designs, the single post hoist may block or prevent the removal of the transmission itself.

The **twin-post lift,** on the other hand, has a design that is especially useful in transmission repair work. The unit shown in Fig. 15-36 has two lifting posts, which are spaced away from the centerline of the vehicle a given distance. This design provides the mechanic with sufficient working space to repair, remove, or install a transmission. Also, the manufacturer locates the lifting arms and racks to each post in such a manner that this hoist will raise the vehicle by the frame, thus permitting the wheels to hang free.

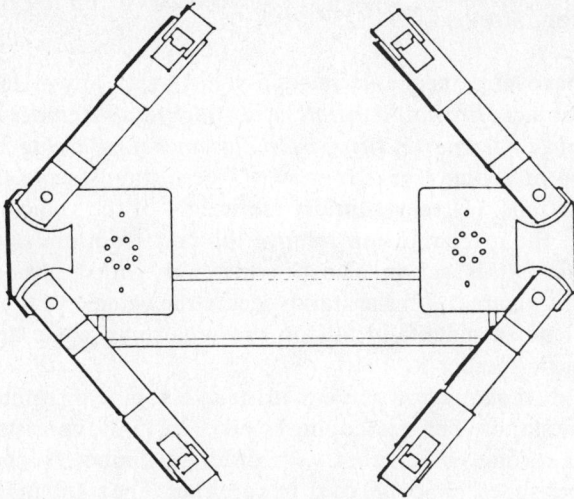

FIGURE 15-36   Typical twin-post lift.

Another common type of hydraulically operated twin-post lift has a design that raises the vehicle by its front and rear suspension systems. With this design, two individually operated lifting posts operate near the center of both the front and rear suspension systems. Attached to each of these posts is a lifting arm and pad that are adjustable to fit the various types and sizes of suspension systems.

Manufacturers equip most lifts with some type of safety device that prevents the vehicle from lowering unexpectedly. Figure 15-37 shows one of these devices; it consists mainly of a rack and gear, a locking dog and release handle, and a trip pin. A rack with teeth attaches to the lower side of each booster pad, which in turn is fastened to each of the lift posts. The teeth of a locking gear mesh with mating teeth located on one of the racks.

The locking dog, when activated by the trip pin, ratchets on the teeth of the rack as it moves upward, but it wedges between the gear teeth if the rack begins to lower. This action prevents the gear from turning on the rack which, in turn, stops the lift from coming down.

The function of the release handle is to pull the dog away from the gear teeth. With the dog lifted, the mechanic can then lower the hoist in the normal manner.

The trip pin, as previously mentioned, activates the locking dog whenever the lift is in the down position. The pin itself connects to a plate bolted to the bolster. As the lift and bolster come down, the trip pin contacts the release handle and moves it forward. This action sets up the dog mechanism to engage in the gear teeth if the hoist should unexpectedly begin to lower.

Another type of safety device stops the lift from lowering only if the hoist has been raised to its maximum height. This unit uses a weighted lock pin that falls into position in a slot or hole machined into the lower section of a reinforced metal tube, attached to the lift post itself. When the mechanic is ready to lower the vehicle, he pulls the pin out of its slot and lowers the hoist in the normal manner. Finally, to protect himself from injury, *the mechanic should install one or more hoist stands under the lift if the unit does not have any safety locking mechanism.*

Hydraulic Jacks

As previously stated, a hydraulic (floor) jack can raise a vehicle so that the mechanic can perform transmission repair work, and mechanics still use this method of lifting a vehicle when a hoist is not available. But there is a great disadvantage in using a floor jack for this purpose. The jack itself cannot raise the vehicle very high; therefore, the mechanic has little working space under the vehicle and must work under the vehicle on his back. This working position slows down any type of repair work, especially

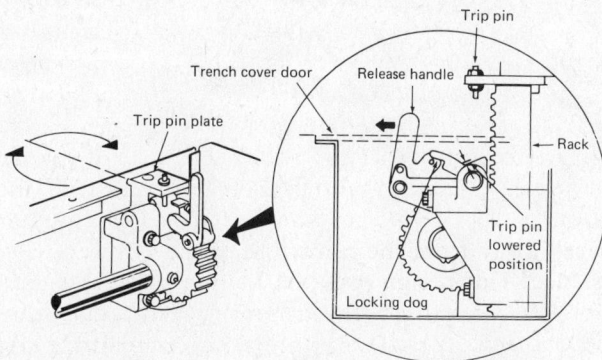

FIGURE 15-37   The safety mechanism located on a twin-post lift.

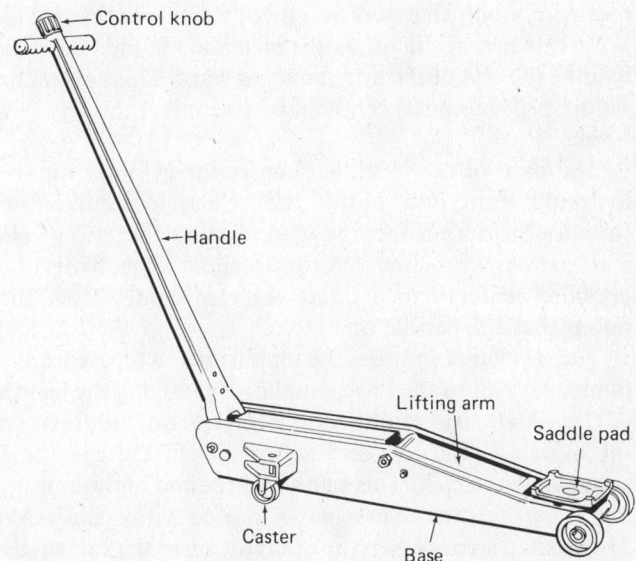

FIGURE 15-38   Common hydraulic floor jack.

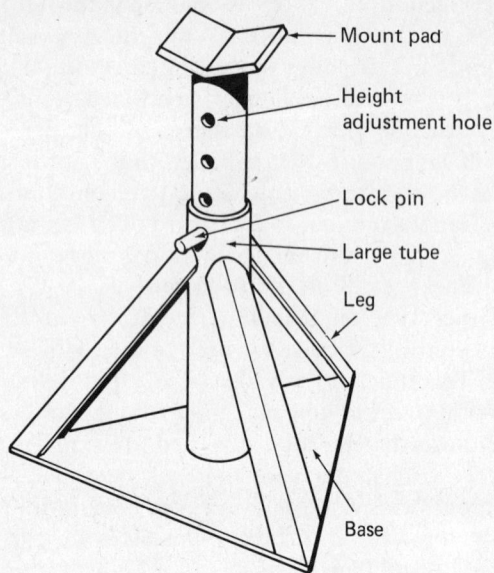

FIGURE 15-39   Design of a typical telescoping stand.

the removal and replacement of the automatic transmission.

The lifting capacity of the average shop floor jack (Fig. 15-38) ranges from two to four tons. This jack design consists of a jack base, lifting arm with saddle pad, hydraulic ram and pump, and a control handle. The jack base is the framework of the jack and comes equipped with steel wheels in the front and casters in the back. The heavy duty wheels and casters support not only the weight of the entire jack but the vehicle as well. They also provide the means by which the mechanic can position the heavy jack under the load—the vehicle.

The lifting arm with its saddle fastens to the base by means of a bolt that acts as a pivot point for these parts. With this arrangement, as the mechanic pumps the jack handle, the end of the arm moves upward. This causes the saddle pad, mounted on the arm, to contact the vehicle's frame.

The base of the jack also houses the cylinder for the hydraulic ram and pump itself. The hydraulic ram supplies the output force needed to move the lifting arm and pad against the vehicle's weight. The hydraulic pressure required to activate the ram comes from the pump that the handle operates.

The mechanic supplies the input force to the hydraulic pump, housed in the base, via the jack handle; the length of this handle will multiply his efforts. Also, mounted on the handle is the release knob, which is the operating control for the jack. This knob is at the end of the handle, and it controls the release valve located within the jack's hydraulic system. When the operator turns the valve fully clockwise, the valve closes; and his pumping of the handle will raise the lifting arm and saddle pad.

Safety Stands

Whenever a mechanic raises a vehicle with any style of floor jack, *he should install safety (jack) stands under the vehicle's frame or suspension system before doing any type of repair work.* In service, these stands serve two functions: (1) they support the weight of the vehicle so that the mechanic can remove the jack from under the vehicle (this action also provides him with additional working area); (2) the stands secure the vehicle so that it will not accidentally fall on the repairman while he is working under it.

Most shops now use either the telescoping or ratchet-type stands. The telescoping stand (Fig. 15-39) consists of two strong, steel tubes with different diameters; these tubes slide (telescope) over one another. The stand manufacturer welds one end of the larger tube directly to a round steel base or directly to three steel legs, which attach to a triangle-shaped base.

Onto one end of the smaller tube, the manufacturer welds a vee-shaped mount pad. This pad will fit under the vehicle's frame or suspension system when the stand is in service. The small tube also has a series of evenly spaced holes machined into it that run almost the full length of the tube. These holes index with a hole cut into the base tube.

These holes provide the stand with height adjustments. The holes accommodate a strong steel lock pin, and the mechanic inserts this pin through the holes in both tubes

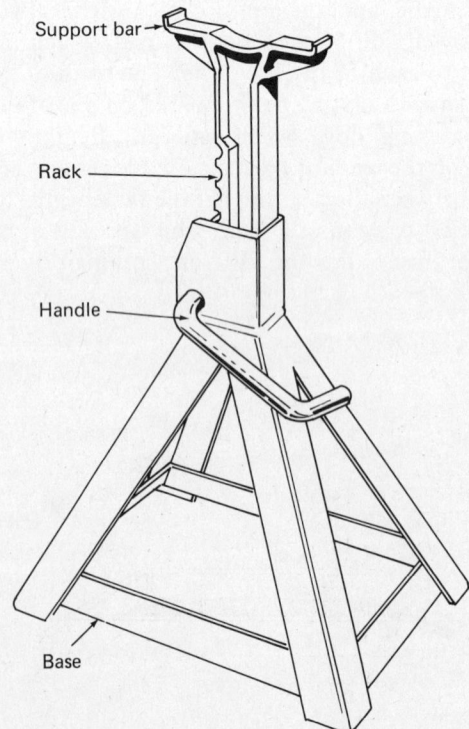

FIGURE 15-40   Design of a ratchet-type stand.

after he adjusts the mount pad to the necessary height.

The ratchet-type stand (Fig. 15-40) consists mainly of a base, a locking dog mechanism, and a rack. This stand manufacturer constructs the square base out of heavy-gauge pressed steel. To increase its strength, the base is rather large in relation to its overall height and has sure-grip type corners.

Built into the upper section of the base is the locking-dog mechanism; this mechanism includes a dog and release handle. The lever portion of the dog engages into the teeth of the rack to lock it to the base at various heights for secure load support. The functions of the handle are to pull the dog away from the rack or release it and to act as a carrying handle for the stand.

The heavy steel rack has a series of teeth cut into it on one side. These teeth, as previously mentioned, index with the lever section of the dog. This design allows the rack to ratchet freely toward its uppermost position, but the dog locks the rack to the base any time the rack attempts to lower by itself. Finally, a sturdy support bar fastens to the top section of the rack. This bar fits under the vehicle's frame to give it positive support.

Transmission Jacks

Automatic transmissions are very heavy and bulky units, and the mechanic should not try to raise, lower, or move them without the aid of a transmission jack. These jacks are available from various manufacturers in different sizes and designs. The overall size and design of the jack will depend on two factors: first, whether the mechanic is going to work under an overhead hoist or beneath a vehicle placed on safety stands; second, the weight of the

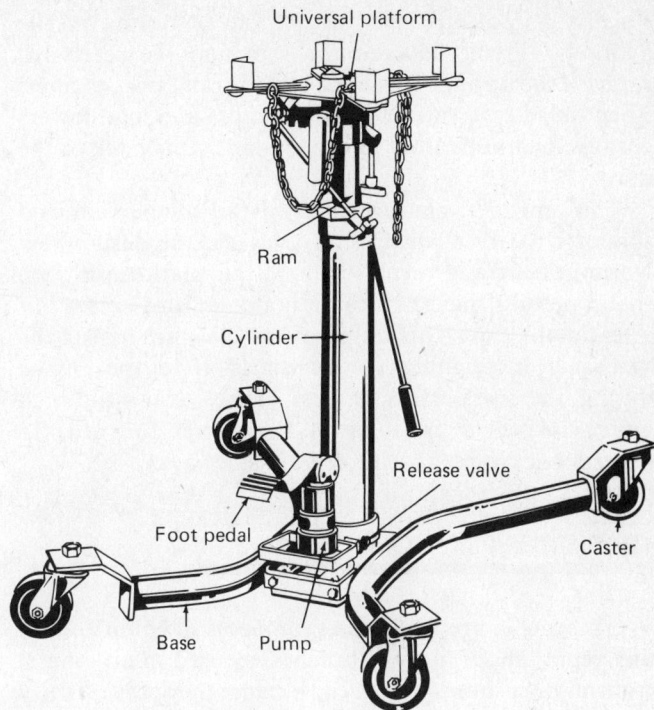

FIGURE 15-42   Design of a hydraulic hoist-type transmission jack.

unit the transmission jack has to support. For example, the heavier units, such as the C-6 and T-400 models, require the use of a hydraulic jack. This jack utilizes hydraulic force to safely raise or lower these large and heavy units, and manufacturers produce hydraulic transmission jacks in both low profile (Fig. 15-41) and hoist models.

Shops that must use the low profile jack because of the unavailability of a hoist often have a jack that operates by a worm-screw drive instead of hydraulics. This device safely raises or lowers the smaller transmissions like the C-4 and Powerglide. But they are not very popular because of the time and effort required to raise and lower the transmission with the worm drive.

A typical hydraulic, hoist-type jack with a 1500-pound lifting capacity is shown in Fig. 15-42. This jack assembly includes such components as a base, hydraulic pump, hydraulic ram, and a universal platform or cradle. The base itself has four legs made of heavy gauge steel. Attached to the end of each leg is a caster; these casters not only support the total weight of the jack and transmission but also provide the means by which the mechanic can position the unit under the vehicle.

The hydraulic pump and cylinder for the ram both attach to the base. The foot-operated hydraulic pump with its reservoir supplies the fluid flow and pressure necessary to activate the ram. The ram that operates inside a cylinder attached to the base produces the output force actually used to raise or support the transmission.

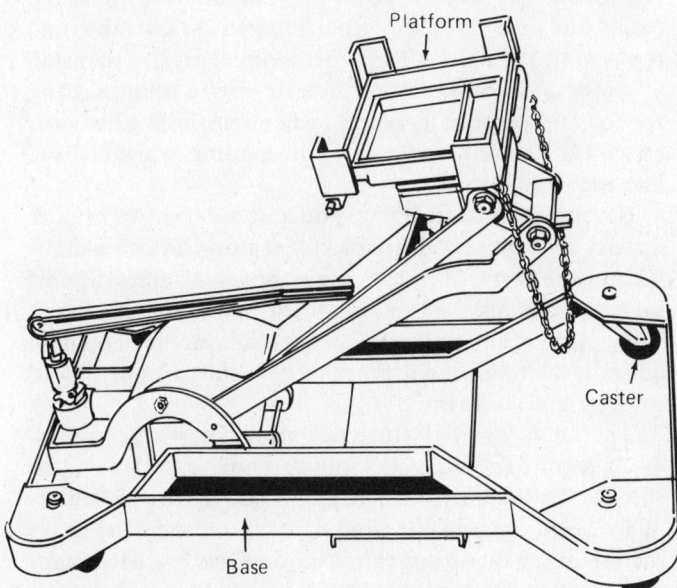

FIGURE 15-41   Design of a hydraulic low profile transmission jack.

Finally, a foot-operated release valve, located in the hydraulic circuit between the ram and the reservoir, controls the up or down position of the ram. For instance, when the release valve is down, the ram and load lowers because hydraulic fluid under the ram vents back to the sump.

The universal platform fits on top of the ram and supports the transmission on the jack; the design of a worm-gear arrangement built into this platform is such that it permits the cradle to tilt not only front to rear but also side to side. This tilting of the platform assists the mechanic in aligning the transmission to the engine during the installation process. The average tilt of a typical platform assembly is 50 degrees forward, 15 degress backward, and 12 degrees sideways.

TRANSMISSION WORK BENCHES

Work benches are very necessary pieces of equipment in any repair shop, but wet benches, as used in automatic transmission shops, are special because they prevent dirty fluid from leaking onto the floor from a transmission the mechanic is tearing down. Therefore, the wet bench assembly reduces the amount of messy clean-up work by collecting dirty fluid and temporarily storing it.

A standard wet-type work bench (Fig. 15-43) consists of the bench, a series of gutters, and a drain-off coupling with an attached sump or container. The manufacturer of these units usually forms this type of bench of heavy-gauge steel with the top (work bench) portion thick

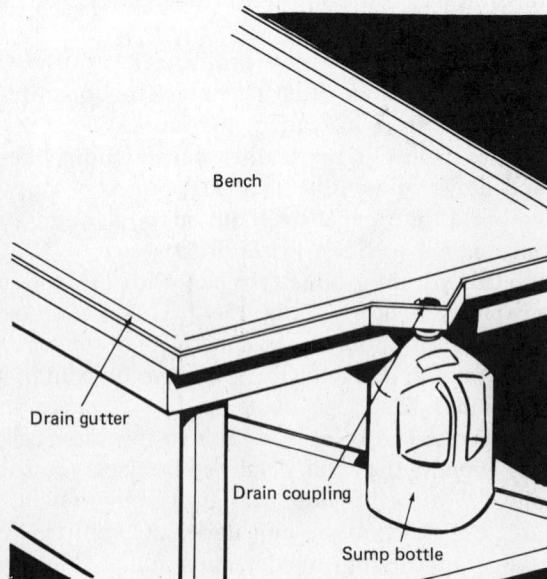

FIGURE 15-43   Typical wet-type transmission work bench.

enough to withstand the heavy weight of the transmission.

Attached to or made as part of the top is a series of drain gutters, which form a fluid trough around the entire outer edge of the top. Any dirty fluid attempting to leak off the bench top and onto the floor collects in these gutters. The gutters along the full length of the bench taper down slightly toward one end to allow the dirty fluid to flow by gravity into the sump. Manufacturers accomplish this by installing the gutters slightly lower from the top of the bench on one side than the other or by making the bench legs slightly longer on one end. This latter action, of course, raises one end of the bench a little higher than the other.

The drain-off coupling, an open tube, fastens into the gutter at its lowest point. This design permits the fluid from all the gutters to flow by gravity toward this lowest point and then drain through the open spout and into the sump bottle which stores it. The bottle itself clamps to the underside of the drain coupling, allowing the repairman to easily remove the container for draining as it becomes full.

TRANSMISSION FIXTURES OR STANDS

A transmission holding fixture, like the one shown in Fig. 15-44, is a very useful piece of equipment in any automatic transmission shop. This device supports the heavy and bulky transmission on top of or to one side of the work bench, and the fixture itself is usually adjustable so that the mechanic can rotate the transmission and fixture to a horizontal or vertical position. This feature makes it easier and faster for the repairman to disassemble and reassemble the unit. Although it is not necessary to install all automatic transmissions in a fixture to rebuild them, the fixture is especially helpful when reinstalling heavy or awkward components in a front-loading transmission, like the T-400.

Several manufacturers produce these fixtures in both factory and universal designs. A factory fixture usually fits only one type of transmission, but some units support several transmissions made by the same manufacturer. The design of the universal stand is adaptable enough to accommodate different types of transmissions produced by different companies.

The factory-style fixture assembly, shown in Fig. 15-44, is supporting a T-400 transmission; Fig. 15-45 now shows the assembly by itself. This fixture assembly includes a base and the holding fixture itself. The lower section of the one-piece cast-iron base has a square shape with four holes machined into it. These holes accom-

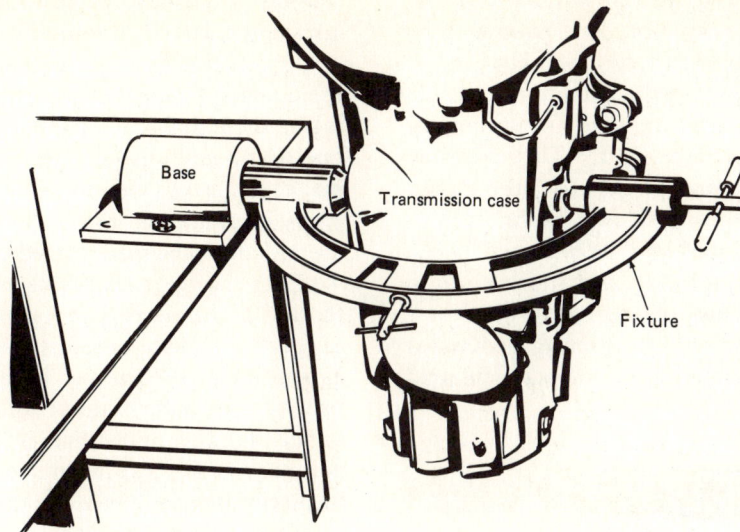

FIGURE 15-44 **An automatic transmission installed in a holding fixture.**

modate bolts that secure the base to the end of the work bench.

The upper part of the base also has a semicircular shape, and the manufacturer machines a large bore and two smaller holes into it. The large bore acts as a guide and support for the shank section of the holding fixture. The threads, cut into the smaller holes, index with the threaded portion of a set screw and thumbscrew.

The set screw fits into the threaded hole nearest the fixture. The purpose of this screw is to prevent the shank and fixture from accidentally sliding out of its bore in the base.

The thumbscrew, on the other hand, threads into the hole farthest away from the fixture. This thumbscrew locks the fixture into either a vertical or horizontal position.

The holding fixture itself attaches to and supports the transmission case in the base assembly; this heavy-duty, cast-iron fixture has a U-shape along with a solid steel shank and two threaded clamp screws. The solid steel shank presses into a bore machined into one end of the fixture. This shank also has a shallow groove cut into it that aligns with the set screw located in the upper section of the base. As previously mentioned, this set screw, when threaded into the base as far as it will go, prevents the fixture from sliding out of the base unexpectedly. Finally, the shank also has a series of placement holes cut into an area near the end of the shank. These holes index with the end of the thumbscrew to secure the fixture in either the horizontal or vertical position.

The two threaded clamp screws are necessary to hold

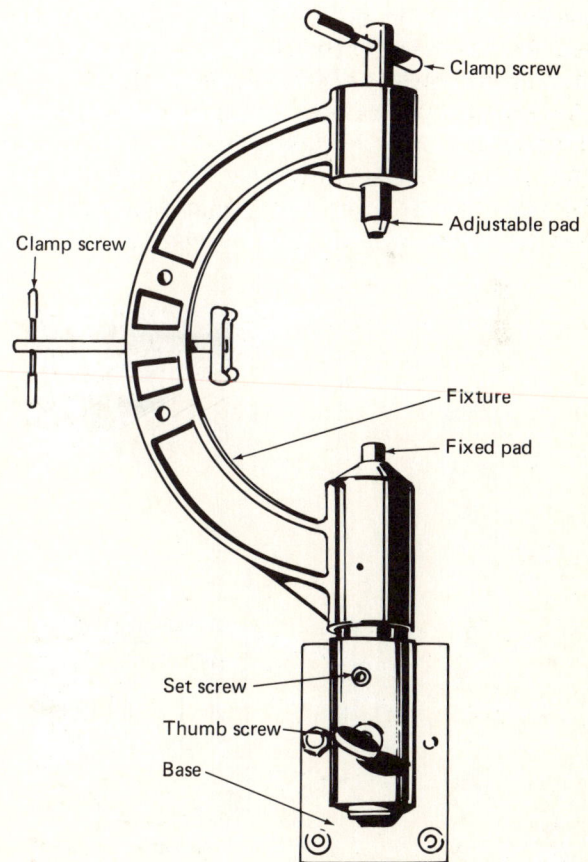

FIGURE 15-45 **Design of a GM transmission holding fixture.**

the transmission case securely into position within the fixture. To understand their function, let's examine how the transmission case attaches to the fixture itself. First of all, the case has two blind mount holes cut into the outside of the case. One of these holes fits over the stationary mount pad on the shank end of the fixture. The other hole accommodates the mount pad fastened to the clamp screw threaded into the opposite side of the fixture.

Now, with the case positioned into the fixture, between the two mount pads, the second (center) screw clamp threads out and bears against the center of the transmission case. This clamp serves as a centering device that maintains the case's centerline almost perpendicular to that of the holding fixture.

THREAD REPAIR TOOL SETS

The mechanic must inspect all threaded holes in the transmission case, front pump, valve body, and extension housing during the rebuilding process for damage and repair them as necessary. If this is not done, the damaged or stripped threads will not be able to withstand the torque applied to the mating threaded fastener, a bolt or screw. As a result, the fastener will not be tight enough, and the components it holds together can work loose,

causing a possible separation of components or an internal or external hydraulic leak.

One method to repair a series of damaged threads is to use a heli-coil thread repair kit (Fig. 15-46). This tool kit includes the heli-coils themselves, a twist drill, heli-coil tap, and installer tool. The heli-coil is a device that the mechanic installs into a specially tapped hole to replace a series of damaged threads. A heli-coil insert is nothing more than a precision formed coil of stainless spring-steel wire; the wire itself forms the external and internal threads of the insert. The coil manufacturer bends the end of the wire across the coil at one end to form a tang, which is necessary to turn the insert during the installation process. The outer diameter of the heli-coil is also slightly larger than the tapped hole it fits into; consequently, the coil has to compress inward somewhat in order to thread into the opening. Once the mechanic installs the insert into its tapped hole, the coil expands because of its elasticity. This elasticity then helps to lock the heli-coil against the tapped bore.

The twist drill of this kit serves two related functions. First, the drill cuts out all the damaged threads from the bore. Second, the bit resizes the hole in order to prepare it for the heli-coil tap. The function of the tap is to machine a special thread into this oversize hole. The newly cut internal thread has the same pitch as the external threads on the heli-coil insert. But, as previously mentioned, the diameter of the heli-coil's thread is slightly larger than the diameter of the new thread within the tapped hole.

The design of the installing tool is such that it supports and rotates the heli-coil during installation. This tool has external threads that mate with the internal threads of the heli-coil. Furthermore, on the one end of the tool is a slot that engages with the tang located on the end of the heli-coil; this locks the coil to the tool. Finally, on the other end of the tool, the manufacturer machines a square drive. This drive accommodates a tap wrench, used by the mechanic to thread the installer tool and its attached heli-coil into the retapped hole.

POWER TOOLS

Automatic transmission specialists use several types of power tools to service and repair transmissions both in or out of the vehicle. These tools save the mechanic time and energy. Power tools—drills and impact wrenches—increase his productivity during a normal work day and, at the same time, reduce his fatigue.

Normally, the classification of these tools is by the type of power used to make them function: electrically powered tools and air-powered (pneumatic) tools. Most popular tool producers now offer power tools that do the same job with either electricity or air as their power source.

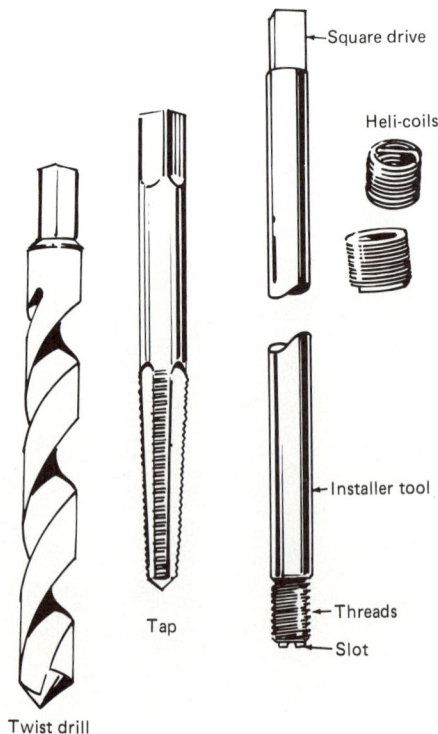

FIGURE 15-46 **Typical heli-coil thread repair kit.**

Drill Motors

The drill motors, shown in Fig. 15-47, are samples of one type of power tool. The transmission technician uses this tool to drive a twist drill, which machines a hole in a given component. For instance, in transmission repair and service work, the drill motor and twist drill together cut the holes necessary to insert heli-coils and alter valve bodies for the installation of shift kits.

Figure 15-47 pictures both the electric and air-type motors. The electric drill motor is not as popular today as the air drill for transmission and general automotive-type repair for several good reasons. For instance, although electricity is probably the most widely used power source today, it can be very dangerous under shop conditions. With the drill motor, normal brush arc can ignite any nearby volatile fumes. Also, if a grounded condition happens to occur within the motor itself and the cord attached to the drill is not the three-wire type with a three-blade plug, the mechanic can receive a severe electrical shock. However, in recent years, drill manufacturers have reduced this risk by forming the motor housing of plastic, which is an insulator.

Air drills, on the other hand, do not have these inherent, negative characteristics and are easier to use than the electric type because they are smaller, lighter, and run cooler. The air drill, with the same overall chuck size and power rating as the electric type, is smaller and quite a bit lighter because of the difference in design between the drive motor found in each unit. Furthermore, since electricity is not the power source to rotate the motor itself, the operating temperature of the air drill, even under loaded conditions, remains relatively low.

Both drill motor designs do have some of the same features. For example, both units have a three-jaw chuck that tightly holds the twist drill in place and a chuck key to tighten and loosen the chuck. Furthermore, manufacturers of both drill designs designate the size of their units by the maximum capacity of the chuck, and the sizes that most automatic transmission shops use have a capacity of $\frac{1}{4}$ inch or $\frac{3}{8}$ inch. Finally, both motor types are available with switches that allow them to operate with varying speeds, which is very helpful when cutting holes in different types of materials.

Impact Wrench

Figure 15-48 illustrates other types of air-powered tools—the impact wrench and the air ratchet. The impact wrench, as its name implies, loosens or tightens most bolts or nuts that hold the transmission itself together. With this device, the mechanic squeezes a trigger to activate the working mechanism of the wrench, which applies partial to full impact force on the fastener to loosen or tighten it. Manufacturers do produce electric impact wrenches, but they are not very popular today for the same reasons that inhibit the use of the electric drill.

A common method of rating air-impact wrenches is by

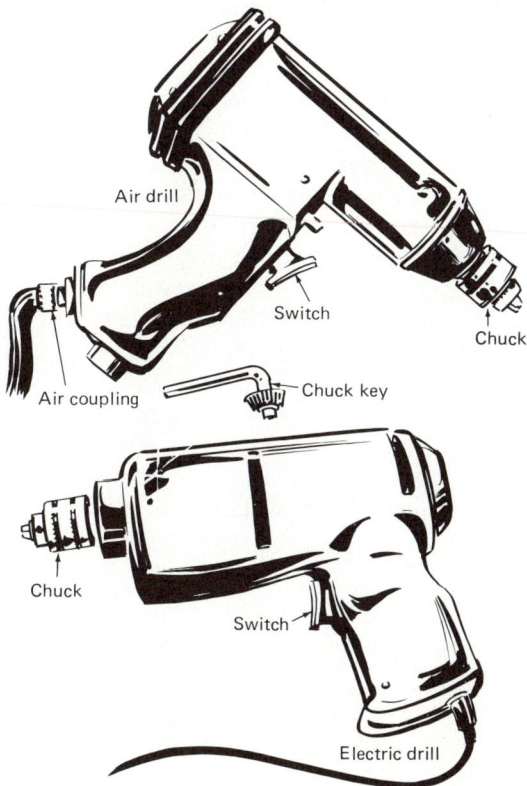

FIGURE 15-47 Typical electric and air drills.

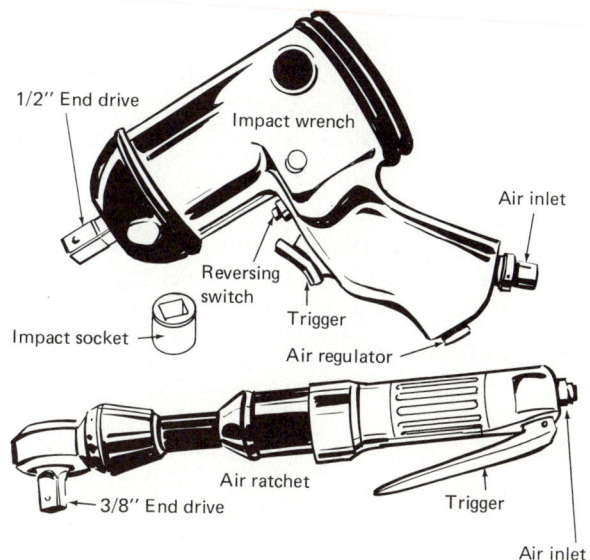

FIGURE 15-48 Air impact wrench and air ratchet.

the size of its end drive. The most common wrenches used by transmission technicians have either a ⅜- or ½-inch end drive, which, in turn, mates with either a ⅜- or ½-inch drive impact socket. These sockets are special in that the manufacturer reinforces them so that the socket can withstand the impact (hammering type) loads placed on them by the air wrench.

Most air wrenches also have some method of regulating input (shop) air pressure to the driving impact mechanism inside the wrench. The position of this regulator then determines not only the speed of the end drive but the torque applied to the impact socket. A typical ⅜-inch air wrench, for example, has a regulator that controls torque output from zero to a maximum of 75 pounds-foot (10.35 kg-m), while the working torque range of a regulated ½-inch impact will be between zero and about 200 pounds-foot (27.6 kg-m).

When using any size air impact wrench, follow these recommendations and the tool will provide long, maintenance-free service along with a reduction in fastener damage:

1. Always use the special impact sockets with an impact wrench.

2. Always use the proper size socket for the nut or bolt.

3. Always put to use with the wrench the simplest assembly of a socket, extension, and universal joint.

4. Where possible, always use a deep socket in place of a standard-length socket and extension.

5. Always hold the wrench in a position so the socket fits squarely on the nut or bolt. As necessary, apply a slight forward pressure on the wrench to hold the socket in place.

6. Once a bolt or nut is tight, never impact it beyond an additional one-half turn on the socket. Continued tightening beyond this point will strip the threads or break the bolt, and *never use the impact wrench to tighten down fasteners that call for a given torque specification.*

7. Never use an impact wrench beyond its rated capacity. If the fastener is not tight enough after impacting with the tool for five seconds, use a larger wrench.

8. Always presoak large, rusty bolts and nuts with penetrating oil before attempting to remove them.

Air Ratchet

The air ratchet, shown in Fig. 15-48, is another type of device that saves the repairman time and energy. This reversible ⅜-inch tool provides a small amount of output torque, about 45 pounds-foot, (6.21 kg-m) for speeding up the removal and replacement of nuts and bolts. When

in service, the mechanic uses the tool as a common ratchet first to break the fastener loose; then he depresses the switch to spin the nut or bolt off. Of course, during the installation of hardware, he reverses this process.

This wrench, due to its unique design, is very versatile. It fits easily into limited work areas because of its smaller size. This increases the mechanic's productivity in removing and disassembling the transmission. Also, because of its low torque output, the serviceman can use the regulated air ratchet to run in smaller fasteners when assembling the transmission. Finally, due to its design and output, this wrench does not require the use of the special impact sockets; therefore, the mechanic can use all the sockets and attachments of a standard ⅜-inch set with this tool.

For proper performance, any type of air-powered tool requires a regulated supply of clean, dry compressed air. The recommended air pressure necessary to operate most common air tools is about 100 psi (7.03 kg/cm²). This rated air pressure is the pressure at the tool while the tool is running under a no-load condition, and it is measurable by attaching a pressure gauge as close to the tool as possible. The pressure that registers on this pressure gauge is the pressure **at the tool** and not the output pressure of the shop's air compressor.

Several problems will occur if the serviceman uses an air tool operated by too low or too high air pressures. If the air pressure is less than recommended, the overall efficiency of the tool is reduced. On the other hand, excessively high air pressure causes the tool to operate above its rated capacity. This action shortens the life of the tool and can damage or break fasteners.

The mechanic should lubricate the air tool each day before using it, by applying three or four squirts of recommended oil into the tool's air inlet, connecting it to a source of air pressure, and operating the tool. This action flushes any moisture, dirt, and gum out of the air motor and lubricates its moving parts. The injected oil removes deposits from the tool by carrying them out the air exhaust.

When performing this lubrication procedure, observe these safety precautions: (1) do not flush out the tool around an open flame; (2) always point the tool's air exhaust port away from your skin or clothing.

PULLERS

Gear Type

The mechanic uses several types of pullers when disassembling an average automatic transmission. These tools, as their names imply, remove gears, seals, or pumps using

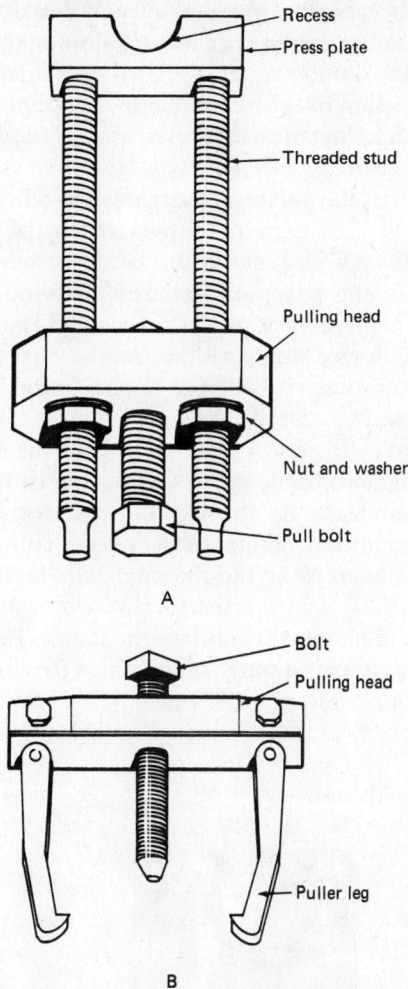

FIGURE 15-49   Two typical speedometer gear pullers.

nuts serve several functions. First, they act as a rigid but adjustable connection between the plate and pulling head. By turning the two nuts one way or the other, the mechanic can shorten or lengthen the distance between the plate and head to match the length of the output shaft (Fig. 15-49).

Second, the technician can use these two studs and their hardware to push or press the gear back onto the output shaft. For a pressing operation, the two nuts fit under the pulling head with the washers against the head itself. Now with the pulling head in position over the output shaft, the repairman can easily press the gear on by turning each nut out a little at a time until the gear moves fully into position on the shaft.

The pulling head and pull bolt also serve two functions. First, the two components together provide the force necessary to remove the gear from the shaft. In this situation, the bolt threads through the head until it contacts the end of the output shaft. Any further rotation of this bolt causes sufficient force on the head, studs, and plate to pull the gear off the shaft.

Second, the head along with the aid of the bolt acts as a reaction area during the gear pressing operation. To accomplish this function, the head must lock to the output shaft. The tool manufacturer accomplishes this by machining a retaining collar into the lower side of the head; this collar, in turn, fits into a groove in the output shaft. Now, with the head attached to the shaft and its bolt threaded in against its end, the head can act as a reaction area (stop) upon which the threaded nut and washers react while pushing the gear in place over the output shaft.

The design of one type of T-400 transmission puller, shown at B of Fig. 15-49, is such that it removes a gear only from its shaft. This tool consists basically of a pair of puller legs, bolts of varying lengths, and the pulling head and bolt. The pointed end of each leg fits under the gear, and they are adjustable to fit any gear size via slots machined into the pulling head.

Two bolts attach the legs to the head. The actual length of the bolts the mechanic will use depends on the length of the shaft the gear is on. In Fig. 15-49, the bolts are short, which brings the two legs in direct contact with the head.

The pulling head and bolt supply the force necessary to remove the gear. The bolt itself threads through the center of the head and bears against the shaft. Now, if the legs are in place around a gear, and the serviceman continues to turn the bolt clockwise, the tool will pull the gear off the shaft.

Seal Pullers

It is also necessary, in most cases, for a mechanic to use another type of puller to remove metal-clad seals from the front-pump or extension housing. The puller normally

either a pulling-type or impact-type force. For example, Fig. 15-49 pictures two pullers used to remove the pressed-on speedometer gear from the output shaft of two different transmission types. Both of these devices use the inclined-plane principle, via the threads on the large bolt, to supply the force necessary to remove the gear.

The Powerglide transmission puller shown at A is a unique device in that the tool not only pulls the gear off, but it also presses the gear back on the output shaft as well. This tool consists of a press plate, two long threaded studs with washers and nuts, and a pulling head and pull bolt. The press plate is nothing more than a piece of steel with a recessed opening and two tapped holes. The opening is large enough so that the plate will fit around the output shaft and under the gear. The recessed area that extends from the opening conforms to the gear's shape and provides adequate support to this area during the removal or installation procedures.

The ends of the two studs thread into the tapped holes in the press plate; these studs along with their washers and

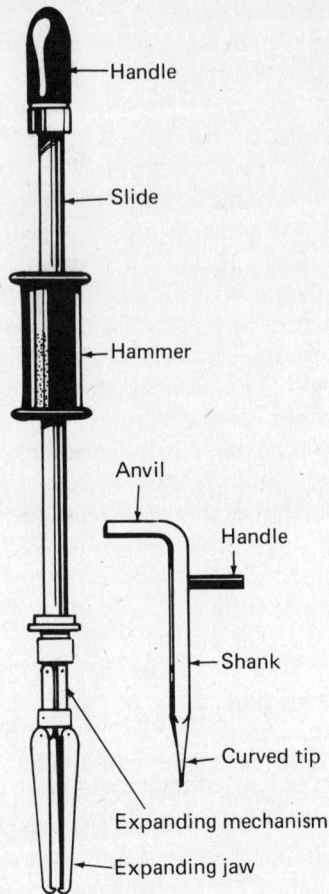

FIGURE 15-50 Two common types of universal seal pullers.

hammer itself, when moved from the jaw end of the slide toward the handle, imposes an impact blow on the handle stop. This blow attempts to force the stop, slide, and jaw set in the direction of the hammer blow. The end result of one or more hammer blows is the removal of the seal from its housing.

The design of the universal tool shown at B, Fig. 15-50, is such that it also uses the impact energy of hammer blows to remove a seal, but in this case, the blows come from a hand-held, ball-peen hammer. The tool itself is made out of square-steel stock with one end of the tool bent at a 90-degree angle to the shank. This end now forms an anvil that receives the impact force from the hammer blows.

On the opposite end of the shank from the anvil is a curved tip made of hardened steel. This tip fits inside the seal and against the lip formed into the steel backing. With this design, any impact force from a hammer blow on the anvil passes from it to the shank and finally to the tip and seal.

The knurled handle threads into the shank. The technician uses this handle to support and guide the tool during a seal pulling operation.

used for this purpose employs impact force to loosen and remove the seal from its housing bore.

Figure 15-50 shows two kinds of universal impact seal pullers. The tool shown in A is a slide-hammer type that includes a set of expanding jaws, a slide and handle, and the hammer itself. The set of jaws fits inside the seal with each jaw bearing against a steel lip formed into the backing. Furthermore, an expanding mechanism threaded onto the end of the slide not only adjusts the set of jaws to fit various seal sizes, but also keeps the individual jaws in firm contact with the steel backing.

The slide with its attached handle acts as a guide for the movable hammer. The hammer fits over the slide and moves back and forth on the unthreaded portion between the expanding mechanism and the stop built into the handle. The handle therefore acts as a stop for the slide hammer and as a place where the mechanic can support the end of the puller.

The heavy slide hammer supplies the tool with the impact force necessary to loosen and remove the seal. The

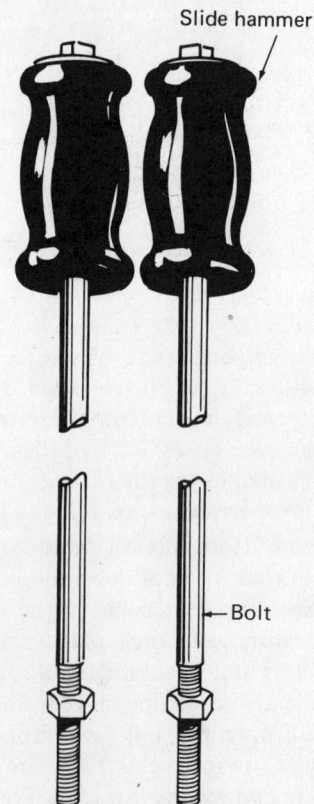

FIGURE 15-51 A set of slide hammers, used to remove the front pump from certain GM transmissions.

Pump Pullers

Certain transmission models also require the use of a puller to remove the pump from the transmission case; this puller is necessary for several reasons. First, the internal component design of the transmission is such that it is impossible to get behind the pump to push it out. Second, the pump's outer seal, which exerts some tension on its bore inside the case along with a stuck pump gasket, usually makes it impossible for the mechanic to remove the pump by hand.

Figure 15-51 shows the most common type of pump puller. This puller consists of two hand-held slide hammers and two long bolts. Each weighted slide hammer fits over one of the bolts, and it slides back and forth on the unthreaded section of the bolt. If the hammer strikes the head of the bolt, its impact energy transmits to the head. This action results in a pulling-type force on the bolt itself in the direction the hammer was moving.

The two bolts, which act as the guides for the hammers, each thread into a tapped hole within the pump housing. Each bolt also has a nut to lock the threaded end of the bolt in place so it cannot come loose during the pulling operation and damage the threads. Now, with these bolts in position, the mechanic can easily remove the pump by firmly pulling both hammers back until they contact the bolt head; the resulting impact force will pull the pump from the case.

SNAP-RING PLIERS

The factory installs several types of snap rings when assembling an automatic transmission to lock or hold a component in a given position. Some snap rings, like those used in a clutch drum, the mechanic can remove easily with a screwdriver. Others require the use of special snap-ring pliers for both their removal and installation.

Snap-ring pliers are of two types, inside and outside. The jaws of the inside snap-ring pliers close and grip an internal snap ring when the mechanic closes the handles. The jaws on the outside snap-ring pliers open to expand an external snap ring as the handles come together.

Figure 15-52 pictures several sizes and types of snap-ring pliers. Pliers A and B are both outside-type tools, which fit plain snap rings. The main difference between these pliers is in the design, shape, and free opening of the jaw tips.

Pliers C and D are outside Truarc pliers, which have a specially designed jaw tip. The manufacturer in this case machines a round pin on the tip of each jaw. These pins fit into the holes formed into the lip ends of a Truarc snap

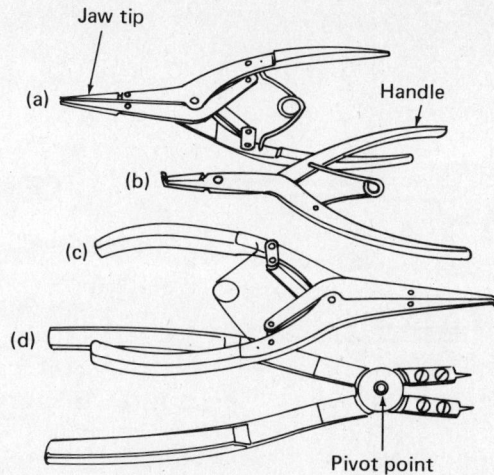

FIGURE 15-52 *Several sizes and designs of snap-ring pliers.*

ring. With this design, there is less possibility of the pliers slipping off the ring during removal or installation.

Some Truarc pliers also have detachable tips, while others have several pivot points. The detachable tip design permits the mechanic, with the help of one pair of pliers and various sizes of interchangeable tips, to service a wide variety of different-sized snap rings. If the tool has

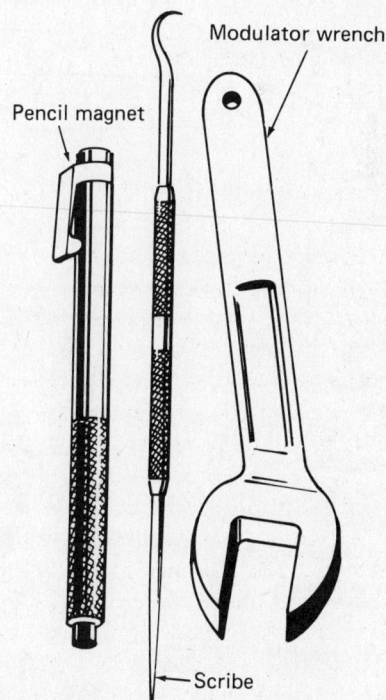

FIGURE 15-53 *Typical scribe, pencil magnet, and modulator wrench.*

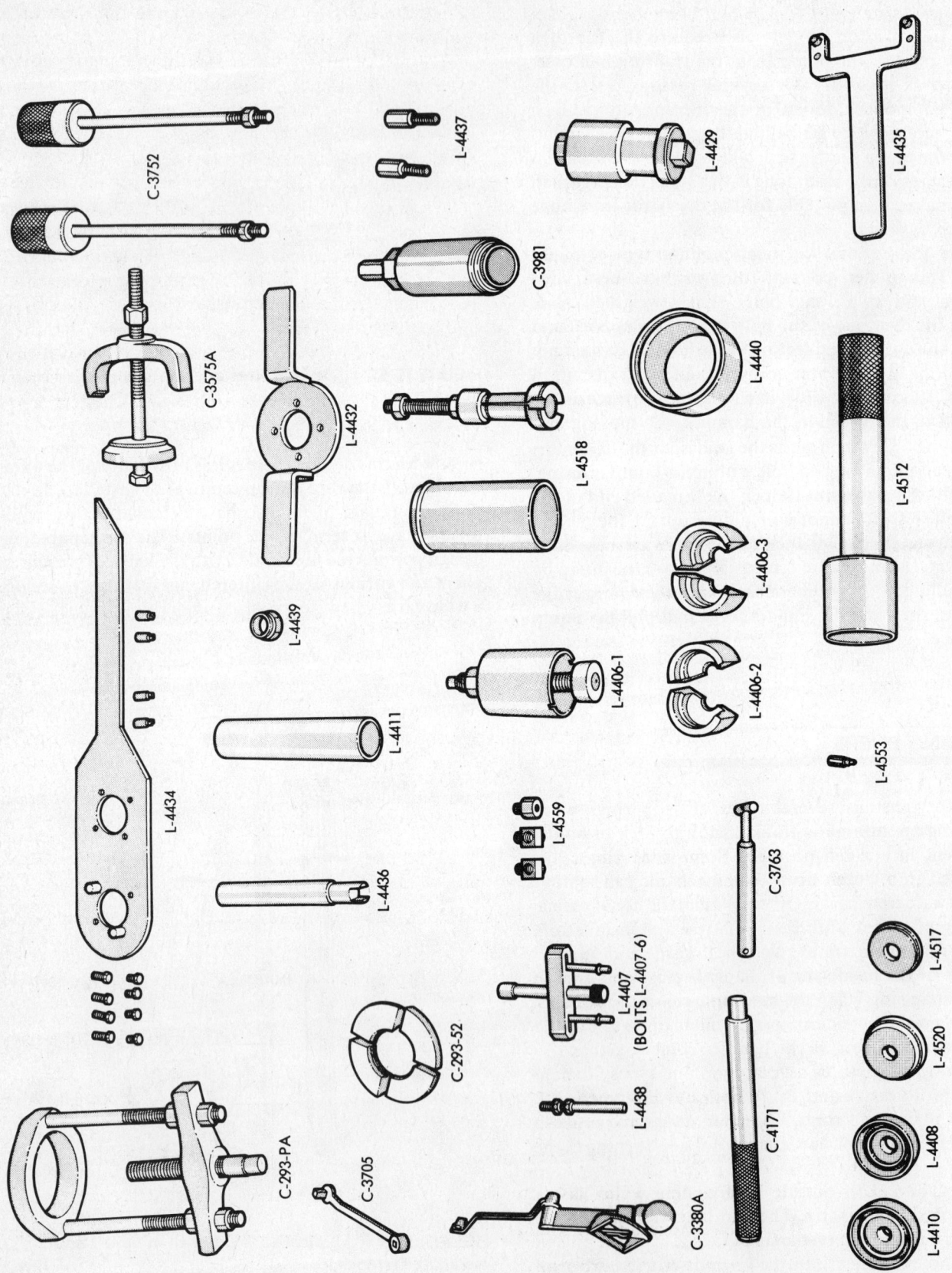

FIGURE 15-54 The special tools recommended by the manufacturer to rebuild an A-415, A-413, and A-470 automatic transaxle. (Courtesy of Chrysler Corp.)

C-3752

L-4437

L-4429

L-4435

C-3981

C-3575A

L-4432

L-4440

L-4518

L-4512

L-4406-3

L-4439

L-4406-2

L-4411

L-4406-1

L-4434

L-4553

L-4436

L-4559

C-3763

L-4517

L-4407
(BOLTS L-4407-6)

C-293-52

L-4520

L-4438

C-293-PA

C-3705

C-3380A

C-4171

L-4408

L-4410

two interchangeable pivot points, the mechanic can convert an inside pair of snap-ring pliers to an outside pair.

OTHER HAND TOOLS AND SPECIAL EQUIPMENT

On the market today are other multi-purpose hand tools that make the automatic transmission mechanic's job a lot easier and faster. Figure 15-53 shows three of these devices, the **scribe,** the **pencil magnet,** and the **modulator wrench.** A mechanic uses the pointed scribe for removing small snap rings or seals in places where a small screwdriver, for example, will not fit. The pencil magnet is very useful in removing check balls from fluid passages in the transmission case or valve body, and it is also quite helpful for removing valves and springs when overhauling a valve body. These two tools have, of course, other

service applications, but the important thing to remember about them is that the technician can use them on all transmission models.

The modulator wrench is necessary, in most cases, to remove and install a vacuum modulator without damaging the housing, but this special tool is not as universal as the scribe and magnet. Most modulator wrenches are made for a specific transmission, others for a family of transmissions, and only a few are universal.

All automatic transmission manufacturers recommend the use of certain other special tools when rebuilding their particular units. All of these tools are nice to own because they make the rebuilding job easier and faster, but they are very expensive, sometimes hard to get, and not always necessary.

Figure 15-54 shows some examples of special tools used to rebuild a transmission. These particular ones are factory tools recommended for use on a Chrysler A-415, A-413, and A-470 automatic transaxle.

REVIEW

This section will assist you in determining how well you remember the material contained in this chapter. Read each item carefully. If you can't complete the sentence, review that portion of the chapter that covers the material.

1. The function of a torque converter flusher is to _____ used converters and eliminate costly comebacks.

2. The flushing machine, described in this section, utilizes the _____ method to circulate the cleaning agent inside the converter.

3. Many sealed converters come from the factory without a _____; this makes servicing the converter much more difficult.

4. The dynamometer _____ _____ the rebuilt or malfunctioning transmission without the need of installing the transmission into the vehicle.

5. The speed-control cylinder of the dynamometer has the responsibility of actually changing the width of the _____.

6. The dynamometer load-application gauge shows the amount of _____ applied to the machine's load section.

7. The dynamometer's 12-volt system tests the _____ circuit on the T-300 and 400 transmissions.

8. The hydraulic gauge set tests the _____ _____ of an automatic transmission that is still in the vehicle.

9. Transmission repair shops can use _____ types of cleaning equipment.

10. The heating coil of a steam cleaner changes tap water into _____.

11. The oscillating jet assemblies of a jet cleaner are _____ the cleaning chamber.

12. The safety-type parts washer, mentioned in this section, has an _____ agitation system.

13. A clutch spring compressor is not usually necessary to compress a _____ type return spring.

14. The factory-type, clutch-spring compressor mentioned in this section uses the principle of the _____ to multiply force.

15. Bushings are a special type of _____ that supports revolving transmission components.

16. A mechanic can use a bushing chisel to remove a bushing for a _____ bore.

17. Because of its design, the _____ lift is not very efficient for transmission repair work.

18. The lifting capacity of the average shop floor jack ranges between _____ and _____ tons.

19. Most shops now use either the _____ or _____ safety stands.

20. The _____-type work bench is a necessary piece of equipment in a transmission shop.

21. A factory-type transmission holding fixture will usually fit only _____ transmission type.

22. One method the mechanic can use to repair damaged threads is through the use of a _____ _____ _____.

23. Mechanics use the air-type drill because it is _____, _____ and runs cooler.

24. Along with the impact wrench, the _____ _____ also saves the repairman both time and energy.

25. Pullers, used on automatic transmissions, primarily utilize _____ or _____ force to remove a gear.

26. The factory installs several types of _____ _____ to lock or hold a component in a given location.

27. The special tool that is very useful in removing valves from the valve body is the _____ _____.

For the answers, turn to the Appendix.

One of the more important jobs of the technician performing service or repair work on the automatic transmission is the taking of accurate measurements. The purpose behind this particular task is to make sure that the transmission itself, along with all of its components, is correctly put together and will operate with given tolerances, called factory specifications. Sometimes this means measuring the torque on bolts or nuts. In other cases, it means checking the clearance in clutch assemblies or the end-play of rotating shafts. But in order to check items against the specifications, the repairman must be familiar with the function of various kinds of measuring tools like the torque wrench, micrometer, feeler gauge, and dial indicator.

CHAPTER 16

Measuring Devices and Fasteners

TORQUE WRENCHES

Function

The torque wrench is a measuring tool that is essential to overhaul the automatic transmission properly. This wrench is basically a measuring device that indicates the amount of torque or twist applied to the fasteners, bolts or nuts, of the transmission during reassembly. *REMEMBER: The automatic transmission manufacturer establishes a given torque value for all threaded fasteners, and these specifications require exact measurements. Therefore, do not depend on "experienced feel" or a regulated impact wrench to torque these fasteners.*

Failure to use a torque wrench or failure to follow factory specifications can cause damage to transmission components or cause the unit to malfunction. For example, over-torquing a valve body cap screw can result in a broken housing, pulled threads, or a warped casting. Under-torquing of the same hardware can cause a hydraulic leak, resulting in a transmission malfunction or total failure.

Types and Sizes

Manufacturers produce torque wrenches in several types and sizes. Figure 16-1 shows several examples of different types and sizes of wrenches. Types A and B are micrometer- or clicker-type wrenches. Wrenches C and D are both beam-type units.

The clicker-type torque wrench has a particular torque value calibrated into it by the mechanic. When the technician tightens a fastener to the preset value with this wrench, it makes a loud click and the wrench handle moves freely a few degrees in the direction of applied torque.

In order to calibrate the torque limit of this wrench, the mechanic loosens the lock nut and rotates the hand grip one direction or the other. The hand grip or handle itself is part of a micrometer mechanism built into that end of the wrench (Fig. 16-2). By twisting the handle and observing the micrometer scale, the mechanic can set whatever torque value he desires into the wrench; it will automatically click and break free at this setting. This provides a very accurate method of torquing nuts or bolts.

The design of the deflecting-beam torque wrench (Fig. 16-3) is such that the amount of twist applied to the fastener registers on a scale as the mechanic pulls on the handle. With this wrench design, the beam itself bends (deflects) in proportion to the amount of torque applied to the fastener. On the free end of the beam is a pointer that moves across the indicator scale mounted on the arm just above the handle. Since the arm does not bend with the beam, the position of the pointer on the scale indicates the amount of applied torque by measuring the amount of bend. Thus, the mechanic can determine the actual amount of twist applied at any given time by simply observing the pointer's position on the scale.

Torque wrench manufacturers produce the clicker and beam wrenches with various sizes of square end drives and overall torque ranges. For example, a common ¼-inch drive micrometer torque wrench measures applied twist from 30 to 200 inch-pounds (34.5 to 230 kg-cm). A ⅜-inch drive wrench of the same design has settings from

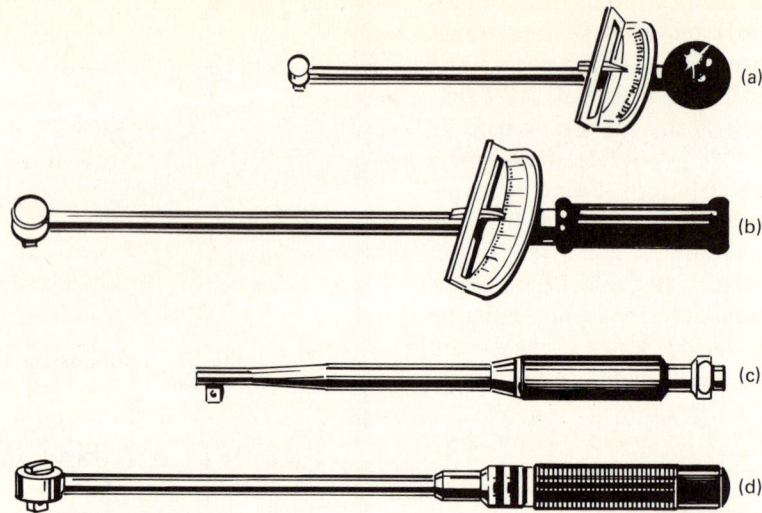

FIGURE 16-1 Examples of different types and sizes of torque wrenches.

5 to 75 foot-pounds (0.69 to 10.35 kg-m). A typical ½-inch drive beam-type wrench has a torque range from 0 to 200 foot-pounds (0 to 27.6 kg-m).

Factors Determining Wrench Size to Use

Three factors determine the size of torque wrench the mechanic should use on a given job. The first factor is the hardness and size of the bolt or screw; the second is the torque specification for the fastener; the third is the torque range of the wrench itself. In most cases, the torque on the bolt or screw relates to the relative hardness and diameter of the fastener. These factors determine its shear point under torque load.

Most nuts, bolts, or screws receive a torque specification value way below their shear point. This reduced specification protects the components attached together by the fasteners from distortion or breakage and the shearing of the hardware itself.

If a torque specification for a given bolt is 30 inch-pounds, a mechanic must use a small torque wrench similar to the ¼-inch drive, mentioned earlier, to tighten the bolt or screw. If he happens to use a larger wrench that cannot accurately measure this low amount, the hardware may break, or the components they fasten together may distort. Also, to prevent damage to the torque wrench itself, the mechanic should select a wrench with a torquing capacity high enough so that the specified torque value will fall near the midpoint on the wrench's scale.

Torque Wrench Extensions or Adapters

When torquing bolts or nuts in obstructed areas, it is sometimes necessary to add certain extensions and/or adapters to the torque wrench in order to provide it with operating clearance. As long as the extension or adapter fits onto the end drive at a 90-degree angle to the wrench handle (Fig. 16-4), the amount of torque, indicated on the scale or set into the clicker-type wrench, will be accurate. In other words, whatever torque reading the mechanic observes on the scale or sets into the wrench will be what it applies to the fastener.

But if the extension or adapter adds length that is parallel to the handle of the wrench (Fig. 16-5), the torque applied to the fastener will be increased. Consequently, the amount of torque showing on the scale or set into the wrench **will no longer be accurate.** Therefore, whenever adding length to the torque wrench, be prepared to cal-

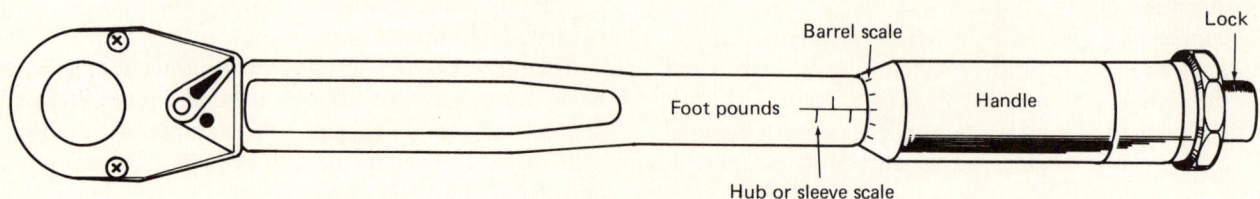

FIGURE 16-2 Micrometer mechanism built into the handle section of the clicker-type torque wrench.

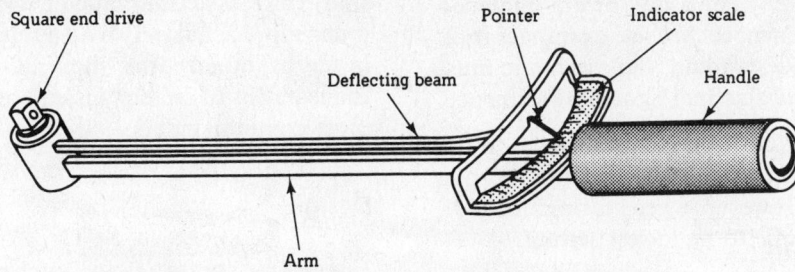

FIGURE 16-3   A deflecting beam torque wrench.

culate a new specification before tightening the hardware.

To determine the correct torque when length extensions are necessary, use the following simple formula,

$$D = \frac{A \times T}{A + B}$$

D is the new dial reading or amount of torque to be preset into the wrench; A is the original length of the wrench from the center of the square drive end to the center of the hand grip; B is the length of the extension from the center of the drive end to the center of the adapter, parallel with the wrench handle; and T is the factory torque specification for the fastener.

Now let's put this information to work on a practical torquing application (Fig. 16-5). Suppose that, for example, a given inch-pound torque wrench has a length (A) of 8 inches. The adapter needed to clear the obstruction (B) has a length of 4 inches; and the specifications call for an applied torque (T) of 45 inch-pounds. Referring to the formula above:

$$D = \frac{8 \text{ inches} \times 45 \text{ inch-pounds}}{8 \text{ inches} + 4 \text{ inches}}$$

$$= \frac{360 \text{ inch-pounds}}{12 \text{ inches}}$$

$$= 30 \text{ inch-pounds}$$

The correct torque reading on the scale or set into the wrench must now be 30 inch-pounds instead of 45 inch-pounds—the amount it would be without the added adapter or extension.

Care and Use of Torque Wrenches

When using a torque wrench, observe the following general rules:

1. A torque wrench is a precision measuring device; do not use it as a general purpose turning tool.

2. Never use the torque wrench to tighten a bolt or nut to a higher value than the maximum capacity of the tool.

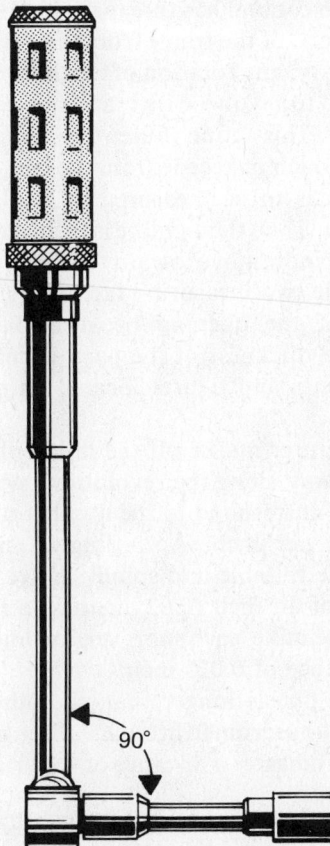

FIGURE 16-4   An extension installed on the drive end of a torque wrench at a 90-degree angle to the wrench handle.

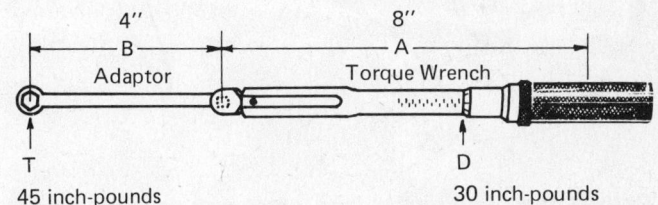

FIGURE 16-5   An adapter installed on the drive end of a torque wrench, parallel to the wrench handle.

3. Never use a torque wrench on a bolt or nut tightened excessively with a wrench, socket set, or impact tool. For an accurate torque reading, the mechanic **must use** a torque wrench for the final tightening sequence.

4. On a clicker-type torque wrench, never position the handle of the micrometer below the minimum torque setting on the scale. However, when storing this tool, **always adjust** the handle to its lowest setting.

5. Always have the clicker-type wrench calibrated at the interval periods specified by the manufacturer.

6. Always clean and lubricate the threads of the fasteners before torquing; this action reduces thread friction that can cause an inaccurate torque reading.

7. Always pull on the torque wrench with an even and steady movement; a fast or jerky motion will cause an incorrect application of torque on the nut, screw, or bolt.

8. Always follow the manufacturer's torque specifications for each fastener used on the transmission.

MICROMETERS

Function

Another kind of measuring device frequently used by transmission mechanics is the micrometer (mike). The technician uses this precision instrument to determine the thickness of such items as selective thrust washers and

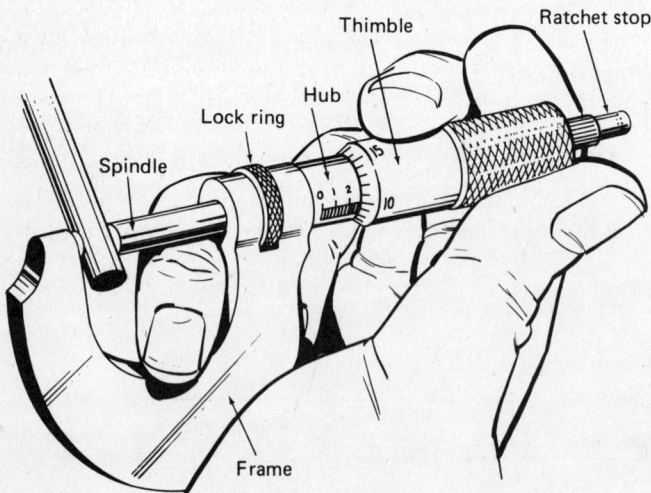

FIGURE 16-6   A typical 0- to 1-inch outside micrometer. The lock ring is fitted into some units to secure the spindle to the frame, after a reading is taken.

snap rings. The micrometer itself measures these parts much more accurately than a ruler because the mike reads in tenths, hundredths, thousandths, and sometimes ten-thousandths of an inch instead of fractions of an inch like most common rulers.

Design

Figure 16-6 shows a 0- to 1-inch micrometer used in automatic transmission service work. This standard mike consists of a frame, anvil, spindle, hub, thimble, and ratchet stop. The frame, as the name implies, is the U-shaped housing or framework of the micrometer. The U-shape allows sufficient access for measuring round objects placed between the anvil and the end of the spindle, while at the same time, provides a convenient place for the mechanic to hold the tool.

The anvil fits into the frame on one end. The highly polished surface of the anvil forms one of the two measuring faces of the tool. The precision measurements taken by the mike are from the face of the anvil to the other measuring face located on the end of the spindle.

The spindle, with its polished face, attaches on the inside to the thimble. The part of this spindle that the thimble and hub conceal has threads, which fit into a nut in the opposite end of the frame from the anvil. Since the frame is stationary, any rotation of the thimble causes the attached spindle to revolve with it and move through the nut in the frame. This action causes the spindle's measuring face to approach or recede from the face of the anvil.

When a part is to be measured, it is placed snugly against the anvil. Then the thimble is turned clockwise so that it and the spindle move toward the anvil, wedging the part between the two measuring faces. Next, the actual measurement of the opening between the anvil and spindle, i.e., the thickness of the part, is determined by observing the lines and figures located on the hub and thimble.

The manufacturer marks off the hub with 40 lines to the inch all the way along the revolution line (Fig. 16-7). These markings correspond to the number of threads on the spindle, 40 per inch. Accordingly, one complete revolution of the thimble and spindle move them longitudinally 1/40 of an inch (0.025 inch). To indicate this movement on the mike, each short vertical line on the hub indicates a distance of 0.025 inch.

Every fourth line is longer than the others and has number designations from 0 through 9. Each of these long numbered lines indicates a distance of 0.100 inch, or one-tenth of an inch.

The beveled end of the rotating thimble also has a series of markings. Each marked line represents 0.001 inch, and there are 25 of these markings. Consequently, every time the mechanic turns the thimble one complete

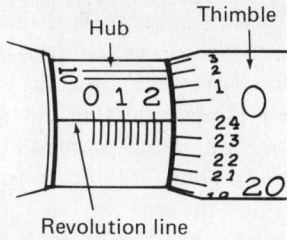

FIGURE 16-7   Hub and thimble markings on a typical micrometer.

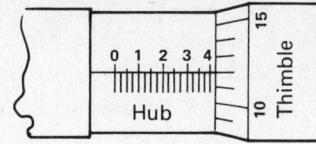

FIGURE 16-8   A 0- to 1-inch micrometer that reads 0.402 inch.

revolution, it and the spindle move exactly 0.025 inch. If he moves it four complete turns, they travel 0.100 inch.

The ratchet stop (Fig. 16-6) is a friction clutch built into the micrometer that allows faster and more accurate readings. By rotating the ratchet stop instead of the thimble, the mechanic applies the correct pressure to the work, which is wedged between the spindle and anvil. In other words, when the force of this wedging action reaches a given point, the ratchet stop will slip. This slippage prevents the spindle from turning any further and, at the same time, indicates to the mechanic the work is held with enough tension to take an accurate reading.

Reading a Micrometer

To properly read the micrometer itself, follow this simple procedure:

1. Multiply the number of vertical divisions (lines) visible on the hub by 0.025 inch. In Fig. 16-8, 16 vertical lines are visible; therefore, the reading on the hub is equal to 16 × 0.025 inch = 0.400 inch.

2. Add to this figure the number of divisions located on the bevel end of the thimble, from 0 to the line that is across from the revolution line on the hub. In Fig. 16-8, this is 0.002 inch.

3. Add the thimble reading to the hub measurement. The total reading in our example is 0.400 inch + 0.002 inch, or 0.402 inch.

Use and Care of a Micrometer

The micrometer is a precision tool and costly to replace if damaged. To avoid damage to this device, the mechanic should follow these general rules concerning its use and care:

1. Never clamp the micrometer down hard on the part being measured. Tighten the thimble only enough to cause the micrometer's measuring faces to drag slightly as you move them over the part, or use the

ratchet stop, if so equipped. Excessive clamping distorts the spindle threads and frame and could ruin not only the part being measured but also the mike.

2. Never open or close the micrometer by holding the thimble and spinning the frame. This action can throw the mike out of adjustment.

3. Never throw the micrometer down or leave it lying unprotected on the work bench. Wipe it clean after every use, and place it in a special drawer or container that protects it from dirt and impact damage from other tools.

FEELER GAUGES

Function and Design

Automatic transmission mechanics use the feeler gauge (Fig. 16-9) for such measuring jobs as checking the operating clearance within clutch assemblies and for the amount of wear in hydraulic pumps. A feeler gauge is a strip or blade of hardened steel or other metal, which is ground or rolled to the proper thickness with extreme accuracy. Then the manufacturer marks on the blade its thickness in thousandths of an inch, thousandths of a millimeter, or both. The blade shown in Fig. 16-9 is a 0.010-inch gauge. Note also the markings for its metric equivalent of 0.25 mm.

For the sake of convenience and compactness, the manufacturer usually assembles a series of individual gauge strips into a set. The measuring range of a typical set is from about 0.0015 to 0.040 inch (0.037 to 1.02 mm). With this arrangement, the mechanic can easily locate

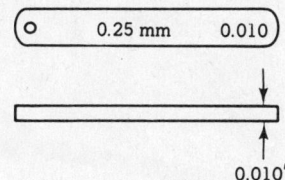

FIGURE 16-9   A 0.010-inch (0.25-mm) feeler gauge strip.

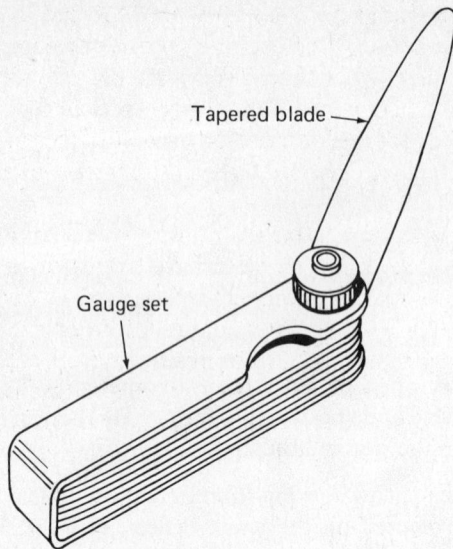

FIGURE 16-10   A typical tapered feeler gauge set.

and pull out the thickness blade he wants to use. Furthermore, he can combine two or more blades together to measure a space for which the set does not contain a gauge of the correct size. In this latter situation, the total thickness of all the blades used to fill the space is the actual measurement between the adjacent surfaces.

Types

Manufacturers also make feeler gauges in several types. The three types most commonly employed by transmission technicians are the tapered, bent, and round. The tapered gauge (Fig. 16-10), as its name implies, has a blade with one end that tapers down nearly to a point. This particular gauge design is very useful for checking the clearances in gears and pumps especially when these components operate in restricted areas.

The bent feeler gauge (Fig. 16-11) has a rounded tip, but the manufacturer bends the metal strip at a given

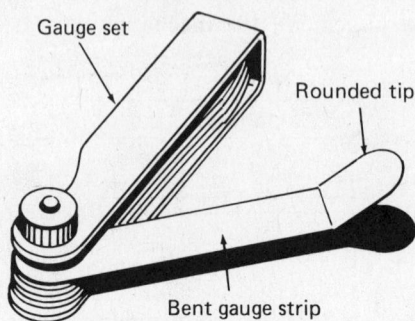

FIGURE 16-11   A typical bent-type feeler gauge set.

angle near the measuring tip. With this design, the mechanic can use the gauge to measure clearances between parts located in recessed locations. For instance, to measure clutch plate operating clearance in many multiple-disc assemblies, the technician must use a bent-type gauge because the plates and their retaining snap ring fit below the level of the opening in the drum, which makes it impossible to insert a flat strip between them.

The round feeler gauge, as its name implies, has a perfectly round cross section. It is made of carefully calibrated steel wire of the proper thickness. This gauge design is especially useful in checking the clearance between a bushing and the shaft it supports.

Use and Care of Feeler Gauges

Feeler gauges are precision measuring devices. In order to preserve their accuracy, the mechanic should follow these general rules regarding use and care:

1. Never force a feeler gauge into an opening that is too small for the size of the strip. Some blades are very thin and this action can easily bend, dent, or tear the gauge strip.

2. Before using a feeler gauge strip, wipe the blade with a clean, oiled cloth. This action removes any dirt from the blade and prevents an inaccurate reading.

3. After using a feeler gauge and before returning it to its

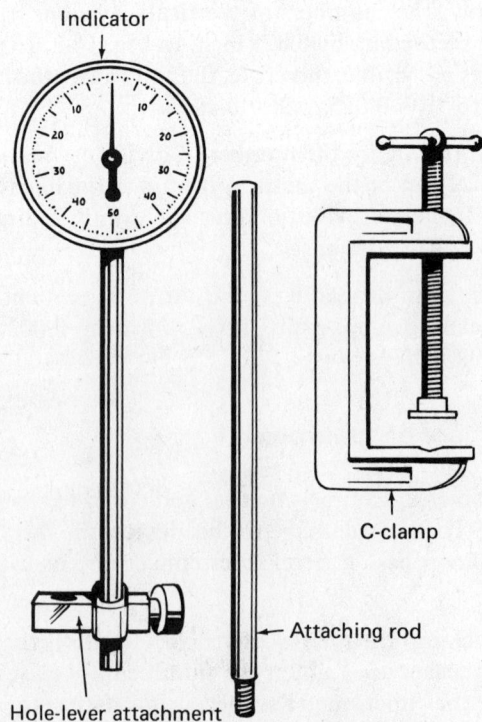

FIGURE 16-12   A universal dial indicator set.

storage area, wipe the blades and the holder with a clean, oiled cloth. The oil from the rag prevents the acids and moisture on your hands from rusting the blades.

DIAL INDICATORS

Design

The dial indicator set shown in Fig. 16-12 includes the indicator itself, C-clamp with attaching rod, and a hole-lever attachment. The indicator mounts in a housing and has a dial face and needle to register the measurements. The capacity of the indicator pictured in Fig. 16-12 is from 0 to 0.050 to 0 inch, in 0.001-inch increments. Therefore, one full revolution of the needle will measure a total distance in travel of 0.100 inch (Fig. 16-13).

A movable, spring-loaded plunger activates the needle. When a shaft or other object contacts this plunger and moves it inward, the needle rotates on the dial face to show the plunger's travel in thousandths of an inch. If the shaft then moves away from the plunger, a spring pushes it outward, and the needle turns in the opposite direction.

Also, the dial face of the indicator is adjustable. This design permits the mechanic to zero the dial after positioning the plunger against the component he is checking. Furthermore, a clamp screw located on the housing locks the dial face in this set position, while the mechanic performs the necessary measurements with the indicator.

The C-clamp and attaching rod, via the hole-lever attachment (Fig. 16-14), are necessary to fasten the indicator housing to the transmission or turbine end-play attachment. The C-clamp itself can fasten directly to a lip or flange on the transmission case if its design permits; but in most cases, the mechanic will tighten down the clamp over a threaded bolt installed in a tapped hole in the hydraulic pump. For example, when checking the end-play of most transmission input shafts, the bolt threads into one of the pump attaching bolts hole that has been tapped out for this purpose. In the case of the turbine end-play attachment, the rod or bolt is a permanent part of the fixture itself.

The hole-lever attachment provides a great deal of flexibility necessary to mount the indicator in different locations. This attachment, when used in conjunction with the C-clamp and bolt, allows the mechanic to turn or twist the indicator to various positions in order to align its plunger with the work. Then, by tightening the thumb screw on the hole-lever attachment, the indicator is solid enough in its mounting to take an accurate measurement.

Indicator Care

The dial indicator is a delicate measuring instrument, and the mechanic must treat it with care. The technician should be careful not to drop or repeatedly push hard on its activating plunger because the internal mechanism is subject to damage or wear if treated roughly. When finished with the indicator, the mechanic should clean and return it and its attachments to their case.

FIGURE 16-13   The scale on the face of a typical dial indicator.

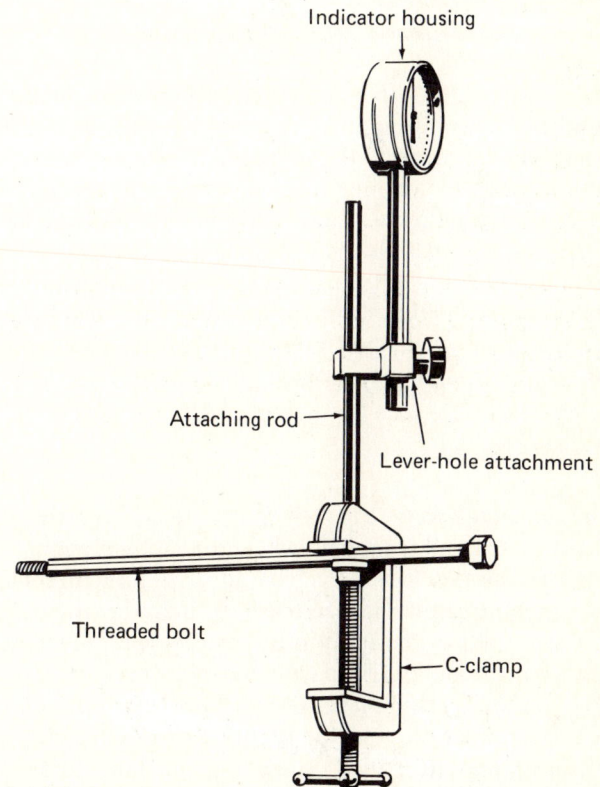

FIGURE 16-14   A C-clamp and attaching rod connected by the hole-lever attachment to the indicator housing.

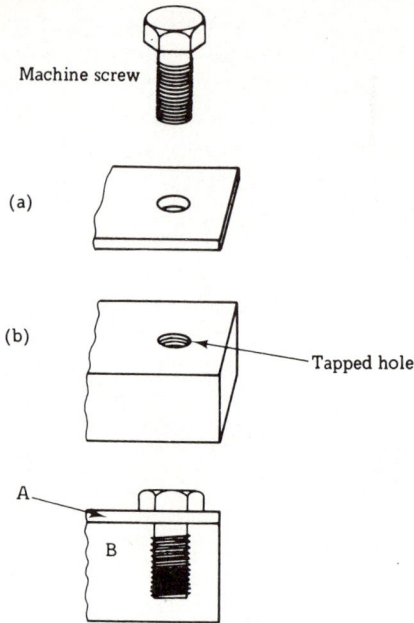

Machine screw

(a)

(b)

Tapped hole

A

B

FIGURE 16-15 **A machine screw attaching part A to part B.**

FASTENING DEVICES

Many types of fasteners are necessary to hold the large number of transmission components together. It is extremely important, therefore, that the mechanic be familiar with the various types, designs, and functions of these devices before actually working on a transmission. The term "fastener" is the name given to any device that secures several parts together. The most common fasteners used in the automatic transmission assembly are the machine screw, bolt and nut, washers, cotter pins, keys, splines, and snap rings.

Machine Screws

The **machine screw** (Fig. 16-15) is a device that threads into a drilled and tapped hole in another component, like the transmission case or valve body. Manufacturers use the machine screw more extensively in automatic transmission buildup than any other single type of fastener. The actual function of the machine screw in practice is to hold another part in place. The part fits between the base of the screw's head and the tapped (threaded) hole.

Since manufacturers utilize the machine screw so extensively in transmission assembly, they use screws with different head designs, sizes, and tensile strengths, in addition to different thread pitches and series to meet cer-

tain structural and load requirements. The most common machine screw head designs, for example, are the Phillips, slotted, Allen, and hexagonal (Fig. 16-16). Manufacturers commonly install the Phillips and slotted screws as fasteners for such items as the valve body and governor assembly. The Allen screw, on the other hand, performs well in holding some types of internal linkages together; but in this type of installation, the screw requires a nut to hold the parts together. Finally, the machine screw that secures the larger transmission components together has a hexagonal (six-sided) head. This head design can withstand larger torque loads on it with less tendency of ruining the areas where the driver, a socket or wrench, normally fits.

Manufacturers designate the size of English or metric screws by a number or by the outside diameter of its threads (D of Fig. 16-17). For instance, using the English system, a certain valve body will have twelve number 10 Phillips machine screws holding it together. A typical servo cover attaches to the transmission case by four 5/16-inch hexagonal machine screws.

English and metric machine screws are also made of different materials, having varied strengths. Figure 16-17 illustrates the head markings of a number of English and metric hexagonal screws; these markings indicate the quality, or tensile strength, of the fastener. **Tensile strength** is the amount of pull in pounds a screw can withstand before it tears apart or breaks. The stronger the screw, the more expensive it is to produce; and the transmission manufacturer only installs them where additional strength is necessary.

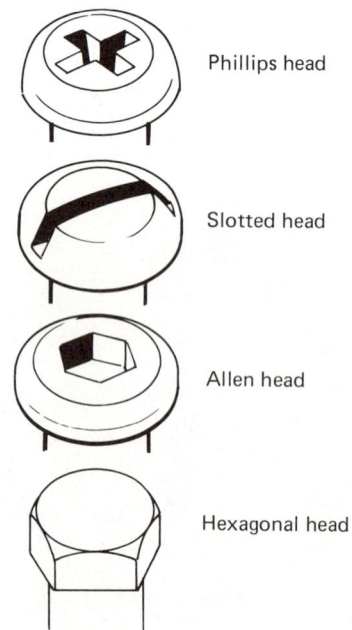

Phillips head

Slotted head

Allen head

Hexagonal head

FIGURE 16-16 **Various head designs for machine screws.**

G- Grade Marking
(bolt strength)
L- Length, (inches)**
T- Thread Pitch
(thread/inch)
D- Nominal Diameter
(inches)

(English) inch system bolt, ½-13 × 1

P- Property Class**
(bolt strength)
L- Length (millimeters)**
T- Thread Pitch (thread width
crest to crest mm)
D- Nominal Diameter
(millimeters)

Metric system bolt, M12.1-1.75 × 25

BOLT STRENGTH IDENTIFICATION

Grade 1 or 2   Grade 5   Grade 8

(English) inch system:

- (English) inch bolts
- Identification marks
  correspond to bolt
  strength
- Increasing number of
  slashes represent
  increasing strength

Metric System:

- Metric bolts
- Identification class
  numbers correspond
  to bolt strength
- Increasing numbers
  represent increasing
  strength
- Common metric fastener
  bolt strength properties
  are 9.8 and 10.9 with the
  class identification on the
  bolt head

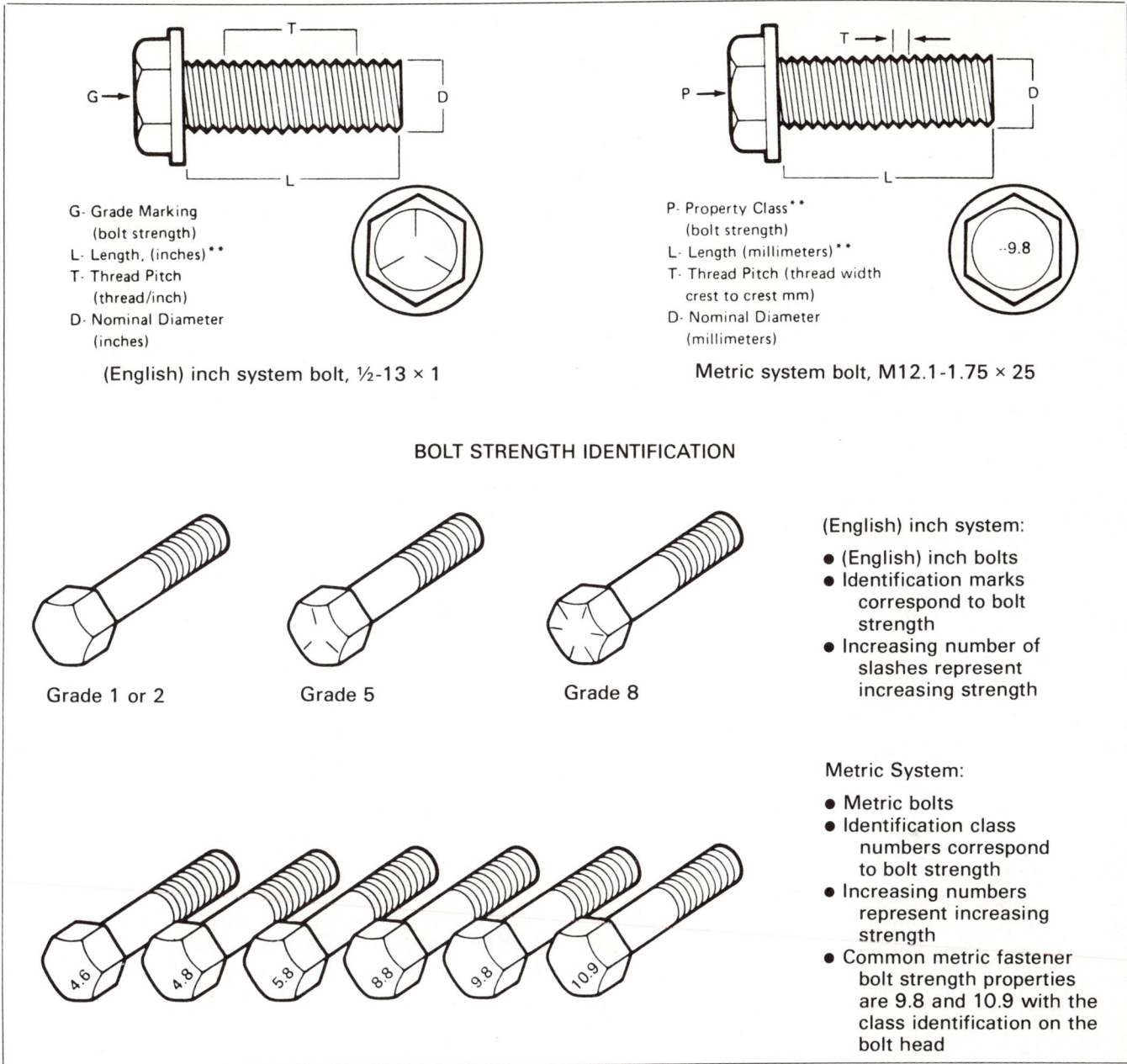

**FIGURE 16-17   Machine screw or bolt terminology.
(Courtesy of Ford Motor Company of Canada Ltd.)**

In the English system, the tensile strength of a screw is identified by the number of radial lines on its head. More lines mean a higher tensile strength. In the metric system, the tensile strength of a screw can be identified by a number on its head. The higher the number, the greater the tensile strength. The important thing to remember here is if a hexagonal machine screw is defective, *always examine its head markings and replace it with one of equal or superior strength.*

Manufacturers use two factors in determining the

length requirements of a given machine screw (Fig. 16-18). The first factor is the thickness of the part the machine screw holds in place. This thickness determines length B of the screw. The second factor is the type and the amount of load place on the thread. If torquing loads are high, the screw should have more pitch, and the threaded section C must be longer as well as the depth of the tapped hole. In other words, the overall length A is a combination of lengths B and C. Finally, the mechanic must consider these facts before choosing a replacement

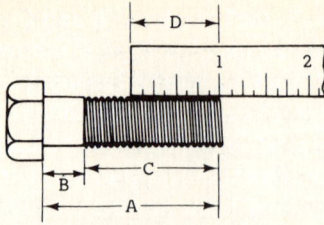

FIGURE 16-18 Measurement for length and pitch on common machine screws and bolts.

machine screw, *or it may fail, resulting in damage to or a malfunction within the transmission.*

Another way to identify a threaded fastener is by its thread pitch or series (Figs. 16-18 and 16-19). **Pitch** in the English system is the actual number of threads per given inch of fastener (D of Fig. 16-18). In the metric system, pitch is the distance in millimeters between two adjacent threads. Thread series, on the other hand, is the coarseness or fineness of the thread itself. This information is important because *thread damage will occur if the mechanic attempts to install a replacement machine screw not mated by thread pitch or series with the original into a tapped hole.*

		Threads per inch	
		Coarse	Fine
Size	Diameter (decimal)	NC	NF
0	0.0600	. . .	80
1	0.0730	64	72
2	0.0860	56	64
3	0.0990	48	56
4	0.1120	40	48
5	0.1250	40	44
6	0.1380	32	40
8	0.1640	32	36
10	0.1900	24	32
12	0.2160	24	28
1/4	0.2500	20	28
5/16	0.3125	18	24
3/8	0.3750	16	24
7/16	0.4375	14	20
1/2	0.5000	13	20

FIGURE 16-19 A chart showing size, diameter, pitch, and series of commonly used English machine screws or bolts.

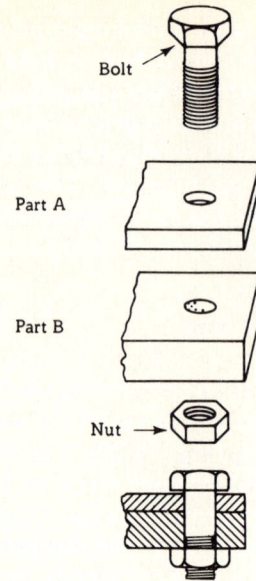

FIGURE 16-20 Several parts held together by a bolt and nut.

Bolts

A **bolt** requires a nut to attach several parts together (Fig. 16-20). The bolt, in this situation, passes through the hole in both parts. A nut threads over the end of the bolt to secure the components together.

Bolts and machine screws are really the same devices; therefore, both have the same classifications for head designs, sizes, tensile strengths, and thread pitch and series. The only practical difference between the two is that the bolt requires a nut; the screw threads into a tapped hole. Finally, it is a general shop practice of mechanics to refer to any hexagonal machine screw as a bolt even if it does, in fact, thread into a tapped hole.

Nuts

Manufacturers produce nuts in many sizes and designs, but the two types of common nuts used in automatic transmission installation or assembly are the plain hex and slotted hex (Fig. 16-21). The factory uses the plain

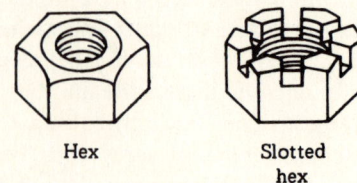

Hex   Slotted hex

FIGURE 16-21 Plain and slotted hex nuts.

hex, usually along with a locking-type washer, more extensively than the slotted hex to hold several parts together tightly. The slotted hex nut, cotter pin, and bolt, on the other hand, connect together linkages or similar components where the parts may require some rotational movement in order to function. The pivot points of throttle and kickdown linkages are good examples of bolt and slotted nut installations.

Washers and Cotter Pins

During assembly of the transmission, the factory can use several types of washers. The two main washers used are the flat and the lock (Fig. 16-22). The flat washer fits under the head of the machine screw or bolt and serves two functions. First, it protects the head surface area from being scratched or marred as the mechanic torques the fastener. Second, the washer extends the head surface contact area, which aids in preventing the head from distorting or tearing the metal surface as the technician tightens the fastener.

Some manufacturers form the flat washer as part of the screw head itself. This design accelerates the assembly process and prevents the loss of the washer when a mechanic services or rebuilds the transmission. An oil pan attaching screw is a good example of a fastener where the factory usually combines the head and washer together into a single unit.

In addition, some flat washers used on the transmission have a gasket or sealing material bonded to one or both of its mating surfaces. The factory installs this type of washer on a screw or bolt to prevent hydraulic leakage. A hydraulic pump-to-transmission case attaching bolt, in many cases, comes equipped with this type of washer to prevent leakage of fluid from the area where the head seats against the pump housing. Whenever the technician removes and replaces a machine screw or bolt equipped with a sealing washer, *he must replace the washer or a leak may develop.*

The factory installs lock washers (B and C of Fig. 16-22) under nuts or the heads of bolts or machine screws to prevent them from working loose from vibration or road shock. This spring-steel washer has a split with edges that cut into the nut or head of the screw or bolt and the surface of the part as the mechanic torques the fasteners. This action keeps the fastener from turning or loosening, and the *mechanic should never reuse a lock washer once the split edges have flattened out because the washer will have lost its locking characteristic.*

The design and installation of the cotter pin (Fig. 16-23) is such that it forms a locking device for a bolt and slotted nut. The bolt, in this situation, has a hole drilled into its threaded end large enough to accommodate the pin. The pin also passes freely through the slots machined into the nut.

To properly install the cotter pin, first thread the slotted nut snugly down on the bolt. Next, tighten or loosen the nut until one of its slots aligns with the machined hole in the bolt. Finally, insert the cotter pin and bend its two exposed legs (ends) over the sides of the nut, so it cannot fall out.

NOTE: *Once a cotter pin has been used, never reinstall it because the ends may later break permitting the pin to fall out and the nut to come off.*

Drive Keys, Balls, and Pins

Woodruff Keys, steel balls, and pins are also forms of fastening devices used to assemble a typical automatic transmission. These special fasteners lock a gear, like the speedometer, to the output shaft of the transmission so they both can rotate together. The **Woodruff Key** (Fig.

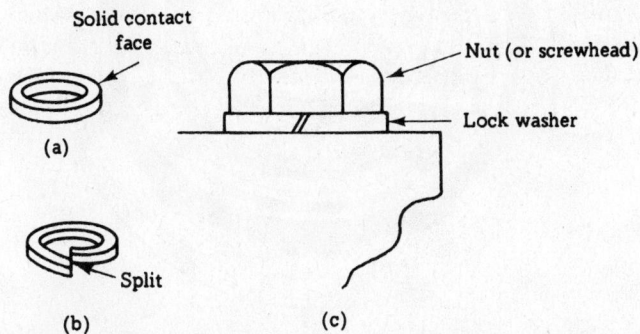

FIGURE 16-22   (a) Typical flat washer; (b) plain lock washer; (c) plain lock washer installed under a nut, screw, or bolt head.

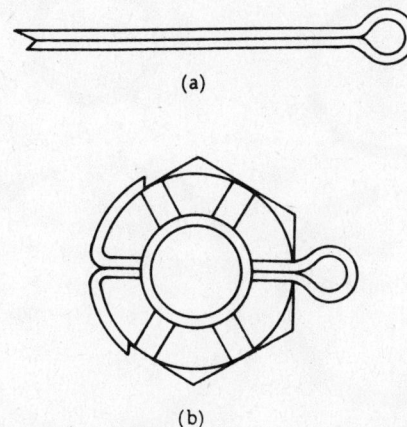

FIGURE 16-23   (a) Cotter pin before installation and (b) its installation through the hole in the bolt and slot in the nut.

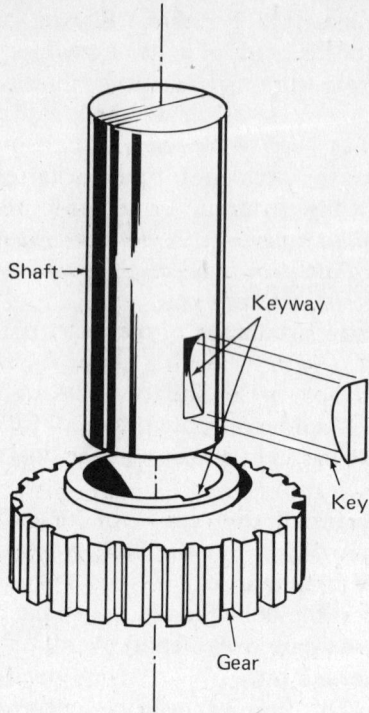

FIGURE 16-24 A key holds a gear to a rotating shaft.

16-24) is nothing more than a half-round piece of hardened steel that fits into a slot (keyway) machined into a shaft.

The gear that the shaft will drive also has a slot or keyway. When the mechanic slides this slotted gear over the Woodruff Key, it extends through the gear, thus locking this part to the shaft.

The round, steel ball has the same basic function as the

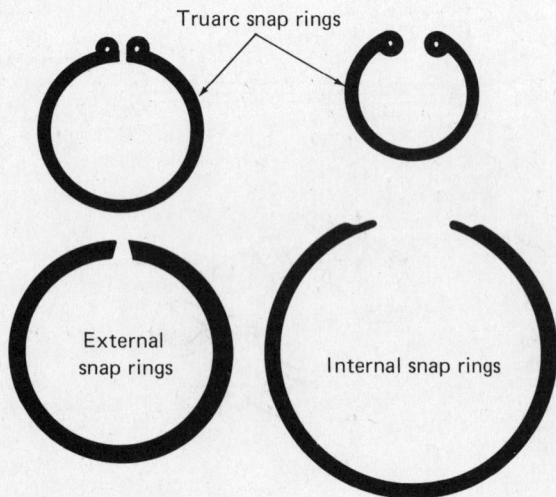

FIGURE 16-25 Typical splined output shaft and slip yoke.

key. The ball, in this situation, also fits between a gear and the shaft. But instead of fitting into a slot in the shaft, the ball rests in a seat machined to conform with the ball's curvature. However, the drive slot in the gear does resemble the one used with a Woodruff Key.

Manufacturers have also used a T-shaped pin to lock the rear pump drive gear to the output shaft on older model transmissions. The shank of this pin fits into a hole drilled into the shaft, leaving the head portion of it resting on the surface of the shaft. The square drive slot of the gear then slides over the head section of the pin during assembly, locking the gear to the shaft.

Splines

Transmission manufacturers also use splines on several shafts of the transmission for assembly purposes. **Splines** are external and internal teeth cut into a shaft and the component the mechanic will install. The splines permit the installed part to move back and forth on the shaft, but

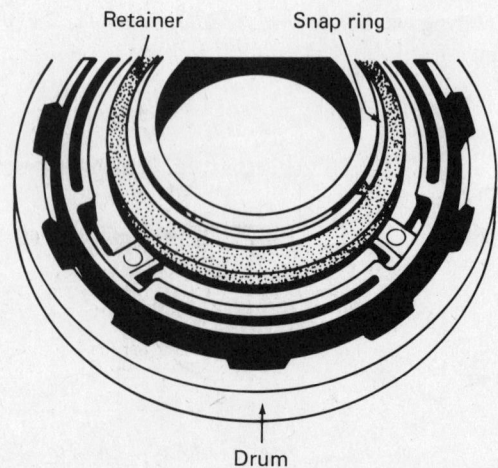

FIGURE 16-26 Common types of snap rings used to assemble a transmission.

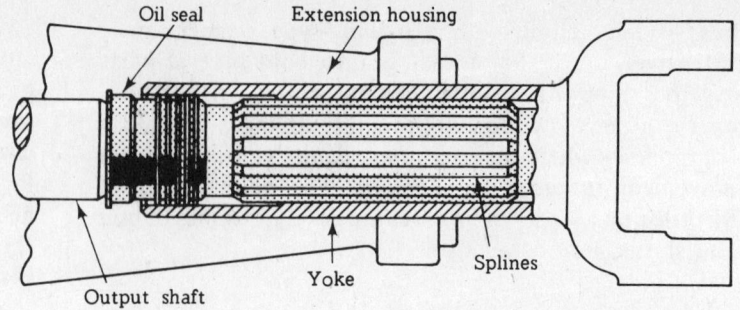

FIGURE 16-27 External-type snap ring installed in its groove in a clutch piston guide.

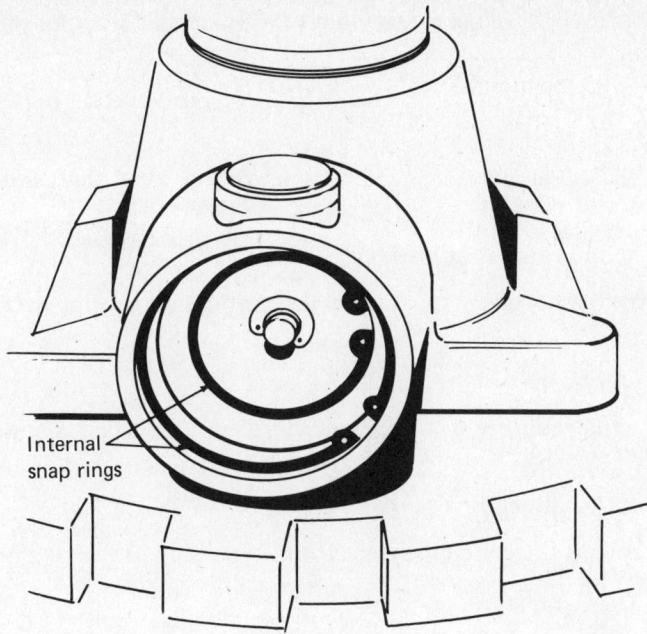

FIGURE 16-28   Internal-type snap ring installed in its groove in a governor housing.

they force the installed part to rotate with the shaft when it turns.

Figure 16-25 illustrates a typical splined output shaft and installed yoke. The external output shaft splines mate snugly with the internal splines of the slip yoke, a part of the drive shaft. The yoke can easily slide over the output shaft, but the splines will force the yoke to turn if the output shaft is rotating.

Snap Rings

The factory installs several types of spring-steel snap or retaining rings when assembling an automatic transmission to lock or hold a component in a given position. The two general types of rings used for this purpose are the **external** and **internal** (Fig. 16-26).

The external snap ring is one type of installation that prevents the end-to-end movement of a gear on a shaft. Another use for this type of snap ring is to secure the return spring retainer to a clutch drum (Fig. 16-27). In either installation, the mechanic has to stretch or expand the ring with special snap-ring pliers before slipping it over the shaft or clutch piston guide and into its groove.

The internal snap ring, on the other hand, holds a shaft or component in a given position (Fig. 16-28). The mechanic, in this case, has to compress or squeeze the ends of the ring together with special pliers before installing it inside its bore, and finally down into a recessed groove.

The **Truarc retaining ring** is a special type of snap ring made for both internal and external applications (Figs. 16-26 and 16-28). This ring has two tangs, each machined with a hole. The holes accommodate the pin ends of a special pair of snap-ring pliers. With this design, there is less possibility of the pliers slipping off the ring during removal or installation.

REVIEW

This section will assist you in determining how well you remember the material contained in this chapter. Read each item carefully. If you can't complete the statement, review the section of the chapter that covers the material.

1. A _____ _____measures how tight a bolt or nut is.

3. Pulled threads can result from improper _____.

3. The wrench that has a micrometer mechanism built into its handle is a _____ type _____wrench.

4. When an extension or adapter adds length to a _____ wrench, the applied torque will be _____.

5. To determine the thickness of a thrust washer, the mechanic should use a _____.

6. The hub section of a _____ has _____ lines to the inch.

7. Each line on the thimble represents _____ inch.

8. The _____ stop is a friction clutch built into the _____

9. To check the operating clearance in a clutch assembly, the mechanic should use a _____ _____ _____.

10. The mechanic should use a _____ _____ to check pump clearances.

11. The mechanic should use a _____ _____ to check input shaft end-play.

12. The _____ face on the indicator is adjustable.

13. The term _____ is the name given to any device that holds several parts together.

14. The device that threads into a tapped hole is the _____.

15. _____ refers to the actual number of threads per inch on a bolt or machine screw.

16. A _____ requires a nut to attach several parts together.

17. The _____ washer protects the contact surface from scratching or marring by the bolt head.

18. A _____ pin is used with a _____ nut.

19. The manufacturer can lock a gear to a shaft through the use of a _____, _____, or _____.

20. The name given to the external teeth machined on a shaft is _____.

For the answers, turn to the Appendix.

Automatic transmission problem diagnosis (trouble-shooting) is probably the greatest challenge to the transmission technician; a challenge even greater, in many instances, than overhauling the unit itself. To understand why this is such a hard task, let us examine for a moment just what the serviceman must know and do to locate the cause of a customer complaint.

OVERVIEW

First of all, before a mechanic can figure out what is wrong with the transmission, he must have a working knowledge of how the transmission and its systems operate. He must know, for example, the function of hydraulically operated units, clutches, and bands; basic hydraulics and the function of the many hydraulic valves and components; transmission powerflow; basic engine operation; engine systems that relate to transmission performance; and basic engine tune-up procedures.

The technician must be familiar with the various tools and charts used in diagnosing transmission malfunctions. These tools include such items as the hydraulic gauge, tachometer, vacuum gauge, and portable vacuum pump. The hydraulic gauge is necessary to measure the amount of transmission internal-hydraulic pressure. The tachometer and vacuum gauge measure engine speed (rpm) and engine vacuum. Finally, the portable vacuum pump is a tool necessary to check a transmission's modulator system and modulator-diaphragm condition.

There are many different types of charts the serviceman may need to refer to when troubleshooting a transmission. These charts provide specifications and procedures for checking stall speed, shift point, control pressure, governor pressure, and throttle pressure. Not all of the charts mentioned above will be available on every transmission model, mainly because the manufacturer either does not desire the technician to do a specific test on the unit or does not provide the unit with a test plug or method of actually performing the check. For instance, many transmission manufacturers do not recommend stall testing their units while others do not provide test points for checking governor or throttle valve pressures.

Regardless of what is available, the main thing that a mechanic **must** always do to be successful in transmission troubleshooting is to follow a logical, specific, and thorough test procedure. Hopefully, if this is done, he will quickly locate and correct the **cause** of the specific transmission problem or malfunction. By following a given procedure (guide), the technician will locate what started the problem (the main cause), and by doing so, he will prevent this particular problem from reoccurring.

Several factors determine the type and scope of any diagnosis procedure. The first is the type of transmission

CHAPTER 17

Automatic Transmission Problem Diagnosis

itself and the actual troubleshooting procedures recommended by its manufacturer. The second factor is the type and severity of the malfunction. Since there is no one simple trouble diagnosis procedure to cover all transmissions and all complaints, this chapter will attempt to provide only an overview of the checks and procedures commonly used by the transmission repair industry with emphasis placed on understanding what the test is for, how to perform it, and what to look for.

PRELIMINARY FLUID LEVEL AND CONDITION CHECKS

Fluid Level

The first and easiest check to perform when there is a complaint of a malfunctioning transmission is the fluid level in the unit. The correct level is of primary importance if the transmission is to function properly. For instance, a low fluid level can cause a number of complaints such as delayed engagement or slippage during the application of a multiple-disc clutch or band. These conditions directly result from air, which is compressible, being pulled into the hydraulic system. The air reduces the apply pressure to the above mentioned friction units.

If a clutch or band slips, the unit **develops** excessively

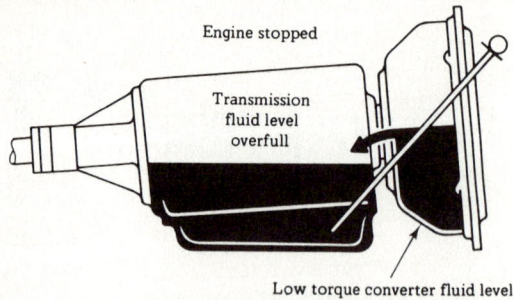

FIGURE 17-1 Fluid level in the transmission will be too high if the oil in the converter drains back due to a defective bushing or valve.

high levels of friction and its operating temperatures exceed safe limits. The increased friction wears out the clutch plate or band friction material prematurely and causes the scoring of steel discs and band drum surfaces. Furthermore, the high temperatures break down the additives in the fluid and varnish forms on internal transmission parts. This can cause hydraulic valves to stick in their bores.

A fluid level that is too high can cause similar problems. The rotating gear train churns up the fluid causing it to aerate and foam. This also results in the introduction of air into the hydraulic system via the fluid, which leads to clutch and band wear, overheating of the fluid, and varnish deposits. Moreover, foaming can cause fluid to be forced out of the transmission vent or filler tube, which may be mistaken for an external fluid leak.

Before examining the actual fluid level checking procedure, it is a good time to mention one particularly misleading problem caused by converter drainback. This condition occurs when fluid drains back from the converter to the transmission case via a faulty valve or worn bushing in the front pump (Fig. 17-1). The problem itself becomes apparent only after the vehicle sits idle for several hours. Then, when the driver starts the engine and shifts the transmission into gear, the vehicle will not move until the hydraulic pump refills the converter. Finally, after the engine has operated for a period of time, the converter and transmission fluid levels stabilize, and the vehicle will move normally in either direction.

The easiest and fastest way to diagnose this malfunction is to check the transmission fluid level after the vehicle has sat for a few hours. If the level on the dipstick is quite high with the transmission cold, an excessive amount of fluid has leaked back from the converter to the case.

This converter drainback generally does not harm either the converter or transmission. However, it is a definite nuisance and appears to most drivers to be a major transmission problem. The situation is correctable only by removing the transmission, repairing or replacing

the valve, installing a new bushing into the pump, and reinstalling the transmission.

To perform a fluid level check, follow this procedure:

1. Position the vehicle on a level floor, and operate the engine at fast idle until the fluid reaches its normal operating temperature.

2. With the engine idling, move the shift selector lever momentarily into each driving range to fill all the clutches and servos.

3. Leave the selector lever in either neutral or park position, as specified by the manufacturer, and with the engine running at idle, take a dipstick reading. (NOTE: Before removing the dipstick, wipe the area around its cap clean to be sure no dirt will get into the transmission fluid.) Then, pull the dipstick out, wipe it off with a clean, lintfree rag, and push it back until the dipstick seats in the tube.

4. Pull the dipstick out again and check to see if the fluid is between the ADD and FULL marks, never lower than ADD or above FULL (Fig. 17-2). If the level is below the ADD mark, slowly pour in the correct fluid until the level on the stick is between ADD and FULL. **Never overfill the transmission with fluid.** (NOTE: If the fluid level was low, perform a fluid leakage check as outlined later in this chapter.)

5. If the fluid level is too high, insert a length of clean hose into the dipstick tube, attach its free end to a suction gun, and pull the excess oil from the transmission (Fig. 17-3). Then, with the engine idling in neutral or park, recheck the fluid level.

Fluid Condition

The condition and color of the fluid can provide the mechanic with clues to specific transmission problems. The fluid should normally be clean and its original color. However, dirty fluid, oil containing solid material, or discolored fluid are good indications that the transmission

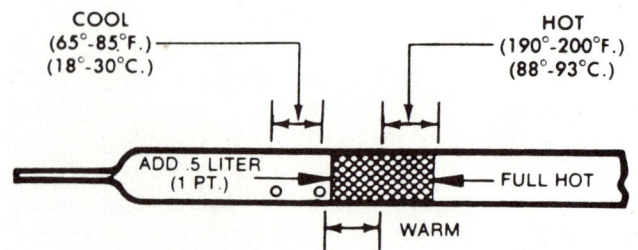

FIGURE 17-2 Reading the fluid on a typical dipstick. (Courtesy of General Motors Corp.)

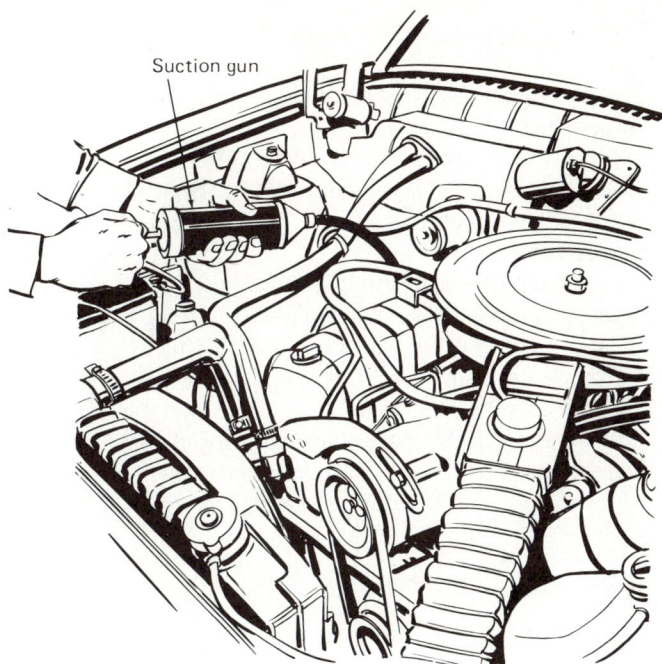

FIGURE 17-3 Removing excess fluid with a suction gun.

may have overheated or has internal damage such as a clutch or band failure.

To check the fluid's condition and color while the engine is operating, pull the dipstick out and wipe it off on a clean, lintless white paper towel or tissue. Once the liquid fluid soaks into the towel or tissue, any solid particles will be clearly visible (Fig. 17-4). If the fluid sample appears discolored, smell it. Fluid that is burnt has quite a distinguishable odor similar to that of a burned-out electrical motor coil.

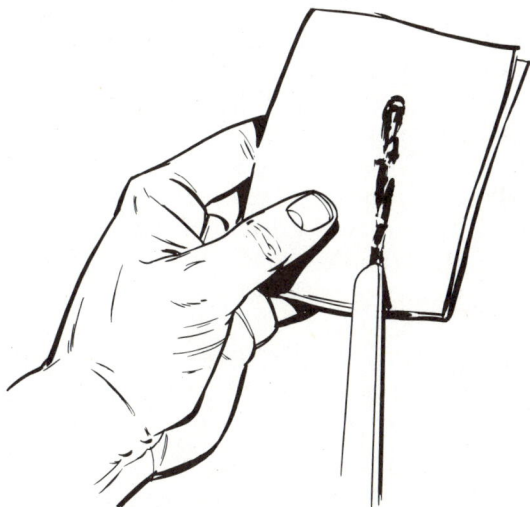

FIGURE 17-4 Checking the fluid's condition using a clean white paper towel or tissue.

As previously mentioned, the results of the fluid condition check will provide a good indication of the general condition of the transmission. Therefore, carefully examine the fluid sample and compare the results with those listed below:

1. The fluid sample appears clean and has a normal color. If the fluid has been in the transmission for some time, this is a good indication that the unit is still in good mechanical condition and the cause of the problem is elsewhere.

2. The fluid sample appears slightly brown, but it has no burnt smell and contains no particles. In this case, the transmission may just be overdue for a fluid change. Change the filter and transmission fluid and inspect the pan for residue.

3. The fluid sample is dark brown or is dirty (contains solid material). This is a very good indication that the transmission has overheated or has some internal damage. Smell the fluid for the burnt odor. Then check the sample for particle content. Dark particles may be from dislodged clutch or band material, while silvery or shining particles may be from a bushing or thrust washer. In either case, remove the pan for further inspection.

4. A dark fluid sample without the burnt odor or presence of solid material and varnish on the dipstick are good indications that there is antifreeze in the transmission oil. When this problem is evident, the transmission cooler requires inspection and repair in addition to a complete transmission fluid change.

5. A black fluid sample that has a strong burnt odor and contains particles of solid material, along with a dark varnish stain on the dipstick, usually indicates prolonged transmission overheating or a burned out clutch or band. In almost all cases, any transmission with fluid such as this will require a complete tear down, inspection, and overhaul.

FLUID LEAKAGE CHECKS

If the first preliminary fluid level check indicated a fluid shortage in the transmission and this condition causes the malfunction, the next logical step is to determine where the oil is going. There are three possibilities: (1) the system was low on fluid to start with, (2) the engine is pulling the fluid through its vacuum system, or (3) the oil is leaking to the outside of the transmission or into the engine cooling system. A quick question to the vehicle owner will verify if the transmission had a fluid change or service recently. It is not uncommon for some transmissions to start mal-

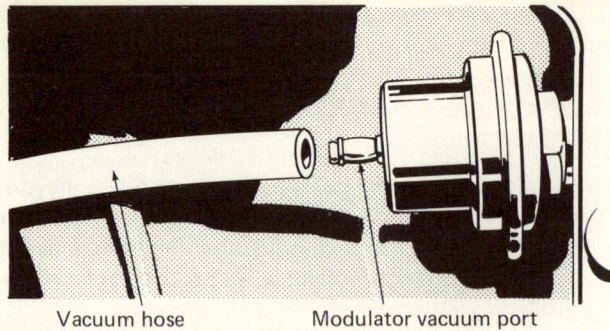

Vacuum hose   Modulator vacuum port

FIGURE 17-5 If evidence of fluid is found in the modulator's vacuum line or inlet port, the diaphragm has a hole in it.

functioning a short time after a fluid change just because it was a quart low on fluid.

Fluid that is being burned inside the engine is another very common cause of fluid loss in transmissions that have a vacuum modulator. This condition is not very hard to detect because there are usually other symptoms other than a loss of fluid. For example, the engine usually smokes excessively, especially when developing high vacuum. Also, engine idle will be erratic and higher than normal. Finally, depending on the type of transmission, the defective modulator will cause either harsh up-and-down shifting, delayed upshifting, or no upshifting at all.

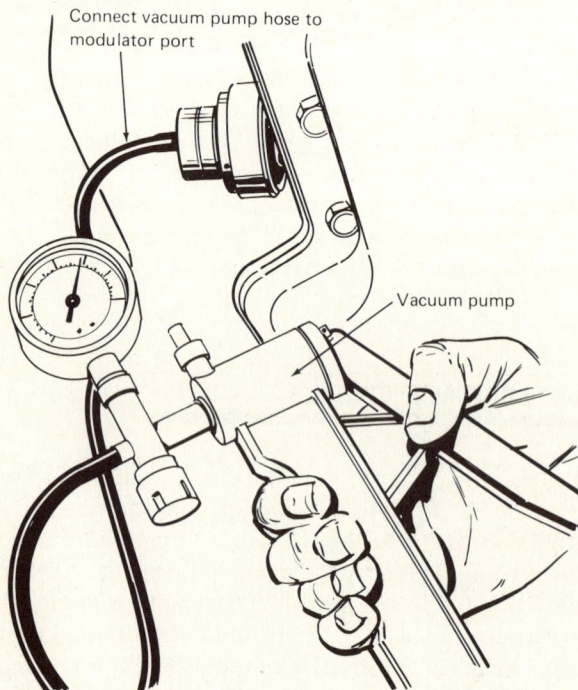

Connect vacuum pump hose to modulator port

Vacuum pump

FIGURE 17-6 Checking a modulator with a hand-operated vacuum pump. (Courtesy of Ford Motor Company)

Testing for Modulator Failure

There are two checks a mechanic can make to verify a defective modulator. The first is to just disconnect the vacuum at the modulator itself. If fluid drips out of the modulator port or the vacuum hose, its diaphragm has ruptured and engine vacuum is pulling the fluid from the transmission (Fig. 17-5).

The second test requires the connecting of the hose from a portable vacuum pump to the port on the modulator and attempting to apply 18 inches Hg to the diaphragm (Fig. 17-6). (NOTE: For this test, the modulator assembly can be removed or remain on the transmission. In either case, if the vacuum pump cannot build up this amount of vacuum in the modulator, or it won't hold it, the diaphragm inside the unit is defective.)

Checking for External Fluid Leaks

The third possibility for fluid loss and unfortunately the most common is through external leakage. This simply means that the fluid is leaking to the outside of the transmission from one or more points. This situation is unfortunate for both the vehicle owner and the mechanic in that it is costly to the owner to have leaks diagnosed and repaired, and it is sometimes very difficult for a technician to locate and correct many types of leaks.

External leak diagnosis can be a very trying experience for a mechanic for several reasons. First, the vehicle may have more than one component leaking fluid into the same general area. For example, there may be an engine or power-steering leak along with a transmission leak. The airstream, passing under the vehicle, compounds the problem by carrying the leaking engine or power-steering oil back where it deposits onto the transmission case and leaks off along with the transmission fluid at various points. Second, a transmission leak itself may also originate at one place, but the force of gravity along with the airstream will cause the fluid to move to another location before it drips off.

With these facts in mind, let's examine the common procedures used to locate the various types of leaks you may have to diagnose. To determine the source of a leak:

1. Position the vehicle on a hoist.

2. Wipe off the underside of the transmission and engine carefully to remove all traces of old fluid leakage. As necessary, steam clean these two areas or use a degreaser and then allow the areas to dry thoroughly.

3. Place a piece of white cardboard or heavy paper on the shop floor under the vehicle (Fig. 17-7). Make sure the cardboard or paper is large enough to encompass the area from the front of the engine to the back of the transmission.

FIGURE 17-7 Placing a piece of clean white cardboard or paper under a vehicle to locate the source of a leak.

4. Start the engine and allow it to run for about 15 minutes.

5. Shut off the engine and examine the cardboard or paper for traces of fluid. The traces will appear on the cardboard at a point approximately below where the fluid has leaked out of the transmission, engine, power-steering component, or cooling system.

6. If under the transmission, check the general area where the leak appears to originate. (NOTE: The leak may have started at a higher point, followed the curvature of the case, and then dripped off the transmission in another location.)

7. Once you have located the general area where the leak originated, it may be necessary to wipe this area off again and observe it carefully with the engine running.

8. If the fluid traces appear on the paper in front of the converter and transmission, examine its color. Engine oil is usually brown unless really dirty; engine coolant is green unless contaminated; and power steering fluid is normally yellow-green, except for those vehicles that use ATF for this purpose. If any of these traces do appear, the leakage is coming from the component indicated by the fluid color and not the transmission.

However, if the leak originated at any of the following areas of the transmission, take note of the usual cause and perform the corrective action as specified.

1. Leakage at the governor housing or cover is usually due to a defective gasket, warped cover, or loose attaching bolts. Replace the defective parts or retorque the bolts as necessary to stop the leak.

2. Leakage at the oil pan usually results from loose attaching bolts, defective gasket, or warped or damaged pan. Retorque the bolts or replace the defective parts in order to prevent further leakage.

3. Leakage from the fluid filler tube is usually caused by a loose fitting, defective seal, or loose attaching bolts. Retorque the fitting or attaching bolts or replace the seal as necessary to stop the leak.

4. Leakage at the cooler lines and fittings usually results from loose or damaged fittings in addition to worn, damaged, or leaking lines. Retorque any loose fitting, or replace any damaged or worn fittings and cooler lines. (NOTE: Refer to Chapter 21 for repairing a leaking cooler line.)

5. Leakage at either the throttle-lever shaft or manual-lever shaft is usually the result of a defective seal. In this case, replace the seal as necessary to correct the leak. (NOTE: Refer to the shop manual for the procedure because it may be necessary to remove and disassemble the transmission in order to replace the defective seal.)

6. Leakage at any of the test plugs usually results from a loose or worn plug. Retorque the plug to specifications; if tightening the unit does not stop the leak, replace it as necessary.

Converter Housing Visual Leak Inspections

Other components than the transmission itself can cause a leak from the converter housing. If the color of the fluid does not indicate its source, there are two methods commonly used by technicians to locate the leak, a visual inspection and the use of a black light.

To locate the source of a leak in the converter housing area by visual inspection, follow these steps:

1. Remove the converter access plate from the transmission.

2. Thoroughly clean off any fluid residue from the top and bottom of the converter housing, front of the transmission case, and rear face of the engine and its oil pan. Clean the areas by washing them with a suitable nonflammable solvent, and blow the areas dry with compressed air.

3. Wash out the inside of the converter housing, the front of the flywheel or flexplate, and the converter drain plug if so equipped. Clean these areas by washing them out using a suitable nonflammable solvent and a squirt-type oil can. Next, blow all of the washed areas dry with compressed air.

4. Start and run the engine until the transmission reaches its normal operating temperature. Shut off the engine, and observe the back of the engine block and top of the converter housing for evidence of engine oil leakage.

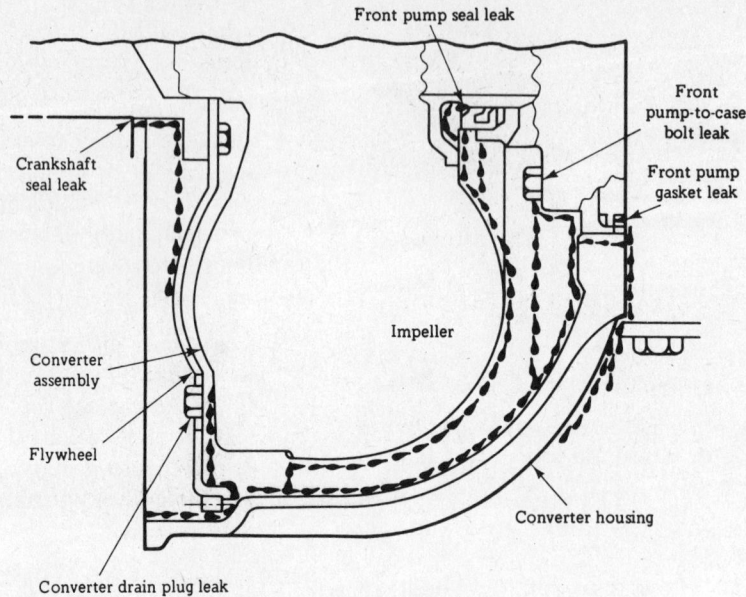

FIGURE 17-8 Sources of fluid leakage from the torque converter housing.

5. If there is no sign of leakage at this point, have an assistant run the engine at fast idle, then at idle, while occasionally shifting to both the drive and reverse ranges to increase the pressure within the transmission's hydraulic system.

6. Shut the engine off and observe the front of the flywheel or flexplate, back of the block (in as far as possible), inside the converter housing, and also the front of the transmission case. (NOTE: Run the engine as long as necessary until oil leakage is evident and the probable source is apparent.)

The actual source of the leak is determined by observing the paths the fluid takes to reach the bottom of the converter housing (as shown in Fig. 17-8):

1. Fluid that leaks by the lip of the front-pump seal tends to move along the pump-drive hub and onto the back of the impeller housing. In the case of total seal failure, fluid leakage by the seal's lip also deposits on the inside of the converter housing itself, near its outside.

2. Fluid leakage past the outside diameter of the front-pump seal will follow the same path as leakage by its lip.

3. Fluid that leaks by one or more front pump-to-case attaching bolts deposits onto the inside of the converter housing only; fluid will not usually be found on the back of the converter.

4. Leakage past the front pump-to-case gasket usually causes fluid deposits on the inside of the converter housing, or it may seep down between the front of the transmission case and the back of the converter hous-

ing. Fluid on the front of the transmission case above the level of the pan gasket is evidence that the front pump-to-case gasket is leaking.

5. Fluid leakage from the converter drain plug appears at the outside diameter of the converter, on the rear face of the flywheel or flexplate, and inside the converter housing only—near the flywheel or flexplate.

6. Engine oil gallery plug leaks will also allow oil to flow down the rear face of the engine block and to the bottom of the converter housing.

7. Leakage by the engine crankshaft rear seal will work its way back to the front of the flywheel or flexplate, and from there into the converter housing.

8. Leakage at the rocker arm covers of the engine may allow oil to flow over the top of the converter housing or to seep down between it and the block, causing oil to be present in or at the bottom of the converter housing.

Black Light Tests

With some transmission designs, it is impossible to perform a good visual inspection of the converter housing area. To determine if the leak coming out of the housing is transmission fluid or not, perform the following black light test **before starting any repair work.**

1. Mix a solution of an oil-soluble fluorescent dye, at the rate of ½ teaspoon of dry powder to ½ pint of transmission fluid, and pour the mixture into the filler tube. (NOTE: This dye is necessary to determine, in some

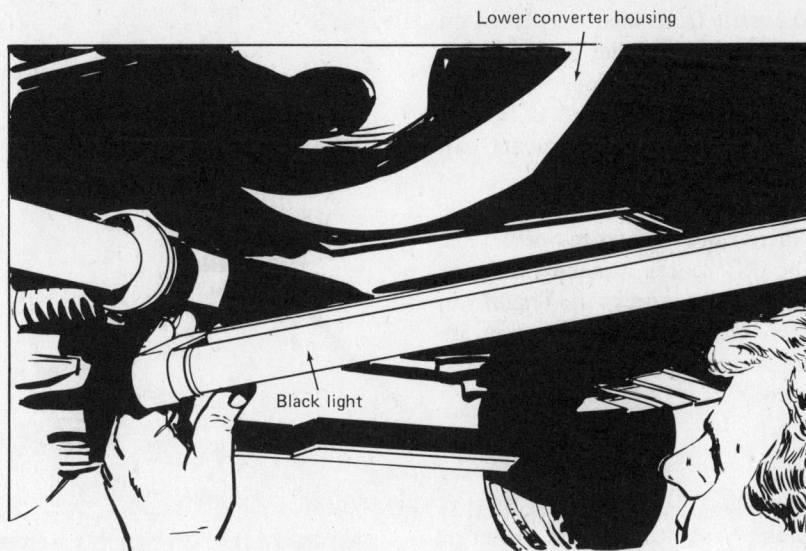

FIGURE 17-9   Checking the converter housing area with a black light to determine if the fluid leaking from this location is from the transmission itself.

cases, if the leaking fluid is from the engine or transmission.)

2. Run the engine for about 10 minutes to mix this solution into the transmission fluid and circulate it throughout the unit.

3. View the converter housing area with a black light (Fig. 17-9). The dye in the transmission fluid, when under a black light, is clearly visible and indicates whether or not the leak from the housing is actually transmission fluid.

Cooler Leakage Tests

Probably the **least** likely cause of low fluid level in the transmission is a loss of oil into the engine cooling system. In this situation, the fluid leak is from a crack or hole in the oil cooler located inside the radiator. As the transmission fluid circulates through this cooler, under pressure, small amounts of it pass into the coolant.

This type of leak creates two problems. First, of course, is the loss of fluid from the transmission itself, which in time causes the unit to malfunction due to the reduced lubricant level. Second, the transmission fluid entering the radiator overfills and contaminates the coolant. This multiple condition can create cooling system deterioration, engine overheating, coolant venting out the radiator overflow tube, and in some cases, the coolant actually being forced into the transmission.

To check for this problem, shut the engine off and allow it to cool sufficiently before removing the radiator cap. Then, check the engine coolant in the radiator. If

transmission fluid is present in the coolant, the cooler assembly inside the radiator is most likely leaking.

To check the cooler assembly, follow this procedure:

1. Disconnect both cooler lines at the radiator and plug one of the cooler fittings.

2. After attaching an air fitting to the other cooler fitting, apply a pressure of 25 to 50 psi to the cooler. CAUTION: *Do not exceed a pressure of 50 psi; higher pressure may damage the cooler.*

3. Check the coolant in the radiator for the presence of bubbles. If the cooler is leaking, remove the radiator for reconditioning or replacement.

OTHER PRELIMINARY CHECKS

If the fluid level and condition are satisfactory and the transmission does not leak, the next items the technician should check are engine idle speed, linkage adjustments, and the vacuum or electrical circuits to the transmission. All of these items influence the manner in which the transmission operates. Consequently, whenever the owner complains of a transmission malfunction, the mechanic should check them **all** as well as perform a stall and road check.

Idle Speed Checks

The idle speed of the engine will cause several problems if not in proper adjustment. If the idle is set too low, the engine will stall or run roughly whenever the transmission is in gear. This condition results from the load placed on the engine by the converter. If, on the other hand, engine idle speed is too high, the vehicle will creep excessively when placed in gear or the initial engagement will be very harsh. Prolonged or repeated harsh engagements can cause a band to break.

There is no one simple method of adjusting engine idle speed. The actual adjustment does vary from one vehicle to another because of the differences in such things as linkages, carburetors, and smog controls. Therefore, when setting engine rpm at idle, use an accurate tachometer and follow the exact procedure set forth in the appropriate service manual.

Linkage Checks

All automatic transmissions have at least one cable- or linkage-type connection to the transmission, while others will have two. This linkage provides control input into the

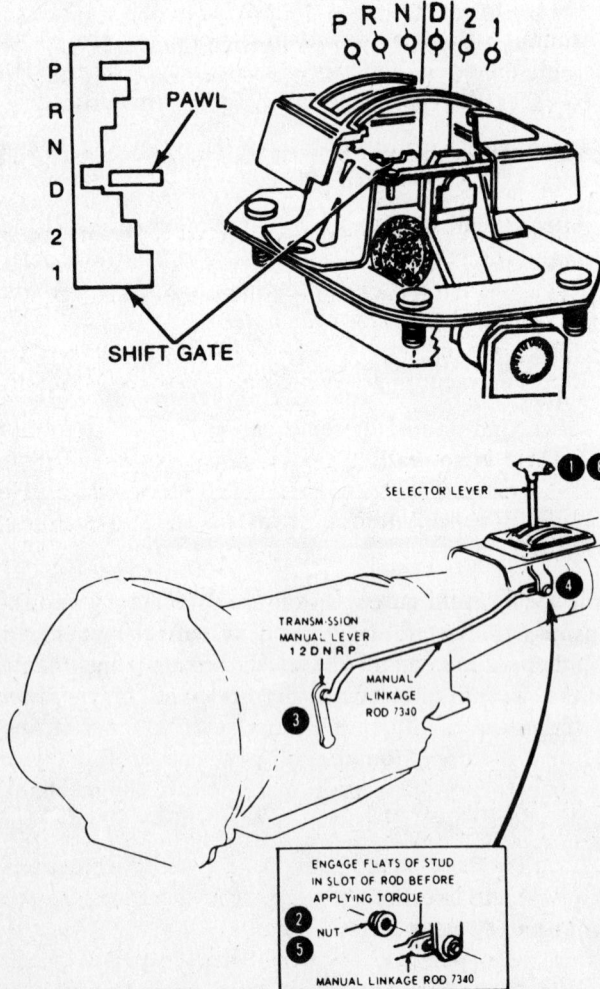

FIGURE 17-11 A typical throttle valve cable and linkage. (Courtesy of General Motors Corp.)

transmission from one or more sources. For instance, there is always a mechanical link between the gearshift-selector lever and the manual valve in the valve body. When the driver moves the selector lever into a given position, the linkage or cable moves the manual valve, which sets up the actual operating conditions within the transmission (Fig. 17-10).

Because of its very important function, proper manual valve linkage adjustment is very important. If this linkage is out of adjustment, the transmission may creep in neutral or park, not engage immediately into any gear, prevent the engine from starting in neutral or park, not lock in park, or have low hydraulic pressure resulting in premature clutch or band wear.

In order for **some** automatic transmissions to respond to the varying loads placed on the engine, there is also some linkage installed from the carburetor down to the transmission (Fig. 17-11). This linkage or cable operates both the throttle and kickdown valves but sometimes only the kickdown valve. The Powerglide, T-200, and the TorqueFlite are all examples where the linkage operates both the throttle and kickdown valves. The Ford C-4 is an example of a transmission in which the linkage only activates the kickdown valve.

Several transmission malfunctions can occur if the combination throttle- and kickdown-valve linkage or cable is out of adjustment. For example, the transmission may upshift too soon or too late in relation to engine load or vehicle speed. Also, a forced downshift or kickdown may not occur when the driver depresses the accelerator pedal enough to move the linkage through the detent position, or the downshift may occur before reaching the detent position. Finally, this premature or no kickdown condition can also occur in a transmission that has only a kickdown linkage if it is out of adjustment.

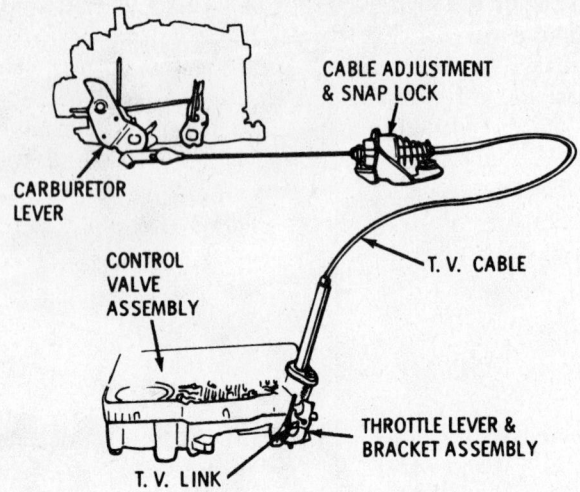

FIGURE 17-10 Typical manual valve linkage. (Courtesy of Ford Motor Company)

The actual procedures for checking and adjusting the manual-valve, throttle-valve, and kickdown-valve linkages sometimes vary greatly between one vehicle manufacturer and another. For this reason, always refer to the linkage adjustment section of the service manual for the make and model of vehicle you are repairing before checking or altering any adjustments. In addition, Chapter 19 of this text provides samples of typical linkage checks and adjustments.

Vacuum System Checks

Some transmissions have a vacuum-operated modulator unit in place of or in addition to a linkage operated throttle-valve system. If the vacuum modulator system replaces the mechanically operated system, the modulator circuit will then vary shift points and control pressure to match engine load. On the other hand, if the transmission has both a modulator and a mechanically operated circuit, the former will vary only control pressure to match engine load.

If the modulator system itself does not function properly or does not receive the correct amount of engine vacuum, the following malfunctions can occur on transmissions without a mechanically operated throttle valve:

1. No upshift

2. Early or late upshifts

3. Slipping

4. Harsh upshifts and downshifts

5. Low or high control pressure

6. Losses of transmission fluid

7. Excessive engine smoking

If the vacuum modulator system fails or does not receive sufficient engine vacuum on a transmission with a mechanically operated throttle valve, the following malfunctions can occur:

1. Harsh upshifts and downshifts

2. Slipping

3. Low or high control pressure

4. Losses of transmission fluid

5. Excessive engine smoking

If a transmission does have a vacuum-operated diaphragm or modulator unit, it must, as previously mentioned, receive sufficient engine vacuum to function properly. To check the amount of engine vacuum at the modulator unit, perform the following procedure:

1. Disconnect the vacuum hose at the diaphragm.

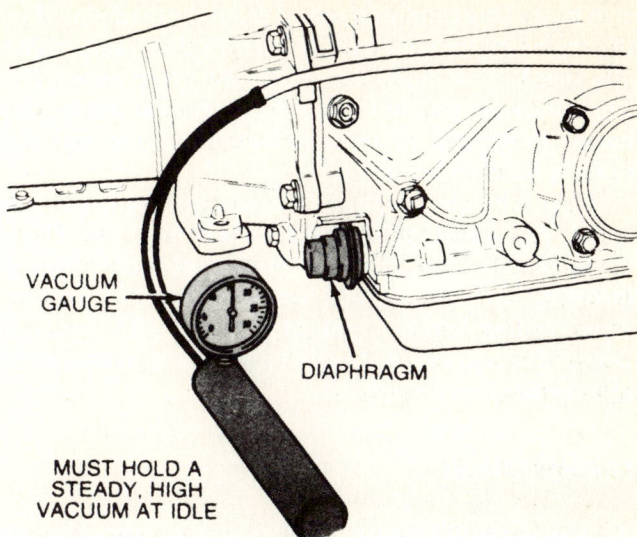

FIGURE 17-12   Checking the vacuum to the modulator. (Courtesy of Ford Motor Company)

(NOTE: If transmission fluid drips out of the hose or modulator, the diaphragm inside the modulator has probably ruptured, and the entire unit will require replacement.)

2. Connect a vacuum gauge to the open modulator hose (Fig. 17-12).

3. Start the engine and check the vacuum reading on the gauge. The reading should be relatively steady and be within the limits specified by the manufacturer.

4. Accelerate the engine briefly and then let it return to idle. The vacuum gauge reading should drop rapidly, then return to a steady idle indication.

5. If there is no reading on the gauge, inspect the hose and line leading from the engine for kinks, restrictions, or breaks. Repair or replace the line or hose as necessary.

6. If the vacuum gauge reading is below specifications, check the engine's condition. *A badly worn engine produces low vacuum, and as previously mentioned, low vacuum can cause abnormal transmission operation.*

Some Ford automatic transmissions come equipped with a dual-port vacuum modulator. The front hose on most of these models connects to the exhaust gas recirculation (EGR) system. The second (rear) port connects via a hose and line directly to the engine's intake manifold and receives normal engine vacuum.

The EGR port, on the other hand, receives ported vacuum. **Ported vacuum,** in this case, means simply that at engine idle there should be no vacuum at this hose connection. However, as the driver opens the throttle to about the ¼ to ½ open position, there should be a specified amount of vacuum routed to the EGR connection on the modulator. To check for EGR-ported vacuum, perform the following steps:

1. Remove the EGR vacuum hose at the modulator unit.

2. Connect a vacuum gauge to the open hose.

3. Start the engine and check the vacuum gauge for a reading. There should not be any at all at idle.

4. Open the throttle until it reaches the ¼ to ½ position. The vacuum gauge should now show a reading. If there is no gauge reading, the EGR valve system is inoperative or the hose or line may have kinks, restrictions, or breaks. (NOTE: When checking any dual-port modulator system, always inspect the hose routing to the modulator itself against the manufacturer's specifications. If the hose connections are incorrect, the transmission will receive the wrong vacuum signal, which will cause it to malfunction and eventually fail.)

Along with checking the condition of the engine, vacuum lines, and hoses, test the modulator itself for leaks. To test the modulator:

1. Connect a hand-operated vacuum pump and gauge assembly to the modulator inlet port (Fig. 17-13).

2. Apply 18 inches of vacuum to the modulator diaphragm.

3. Note the reading on the gauge. The reading should

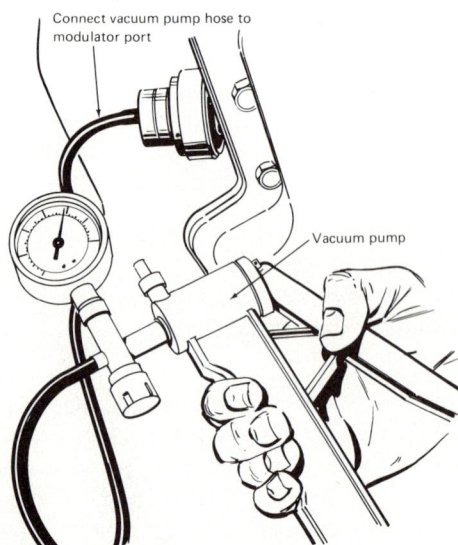

Connect vacuum pump hose to modulator port

Vacuum pump

FIGURE 17-13 **Checking the vacuum modulator in the vehicle for leakage using a hand-operated vacuum pump and gauge assembly.**

hold steady for at least 30 to 60 seconds. If the vacuum bleeds down, the diaphragm inside the unit is leaking, and the entire assembly requires replacement.

Checking Electrical Transmission Circuits

Some automatic transmissions have one or more electrical leads connected to switches or terminals on the transmission case. These wires, switches, or terminals are parts of electrical circuits that function during various phases of transmission operation. As the transmission operates at different times, these units will complete an electrical circuit, through the assistance of the transmission, to such items as the back-up lights, starter, kickdown system, and the transmission controlled spark system (TCS).

Because of the differences in the way manufacturers construct their vehicles, these circuits and their components all vary slightly and so do the testing and servicing procedures. Therefore, if one of these circuits malfunctions, always refer to the appropriate service manual for the make and model of the vehicle you are repairing for the proper troubleshooting procedure.

STALL SPEED TESTS

Now that the preliminary inspections are out of the way, the next usual step in diagnosing transmission malfunctions is a series of operational tests. These include the stall speed, road, and hydraulic pressure.

Function

A technician uses the stall test to find the highest engine speed (rpm) at wide open throttle with the wheels locked and the transmission in gear. This test quickly checks engine performance, stator one-way clutch operation, and the holding ability of bands and clutches.

Some manufacturers **do not** specify the stall test as part of their recommended diagnostic procedures. The reason for this is twofold. First, if a clutch or band is marginal before a stall test, or if the design of the transmission is such that controlled slippage of a clutch or band is normal, the strain of the test can ruin them. Second, if the technician does not follow the factory procedures, the test can damage the transmission, engine, and brakes. Consequently, before attempting to perform a stall speed test on a given transmission, always refer to the vehicle or transmission service manual to see if the test is permissible and, at the same time, review the recommended procedure.

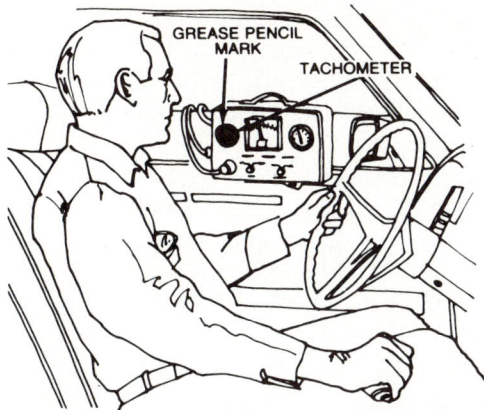

RANGE	SPECIFIED RPM	HIGH, LOW OR OK
D		
2		
1		
R		

FIGURE 17-14   A tachometer positioned inside a vehicle in preparation for a stall test. (Courtesy of Ford Motor Company)

Typical Procedures

To conduct a typical stall speed test, follow these steps:

1. Check the engine coolant level, transmission fluid, and band adjustment.

3. Connect an accurate tachometer to the engine. Position the gauge so that you can easily view it from the driver's seat (Fig. 17-14).

3. Mark on the gauge with a grease pencil the maximum stall speed (rpm) for each driving range listed in the service manual for the particular engine and transmission configuration you are testing (Fig. 17-15). This will provide you with a quick reference during the test.

4. Start the engine and operate it at fast idle until the engine reaches its normal operating temperature.

Selector lever position	Clutch applied	Band applied	Stall speed (rpm)
D2	Front	Front	1350-1700
D1	Front	One-way clutch	1350-1700
L	Front	Rear	1350-1700
R	Rear	Rear	1350-1700

FIGURE 17-15   A typical stall test chart showing the maximum stall speed for each transmission operating range.

5. Apply fully the service and parking brakes. (NOTE: *Never allow anyone to stand in front of the vehicle during the test. If the vehicle should break free during the stall test, the impact could injure him severely.*)

6. Shift the gearshift-selector lever into the range being tested. Then, push the accelerator pedal rapidly to the floor, and hold it there just long enough until a stable rpm reading registers on the tachometer. CAUTION: *Do not keep the accelerator pedal depressed longer than 5 seconds because this will severely overheat the transmission.*

7. Release the accelerator pedal immediately if engine rpm goes over the specified amount previously marked on the tachometer; this is a good indication a clutch or band is slipping.

8. Record the maximum rpm achieved during the test.

9. Run the engine at fast idle (about 1000 rpm) for 60 seconds with the transmission in neutral to cool its fluid.

10. Repeat step 6 for each driving range and record the results. Cool the fluid after each check as outlined in step 9.

11. Compare all the maximum stall speed readings taken during the test with manufacturer's specifications. If the readings are not within tolerances, follow the diagnosis chart in the appropriate service manual to locate the cause of the problem. Listed below are typical causes of high or low stall speed readings, along with what a normal reading indicates.

Results and Indications

High Stall Speed.   If the stall speed is more than 200 rpm above the manufacturer's specifications, this indicates a possible clutch or band slippage. For instance, if this situation occurs in a C-4 transmission in **all** forward driving ranges, the forward clutch is probably slipping because it is the only friction-type one applied in these ranges. Consequently, a high stall reading in **all** forward-drive ratios is a good indication of forward clutch slippage.

If, on the other hand, the high stall speed only occurs in second ratio of a C-4 transmission, the clutch is probably not at fault. In this range, the intermediate band and forward clutch are both applied. Therefore, if the excessive stall speed is obvious only in second ratio, the intermediate band is slipping.

Low Stall Speed.   If the stall speed is about 250 to 400 rpm below manufacturer's specifications and the engine is in tune, the torque converter is usually at fault. Within the converter, the stator's one-way clutch has failed and is no longer holding it against counterclockwise rotation. This causes the returning turbine fluid to strike the forward faces of the impeller blades, which slows them and the engine down. The end result of this failure is a low rpm reading, indicating an early stall condition.

You can also confirm this condition during the road test. The vehicle should operate satisfactorily at cruising speeds, but it will have poor low speed acceleration. If the road test confirms a failure of the one-way clutch, replace the converter.

Normal Stall Speed.   A normal stall speed does provide a positive indication of the serviceability of clutches and bands, but the converter can still be malfunctioning. Therefore, a road test should follow the stall test to check stator operation during cruise conditions. If low speed acceleration is good but the vehicle drags or requires a high throttle opening at highway speeds, the stator has seized or locked up. In this situation, the transmission will also usually run hotter than normal.

However, before replacing the converter to correct this problem, check the vehicle's exhaust system. A partially blocked exhaust system produces the same symptoms as a seized stator clutch. An almost completely restricted exhaust system will affect both low speed and high speed vehicle operation.

ROAD TESTS

Purpose

A properly conducted road test, a test in which the technician checks the operation of the transmission under various vehicle driving conditions, is a valuable diagnostic tool for several good reasons. The road test, for example, verifies the owner's complaint; this is very important in locating and repairing the **cause** of the problem. There will be times when the owner's explanation of what the transmission is doing or not doing will be unclear just because he cannot explain in technical terms just what the problem really is. In some situations, there will be no real transmission defect at all, just a case of improper driving techniques or poor engine performance.

Furthermore, the road test verifies the results of a stall test, and at the same time, it checks items that the stall test could not. For instance, suppose that while stall testing a Ford FMX transmission, the stall speed was too high in reverse **only**, indicating a possible slipping low reverse band or rear clutch. The stall test, in this case, cannot really tell you which of the two units is slipping. The

reason for this is that while the band holds the carrier by itself in reverse, the band along with a one-way clutch holds the carrier in manual low. Consequently, if the band slips in low, the one-way clutch automatically stops and holds the carrier.

The rear clutch, on the other hand, also engages in third. Therefore, a road test should certainly verify a slipping clutch condition when the transmission upshifts to third ratio. Finally, a well-executed road test also provides a good indication of how the throttle, governor, and shift valves are functioning.

Preliminary Steps

Before road testing any vehicle for a transmission malfunction, perform the following steps:

1. Check the engine's coolant and oil levels along with the transmission fluid level.

2. Check engine performance. The engine has to idle properly and otherwise perform relatively well, or the transmission can operate erratically.

3. Look up the manufacturer's shift-speed specifications in the service manual (Fig. 17-16). The shift points will differ according to the type of engine and transmission, axle and tire sizes, and model-year combinations. Write these specifications down on a work sheet for easy reference during the test.

4. As needed, connect a tachometer and vacuum gauge to the engine in addition to a fluid pressure gauge to the transmission. Position the gauge set inside the vehicle at a convenient place so you can view them when necessary during the test (Fig. 17-17).

Conducting the Road Test

When performing the road test, check the transmission for the following:

1. **Slippage**—Operate the transmission in all ranges to test for variation or signs of slippage during all the shifts. (NOTE: A slipping condition is verifiable by observing the tachometer. If the transmission is slipping, the gauge will register an engine overspeed resulting from a defective clutch or band.)

2. **Shift quality**—Note the quality of each upshipshifts, closed throttle downshifts, and engine braking in manual low and intermediate ranges.

3. **Premature, late, or erratic shift timing**—Check and note the actual speeds at which all automatic upshifts and downshifts take place. Check light throttle upshifts, full throttle upshifts, closed throttle downshifts, and engine braking in manual low and intermediate ranges.

Engine cu. in.	225	318	360-4	360-2	400-2
Axle ratio	2.76	2.45	3.21	2.45	2.71
Tire size	6.95 × 14	E78 × 14	H78 × 14	GR78 × 15	HR78 × 15
Throttle minimum					
1-2 Upshift	9-16	8-16	8-15	9-16	9-16
2-3 Upshift	15-25	15-25	15-23	17-25	15-25
3-1 Downshift	8-13	9-14	8-13	9-14	8-13
Throttle wide open					
1-2 Upshift	31-43	39-54	43-56	41-57	37-52
2-3 Upshift	63-76	79-95	78-93	83-100	77-92
Kickdown limit					
3-2 WOT downshift	60-73	76-92	75-90	79-96	73-89
3-2 Part throttle downshift	46-61	30-56	34-57	31-58	30-56
3-1 WOT downshift	28-35	30-44	34-47	31-46	29-43

FIGURE 17-16   A chart showing shift speed specifications.

Perform the initial engagement checks as follows to determine if the band and clutch engagements are smooth:

1. Run the engine until it reaches normal operating temperature.

2. With the engine at the correct idle speed, shift the selector lever from neutral to drive, neutral to intermediate, neutral to low, and neutral to reverse. Band and clutch engagements should be smooth in all positions.

After the initial engagements check is complete, begin the road test from a stationary position with the gearshift selector in drive. Check the light throttle upshifts as the transmission starts the vehicle off in first, then upshifts to second, and finally to third. Note the results of this check.

Again start the vehicle off in drive, allowing the transmission to normally upshift to direct. While the transmission is in direct drive, depress the accelerator pedal to the floor (through the detent). The transmission should shift down to the next lower gear ratio, depending on the vehicle's speed. Note the results of this check.

Accelerate the engine once more until the transmission upshifts to its highest gear ratio—third or fourth ratio. Then check the closed throttle downshift from the highest to the lowest gear ratio by allowing the vehicle to slowly coast down from about 30 mph. The downshifts should occur as specified in the service manual. NOTE: You may experience a 4-3-2-1 or 3-2-1 downshift (depending on transmission type) under these conditions; this is not abnormal. Note the results of this check.

Check the partial throttle downshifts in drive using the service brakes as a load. To do this, accelerate the engine until the transmission upshifts into third or fourth gear and vehicle speed reaches 30 mph or the speed specified in the manual. Then depress the accelerator pedal to the half-throttle position while applying the service brakes to the point where road speed slowly reduces. The transmission should begin to downshift into its lower gear ratios as vehicle speed decreases. Note the results of this test.

Accelerate the vehicle from a stationary position with the selector level in drive two, 2, or intermediate range. NOTE: Many transmissions will start the vehicle off in first ratio as they would normally do in drive, and then upshift to second. Check and note the vehicle speed as the 1-2 upshift occurs. There will be no 2-3 upshift in this driving range. In other words, these transmissions should remain in second ratio regardless of engine or vehicle speed and provide engine braking on deceleration.

Vacuum gauge   Tachometer   Fluid pressure gauge

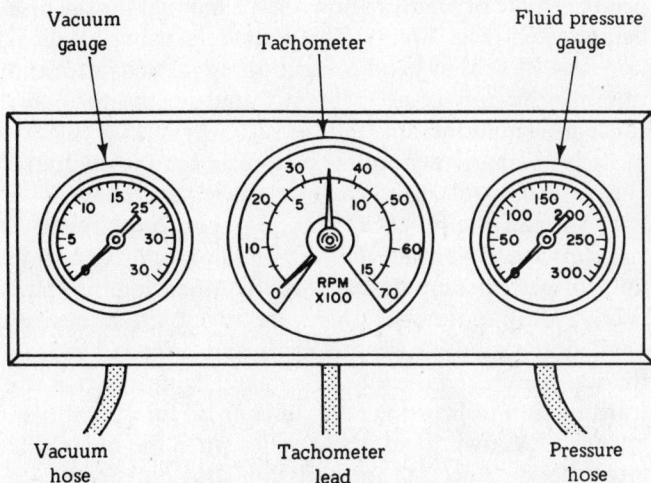

Vacuum hose   Tachometer lead   Pressure hose

FIGURE 17-17   A tachometer, vacuum, and fluid pressure gauge set positioned inside a vehicle in preparation for a road test.

Selector-lever position Drive ratio	Front clutch	Rear clutch	Front (kickdown) band	Rear (low-rev) band	Overrunning clutch
N–Neutral	Disengaged	Disengaged	Released	Released	No movement
D–Drive (first) 2.45 to 1	Disengaged	Engaged	Released	Released	Holds
(second) 1.45 to 1	Disengaged	Engaged	Applied	Released	Over runs
(direct) 1.00 to 1	Engaged	Engaged	Released	Released	Over runs
Kickdown (to second) 1.45 to 1	Disengaged	Engaged	Applied	Released	Over runs
(to low) 2.45 to 1	Disengaged	Engaged	Released	Released	Holds
2–Second 1.45 to 1	Disengaged	Engaged	Applied	Released	Over runs
1–1 Low 2.45 to 1	Disengaged	Engaged	Released	Applied	Partial Hold
R–Reverse 2.20 to 1	Engaged	Disengaged	Released	Applied	No movement

FIGURE 17-18   A typical clutch and band application chart.

However, some transmission types will start the vehicle from a stationary position in second gear with the gearshift-selector lever in drive two, 2, or intermediate. On some models, there will be a 2-3 upshift at a given road speed and throttle setting, while on still others the transmission will remain in second gear. But in either case, the transmission will provide braking on deceleration if working properly. Note these results.

Accelerate the vehicle from the stationary position with the gearshift-selector lever in manual low. The transmission must start the vehicle off in first ratio and remain in this gear range. (NOTE: Some transmissions will upshift to second at a given road speed; this protects the engine from operating at an excessively high rpm.) After accelerating the vehicle for a few moments, release the throttle pedal. The transmission should provide engine braking on deceleration. Note the results of this test.

With the transmission in drive and upshifted into third or direct, move the selector lever back into manual low at about 25 mph. The transmission should downshift from third or direct to first and provide engine braking in manual low. NOTE: At speeds above 25 mph, most transmissions when manually downshifted from drive to manual low will downshift into second and provide engine braking before finally dropping into low. Note the results of this check.

With the vehicle stationary, place the gearshift-selector lever into reverse. Release the brakes, and check transmission operation in reverse under part- and full-throttle settings. Note the results of this check.

Results and Indications

After completing the above tests and noting the results, it is time to analyze the results and what they indicate. A handy reference for analyzing the results of the road test is a clutch-and-band application chart, found in the appropriate vehicle or transmission service manual for the unit being tested (Fig. 17-18). This chart tells which application unit, clutch, or band is functioning in each ratio; and this information is necessary to analyze the probable cause of a malfunction.

Let's compare the results of a typical road test with the clutch-and-band application chart to analyze a transmission malfunction. Suppose you have tested a vehicle with a TorqueFlite transmission and found the unit slipped in second ratio only. By observing the chart (Fig. 17-18), note that there are two friction devices applied in both drive, second ratio, and 2, second ratio—the rear clutch and the kickdown band. Note also that the transmission utilizes the rear clutch in all forward ratios, but the kickdown band only applies when the unit shifts into second. Since the transmission slips only in second gear, in drive and 2, and in no other forward ratios, the rear clutch is functioning satisfactorily; so the kickdown band is most likely the cause of the slipping condition.

PRESSURE TEST

Purpose

The stall and road test can assist you in determining which clutch or band is not functioning properly. These tests, however, cannot identify the **actual cause** of the problem. The cause may be mechanical, a worn out or burned out clutch or band; or it may be due to problems within the transmission's hydraulic system such as an internal leak, worn pump, sticky valve, or an inoperative modulator system.

A pressure test will indicate any major hydraulic malfunction and thereby eliminate the system as the cause of a particular problem. In other words, the pressure test eliminates guesswork as to the exact cause of a transmission problem, and this test should be part of your diagnosis procedure **before** tearing down the unit for visual inspection and replacement of worn parts.

In order to perform a hydraulic pressure test, you must have a pressure gauge set, a vacuum pump and gauge, and an accurate tachometer. The vacuum pump and gauge will not be necessary if the transmission does not have a modulator system. In addition, you must have access to the hydraulic pressure charts for the particular transmission you are testing.

Preliminary Steps

To prepare to pressure test a transmission still in the vehicle, perform the following steps:

1. Attach a tachometer to the engine, following the manufacturer's instructions.

2. If the transmission has a modulator, connect the vacuum pump and gauge assembly to the unit (Fig. 17-19), and plug its vacuum hose.

3. With a pressure hose and fittings, attach a pressure gauge to the control pressure outlet port on the transmission (Fig. 17-20).

4. Have an assistant available to start the engine and operate the transmission in its various ranges, while applying the vehicle's service brakes.

5. Adjust the engine's idle speed to the specified rpm in drive. NOTE: If engine idle speed cannot be brought to within limits by adjustment at the carburetor's idle speed adjustment screw, check the throttle or downshift linkage for binding. If the linkage is satisfactory, check for vacuum leaks in the connecting lines or hoses to the modulator if so equipped. Check as necessary for leaks in other vacuum-operated units such as the power brakes or distributor advance mechanism.

Test Procedures

The actual test procedures do vary from one transmission model to another. Therefore it will be necessary to check the service manual for exact instructions. However, listed below are three tests, performed on a given transmission, which has a vacuum modulator.

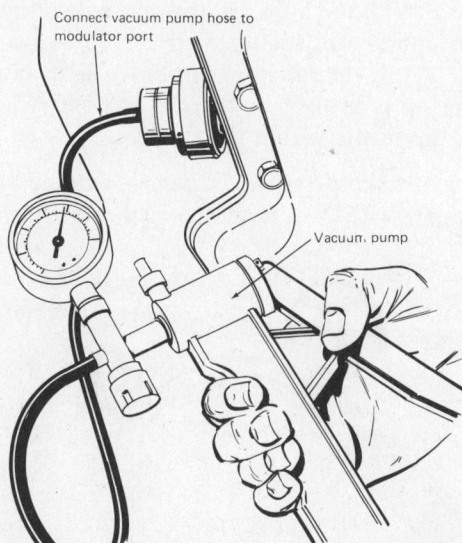

FIGURE 17-19   A vacuum pump and gauge connected to a modulator in preparation for checking control pressure.

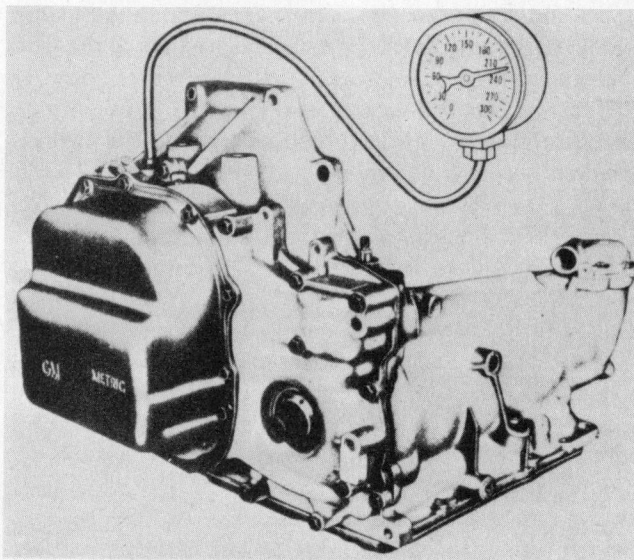

FIGURE 17-20   GM-125C pressure test hookup. Use your oil pressure gauge in this manner during the preliminary checking procedure. (Courtesy of General Motors Corp.)

To perform **hydraulic test number one**, do the following:

1. Apply 18 inches of vacuum to the modulator by means of the hand-operated pump.

2. Start the engine and allow it to warm up.

3. Operate the transmission in **all** operating ranges. Note and record the pressure reading during each phase of transmission operation.

To perform **test number two:**

1. Apply 13 to 16 inches of vacuum to the modulator.

2. Operate the transmission again in **all** its operating ranges. Note and record the pressure reading during each phase of transmission operation. These readings should be higher than the ones taken during test one.

To perform **test number three:**

1. Apply no more than 1.5 inches of vacuum to the modulator.

2. Operate the transmission for the final time in **all** its operating ranges. Note and record the pressure reading during each phase of transmission operation. NOTE: During this test, it may be necessary to run the engine between 1800 and 2000 rpm in order for the hydraulic pump to produce the specified pressure.

Results and Indications

If hydraulic pressure at engine idle is below factory specifications in **all** selector positions during test one, this is an indication of problems other than the vacuum modulator unit. Check for excessive leakage in the front hydraulic pump, transmission case, and valve body.

If hydraulic pressure at engine idle is above factory specifications in **all** selector lever positions, the problem may be the vacuum modulator system. To check the

system with the engine stopped, perform these steps:

1. Remove the modulator and push rod.

2. Inspect the push rod for a bent condition and for corrosion.

3. Install the modulator assembly into the transmission case to prevent fluid loss, but leave the push rod out. (NOTE: With the push rod removed, the modulator diaphragm cannot affect hydraulic pressure; therefore, it should remain constant in all operating conditions.)

4. Start the engine and check hydraulic pressure again in all selector lever positions. If hydraulic pressure is still too high, the trouble is usually in the control system. If, on the other hand, the hydraulic pressure is now within limits, the modulator was not functioning correctly and should be checked.

To check the assembly for diaphragm leakage and the condition of the spring:

a. Remove the modulator from the transmission.

b. Connect the vacuum hose from the vacuum pump and gauge to the unit's inlet port (Fig. 17-21).

c. With the pump, apply 18 inches of vacuum to the modulator. The vacuum modulator should hold the 18 inches for at least 30 seconds. If it does not, the diaphragm inside the unit is leaking.

d. Install the push rod into the modulator.

e. With 18 inches of vacuum showing on the gauge, remove the hose from the modulator, while holding your finger over the end of the control rod. With the hose off, the internal spring within the modulator should push the rod outward (Fig. 17-22). If the rod does not move as specified, the spring is broken or there is internal binding of components within the modulator assembly.

If the hydraulic pressure did not rise in the forward-driving ranges with the modulator vacuum setting of 13 to

TRAP 18 INCHES
VACUUM

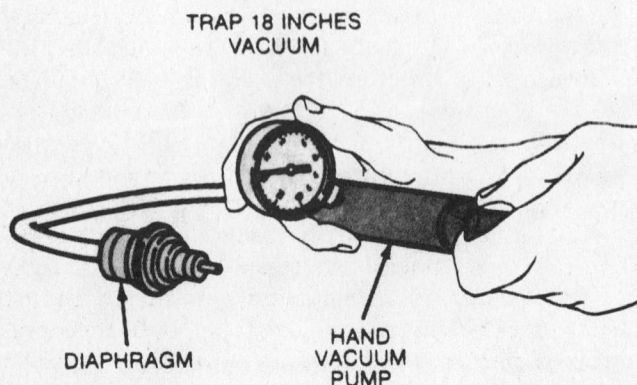

FIGURE 17-21   Checking the modulator's diaphragm with a hand-operated vacuum pump. (Courtesy of Ford Motor Company)

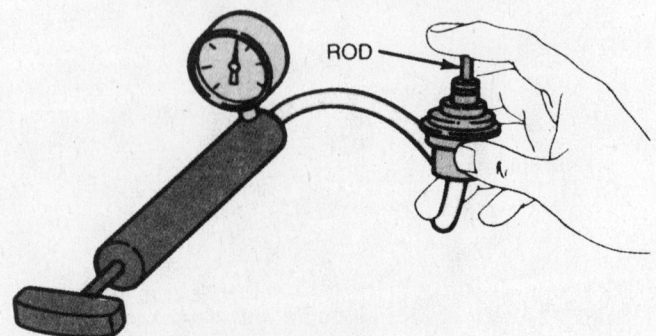

FIGURE 17-22   Checking the diaphragm rod for binding. (Courtesy of Ford Motor Company)

16 inches, again check the transmission's pressure rise capacity by operating the unit in reverse. If pressure rise is normal in reverse, check the modulator as outlined under "test one results and indications." If the modulator is serviceable, the pressure problem is in the hydraulic circuits to the clutches or band servos.

If idle pressures are normal in both tests one and two but test three pressure increases are not to specifications in **all** operating ranges, this indicates excessive hydraulic circuit leakage, low pump capacity, or restricted fluid screen or filter.

If pressures are not within specifications for specific driving ranges only, this indicates excessive hydraulic leakage in the clutch or servo circuits used in those affected ranges.

If control pressure is extremely erratic, check the modulator and push rod as mentioned earlier. Replace the unit and push rod as necessary and retest the transmission. If the pressure is still extremely erratic, clean and inspect the pressure regulator valve train and other valve body components.

ADJUSTING THE CONTROL PRESSURE

Hydraulic control pressure is adjustable on some automatic transmissions by means of an adjustment screw on the modulator or selective push rods. These devices are sometimes necessary to correct a soft or harsh shift condition; however, *the manufacturer does not produce these components to raise low hydraulic pressure caused by worn or leaking components.*

The adjustable vacuum modulator replaces the factory-installed nonadjustable unit. This modified unit has an adjustment screw in the vacuum inlet port (Fig. 17-23). By turning the screw in or out, the control pressure increases or decreases to correct the poor shift condition.

The use of selective push rods with a nonadjustable modulator will have the same effect. When installed, a longer than standard push rod raises hydraulic pressure.

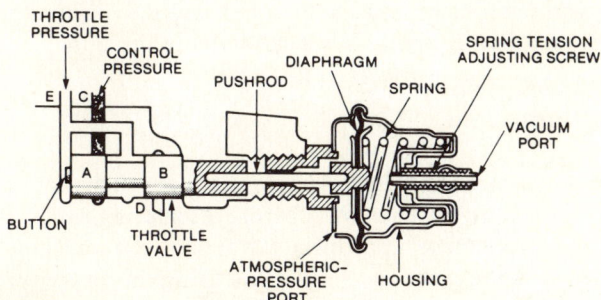

FIGURE 17-23 A typical adjustable modulator assembly with an adjustment screw, located in the inlet port.

A shorter rod decreases control pressure.

To adjust control pressure, perform the following steps:

1. Before installing an adjustable modulator or selective push rod, perform pressure tests one through three with the original nonadjustable modulator installed on the transmission. This action ensures that pressures are all within limits and that the shift problem is not due to other items within the transmission or modulator system.

2. If pressures are within specifications, install the adjustable modulator and operate the transmission with 10 inches of vacuum applied to its diaphragm. NOTE: It may be necessary to make an initial adjustment on this new assembly to provide the specified pressure at 10 inches of vacuum. Once this initial adjustment is complete, make any further adjustments on the unit to overcome the shift feel problem.

3. If the shifts are harsh, reduce hydraulic pressure by turning the adjustment screw counterclockwise. NOTE: Installing a shorter selective push rod with the original modulator will have the same effect.

4. If the shifts are too soft, increase control pressure by turning the adjusting screw clockwise. NOTE: Installing a longer selective push rod with the original modulator will have the same effect.

5. Perform hydraulic pressure tests one through three again. The pressures obtained should still be within the range of those specified in the service manual.

DIAGNOSIS GUIDES

All vehicle manufacturers provide diagnosis guides for their automatic transmissions (Fig. 17-24). These guides list the most common symptoms of problems and list the items a technician should check to find the cause of the malfunction. The guides arrange these check items in a logical sequence, which the serviceman should follow for quickest results. Therefore, the guides are very useful to use during a stall or road test to accelerate the diagnosis procedure.

Let's now examine one of these guides and see how to use it in diagnosing a particular problem. The particular chart, shown in Fig. 17-24, begins at the top center of the page with the problem "no upshift." Moving straight down the guide from the symptom, you will see the first item to check—fluid level and condition. If the fluid level and condition are satisfactory, you continue to move down the list of check items, one at a time, until the cause of the problem is located.

Note also that the check list is split into two groups beginning after the check for fluid level and condition.

```
                    ┌─────────────────┐
                    │   No upshift    │
                    └────────┬────────┘
                             │
                             ▼
                 ┌────────────────────────┐
         ┌───────┤ Check fluid level and  ├────────┐
         │       │      condition         │        │
         │       └────────────────────────┘        │
         ▼                                          ▼
┌──────────────────────┐              ┌──────────────────────────┐
│ Perform all linkage  │              │ Perform hydraulic        │
│     adjustments      │              │    pressure tests        │
└──────────┬───────────┘              └──────┬────────────┬──────┘
           │                                 │            │
           ▼                                 ▼            ▼
┌──────────────────────┐        ┌──────────────────┐  ┌──────────────┐
│   Check kickdown     │        │  Check governor  │  │ Inspect oil  │
│   band adjustment    │        │ assembly including  │ filter:      │
└──────────┬───────────┘        │   seal rings:    │  │ replace if   │
           │                    │ replace if necessary│ necessary    │
           │                    └────────┬─────────┘  └──────────────┘
           │                             │
           │                             ▼
           │                    ┌──────────────────┐
           │                    │ Check valve body:│
           │                    │ disassemble,     │
           │                    │ clean, inspect   │
           │                    └────────┬─────────┘
           │                             ▼
           │                    ┌──────────────────┐
           │                    │ Check kickdown   │
           │                    │ servo:           │
           │                    │ disassemble,     │
           │                    │ clean, inspect   │
           │                    └────────┬─────────┘
           │                             ▼
           │                    ┌──────────────────┐
           │                    │ Check accumulator:│
           │                    │ disassemble,     │
           │                    │ clean, inspect   │
           │                    └────────┬─────────┘
           │                             ▼
 In-vehicle│                    ┌──────────────────┐
  checks   │                    │ Perform air      │
     ▲     │                    │ pressure test    │
     │     │                    └────────┬─────────┘
     └─────●─────────────────────────────●───────────┐
┌──────────────────────┐                        Requires
│ Check trans. oil pump:│                       transmission
│ disassemble, clean,  │                          removal
│     inspect          │
└──────────────────────┘
                    ┌──────────────────┐
                    │ Check front clutch:│
                    │ disassemble,     │
                    │ clean, inspect   │
                    └────────┬─────────┘
                             ▼
                    ┌──────────────────┐
                    │ Check kickdown   │
                    │     band         │
                    └──────────────────┘
```

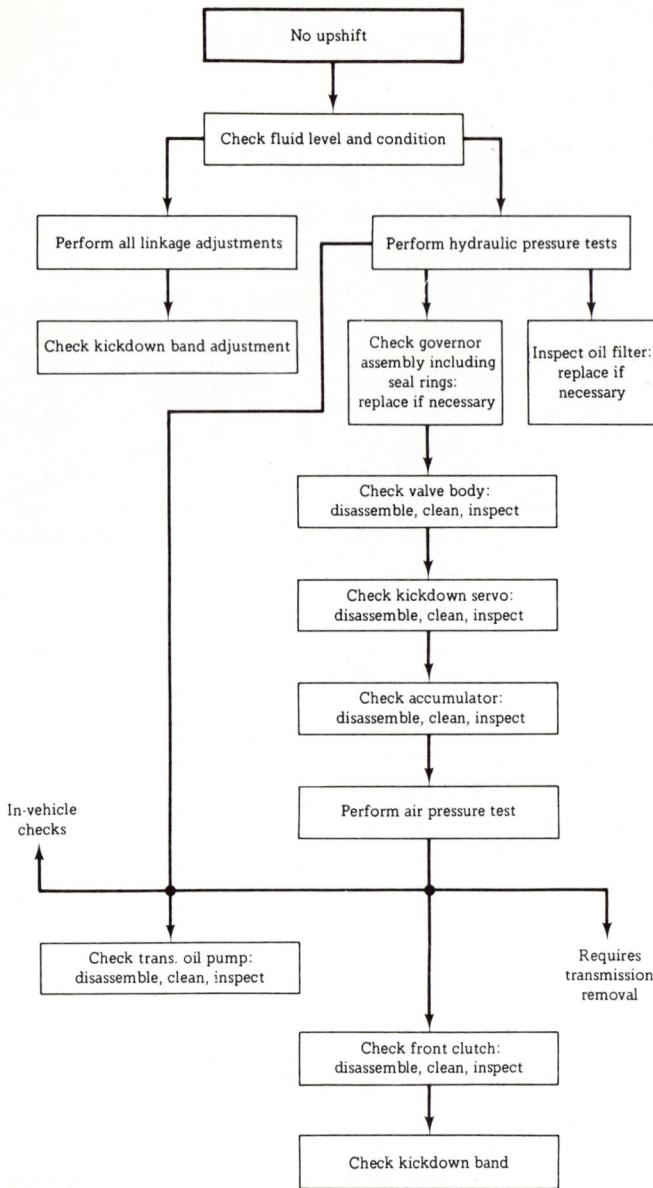

FIGURE 17-24   A typical automatic transmission diagnosis guide.

The check list on the left consists of mechanical things such as linkage and band adjustments; the list on the right includes hydraulic system components.

Note the horizontal line near the bottom of the page with the arrow on the left end pointing upward, and the one on the right pointing downward. This represents the dividing line between the checks that are performable with the transmission installed in the vehicle from those that require transmission removal. In other words, all check items above the line are accessible with the transmission still in the vehicle; those below the line are not.

There are a few things to remember in reference to these guides. First, there may not be a guide for the exact transmission malfunction you are attempting to diagnose. Remember that several transmission problems can cause a combination of several different symptoms or malfunctions. Second, the guide itself is only as good as how well you follow it.

AIR CHECKS

Purpose

One of the check items on the diagnosis guide just discussed is an air pressure test. The air test, as its name implies, is the checking of certain components with compressed air. In transmission work, the air is a substitute for fluid pressure and when directed into given hydraulic ports, it will indicate if a clutch, servo, or governor is functioning properly.

The air pressure test basically indicates if certain hydraulic components are leaking excessively. It is very possible, in certain cases, to have an internal hydraulic leak inside the transmission that will not be detectable using the fluid pressure tests mentioned earlier in this chapter. The reason for this is that the transmission's hydraulic pump has a potential of producing high pressure and volume when in operation. Therefore, the fluid pressure test may not detect a small, internal hydraulic leak because the pump has the ability to maintain reasonable system pressure even when it has a leaking hydraulic component. The air pressure test, on the other hand, can pinpoint the location of small and large internal leaks.

The air pressure test is performable with the transmission still in the vehicle or on the work bench. It it is done with the transmission in the vehicle, the test will be part of a diagnostic procedure used to supplement a stall, road, or hydraulic pressure test. However, it is more common to air check a transmission during the overhaul procedure either to verify a suspected leak before disassembling the transmission or to test clutches and servos to ensure that they function properly when assembling the unit.

Test Procedure

Not all automatic transmission designs are such that you can air test them. Therefore, before attempting to air test a unit, check the appropriate transmission or service manual to see if the unit is testable using this procedure, and what the approved instructions for performing the test are.

To perform a typical air test on a transmission still

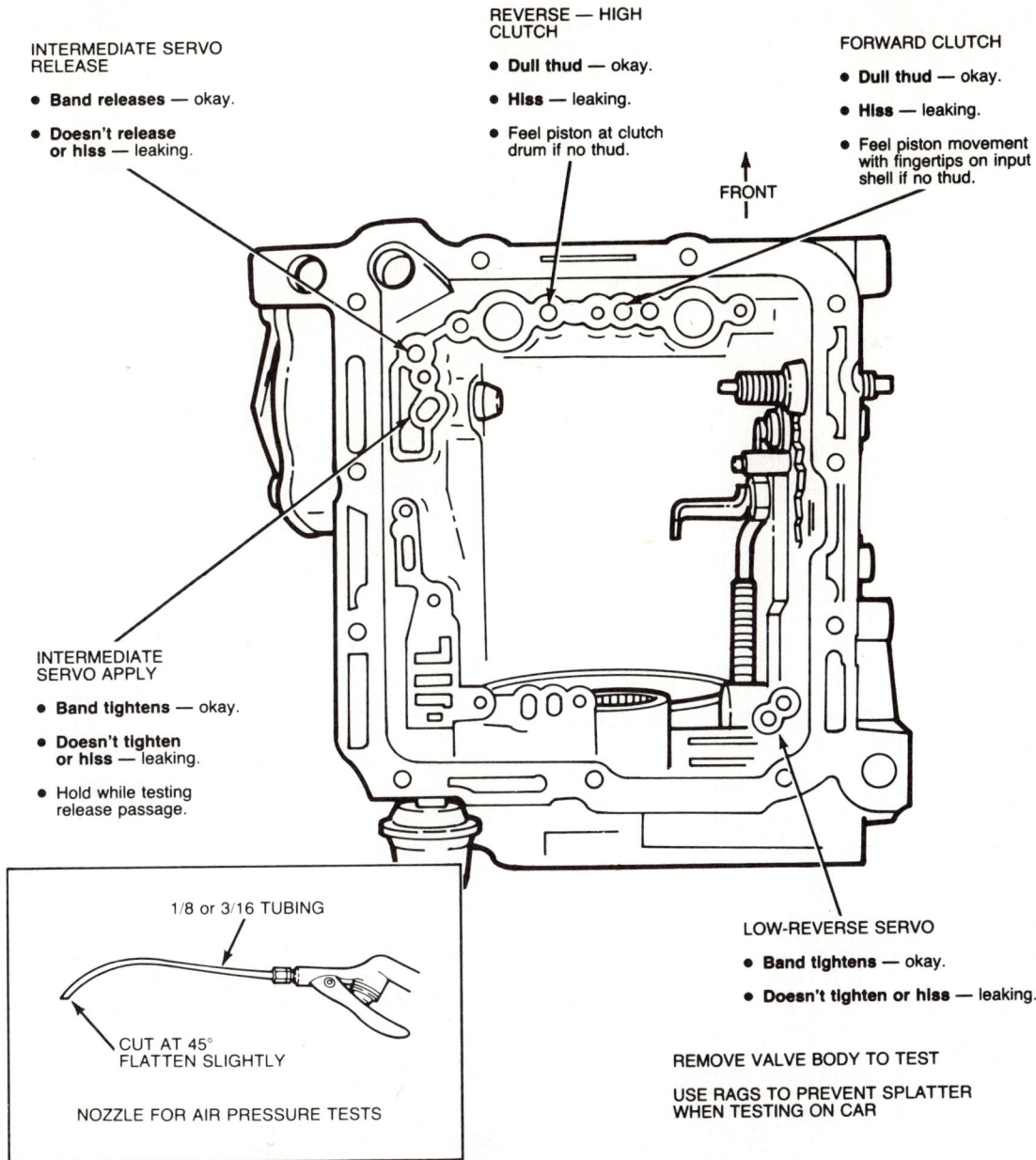

INTERMEDIATE SERVO
RELEASE

- **Band releases** — okay.
- **Doesn't release or hiss** — leaking.

REVERSE — HIGH
CLUTCH

- **Dull thud** — okay.
- **Hiss** — leaking.
- Feel piston at clutch drum if no thud.

FRONT

FORWARD CLUTCH

- **Dull thud** — okay.
- **Hiss** — leaking.
- Feel piston movement with fingertips on input shell if no thud.

INTERMEDIATE
SERVO APPLY

- **Band tightens** — okay.
- **Doesn't tighten or hiss** — leaking.
- Hold while testing release passage.

1/8 or 3/16 TUBING

CUT AT 45°
FLATTEN SLIGHTLY

NOZZLE FOR AIR PRESSURE TESTS

LOW-REVERSE SERVO

- **Band tightens** — okay.
- **Doesn't tighten or hiss** — leaking.

REMOVE VALVE BODY TO TEST

USE RAGS TO PREVENT SPLATTER
WHEN TESTING ON CAR

FIGURE 17-25   An automatic transmission fluid-passage identification chart. (Courtesy of Ford Motor Company)

installed in a vehicle, complete the following steps:

1. Raise the vehicle on an overhead hoist or with a jack to a suitable working height. *If raised with a jack, lower the vehicle on safety stands before working under it.*

2. Place a drain container under the transmission, and remove the fluid from the converter and transmission as outlined in Chapter 18.

3. Remove the fluid pan, gasket, and screen or filter as outlined in Chapter 18.

4. Loosen all the valve body attaching bolts and carefully remove the valve body. Place the valve body on a clean work bench.

5. Check the transmission or vehicle service manual for the appropriate passage identification chart (Fig. 17-25).

6. Apply 25 to 30 psi air pressure to all the designated passages. Hold the air nozzle on each port opening for several seconds and listen for a hissing noise, indicating a leak.

7. Listen also for a dull thud or clunk as the clutches and servos apply. Watch for the operation of the bands by observing them tighten and loosen around their respective drums as you apply and release the air pressure.

8. After air checking the unit, reinstall the valve body and torque its attaching bolts to specifications.

9. Reinstall the filter or screen.

10. Reinstall the fluid pan, using a new gasket. Torque all of its attaching bolts to specifications.

11. Lower the vehicle and refill the transmission with fluid using the procedure outlined in Chapter 18.

NOTE: When air testing a transmission on the work bench, you will need to follow only steps 5 through 10.

Results and Indications

If during the air test the following problems are noted, remove and disassemble the transmission and its components for inspection and service:

1. **A hissing sound.** This indicates a leak usually caused by defective rubber or metal sealing rings.

2. **No indication of clutch or band operation.** This is also an indication of possible defective sealing rings, a binding clutch or servo piston, or broken piston return spring or springs.

3. **No whistling, clicking, or buzzing from the governor assembly.** This is a definite sign of a seized governor valve.

NOISE DETECTION

Noise can be one of the hardest problems to diagnose because it can travel from one area of the vehicle to another through the vehicle's metal parts. In addition, a vibration often accompanies a noise; and when one or both are present, the problem can be in the tires, suspension, engine and its accessories, drive line, or the transmission. There are, of course, certain things you can do to locate the exact **cause** of the noise.

Preliminary Steps

The first thing to do is to isolate the general area where the noise originates. To do this, road test the vehicle and look and listen for the following:

1. The speed at which the noise or vibration occurs.

2. The engine rpm, measured with a tachometer, at which the noise and/or vibration occurs.

3. Noise levels that change with different selector lever positions or gear ratios.

4. Noise levels that increase or decrease with engine temperature.

5. Harshness of vehicle ride. In this case, check tire inflation.

6. Consistency of the noise. If the noise changes intensity several times a minute, there could be more than one problem.

7. Changes in the noise level during acceleration and deceleration.

8. Presence of noise when the vehicle is coasting or standing stationary with the engine idling.

Hoist Test

After the road test is complete, further diagnosis is performable with the vehicle on a hoist. With the wheels free to revolve, operate the vehicle at the road speed where the noise is noticeable. Use a stethoscope (Fig. 17-26) to assist in pinpointing the source of the noise by placing the tip of its probe as necessary on the converter housing, transmission case, extension housing, differential housing, axle housing, and the various engine components.

The following conditions are examples of noise or vibration problems and their causes:

1. A vibration at 1000 to about 2000 rpm after an engine or transmission overhaul is a good indication of a crankshaft, flywheel, or torque converter unbalance.

FIGURE 17-26 *Using a stethoscope to pinpoint the source of a noise.*

2. A high-pitched whine coming from the converter housing area can result from either a low fluid level, clogged filter, or a worn or defective pump. (NOTE: If a similar type of noise is heard only during a stall test or on vehicle acceleration, the stator's one-way clutch is probably defective, which necessitates converter replacement.)

3. A grinding noise in neutral only is usually due to a very worn planetary gear train.

4. A noise in the lower gear ratios but not in third or direct drive is a good indication of worn planetary gears, thrust washers, or needle bearings.

5. Noise in high gear only. This usually means a problem other than the transmission like a worn drive shaft, center-support bearing, differential bearing, carrier bearing, or axle bearing.

6. Vibration and possible noise on vehicle acceleration up to 45 mph are usually the result of a bent drive line or worn universal joints.

7. Noise or a howl when the vehicle is stationary with the engine running can indicate defective bearings in the air conditioner compressor, alternator, air pump, or power-steering pump.

DYNAMOMETER TESTS

Purpose

A number of automatic transmission speciality shops now use a special dynamometer to test rebuilt transmissions or those that have malfunctions. The dynamometer itself simulates the operating conditions the transmission encounters while in service in a motor vehicle without the need of the unit being in the vehicle itself. The machine bench checks the rebuilt or malfunctioning transmission for various problems with the unit operating under various load conditions, and the technician can spot and repair defects in the transmission before installing it in the vehicle.

Setup Procedures

To set up and test a typical transmission on the dynamometer mentioned in Chapter 15, perform the following steps:

1. On the bench, bolt the appropriate mount plate to the transmission converter housing, using at least four bolts (Fig. 17-27). The converter housing dowel

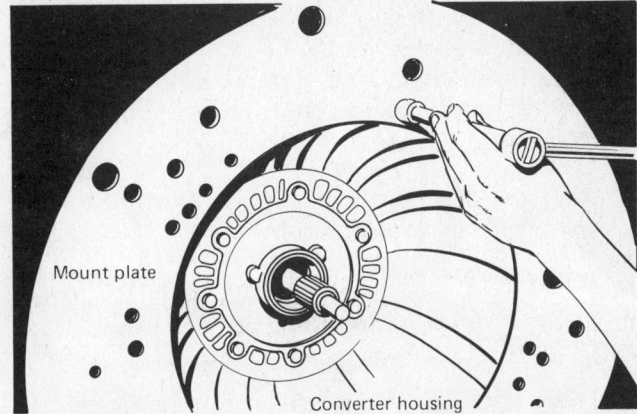

FIGURE 17-27 Bolting the mount plate onto the transmission's converter housing.

pins or those supplied with the machine will assist in locating the mount plate onto the housing.

2. Remove all the test pipe plugs from the transmission case, and install the quick-disconnect fitting into their tapped holes (Fig. 17-28).

3. Attach a bypass hose between the two cooler fittings (Fig. 17-29). (NOTE: On air-cooled transmissions, this step is not necessary.)

4. Install two impact risers over the converter drive studs 180 degrees apart (Fig. 17-30).

5. Using an overhead crane and chain fall, transfer the transmission from the work bench to the dynamometer.

6. Install the two, centering dowel bolts and one top attaching bolt, and tighten them securely.

7. Install the appropriate drive fingers on the converter driver, and insert the correct size bushing into the

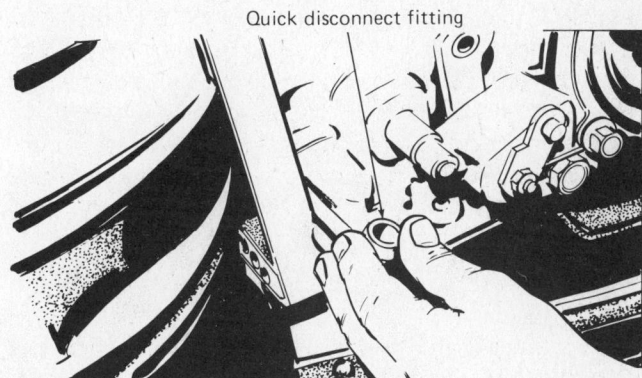

FIGURE 17-28 Installing a quick-disconnect fitting into a pressure test port.

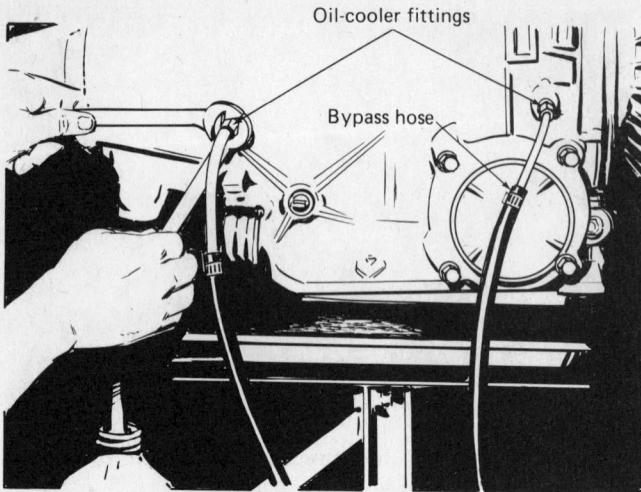

FIGURE 17-29 Installing a bypass hose to the transmission's two cooler fittings.

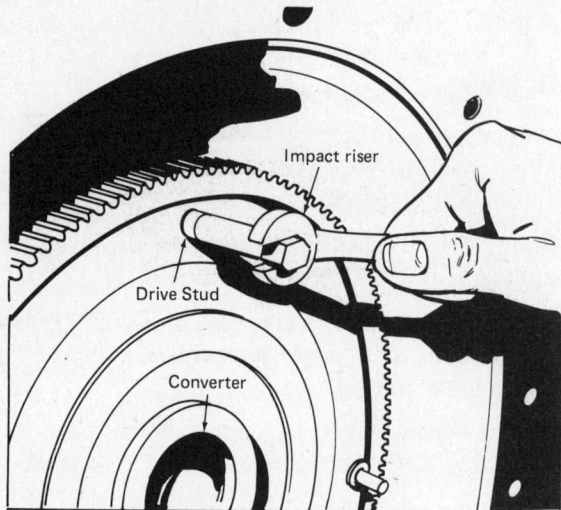

FIGURE 17-30 Installing the impact risers to two converter drive studs.

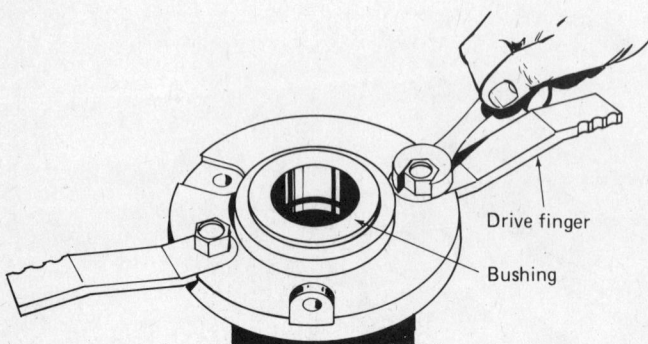

FIGURE 17-31 Installing the bushing into and the drive fingers onto the converter driver.

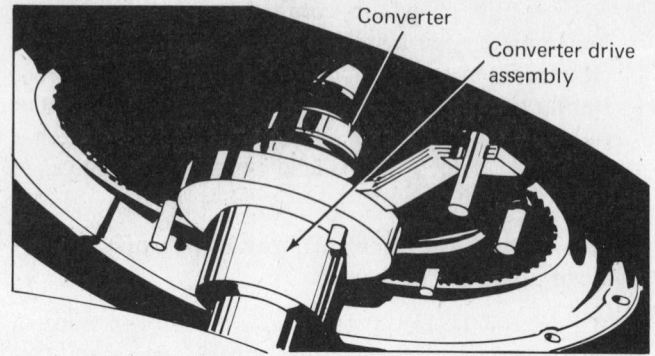

FIGURE 17-32 Engaging the converter with the converter drive assembly.

center of the driver (Fig. 17-31). Position the completed converter assembly into the input shaft socket.

8. Release the cradle mount slide lock and move the cradle mount forward to fully engage the converter with the converter drive assembly. Make certain that the drive fingers make contact with the impact risers on their driving sides (Fig. 17-32). Lock the cradle mount in this position.

9. Connect the appropriate number of pressure gauge hoses to the quick-disconnect fittings (Fig. 17-33).

10. Slide the correct splined sleeve onto the transmission's output shaft (Fig. 17-34), and install the front U-joint into the sleeve.

11. Release the lock on the load section, and loosen the dynamometer's output shaft set screw (Fig. 17-35). Bring both units forward until the drive shaft engages

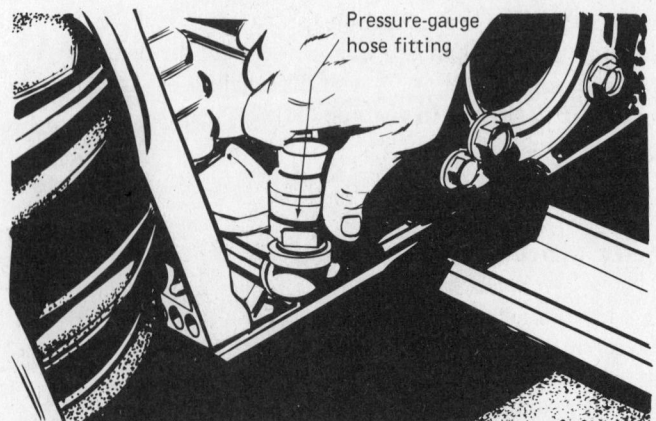

FIGURE 17-33 Connecting the machine's pressure hose to the quick-disconnect fitting.

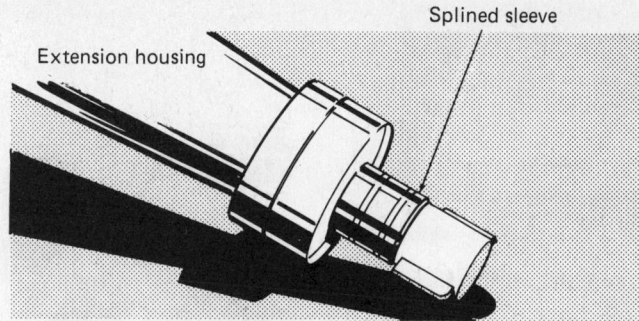

FIGURE 17-34 *Installing the splined sleeve over the output shaft.*

into the U-joint. Move the load section back until about two inches of splines are showing on the output shaft. Tighten the drive shaft set screw and load section lock securely (Fig. 17-36).

12. Position and lock the drive shaft safety shield in place (Fig. 17-37).

13. Position the torque converter safety shield into place (Fig. 17-38).

14. Connect the machine's 12-volt test leads to the transmission electrical kickdown system if so equipped.

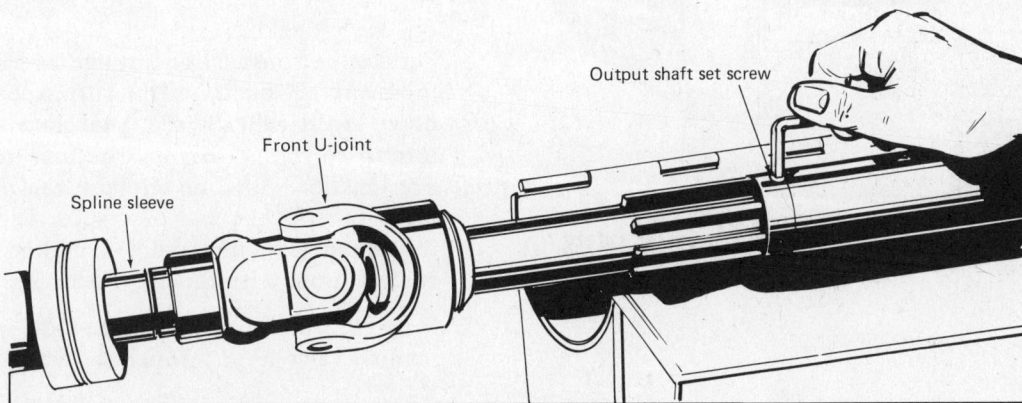

FIGURE 17-35 *Loosening the output shaft set screw.*

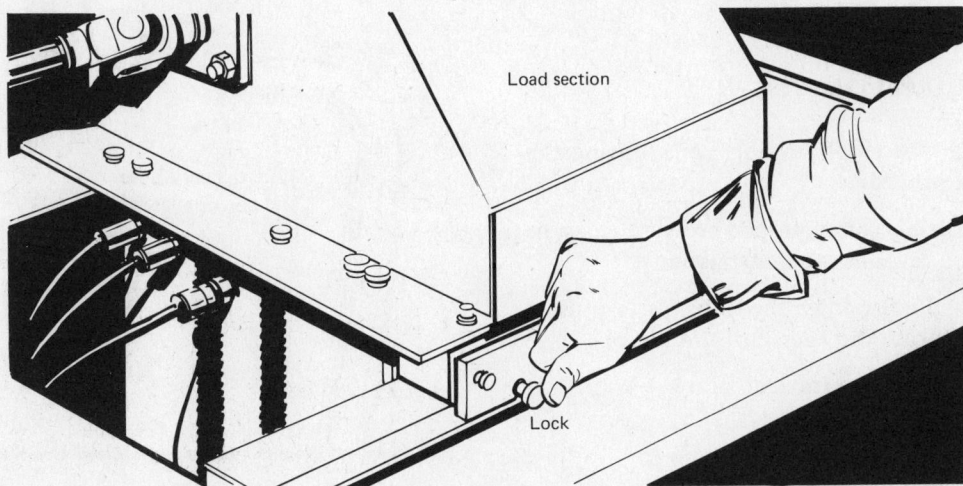

FIGURE 17-36 *Tightening the load section lock.*

FIGURE 17-37   Locking the drive shaft safety shield in position.

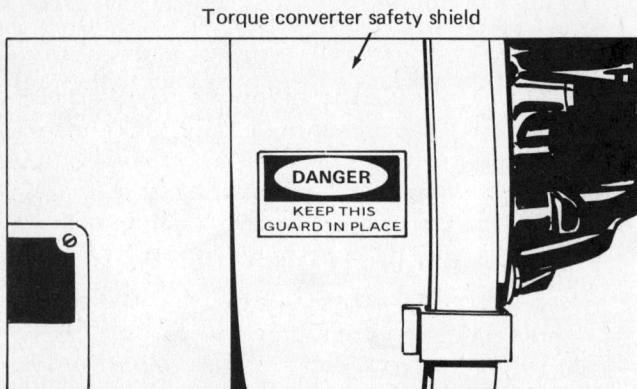

FIGURE 17-38   Positioning the torque converter safety shield into place.

15. Install the machine's vacuum gauge hose over the transmission modulator's inlet port if so equipped (Fig. 17-39).

16. Connect the filler tube to the transmission.

Items to Observe During the Test

You should watch for the following items during the dynamometer test procedure:

1. Pressure fluctuations or improper pressure (may be due to low fluid level in the transmission).

2. Loud or unusual noise from the transmission (use a stethoscope to locate the source of the noise).

3. Unusual vibration in the transmission.

4. External leakage.

5. Missing of upshifts and downshifts.

6. Signs of slippage.

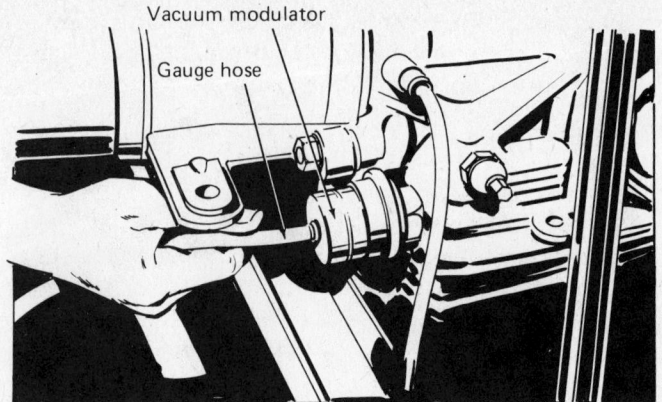

FIGURE 17-39   Connecting the vacuum hose to the modulator.

Testing Procedures

To carry out the dynamometer test, perform the following steps:

1. Service the transmission through the filler or dipstick tube with six quarts of fluid from the fluid supply hose (Fig. 17-40). Start the machine and allow the torque converter to fill. Continue to service the transmission with fluid until the level on the dipstick reads full. At the end of this servicing procedure, transmission fluid pressure should be stable and to specifications with the unit operating in neutral.

2. With the transmission turning at idle speed, shift the manual valve lever through the drive, intermediate,

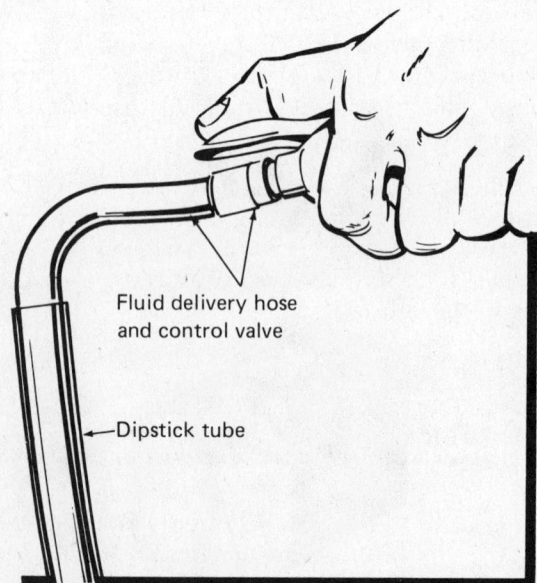

FIGURE 17-40   Servicing the transmission with fluid from the delivery hose.

low, and reverse ranges, while noting the control pressure of the unit during each operating phase. The pressure readings should be to specifications.

3. Return the manual valve lever to the drive position. With the speed control handle, slowly increase input shaft speed while observing both the input and output tachometers. The upshift from low to intermediate should occur at about 900 rpm, and the shift from second to third should occur at about 1400 rpm (input shaft speed) on a typical three-speed transmission.

4. With the input speed set at 1600 rpm, remove the vacuum hose from the modulator if so equipped. Control pressure should now increase to about 150 psi, and the transmission should downshift to second.

5. Reconnect the modulator vacuum hose. The transmission should upshift again to third, and the hydraulic pressure should drop back to normal, about 60 psi.

6. Apply a dynamometer load of about 1200 psi to the transmission, and check the zero vacuum downshift (step 4) again. The downshift should now occur much more abruptly. Reconnect the vacuum hose as in step 5. The transmission should again upshift to third, but the shift should be sharper and more distinct than it was during step 5.

7. With the dynamometer load still applied, reduce the input shaft speed to idle, observing the downshift of the transmission into second and finally to low as input speed decreases. Then, increase the input speed again, observing the transmission's upshift pattern as the input speed goes up. The upshifts should occur at about the same points as they did with no load. Next, reduce input speed to idle and cut the dynamometer load on the transmission.

8. Set the disc brake lever to stall the drive shaft. Then, increase the input speed to about 750 rpm, allowing the transmission to operate at this speed for several minutes. This action will cause the internal temperature of the unit to rise.

9. Increase the input speed to about 1100 to 1200 rpm. At this speed, the drive shaft should attempt to break loose from the disc brake, and the input motor will be laboring. If these characteristics **are not** present, try to increase input shaft speed. It should not be possible to reach 1300 rpm with the drive shaft locked unless internal slippage is present within the transmission itself. Reduce the input speed to idle, release the disc brake, and allow the transmission to idle in drive range for about one minute to cool down.

10. Apply a full dynamometer load (1600 psi) to the transmission, and adjust the vacuum to the modulator to 10 inches. Next, increase the input speed rapidly; the transmission should upshift to third at about 2200 rpm.

11. With the transmission operating in third at 2200 rpm, move the kickdown lever to the detent position; the transmission should downshift to second. Release the kickdown lever; the transmission should again upshift to third. Finally, reduce the dynamometer load and reduce the input speed to idle.

12. When the overall performance of the transmission satisfies all requirements, drain the torque converter and pan into the tray. Remove the transmission from the machine and return it to the work bench. Remove all the quick-disconnect fittings, replace them with plugs, and remove the mount plate and drive accessories.

Post-Test Observations

When draining the transmission and torque converter, observe the fluid as it comes out:

1. Does the fluid contain particles of friction material or metal? This is a good indication of either internal failure or misfitting of components.

2. Is the fluid sharpy discolored in relation to its color before the test? This indicates that the fluid overheated due to internal transmission slippage or to improper test procedures.

REVIEW

This section will assist you in determining how well you can remember the material in this chapter. Read each item carefully. If you cannot complete the statement, review the section in the chapter that covers the material.

1. To be successful in troubleshooting a transmission, the mechanic must follow a _____, _____, and _____ test procedure.

2. The first check to perform for any transmission malfunction is fluid _____ and _____.

3. Excess fluid can be pulled out of an automatic transmission using a _____ _____.

4. Fluid being burned inside the engine is a common cause of fluid loss from a transmission that has a _____ _____.

5. To help locate the source of a leak, place a piece of white _____ or _____ under the vehicle.

6. In order to determine the source of a leak from the converter housing, visually inspect the housing area, or use a _____ _____ to determine if the fluid is coming from the transmission.

7. Fluid leakage into the engine cooling system usually results from a crack or hole in the _____ inside the radiator.

8. There is always a mechanical link between the gearshift-selector lever and the _____ _____ in the valve body.

9. If a transmission has a vacuum modulator system instead of a mechanically operated throttle valve system, the modulator will then vary _____ _____ and _____ _____ to match engine load.

10. To test a modulator diaphragm, connect a vacuum gauge and _____ _____ to the unit's inlet port.

11. The stall test finds the highest engine speed at _____ _____ with the wheels locked and the transmission in gear.

12. To verify an owner's complaint, _____ _____ the vehicle.

13. The initial engagement part of the road test determines how smooth a _____ or _____ applies.

14. A handy reference for analyzing the results of a road test is a _____ and _____ chart.

15. A _____ test indicates any major hydraulic system malfunction.

16. During pressure test number three, apply no more than _____ inches of vacuum to the modulator.

17. If a vacuum modulator has a ruptured diaphragm, hydraulic pressure readings will be _____ during pressure tests one and two.

18. Hydraulic pressure is adjustable on some automatic transmissions by means of a modified _____ or _____ push rods.

19. _____ guides list the most common trouble symptoms and provide the items a technician should check to find the cause of the malfunction.

20. An _____ test will determine if a clutch, servo, or governor is functioning properly.

21. It is more common to perform an air check during transmission _____.

22. An internal hydraulic leak is usually due to a defective _____ or _____ sealing ring.

23. A _____ is a handy tool in locating the source of a noise.

24. Many transmission shops use a _____ to test rebuilt or malfunctioning units.

25. The dynamometer's 12-volt system connects to a transmission's electrical _____ system.

For the answers, turn to the Appendix.

CHAPTER 18

Changing the Transmission Fluid and Filter

Changing fluid is one type of service work done on an automatic transmission. Mechanics usually change the fluid within the transmission for one of two reasons. The first and most pressing reason is that the fluid has become oxidized or has lost part or all of its lubricating properties. Chapter 17 of this text explains how to check the fluid for this condition. The second practical reason is for preventive maintenance, which includes changing the fluid and filter to comply with factory recommendations or specifications.

The majority of automatic transmission manufacturers recommend fluid changes in their respective units at given intervals, but these interval periods differ from one manufacturer to another. A given manufacturer may specify the interval period in chronological time while another will use vehicle mileage. In either case, the owner's manual or service manual for the vehicle contains the recommended fluid change intervals.

Interval Period

A typical mileage interval may be as low as 20,000 miles or as high as 60,000 miles. Under normal driving conditions, the owner should adhere to these specifications. If the owner does not follow them and change the fluid at the specified time, the oil can lose its ability to function as a lubricant and the transmission will wear out prematurely.

However, under certain driving situations, the owner may need to have the transmission fluid changed more often. The driving conditions that reduce the useful life of the fluid are the constant operation of a vehicle in hot or dusty climates, the frequent use of the vehicle to haul heavy loads, and the constant use of the vehicle in stop-and-go driving such as in heavy traffic. These conditions can reduce the life of the fluid considerably and require the owner to replace it at 6,000-mile intervals or less.

A total fluid change of an automatic transmission requires a lot more than just draining and refilling the reservoir or pan. For example, every fluid change should include the removal of the pan and the cleaning or replacement of the filter or screen. Furthermore, the mechanic should drain or pump the dirty fluid from the torque converter.

DRAINING THE FLUID FROM THE TRANSMISSION PAN

To drain the fluid in a typical automatic transmission, proceed as follows:

1. Raise and level the vehicle using either a hydraulic lift or suitable jacks so that the fluid pan and converter access plate are easy to reach. When using jacks for this purpose, *always insert safety stands under the frame at appropriate locations*. These stands prevent the vehicle from falling on the mechanic as he works underneath the transmission, *and they should remain in place until he completes the repair job.*

2. Inspect the transmission pan for a drain plug (Fig. 18-1). Most of the current model transmissions **will not** have a drain plug. Transmission manufacturers for the most part have deliberately eliminated this plug to assure a more complete fluid change and filter service. Consequently, the mechanic must remove the pan to drain the fluid from the transmission.

3. Inspect the transmission for a pan-mounted dipstick. The mechanic can also drain some transmissions by removing the dipstick (fluid filler) tube from the side of the pan (Fig. 18-2). Tube fittings secure these types of filler tubes to the side of the pan.

4. Place a drain pan under the transmission. If the transmission pan has a drain plug or detachable filler tube, remove either one and permit the dirty fluid to flow into the container.

5. To inspect, clean, or change the filter, remove the transmission pan's attaching bolts and lower the pan. (BE CAREFUL: *The pan may still be very hot.*)

6. If the pan has no drain plug or detachable filler tube,

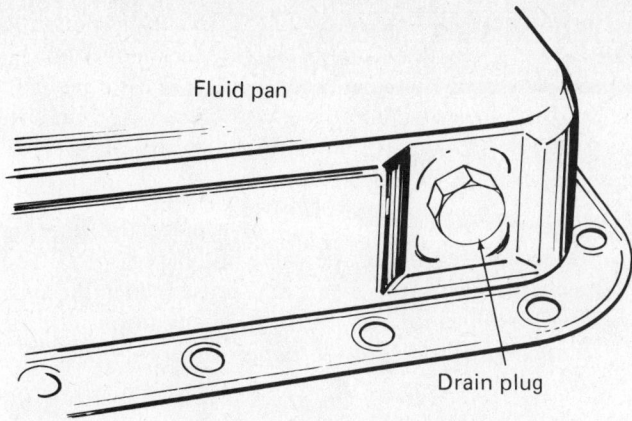

FIGURE 18-1   Some transmission pans have a drain plug.

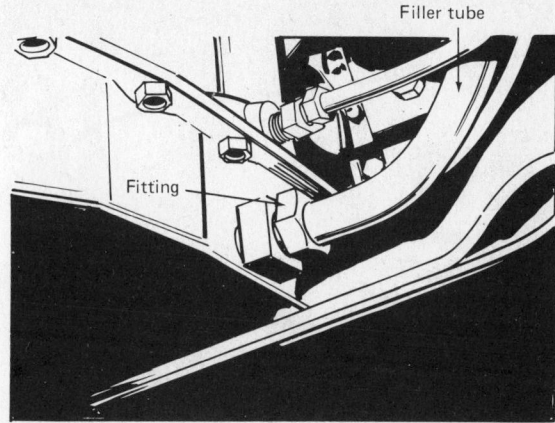

FIGURE 18-2   A dipstick filler tube located on the side of the pan.

remove the pan using the following procedure: Install a drain pan under the transmission. From the rear of the pan, remove all the attaching bolts (Fig. 18-3). Then, one at a time, begin to take out a bolt from first one side and then the other. This process permits the rear end of the fluid pan to slowly tip downward for easy draining.

If the mechanic does not follow this procedure, the hot oil will gush out. This makes it very difficult to catch all the dirty fluid in the drain pan and usually causes a mess on the floor. In addition, the hot oil rushing out of the pan can burn the mechanic's hands as he works under the unit.

7. As soon as the draining fluid has slowed down to a trickle, remove the remaining attaching bolts. Then, remove the pan and place it on a work bench.

INSPECTING AND SERVICING THE PAN

With the pan removed, check and service it as follows:

1. Carefully inspect the inside of the pan for the presence of metal or friction particles (Fig. 18-4). The presence of a few particles of these materials is normal, but large quantities indicate a transmission with excessive wear. In this situation, the mechanic should tear the transmission apart for further inspection and repair.

2. Inspect the pan and its mating surfaces on a flat surface for distortion. Check the bolt holes on the gasket side of the pan to see if prior torquing of the attaching bolts has raised (dimpled) the metal surface (Fig. 18-5).

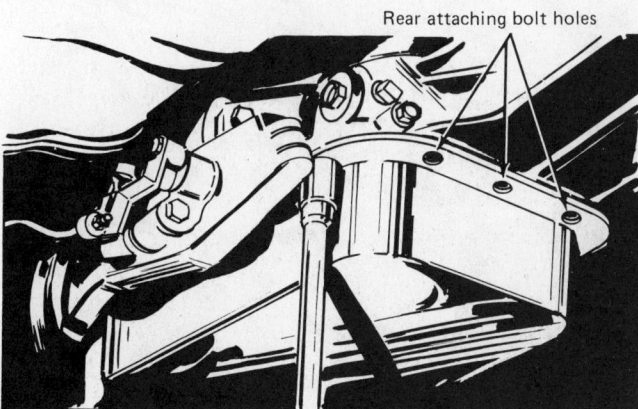

FIGURE 18-3   A mechanic removing the bolts from the rear section of the pan.

FIGURE 18-4   Inspection of the inside of the pan for the presence of metal and friction particles.

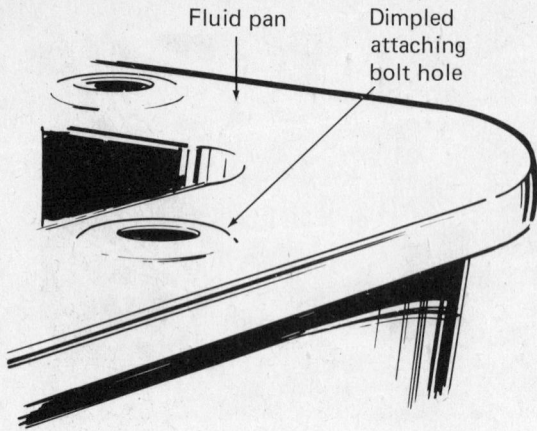

FIGURE 18-5 A pan with dimpled bolt holes.

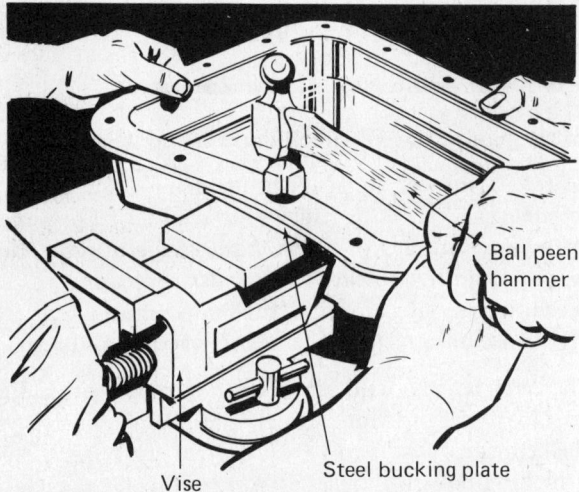

FIGURE 18-6 A mechanic flattening a dimpled pan with a ballpeen hammer.

To remove these dimples, place a flat metal bar or plate under each hole, on the bolt-head side of the pan, and carefully flatten the dimples, using a ballpeen hammer (Fig. 18-6). CAUTION: *Be careful not to distort the pan by striking the surface with excessive force.* If this process does not successfully flatten the dimples, or the pan is distorted, replace it.

3. After removing all traces of the old gasket, wash the pan in clean solvent, using a brush (Fig. 18-7). Then, using low-pressure compressed air, blow the pan dry. CAUTION: *Never use a cloth to dry the pan because any rag lint left behind may enter the hydraulic system and cause valves to stick.*

FILTER OR SCREEN REMOVAL, SERVICE, AND INSTALLATION

Removal and Service

1. While the pan is off, clean or replace the filtering device, located under the valve body (Fig. 18-8). The mechanic can often clean and reuse a metallic-type filter screen as long as it is in good condition. *But the technician must replace a paper or fabric filter.*

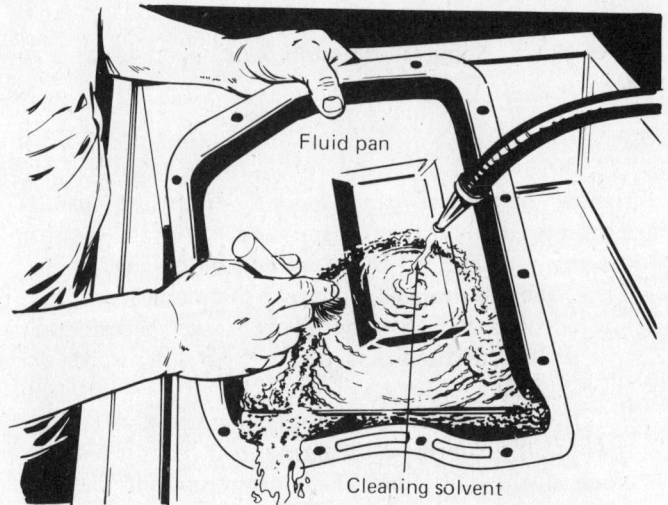

FIGURE 18-7 Washing a pan in clean solvent.

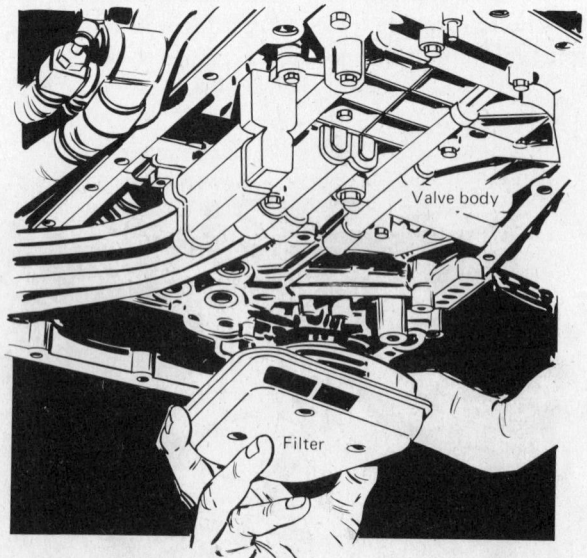

FIGURE 18-8 A typical filter location under the valve body.

2. To actually remove a filter, remove the screws, bolts, or spring clips that secure the filter to the valve body and discard the gasket when used. Be careful when removing the screen from 1970 and some later C-4 transmissions because a tab section of the filter's housing holds the throttle-pressure limit valve into the valve body (Fig. 18-9). When working on this type of transmission, therefore, carefully lower the screen so that you can reach over the tab and grab the valve and spring before they fall out.

3. Wash the wire screen thoroughly with clean solvent and a brush (Fig. 18-10). If solvent will not dissolve the varnish built up on the surface of the screen, soak the filter in a good grade of carburetor cleaner for 15 to 30 minutes and rinse it out thoroughly with water.

4. Blow the filter screen dry with low-pressure compressed air. Never dry a filter screen with any type of rag. *A piece of lint from the rag, trapped in the*

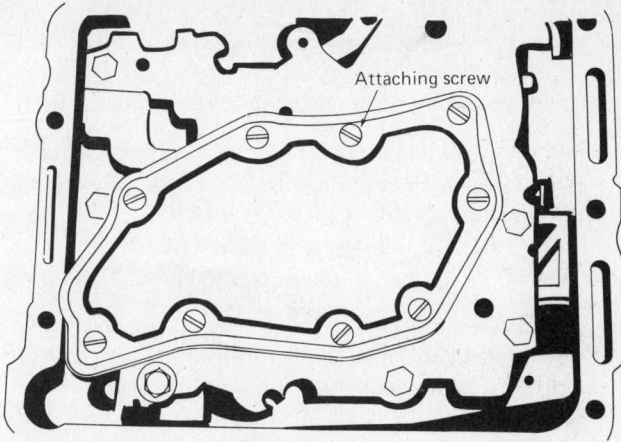

FIGURE 18-11   The attaching screw pattern of a C-4 filter screen.

screen, can work its way into the valve body and cause a valve to stick in its bore.

5. Inspect the screen after cleaning. If the cleaning job does not remove all the varnish, replace the assembly. Varnish buildup will reduce the fluid flow through the screen and cause problems within the transmission due to oil starvation. Lastly, if the screen is bent excessively or torn open, replace it.

Installation

1. Replacing a filter or reinstalling a screen is usually not a hard job because they fit onto the valve body only one way. There are several reasons for this. First, the filter or screen manufacturer forms the unit a certain shape, either square or rectangular, and its attaching screw holes are machined so that the filter must fasten to the valve body a given way (Fig. 18-11). Second, to make it easy for the mechanic to quickly locate the filter on the valve body, some manufacturers stamp the word "front" on the filter itself. This indicates to the mechanic the proper position of the filter relative to the valve body.

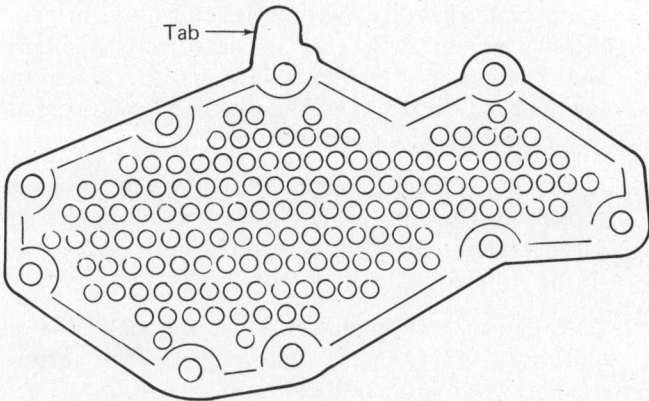

FIGURE 18-9   Tab section of a C-4 filter screen assembly.

2. When installing the filter or screen, always replace the sealing device—seal or gasket—between the filter and the valve body. If the screen or filter has a mounting tube and it has an O-ring seal, be sure to replace it also. Then, push the screen assembly straight up into position until the tube seats in its bore in the transmission case.

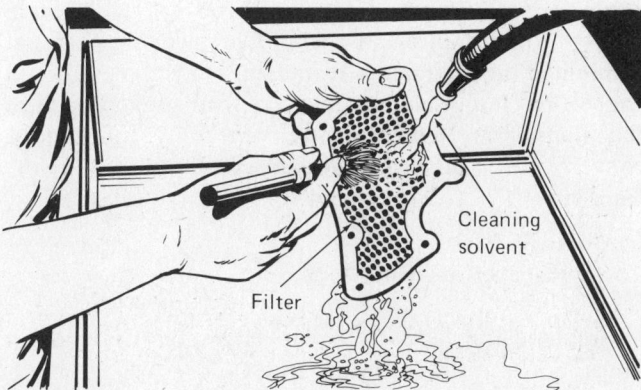

FIGURE 18-10   Washing a screen-type filter in cleaning solvent.

3. Install any screws, bolts, or attaching straps that secure the filter to the valve body. Tighten all attaching screws or bolts to factory specifications.

PAN INSTALLATION

To install the oil pan after servicing or replacing the filter:

1. Place a new gasket on the pan's mounting flange (Fig. 18-12). To hold the gasket in place, coat the flange with petroleum jelly or grease. CAUTION: *Do not use gasket cement or sealer unless there has been a prior problem of gasket slippage during the tightening process.*

2. Reinstall the pan onto the transmission case, making sure the gasket does not slip out of alignment or become pinched. Install all the pan's attaching bolts and washers if used. Tighten all bolts finger tight (Fig. 18-13).

3. With a suitable torque wrench, tighten the pan bolts to factory specifications, using a crisscross pattern. It is sometimes advisable to set the initial tightening torque

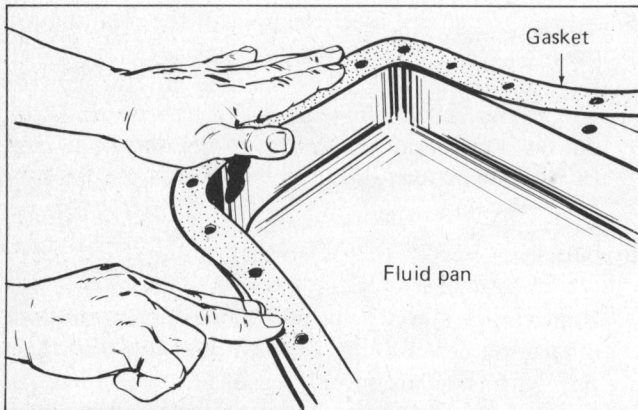

FIGURE 18-12   Installing the gasket on the pan.

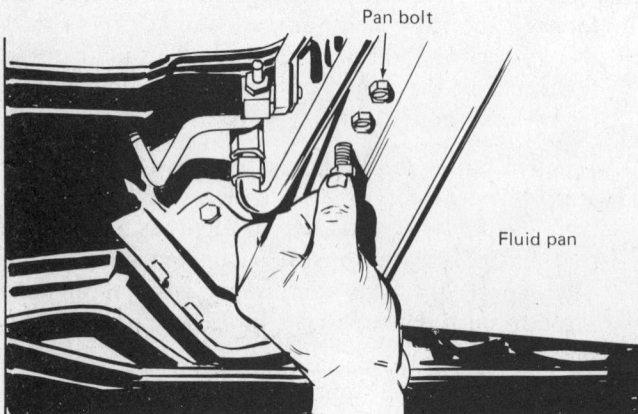

FIGURE 18-13   Installing the pan's attaching bolts finger tight.

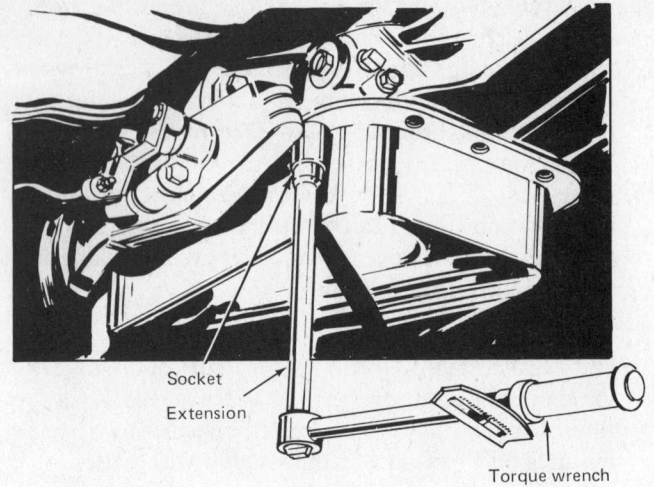

FIGURE 18-14   Torquing the pan's attaching bolts.

to one-half of normal factory specifications to compress the gasket slowly. Next, retorque all bolts to the full amount (Fig. 18-14). The crisscross pattern and this torquing sequence prevent the pinching, breaking, or distorting of the gasket out of shape as well as dimpling of the pan's bolt holes.

4. If the transmission has a filler tube that attaches directly to the pan, connect the tube and tighten its fitting finger tight. Using a torque wrench, tighten the fitting to factory specifications.

5. If the pan has a drain plug, install a new gasket over it and thread it into the pan finger tight. Then, torque the plug to factory specifications.

DRAINING THE CONVERTER

Removing the dirty fluid from a torque converter can be a problem because many converters do not have drain plugs. For instance, in 1978, Chrysler Corporation eliminated the drain plugs from their converters. General Motors has not used plugs in any of their converters since the mid-1960s. Consequently, the only automotive converters that the mechanic can actually "drain" are the Ford, pre-1978 Chrysler, and some early GM units.

Converters with Drain Plugs

To remove the fluid from converters that have drain plugs, perform the following steps:

1. Remove the converter inspection or access panel from the housing (Fig. 18-15).

2. Using a remote button or the starter switch, turn the

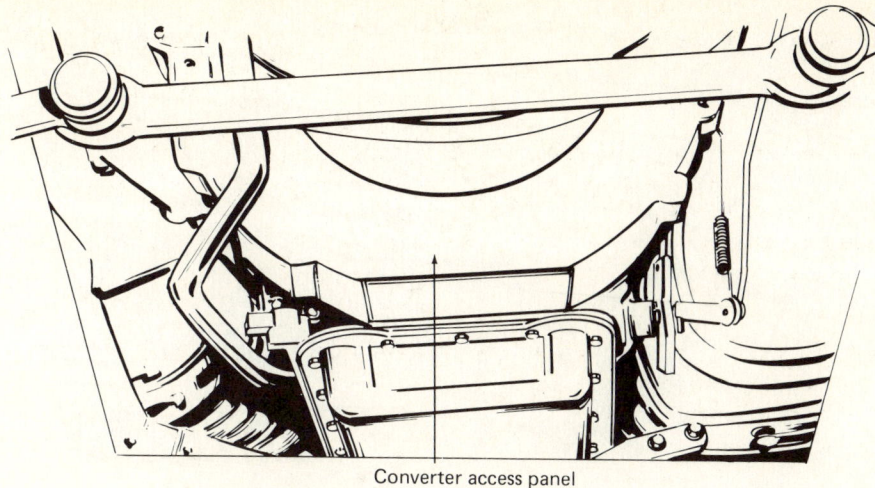

FIGURE 18-15   A typical converter access panel.

engine over until the converter drain plug is in its 6 o'clock position (Fig. 18-16). CAUTION: *Never pry on the starter ring gear or flexplate with a screwdriver or use a wrench on the converter mount bolt in order to turn the engine. This action could distort the ring gear or flexplate, or damage the converter bolts.*

3. Remove the plug and allow the dirty fluid to flow into a drain pan (Fig. 18-17).

4. After the old fluid has drained out of the converter, replace and torque the plug.

5. Install the access panel back onto the converter housing, torquing all attaching bolts to specifications.

Pumping the Fluid from the Converter

If the torque converter has no drain plug, a mechanic can pump the majority of the dirty fluid out of the unit by following this technique:

1. At the radiator, disconnect the inlet-pressure line fitting at the transmisson cooler and remove the tube (Fig. 18-18).

2. Slip a suitable snug-fitting drain hose over the end of the line and place the free end of the hose into a pan (Fig. 18-19). If in doubt as to which line is the inlet, disconnect both lines and attach drain hoses to both.

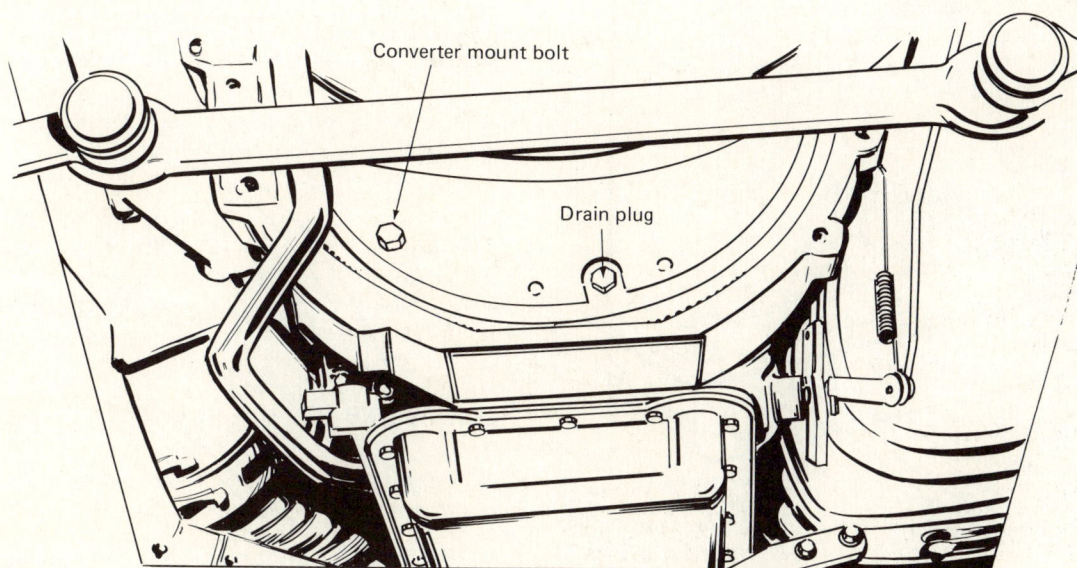

FIGURE 18-16   The converter drain plug, positioned for removal.

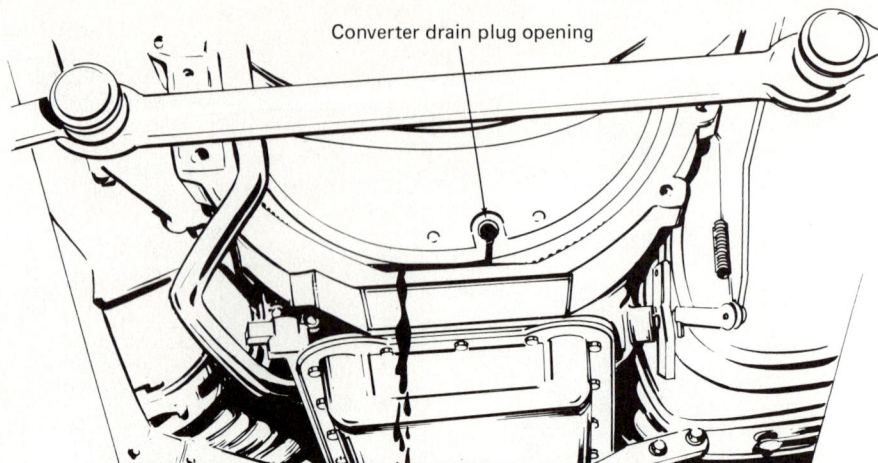

FIGURE 18-17 **Fluid drains easily from a converter with the plug removed.**

3. After cleaning the pan and servicing or replacing the filter, add four to five quarts of the correct type ATF to the transmission through the filler tube.

4. While observing the hoses in the drain pan, have an assistant start the engine. Dirty fluid should begin to flow out of the inlet hose.

5. Keep the engine running at idle rpm until the fluid changes color, indicating new fluid is now in the converter and cooler line, or until air bubbles appear (Fig. 18-20). If air bubbles appear before the fluid changes color, stop the engine and add two additional quarts of fluid to the transmission. Then, restart the engine and continue the process until all the dirty fluid is out of the converter. (NOTE: *Always stop the engine immediately when air bubbles appear, because this indicates the pan is empty of fluid and the pump is sucking in air.*)

6. When clean fluid begins to pour from the inlet line, shut the engine off and reconnect the cooler lines to the radiator. Torque the fittings to factory specifications.

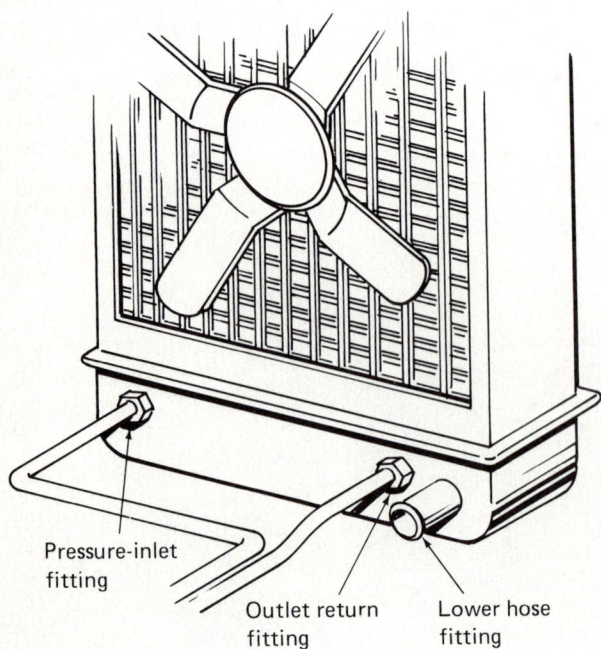

FIGURE 18-18 **Typical cooler inlet-pressure and return lines.**

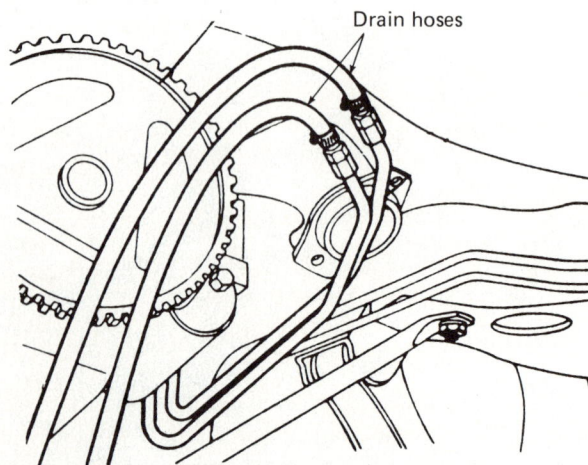

FIGURE 18-19 **Two drain hoses from the cooler lines leading down to a drain pan.**

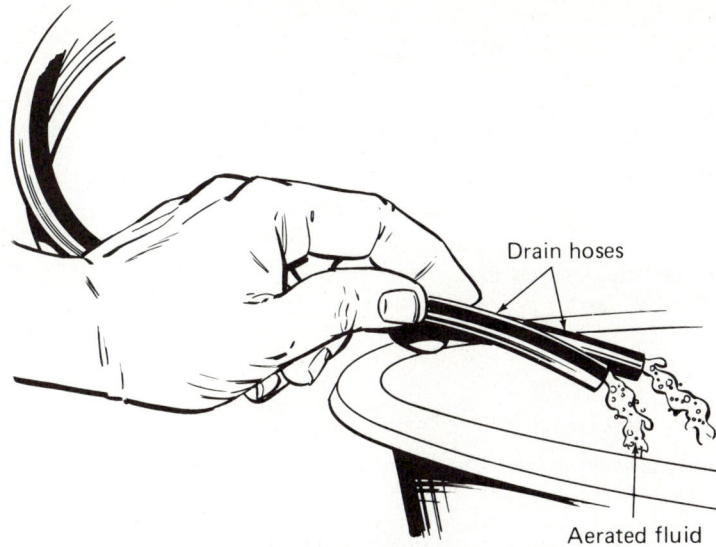

FIGURE 18-20   Air bubbles coming out of the drain hose indicate the pump is sucking air because the pan is empty.

FILLING THE TRANSMISSION WITH NEW FLUID

After draining the fluid from the transmission and converter, service the unit with new fluid as follows:

1. Look up the total fluid capacity of the transmission and converter in the shop manual. The actual amount of fluid to initially add and the quantity necessary to completely fill the unit will depend on two factors: (1) the total capacity of the converter and transmission, and (2) whether the mechanic drained or pumped the dirty fluid out of the converter. If the transmission and converter hold 10 quarts and you drained the pan and converter, add 4 quarts of ATF to the transmission before starting the engine (Fig. 18-21). If, on the other hand, you pumped the fluid out of the converter until it changed color, add 2 to 3 quarts to the transmission before restarting the engine. Finally, if you drained the pan by itself, add 1 quart **less** than pan capacity before starting the engine.

2. With the gearshift-selector lever in park, start the engine and allow it to operate at idle speed. *If you drained the converter via the plug, immediately add 4 quarts of fluid to the transmission.*

3. With the parking brake set and the brake pedal held down, operate the transmission in all its operating ranges, stopping in each range until the transmission fully engages.

4. Return the gearshift-selector lever to neutral or park

as specified by the manufacturer and check the fluid level.

5. Add more fluid as required to raise the level up, but never higher than the ADD mark on the dipstick (Fig. 18-22).

FIGURE 18-21   Servicing the transmission with fluid.

COOL
(65°-85°F.)
(18°-30°C.)

HOT
(190°-200°F.)
(88°-93°C.)

ADD .5 LITER
(1 PT.)

FULL HOT

WARM

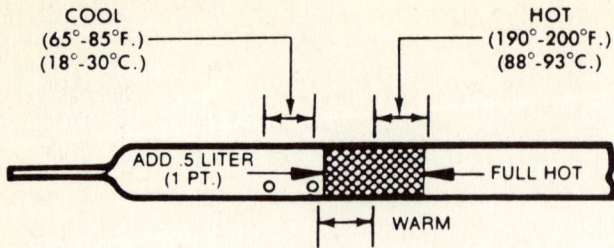

FIGURE 18-22   The ADD mark on a typical dipstick.

6. Permit the transmission and engine to operate until they reach normal operating temperature and then recheck the fluid level. Now, bring the fluid level up to the FULL mark on the dipstick by adding fluid carefully. CAUTION: *Never overfill the transmission.*

7. Inspect the cooler fittings, drain plug, pan gasket, and filler tube installations for external leakage. Perform any repair necessary to stop any observed leak.

REVIEW

This section will assist you in determining how well you remember the material in this chapter. Read each item carefully. If you can't complete the statement, review the section in the chapter that covers the material.

1. If the transmission fluid becomes _____, it should be changed immediately.

2. Fluid change intervals for a given automatic transmission can be found in the _____ _____ or the _____ _____.

3. A total fluid change of an automatic transmission includes draining the _____, _____, and servicing the filter.

4. Some fluid pans are drained by removing the _____ _____ attached to its side.

5. To remove a pan that has no drain plug, remove the _____ attaching bolts first.

6. A large amount of metal or friction particles found in the pan is a good indication that the transmission has excessive _____.

7. Excessive torque can cause the pan's bolt holes to _____.

8. The mechanic should never use rags to dry a pan or filter because _____ may get into the hydraulic system.

9. The mechanic should never attempt to clean a _____ type filter.

10. To hold the gasket in place on the pan, coat the mounting flange with _____ _____ or _____.

11. The mechanic should tighten the pan's attaching bolts with a _____ _____ to _____ _____.

12. To remove the converter's drain plug, rotate the engine over until the plug is in its _____ _____ position.

13. If the converter has no drain plug, the mechanic can _____ the majority of the fluid out of it.

14. If air bubbles appear in the fluid flowing from the converter drain hose, _____ the engine immediately.

15. The mechanic can look up the fluid capacity of a given automatic transmission in the _____ _____ for the unit.

For the answers, turn to the Appendix.

Along with fluid and filter changes, other important types of service work that a mechanic can perform while a transmission is still in a vehicle are band and linkage adjustments. The adjustment of these components must be correct (to factory specifications), or the transmission can not operate properly. In other words, many transmission malfunctions can result from incorrect band or linkage adjustments. However, these adjustments obviously will not correct all transmission problems as many people believe; the material contained in Chapter 17 of this text pointed this out rather clearly.

BAND ADJUSTMENTS

Effects of Improper Band Adjustments

In order for a band to perform its task of holding a planetary member stationary, it must be adjusted properly. This establishes the correct amount of open or operating clearance between the band's lining and the drum surface. If, for example, a band is too loose or there is **excessive clearance** between this lining and the drum it must stop, the servo piston will not be able to firmly lock the band around the drum before the piston bottoms in its bore. As a result, the band permits the drum and its attached planetary member to slip, which causes the engine to overspeed. In other words, the transmission will slip in any driving range where this excessively loose band applies.

If, on the other hand, the band is too tight or there is **insufficient clearance** between the lining and the drum, the band may drag or not release its hold on the drum at all. This will cause a total band failure or the planetary gear train to seize up, depending on how tight the adjustment is. If, for instance, the band adjustment is such that the lining constantly drags on the drum, the band will prematurely burn out from excessive friction caused by a lack or reduction in fluid lubrication between the two components.

If the adjustment is so tight that the lining will not permit the drum to move at all, the planetary gear train will, most likely, seize as it attempts to change from one ratio to another. This seizure results from the inability of the band-controlled planetary member to freewheel when it should. In other words, the planetary would, technically speaking, be trying to operate in two different gear ratios at the same time; this is what causes the unit to seize, or bind up.

As previously stated, many people still think that a band adjustment will correct most transmission problems; this is not really the case. While a band adjustment does play an important part in the overall operation of the transmission, it may not always be the cause of a

CHAPTER 19

Band and Linkage Adjustments

slipping condition. For example, low system pressure on the servo piston that applies the band or a defective friction or overrunning clutch can also cause an automatic transmission to slip. Consequently, the adjustment of the band, in an attempt to correct a slipping condition caused by any of the other conditions mentioned, would be a waste of time and effort.

Frequency of Band Adjustments

Each transmission manufacturer recommends a given interval period between band adjustments. This interval may be in vehicle mileage such as every 20,000 miles, or it may be at every fluid change, or only during transmission overhaul. In any case, the manufacturer makes these recommendations for a reason, and the vehicle owner and mechanic should adhere to them.

The main purpose behind adjusting a band at a given mileage interval or fluid change is to compensate for lining wear. The manufacturer does expect a given amount of lining wear over a certain period of time. Consequently, the owner has to have someone perform the adjustment to compensate for this wear, or the band will begin to slip after a while.

Manufacturers recommend band adjustments during unit overhaul for several reasons. First, the band adjustment procedure, if followed correctly, centers the band properly around the drum it will hold. Second, the adjust-

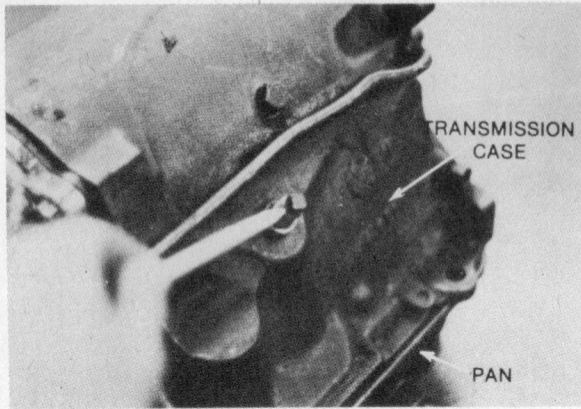

FIGURE 19-1 Typical external-type band adjuster.

ment procedure inserts the specified lubrication clearance between the lining and the drum, which prevents premature band failure due to friction.

Types of Band Adjustments

Basically speaking, there are two ways of making band adjustments—externally or internally. If the transmission design is such that the mechanic can adjust a band externally, the unit will have a threaded adjuster that screws into the side of the case (Fig. 19-1). Since the head of the adjuster is on the outside of the transmission case, it provides the means by which the technician can perform an external adjustment.

However, if the transmission design is such that the adjuster is not accessible from the outside of the case, the mechanic must first remove the oil pan in order to make an internal adjustment. Usually when a transmission has internal adjusters (Fig. 19-2), the manufacturer recommends a band adjustment only during a fluid

FIGURE 19-2 An internal-type band adjuster.

change. The reason for this, of course, is to reduce the number of times the oil pan has to be pulled off for maintenance.

In some transmissions, the manufacturer does not provide a means to adjust the bands by the use of a threaded adjustment screw. For instance, the General Motors T-400 transmission has two bands that are not adjustable through this method. Instead, the band-to-drum clearance is a factor determined by the diameter of the drum, the thickness of the band and lining, and the stroke of the servo piston. With this arrangement, when the band-to-drum clearance becomes excessive enough through the effects of wear to cause slippage, the mechanic has no alternative but to replace the band.

If and when a mechanic overhauls a transmission with nonadjustable bands, he must check the band-to-drum clearance with a special gauge (Fig. 19-3). If the band is in

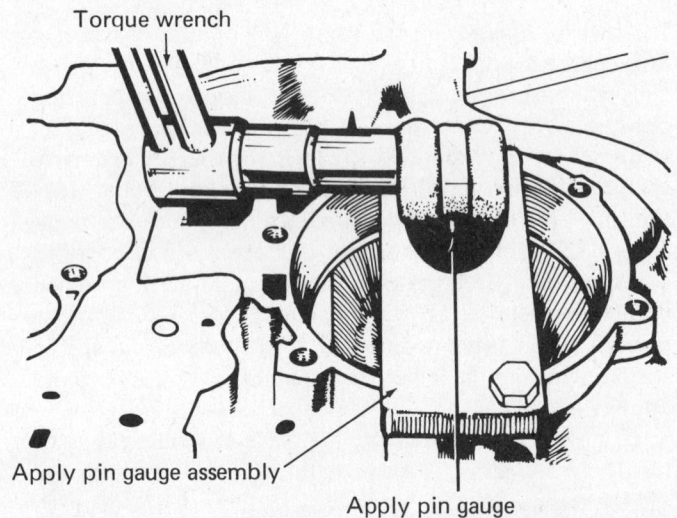

FIGURE 19-3 Checking the rear-servo apply-pin length of a T-400 transmission using a special gauge assembly.

good condition but the clearance is not to specifications, the technician must change the selective servo apply pin. In other words, this transmission manufacturer provides (selective) servo apply pins of various lengths in order to adjust band clearance during the overhaul process.

Common Band Adjustment Procedures

Although band adjustment procedures vary from one transmission type to another, there are certain service operations common to them all: torquing the adjuster and then either tightening or loosening the screw a specified number of turns. The first important service operation is the torquing of the adjuster itself to a given specification. This action loads (tightens) the band around the drum,

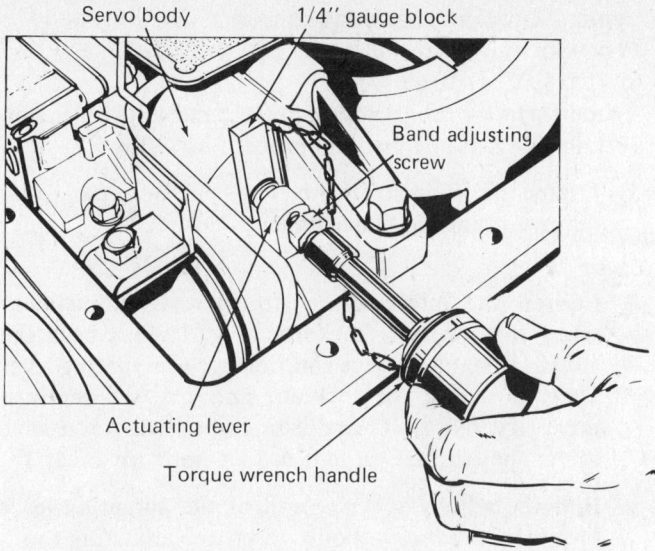

Servo body 1/4" gauge block Band adjusting screw Actuating lever Torque wrench handle

FIGURE 19-4 The special gauge block and wrench used to adjust the front (intermediate) band on an FMX transmission.

and if the mechanic does not use an accurate torque wrench or follow the tightening specifications, the second operation or part of the procedure will not be correct either.

The second, and equally important operation, is the actual establishment of the correct band-to-drum clearance. The mechanic accomplishes this by turning the

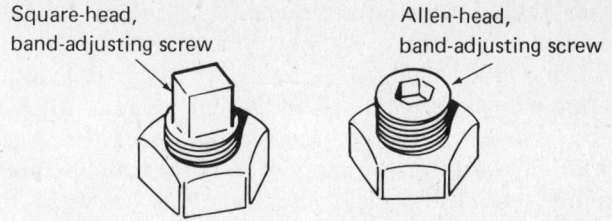

Square-head, band-adjusting screw Allen-head, band-adjusting screw

FIGURE 19-5 Head designs of typical band adjustment screws.

adjuster either in or out a specified number of turns. The proper clearance, in other words, will exist between the band and drum if the technician first torques the adjuster and then turns it in or out the number of turns specified by the manufacturer.

Equipment and Tools Necessary to Adjust Bands

As previously mentioned, some transmissions require the use of a special piece of equipment to check and/or set band-to-drum clearance (Fig. 19-3). Figure 19-4 shows another piece of equipment—a special gauge block and torque wrench—used to adjust the front (intermediate) band on an FMX transmission. The 1/4-inch gauge block fits between the intermediate, servo-piston stem, and the adjuster screw. A torque wrench with a given setting is built into the handle of the tool; this handle breaks or slips

Torqueflite transmission
band adjusting chart

Kickdown band

Transmission model	Engine size	Adjuster torque	Number of turns
A-904	225 and 318	72 inch pounds	2
A-904 LA	360 (1)	72 inch pounds	2
A-904 LA	225, 318, 360 (2)	72 inch pounds	2-1/2
A-727	360, 400, 440	72 inch pounds	2-1/2

Low and reverse band

A-904	225	41 inch pounds	7
A-904	318	72 inch pounds	4
A-904 LA	360 (1)	72 inch pounds	4
A-904 LA	225, 318, 360 (2)	72 inch pounds	2
A-727	360, 400, 440	72 inch pounds	2

(1) Transmissions that have five clutch discs in the reverse-high clutch pack.

(2) Transmissions that have three clutch discs in the reverse-high clutch pack.

FIGURE 19-6 A typical chart listing band adjustment specifications.

when the adjuster screw reaches a torque of 10 inch-pounds.

Some additional tools necessary to adjust other common types of bands include such items as an Allen or 8-point socket and an accurate torque wrench. The Allen or 8-point socket is necessary because the manufacturers use these head designs for the adjustment screws to differentiate them from other transmission fasteners (Fig. 19-5). The torque wrench, remember, is necessary to tighten the adjustment screw a given amount, either in inch- or foot-pounds.

Band Adjustment Specifications

Before a mechanic performs a band adjustment on any transmission, it is important that he look up the specifications as to the initial torque on the adjustment screw and the number of turns it must be rotated out or in (Fig. 19-6). However, when looking up these specifications in a manual, the mechanic may find that the manufacturer may require some specific information from the transmission itself. For example, it may be necessary for the technician to check the data tab or transmission case for the unit's serial number.

The transmission serial number now, in most cases, not only identifies the vehicle the unit fits into but also the engine type and displacement. The knowledge of the engine type and displacement is necessary because manufacturers install the same transmission model with several engine styles. When the manufacturer does this, the internal components of the transmission are usually somewhat different and so are the adjustments. Therefore, a mechanic should always check the serial number on the transmission before looking up the specifications in the manual for band adjustments and special information required for other service procedures including a complete overhaul.

Typical Internal Band Adjustment Procedures

To perform a routine intermediate band adjustment on certain FMX transmissions, proceed as follows:

1. Drain the fluid from the transmission pan, and remove it and the fluid screen as outlined in Chapter 18.

2. Loosen the intermediate (front) servo adjustment screw locknut two full turns (Fig. 19-7). Check the adjusting screw for free rotation in the actuating lever after loosening the locknut, and free the screw if necessary. (NOTE: The adjusting screw will not receive the proper torque value if it does not turn freely.)

3. Pull the adjusting screw end of the actuating lever away from the servo body. Insert the adjusting tool's ¼-inch gauge block between the servo-piston stem and the adjusting screw (Fig. 19-8).

4. Tighten the adjusting screw with the tool wrench handle until it overruns the screw. If this special wrench is not available, torque the screw to 10 inch-pounds with a common torque wrench.

5. Remove the gauge block and either tighten or loosen the adjustment screw the specified number of turns.

6. Hold the adjusting screw stationary and torque the locknut to 20 to 25 foot-pounds.

7. Reinstall the filter and oil pan as outlined in Chapter 18.

8. Refill the transmission to the specified level with the correct type of ATF.

The adjustment of the low-reverse band of a Torque-Flite transmission does not require the use of a special gauge. To adjust this band, follow these instructions:

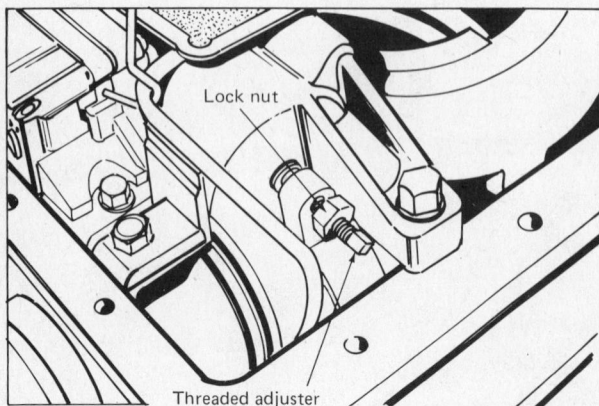

FIGURE 19-7 An FMX intermediate band adjustment screw.

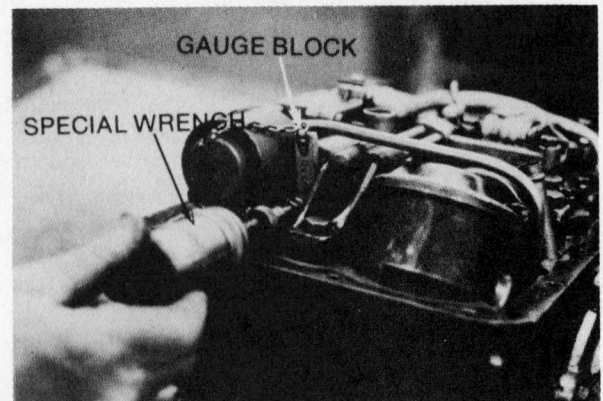

FIGURE 19-8 Adjusting the FMX intermediate band using the special gauge block and wrench.

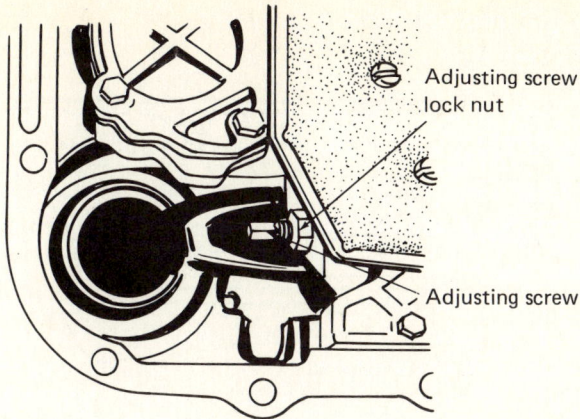

FIGURE 19-9   *The TorqueFlite low-reverse band adjuster.*

1. Raise the vehicle on a hoist, drain the transmission fluid, and remove the pan as outlined in Chapter 18.

2. Loosen the adjusting screw locknut and back it off about five turns (Fig. 19-9). Then, check the adjusting screw itself for free movement within the lever.

3. Torque the adjusting screw to specifications (Fig. 19-10).

4. Back out the adjusting screw the specified number of

turns. Next, while holding the adjusting screw, tighten the locknut to specifications.

5. Reinstall the oil pan and refill the transmission with the correct type of fluid as outlined in Chapter 18.

Typical External Band Adjustment Procedures

To adjust the bands on a typical C-4 transmission, proceed as follows:

1. Clean all the dirt away from the intermediate band adjusting screw area (Fig. 19-11).

2. While holding the adjusting screw, loosen, remove, and discard the locknut. This type of locknut has a seal bonded to its inside surface so it will no longer prevent leakage if reused.

3. Install a new locknut loosely onto the adjusting screw.

4. Torque the adjusting screw to specifications (Fig. 19-12).

5. Back off the adjusting screw the number of turns specified by the manufacturer.

6. While holding the adjusting screw, torque the locknut to 35 to 45 foot-pounds.

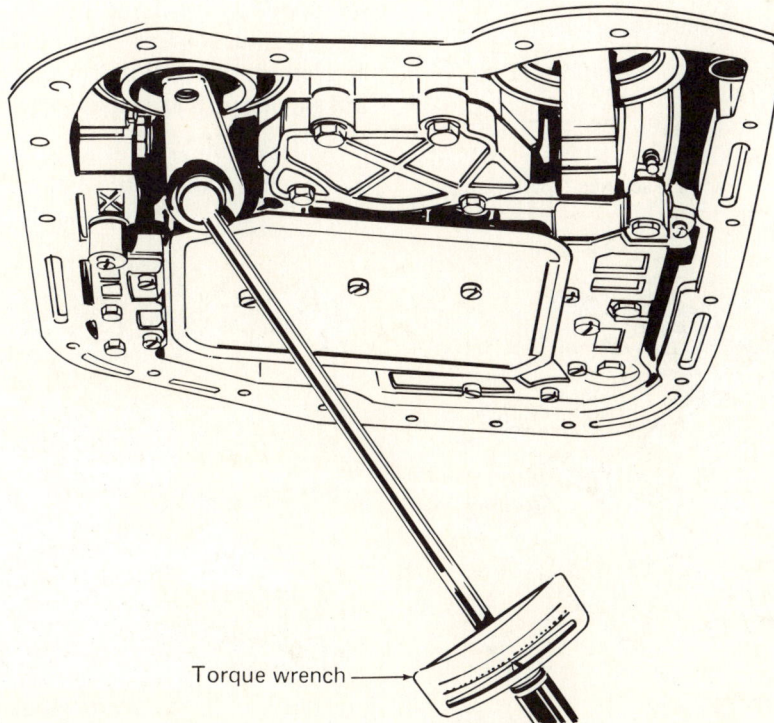

FIGURE 19-10   *Torquing the low-reverse band adjuster.*

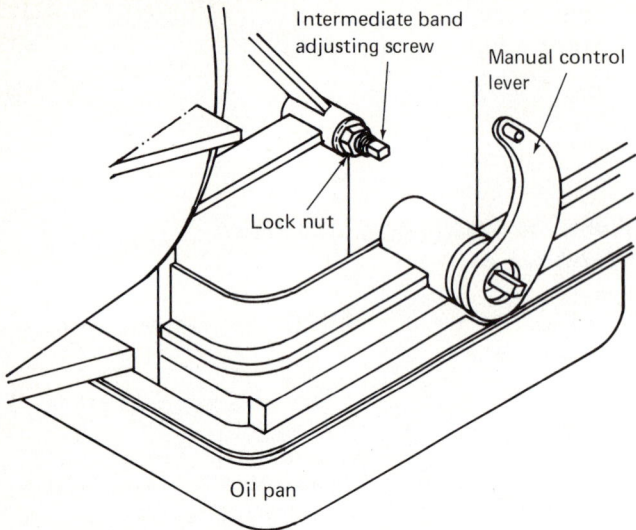

FIGURE 19-11   The C-4 intermediate band adjuster.

To adjust the low-reverse band of a typical C-4 transmission:

1. Clean away any accumulated dirt from around the band adjusting screw area (Fig. 19-13).

2. While holding the adjusting screw, loosen, remove, and discard the locknut. This locknut has a seal bonded to its inside surface so it will no longer prevent leakage if reused.

3. Install a new locknut loosely onto the adjusting screw.

4. Torque the adjusting screw to specifications (Fig. 19-14).

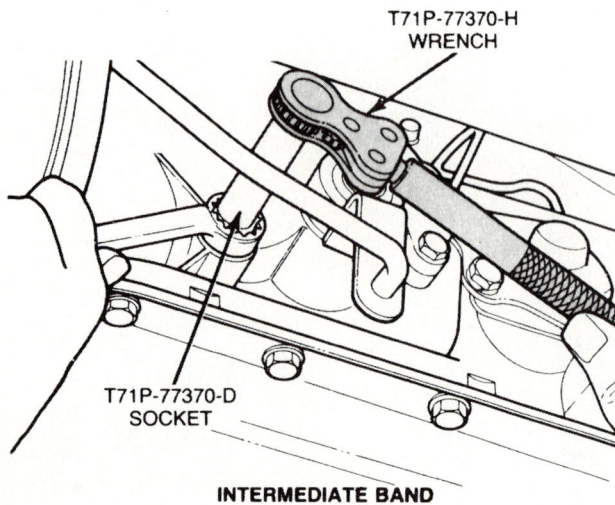

FIGURE 19-12   Torquing the C-4 intermediate band adjuster. (Courtesy of Ford Motor Company)

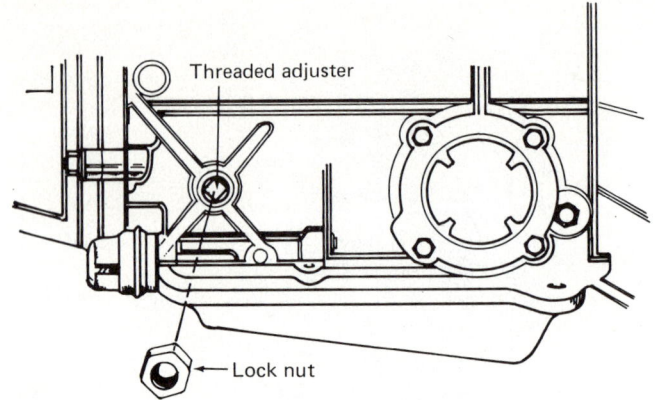

FIGURE 19-13   C-4 low-reverse band adjusting screw.

5. Back out the adjusting screw the number of turns specified in the service manual.

6. While holding the adjusting screw, tighten the locknut to 35 to 45 foot-pounds.

LINKAGE ADJUSTMENTS

As mentioned in an earlier chapter, correct linkage adjustments play a very important role in transmission operation. The reason for this is fairly simple when you consider what the linkage really does. The linkage itself provides control input into the automatic transmission from various sources, the gearshift-selector lever, and the accelerator pedal. With the gearshift-selector lever and its linkage, the driver manually selects a given driving range

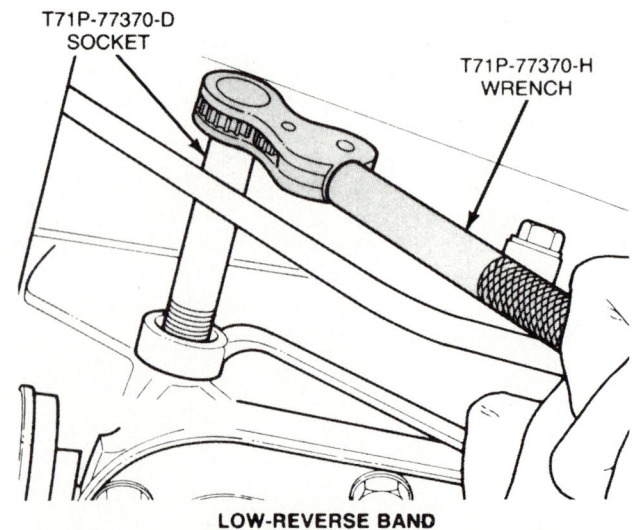

FIGURE 19-14   Torquing the C-4 low-reverse band adjusting screw. (Courtesy of Ford Motor Company)

within the transmission. The accelerator pedal, with its attached linkage to the transmission, mechanically signals the unit as to the load on the engine at any given time or when the transmission should downshift to a lower gear ratio. With these facts in mind, it should be easy to figure out what could happen if one or more of these linkages are out of adjustment.

The four types of linkages that this chapter will discuss are the (1) gearshift (manual valve) linkage, (2) the accelerator pedal, (3) throttle valve, and (4) downshift linkages. Not every vehicle has all four types of linkages; their use, therefore, depends on the type of transmission and the motor vehicle. Consequently, when in doubt as to the types of linkages used in a given vehicle, check the appropriate service manual for the type, location, and the adjustment.

Gearshift (Manual Valve) Linkage

The gearshift linkage moves the manual valve within its bore in the valve body. The manual valve, in turn, sets up the actual operating conditions within the transmission by directing fluid to apply the clutches and/or bands. If this linkage is out of adjustment, the transmission can malfunction because the manual valve will not be in its correct position, thus permitting clutch-apply or band-apply fluid to leak past the valve and back to the sump.

Certain conditions can cause gearshift linkage to become out of adjustment. For instance, if the mechanic removes the transmission or the gearshift linkage to perform maintenace, he must check and readjust the linkage as required. In addition, linkage adjustment may be necessary because of wear caused by constant use, or because of loose linkage fasteners.

In practice, two components control the actual position of the manual valve within its bore in the valve body: a detent pawl and a shift gate. The detent pawl is part of the manual valve lever arrangement mounted on the valve body (Fig. 19-15). An arm on this lever fits into a groove on one end of the manual valve itself. With this design, whenever the lever moves due to the action of the gearshift-selector lever, the manual valve moves in its bore.

Another section of this very same lever forms the detent pawl that has notches for each gearshift-selector position. A spring-loaded detent ball that fits into a special bore in the valve body engages into these notches, one at a time, as the manual valve moves from position to position. With this arrangement then, the detent ball and pawl hold the manual valve into one selected position at a time.

The shift gate is part of the gearshift-selector lever mechanism, built either into the column or console. Figure 19-16 shows a typical column-shift gate with pawl. The shift gate itself has a series of notches or stops that

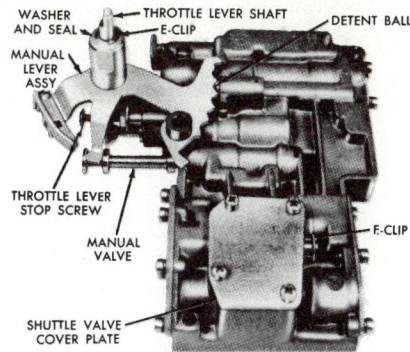

FIGURE 19-15   A detent pawl and ball mounted on a valve body. (Courtesy of Chrysler Corp.)

correspond to the notches in the pawl section of the manual valve lever. The gate pawl moves with the gearshift-selector lever, and it engages into the various stops. When the gearshift-selector pawl and gate-stop position synchronize with the pawl and notch position on the manual valve lever, the shift linkage correctly positions the manual valve.

The following procedure for checking linkage adjustment is typical of those commonly used in the automotive industry. However, because the actual process varies somewhat from one vehicle to another, always refer to the appropriate service manual before checking or altering linkage adjustment.

To check the linkage adjustment:

1. Position the gearshift-selector lever into N (neutral).

2. Lift upward on the gearshift-selector lever and move it toward the D (drive) position until you feel the manual-valve pawl detent drop into place in the drive position (Fig. 19-16).

3. Lower the gearshift-selector lever until the pawl stops against the shift gate.

4. Without raising the gearshift-selector lever, attempt to move the lever into the 2 position. You should feel little or no movement in the lever. If excessive movement exists, the linkage requires adjustment.

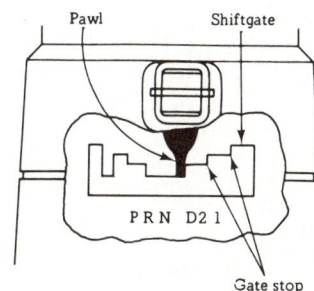

FIGURE 19-16   A common type of shift gate and pawl.

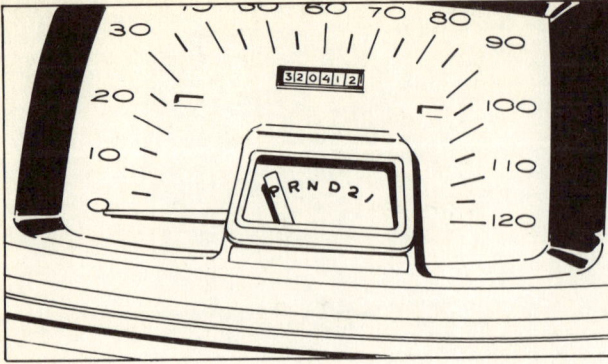

FIGURE 19-17   *The gearshift-selector lever positioned in park (P).*

To adjust the linkage:

1. Place the gearshift-selector lever into P (park) position (Fig. 19-17).

2. Raise the vehicle on a hoist or with a jack to a suitable working height. If a jack is used, *lower the vehicle down on safety stands before working under it.*

3. Locate and loosen the control-rod swivel-clamp screw a few turns (Fig. 19-18).

4. Move the transmission control lever all the way to the rear, the P (park) position.

5. With the transmission control lever in P detent position and the gearshift-selector lever in park position, tighten the swivel-clamp screw securely.

6. Recheck all the gearshift-selector lever detent positions, and then lower the vehicle and road test.

Accelerator and Throttle Valve Linkage

In order for the automatic transmission to respond to the varying loads placed on the engine, the linkages from the accelerator pedal to the carburetor and from the carburetor to the transmission have to be adjusted correctly. This is especially true for vehicles that use such transmissions as the Powerglide, T-200, and TorqueFlite because these units utilize mechanical linkage or a cable from the carburetor to operate both the throttle and kickdown valves. This is also true to a lesser degree for transmissions that use linkage to control only the kickdown valve.

Several transmission malfunctions are evident if the combination throttle and kickdown linkage or cable is out of adjustment. The transmission may upshift too soon or too late in relation to the engine load and vehicle speed. Also, a forced downshift or kickdown may not occur when the driver depresses the accelerator enough to move the linkage through the detent position, or the downshift

may even occur before reaching the detent position. This premature or no kickdown condition can also occur in a transmission that has only a kickdown valve linkage if it is out of adjustment.

Although the actual procedures for checking and adjusting accelerator, throttle, and kickdown linkage may differ between vehicle manufacturers, in practice, there are three important rules to **always** follow: **One,** refer to the linkage adjustment section of the service manual for the make and model of the vehicle you are repairing before checking or altering any adjustments. **Two,** make sure that all linkages are intact and serviceable. **Three,** make certain that when the accelerator pedal is all the way to the floor, the carburetor throttle plate is wide open with the throttle lever against the wide-open stop. And equally important is that, as you release the accelerator pedal, the linkage returns freely and immediately to the idle position.

To check and adjust a typical throttle and carburetor linkage, follow these instructions:

1. Disconnect the automatic choke at the carburetor and block the choke valve in the full-open position with a screwdriver (Fig. 19-19). Open the throttle slightly to release the fast idle cam. Return the carburetor linkage to the curb idle position.

2. Remove the spring, cotter or retaining pin, washer, and slotted throttle-rod adjuster from the bellcrank lever pin.

3. By means of the transmission throttle rod, hold the transmission lever forward against its stop (rod or lever must not move vertically while you move the lever against the stop). Adjust the length of the transmission rod, with the threaded adjuster at its upper end. After adjustment, the rear end of the slot should just contact the bellcrank lever pin without exerting any additional forward force on the throttle rod.

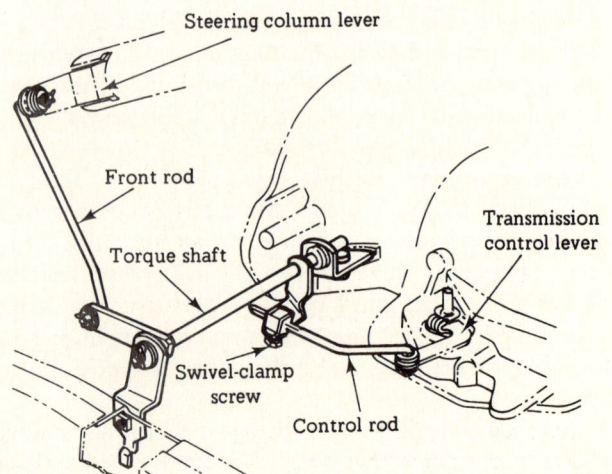

FIGURE 19-18   **Adjustment of the manual valve linkage.**

Choke valve

Accelerator cable

Cable-clamp nut

Slotted throttle-rod adjuster

Housing ferrule

Throttle-valve lever

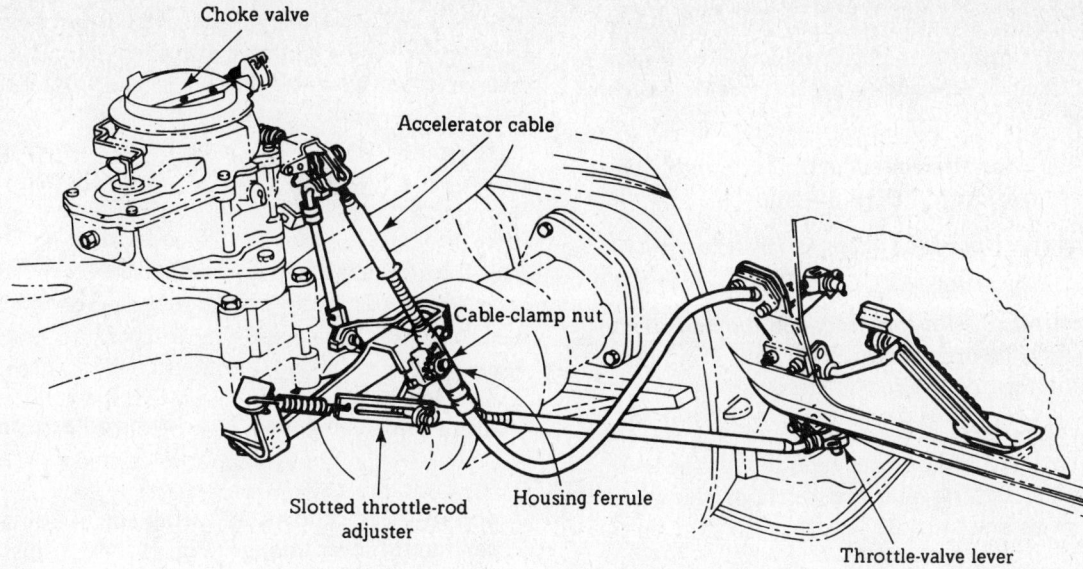

FIGURE 19-19   Typical accelerator and throttle valve linkage adjustments.

4. Lengthen the throttle rod by turning the adjuster one full turn.

5. Assemble the slotted adjuster to the bellcrank lever pin and reinstall the washer and retainer pin. Position the transmission linkage return spring in place.

6. Check the transmission throttle rod linkage for freedom of operation by moving the slotted adjuster to its full rearward position. Allow it to return slowly, making sure the adjuster returns to its full-forward position.

To adjust the accelerator cable shown in Fig. 19-19:

1. Loosen the cable clamp nut. Then, adjust the position of the cable housing ferrule in the clamp so that you remove all the slack from the cable with the carburetor at curb idle. To remove the slack from the cable, move the ferrule, in the clamp, in a direction **away** from the carburetor lever.

2. Back off the ferrule ¼ inch. This action provides ¼-inch cable slack at curb idle. Finally, tighten the cable clamp nut securely.

3. Route the cable so that it does not interfere with the transmission throttle rod throughout its full range of travel.

4. Reconnect the automatic choke rod, and remove the screwdriver that blocks open the choke valve or plate.

5. Road test the vehicle and check shift points and kickdown valve operation.

Throttle Valve Cable

If a transmission has a throttle valve cable, it serves the same function as the throttle rod just described. In other words, the cable connects the carburetor to the valve body; and the cable transfers any carburetor throttle plate movement directly to the throttle lever and bracket assembly, attached to the valve body. This assembly then operates the throttle valve within its valve body bore.

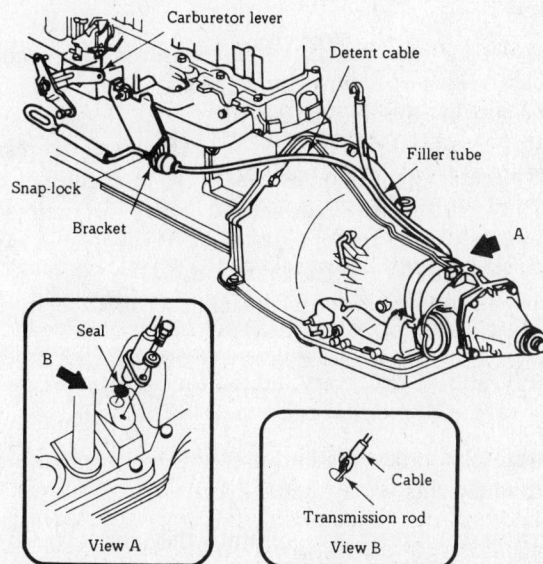

Carburetor lever

Detent cable

Filler tube

Snap-lock

Bracket

A

Seal

B

View A

Cable

Transmission rod

View B

FIGURE 19-20   Adjusting a common-type throttle valve cable.

A broken, sticking, or misadjusted cable will also interfere with normal transmission performance. To see if the cable and its attached components are serviceable, follow this procedure:

1. Check the cable for attachment at the carburetor lever and the transmission rod (Fig. 19-20).

2. Remove any sharp bends as necessary by rerouting the cable.

3. Inspect the transmission rod between the end of the cable and the throttle valve lever for alignment. Straighten or replace the rod as necessary.

4. If necessary, remove the oil pan and inspect the throttle valve lever and bracket assembly. If the lever, bracket, and retaining pin are worn or damaged, replace them as necessary.

To adjust the throttle valve cable, proceed as follows:

1. Unlock or disengage the snap lock on the cable (Fig. 19-20).

2. Turn the cable lever on the carburetor to its wide-open position.

3. While holding this lever in its wide-open position, push the snap lock downward until its top is flush with the rest of the cable. Engage the snap lock.

4. Slowly return the carburetor lever to its closed position.

5. Road test the vehicle to check the upshift points and through-detent downshifts.

Kickdown Linkage

As previously mentioned, some transmissions require only a kickdown or downshift rod. This rod that operates by accelerator linkage causes a forced transmission downshift at wide-open throttle or as the driver depresses the accelerator pedal to the floor. To accomplish the forced downshift, the kickdown rod (Fig. 19-21) activates the downshift valve in the valve body; this valve, in turn, actually downshifts the transmission.

To check and adjust a typical kickdown linkage, follow these steps:

1. Check and, if necessary, adjust the accelerator pedal and carburetor linkages.

2. Rotate and block the carburetor throttle valve open and against its wide-open stop.

3. Depress the kickdown rod until the kickdown valve bottoms against its stop within the valve body.

4. While holding the kickdown rod in this position, check the clearance between the throttle lever and the

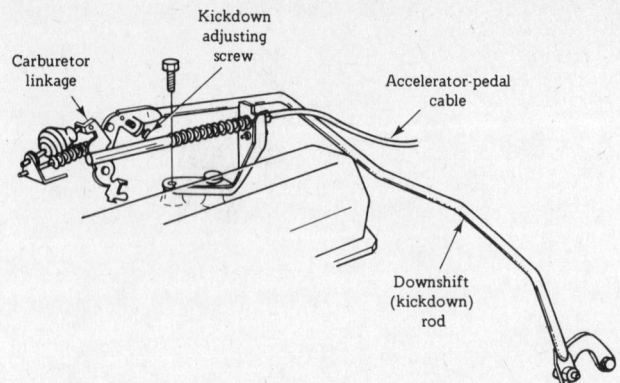

FIGURE 19-21   A typical kickdown rod adjustment.

end of the kickdown adjusting screw, located on the carburetor lever linkage (Fig. 19-21). Adjust the screw as required to obtain the specified clearances.

5. Release the kickdown rod and allow it to return to its normal position. Unblock and allow the throttle valve to return to the idle position.

6. Road test the vehicle. A detent downshift should now occur at the appropriate road speed.

ELECTRICAL KICKDOWN SWITCHES

Other transmissions, like the T-300 and 400 units, have an electrical kickdown (detent) switch (Fig. 19-22). Wide-open travel of the accelerator pedal operates the switch, and it, in turn, directs an electrical signal to a downshift solenoid inside the transmission. This solenoid then causes a hydraulic detent and kickdown valve to operate, which downshifts the transmission.

Typical Switch Adjustment

When kickdown shift points are not within specifications, the switch position should be checked and adjusted as follows:

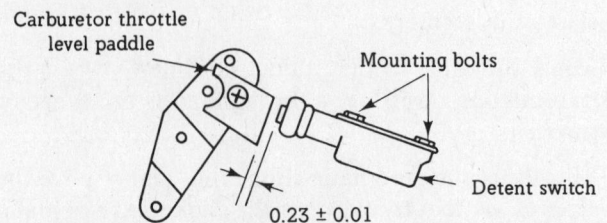

FIGURE 19-22   Adjusting a common-type electrical kickdown (detent) switch.

1. Check and adjust accelerator pedal and carburetor linkages as necessary.

2. Loosen the switch mounting bolts (Fig. 19-22).

3. With the automatic choke and accelerator linkage blocked in the wide-open position, depress the detent switch plunger until it bottoms in the switch.

4. While holding the plunger in, move the switch toward the throttle lever paddle until you obtain a clearance of 0.23 inch between the paddle and plunger.

5. Tighten the mount bolts, and unblock the choke and accelerator linkages.

6. Road test the vehicle to check the detent downshift. If the downshift still does not occur properly, the solenoid, detent, or downshift valves may be at fault and should be checked one at a time.

REVIEW

This section will assist you in determining how well you remember the material contained in this chapter. Read each item carefully. If you can't complete the statement, review the portion of the chapter that covers the material.

1. The task of a band is to hold a _____ _____ stationary.

2. A band will burn out if its adjustment is too _____.

3. The main reason why a band is adjusted at a given mileage interval is to compensate for _____ wear.

4. If the head of the band adjuster is on the outside of the transmission case, the mechanic can perform an _____ adjustment.

5. In transmissions with nonadjustable bands, the manufacturer sometimes supplies _____ servo apply pins of various lengths to adjust band clearance during an overhaul.

6. When a mechanic torques a band adjuster, he _____ the band around the drum.

7. To tighten most band adjusters, the mechanic must use either a _____ or _____ socket.

8. Before adjusting any band, look up the _____ in the service manual.

9. In order to make an internal band adjustment, you must first drain the _____ and _____ the oil _____.

10. The linkages provide _____ _____ into the transmission.

11. The gearshift linkage moves the _____ _____ within the valve body.

12. The _____ _____ and the _____ _____ control the position of the manual valve within the valve body.

13. On some transmissions, linkage from the carburetor to the transmission controls the operation of both the _____ and the _____ valves.

14. Before making any linkage adjustments, look up the procedures in the _____ _____ for the make and model of vehicle you are repairing.

15. Instead of a throttle rod, some transmissions use a _____.

For the answers, turn to the Appendix.

CHAPTER 20

Transmission and Seal Removal and Installation

It may be necessary for a mechanic to remove the automatic transmission for a number of different reasons. For example, the transmission has to be removed in order to replace the pump seal or bushing if a leak develops around the converter hub. Also, the technician may even have to pull the transmission to perform repair work on the engine. Finally, the most common reason to remove the transmission is, of course, to rebuild the unit.

There is no one single, detailed procedure that a mechanic can use to remove and reinstall all automatic transmissions. Each procedure is slightly different because of the many model years, body styles, and engine transmission configurations now in use. All these things determine the location of exhaust system components, crossmembers, underbody parts, and linkages that affect the manner or procedure by which the mechanic will remove and reinstall the transmission. Consequently, the technician will find it necessary to follow the specific instructions found in the vehicle service manual. However, the following general instructions apply to most makes and models of domestic vehicles.

GENERAL INSTRUCTIONS

1. Open the hood and visually inspect the radiator hoses, fan, and other parts located at the front of the engine that might sustain damage during the removal or installation process.

2. On vehicles in which the engine is **not** transverse mounted, visually inspect all components at the rear of the powerplant for working clearance between them and the firewall. This clearance is sometimes necessary to permit the engine to tilt downward at the back during transmission removal or installation. If the vehicle has a V-8 engine with a rear-mounted distributor, remove the cap and rotor before removing the transmission, or they may be broken in the process.

3. In most cases, it is mandatory that the mechanic remove the transmission and torque converter as an assembly; otherwise, damage will occur to the converter drive plate, pump bushing, or oil seal. Furthermore, the drive plate itself will not support a load; therefore, the mechanic cannot permit any of the weight of the transmission to rest on this part during the removal process.

4. There are a number of methods of turning the engine over to either drain the converter or remove its attaching bolts or nuts. For instance, one way is to rotate the engine by using a wrench on the crankshaft front pulley bolt. Another and equally safe method is to turn the starter ring gear on the converter or flexplate with a flywheel wrench. A third method, which is not as safe as the other two but is necessary on some vehicle configurations, employs the use of a remote starter button and the engine starter to crank the unit over. However, many manufacturers do not recommend this method because of the possible damage to the torque converter or drive plate that may occur due to the starter motor torque, once the mechanic loosens the converter-to-drive plate mount bolts. (CAUTION: Never use a screwdriver on the ring gear teeth, the flex drive plate, or the converter itself; and never use a wrench on the converter drive bolts to turn the engine over. This action can damage either the ring gear teeth, flex drive plate, or the converter mount bolts.)

TRANSMISSION REMOVAL

With these general instructions in mind, let's turn our attention to the actual procedure used to remove a typical transmission. As an example, we will use the A-415, A-413, or A-470 automatic transaxle, as used in Chrysler-built vehicles. During this procedure, note especially the recommended method by turning the engine over in order to remove the converter-to-drive plate attaching bolts.

To remove the transaxle and torque converter as an assembly:

1. Open the hood and disconnect the negative battery

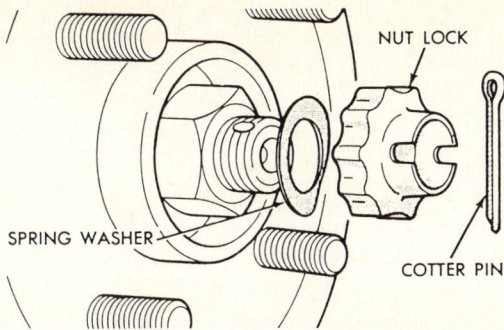

FIGURE 20-1   Removing the cotter pin and nutlock from a front wheel. (Courtesy of Chrysler Corp.)

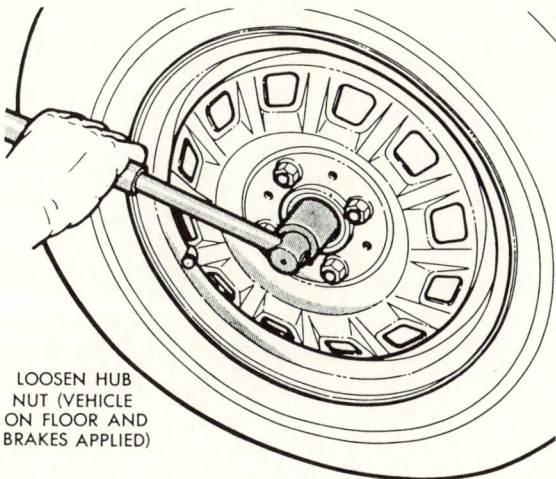

FIGURE 20-4   Installation of the engine support fixture. (Courtesy of Chrysler Corp.)

cable along with the throttle and shift linkage from the transaxle.

2. From each front wheel, remove the cotter pin and nutlock (Fig. 20-1). Then, apply the brakes, and loosen but **do not** remove the front wheel hub nuts (Fig. 20-2).

3. Remove the upper and lower oil cooler lines (Fig. 20-3).

4. Install the engine support fixture (Fig. 20-4); adjust it in order to take the engine weight off the mounts.

5. From inside the engine compartment, loosen and remove all the upper bell housing-to-transaxle case bolts (Fig. 20-5).

6. Raise the vehicle on a hoist to a comfortable working height.

7. Remove the hub nut, washer, and the front wheel and tire from each side of the vehicle (Fig. 20-6).

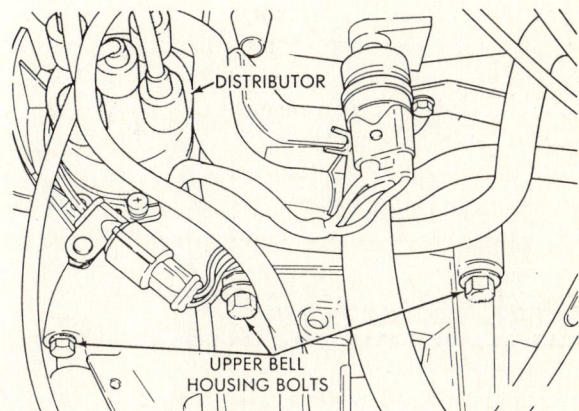

FIGURE 20-2   Loosening a front wheel hub nut. (Courtesy of Chrysler Corp.)

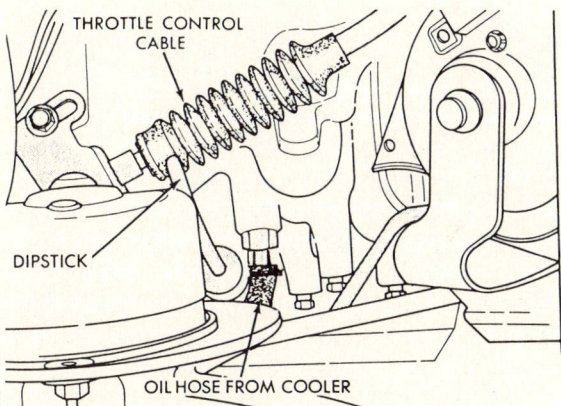

FIGURE 20-3   Removing both oil cooler lines from the engine compartment. (Courtesy of Chrysler Corp.)

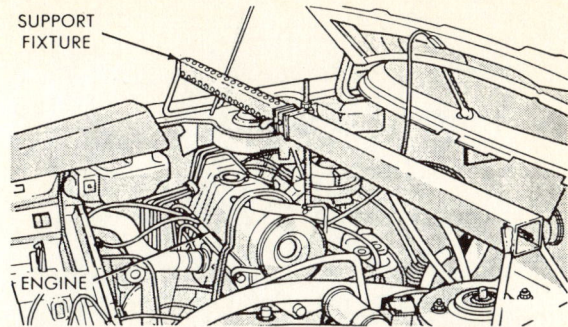

FIGURE 20-5   Removing the upper bell housing bolts. (Courtesy of Chrysler Corp.)

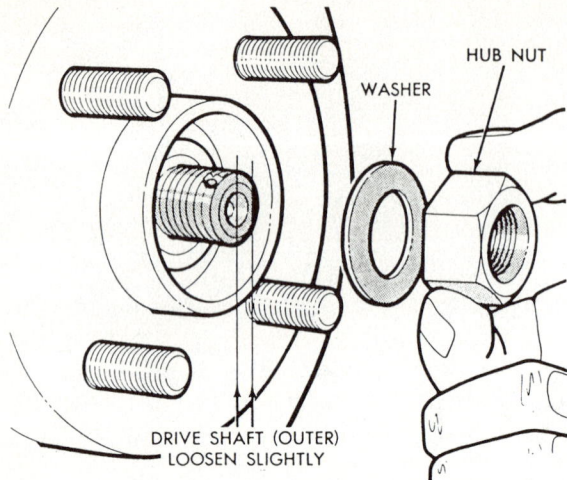

FIGURE 20-6   Removing the nut, washer, front wheel, and tire. (Courtesy of Chrysler Corp.)

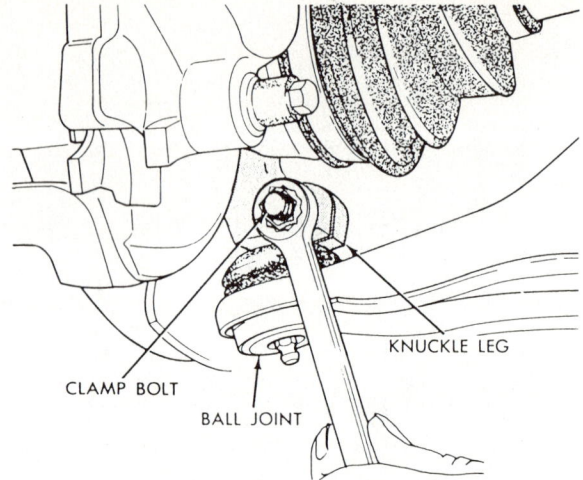

FIGURE 20-8   Removing the ball joint-to-steering knuckle clamp bolt. (Courtesy of Chrysler Corp.)

8. From underneath the vehicle, remove the speedometer pinion from the right drive shaft extension (Fig. 20-7).

9. Loosen and remove the clamp bolt that secures the ball joint studs into the steering knuckles (Fig. 20-8).

10. Separate the ball joint studs from the steering knuckles by prying against the knuckle leg and control arm (Fig. 20-9). (CAUTION: Be careful not to damage either the ball joint or C/V joint boots.)

11. Separate both the outer C/V joint splined shafts from the hubs by holding the C/V housing and moving the steering knuckle assembly outward (Fig. 20-10). (CAUTION: Do not pry on or otherwise damage either wear sleeve on the outer C/V joints.)

12. Support the right driveshaft assembly at both the

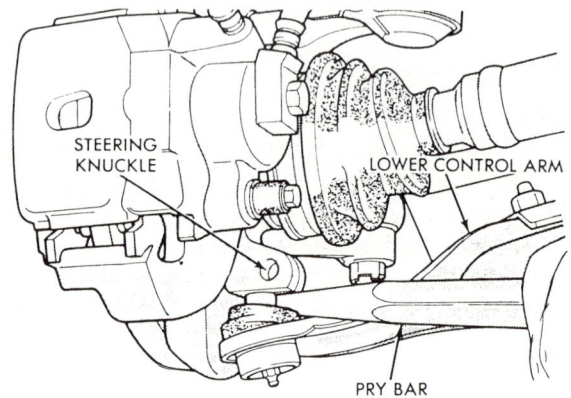

FIGURE 20-9   Separating the ball joint from the steering knuckle. (Courtesy of Chrysler Corp.)

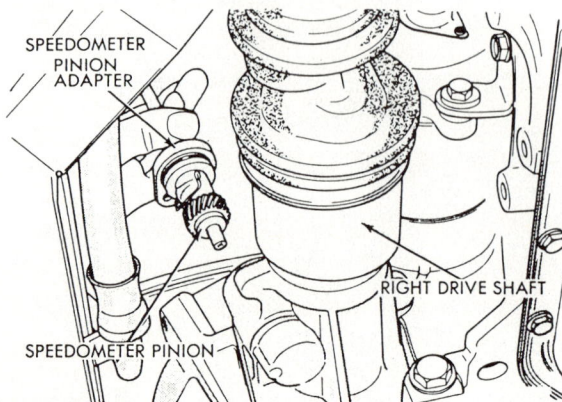

FIGURE 20-7   Removing the speedometer pinion clamp. (Courtesy of Chrysler Corp.)

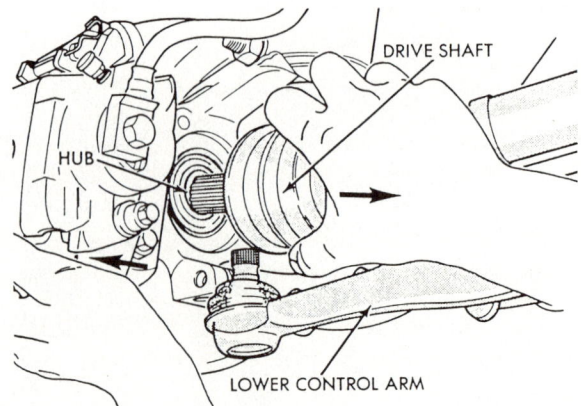

FIGURE 20-10   Separating the outer C/V joint shaft from the knuckle assembly. (Courtesy of Chrysler Corp.)

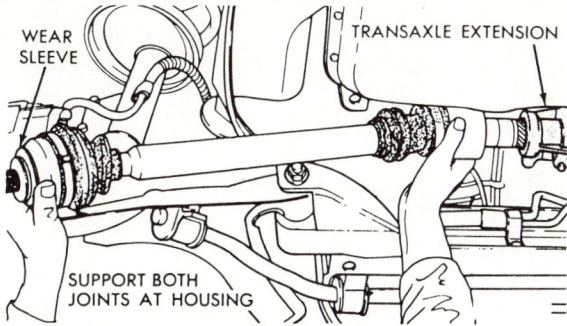

FIGURE 20-11 Removing the driveshaft assembly. (Courtesy of Chrysler Corp.)

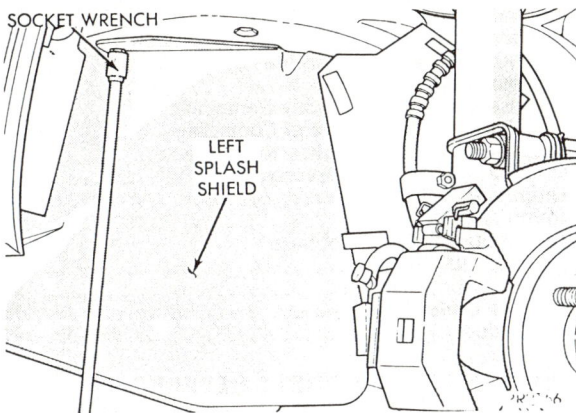

FIGURE 20-12 Removing the left splash shield. (Courtesy of Chrysler Corp.)

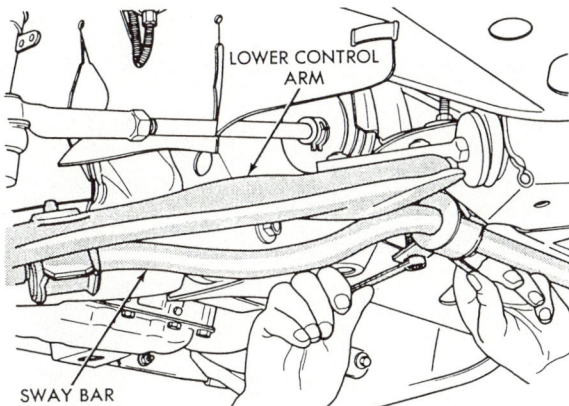

FIGURE 20-13 Disconnecting the sway bar from a lower control arm. (Courtesy of Chrysler Corp.)

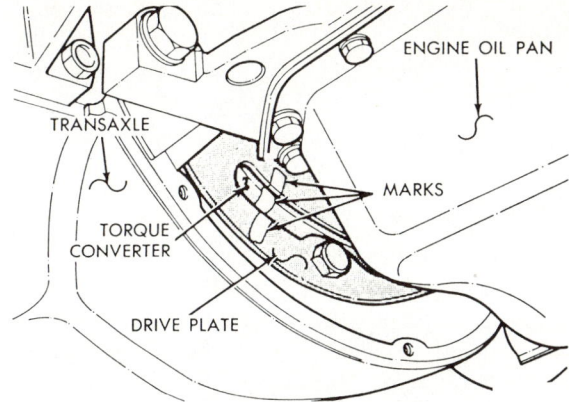

FIGURE 20-14 Inserting alignment marks on the converter and drive plate. (Courtesy of Chrysler Corp.)

C/V housings (Fig. 20-11). Then, remove the entire assembly by pulling outward on the inner joint housing.

13. Support the left driveshaft assembly, and remove it by pulling outward on the inner joint housing.

14. Loosen and take out the left splash shield attaching bolts (Fig. 20-12). Next, remove the shield assembly.

15. Loosen and then remove the nuts, bolts, and retainers that hold the sway bar to the control arms (Fig. 20-13).

16. Loosen and take out the bolts from the crossmember clamps, and remove the clamps and sway bar.

17. Remove the torque converter dust cover.

18. Scribe or paint a suitable mark on the torque converter and drive plate (Fig. 20-14). This will assist you in aligning these two parts during transaxle installation.

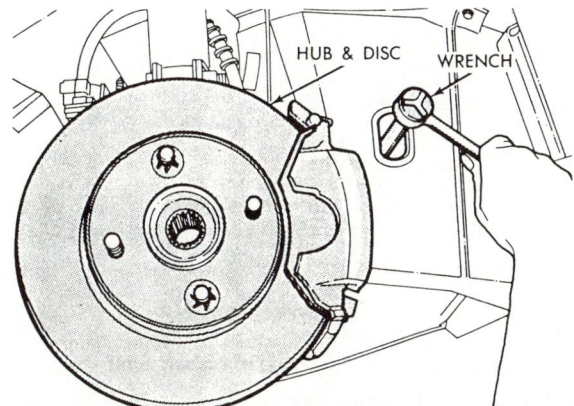

FIGURE 20-15 Turning the engine over by inserting a wrench through the crankshaft access plug. (Courtesy of Chrysler Corp.)

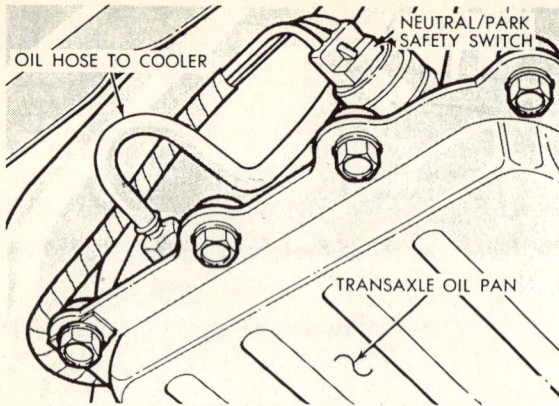

FIGURE 20-16 Removing the electrical connector at the neutral safety switch. (Courtesy of Chrysler Corp.)

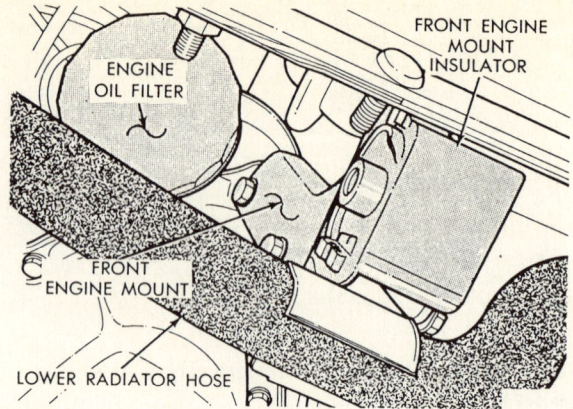

FIGURE 20-18 Removing the front mount attaching hardware. (Courtesy of Chrysler Corp.)

19. Remove the crankshaft access plug in the right splash shield (Fig. 20-15).

20. Insert a wrench through the access hole to rotate the crankshaft.

21. Remove the torque converter-to-drive plate mount bolts. (NOTE: Turn the engine over as necessary to bring each bolt in turn into the converter dust cover area, prior to removal.)

22. Remove the electrical connector to the neutral safety switch (Fig. 20-16).

23. Loosen and remove the bolts and nuts from the front engine mount bracket (Fig. 20-17). Next, take the bracket off the crossmember.

24. Loosen and remove the front mount insulator through bolt and bell housing bolts (Fig. 20-18). Then, take off the front engine mount.

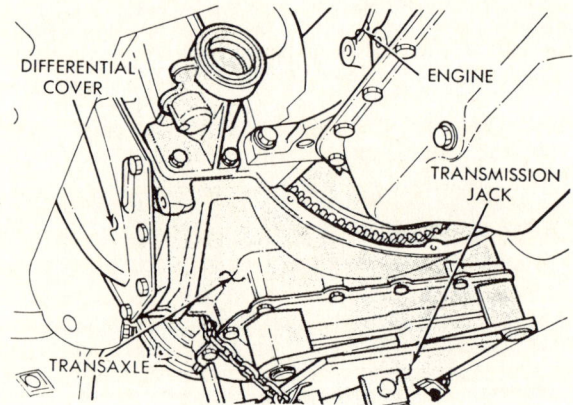

FIGURE 20-19 Positioning a transmission jack under the transaxle. (Courtesy of Chrysler Corp.)

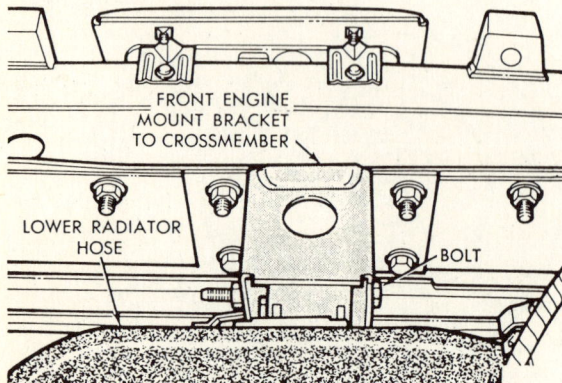

FIGURE 20-17 Removing the engine mount bracket from the crossmember. (Courtesy of Chrysler Corp.)

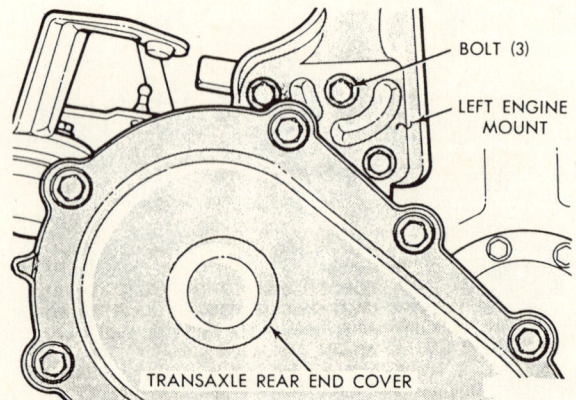

FIGURE 20-20 Removing the three left engine mount bolts. (Courtesy of Chrysler Corp.)

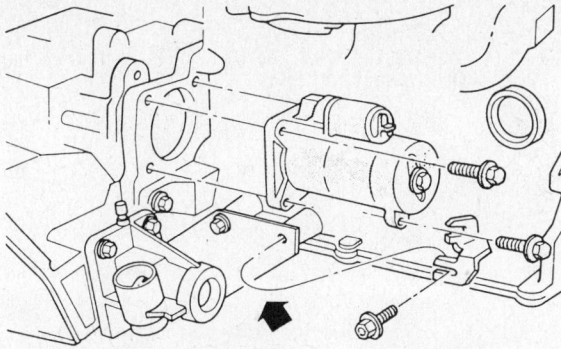

FIGURE 20-21 Removing the starter assembly. (Courtesy of Chrysler Corp.)

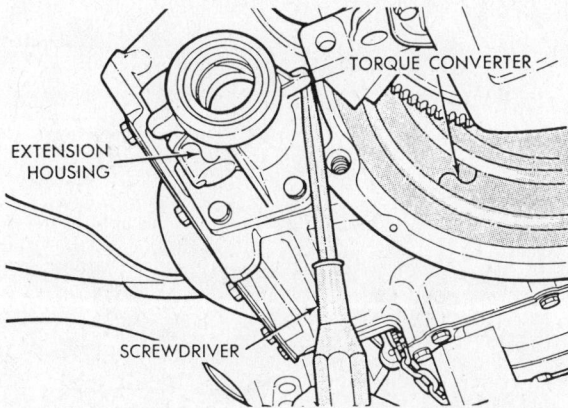

FIGURE 20-22 Obtaining clearance between the engine and transaxle. (Courtesy of Chrysler Corp.)

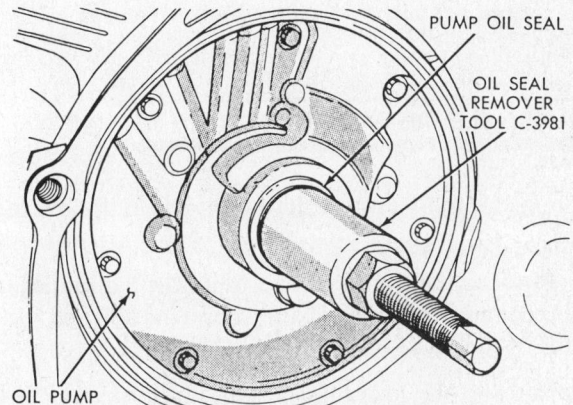

FIGURE 20-23 Lowering the transaxle from under the vehicle. (Courtesy of Chrysler Corp.)

25. Position a hydraulic transmission jack under the transaxle and install its safety chain (Fig. 20-19). Raise the jack to place some tension on the transaxle.

26. Loosen and take out the three left engine mount bolts (Fig. 20-20).

27. Remove the starter assembly (Fig. 20-21).

28. Loosen and remove the lower bell housing attaching bolts.

29. With a bar, pry between the engine and transaxle to provide some clearance between the two (Fig. 20-22).

30. Lower and tilt the jack as necessary to remove the transaxle from the vehicle (Fig. 20-23).

PUMP OIL SEAL REPLACEMENT

The pump oil seal can be replaced without removing the pump and reaction shaft assembly from the transaxle. To remove and replace the seal:

1. Disconnect the safety chain from around the transaxle, and place it on a suitable work bench.

2. Carefully remove the torque converter from the front of the transaxle. Pull the converter straight out from the pump; this will prevent scoring or damage to the pump bushing and seal. Then, place the converter on an unused portion of the work bench.

3. Thread the remover tool (Fig. 20-24) into the seal. Next, tighten down the threaded shaft portion of the tool to withdraw the seal.

4. With a flashlight, inspect the pump bushing for scoring, wear, or damage. If it is worn excessively,

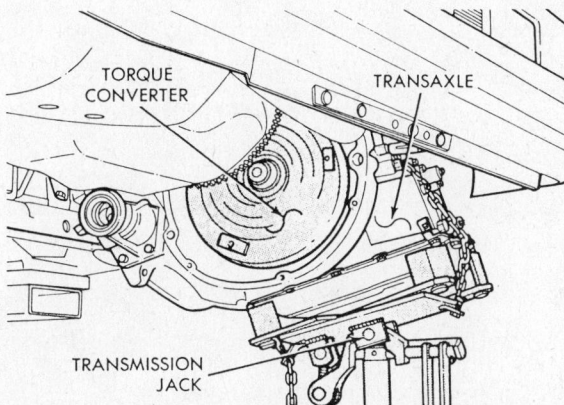

FIGURE 20-24 Removing the front pump seal. (Courtesy of Chrysler Corp.)

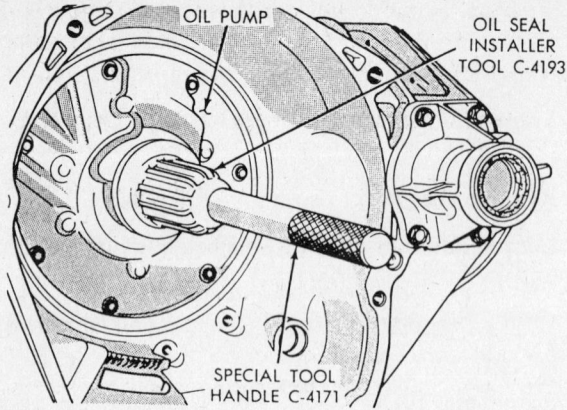

FIGURE 20-25 Installation of the pump seal. (Courtesy of Chrysler Corp.)

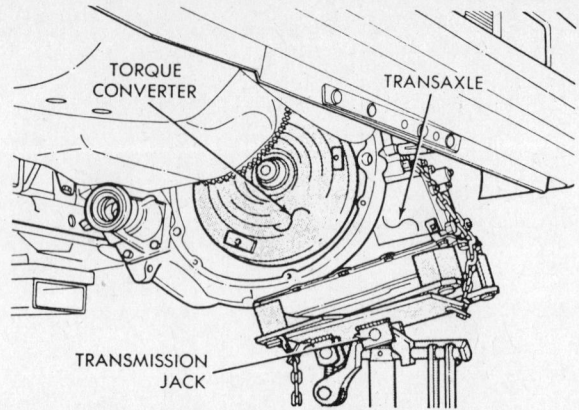

FIGURE 20-26 Positioning the jack and transaxle under the vehicle during its installation process. (Courtesy of Chrysler Corp.)

remove the pump and replace the bushing as outlined in Chapters 21 and 23. (NOTE: *Excessive bushing wear can cause a leak at the pump seal.*) If there is any doubt as to bushing condition, replace it or a leak may develop.

5. Before installing the new seal, coat the outer circumference of its backing with a nonhardening sealer. This procedure is unnecessary if the seal backing has a rubber or resin coating.

6. To install the new seal, place it in the opening of the pump housing with the lip facing inward or toward the transaxle. Then, with the tool shown in Fig. 20-25, drive the seal into the housing bore until it bottoms.

7. Lubricate the converter hub with transmission fluid. Then, carefully reinstall the converter into the front of the transaxle by sliding it over the stator support and into the pump. CAUTION: *Severe damage can occur to the pump, converter, or drive plate if you attempt to install the transaxle with the converter hub not engaged properly into the pump drive gear.*)

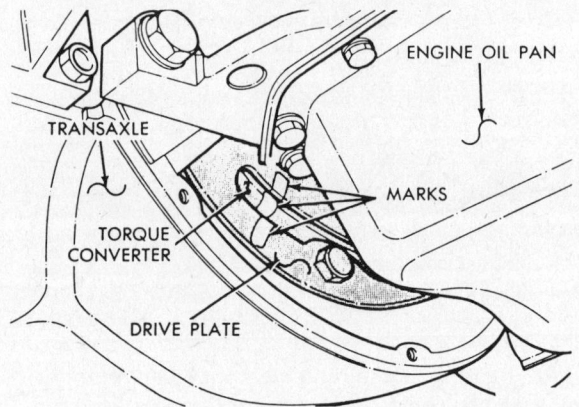

FIGURE 20-27 Aligning the converter and drive plate markings. (Courtesy of Chrysler Corp.)

TRANSAXLE INSTALLATION

To reinstall the transaxle into the vehicle, follow these steps:

1. Position the transaxle back onto the hydraulic jack; reconnect the safety chain around the unit and secure it to the jack.

2. Position the hydraulic jack and transaxle under the vehicle for installation. Then, raise and/or tilt the platform as necessary until the transaxle aligns with the engine (Fig. 20-26).

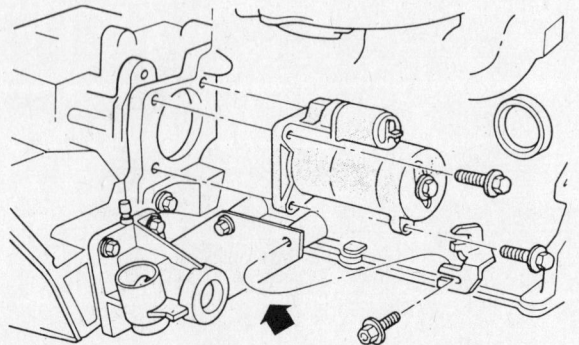

FIGURE 20-28 Installing the starter assembly. (Courtesy of Chrysler Corp.)

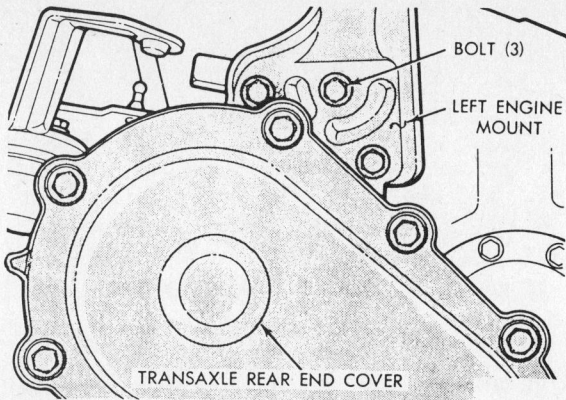

FIGURE 20-29   **Installing the left engine mount bolts.** (Courtesy of Chrysler Corp.)

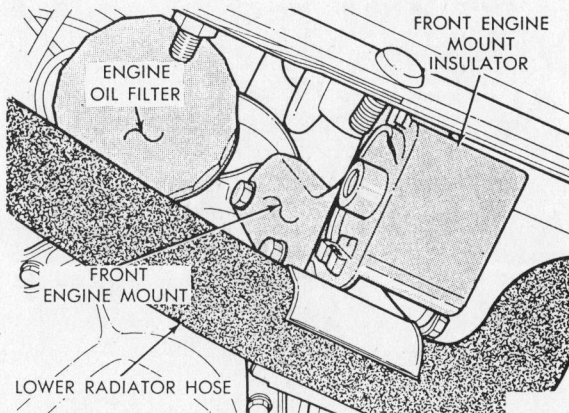

FIGURE 20-30   **Installing and torquing the front mount attaching hardware.** (Courtesy of Chrysler Corp.)

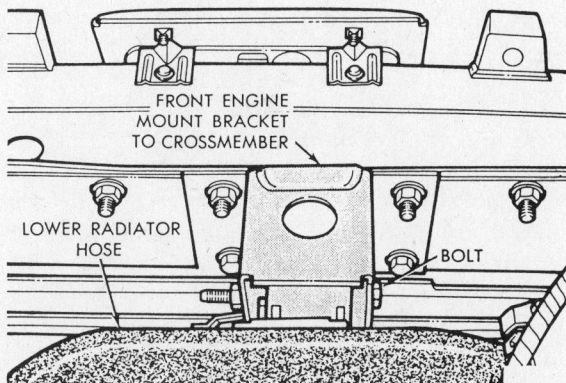

FIGURE 20-31   **Installing the front mount bracket onto the crossmember.** (Courtesy of Chrysler Corp.)

FIGURE 20-32   **Installing the connector onto the neutral safety switch.** (Courtesy of Chrysler Corp.)

3. Rotate the converter so that the mark on it, made during transaxle removal, will align with the mark on the drive plate (Fig. 20-27).

4. Carefully work the transaxle assembly forward into alignment with the engine. The converter front hub should now freely enter the opening in the crankshaft.

5. Once the transaxle is in the proper position, install and torque the lower bell housing bolts.

6. Install the starter assembly (Fig. 20-28).

7. Reinstall and torque the three left engine mount bolts (Fig. 20-29).

8. Position the front engine mount into place. Next, install and torque the front mount insulator through bolt and bell housing bolts (Fig. 20-30).

9. Remove the hydraulic jack out from under the transaxle.

10. Position the front engine mount bracket into place on the crossmember. Then, install and torque its bolts and nuts (Fig. 20-31).

11. Install the electrical connector onto the neutral safety switch (Fig. 20-32).

12. Rotate the crankshaft as necessary, and install and torque the converter-to-drive plate attaching bolts.

13. Install the crankshaft access plug in the right splash shield.

14. Install the torque converter dust cover.

15. Position the sway bar and clamps into place on the crossmember. Next, install and torque the clamp bolts.

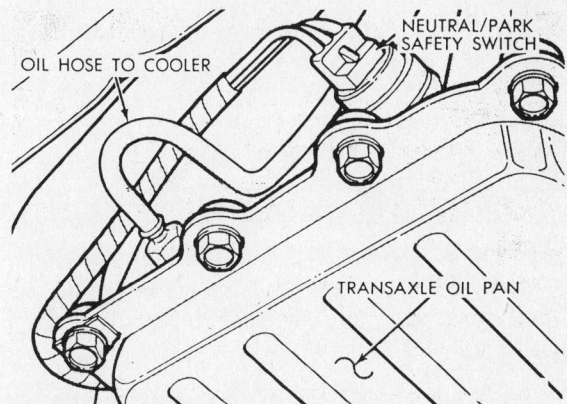

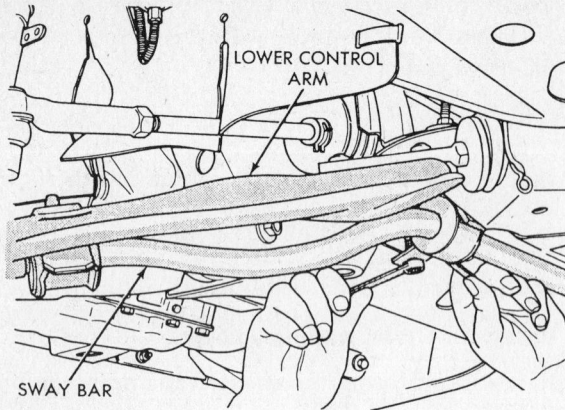

FIGURE 20-33   Connecting the sway bar to a lower control arm. (Courtesy of Chrysler Corp.)

16. Reinstall and torque the nuts, bolts, and retainers that hold the sway bar to the control arms (Fig. 20-33).

17. Position the left splash shield into place; install and torque its attaching bolts.

18. Support the left driveshaft assembly at both C/V housings, and insert its inner joint's splined shaft into the transaxle (Fig. 20-34).

19. Support the right driveshaft assembly at both C/V housings, and insert its inner joint's splined shaft into the transaxle extension (Fig. 20-35).

20. Reinstall both driveshaft outer splined shafts into their respective hubs. To accomplish this, move each steering knuckle outward, while inserting the shaft into its hub splines (Fig. 20-36).

21. Insert the lower ball joint studs into the steering knuckles (Fig. 20-37).

22. Reinstall both lower ball joint-to-steering knuckle clamp bolts, and torque them to specifications (Fig. 20-38).

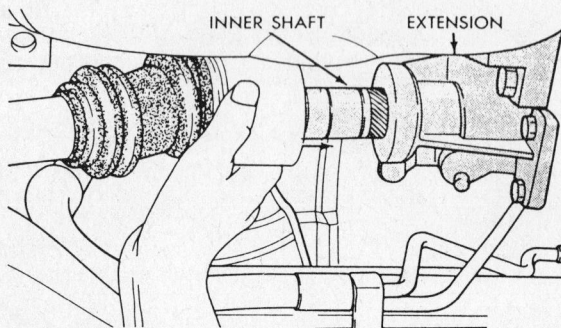

FIGURE 20-34   Installing the inner shaft into the transaxle. (Courtesy of Chrysler Corp.)

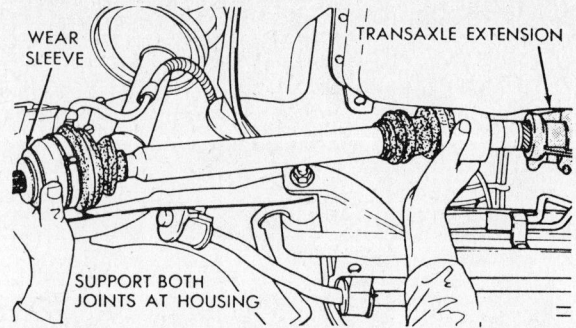

FIGURE 20-35   Installing the right driveshaft assembly. (Courtesy of Chrysler Corp.)

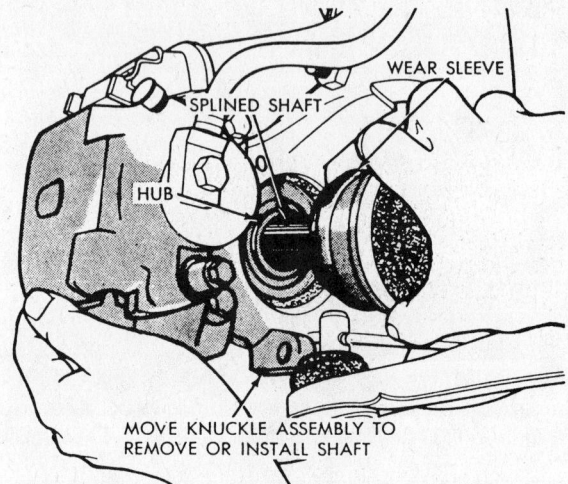

FIGURE 20-36   Installing the outer shaft splines into the hub. (Courtesy of Chrysler Corp.)

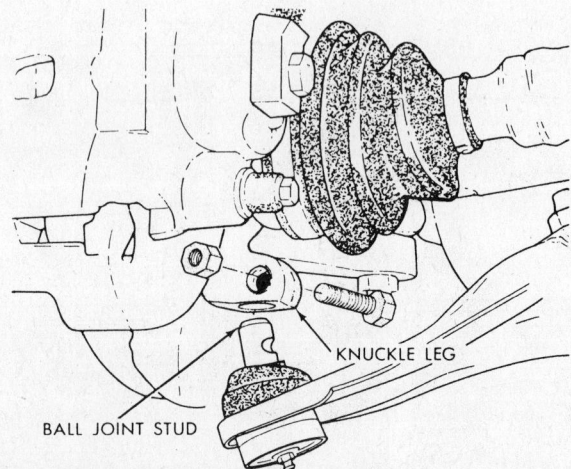

FIGURE 20-37   Installing the ball joint stud into the knuckle leg. (Courtesy of Chrysler Corp.)

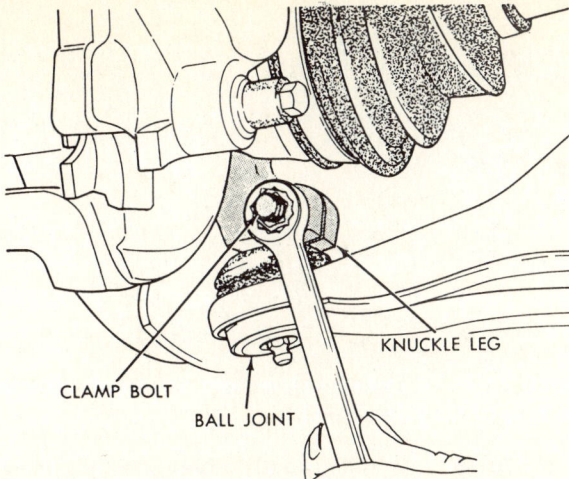

FIGURE 20-38 Torquing the ball joint-to-steering knuckle clamp bolt. (Courtesy of Chrysler Corp.)

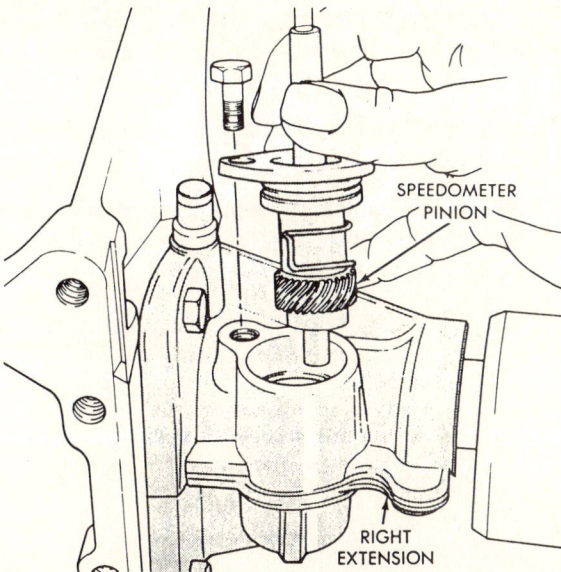

FIGURE 20-39 Installing the speedometer pinion and adapter. (Courtesy of Chrysler Corp.)

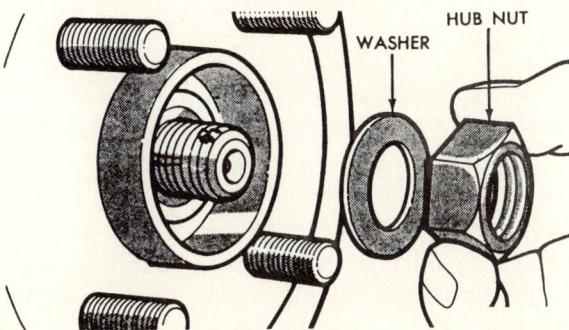

FIGURE 20-40 Installing the hub washer and nut. (Courtesy of Chrysler Corp.)

FIGURE 20-41 Torquing the wheel hub nuts. (Courtesy of Chrysler Corp.)

23. Install the speedometer pinion into the right drive-shaft extension (Fig. 20-39).

24. Install both front wheels along with their hub washers and nuts (Fig. 20-40).

25. With the brakes applied, tighten each hub nut to specifications (Fig. 20-41).

26. Install the hub nutlock, spring washer, and a **new** cotter pin. Wrap the cotter pin prongs tightly around the nut as shown in Fig. 20-42.

27. From inside the engine compartment, install and torque the upper bell housing-to-transmission case bolts (Fig. 20-43).

28. Disconnect and remove the engine support fixture (Fig. 20-44).

29. Install the upper and lower oil cooler lines (Fig. 20-45).

30. Reinstall both the throttle and shift linkages; adjust both to specifications.

31. Install the negative battery cable.

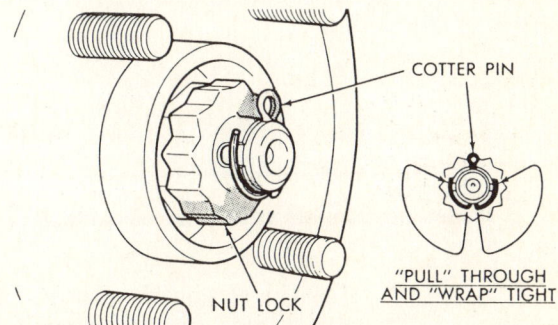

FIGURE 20-42 The correct installation of the hub nutlock, spring washer, and cotter pin. (Courtesy of Chrysler Corp.)

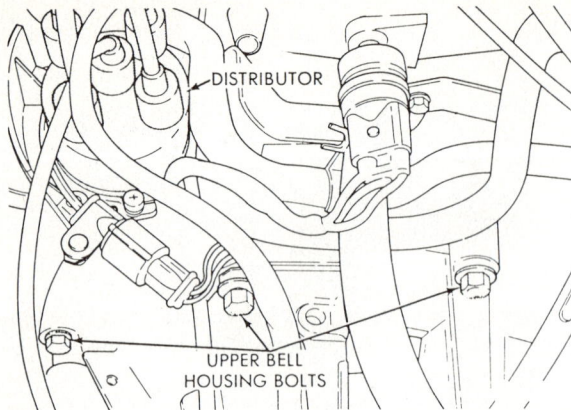

FIGURE 20-43 Installation of the upper bell housing-to-transmission case bolts. (Courtesy of Chrysler Corp.)

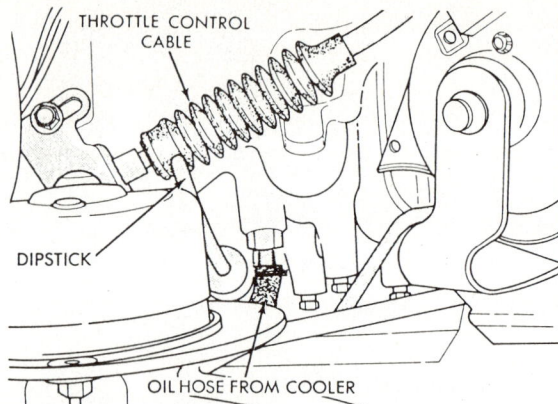

FIGURE 20-45 Installing both cooler lines. (Courtesy of Chrysler Corp.)

32. Check the fluid level in the transaxle. If it is low or has been drained, pour in sufficient fluid to bring the level into the ADD zone on the dipstick (Fig. 20-46).

33. Start the engine and allow it to idle for at least one minute. Then, with the parking and service brakes applied, move the selector lever momentarily to each position, ending in park or neutral.

34. Add sufficient fluid to bring the level to $\frac{1}{8}$ inch below the ADD mark. Recheck the fluid level after the transaxle is at normal operating temperature; the level should be in the HOT region on the dipstick.

35. Visually inspect the transaxle, converter area, and cooling lines for leakage.

36. Road test the vehicle to check the operation of the transaxle. Make any necessary adjustments to the linkages, and recheck the fluid level.

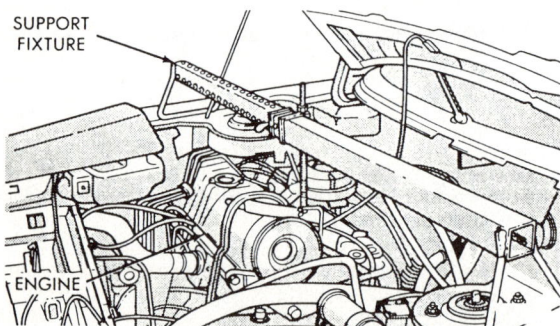

FIGURE 20-44 Removing the engine support fixture. (Courtesy of Chrysler Corp.)

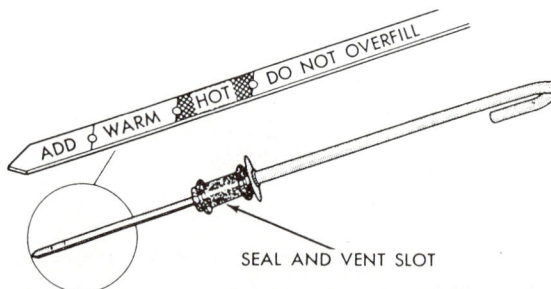

FIGURE 20-46 The level markings on the dipstick. (Courtesy of Chrysler Corp.)

REVIEW

This section will assist you in determining how well you remember the material contained in this chapter. Read each item carefully. If you cannot complete the statement, review the section in the chapter that covers the material.

1. After installing the transaxle, adjust all the _____.

2. Severe damage can occur during transaxle installation if the converter hub does not engage properly in the _____ _____ of the pump.

3. Install the pump seal with its lip facing _____ .

4. Excessive _____ wear can cause the pump seal to leak.

5. Marking the converter and drive plate is an aid to _____ both units during transaxle installation.

6. To turn the engine over to remove the converter bolts, insert a wrench through the _____ _____ in the right splash shield.

7. It is necessary to remove the transaxle to replace the pump _____ .

8. In order to remove the driveshafts, it is necessary to first separate the _____ _____ from the steering knuckles.

9. Before beginning to remove the transaxle, first install an _____ _____ fixture.

10. To take the old seal out of the pump, use a _____ tool.

For the answers, turn to the Appendix.

CHAPTER 21

Torque Converter and Hydraulic Pump Inspection, Testing, and Service

The torque converter and hydraulic pump are two of the more expensive components of the automatic transmission. When the mechanic rebuilds or services the transmission, it therefore makes good sense to do everything practical to save these components for reuse. With this in mind, let's examine the methods of checking whether the converter and pump are still serviceable, and the cleaning and repair operations necessary to restore them both to good condition.

The modern torque converter is a sealed assembly. Consequently, the mechanic can perform only a limited number of service operations on it. The most important of these is to clean the converter thoroughly with a mechanically agitated cleaner whenever the technician rebuilds or services the transmission. A special item of equipment, the torque converter flusher, is necessary to perform this operation effectively and automatically. The mechanic should always use this machine to clean the converter, provided the converter passes a few simple inspection checks for serviceability.

CONVERTER INSPECTION

The mechanic should perform the serviceability checks before he installs the converter on the cleaning machine since they take only a few minutes, and there is no point in cleaning a converter that is no longer serviceable. The serviceability checks include drive stud, lug, or flange inspection; converter hub inspection; stator-to-impeller interference test; stator-to-turbine interference test; stator one-way clutch test; turbine end-play test; and converter leakage test.

Drive Stud, Lug, or Flange Inspections

Torque converters may have either drive studs (Fig. 21-1), welded drive nuts (lugs), or a drilled drive flange. These parts mate the converter to the flexplate, permit the engine to drive the converter, and in many cases, align the converter so that it will run true with the crankshaft of the engine. To perform an inspection of these parts, place the converter on the work bench, pump hub facing down, and check the drive studs for the following conditions:

1. Tightness in the converter itself and broken support welds.

2. Damaged or worn threads.

3. On Ford converters, damaged or worn pilot shoulders (Fig. 21-2). (NOTE: These areas pilot the converter so that it runs true with the drive or flex plate, which fastens to the engine crankshaft. Any converter misalignment caused by worn or damaged shoulders causes the hub on the converter, which drives the pump, to operate eccentrically; this would quickly destroy the pump bushing or cause a possible vibration. If these shoulders are worn or damaged, replace the converter with a new or rebuilt unit.)

Check a converter that has drive nuts or lugs for the following:

1. Loose lugs or cracked or broken support welds (Fig. 21-3).

2. Worn or damaged threads. If the lugs are no longer serviceable, repair them as necessary or replace the converter.

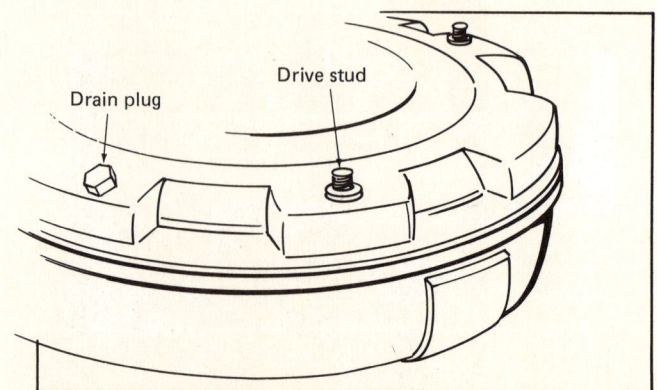

FIGURE 21-1   Drive studs of a typical converter.

INSPECT DRIVE STUDS

- Studs pilot converter to run true with flywheel.

NOTE: C-3 converters have welded drive nuts in place of drive studs. Check these for damage and for cross threading. Replace drive bolts after each disassembly.

1 Check STUDS for:
- Tightness
- Good threads

2 Stud shoulders must not be damaged.

3 Raised or lowered shoulder causes:
- Misalignment
- Pump drive hub eccentric
- Pump bushing damage

DRIVE PLATE

RESULTS
- **Studs or welded drive nuts damaged or loose** — replace converter.
- **Shoulder damaged** — clean up burrs; inspect pump body bushing.
- **Okay** — check drive hub (step E).

FIGURE 21-2   Inspection of a drive stud with a pilot shoulder. (Courtesy of Ford Motor Company)

Inspect the converter equipped with drilled drive flanges for the following:

1. Damaged or elongated holes that accommodate the drive bolts.

2. Cracked or broken drive flanges or mount welds. (NOTE: The mechanic should replace the converter if the flanges or drive holes are worn or damaged.)

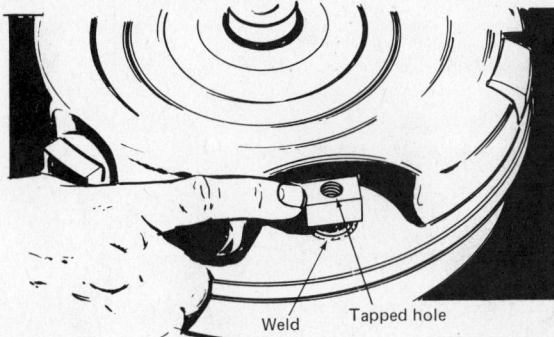

Weld   Tapped hole

FIGURE 21-3   Inspection of a typical converter with a threaded drive lug.

Converter Hub Inspections

Examine the converter crankshaft hub for burrs and nicks. Then, turn the converter over and take a close look at the outside of the pump hub (Fig. 21-4). There are opportunities for hub wear at two points: (1) where the hub operates inside the pump seal and (2) where it operates inside the pump bushing. If the hub has score or wear marks, this indicates that the seal or bushing or both are also unserviceable, and each should be replaced as necessary, according to the location of the wear marks.

Frequently, the score marks on the hub appear severe to the eye, but they may in fact be very shallow. To check whether the score marks are deep, run a fingernail across them. If it catches in the score indentations, this is a good indication that the wear is excessive, which necessitates converter replacement.

If the wear marks are not too deep, polish them out using this procedure:

1. Cover the hub opening to prevent particles from entering the converter itself during the polishing operation.

INSPECT DRIVE HUB AND PILOT HUB
- Scoring may not be as deep as it appears to the eye.

CHECK FOR DAMAGE
- **Burrs or nicks** — clean up.

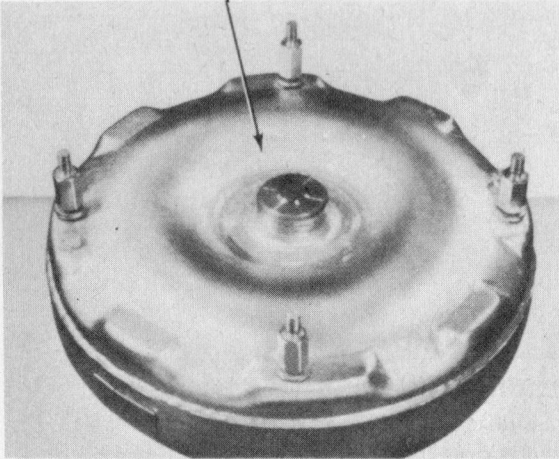

CHECK WITH TIP OF FINGERNAIL.
- **Light scoring** — remove as shown below.
- **Deeply scored** — replace converter.
- Cover impeller hub to prevent dirt from entering.

FIGURE 21-4   Inspection of a converter hub. (Courtesy of Ford Motor Company)

REMOVE LIGHT SCORING
- Check and replace pump seal and bushing as required.

POLISH AS SHOWN
(CLEAN HUB THOROUGHLY
AFTER POLISHING)

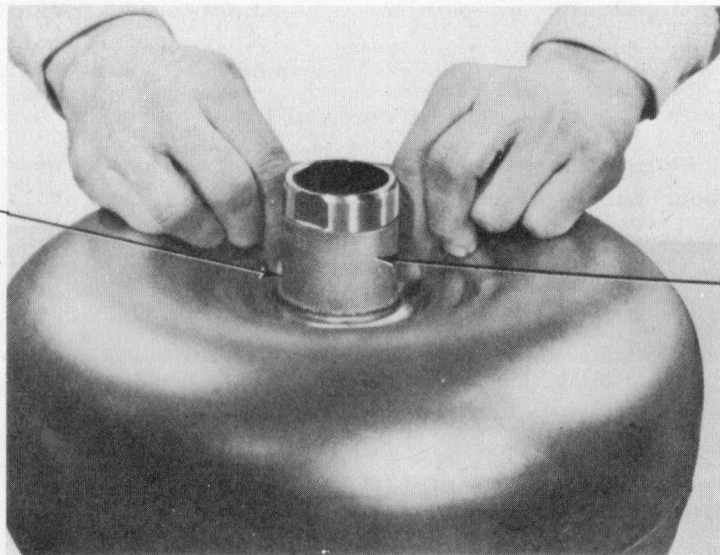

LIGHT CROCUS CLOTH
(GRIT #600)

If scoring cannot be removed by light polishing, the converter must be replaced.

FIGURE 21-5   Polishing the converter hub with emery cloth. (Courtesy of Ford Motor Company)

CHECK FOR STATOR-TO-IMPELLER INTERFERENCE

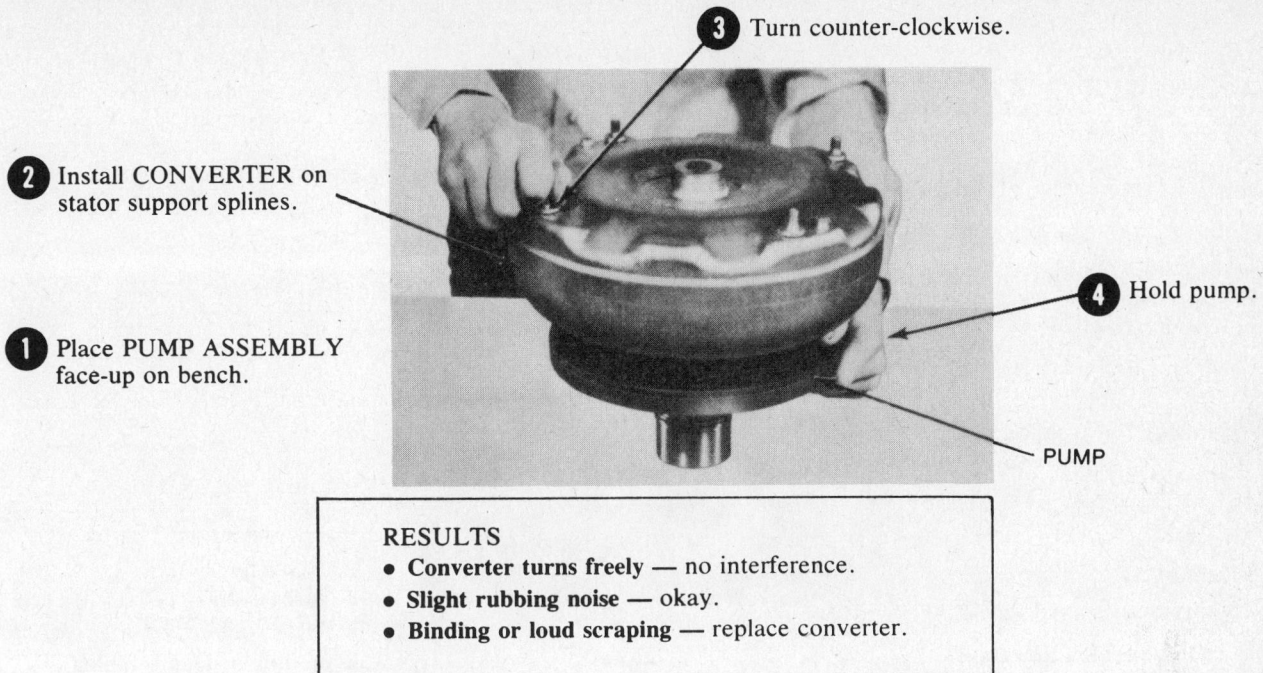

3 Turn counter-clockwise.

2 Install CONVERTER on stator support splines.

4 Hold pump.

1 Place PUMP ASSEMBLY face-up on bench.

PUMP

RESULTS
- **Converter turns freely** — no interference.
- **Slight rubbing noise** — okay.
- **Binding or loud scraping** — replace converter.

FIGURE 21-6 *Conducting a stator-to-impeller interference test.*

2. Wrap a piece of 600-grit crocus cloth around the hub and polish the marks away (Fig. 21-5).

3. Clean the hub thoroughly after polishing.

Stator-to-Impeller Interference Test

This test, as its name implies, indicates whether the stator assembly is touching or interfering with the normal rotation of the impeller. To perform such a test follow these steps:

1. Place a hydraulic pump assembly on a clean work bench with the splined end of the stator shaft pointed upward (Fig. 21-6).

2. Install the torque converter over the pump assembly in such a manner that the splines on the stator one-way clutch inner race engage with the mating splines of the stator support, and the converter hub drive slots engage the pump drive gear.

3. While holding the pump stationary, attempt to turn the converter counterclockwise. The converter should rotate freely without any signs of scraping or interference from within the unit itself.

4. If there is an indication of scraping or binding, the tailing edges of the stator blades are probably interfering with the leading edges of the impeller blades. If this occurs, replace the converter with a new or rebuilt unit.

Stator-to-Turbine Interference Test

This check will indicate whether the stator is striking or interfering with the normal rotation of the turbine. The perform this test, follow these directions:

1. Place the torque converter assembly on a clean work bench with its front side facing down (Fig. 21-7).

2. Install the input shaft and engage its splines into the mating ones within the turbine hub.

3. While holding the pump and converter stationary, try to turn the turbine with the input shaft. The turbine should turn freely in both directions without any audible signs of interference or scraping.

4. If interference does exist, the stator's front thrust washer is probably worn out, allowing the stator to hit

CHECK FOR TURBINE INTERFERENCE

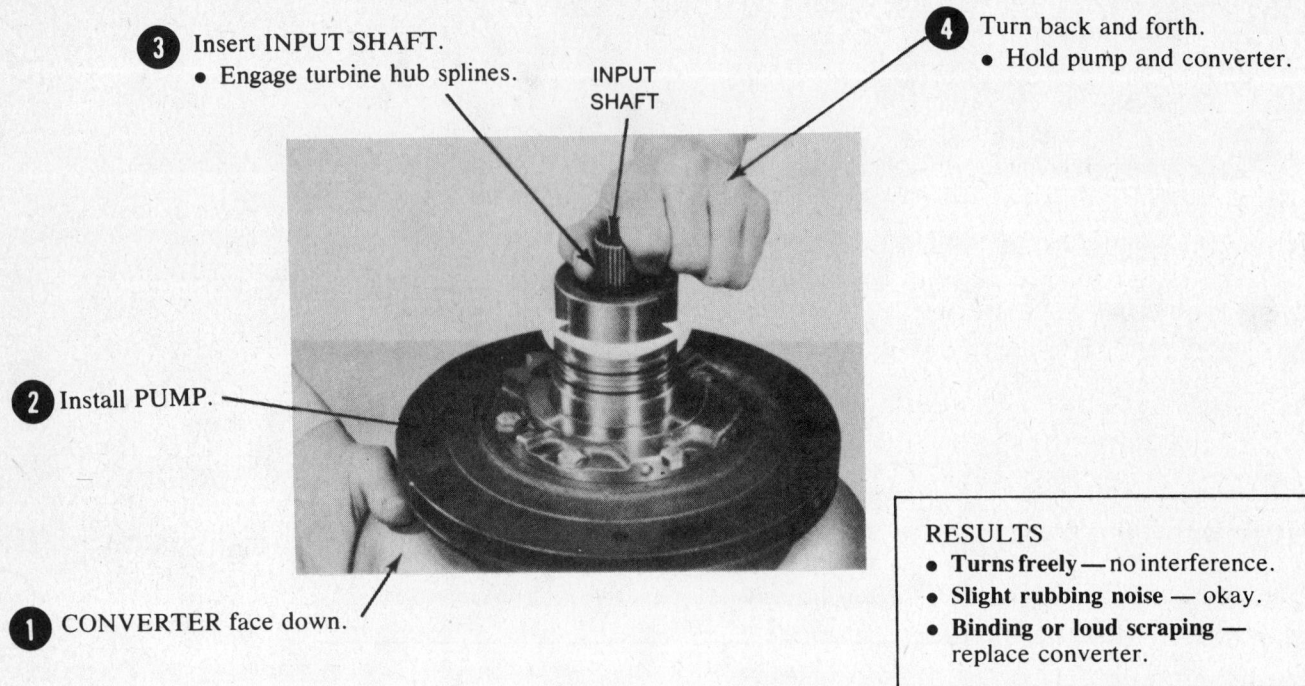

3 Insert INPUT SHAFT.
  • Engage turbine hub splines.

INPUT SHAFT

4 Turn back and forth.
  • Hold pump and converter.

2 Install PUMP.

1 CONVERTER face down.

RESULTS
  • **Turns freely** — no interference.
  • **Slight rubbing noise** — okay.
  • **Binding or loud scraping** — replace converter.

FIGURE 21-7   Performing a stator-to-turbine interference test. (Courtesy of Ford Motor Company)

the turbine. In this situation, replace the converter with a new or rebuilt unit.

Stator One-way Clutch Test

This test indicates whether or not the stator's one-way clutch will hold the stator against counterclockwise rotation. To perform this test accurately, a special tool set, like the one shown in Fig. 21-8, is necessary; this set consists of a holding tool, gauge post, and a guide collar.

To use this tool set to perform the test, proceed as follows:

1. Install the end of the outer-race holding tool into one of the four holes provided in the stator itself. This tool must engage into one of these holes to prevent the stator from turning during the test.

2. Insert the post tool so that its splines engage into those of the one-way clutch inner race.

3. Slide the guide collar over the post, hub, and holding tool.

4. While holding the guide collar and holding tool stationary, turn the post tool in both directions with a torque wrench. The tool should freely turn clockwise,

but it should hold at least a torque of 10 foot-pounds counterclockwise. Try the clutch for lock-up and hold in at least five different locations around the converter.

5. If the one-way clutch fails to lock up and hold the 10 foot-pounds torque, replace the converter with a new or rebuilt unit.

NOTE: If the tool set shown in Fig. 21-8 is not available and you suspect a total one-way clutch failure, check it using this simple method: Insert one finger into the splined, one-way clutch inner race and attempt to turn the race in both directions. The race should turn freely in the clockwise direction, but it should lock up when turned counterclockwise. This is not an accurate test of clutch reliability, of course, but it will indicate a total failure of the one-way assembly.

Turbine End-Play Test

This check actually measures the amount of wear on the thrust washers or bearings separating the internal components within the torque converter. If excessive wear does exist, the converter turbine can lose some of its operating efficiency or produce varying levels of noise. In

CHECK ONE-WAY CLUTCH
- Must hold stator for torque multiplication at low speeds.
- Must "free-wheel" at high speeds (coupling phase).

1 Use these SPECIAL TOOLS for converter checks.

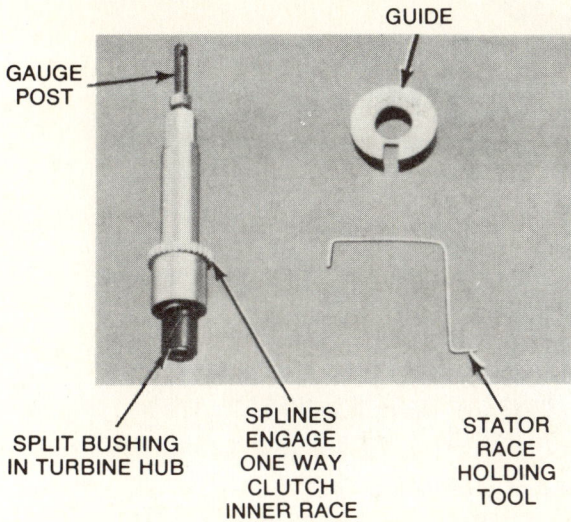

GUIDE

GAUGE POST

SPLIT BUSHING IN TURBINE HUB

SPLINES ENGAGE ONE WAY CLUTCH INNER RACE

STATOR RACE HOLDING TOOL

2 Insert HOLDING TOOL in groove in stator.
- Prevents stator from turning.

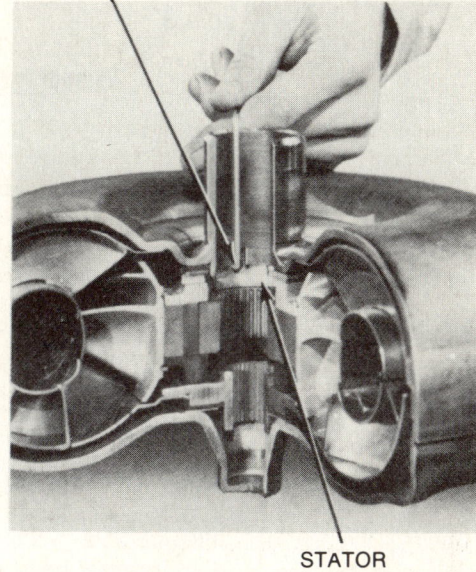

STATOR

3 Drop GAUGE POST into converter.
- Use to turn inner race.

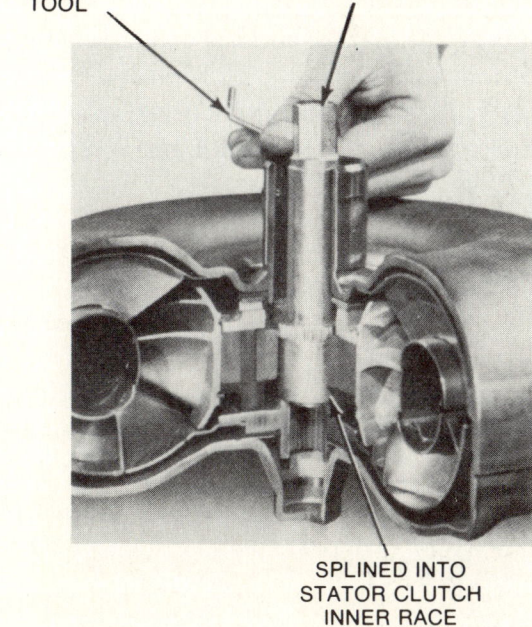

STATOR HOLDING TOOL

SPLINED INTO STATOR CLUTCH INNER RACE

4 Install GUIDE to lock holding tool.

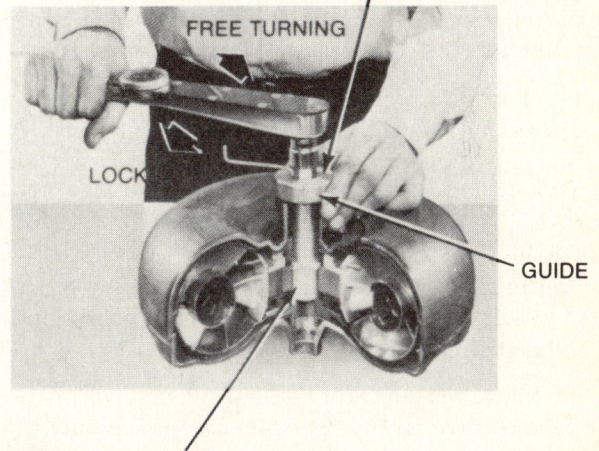

FREE TURNING

LOCK

GUIDE

5 Turn with torque wrench.
- Check in five positions.

RESULTS REQUIRED

- **Clockwise** — turn freely.
- **Counter-clockwise** — lock up with a 10 foot-pound pull.

REPLACE CONVERTER IF NOT AS REQUIRED.

FIGURE 21-8   *Performing a torque converter one-way clutch test. (Courtesy of Ford Motor Company)*

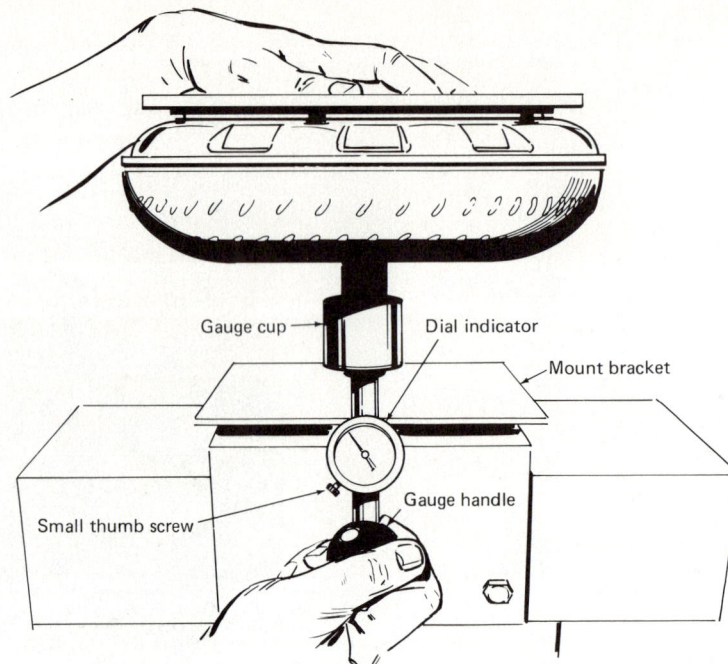

FIGURE 21-9   Performing a turbine end-play test using the universal gauge.

order to accurately test the turbine's end-play, special equipment is necessary. This may be in the form of universal- or factory-type equipment.

To check the turbine end-play of any converter using the universal gauge, mentioned in Chapter 15, proceed as follows:

1. Slide the gauge assembly into the slot in the bracket mounted on the converter flushing machine.

2. With the gauge (Fig. 21-9) installed into the mount bracket, set the converter hub down into the gauge cup.

3. Push up on the gauge handle until you feel the tip make contact with the turbine splines inside the converter.

4. Adjust the indicator assembly up or down on the shaft until the dial indicator stem makes contact with the gauge cup. Lock the dial indicator in place with its large thumb screw.

5. Push up on the gauge handle lightly until you meet some resistance. Loosen the small thumb screw on the dial indicator face and zero the gauge.

6. Push up on the gauge handle once more and read the dial indicator. If the reading is not to specifications, replace the torque converter with a new or rebuilt unit.

To test a converter for turbine end-play with the factory-type tool shown in Fig. 21-10, follow these steps:

1. Insert the guide post into the converter hub until it bottoms.

2. Install the guide over the post and onto the converter hub.

3. Expand the guide post bushing in the turbine hub by tightening the nut on the threaded inner post.

4. Mount a dial indicator on the post.

5. Position the dial indicator button on the converter hub, and adjust the dial face to zero.

6. Insert a rod through the gate post eye for use as a handle.

7. Lift the handle upward as far as it will go, and note the indicator reading, which represents the total end-play of the turbine.

8. If the reading exceeds specifications, replace the converter.

Converter Leakage Test

The purpose of this test is to determine if the welds around the torque converter housing are leaking. A technician will usually make this test only if (1) there has been an undetectable leak in the converter housing area or (2) the serviceman has cut open the converter, rebuilt its internal components, and rewelded the housing back together.

CHECK STATOR END PLAY

1 Use these SPECIAL TOOLS for converter checks.

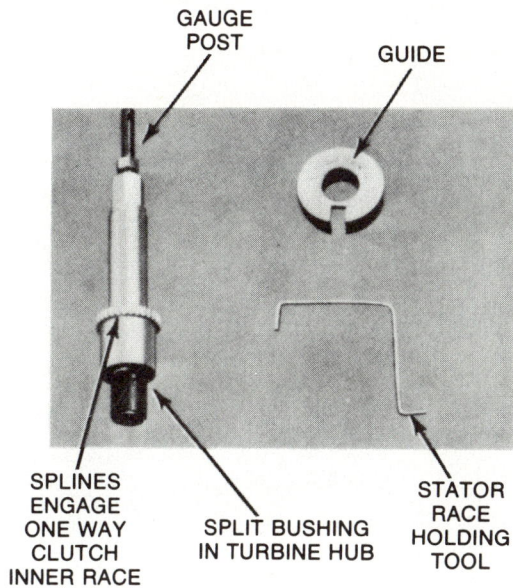

GAUGE POST

GUIDE

SPLINES ENGAGE ONE WAY CLUTCH INNER RACE

SPLIT BUSHING IN TURBINE HUB

STATOR RACE HOLDING TOOL

4 Tighten NUT firmly.
  • Split bushing will expand in turbine hub.

3 Install GUIDE over tool onto hub.

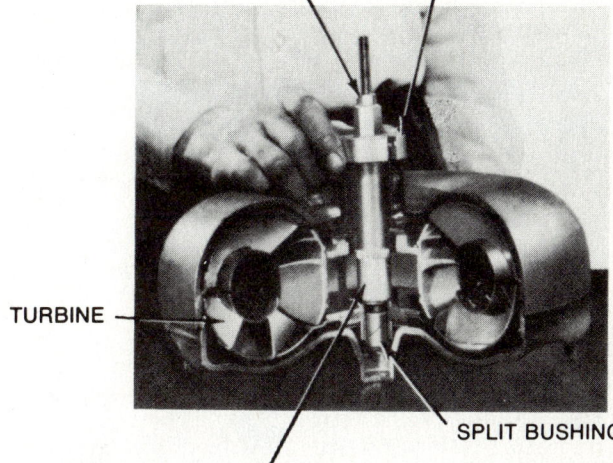

TURBINE

SPLIT BUSHING

2 Drop GAUGE POST TOOL to bottom in converter.

5 Mount DIAL INDICATOR on gauge post.

6 Set indicator tip on hub.

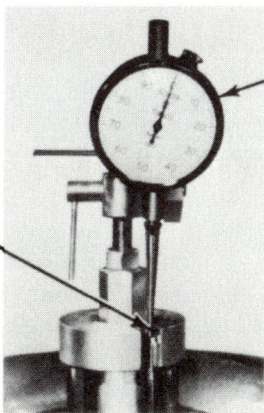

7 Zero the dial.

8 Insert rod through gate post eye for handle and lift to stop.

9 Read end play.

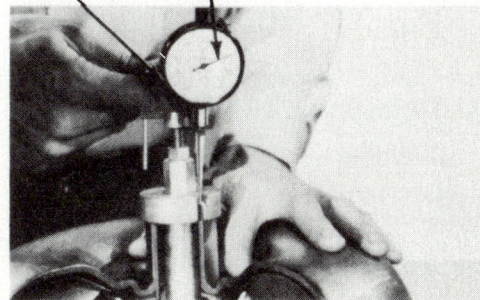

FIGURE 21-10 *Performing the turbine end-play test with factory equipment. (Courtesy of Ford Motor Company)*

Regardless of the reason for the test, it does require certain special tools in order to perform it. If the shop does not already have a leakage test tool, build a substitute with the commonly available parts as shown in Fig. 21-11. When properly constructed, this device can test almost any converter with a drain plug. In order to use this tool on a converter without a plug, you will have to drill and tap the housing for one, or use another type of testing tool that has an air valve built into the expanding plug assembly.

TOOLS FOR BENCH LEAKAGE CHECK

STANDARD 1/8" FITTING — 87971-S FOR
RETAPPED DRAIN PLUG THREADS — USE
1/4" OVERSIZE FITTINGS — 87973-S

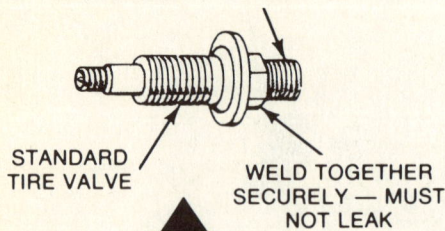

STANDARD
TIRE VALVE

WELD TOGETHER
SECURELY — MUST
NOT LEAK

1 VALVE ASSEMBLY is
used to introduce and
hold air pressure for
leak test.

2 PLUG ASSEMBLY is
used to seal converter
drive hub opening.

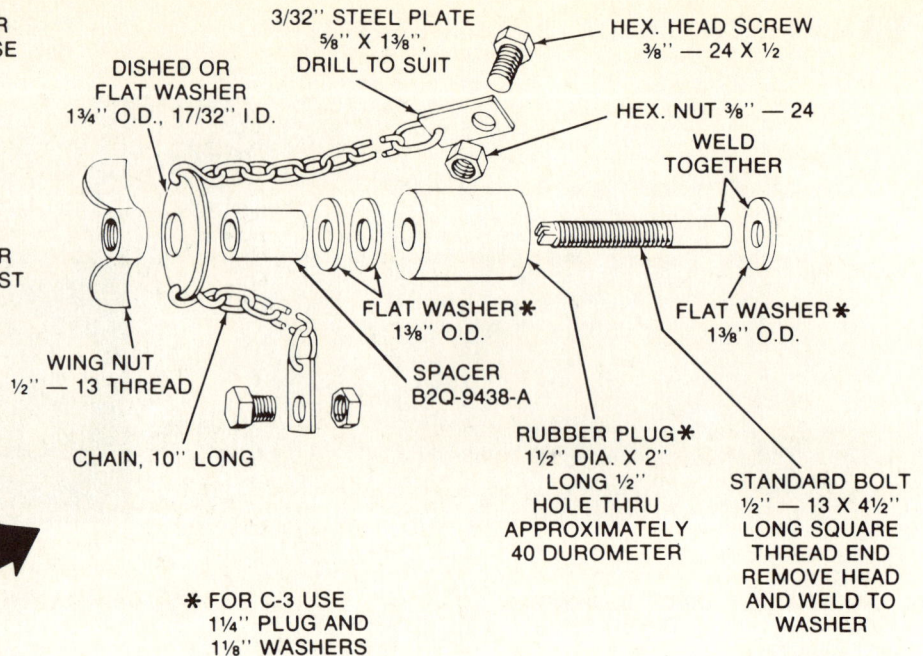

3/32" STEEL PLATE
5/8" X 1 3/8",
DRILL TO SUIT

DISHED OR
FLAT WASHER
1 3/4" O.D., 17/32" I.D.

HEX. HEAD SCREW
3/8" — 24 X 1/2

HEX. NUT 3/8" — 24
WELD
TOGETHER

FLAT WASHER *
1 3/8" O.D.

FLAT WASHER *
1 3/8" O.D.

WING NUT
1/2" — 13 THREAD

SPACER
B2Q-9438-A

CHAIN, 10" LONG

RUBBER PLUG *
1 1/2" DIA. X 2"
LONG 1/2"
HOLE THRU
APPROXIMATELY
40 DUROMETER

STANDARD BOLT
1/2" — 13 X 4 1/2"
LONG SQUARE
THREAD END
REMOVE HEAD
AND WELD TO
WASHER

* FOR C-3 USE
1 1/4" PLUG AND
1 1/8" WASHERS

FIGURE 21-11 A universal torque converter leakage
tester can be made using these parts. (Courtesy of Ford
Motor Company)

To perform a test using the tool shown in Fig. 21-11,
follow these steps:

1. Remove one of the converter drain plugs and remove
all the old fluid from the converter.

2. Clean the outside of the converter with solvent and a
clean rag.

3. Install the air valve in the open drain plug hole and
tighten securely (Fig. 21-12).

4. Install the expandable rubber plug assembly into the
converter hub opening, and expand it by tightening
the wing nut securely. Then, install the safety chain
around the converter.

5. Introduce air pressure into the converter housing via
the air valve. Check the air pressure with a tire gauge,
and adjust it to 20 psi.

6. Place the torque converter into a tank of water for 5 to
10 minutes, and then observe all the welded areas for
signs of bubbles. If you observe no bubbles, assume
that the welds are not leaking.

TORQUE CONVERTER SERVICE

Converter Flushing

If the torque converter has passed all the above-
mentioned tests, assume that it is still serviceable and is
ready for an internal cleaning or flushing. In order to
properly clean a sealed converter, a flushing machine like
the one shown in Fig. 21-13 is necessary. As previously
mentioned in Chapter 15, this machine, when in
operation, cleans a converter by pumping solvent through
the unit while at the same time rotating the turbine.

To set this machine up to clean a typical converter,
perform the following procedures:

1. Check the converter for an available drain plug.
Remove the plug and completely drain the old fluid
out of the converter into a waste oil container. If the
converter has no drain plug, drill a hole in it for a rivet
or drill and tap the converter for a plug as explained
later in this chapter. Having a drain plug in the

CHECK FOR LEAKAGE ON BENCH

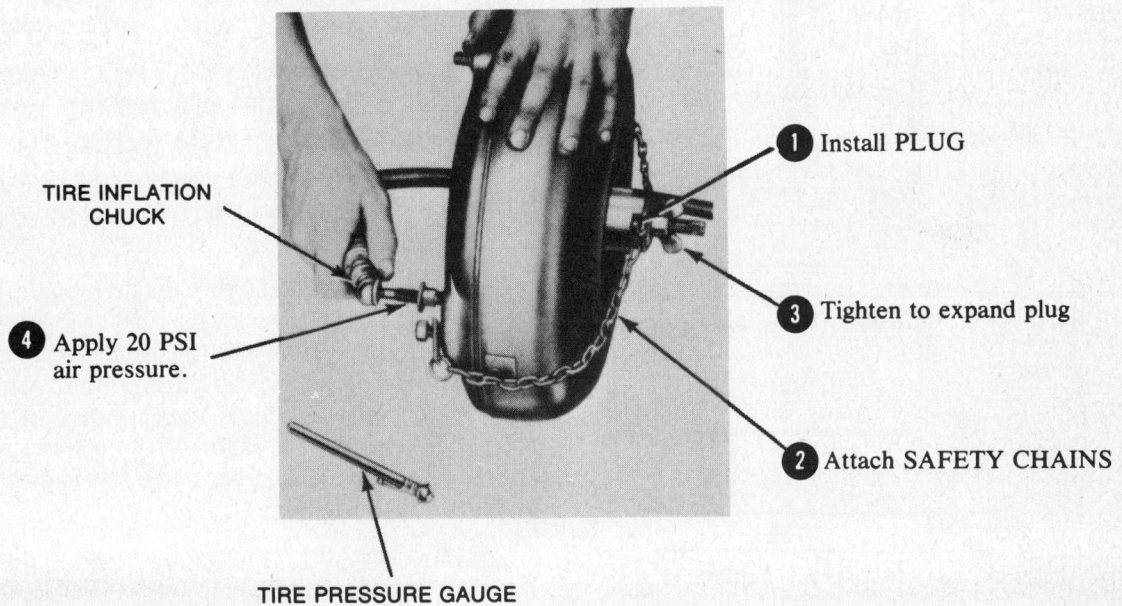

TIRE INFLATION CHUCK

4 Apply 20 PSI air pressure.

1 Install PLUG

3 Tighten to expand plug

2 Attach SAFETY CHAINS

TIRE PRESSURE GAUGE

5 Check for bubbles at possible leak points—
- Wait 5-10 minutes for small leaks
- May not show with converter cold

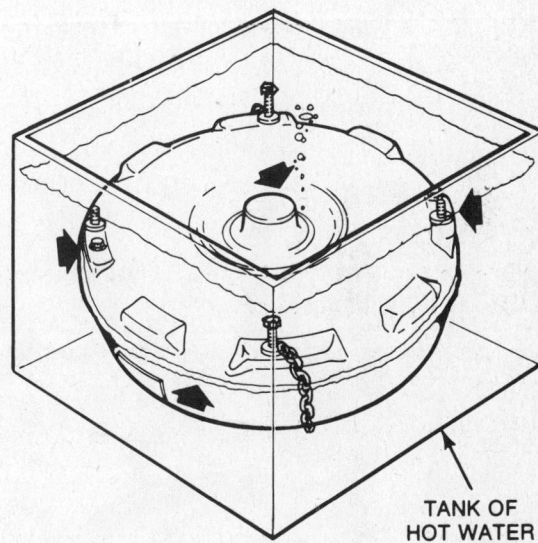

TANK OF HOT WATER

RESULTS

- **Leakage at Weld** — replace converter.

- **Leakage at drain plug** — tighten.

- **No leakage** — check pump and other causes in converter area

FIGURE 21-12   Testing a converter for leakage.
(Courtesy of Ford Motor Company)

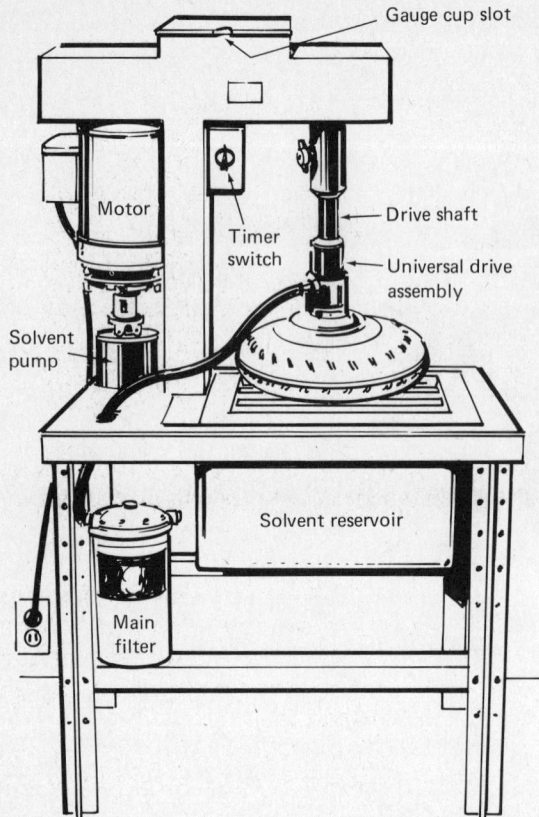

FIGURE 21-13   A typical torque converter flushing machine.

converter makes it easier to remove not only the dirty fluid but, later on, the cleaning solvent as well. Any residual dirty oil left in the converter will contaminate the machine's cleaning solvent, and any solvent left in the converter can circulate through the overhauled transmission, where it can harm the seals.

2. After completely draining the converter, place it on the flushing machine supports over the sump tank with the pump hub side facing up.

3. Insert the drive assembly over the converter hub, and push the hub adapter down as far as it will go (Fig. 21-14).

4. Push down and turn the drive shaft in order to engage the drive tip with the turbine splines. (NOTE: Always make certain that this shaft turns freely but with an inertia effect, caused by the weight of the revolving turbine inside. If the turbine does not feel like it is revolving, or if the drive tip seems as if it's skipping, remove the drive assembly and visually check the turbine splines.)

5. Place the converter assembly directly under the motor-driven shaft. Lower the drive sleeve and align it with the drive shaft of the drive assembly. Lock the drive sleeve in position with the locking knob (Fig. 21-15).

6. Attach the inlet pressure hose to the quick-disconnect fitting on the drive assembly (Fig. 21-16). Make sure that the drive assembly outlet hose is in position inside

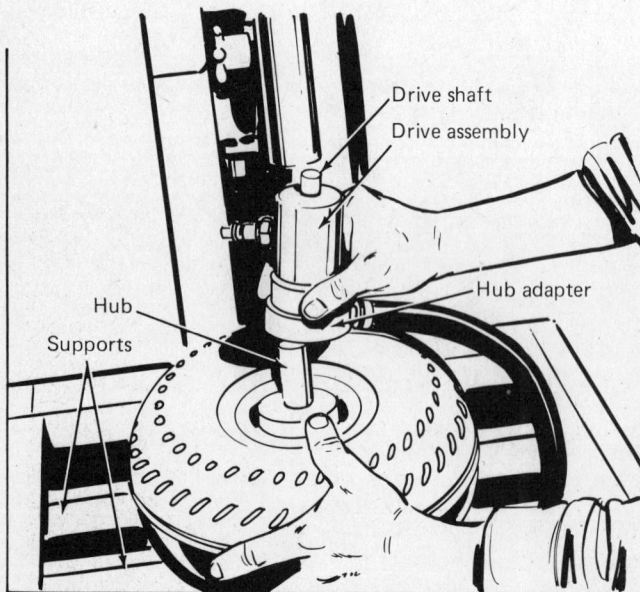

FIGURE 21-14   Installation of the drive assembly onto the converter hub.

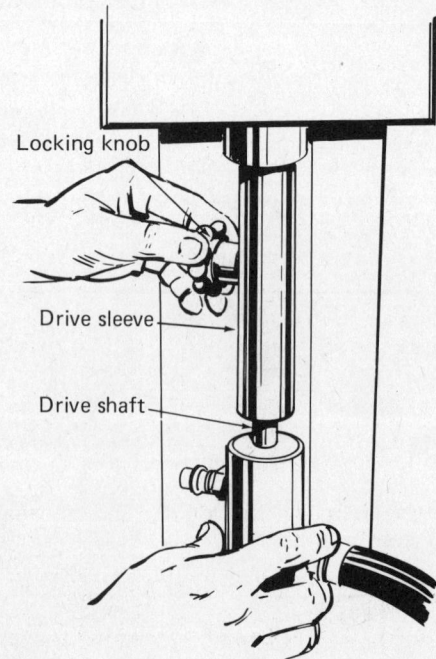

FIGURE 21-15   Locking the drive sleeve in position with the knob.

the sump tank. (NOTE: When flushing a converter with a lock-up feature, it is a good idea to position the outlet hose over a fine wire mesh screen so you can observe the fluid coming out of the unit for signs of friction particles, which indicate failure of the lining on the lock-up piston.)

7. Set the timer to the desired time, 10 minutes for a moderately dirty converter and 30 minutes or more for an extremely contaminated one (Fig. 21-17). (NOTE: If the converter makes a racheting sound while the drive is turning, stop the machine and recheck the drive shaft to make sure it is securely held down in position. If the noise continues, check the splines of the turbine. If, on the other hand, the drive will not rotate the turbine at all, replace the converter with a new or rebuilt unit.)

8. After flushing the converter, loosen the lock knob, raise the drive shaft, and disconnect the inlet hose from the drive assembly.

9. Tilt the converter to a vertical position with the drain plug hole at the bottom. This facilitates a complete draining of the cleaning solvent from the converter. Install the rivet or drain plug and torque the latter to specifications.

Drain Plug Installation

As previously mentioned, some converters do not have factory-installed drain plugs. Consequently, changing the fluid with the transmission and converter still in the

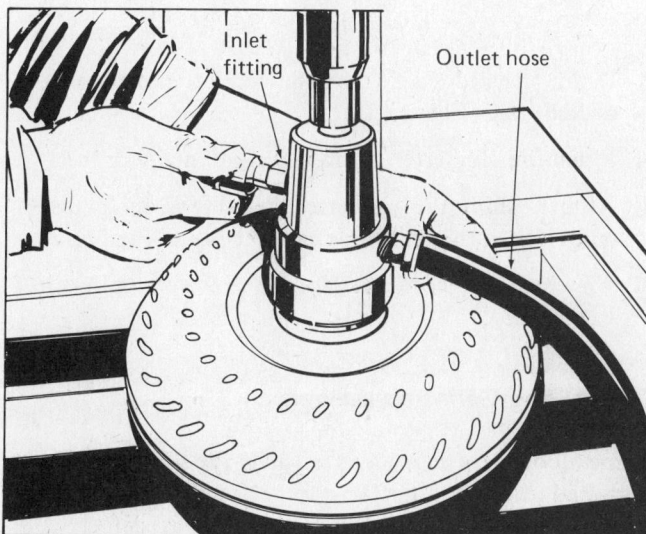

FIGURE 21-17   Setting the timer to the desired position to start the machine and keep it operating for a given period.

vehicle and flushing the converter is more difficult and time consuming. In the long run, then, it will save someone time and effort if whenever one of these converters is out of the vehicle, a mechanic drills, taps, and installs a drain plug in the unit.

Using the drill and tapping equipment referred to in Chapter 15, a technician can install a plug in a converter following these instructions:

1. Clean the outside of the converter with solvent and a clean rag; then place the converter on a clean, wet-type work bench.

2. Install the mount bracket on the converter drive lugs or flanges with the hardware provided with the kit (Fig. 21-18). Tighten the mounting hardware securely.

3. Install the drill guide insert into its receptacle on the mount bracket. Lubricate the guide with oil.

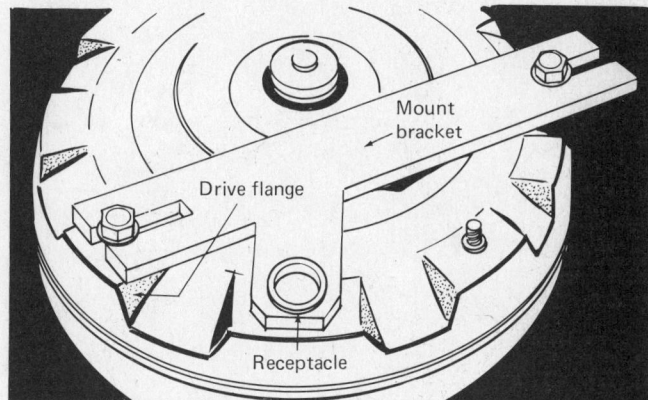

FIGURE 21-16   Connecting the pressure inlet hose to the drive assembly.

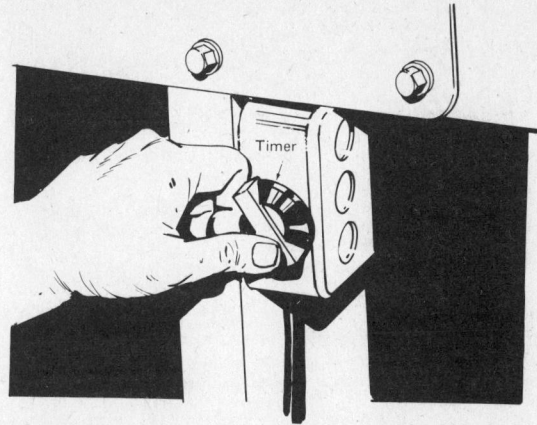

FIGURE 21-18   Installation of the bracket onto the converter.

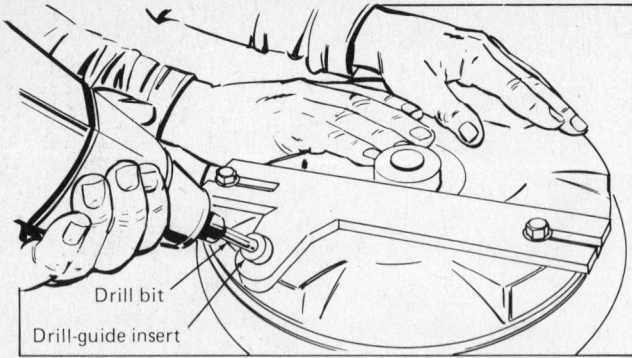

FIGURE 21-19   Drilling the hole into the converter.

4. Install the special drill in a drill motor. Then, bore a hole through the outer converter housing (Fig. 21-19). (NOTE: As the drill begins to cut through the last few layers of metal, ease off your pressure on the motor; otherwise, the drill may strike and damage the turbine as it breaks through the remaining layers of housing material.)

5. With low-pressure compressed air, blow the chips away from the hole and drill **before** removing it from the hole and guide. This procedure prevents excess metal chips from entering the converter.

6. Remove the drill guide insert and install the tap guide insert into the bracket receptacle.

7. Using the ⅛-inch pipe tap supplied with the kit and a tap wrench, cut the plug threads into the drilled hole (Fig. 21-20). (NOTE: To prevent tap or thread damage, it may become necessary to break the chips up by occasionally reversing the direction of the tap wrench. If this becomes necessary, reverse the direction of the tap only one full revolution of the wrench at any given time.)

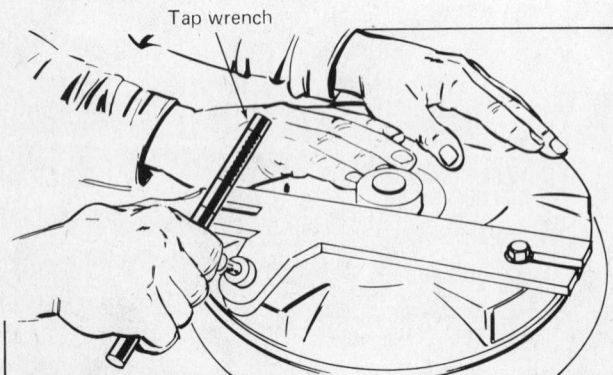

FIGURE 21-20   Cutting the plug threads into the converter using a tap and wrench.

8. Before removing the tap completely from the hole and guide, again blow any chips away from the area using low-pressure compressed air.

9. Remove the hardware and bracket from the converter. Next, stand the converter up with the new drain opening down and permit any remaining fluid to flow out of it.

10. Thoroughly flush the converter as outlined earlier in this chapter. Then, install a new drain plug, and torque to specifications.

Drilling a Drain Hole and Installing a Rivet

Another method sometimes used to drain a converter that has no plug before flushing it is to drill a ⅛-inch hole in the unit and then close the opening later with a rivet. To perform this operation:

1. Center punch and drill the ⅛-inch hole close to the converter weld bead (Fig. 21-21).

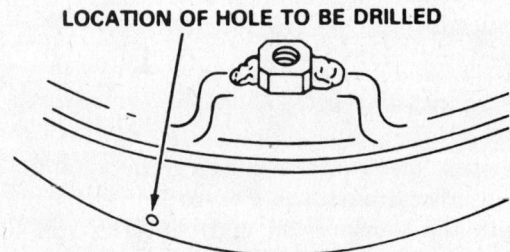

FIGURE 21-21   Location of the drilled hole in the converter. (Courtesy of General Motors Corp.)

2. Drain the old fluid from the converter through the drilled hole (Fig. 21-22).

3. Flush the converter as previously explained.

4. After flushing the converter, install a ⅛-inch closed-end rivet in the drill hole (a) and (b) of Fig. 21-23.

5. Air check the converter for leaks as shown in (c) of Fig. 21-23.

Starter Ring-Gear Replacement

Some modern torque converters have the starter ring gear mounted directly on the outer circumference of the converter front cover. When this gear is defective, a mechanic can remove and replace it. It is not necessary, then, to replace the converter unless it is also defective or leaking.

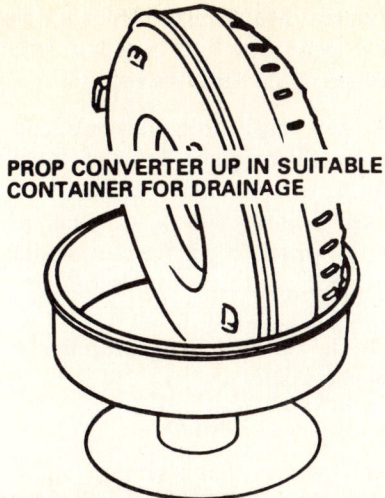

PROP CONVERTER UP IN SUITABLE
CONTAINER FOR DRAINAGE

DRAINING CONVERTER AFTER
DRILLING 1/8" HOLE

FIGURE 21-22 Draining the old fluid from the ⅛-inch hole. (Courtesy of General Motors Corp.)

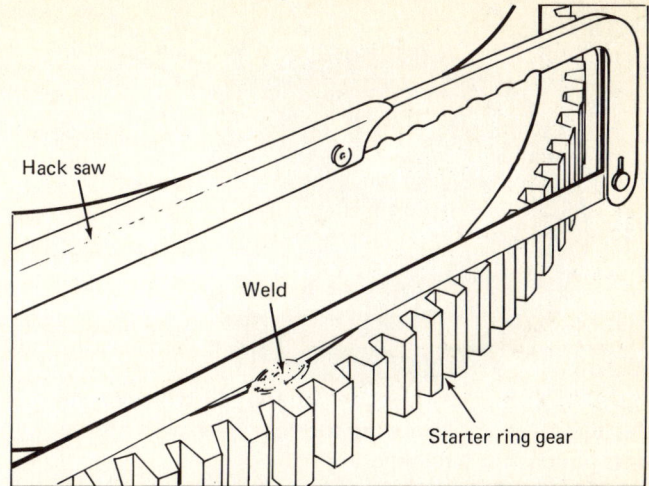

Hack saw

Weld

Starter ring gear

FIGURE 21-24 Cutting through the ring gear weld material.

With the converter removed from the vehicle, take off the gear following these general procedures:

1. Cut through the weld material at the rear side of the ring gear, using a hack saw or grinding wheel (Fig. 21-24). Be very careful not to cut or grind into the front cover.

2. Scribe a heavy line on the front cover next to the front face of the ring gear as an aid in locating the new ring gear.

3. Support the converter on the bench with the four drive lugs or studs resting on blocks of wood. (NOTE: *The converter must not rest on the front cover during this operation.*)

4. Using a blunt chisel or drift punch and hammer, tap

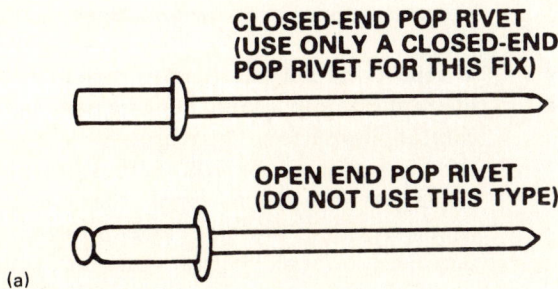

CLOSED-END POP RIVET
(USE ONLY A CLOSED-END
POP RIVET FOR THIS FIX)

OPEN END POP RIVET
(DO NOT USE THIS TYPE)

(a)

POP RIVETS

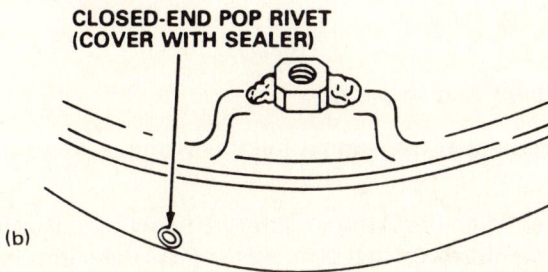

CLOSED-END POP RIVET
(COVER WITH SEALER)

(b)

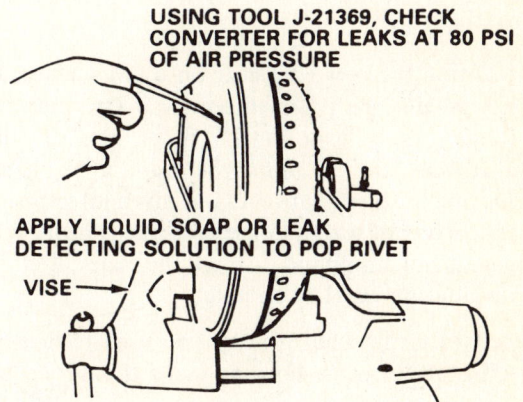

USING TOOL J-21369, CHECK
CONVERTER FOR LEAKS AT 80 PSI
OF AIR PRESSURE

APPLY LIQUID SOAP OR LEAK
DETECTING SOLUTION TO POP RIVET

VISE

(c)

FIGURE 21-23 Closing the hole with a rivet and pressure testing the converter. (Courtesy of General Motors Corp.)

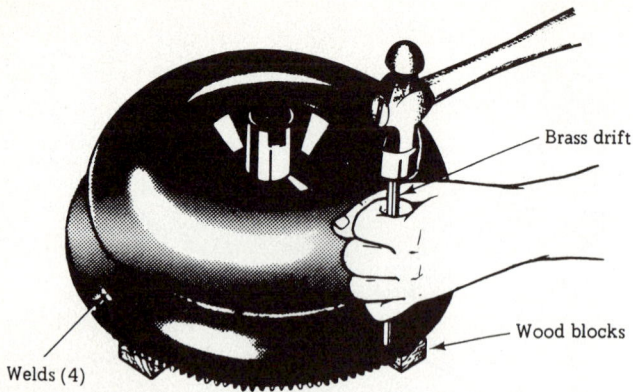

FIGURE 21-25 Removing the rear gear with a brass drift punch and a hammer.

downward on the ring gear near the welded areas in order to break away the remaining weld material (Fig. 21-25). Tap around the gear until it comes completely off the converter housing.

5. Smooth off the welded areas on the cover, using a suitable file.

Using any one of the following methods, heat and expand the starter ring gear before installation on the torque converter. (NOTE: *To prevent injury, wear heavy gloves when handling the hot ring gear.*)

1. Place the new ring gear in an electric or gas oven and set the oven temperature to 200° F. Keep the gear in the oven for 15 to 20 minutes, or

2. Position the new ring gear in a shallow container, add water, and heat the water until it reaches its boiling point. Continue to heat the ring gear, using the boiling water, for about 8 minutes, or

3. Place the new ring gear on a flat surface, and then direct a steam flow around the gear for about 2 minutes, or

4. Position the new ring gear on a flat surface. Using a gas welder with a medium-sized tip, direct a slow flame evenly around the inner rim of the gear. CAUTION: *Do not apply the flame to the gear teeth.* During the heating process, at given intervals, apply a few drops of water to the face of the gear. When the gear is hot enough to just boil the water, it is ready for installation on the converter.

Regardless of what process you used to heat the ring gear, install it on the housing using this procedure:

1. Place the ring gear in position on the converter front cover. Then, tap the gear onto the cover evenly with a plastic or rawhide mallet until the gear's face is even with the scribed line (made during gear removal) on

the front cover. Make certain that the gear's face is even with the scribed line, running around the full circumference of the front cover.

2. Reweld the new ring gear to the torque converter front cover. Be careful to place, as nearly as possible, the same amount of weld material in exactly the same location as the original weld. This procedure is necessary in order to maintain torque converter balance. Also, place the welds alternately on opposite sides of the converter to minimize the possibility of distortion. The following suggestions may assist you in making the weld:

   a. Do not use a gas welder.

   b. Use a DC welder that is set at straight polarity, or an AC unit if the correct type of electrode is available.

   c. Use a ⅛-inch diameter welding rod, and set the welding current to 80 to 125 amps.

   d. Direct the arc at the intersection of the ring gear and front cover from an angle of 45 degrees from the rear face of the ring gear.

3. Inspect all the gear teeth and remove all nicks, raised metal, or weld-metal splatter. This will assure longer gear life and quieter starter operation.

HYDRAULIC PUMP INSPECTION AND SERVICE

Whenever an automatic transmission is overhauled, the hydraulic pump should be disassembled, cleaned, and carefully inspected. The reason for this is twofold: First, the pump gears or rotors must operate with given clearances in order to supply the transmission with the necessary fluid volume and pressure. If clearances become too large, the transmission can malfunction or even fail because of a decrease in fluid flow. Second, as previously stated, the pump is also a very expensive component; the technician, then, should do everything possible to reuse it.

Teardown and Inspection

To determine if the pump is still serviceable, follow these steps:

1. Remove the stator support-to-pump housing attaching bolts and separate the pump assembly into three sections: the pump housing, pump gears or rotors, and the stator support (Fig. 21-26). (NOTE: Before removing the pump gears or rotors from the pump housing, check them both for identification

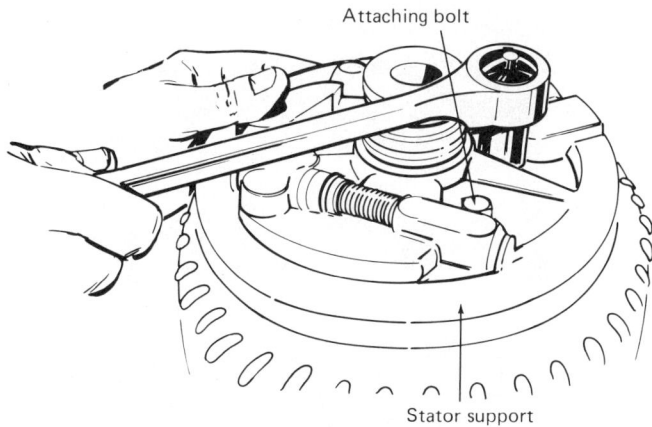

FIGURE 21-26   Removing the stator support-to-housing attaching bolts in order to separate the main components.

marks. If there are none, make an alignment mark across both gears or rotors and the housing with blue layout fluid or an indelible marking pen (Fig. 21-27). *Never use a center punch or metal scribe.* These marks will assist in reinstalling the gears or rotors in the same relative position when you reassemble the pump components.)

2. With the tool shown in Fig. 21-28, remove and discard the pump seal. Remove and discard the pump housing-to-case squarecut seal.

3. Clean the three components in a suitable solvent and blow dry with low-pressure compressed air. CAUTION: *Make sure you thoroughly blow out the fluid passages: any clogged or restricted passage can cause a malfunction.*

4. Check all the parts for burrs and light scoring; if you find any, use crocus cloth to clean and polish the surfaces. Then wash these parts again to remove any grit.

5. Inspect the gears or rotors carefully for broken teeth or lobes, which require replacing the pump.

6. Inspect the inner and outer circumferences of the gears or rotors and also the teeth or lobes for excessive wear. (NOTE: Moderate wear does not necessitate using a new pump. In fact, if the hydraulic pressure was normal during the entire diagnosis procedure, the pump very likely has enough output for reuse.)

7. Inspect the mating surfaces of the gears or rotors, pump housing, and stator support for abnormal wear. Don't be concerned if the black lubrite coating has worn off during previous pump use; this is a normal tendency. Pay special attention to the inside of the pump body (the counterbore) where the gears or rotors operate (Fig. 21-29). Check the depth of any scoring with your fingernail; if there is excessive wear in this area and the pump pressure was low during diagnosis, replace the pump.

8. If everything has checked out so far, reinstall the gears or rotors in the pump body, aligning the marks made during disassembly.

9. Measure the pump body-to-gear or rotor end-play clearance using a straight edge and a feeler gauge (Fig. 21-30). This measurement must be within specifications.

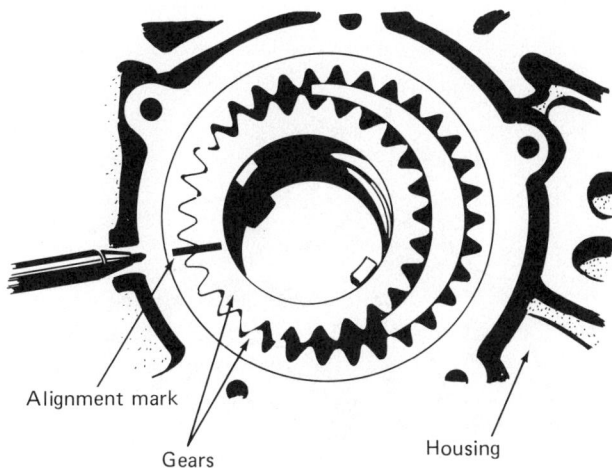

FIGURE 21-27   Marking the gears and housing using blue layout fluid.

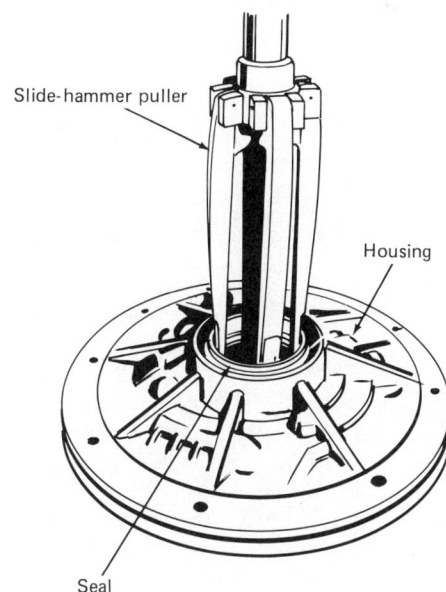

FIGURE 21-28   Removing the pump seal with a puller.

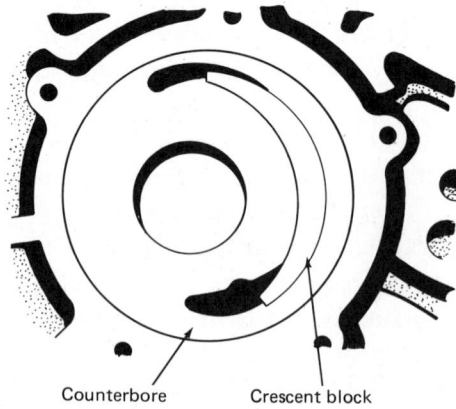

FIGURE 21-29   Checking the pump counterbore for wear.

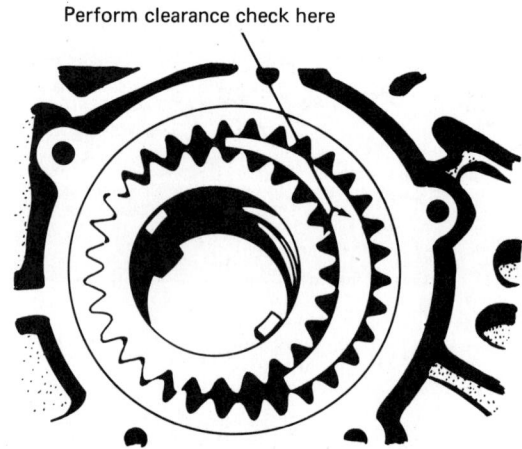

FIGURE 21-32   Locations where you measure the clearance between both gears and the crescent.

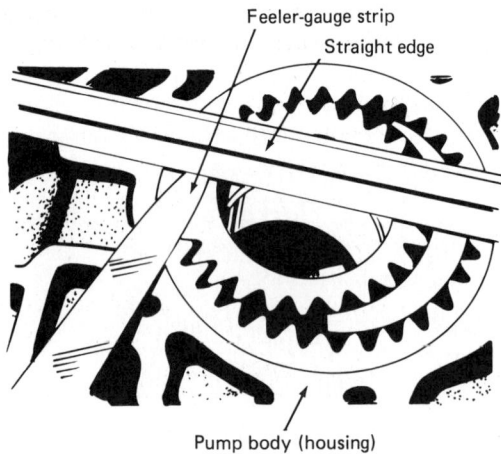

FIGURE 21-30   Measuring the pump body-to-gear end clearance.

10. Using a feeler gauge of the specified thickness, measure the clearance between the outer gear or rotor and the pump housing bore (Fig. 21-31).

11. On the gear-type pump only, measure, with the specified size feeler gauge, the clearance between the teeth of each gear and the pump crescent (Fig. 21-32).

12. On the rotor-type pump only, measure, with the specified size feeler gauge, the clearance between the tips of the inner and outer rotors (Fig. 21-33). (NOTE: *If any of these feeler gauge clearance checks are not to specifications, replace the pump assembly.)*

13. Remove all the sealing rings from the stator support. Inspect the support for worn or damaged ring grooves, clutch drum bearing surfaces, thrust washer surfaces, or splines. If wear or damage exists in any of these areas, replace the stator support or pump assembly.

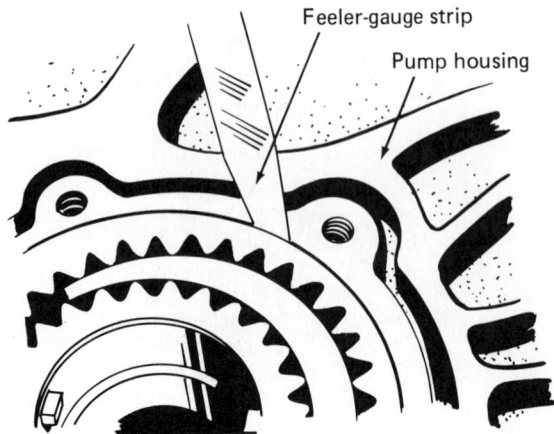

FIGURE 21-31   Measuring the clearance between the outer gear and the pump housing.

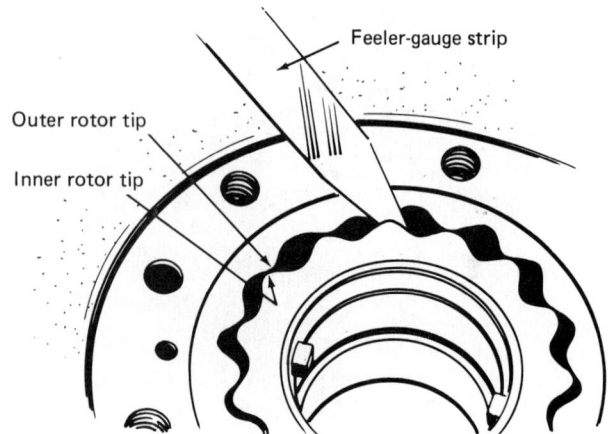

FIGURE 21-33   Measuring rotor tip clearance.

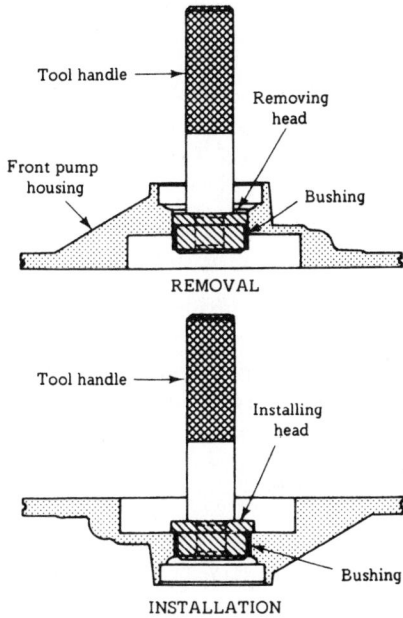

FIGURE 21-34   Removing and replacing a pump converter bushing.

Service and Reassembly

After cleaning and inspecting the pump assembly, service its components and reassemble them as follows:

1. Again remove the gears or rotors from the pump housing.

2. With the tools shown in Fig. 21-34, replace the pump converter bushing. (NOTE: For installation of the bushing, always refer to the transmission service

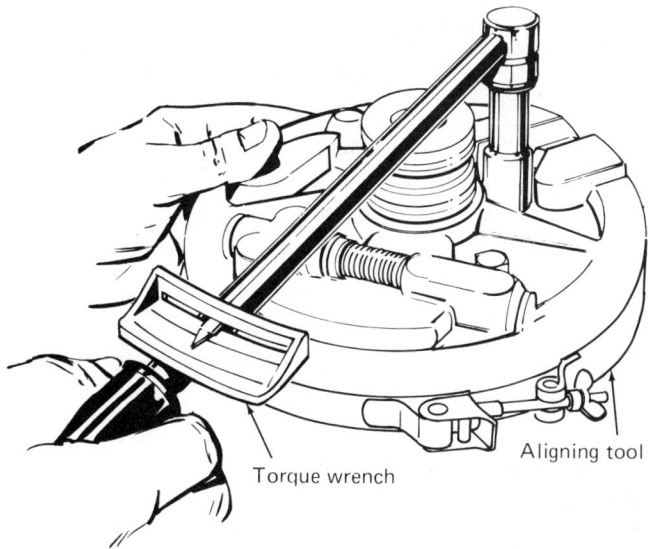

FIGURE 21-36   A typical pump aligning tool installed around the housing assembly, during the torquing procedure.

manual for staking procedures, when used, and the bushing's proper location in the bore. Some bushings have grooves, slots, or oil passages, which have to be located in a given area of the bore for proper bushing lubrication.)

3. Install a new pump seal using the tool shown in Fig. 21-35. (NOTE: If the new seal does not have a rubber or resin-coated backing, spread a thin layer of nonhardening sealer around the contacting edges before installation. Also, make sure the sealing lip faces inward toward the gears or rotors.)

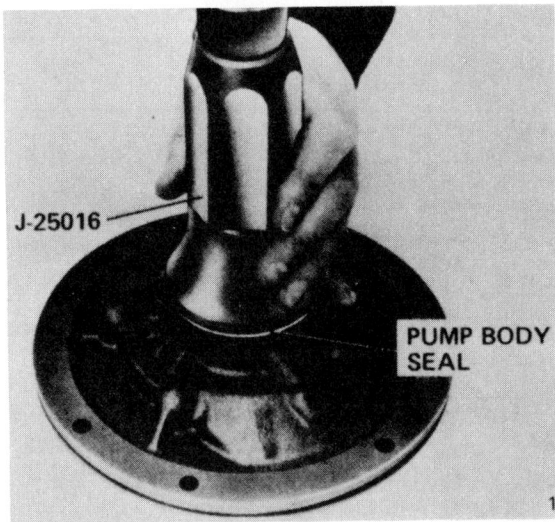

FIGURE 21-35   Installing the pump seal. (Courtesy of General Motors Corp.)

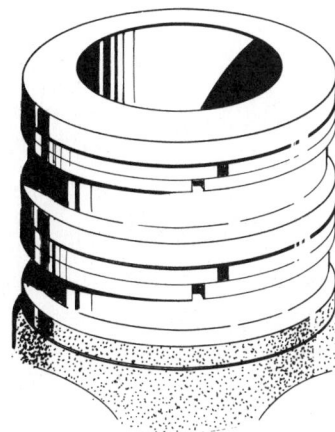

FIGURE 21-37   Replacing the stator support sealing rings.

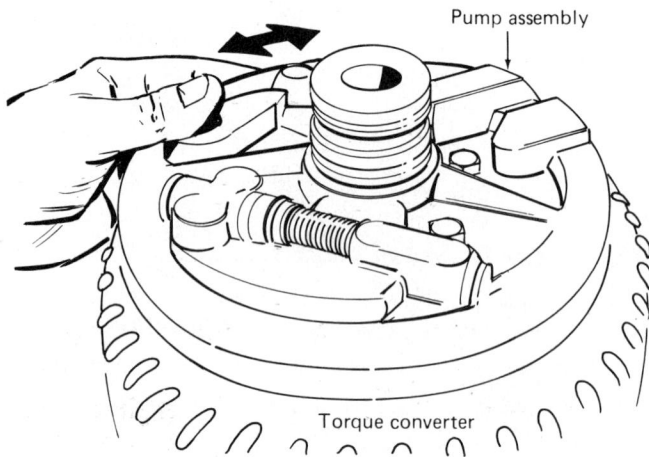

FIGURE 21-38 Checking the pump gears or rotors after pump assembly for free movement.

4. Lubricate the gears or rotors with clean hydraulic fluid and reinstall them into the pump housing, aligning their locating marks.

5. Position the stator support over the pump housing and install the stator support-to-housing attaching bolts. Tighten these bolts finger tight.

6. When specified by the manufacturer, install the housing-to-stator support aligning tool (Fig. 21-36), and torque the stator support-to-housing bolts to specifications.

7. Coat the selective thrust washer with petroleum jelly and install on the stator support.

8. Install the new sealing rings in their proper locations on the stator support (Fig. 21-37). BE CAREFUL: *These rings may be of different diameters. If in doubt as to their proper locations, always refer to the appropriate transmission service manual.* After installation, check the rings for freedom of movement in their respective grooves.

9. Place a torque converter on the bench with its pump drive hub facing upward. Then, position the pump assembly over the converter hub, engaging the hub with the pump drive gear or rotor. Rotate the pump housing in both directions to check the pump gears or rotors for freedom of movement within the pump housing and stator support (Fig. 21-38).

REVIEW

This section will assist you in determining how well you remember the material contained in this chapter. Read each item carefully. If you can't complete the statement, review the section of the chapter that covers the material.

1. It makes good sense to do everything practical to save the converter and hydraulic pump for reuse because they are very _____ components.

2. The most important service operation performed on a converter is to _____ it.

3. The mechanic should perform the serviceability checks _____ using the flusher on the converter.

4. Pilot shoulders permit the converter to run _____ with the engine crankshaft and transmission.

5. There can be converter hub wear at _____ points.

6. Use your _____ to check the depth of the score marks on the converter hub.

7. If the score marks are not too deep on the hub, _____ them away using _____ cloth.

8. If there is any scraping or binding felt during the stator-to-impeller interference test, _____ the converter.

9. The stator one-way clutch test indicates whether or not the clutch will hold against _____ rotation.

10. The turbine end-play test measures the amount of wear on _____ or _____ _____ .

11. The purpose of the converter leakage test is to see if the _____ around the housing are leaking.

12. To properly clean the inside of a sealed converter a _____ _____ is necessary.

13. If the torque converter does not have one, install a _____ _____ before flushing the unit.

14. It is not necessary to replace the converter if the starter _____ _____ is defective.

15. If the hydraulic pump is worn excessively, it may not be able to supply the _____ and _____ necessary to properly operate the transmission.

16. Before removing the gears or rotors from the pump housing, mark them with _____ _____ or an _____ _____.

17. When cleaning the pump housing and stator support, make sure you blow out all fluid _____.

18. Moderate teeth or lobe wear does not always necessitate using a _____ _____.

19. A feeler gauge and a _____ are necessary to perform the body-to-gear or rotor end-play test.

20. When specified by the manufacturer, use an _____ _____ to hold the pump housing and stator support together while you torque their attaching bolts.

For the answers, turn to the Appendix.

CHAPTER 22

Subassembly Cleaning, Inspection, and Service

In addition to the torque converter and hydraulic pump, the automatic transmission contains may other hard and soft parts that the mechanic must clean, inspect, service, or replace during the overhaul to make the job successful. The **hard parts** include such items as the transmission case, extension housing, drums, one-way (overrunning) clutches, servos and accumulators, governors, planetary gear trains, shafts, thrust washers, bearings, bushings, and the valve body. The **soft (commonly replaced) parts** include the clutch plates; bands; metal, Teflon, and rubber sealing rings; metal-clad seals; and gaskets.

Since it would be impossible in the space available to cover all the varied service techniques used by all the transmission manufacturers, this chapter will present only an overview of the commonly used methods for cleaning, inspecting, and repairing or replacing the transmission's hard and soft parts along with the cooler lines and cooler. If you are ever in doubt as to how to perform similar procedures on any particular make of transmission, always refer to the appropriate service manual.

TRANSMISSION CASE

The **transmission case** (Fig. 22-1) forms the main framework or housing for the unit's internal components. Furthermore, its many drilled passages make up a major portion of the transmission's hydraulic circuits. Therefore, it is very important that the mechanic pay special attention to this assembly during an overhaul procedure.

Cleaning the Case

To clean the transmission case properly, the process will have to remove many types of contamination such as oil, grease, road dirt, varnish, or shellac. Of the five, the mechanic can easily remove the first three kinds by steam cleaning or washing the case thoroughly in a safety-type, solvent-filled parts washer. Varnish and shellac deposits are much harder to dissolve.

There are two general methods used to remove varnish or shellac from the case. The first method is to place the case in a jet-type cleaner (Fig. 22-2). This device sprays a solution of hot water mixed with a chemical that dissolves the contaminants without damaging the metal, while the case revolves inside the machine. The second method is to submerge the case in a cold or hot tank filled with a chemical that also cuts through the varnish and shellac without harming the metal itself. Cold tanks normally contain a chemical similar to carburetor cleaner, while a hot tank holds a solution much like that used in a jet cleaner.

No matter what chemical treatment the technician uses to clean off the varnish or shellac built up inside a case, he should follow it up by rinsing and air drying the case. The serviceman can rinse a case by either steam cleaning it again or by washing it thoroughly with large quantities of water—perferably hot. After the rinsing process, he should then air dry the case with low-pressure compressed air, *making sure to blow out all passageways completely.*

Case Inspection

With the case clean and dry, inspect it for the following problems:

1. Damaged or stripped threads.

2. Plugged or restricted fluid or ventilation passages.

3. Worn or damaged bushings.

4. Worn or damaged manual valve and parking pawl linkages and seal.

5. Worn or damaged clutch plate slots or lugs and snapring grooves.

6. Worn or damaged governor, band anchor pin, modulator valve, and speedometer gear bores.

7. Damage, cracks, or case porosity. If the case has damage, cracks, or porosity, it may be better to replace the unit at this point because repairs for these

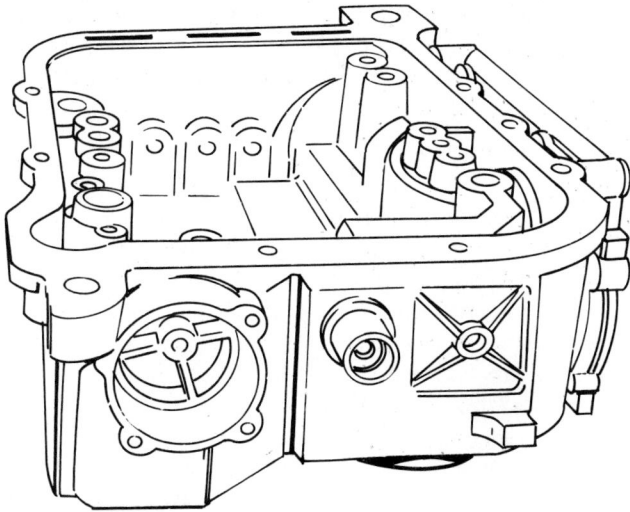

FIGURE 22-1   A typical transmission case.

defects are not always successful. But this chapter, later on, will cover a repair for case porosity.

Thread Repair

To repair damaged threads with the Heli-coil kit, referred to in Chapter 15, follow this simple procedure:

1. With the specified drill bit and motor, machine out all the damaged threads (Fig. 22-3).

2. With a tap wrench and the special tap included in the kit, cut the threads for the Heli-coil into the drilled hole.

3. Thread the installation tool into the Heli-coil. Make certain that the coil's driving tang fully engages with the slot in the end of the installation tool.

4. Blow out all the metal chips from the tapped hole, and

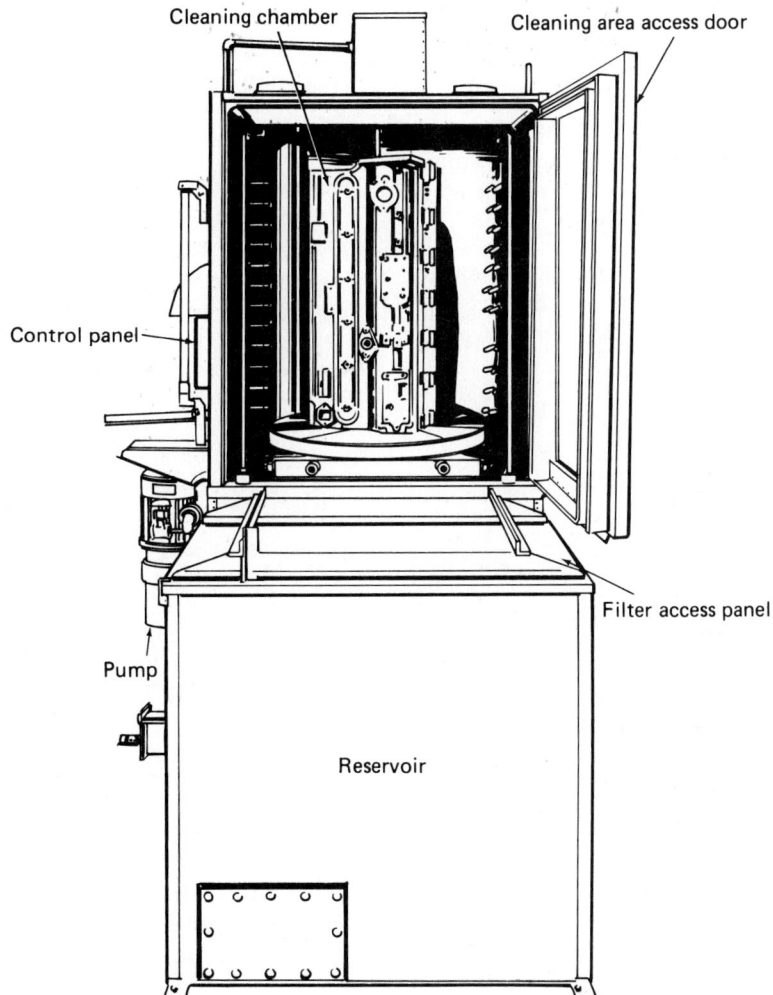

FIGURE 22-2   A jet cleaner used to remove varnish or shellac from the transmission case.

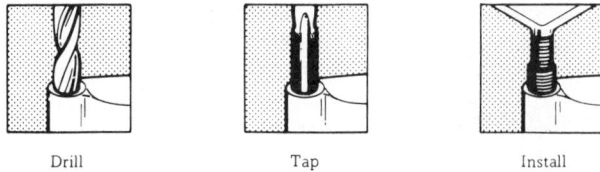

Drill　　　　　　Tap　　　　　　Install

FIGURE 22-3　Removing damaged threads using a Heli-coil repair kit.

thread the Heli-coil into it, using the installation tool. Stop the coil installation when the insert's top is ¼ to ½ turn below the metal surface of the hole.

5. Remove the installation tool from the coil, and break the tang off the coil at its notched point with a square-end punch or pliers.

Bushing Replacement

A transmission case bushing supports and guides the output shaft and (in some designs) the input shaft as well. This bushing must be in good condition or the affected shaft will wobble in the case, causing noise and possible vibration. If upon inspection the bushing shows damage or excessive wear, replace it using the equipment mentioned in Chapter 15 and this procedure:

1. Using a bushing chisel, cut along the bushing's seam until the tool breaks through the wall of the part (Fig. 22-4). Pry up the loose ends of the old bushing with an awl or needle-nose pliers and remove the part, or

2. With the handle and correct-size driving head, press or drive the old bushing from the case (Fig. 22-5).

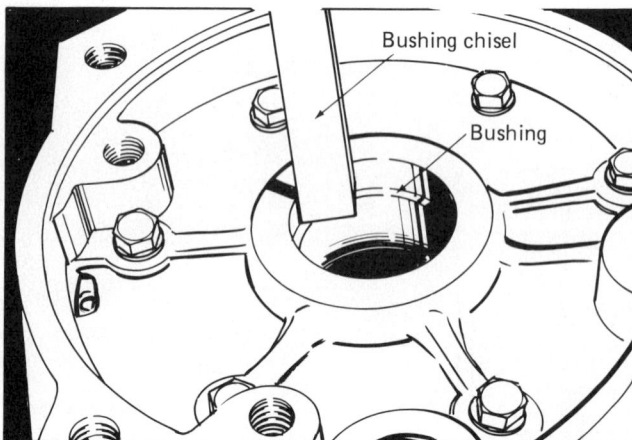

FIGURE 22-4　Removing a transmission case bushing using a cape or bushing chisel.

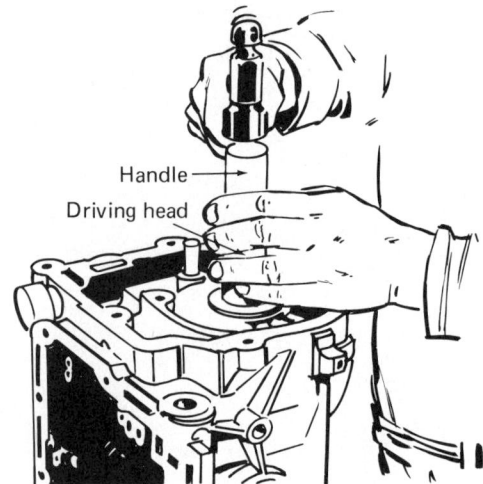

FIGURE 22-5　Using a driving head and handle to remove a bushing from the transmission case.

Install the new bushing using the following procedures:

1. Inspect the bushing bore for cracks, wear, or damage.

2. Position the new bushing into its bore. Be sure to align, where used, any bushing lubrication holes or grooves with the corresponding passages or openings in the case. Always start a new bushing into the chamfered side of the bore. This chamfer helps prevent the bushing from cocking in the bore as you press or drive it into place.

3. Using a press or hammer, the handle, and the correct-size installing head, insert the bushing into its case bore. Be sure to press or drive the bushing in sufficiently so its ends are the same distance from each end of the case bore, or the distance specified by the manufacturer (Fig. 22-6).

4. Carefully insert the shaft that rides in the bushing, and check it for free rotation within the bushing. If the shaft will not go into place or is hard to turn, the pressing operation damaged the bushing, and you will most likely have to replace it.

Seal Replacement

The **shifter shaft seal** prevents external fluid leakage around where both the throttle and manual valve shafts protrude through the case. The mechanic should always change this seal during a transmission overhaul because a leak may develop later on.

In order to replace a shifter shaft seal in the side of a typical transmission case, follow this general procedure:

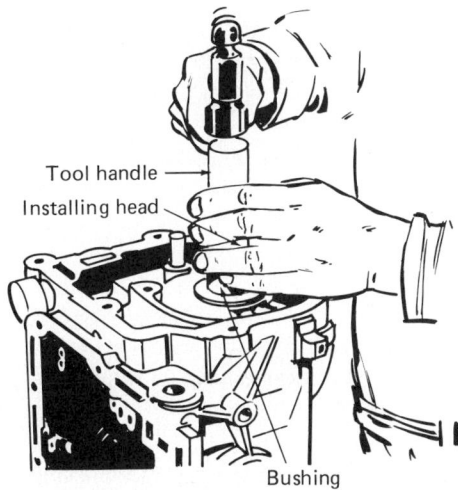

FIGURE 22-6   *Installing a typical transmission case bushing.*

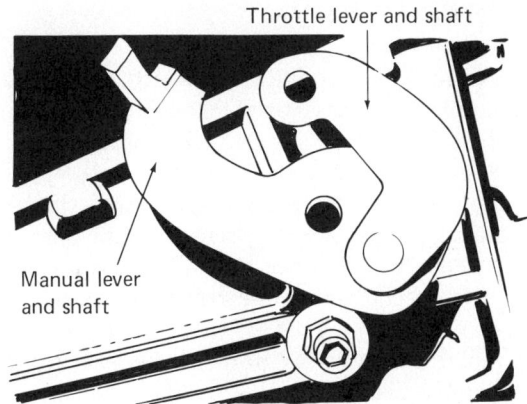

FIGURE 22-8   *After removing the throttle- and manual-valve levers, just pull the two shafts from the case.*

Use the following service steps to install the new seal and replace the shafts and levers:

1. From inside the case, loosen the throttle lever attaching bolt, and remove the lever from its shaft (Fig. 22-7).

2. Also from inside the case, loosen the manual-valve lever attaching bolt, and remove this lever from its shaft.

3. From the outside of the case, remove both the manual and throttle-lever shafts from the case (Fig. 22-8).

4. With a suitable puller or screwdriver, pull or pry the seal from its case bore.

1. Using a hammer, handle, and the correct-size installing head, drive the new seal into the case (Fig. 22-9). CAUTION: *Be certain that the seal firmly seats into the case counterbore, and its lip faces inward or toward the inside of the case.*

2. Lubricate both the manual- and the throttle-valve shafts, and carefully reinstall them through the seal and into their case bore.

3. From inside the case, reinstall the manual-valve lever over the manual lever shaft, and torque its attaching bolt to specifications.

4. Also from the inside of the case, reinstall the throttle-valve lever over the throttle lever shaft, and torque its attaching bolt to specifications.

FIGURE 22-7   *Removing the throttle- and manual-valve levers from the inside of the transmission case.*

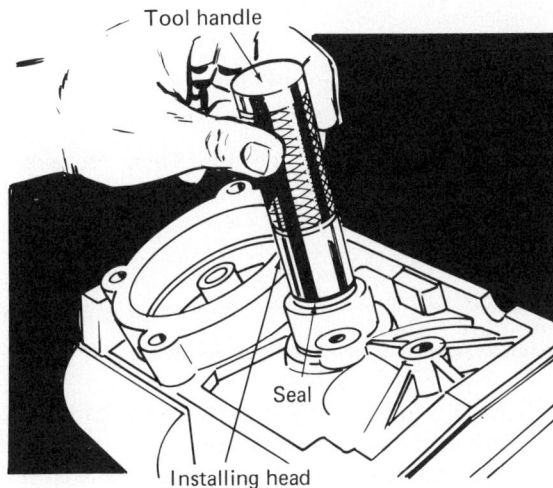

FIGURE 22-9   *Installing the shifter shaft seal.*

Repair of Case Porosity

Case porosity is a condition in which a pressurized or static fluid leak has developed through the outside of the metal case in certain places. It is possible, in some situations, to repair a leak of this nature, without removing and/or replacing the case itself, through the application of a special epoxy cement to the porous area. While this method does not always work, it is sometimes worth attempting *before the transmission is removed from the vehicle.*

To repair a case with a porous-type leak, follow this recommended procedure:

1. Road test the vehicle in order to bring the transmission's temperature up to operating range, approximately 180° F.

2. With another mechanic in the vehicle, raise it with an overhead hoist or floor jack. If you use a floor jack, position jack stands under the frame at suitable locations, and lower the vehicle down on them.

3. Locate the leak by having the other mechanic start the engine and operate the transmission in all its driving ranges. The use of a mirror and flashlight is helpful in finding a leak of this nature.

4. Shut the engine off and thoroughly clean the porous area with a cleaning solvent and brush. Then blow the area dry with compressed air.

5. Using the manufacturer's instructions, mix a sufficient amount of epoxy to make the repair. *Observe all the precautions set forth by the manufacturer while handling this material.*

6. While the transmission case is still hot, apply the epoxy to the area under repair. A clean, dry, soldering brush works well to not only clean but apply the epoxy cement to the area. CAUTION: *Make sure to completely cover the porous area.*

7. Allow the cement to cure for three hours before starting the engine to perform a leak test.

8. Lower the vehicle and road test. Then, recheck the repaired area for leakage. If the case still leaks, remove the transmission and replace the case.

EXTENSION HOUSING

The **extension housing** itself attaches to the back end of the transmission case. Its function is to enclose and protect the components that extend from the rear of the case such as the output shaft, governor, and speedometer drive assembly (Fig. 22-10). If the mechanic removes this

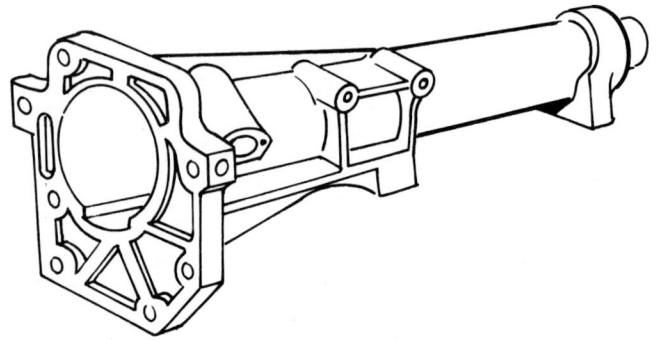

FIGURE 22-10   A typical extension housing.

housing from the case as during a transmission overhaul, he should clean, inspect, and service it before reinstallation.

Cleaning

Before cleaning the extension housing with the **same** procedures used on the transmission case, the technician should first remove the extension housing oil seal. The reason for this is that most solvents or cleaning agents used to dissolve varnish or shellac will deteriorate this seal. But if the seal is in good condition and the mechanic is not going to change it, *he can wash the housing in a cleaning detergent that will not harm the seal,* and then blow the assembly dry with compressed air.

Inspection

With the extension housing removed and cleaned, inspect it for the following defects:

1. Cracks or signs of other visible damage.

2. Worn or hard lip on the oil seal.

3. Worn or damaged bushing.

4. Stripped or damaged threads.

5. Warpage at its mating surface.

Bushing and Oil Seal Replacement

The **extension housing bushing** supports the output shaft and the slip yoke of the drive shaft. This bushing is a precision-type bearing that requires no reaming or finishing after installation.

To replace a worn or damaged bushing along with the oil seal, follow these directions:

1. With a chisel or suitable puller, remove the old seal from the housing (Fig. 22-11).

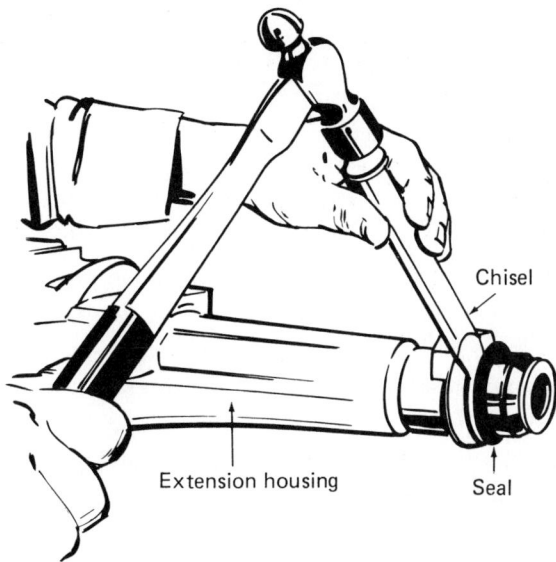

FIGURE 22-11 **Using a hammer and chisel to remove the extension housing seal.**

2. With a bushing chisel, cut along the bushing's seam until the tool breaks through its wall. Next, pry the loose ends of the old bushing up with an awl or needle-nose pliers and remove it from the housing, or,

3. With the handle and correct-size head, press or drive the old bushing from the housing (Fig. 22-12).

To install the new bushing:

1. Inspect both the bushing and seal bores for cracks, wear, or damage.

2. Position the new bushing into its bore. Be sure to align, where used, any bushing lubrication holes or grooves with the corresponding passages or openings within the housing.

3. Using a press or hammer plus the tool handle and the correct-size installing head, drive the bushing into its housing bore (Fig. 22-13). Be certain to install the bushing in far enough so that its end is flush with the top of the bore, or the distance specified by the manufacturer.

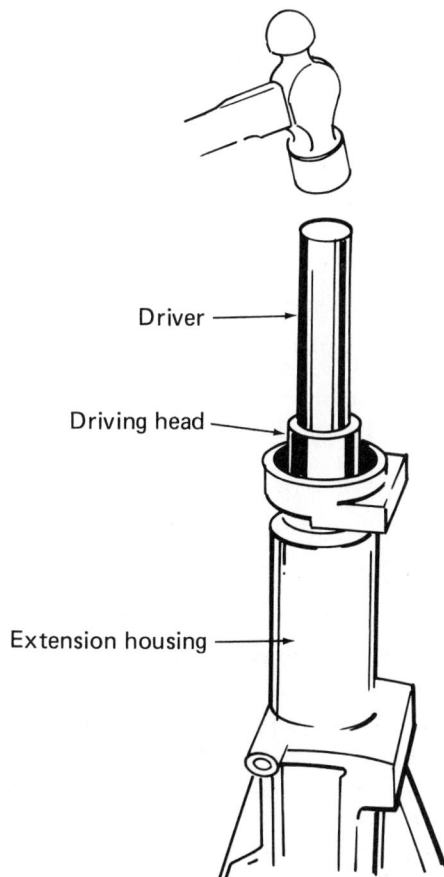

FIGURE 22-12 **Removing the extension housing bushing.**

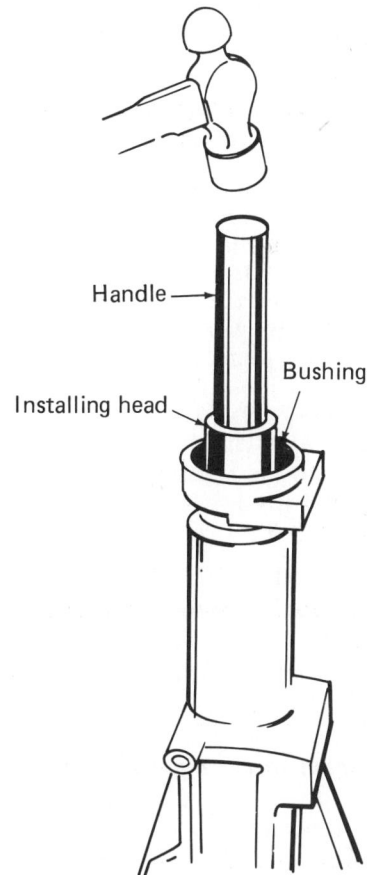

FIGURE 22-13 **Installing the extension housing bushing.**

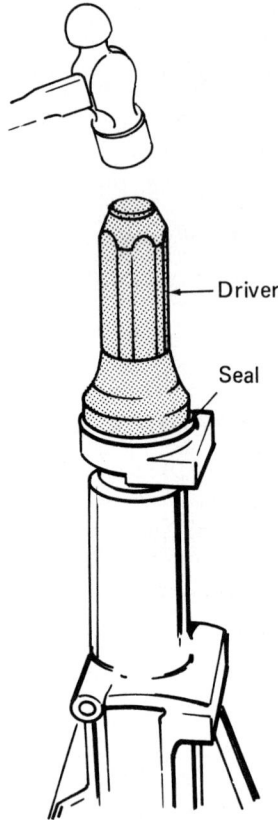

FIGURE 22-14   Installing the extension housing metal-clad seal.

4. Carefully insert the slip yoke that rides in the bushing, and check it for free rotation within the housing. If the yoke will not go into the bushing or is hard to turn, the installation process damaged the bushing, and you will most likely have to replace it.

5. Remove the slip yoke. Using a hammer and the driver, install a new seal into the housing. Be sure that the seal firmly seats into the housing counterbore, and *its lip faces inward or toward the bushing* (Fig. 22-14).

DRUMS

A drum may serve several functions within the automatic transmission. For example, it can act as the housing for a multiple-disc clutch assembly used to connect a planetary gear train member to the input shaft. Or along with a brake band, a drum can hold a planetary gear train member stationary.

Since the drum can serve several different functions, each independent of the other, its cleaning, inspection, and service requirements will also vary slightly. To differentiate the various drum usages and their service

requirements, this chapter will refer to them as either a clutch drum or brake drum.

Clutch Drum Disassembly

Before cleaning and inspecting a typical clutch drum, the mechanic must first remove its internal components. To accomplish this, he should follow this procedure:

1. Remove the retainer snap ring with a screwdriver, and lift the cover or flange assembly from the clutch drum (Fig. 22-15).

2. Remove the clutch hub's outer thrust washer if so equipped.

3. Lift out the clutch hub. Remove the clutch pack and the hub's inner thrust washer, if the unit has one. Then, tie all these components together with a piece of fine wire as an aid to reassembly.

4. Install a spring compressor (refer to Chapter 15) over the clutch spring retainer, and slowly compress the springs until you can easily remove the snap ring with a scribe, screwdriver, or pliers (Fig. 22-16). Carefully release the tool's pressure on the springs, and remove the retainer and springs.

5. Lift up on the piston with a twisting motion to remove it from the drum. (NOTE: On some clutch assemblies, it is necessary to direct air pressure to the clutch apply port in order to force the piston from its bore. Then, remove the inner piston seal from the drum guide and the outer seal from the piston.)

Clutch Drum and Component Cleaning

To prepare the clutch drum and its various components for inspection, wash or clean all the metal parts as outlined under the section on case cleaning. CAUTION: *Do*

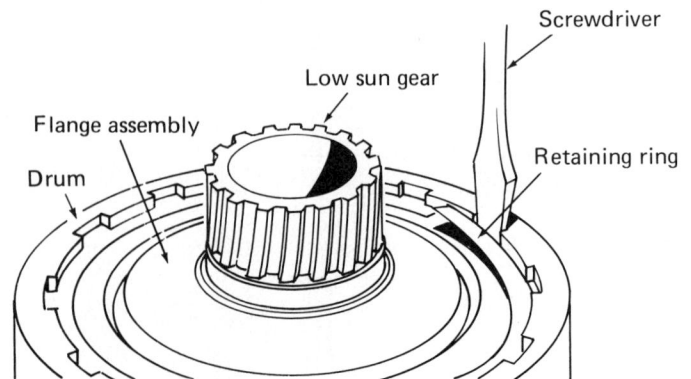

FIGURE 22-15   Removing the retainer snap ring from a typical clutch drum.

not wash or submerge any rubber seals or friction plates in solvent or a varnish dissolving agent. These materials will deteriorate the rubber seals and the lining on the plates.

Clutch Drum and Component Inspection

With all the components clean and dry, inspect them for the following:

1. The drum bushing for excessive wear or scoring.

2. The steel residual check ball in the clutch drum. Be sure that this ball is free to move in its bore and the orifice leading to the front of the drum is open. (NOTE: *On some clutch assemblies, the check ball's location is in the piston itself.)* In either case, if the check ball is loose enough to come out or not loose enough to rattle, replace the clutch drum or the piston assembly. CAUTION: *Do not attempt to replace or restake the check ball.*

3. The fit of the clutch flange (cover) in the drum slots. There should be no appreciable radial play between the two components.

4. The drum for burred, scored, or damaged thrust surfaces.

5. The inside of the drum for rough or badly worn clutch plate slots or grooves.

6. The drum for nicked or deeply worn sealing ring bearing surfaces and for obstructed or restricted fluid passages.

7. The piston bore in the drum for wear or scoring.

8. The piston for cracks, wear, or damage.

9. The piston return springs for wear, damage, or signs of overheating.

10. The friction plates for excessive wear; cracked lining material; charred, burned, or glazed lining; pitted or scored lining; or distortion.

11. The steel plates for distortion, surface scuffing or scoring, and damaged drive lugs.

12. If the clutch drum or any of its various parts are defective, replace them as necessary before reassembling the unit.

Clutch Drum Reassembly

To put the clutch assembly back together, follow these steps:

1. Prepare the steel and friction plates as outlined in this chapter under the section on soft parts—inspection and service.

2. Install a new piston inner seal over the hub of the clutch drum.

3. Install a new outer seal in the groove located on the clutch piston. Refer to the portion of this chapter entitled "Transmission Soft Parts" for further details on the installation of both the inner and outer piston seals.

4. Lubricate the seals and the piston bore in the drum generously with clean transmission fluid. Reinstall the piston in the clutch drum using a twisting motion. (NOTE: If the inner and outer seals are the lip type, it may also be necessary to use a feeler gauge strip as an aid in installing the piston. If this is necessary, slowly work the gauge strip around the circumference of the

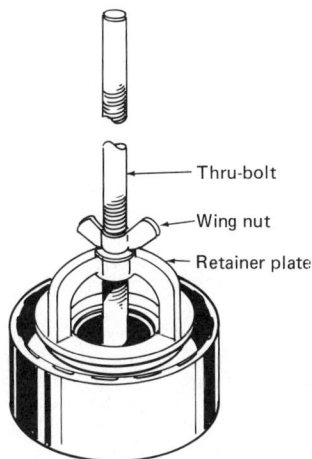

FIGURE 22-16   *Using a compressing tool to relieve the tension on the piston return springs.*

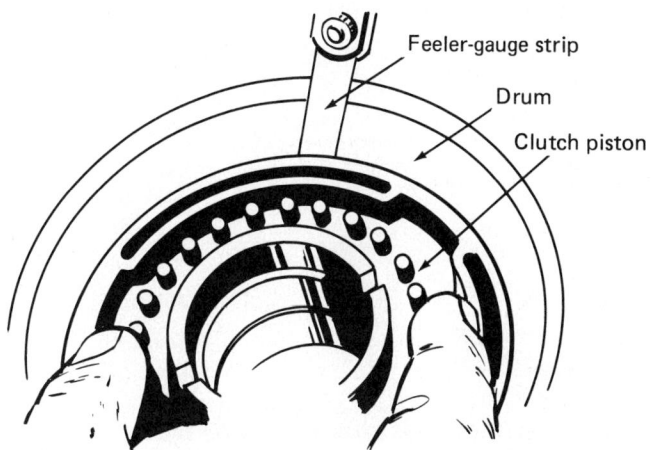

FIGURE 22-17   *Using a feeler gauge as an aid in installing a clutch piston.*

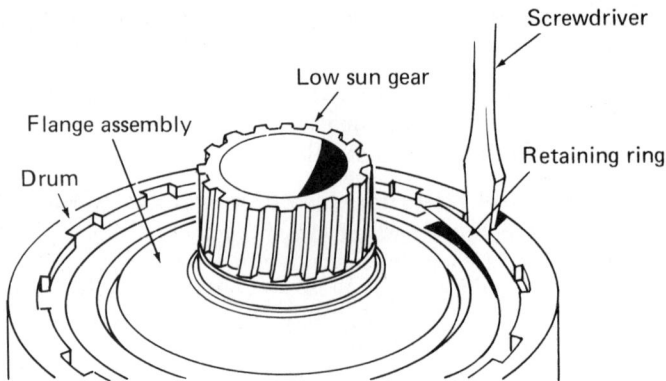

FIGURE 22-18   Installing a clutch drum retainer snap ring.

piston to avoid damage to the seal, while applying some slight downward pressure on the piston itself [Fig. 22-17]).

5. Position all the springs onto the installed piston. Then, place the retainer in position on the springs.

6. Using the compressor tool shown in Fig. 22-16, slowly depress the retainer plate and the springs far enough to permit the installation of the snap ring into its groove in the clutch hub guide.

7. Reinstall the clutch hub inner washer, if used. Next, install the clutch hub over the washer.

8. Install the steel clutch plates and the faced friction discs alternately, beginning with a steel plate.

9. Install the outer, clutch hub washer, if used, with its flange upward or toward the cover flange. Then, reinstall the cover flange assembly, and secure it with the snap ring (Fig. 22-18). (NOTE: When installed, the

opening in the snap ring should be across from one of the flange slots in the clutch drum.)

10. Check the clutch assembly by turning the clutch hub to be certain that it is free to rotate. (NOTE: On some types of clutch assemblies, it is also necessary to check clutch plate operating clearance. When in doubt as to whether this check is necessary, refer to the appropriate transmission service manual.)

BRAKE DRUMS

Cleaning and Inspection

Before inspecting a brake drum for defects, clean it as outlined in the section on case cleaning. With the drum clean and blown dry with low-pressure compressed air, inspect it for the following defects:

1. Worn, scored, or damaged thrust washers.
2. Worn or damaged splines.
3. Worn or damaged drive lugs.
4. Worn, scored, or damaged band application surface.
5. Worn or scored bushing.
6. Cracked or damaged drum surfaces.

Brake Drum Service

If the drum is still usable, perform the following service operations on it:

1. Remove and replace the drum bushing, if so equipped, using the tools and procedures as outlined in this

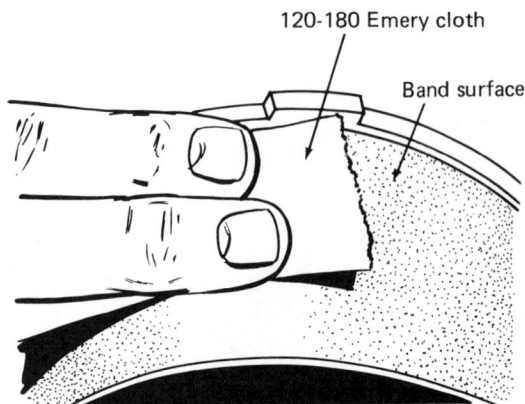

FIGURE 22-19   Deglazing a drum that uses a paper-lined band.

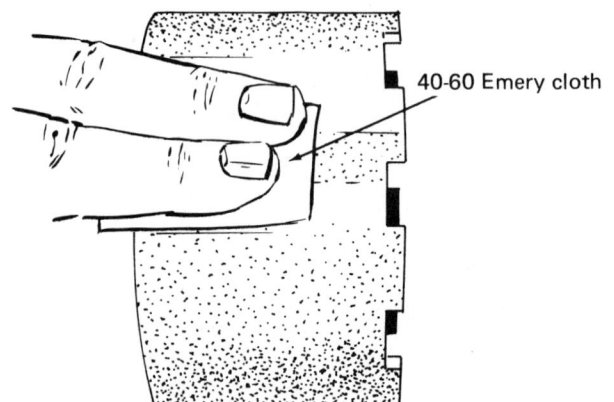

FIGURE 22-20   Deglazing a drum that uses an asbestos-type band.

chapter under the section on transmission case or extension housing service.

2. Deglaze the portion of the drum where the band rides to restore its friction characteristic. For drums that use a paper-lined band, deglaze the band with a 120-180 grit sandpaper or emery cloth by sanding around the drum, but never in a front to back direction (Fig. 22-19). For drums that have an asbestos-lined band, deglaze with 40-60 grit emery cloth or sandpaper by sanding the drum front to back (Fig. 22-20).

ONE-WAY (OVERRUNNING) CLUTCHES

Automatic transmissions have one or more one-way clutches for the main purpose of holding a component stationary in one direction; but when necessary, this same device allows the same part to rotate freely in the opposite direction. The stator one-way clutch, within the torque converter, is a good example of one practical use for this device. In this situation, the clutch locks the stator against any counterclockwise rotation, but it permits the stator assembly to turn freely in a clockwise direction.

Within the transmission itself, this clutch acts as a holding and freewheeling mechanism for a given planetary gear train member. In other words, the one-way clutch holds a planetary member against any counter-clockwise rotation, but it permits the same member to spin freely in the other direction.

There are two basic types of one-way clutches used to hold a planetary member, the sprag and roller. Although these clutches each have a different design, both of them do the same job in a slightly modified manner. Due to

their somewhat different design, each one will have special inspection and service requirements.

Cleaning

Because of the size and, in the case of the roller-type clutch, the number of individual parts, you must follow certain precautions when cleaning the unit to prevent the loss of parts. With this in mind, take the disassembled clutch components and do the following:

1. Place them in a small parts basket and submerge them in an agitated solvent tank or cold tank solution.

2. After soaking the parts for a period of time in the solvent tank, blow them dry, using low-pressure compressed air. (NOTE: The parts cleaned in a cold tank require neutralizing with water before blowing them dry with compressed air.)

Inspection: Sprag Type

To inspect a sprag-type clutch (Fig. 22-21), check for these conditions:

1. Scored or damaged surfaces on both the inner and outer races.

2. Excessively worn or damaged sprag segments.

3. Bent or damaged sprag spring retainer.

Inspection: Roller Type

To inspect a roller-type clutch (Fig. 22-22), check for these defects:

1. Scored or damaged surfaces and signs of brinelling on the inner race.

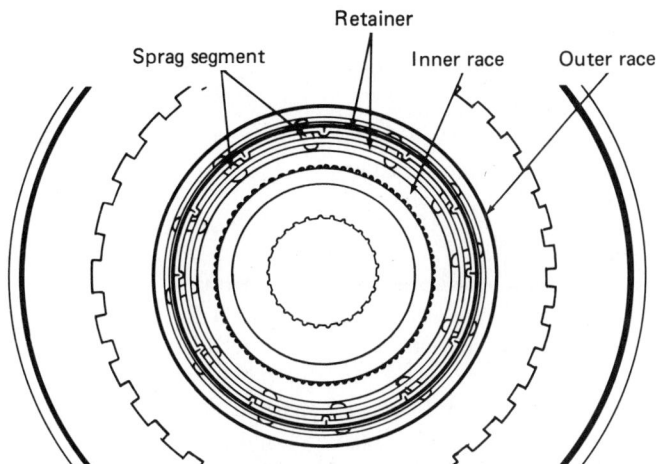

FIGURE 22-21 Inspecting the components of a typical sprag clutch.

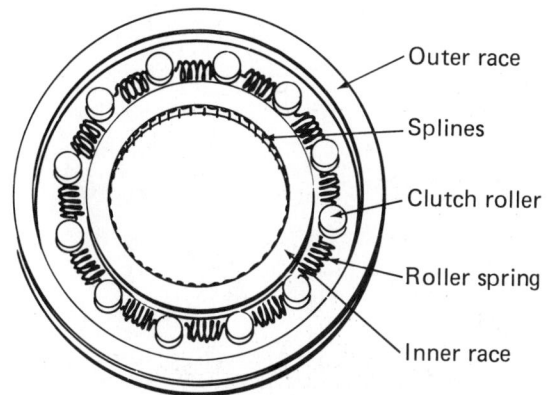

FIGURE 22-22 Inspecting the components of a common, roller-type overrunning clutch assembly.

2. Scored or damaged roller ramp surfaces on the outer race.

3. Flat spots, chipped edges, or signs of excessive wear on each of the clutch rollers.

4. Bent or damaged spring retainer.

5. Worn, bent, or damaged roller springs.

6. Wear or damage to splined areas.

7. Stripped or damaged threads in any tapped holes in the races.

SERVOS AND ACCUMULATORS

The servo and accumulator assemblies in many transmissions look very much alike, but they each have different functions (Fig. 22-23). The **servo** is responsible for applying a band around a drum. The **accumulator** cushions the application of a band or clutch by absorbing servo or clutch-apply pressure.

Not only do the servo and accumulator components resemble one another, their service requirements are also about the same. In other words, the service procedures that apply to a servo assembly for the most part also apply to the accumulator as well.

Cleaning

Because of the size and number of smaller components, you must follow certain precautions when cleaning servo and accumulator parts to prevent their loss. With this in

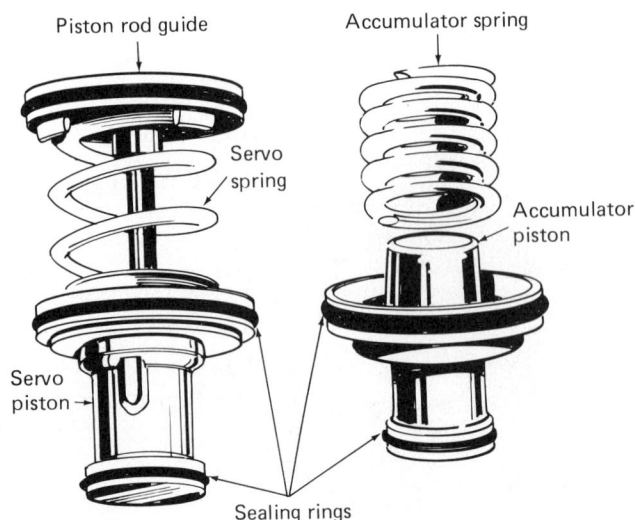

FIGURE 22-23   Typical servo and accumulator assemblies.

mind, take the disassembled servo or accumulator components and perform the following steps:

1. Remove all rubber seals from them because most cleaning agents will cause their deterioration.

2. Place all the pieces in a small parts basket and submerge them in an agitated solvent tank or a cold tank solution.

3. After soaking the parts for a period of time in the solvent bath, blow them dry, using low-pressure compressed air. (NOTE: Any parts cleaned in a cold tank require neutralizing with water before blowing them dry with compressed air.)

Inspection

After cleaning and drying all the components along with their bores, inspect them for the following problems:

1. Cracked, scored, or worn servo or accumulator bores.

2. Worn, damaged, or hard seals. Discard all the seals after this inspection, and install new ones on all the components as outlined later in this chapter.

3. Cracked, broken, distorted, or worn return springs.

4. Worn or damaged pistons and piston rods.

5. Restricted piston movement in its bore.

6. Damaged or distorted servo- or accumulator-cover mating surfaces.

7. Closed or obstructed case or housing fluid passages.

8. Frozen or stiff movement of any servo apply linkage or lever. If necessary, remove the linkage or lever and inspect the assembly for worn or frozen support bearings, worn support shaft, or worn or damaged lever. Replace worn or defective parts and reassemble the linkage.

9. Worn or damaged band adjusting screw, adjusting screw threads, and adjusting screw lever threads. Before reassembly into their respective bores, replace any servo or accumulator parts found to be defective.

GOVERNORS

The **governor assembly** is a device that produces a hydraulic pressure signal that is in proportion to vehicle speed. This signal is responsible for automatically upshifting the transmission at a given road speed. At a given road speed, the pressure signal from the governor valve will be great enough to open a shift valve that directs fluid to a clutch or band servo to upshift the transmission.

It should be obvious then that whenever the mechanic removes this device during a transmission overhaul, he must give it very special attention, or the unit may not upshift properly.

The location of the governor is in the rear section of the transmission usually inside the extension housing. On some transmission styles, the assembly mounts to and rotates with the output shaft; while on still others, the transmission case supports the governor assembly, and a gear on the output shaft spins it.

Because of the different methods of mounting and driving governor assemblies, their construction along with the inspection and service requirements will vary among transmission manufacturers. Therefore, when servicing a given governor assembly, it may be necessary to refer to the transmission service manual for specific instructions.

Cleaning the Typical Governor Assembly

With the governor removed from the transmission and disassembled, do the following:

1. Place all of its components into a small parts basket and submerge them in an agitated solvent tank or a cold tank solution.

2. After soaking the parts for a period of time in the solvent bath, blow them dry, using low-pressure compressed air. (NOTE: Any parts cleaned in a cold tank require neutralizing with water before blowing them dry with compressed air.)

Component Inspection

With all the governor parts clean and dry, inspect them for the following defects (Fig. 22-24):

1. Scored, worn, or corroded valve housing bore.

2. Scored, nicked, or rusted governor valve spool.

3. Binding or sticking of the governor valve in its bore. With the valve clean and dry, it must slide freely back and forth within its bore.

4. Scored, nicked, or rusted governor weights.

5. Binding or sticking of the governor weights in their bores. With the weights clean and dry, they must slide freely back and forth in their bores.

6. Broken or distorted governor spring.

7. Broken snap rings and damaged snap ring grooves.

8. Damaged or distorted governor housing mating surfaces.

9. Damaged or restricted governor housing screen. (NOTE: The location of this screen may also be in either the valve body or main transmission case.)

10. Worn or broken sealing rings.

11. Plugged or restricted governor fluid passages in the governor housing, output shaft, and transmission case.

12. Replace all defective parts before reassembling the governor.

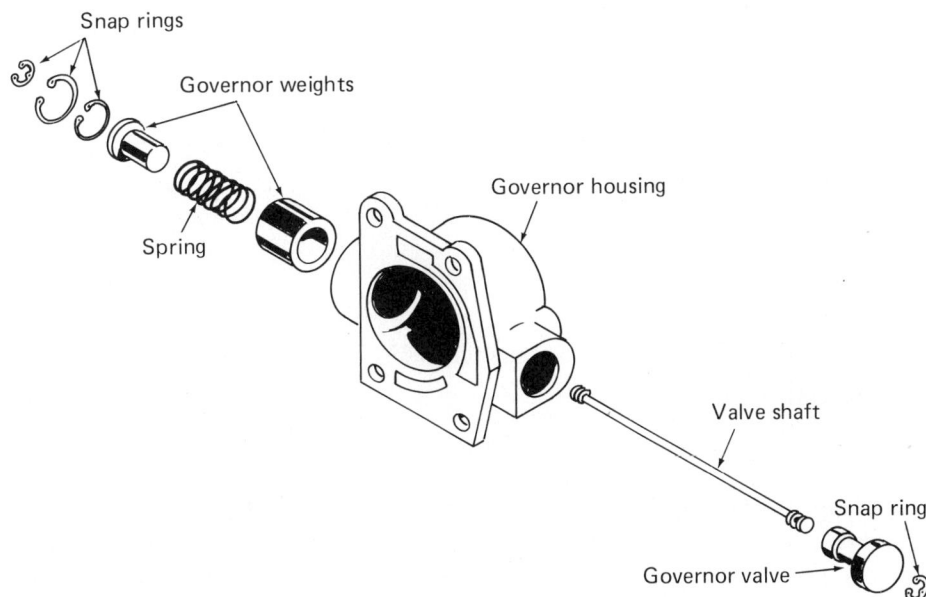

FIGURE 22-24 **Inspecting the components of a typical governor assembly.**

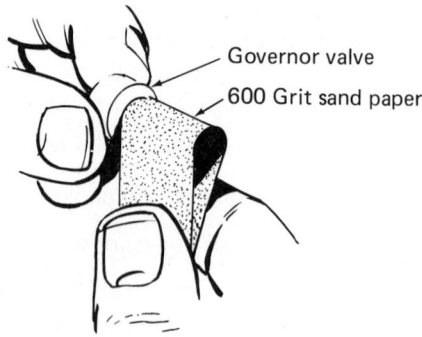

FIGURE 22-25 Polishing a governor-valve spool with sandpaper.

Component Service and Governor Reassembly

Before reassembling the governor, service the components as follows:

1. Polish a sticky governor valve or weight using 600-grit sandpaper or emery cloth (Fig. 22-25). Polish evenly around the circumference of the valve spool or weight. *Do not round off any square edges or corners.* These edges are necessary to cut through any varnish or shellac that may build up within the valve's or weight's bore.

2. Clean thoroughly any polished valve or weight and blow dry. Wet the governor valve and weight with clean transmission fluid prior to their installation in the bores of the housing.

3. Install all snap rings or C-clips making sure to secure them tightly into their respective grooves.

PLANETARY GEAR TRAINS

All automatic transmissions have some form of planetary gear train. This device is responsible for actually providing the vehicle with its various forward gear ratios and a reverse. To provide these ratios, the gear train will consist of one or more ring gears, carriers, and sun gears.

Because of its design, function, and the number of components, the planetary gear train is a very expensive transmission hard part. Consequently, the technician should do everything possible to reuse it when overhauling the transmission. But like so many other transmission components, there are various types of planetaries with each having its own inspection and service requirements. Therefore, this chapter will only attempt to explain to the reader the inspection and service techniques for one type (Fig. 22-26). If in doubt as to how to service another kind of gear train, always refer to the appropriate transmission service manual.

Cleaning

1. Wash or soak the parts as outlined under the section of the chapter on case cleaning, and rinse them off with water as necessary.

2. Blow all parts dry using low-pressure compressed air. CAUTION: *Do not allow the compressed air to spin the pinion gears; this will damage their support bearings.*

Inspection

With the gear train components clean and dry, check them for the following:

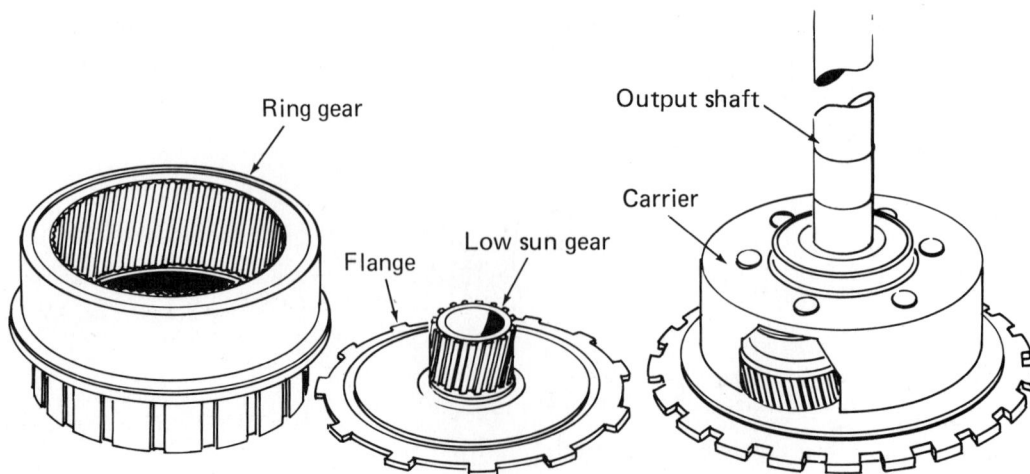

FIGURE 22-26 Inspecting a typical planetary gear train assembly.

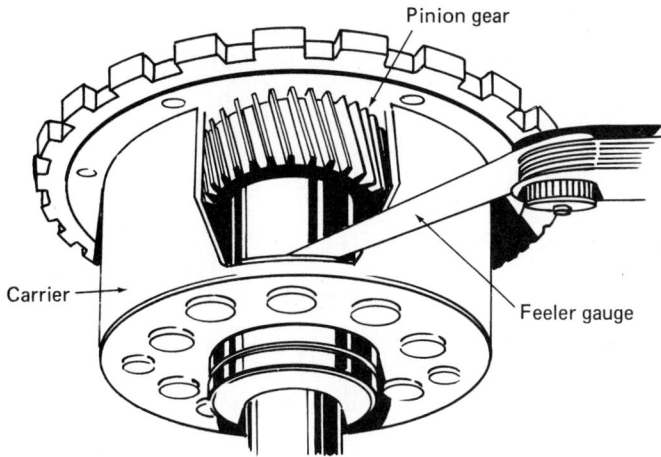

FIGURE 22-27 To check pinion-gear end-play, insert a feeler gauge between the end of the gear and its thrust washer.

1. All the planet pinions for wear, nicks, tooth damage, and binding or looseness in their support bearings.

2. Check the end-play clearance of each planet pinion (Fig. 22-27). For this unit, the clearance should be 0.006 to 0.030 inch.

3. The input sun gear for wear, nicks, or tooth damage. Also, the input sun gear's thrust washer for damage.

4. The low sun gear for wear, nicks, or tooth damage.

5. The ring gear for wear, nicks, or tooth damage.

6. The ring-gear clutch-hub splines for excessive wear or damage.

7. The carrier for cracks or damage. Also, its attached output shaft for worn or damaged splines, obstructed lubrication passages, and worn or nicked bearing surfaces. (NOTE: Individual parts on most planet carriers are now no longer serviceable. Therefore, if the carrier, pinions, thrust washers, or bearings do not pass inspection, replace the entire assembly.)

Service

Before reinstalling the planetary gear train into the case, follow these service procedures:

1. Replace any worn or damaged ring or sun gears, thrust washers, bearings, or the carrier.

2. Coat all thrust washers or bearings with petroleum jelly to hold them in place.

3. Lubricate all other components with clean transmission fluid.

TRANSMISSION SHAFTS

While the average automatic transmission has two rotating shafts, others will have as many as four. But no matter how many shafts the unit has, they all do the same thing—they carry engine torque and power through the various gear train members and then to the drive shaft. An **input shaft** brings the torque and power into the transmission and to the planetary gear train. The **output shaft** delivers the torque and power from the planetary to the drive shaft (Fig. 22-28).

These shafts, which bushings support within the transmission, operate independently of one another. And under normal conditions, they never touch each other. Furthermore, splines mate each independent shaft with the appropriate gear train member or the drive shaft. These splines provide not only a positive method of locking the two components together but also provide a way to quickly separate the two during service operations.

Cleaning and Inspection

With the shafts removed from the transmission, clean and dry them following the same procedure as outlined previously for other large hard parts, like the case and planetary gear train. Then, inspect the shafts for these conditions:

1. Worn or damaged shaft splines.

2. Shaft distortion or cracks.

3. Burrs, scores, or other damage to the bearing or thrust washer surfaces.

4. Worn, broken, or damaged sealing rings, if so equipped.

5. Restricted or plugged fluid passages.

6. Worn or damaged snap-ring grooves.

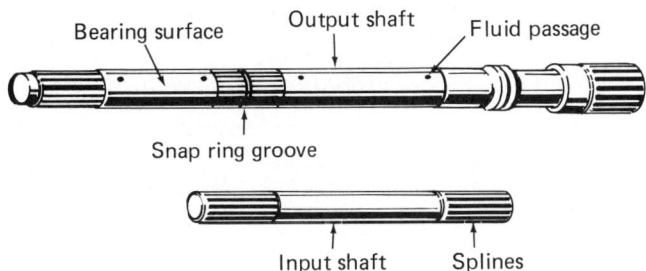

FIGURE 22-28 Inspecting a typical input and output shaft.

7. Worn or damaged speedometer- or governor-drive gear teeth.

8. Scored, worn, or damaged bushings.

Service

Before reinstalling a shaft, perform these service steps:

1. Replace any shaft that does not pass inspection.

2. Remove and replace any worn or damaged support bushings.

3. Replace all sealing rings, if so equipped.

4. Lubricate the shaft generously with clean transmission fluid before reinstalling it into the unit.

THRUST WASHERS

The automatic transmission also usually contains a number of thrust washers (Fig. 22-29), which separate the many moving parts within the assembly. The moving components rotate against the thrust washers; consequently, they absorb any endwise movement of the parts.

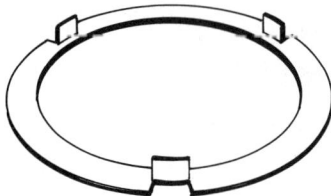

FIGURE 22-29   A typical transmission thrust washer.

Cleaning and Inspection

With the transmission disassembled, clean and dry all thrust washers as you would any small component like the overrunning clutch or governor. With the thrust washers clean and dry, inspect them for the following defects:

1. Nicked, burred, or scratched bearing surfaces on gears, shafts, or clutch drums.

2. Nicked, burred, or scratched washer thrust surfaces.

3. Worn thrust washers. Use a 0- to 1-inch micrometer to measure the thickness of the unit (Fig. 22-30). Compare the reading to the manufacturer's specifications for each thrust washer.

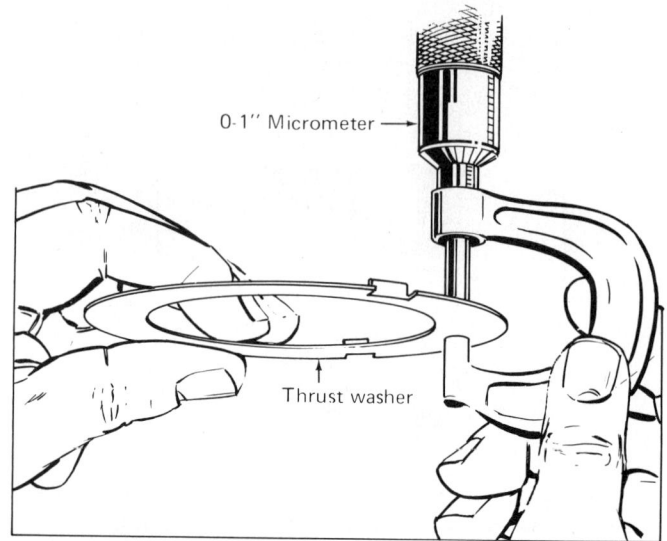

FIGURE 22-30   Measuring the thickness of a thrust washer with a micrometer.

Service

Before reinstalling the thrust washer in the transmission, perform these service steps:

1. Replace any worn or damaged thrust washers and, if necessary, the component it fits against.

2. To hold the thrust washer in place during transmission assembly, coat its backing with petroleum jelly.

3. Coat all other thrust washer bearing surfaces generously with clean transmission fluid.

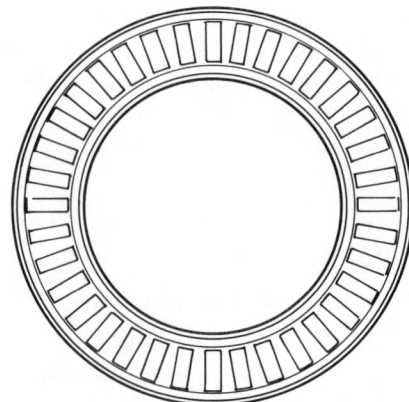

Roller-type bearing thrust washer

FIGURE 22-31   A typical roller-type thrust washer.

BEARINGS

The automatic transmission may also contain one or more roller- or ball-type bearings (Fig. 22-31). These devices serve several functions. First, a bearing assembly can act as a thrust washer to absorb endwise movement of parts. Second, a bearing can support a rotating shaft, like the output.

Cleaning and Inspection

No matter which function the bearing serves, clean it as you would a thrust washer, and blow it dry with low-pressure compressed air. CAUTION: *Be careful not to allow the air blast to spin any of the bearings races at high speeds; this can damage the unit's rollers or balls.* With the bearing clean and dry, check it for the following problems:

1. Binding. Hold one of the bearing races and rotate the other; you should **not** feel any catching or binding between the two. If you do, the bearing is still dirty or is defective.

2. Missing or damaged rollers or balls.

3. Badly worn thrust surfaces.

4. Worn, damaged, or distorted races and bearing cage.

Service

Before reinstalling a bearing, perform these service steps:

1. If the bearing is worn or defective, replace it.

2. Lubricate a shaft support bearing before installation with clean transmission fluid.

3. To hold roller-type thrust bearings in position during transmission assembly, coat them with petroleum jelly.

BUSHINGS

The bushing serves the same function as a shaft supporting roller or ball bearing (Fig. 22-32). In other words, it also supports rotating shafts and other transmission components such as clutch drums. However, the bushing has a much different design that necessitates modified inspection and service procedures.

The **bushing** itself is a precision nonadjustable bearing with a steel backing, lined with either copper or a babbit-type material. Once the technician presses it into a bore, there is a given clearance established between its lining

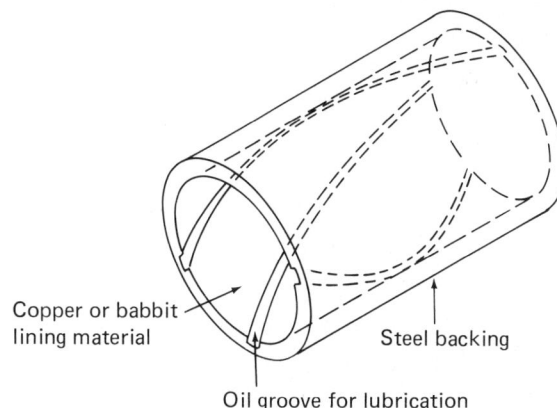

Copper or babbit lining material

Steel backing

Oil groove for lubrication

FIGURE 22-32   A common-type transmission bushing.

and the shaft or component the bushing supports. Over a period of time, the lining can wear or become damaged, necessitating a bushing replacement.

Cleaning and Inspection

Since the bushing presses into another component, the mechanic cleans it at the same time he does the component itself. Once the bushing is clean and dry, inspect it for scoring, pits, and wear. If the bushing requires replacement, inspect the shaft or component that rides in it. This part may also show signs of scoring or wear.

In some situations, shaft-mounted Teflon or metal sealing rings ride inside the bushing. Wear on these sealing rings can also damage the bushing. Therefore, check the bushing closely for ring grooving; if this condition exists, check the shaft seal-ring lands for wear and damage. Replace the bushing, seals, or shaft as necessary.

Service

If a bushing is worn or damaged, it should be replaced using the tools and equipment mentioned in Chapter 15. The choice of tool to remove and replace a bushing depends, of course, on its location. To remove a bushing in a closed or blind bore, follow this typical procedure:

1. Use a cape or bushing chisel and cut along the bushing's seam until the tool breaks through the component's wall (Fig. 22-33). Next, pry the loose ends of the old part up with an awl or needle-nose pliers and remove it.

2. To pull a bushing using a threaded remover and cup assembly:

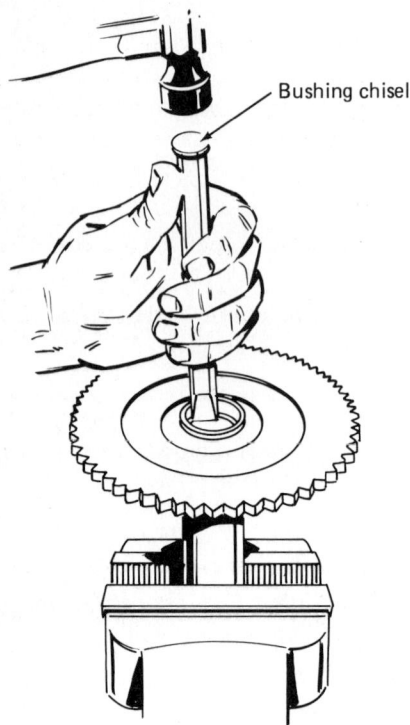

FIGURE 22-33 Using a chisel to cut through a bushing's wall.

a. Thread the remover into the bushing as far as possible by hand (Fig. 22-34).

b. Using a wrench, screw the remover into the bushing three to four additional turns to firmly engage its threads into the unit.

c. Install the cup assembly over the remover and down onto the component that houses the bushing.

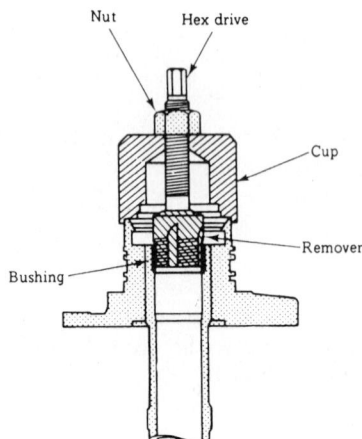

FIGURE 22-34 Pulling a bushing using a threaded remover.

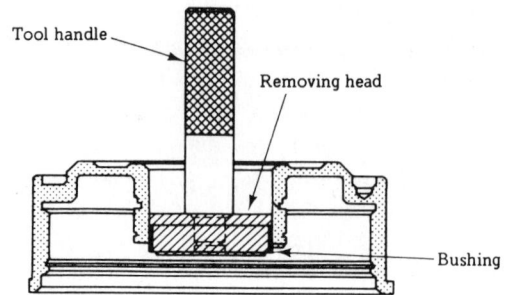

FIGURE 22-35 Removing a bushing using a tool handle and the correct-size head.

d. Thread the hex nut down on the remover until it contacts the cup.

e. Tighten the hex nut with a wrench to pull the bushing from its bore.

3. To press or drive a bushing from its bore using a tool handle and removing head:

a. Place the correct-size removing head into the bushing, and install the tool handle into the opening in the head (Fig. 22-35).

b. Drive or press the bushing straight down and out of its bore. Be careful not to cock the tool while it is in the bore.

4. To install a bushing in either a closed or open bore, clean and check the opening for damage; then, follow these steps:

a. Position the new bushing onto the proper-size installing head.

b. Install the tool handle into the head.

c. Start the new bushing into its bore, making certain to align all oil holes or grooves in the bushing with the corresponding oil passages in the bore.

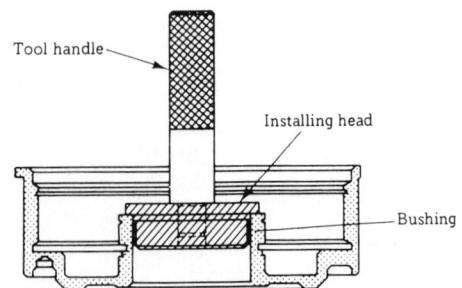

FIGURE 22-36 Installing a bushing using the tool handle and the correct-size head.

d. Press or drive the bushing into the bore (Fig. 22-36). CAUTION: *Be sure the bushing bottoms in its bore, or drive it into the position specified by the transmission manufacturer.*

5. Lubricate the bushing thoroughly with clean transmission fluid before installing the shaft or component it supports. (NOTE: If the component will not go into the bushing or it is hard to turn, the installing process damaged the bearing, and it will most likely have to be replaced.)

VALVE BODIES

The inside of the average valve body (Fig. 22-37) contains a great number of springs and valves. Each of these many parts performs a small but important function during the operation of the transmission. When working together, they convert the valve body into a hydraulic computer that is responsible for upshifting or downshifting the transmission according to the will of the driver, vehicle speed, and engine load.

Because of its many precision components and the very important job it performs, the mechanic must **carefully** and **thoroughly** clean and inspect the valve body as part of every transmission overhaul, or a malfunction may develop. This procedure is usually the most difficult of all overhaul tasks because of the wide variety of valve body designs and the variation in internal component structure. Therefore, when servicing the valve body for any particular transmission model, the mechanic must follow these important general rules:

Disassembly, Cleaning, and Inspection

1. Make certain your hands, tools, and work bench are clean before working on a valve body.

2. Use low-pressure compressed air to clean or dry a component. CAUTION: *Never use rags for this purpose. A single particle of lint can cause a valve to stick in its bore.*

3. Be careful never to drop or nick a valve spool during the service work. A nick can also cause a valve to stick in its bore.

4. When disassembling the valve body, lay the parts out on a clean surface in the order in which you remove them.

5. Whenever possible, obtain an illustrated parts breakdown for the valve body you are servicing to assist you in reassembling the unit (see Fig. 22-37).

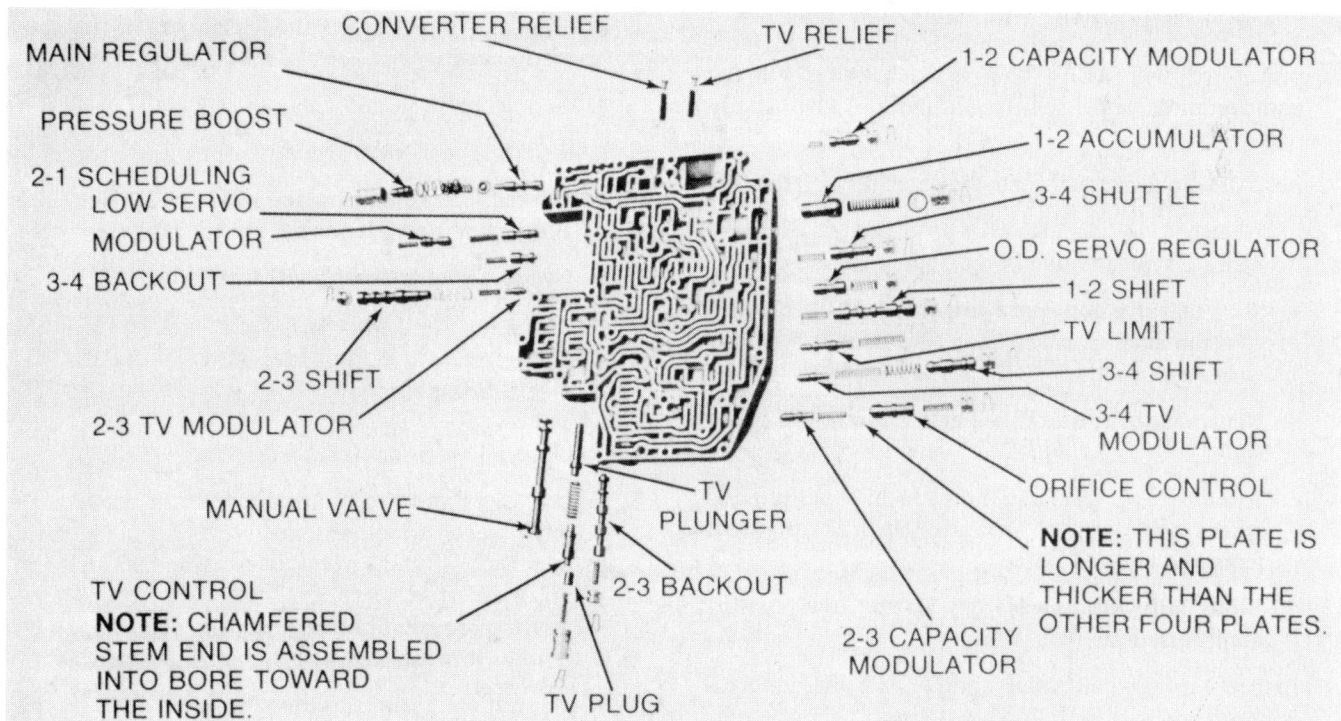

FIGURE 22-37 The illustrated parts-breakdown of a typical valve body. (Courtesy of Ford Motor Company)

Spring holder

Spring-holder cover

Disassembled valve body

Parts breakdown

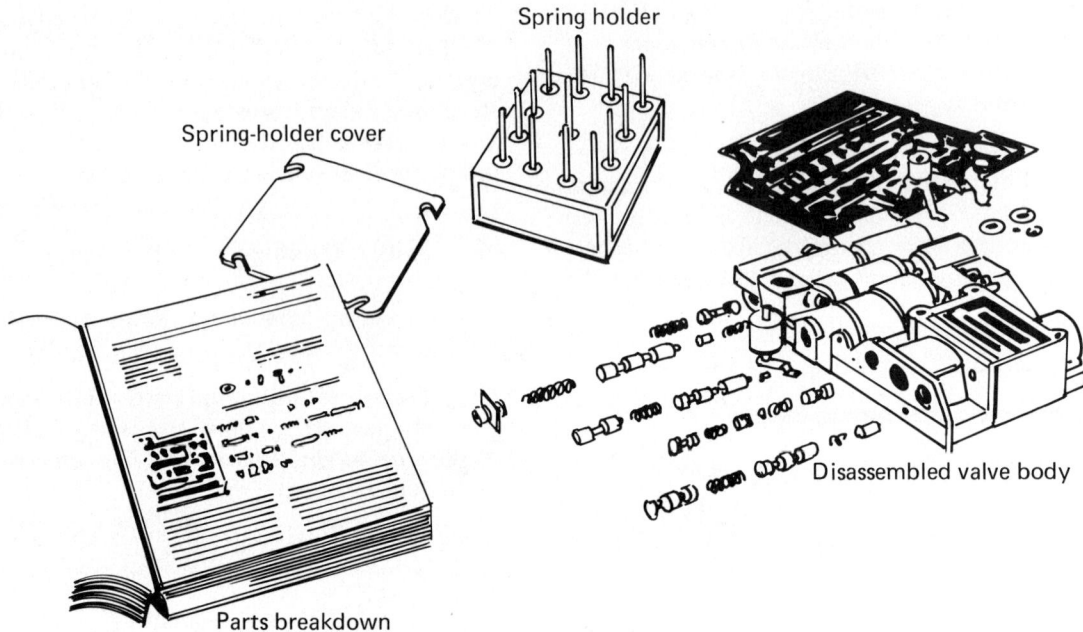

FIGURE 22-38 *Before cleaning the springs, install them on a numbered-peg, valve-body spring holder.*

6. Using the parts breakdown, identify all the disassembled parts; number each spring on the diagram; and install each individual unit on a numbered peg of a spring holder (Fig. 22-38). This procedure will assist you in identifying all the components and springs during valve body reassembly.

7. Save the old gaskets, if used, to match up with the new ones.

8. To clean a typical valve body, follow these steps:

   a. Soak all valve body parts in a cold tank filled with a solution such as carburetor cleaner that dissolves varnish or shellac.

   b. Neutralize or rinse all the parts in quantities of hot water.

   c. After rinsing, immerse the parts in clean mineral spirits or solvent to separate the drops of water from the cleaned parts.

   d. Blow all the parts dry with low-pressure compressed air.

9. Inspect a typical valve body and its internal components for the following defects:

   a. A bent manual valve.

   b. A bent or damaged separator plate.

   c. A cracked valve body casting and distorted or damaged mating surfaces.

   d. Plugged or restricted fluid passages in the valve body casting.

   e. Scored or corroded valve bores.

   f. Broken, bent, or worn valve springs.

   g. Scored, cracked, or burred valves, valve sleeves, or plugs, and valve lands for shiny areas.

   h. Stuck or worn check valves or balls.

Service and Reassembly

1. Replace all worn or damaged parts.

2. Shiny valve land areas indicate friction between the affected valve spool and its bore. To correct this problem, polish these areas away using 600-grit sandpaper or emery cloth (Fig. 22-39). CAUTION: *Be careful not to round off any edges of the valve land when polishing.* Rewash and dry the polished valve.

3. Match the old gaskets to the new ones to ensure they are exact replacements.

4. Check each clean valve in its bore. It should slide freely back and forth in its bore due to its own weight

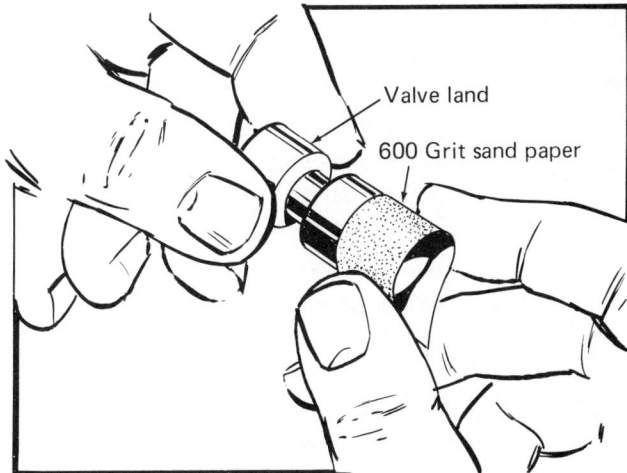

FIGURE 22-39 *Polishing a valve land with fine sandpaper.*

without sticking or hesitating. If a valve fails this test, polish or replace it.

5. Before final installation, lubricate each valve with clean transmission fluid.

6. Make sure the springs and check balls are in their proper locations.

7. Always follow the manufacturer's torquing sequence and specified torque values when assembling the valve body.

TRANSMISSION SOFT PARTS

The **soft parts** of the automatic transmission are those less expensive components that the technician usually replaces during a complete transmission overhaul. Soft parts include such items as the clutch plates; bands; metal, Teflon, and rubber sealing rings; metal-clad seals; and gaskets. Soft parts are those that usually wear out or deteriorate and cause transmission malfunctions.

The only parts mentioned above that are sometimes reused are the clutch steel plates and the bands. But as a rule, it requires more time and effort to prepare steel plates for reuse in the rebuilt transmission than it is worth. And as far as the bands are concerned, the general rule is to replace all paper-lined bands, but reuse an asbestos-lined one if it is still in good condition. Therefore, with the exception of the steel plates and asbestos-lined bands, the mechanic will usually not clean and inspect the soft parts; he will just replace them. The only time a technician will waste the time examining these items is if he is looking closely for the cause of a certain malfunction, which up to this point has eluded him.

CLUTCH PLATES

Steel Clutch Plate Service

As previously mentioned, it is far less expensive to use new steel clutch plates because of the time it takes to prepare the old ones for reuse. But if the technician is going to reuse the old plates, he should bead blast or refinish them on both sides with 120 to 320-grit emery cloth or sandpaper (Fig. 22-40). When sanding the plates, rub until sanding marks are visible over the entire area on both sides of each plate. This bead blasting or sanding process removes polished surfaces and restores new roughness to the plate, which is necessary for proper clutch application and break-in. After servicing the plates, clean and then lubricate the steel plates with transmission fluid before installation.

Friction Clutch Disc Service

The technician should presoak all new friction-type (lined) clutch discs in clean automatic transmission fluid for no less than 30 minutes before installation for two reasons: (1) The composition lining on the discs acts as an insulator that holds frictional heat on the plate surface, causing premature clutch failure due to surface glazing or burning; and (2) a thoroughly presoaked disc quickly dissipates surface heat through the fluid retained in the composition lining. (NOTE: *If for any reason friction-type discs are reused, they should also be soaked in fluid for a few minutes before installation to prevent damage caused by friction.*)

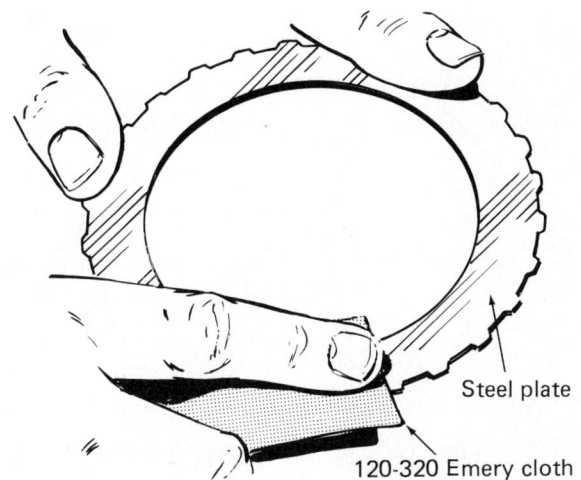

FIGURE 22-40 *Deglazing a typical steel clutch plate.*

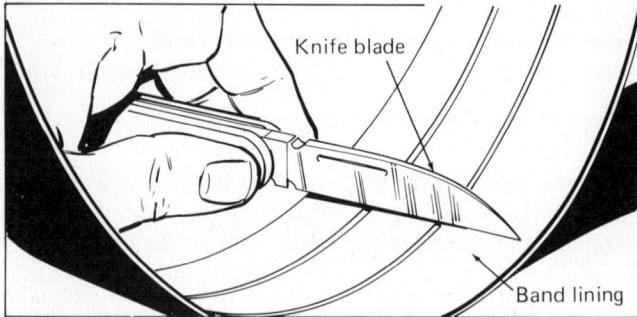

FIGURE 22-41   Scraping the carbon and glaze off an asbestos-lined band.

BANDS

Cleaning and Inspection

If a band appears to be serviceable at first glance, wash it in a cleaning detergent and blow it dry with compressed air. CAUTION: *Do not soak the band in solvent or cold tank solution because these chemicals can deteriorate the lining material or its bonding agent.* Then, inspect the band for these conditions:

1. Distortion.

2. Cracked anchor ends.

3. Excessive or uneven lining wear.

4. Burned, charred, or glazed lining.

5. Loose lining.

6. Flaking or pitted lining.

7. Replace a band that shows any sign of these defects.

Service

Before installing a new or used band, follow these instructions:

1. For quick surface heat dissipation, soak all new paper- and asbestos-lined bands in clean transmission fluid for at least 30 minutes.

2. Before reusing an asbestos-lined band, scrape the lining with a knife or bearing scraper to remove glaze or carbon (Fig. 22-41).

METAL AND TEFLON SEALING RINGS

Metal and Teflon sealing rings, when installed on rotating shafts and pistons, prevent excessive losses in fluid pressure and flow (Fig. 22-42). Manufacturers install this type in high temperature areas of the transmission where a positive seal is not necessary, or where a seal is necessary on rotating units.

Metal Seal Installation

When installing a metal seal, follow these procedures:

1. Compare the old sealing ring with the new one to make sure you have the correct size.

2. Never expand a sealing ring any further than necessary to install it into its groove.

3. Make sure the installed ring is free to rotate in the groove.

4. When installing the locking end sealing ring, make sure that the ring ends properly lock together.

5. Always lubricate the sealing rings with clean transmission fluid before installing the rings over or into the component they seal.

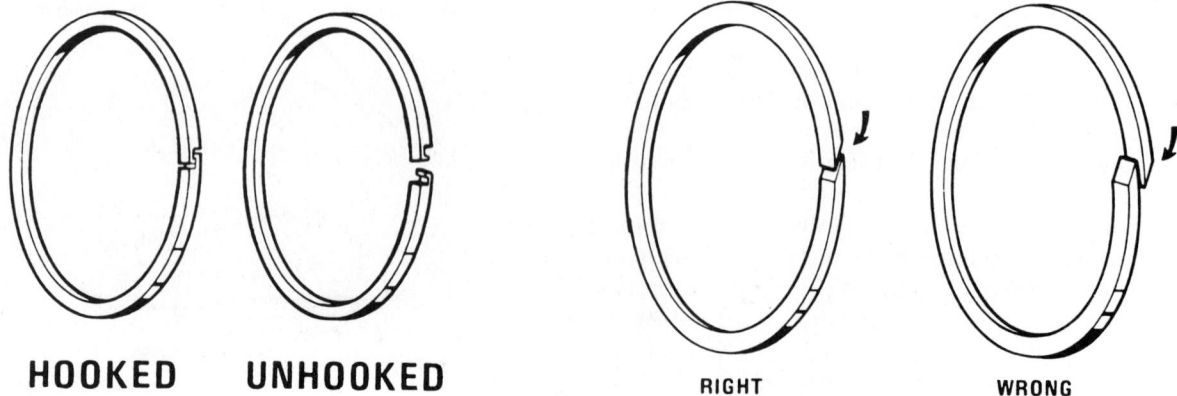

HOOKED   UNHOOKED

RIGHT   WRONG

FIGURE 22-42   Typical metal and Teflon sealing rings.

Teflon Seal Installation

1. Always replace Teflon sealing rings with the metal type when overhauling the transmission.

2. While replacing the Teflon rings with the metal ones, use the same procedures as outlined in the section on metal rings.

RUBBER SEALING RINGS

The mechanic will encounter three types of Neoprene rubber sealing rings in automatic transmissions: the O-ring, lathe-cut, and lip seal (Fig. 22-43). All of these devices prevent fluid pressure and flow losses around such components as clutch, servo, and accumulator pistons. Although the service procedures for all the seals are nearly the same, the lip type requires special attention during its installation.

O-Ring and Lathe-Cut Seal Installation

When installing an O-ring or a lathe-cut seal, follow these general instructions:

1. Compare the old seal with the new one to make certain you have the same size and type.

2. Never expand the seal any farther than necessary to install it into its groove.

3. Make certain the seal fully seats in its groove.

4. Always generously lubricate the seal and the bore it fits into with transmission fluid before attempting to install the sealed component into the bore. (NOTE: Some mechanics prefer to use petroleum jelly, mineral oil, or Door-Ease on the seal instead of fluid as an aid in installation.)

Lip-Seal Installation

When installing lip seals be sure to follow these instructions:

1. Always adhere to the installation procedures set forth earlier for O-ring and lathe-cut seals.

2. **Always install the seal so that the lip faces toward the fluid source.** Some clutch assemblies will use two of these seals, one inside the drum and the other around the outer circumference of the piston. The lips of these two seals must face down into the drum or toward the fluid source that will apply the clutch piston. On another type of clutch assembly that has three lip seals, the fluid source enters the inside of the drum between the lower and two upper seals. Therefore, the lowest seal's lip faces upward toward the fluid source, and the lips of the inner and outer piston seals face downward.

METAL-CLAD SEALS

The metal-clad seal is nothing more than a Neoprene lip seal encased in a metal housing (Fig. 22-44). This seal design also controls fluid leakage around rotating components such as the converter hub and output shaft.

Installation

To install a metal-clad seal properly, follow these general rules:

1. Always install its lip facing toward the fluid source.

2. If the seal's backing does not have a rubber or resin coating, lightly cover the steel backing, where it contacts the case or housing, with nonhardening sealer just prior to installation.

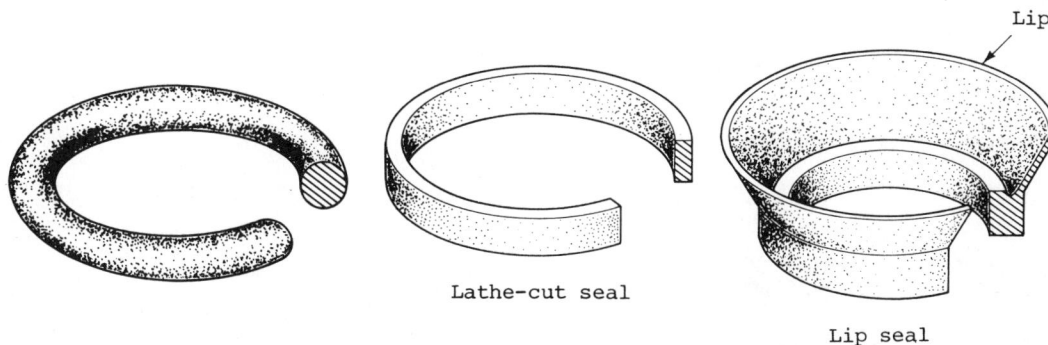

FIGURE 22-43 Common-type O-ring, lathe-cut, and lip seals.

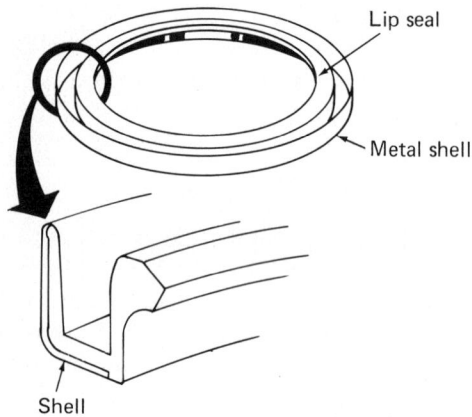

FIGURE 22-44   A typical metal-clad seal.

3. Always use a seal driver to install the seal into its bore.

4. After installation, always lubricate the seal's lip with clean transmission fluid.

GASKETS

Gaskets are a necessary sealing device, found in many locations, within the automatic transmission. These devices provide a positive seal between two components, thus preventing losses of fluid pressure and flow.

When working with transmission gaskets, follow these simple rules:

1. Never reuse an old gasket.

2. Always compare the old gasket to the new one to make sure you have the correct one.

3. Never use any form of sealer on any transmission gasket unless specified by the unit's manufacturer.

4. Always clean off the gasket mating surfaces thoroughly.

TRANSMISSION FLUID COOLING SYSTEM

Every automatic transmission has some form of fluid cooling system. This system, as its name implies, reduces the temperature of the transmission fluid, thereby increasing its overall life. Therefore, if the system should fail to function properly, the fluid's life is cut short, which can result in severe damage to the transmission.

There are three basic kinds of fluid cooling systems: a water-cooled type; an air-cooled type; or an after-market, air-cooled type. The water- or coolant-type system (Fig.

22-45) includes two cooler lines and a fluid cooler located in the lower or side tank of the radiator. The air-cooled design consists of a finned converter and a ducted or drilled converter housing (Fig. 22-46). The after-market, air-cooled system uses a special finned, auxiliary cooler (Fig. 22-47) connected to the transmission by two cooler lines. Manufacturers also produce special finned oil pans to increase the fluid capacity of the transmission and cool the lubricant normally contained in the pan.

Before a mechanic reinstalls the rebuilt transmission into the vehicle, he should inspect and service the transmission's cooling system. The reason for this is twofold. First, the system may now contain foreign material, metal and friction plate particles, carried into the system by the fluid. These particles, if allowed to remain inside the cooling system, will damage the newly rebuilt transmission as the new fluid carries them back to the assembly. Second, the system may be partially or completely restricted, which would cause the new fluid to overheat and eventually result in damage to the transmission because of improper lubrication. Furthermore, if the water-cooled system has internal or external leakage, this can damage the transmission by allowing coolant to enter the transmission or by permitting transmission fluid to leak out.

COOLING SYSTEM INSPECTION

Water-Cooled Type

Referring to Fig. 22-45, inspect this system for the following defects:

1. Kinked cooler lines.

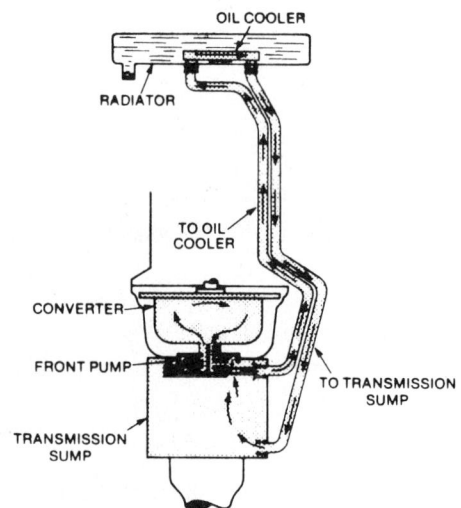

FIGURE 22-45   A water-type, fluid-cooling system.

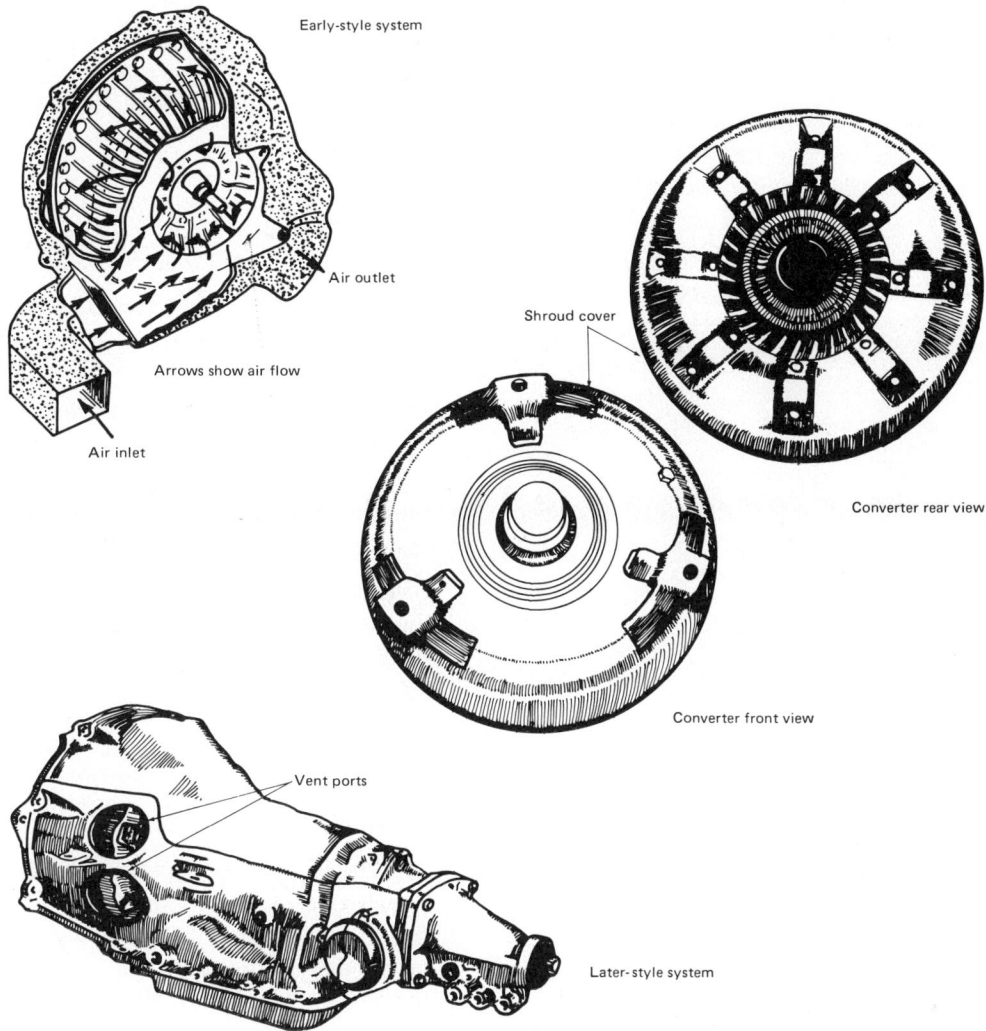

FIGURE 22-46 An air-cooled, fluid-cooling system.

2. Cracked, broken, or leaking lines.

3. Deteriorated flexible cooler hoses and broken clamps.

4. External leakage in the radiator.

5. Internal leakage in the transmission cooler. NOTE: If the cooler lines or transmission, upon teardown for overhaul or at any time, contain **any** radiator coolant, perform this cooler leakage test:

   a. Remove the radiator cap **slowly** to release cooling system pressure on the cooler itself.

   b. Disconnect both cooler lines at the radiator, and cap one of them.

   c. Attach an air coupling or valve to the other cooler fitting.

   d. Apply 25 to 50 psi air pressure to the cooler, and check the coolant for air bubbles.

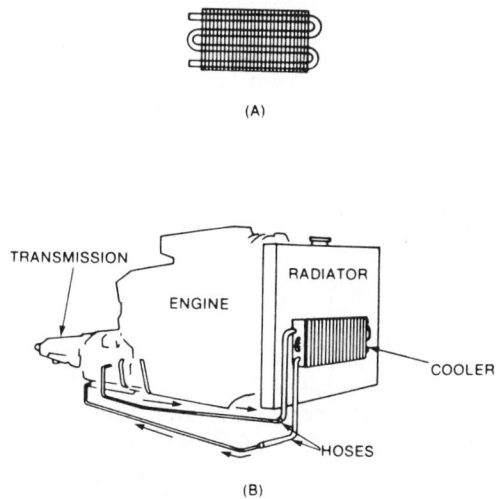

FIGURE 22-47 An auxiliary cooler is shown at A. B shows the cooler connected to the transmission's cooler lines.

e. If the cooler will not hold air pressure or the coolant shows the presence of bubbles, replace or repair the cooler.

6. Restricted or plugged cooler lines or cooler. (NOTE: Make this check while you are flushing the system, using the procedure outlined under cooler flushing.)

Air-Cooled Type

Referring to Fig. 22-46, inspect this system for the following defects:

1. Cracked, loose, or damaged air fins or the shroud cover on the torque converter.

2. Plugged or restricted air ducts or ports in the converter housing.

After-Market Type

Referring to Fig. 22-47, inspect this system for the following problems:

1. Kinked cooler lines.

2. Cracked, broken, or leaking lines.

3. Deteriorated flexible cooler hoses and broken clamps.

4. External cooler for leaks.

5. External cooler fins for obstructions.

6. Restricted or plugged cooler lines or external cooler. (NOTE: Check this type of system for restriction also while flushing it out.)

COOLING SYSTEM SERVICE

Line Repair

If a cooler line becomes kinked or badly damaged, it is better to replace it whenever possible. But a small pinhole leak can be repaired by using a connector fitting or a piece

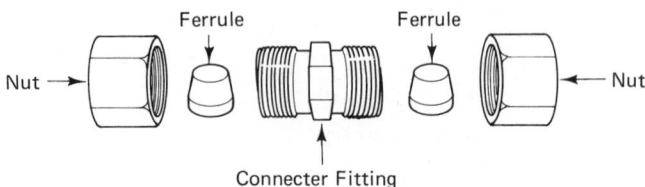

FIGURE 22-48 A connector assembly used to repair a leak in a transmission cooler line.

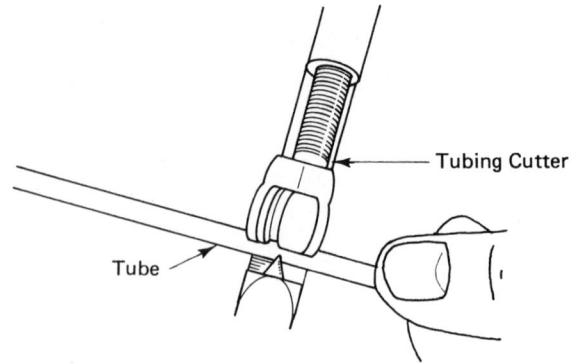

FIGURE 22-49 Cutting out the damaged section of a cooler line using a tubing cutter.

of high-pressure hydraulic hose. To repair a leaky line using a connector (Fig. 22-48), follow these steps:

1. Using a tubing cutter, cut away the damaged section of the pipe (Fig. 22-49). If the damaged area of the tube is longer than the distance inside the connector where the cut-off end of each pipe will bottom in it, use a pressure hose to repair the problem. Otherwise, the repaired cooler line will no longer be the correct length.

2. Install a nut and ferrule over each cut-off end of the tubing.

3. Install one of the tubing ends into the connector, making sure to push it in as far as it will go. Slide the ferrule against the connecter, and run the nut down on the connecter's threads finger tight.

4. Using two wrenches, tighten the nut securely. Position one of the wrenches on the connecter and the other on the nut while tightening the latter to prevent damage to the line or fitting (Fig. 22-50).

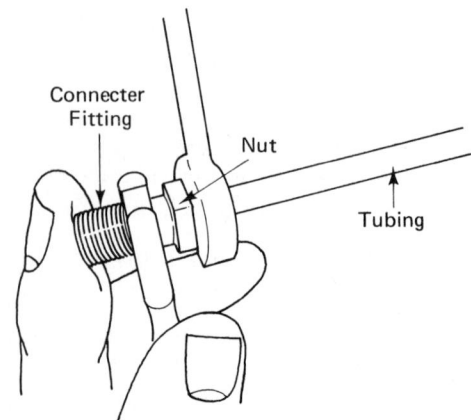

FIGURE 22-50 Using two wrenches, tighten the connecter nut securely.

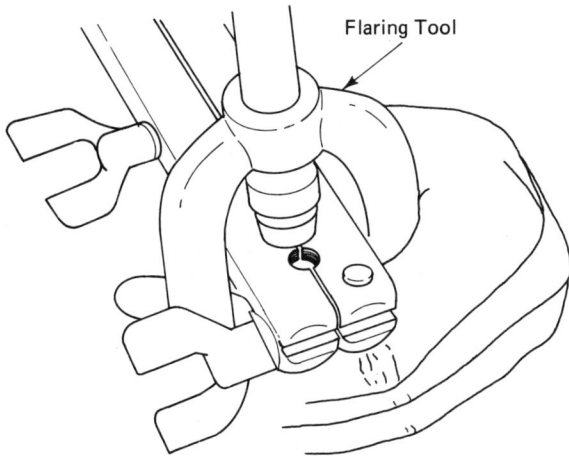

FIGURE 22-51 Flaring the end of a cooler line, using a flaring tool.

5. Repeat the same process on the other portion of the line.

6. Check the line and connector for leaks by operating the transmission.

To repair a damaged cooler line using the high-pressure hydraulic hose:

1. Using a tubing cutter, remove the damaged section of line.

2. With the tools shown in Fig. 22-51, double-flare each cut-off end of the tubing. The flare will prevent the hose from slipping off the line after the clamp is tight.

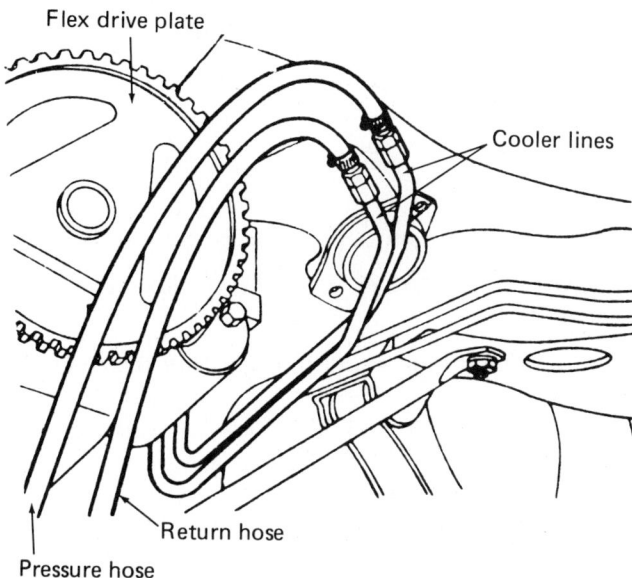

FIGURE 22-52 Connecting the flushing machine's pressure and return hoses to the cooler lines.

3. Install a clamp loosely over each end of the tubing.

4. Cut a piece of hydraulic hose to a length approximately 2 inches longer than the section of damaged pipe.

5. Insert each end of the hose over the double-flared ends of the line. Make certain to push about 1 inch of hose onto each tubing end.

6. Slip the clamps to within $\frac{1}{4}$ to $\frac{1}{8}$ inch from the ends of the hose and tighten them securely.

7. Check the hose and line for leaks by operating the transmission.

Cooler and Line Flushing

With the flushing machine mentioned in Chapter 15, flush out the cooler and lines using this procedure:

1. Roll the machine to the vehicle. Be careful not to tip the machine over or slosh the solvent from the unit.

2. With the two auxiliary hoses and adapters supplied with the machine, connect the unit's pressure hose to either of the two cooler lines (Fig. 22-52). Then,

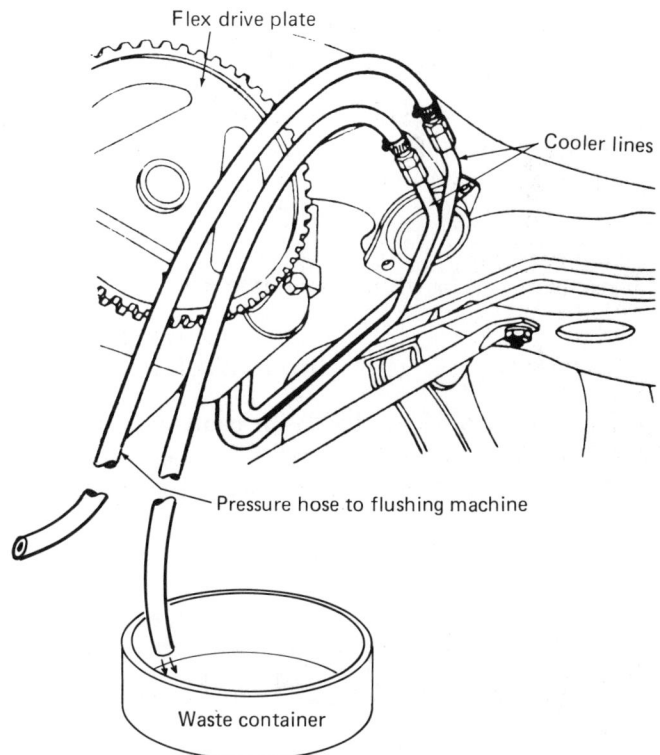

FIGURE 22-53 Pumping a few quarts of solvent through the system and into a waste container.

connect the long auxiliary return hose to the remaining cooler line.

3. Place the return hose in a waste oil container.

4. Turn the machine's timer on for 2 minutes to pump a few quarts of solvent through the system and into the waste container (Fig. 22-53). This procedure prevents excessive amounts of dirty fluid, particles, or sludge from entering the machine's sump.

5. Remove the drain hose from the waste container, and place it into the machine's sump.

6. Set the timer for a 5 to 10 minute flushing and occasionally check the sump return hose to see if solvent is returning freely from the system. *If the solvent is not flowing freely, the system, cooler, or lines have a restriction.*

7. Disconnect the pressure line from the machine and remove the auxiliary hose from the cooler line. Place the return hose once more into a waste container.

8. Using low-pressure compressed air, **less than 50 psi,** blow the remaining solvent from the system. Apply the air pressure to the cooler line, where the machine's auxiliary hose had previously been connected.

9. Disconnect the auxiliary return hose from the cooler line, and return the machine to its storage area.

REVIEW

This section will assist you in determining how well you remember the material contained in this chapter. Read each item carefully. If you cannot complete the statement, review the portion of the chapter that covers the material.

1. The transmission case is the main _____ of the transmission.

2. _____ or _____ deposits are much harder to dissolve than other types of contaminants from the transmission case or its components.

3. To dry the transmission case after cleaning it, use _____ _____ _____ .

4. If the transmission case has cracks or porosity, it is better to _____ it.

5. Damaged threads are repairable using a _____ _____ .

6. A _____ in the transmission case usually supports the output or input shaft.

7. A mechanic can press a bushing out or _____ it in two with a chisel.

8. If leakage occurs around the shift linkage shafts, replace the _____ _____ _____ .

9. Case porosity is a condition where a _____ _____ develops through the metal of the transmission case.

10. A(n) _____ _____ is necessary to repair case porosity.

11. The component that encloses the parts directly behind the transmission case is the _____ _____ .

12. An extension housing _____ usually supports both the output shaft and the slip yoke.

13. When installing the extension housing seal, make sure its lip faces _____ or toward the bushing.

14. A drum along with a multiple-disc clutch connects a planetary gear train member to the _____ _____ .

15. After removing the clutch pack, tie its components together with _____ _____ .

16. It is sometimes necessary to use air pressure to force the _____ _____ from its bore in the drum.

17. The location of the residual check ball will be either in the clutch _____ or _____.

18. If the lip seal makes it difficult to reinstall a piston in a clutch drum, work a _____ _____ around the circumference of the piston.

19. When deglazing a drum that has an asbestos-lined band, sand it with _____ grit emery cloth or sandpaper.

20. An overrunning clutch holds a component stationary in _____ _____ .

21. The _____ cushions the application of a band or clutch.

22. The _____ is a device that produces a pressure signal in proportion to vehicle speed.

23. When polishing a valve spool, do not round off its _____ _____ or corners.

24. Transmission _____ carry engine torque through the various gear train members and then to the drive shaft.

25. _____ _____ separate the many moving parts within the transmission.

26. A _____ can also support a shaft or act as a thrust washer.

27. The device that has a steel backing lined with a copper or babbit-type material is a _____.

28. To remove a bushing from a blind bore, use either a bushing _____ or a threaded _____ .

29. The hydraulic computer of the transmission is the _____ _____ .

30. When cleaning any transmission component, never use a _____ to dry it.

31. Use _____ grit emery cloth or sandpaper to sand a steel plate before reusing it.

32. Soak all new friction clutch discs in clean automatic transmission fluid for _____ minutes before installation.

33. Before reusing an asbestos-lined band, _____ the lining with a knife or bearing scraper.

34. Always replace _____ seals with the metal type.

35. Always install a lip seal so that the lip faces _____ the fluid source.

For the answers, turn to the Appendix.

CHAPTER 23

Transmission Overhaul

The complete overhauling or rebuilding of an automatic transmission is quite a complicated process for several reasons. First, an overhaul involves the restoration of the transmission to a condition of factory newness. This procedure can include most if not all the various service procedures, presented in the previous chapters of Section 2. Second, it should be obvious that there exist significant differences in automatic transmission design that result in variations in the overhaul procedures.

For this reason, along with limited space, this chapter only provides the step-by-step procedure for the overhaul of one transmission type, a Chrysler A-415, A-413, or A-470 transaxle. However, the repair procedures or sequences presented here are typical of those used on other units. As you proceed through these pages and perform the various service steps, it will be helpful to refer back to the various other chapters of this service section for further explanation of the many pieces of equipment, tools, and procedures presented here.

GENERAL INSTRUCTIONS

During the teardown, repair, and assembly of the many subassemblies of this transaxle, the mechanic must follow certain general instructions. By following these instructions carefully, the mechanic will avoid damaging

transaxle components and the unnecessary repetition of certain overhaul operations.

1. After removing the transaxle from the vehicle, plug all of its openings and steam clean the outside of the case thoroughly. This action prevents dirt from entering the mechanical parts during disassembly. **Cleanliness through the entire disassembly and reassembly process cannot be overemphasized.**

2. Handle all transaxle parts carefully to avoid nicking or burring the bearing or mating surfaces.

3. Lubricate all internal transaxle parts during assembly with clean Dexron fluid. *Do not use any other lubricants except on thrust washers. The mechanic can coat these with petroleum jelly to facilitate assembly.*

4. Always install new gaskets, sealing rings, and seals when rebuilding the transaxle.

5. Always tighten all bolts and screws to manufacturer's specifications (see Table 23-1).

6. When rebuilding this transaxle, the mechanic must remove and reinstall 11 thrust washers. It is very important that each of these washers be in its correct position during the assembly process. To properly locate and identify the washers and their locations in the upcoming illustrations and text, the washers are numbered from 1 to 11. For example, No. 1 is the thrust washer located on the oil pump; the last washer —location No. 11—is at the rear of the transaxle case in front of the rear annulus gear.

7. With the exception of the bands and friction clutch plates, wash all parts with a suitable, clean solvent.

8. Do not dry parts with a rag; instead, use low-pressure compressed air.

TRANSAXLE DISASSEMBLY

To disassemble the transaxle on the work bench:

1. Loosen and then remove the oil pan attaching bolts (Fig. 23-1).

2. Remove the oil pan from the transaxle case, and discard the gasket (Fig. 23-2). Next, clean the old R.T.V. sealant from the pan and case mating surfaces.

3. Loosen and remove the oil filter attaching screws (Fig. 23-3).

4. Remove the oil filter from the valve body (Fig. 23-4).

TABLE 23-1   Torque Specifications for the A-415, A-413, and A-470 Automatic Transaxle

Item	Qty.	Thread Size	Driver	Torque Newton-meters	Torque Inch-Pounds
A-415, A-413, and A-470 Automatic Transaxle:					
Bolt—Bell Housing Cover	3	9.8-M6-1-10	10mm Hex	12	105
Bolt—Flex Plate to Crank (A-415)	6	10.9-M10-1-16†‡	16mm Hex	68	50*†‡
Bolt—Flex Plate to Torque Converter (A-415)	3	10.9-M10-1.5-11#	18mm Hex	54	40*#
Screw Assy. Transaxle to Cyl. Block	3	9.8A-M12-1.75-65	18mm Hex	95	70*
Screw Assy. Lower Bell Housing Cover	3	9.8-M6-1-10	10mm Hex	12	105
Screw Assy. Manual Control Lever	1	9.8A-M6-1-35	10mm Hex	12	105
Screw Assy. Speedometer to Extension	1	9.8A-M6-1-14	10mm Hex	7	60
Connector, Cooler Hose to Radiator	2	1/8-27 NPTF	12mm Hex	12	110
Bolt—Starter to Transaxle Bell Housing	3	M10-1.5-30	15mm Hex	54	40*
Bolt—Throttle Cable to Transaxle Case	1	M6-1.0-14	10mm Hex	12	105
Bolt—Throttle Lever to Transaxle Shaft	1	M6-1-25	10mm Hex	12	105
Bolt—Manual Cable to Transaxle Case	1	M8-1.75-30	13mm Hex	28	250
Bolt—Front Motor Mount	2	M10	15mm Hex	54	40*
Bolt—Left Motor Mount	3	M10-1.5-25	15mm Hex	54	40*
Dress Up:					
Connector Assembly, Cooler Line	2	M12-1.75-122	17.5mm Hex	28	250
Plug, Pressure Check	7	1/16-27NPTF	5/16 in. Hex	5	45
Switch, Neutral Safety	1	3/4-16UNF	1.0 in. Hex	34	25*
Differential Area:					
Ring Gear Screw	8	12.9-M13-1.5-25	Tool C-4706	95	70*
Bolt, Extension to Case	4	9.8-M8-1.25-28	13mm Hex	28	250
Bolt, Differential Bearing Retainer to Case	6	9.8-M8-1.25-28	13mm Hex	28	250
Screw Assy., Differential Cover to Case	10	9.8-M8-1.25-16	13mm Hex	19	165
Transfer & Output Shaft Areas:					
Nut, Output Shaft	1	M20-1.5	30mm Hex	271	200*
Nut, Transfer Shaft	1	M20-1.5	30mm Hex	271	200*
Bolt, Gov to Support	2	9.8-M5-0.8-20	7mm Hex	7	60
Bolt, Gov to Support	1	9.8-M5-0.8-30	7mm Hex	7	60
Screw Assy., Governor Counterweight	1	M8-1.25-35	13mm Hex	28	250
Screw Assy., Rear Cover to Case	10	9.8-M8-1.25-16	13mm Hex	19	165
Plug, Reverse Band Shaft	1	1/4-18-NPTF	1/4 in. Sq. Skt.	7	60
Pump & Kickdown Band Areas:					
Bolt, Reaction Shaft Assembly	6	9.8-M8-1.25-19	13mm Hex	28	250
Bolt Assy., Pump to Case	7	9.8-M8-1.25-25	8mm 12 Pt.	31	275
Nut, Kickdown Band Adjustment Lock	1	M12-1.75	18mm Hex	47	35*
Valve Body & Sprag Areas:					
Bolt, Sprag Retainer to Transfer Case	2	9.8-M8-1.25-23	13mm Hex	28	250
Screw Assy., Valve Body	16	9.8A-M5-0.8-11	Torx, T25	5	40
Screw Assy., Transfer Plate	16	9.8A-M5-0.8-25	Torx, T25	5	40
Screw Assy., Filter	2	9.8A-M5-0.8-30	Torx, T25	5	40
Screw, Transfer Plate to Case	7	9.8-M6-1-30	10mm Hex	12	105
Screw Assy., Oil Pan to Case	14	9.8-M8-1.25-16	13mm Hex	19	165
Nut, Reverse Band Adjusting Lock	1	M8-1.25	13mm Hex	14	120

*foot-pounds

† A-413 =	M10 x 1.5 x 18	17mm Hex	88 N·m	65 ft. lbs.	
‡ A-470 =	M12 x 1.25 x 21	19mm Hex	136 N·m	100 ft. lbs.	
# A-413 and A-470 =	M10 x 1.5 x 11.7	18mm Hex	54 N·m	40 ft. lbs.	

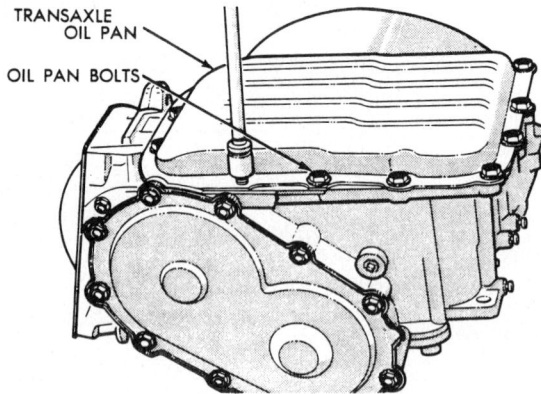

FIGURE 23-1 Removing the oil pan attaching bolts. (Courtesy of Chrysler Corp.)

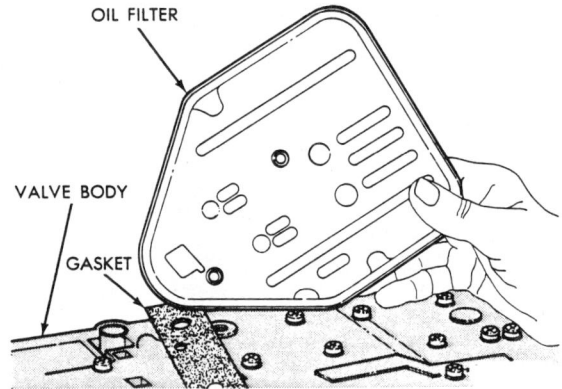

FIGURE 23-2 Removing the transaxle oil pan. (Courtesy of Chrysler Corp.)

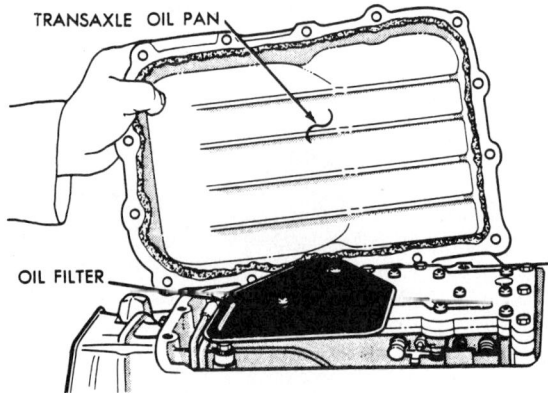

FIGURE 23-3 Removing the filter screws. (Courtesy of Chrysler Corp.)

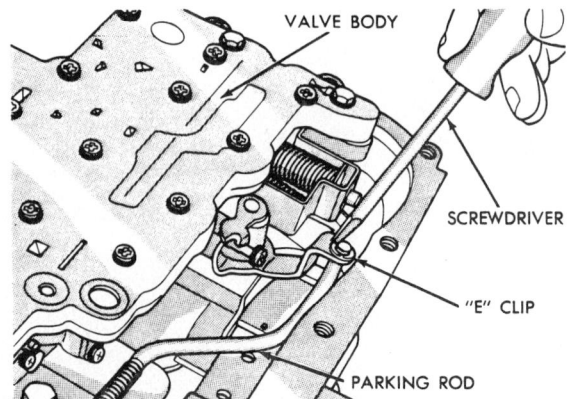

FIGURE 23-4 Removing the oil filter. (Courtesy of Chrysler Corp.)

5. Loosen and remove the neutral starting and back-up lamp switch from the transaxle case.

6. Remove the "E" clip from the parking rod (Fig. 23-5).

7. Take out the parking rod (Fig. 23-6).

8. Loosen and remove the seven valve body-to-case attaching bolts (Fig. 23-7).

9. Carefully lift the valve body and governor tubes away from the transaxle case (Fig. 23-8).

10. Attach a dial indicator onto the transaxle bell housing with its plunger resting against the end of the input shaft (Fig. 23-9).

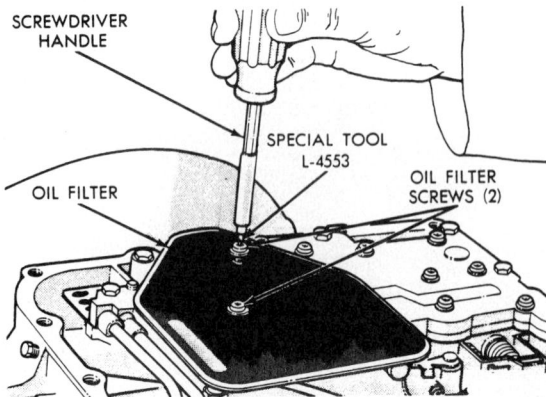

FIGURE 23-5 Removing the parking rod "E" clip. (Courtesy of Chrysler Corp.)

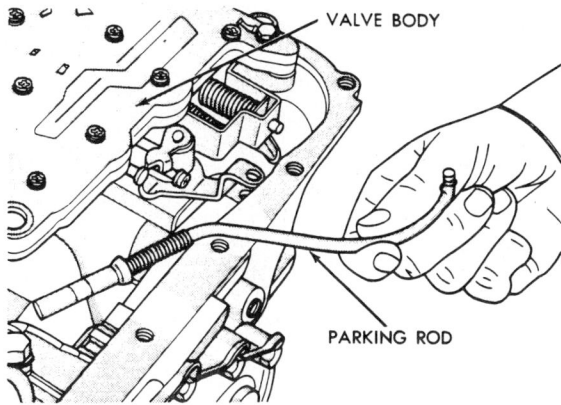

FIGURE 23-6 Taking out the parking rod. (Courtesy o Chrysler Corp.)

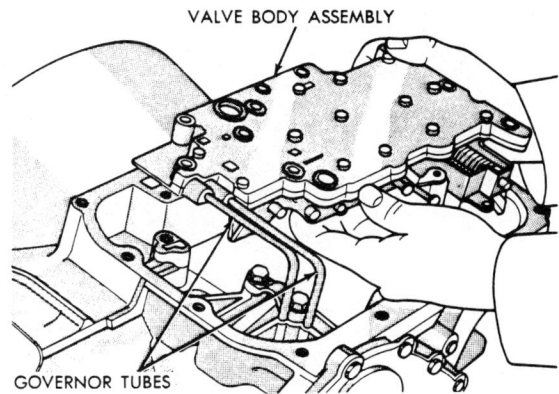

FIGURE 23-8 Taking off the valve body and governor tubes. (Courtesy of Chrysler Corp.)

11. Move the input shaft in and out to obtain an end-play reading on the dial indicator. The end-play specification is 0.007 to 0.073 inch. Then, record the indicator reading for reference when assembling the transaxle.

12. Loosen the locknut and tighten down the kickdown band adjusting screw (Fig. 23-10).

13. Loosen and take out the pump-to-case attaching bolts (Fig. 23-11).

14. Into the two tapped holes in the pump housing, thread the two pump puller adapters (Fig. 23-12).

15. Pull back firmly on the two handles of the pullers until the oil pump assembly and the No. 1 thrust

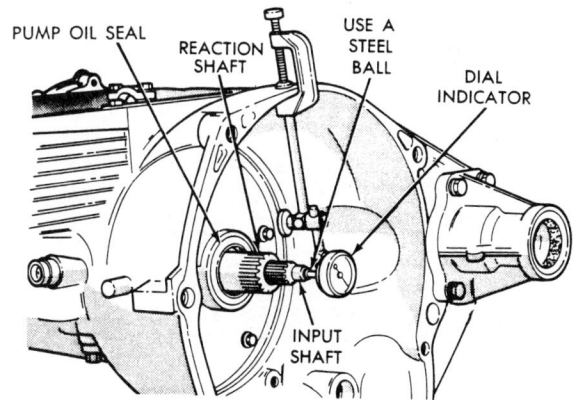

FIGURE 23-9 Measuring input shaft end-play with a dial indicator. (Courtesy of Chrysler Corp.)

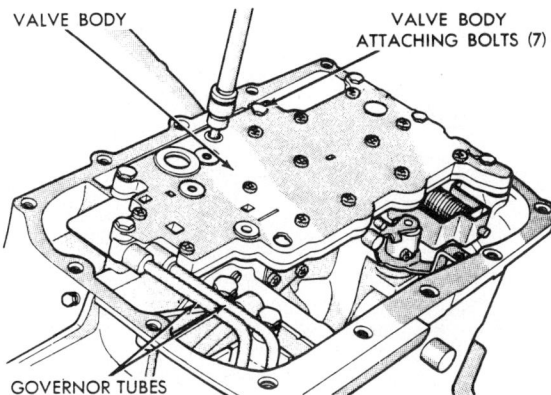

FIGURE 23-7 Removing the valve body attaching bolts. (Courtesy of Chrysler Corp.)

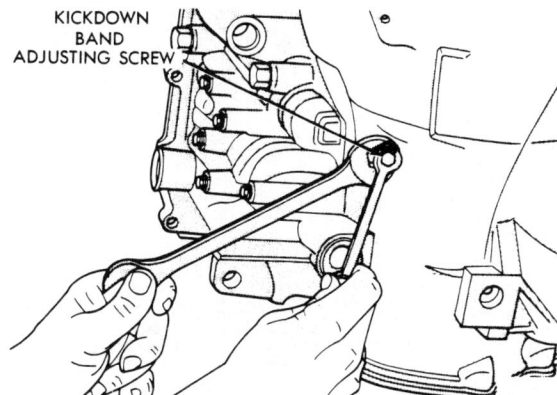

FIGURE 23-10 Tightening down the kickdown band adjusting screw. (Courtesy of Chrysler Corp.)

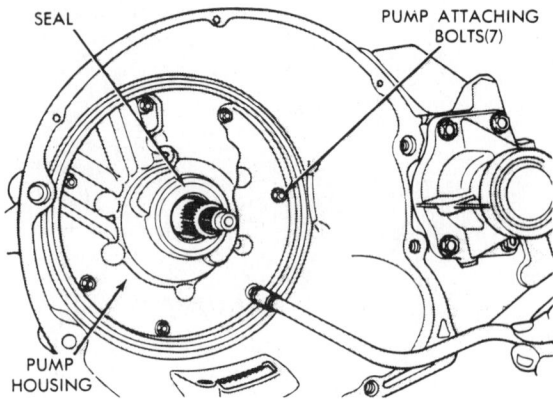

FIGURE 23-11 Removing the oil pump attaching bolts. (Courtesy of Chrysler Corp.)

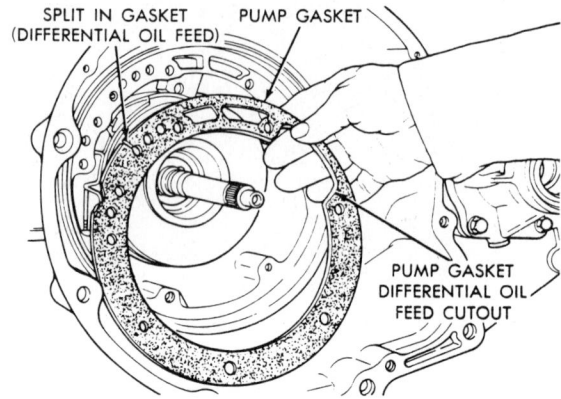

FIGURE 23-14 Taking off the old pump gasket. (Courtesy of Chrysler Corp.)

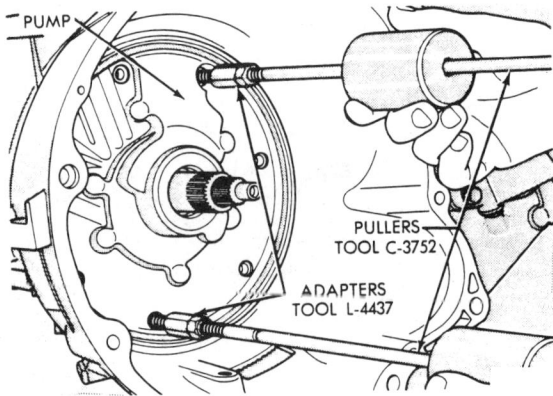

FIGURE 23-12 Threading the puller adapters into the pump housing. (Courtesy of Chrysler Corp.)

washer come loose from the transaxle case (Fig. 23-13).

16. Remove and discard the old pump gasket (Fig. 23-14).

17. Loosen and then back off the kickdown band adjusting screw (Fig. 23-15).

18. Remove the kickdown band and strut from the front of the transaxle case (Fig. 23-16).

19. Carefully lift out the front clutch assembly (Fig. 23-17).

20. Remove the No. 2 thrust washer and the rear clutch assembly (Fig. 23-18).

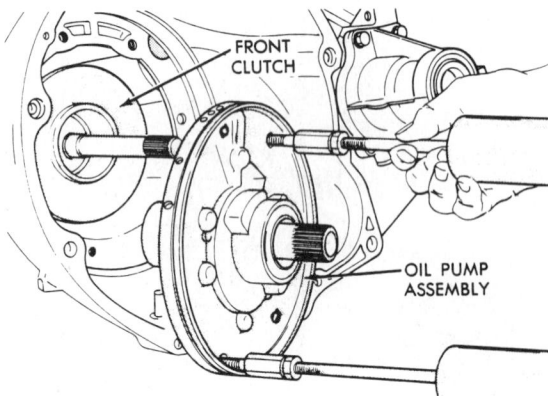

FIGURE 23-13 Removing the pump and No. 1 thrust washer. (Courtesy of Chrysler Corp.)

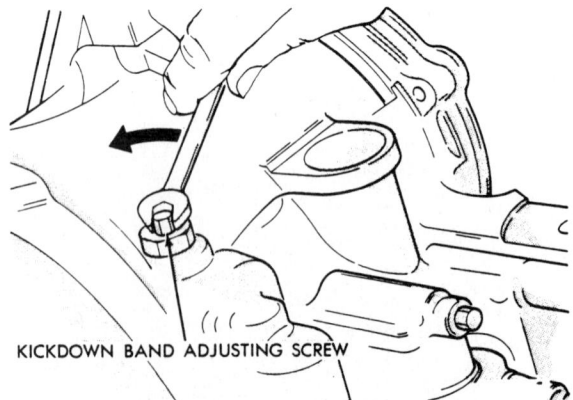

FIGURE 23-15 Loosening the kickdown band adjusting screw. (Courtesy of Chrysler Corp.)

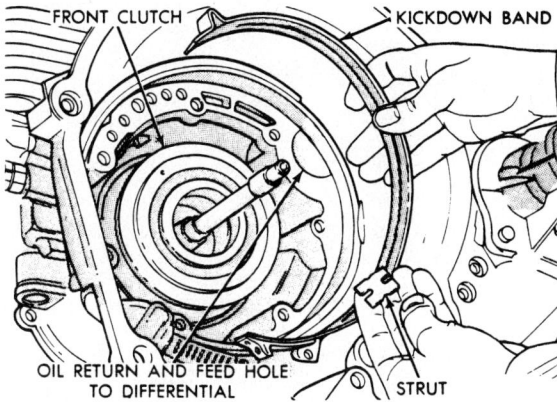

FIGURE 23-16   Removing the kickdown band and strut. (Courtesy of Chrysler Corp.)

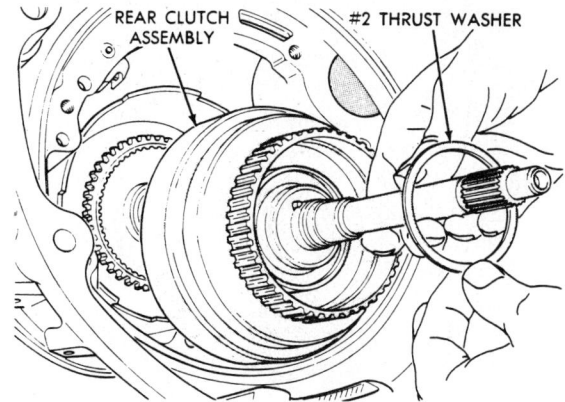

FIGURE 23-18   Removing the No. 2 thrust washer and rear clutch assembly. (Courtesy of Chrysler Corp.)

21. From the end of the output shaft, take off the No. 3 selective thrust washer (Fig. 23-19).

22. Using the pliers shown in Fig. 23-20, remove the front planetary snap ring.

23. Carefully take out the front planetary gear assembly (Fig. 23-21).

24. From inside the sun gear driving shell, remove the No. 6 thrust washer (Fig. 23-22).

25. Take out the sun gear driving shell (Fig. 23-23).

26. From in front of the rear planetary gear assembly, remove the No. 9 thrust washer (Fig. 23-24).

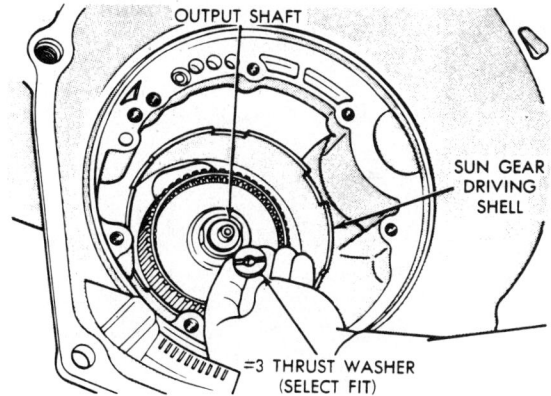

FIGURE 23-19   Taking off the No. 3 thrust washer. (Courtesy of Chrysler Corp.)

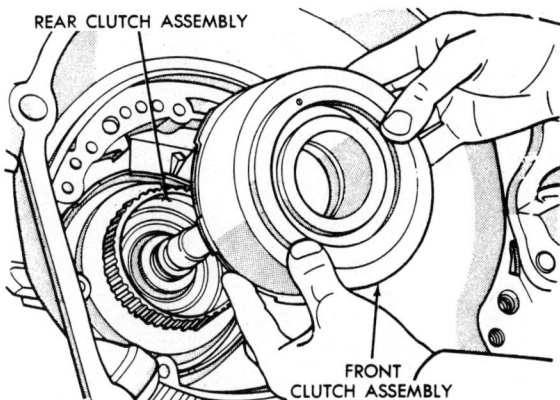

FIGURE 23-17   Lifting out the front clutch assembly. (Courtesy of Chrysler Corp.)

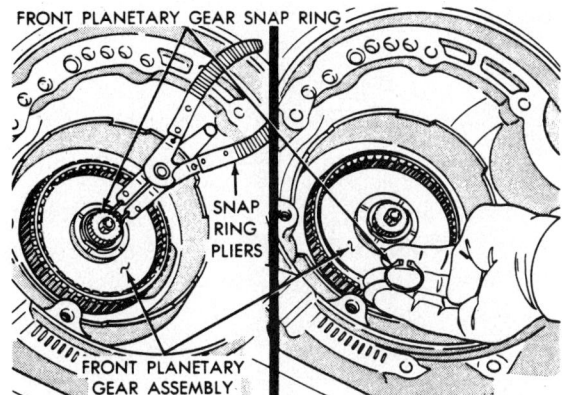

FIGURE 23-20   Using pliers to remove the front planetary snap ring. (Courtesy of Chrysler Corp.)

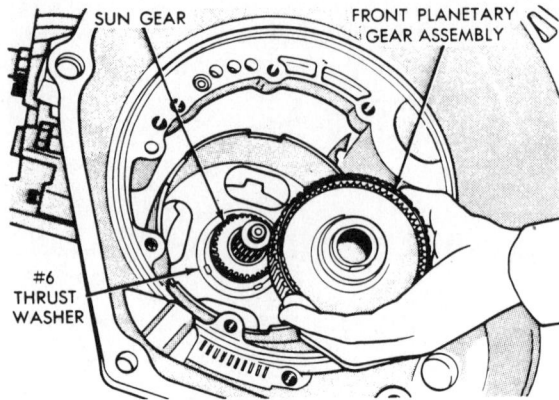

FIGURE 23-21 Removing the front planetary gear assembly. (Courtesy of Chrysler Corp.)

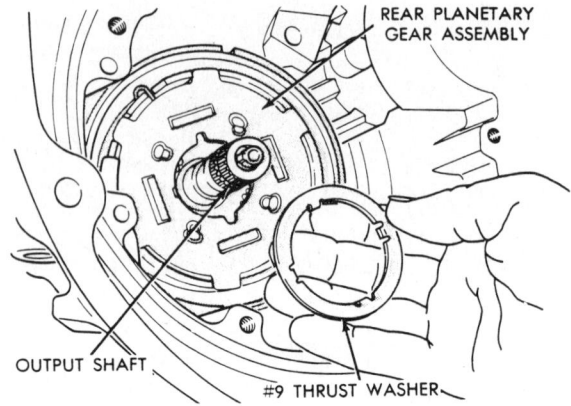

FIGURE 23-24 Removing the No. 9 thrust washer. (Courtesy of Chrysler Corp.)

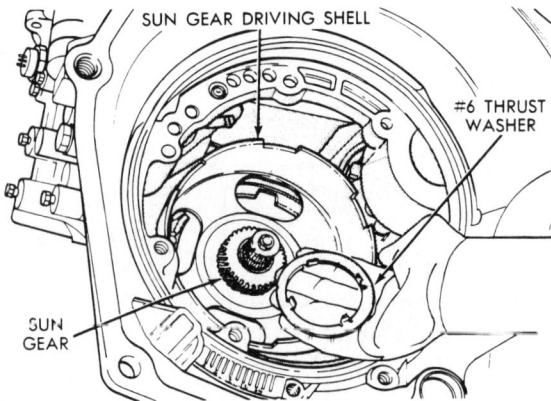

FIGURE 23-22 Taking out the No. 6 thrust washer. (Courtesy of Chrysler Corp.)

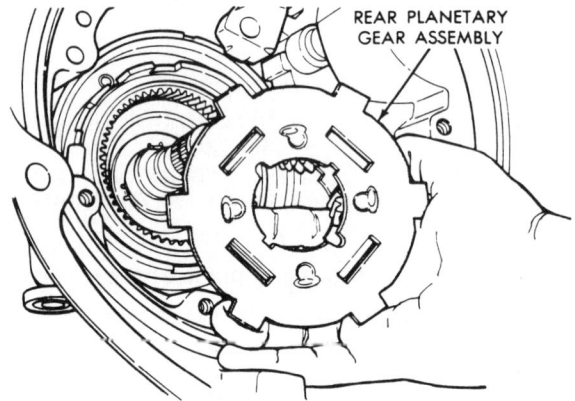

FIGURE 23-25 Lifting out the rear planetary gear assembly. (Courtesy of Chrysler Corp.)

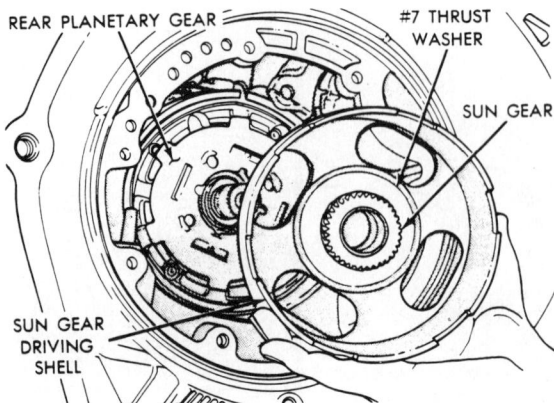

FIGURE 23-23 Lifting out the sun gear driving shell. (Courtesy of Chrysler Corp.)

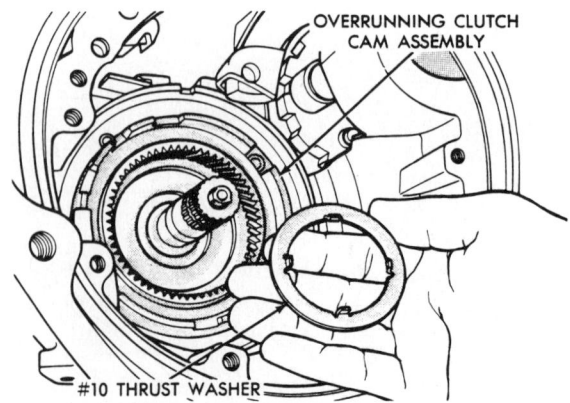

FIGURE 23-26 Removing the No. 10 thrust washer. (Courtesy of Chrysler Corp.)

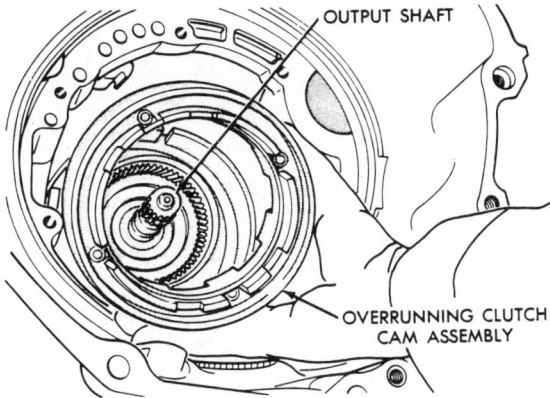

FIGURE 23-27 Taking out the overrunning clutch cam assembly. (Courtesy of Chrysler Corp.)

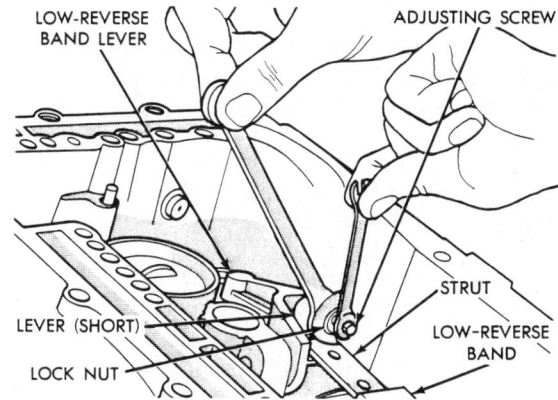

FIGURE 23-29 Loosening the low-reverse band adjuster. (Courtesy of Chrysler Corp.)

27. Remove the rear planetary gear assembly (Fig. 23-25).

28. Take out the No. 10 thrust washer (Fig. 23-26).

29. Carefully remove the overrunning clutch cam assembly (Fig. 23-27).

30. Take out from inside the case the eight overrunning clutch springs and rollers (Fig. 23-28). (NOTE: This step is only necessary if these parts came out of the clutch during its removal.)

31. Loosen off the low-reverse band adjusting screw (Fig. 23-29).

32. Remove the low-reverse band and strut from the case (Fig. 23-30).

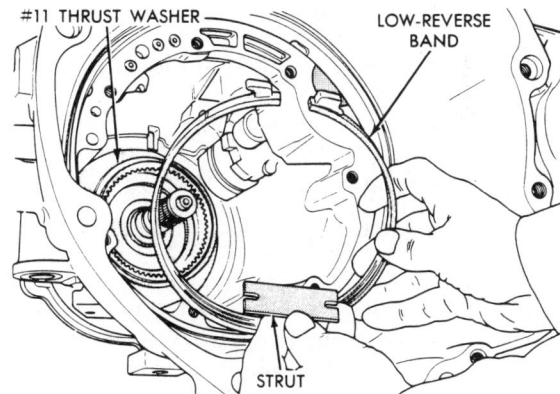

FIGURE 23-30 Lifting out the low-reverse band and strut. (Courtesy of Chrysler Corp.)

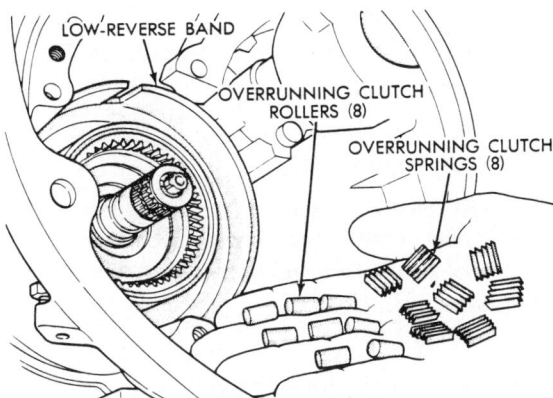

FIGURE 23-28 Removing the eight overrunning clutch springs and rollers. (Courtesy of Chrysler Corp.)

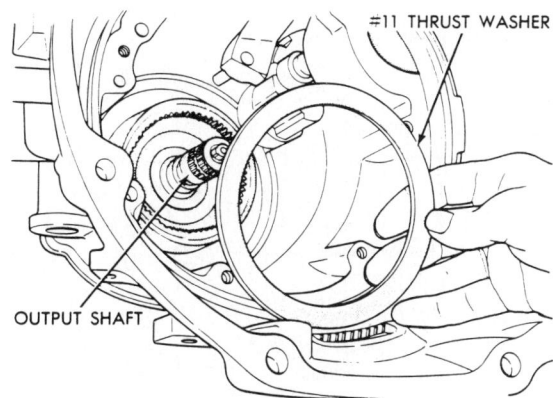

FIGURE 23-31 Removing the No. 11 thrust washer. (Courtesy of Chrysler Corp.)

33. Take out the No. 11 thrust washer (Fig. 23-31). (NOTE: The output shaft and rear annulus gear can remain in the case unless the assembly must be removed to repair the transfer shaft or gears.)

HYDRAULIC PUMP RECONDITION

During the overhaul procedure, the hydraulic pump should be reconditioned before reinstallation on the transaxle. The reason for this is twofold. First, the pump gears must operate with given clearances in order for them to provide the transaxle with the necessary fluid flow and pressure. If the clearances become too large, the transaxle can malfunction or even fail due to a loss in fluid flow. Second, the pump is a very expensive component; therefore, the technician should do everything possible to reuse it.

Teardown and Inspection

To determine if the pump is still serviceable, follow these steps:

1. From the pump housing, loosen and remove the six reaction shaft attaching bolts (Fig. 23-32).

2. Take the reaction shaft support off the pump housing (Fig. 23-33).

3. Check both gears for identification marks. If there are none, make an alignment mark across both gears and the housing with layout fluid or an indelible pen (Fig. 23-34). This mark will assist in reinstalling the gears in the same relative position when you assemble the pump. (CAUTION: *Never use a center punch or metal scribe to mark these components.*)

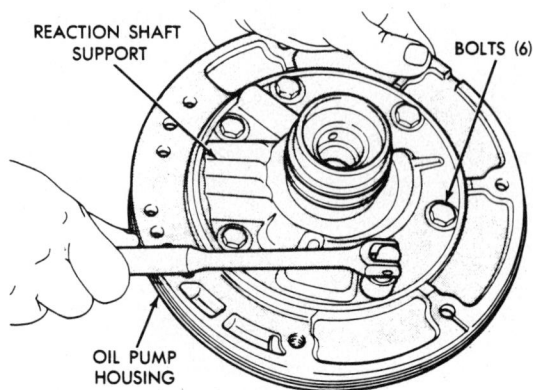

FIGURE 23-33 Separating the reaction shaft support from the oil pump housing. (Courtesy of Chyrsler Corp.)

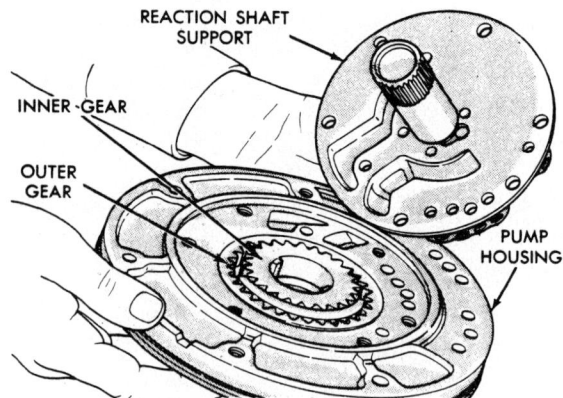

FIGURE 23-34 Marking the gears and housing with layout fluid.

4. Remove the inner and outer gears from the pump housing (Fig. 23-35).

5. With the tool shown in Fig. 23-36, remove and discard the pump seal.

6. Clean all the components in clean solvent, and blow them dry with low-pressure compressed air. CAUTION: *Make sure you thoroughly blow out all the fluid passages; clogged or restricted passages can cause a malfunction.*

7. Check all the parts for burrs and light scoring; if you find any, use crocus cloth to clean and polish the surfaces. Next, rewash these parts again to remove any grit.

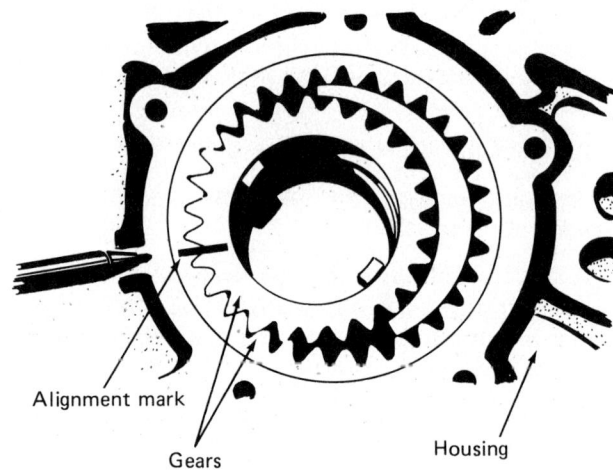

FIGURE 23-32 Removing the reaction shaft support attaching bolts. (Courtesy of Chrysler Corp.)

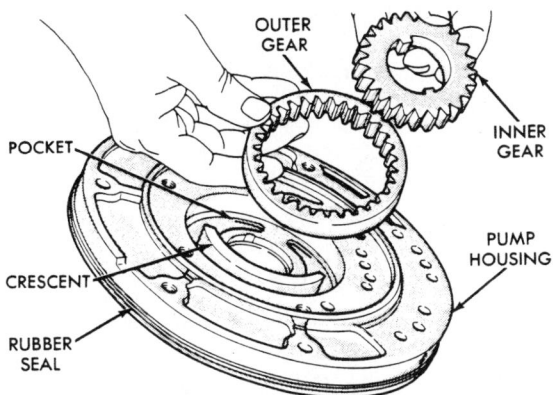

FIGURE 23-35 Removing both gears from the pump housing. (Courtesy of Chrysler Corp.)

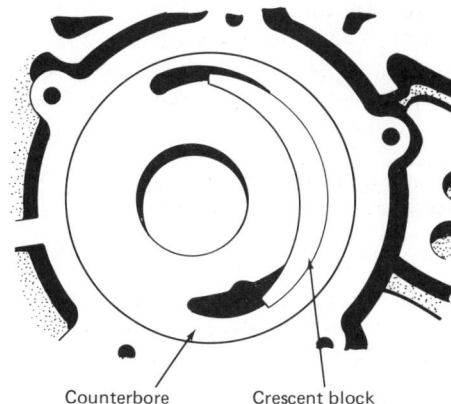

FIGURE 23-37 Checking the pump's counterbore for wear.

8. Inspect the gears carefully for broken teeth. If there are any, replace the pump.

9. Inspect the inner and outer circumference of the gears along with the teeth for excessive wear. (NOTE: Moderate wear does not necessitate replacing the pump. In fact, if the hydraulic pressure was normal during the entire diagnosis procedure, the pump very likely has enough output for reuse.)

10. Inspect the mating surfaces of the gears, pump housing, and reaction shaft support for abnormal wear. Don't be concerned if the black lubrite coating

has worn off during previous pump operation; this is a normal tendency. However, pay special attention to the inside of the pump body (the counterbore), where the gears operate (Fig. 23-37). Check the depth of any scoring with your fingernail; if there is excessive wear in this area and pump pressure was low during diagnosis, replace the pump.

11. If everything has checked out so far, reinstall the gears in the pump housing, aligning the marks made during disassembly.

12. Measure the pump housing-to-gear clearance using a straightedge and a feeler gauge (Fig. 23-38). This measurement must be within specifications.

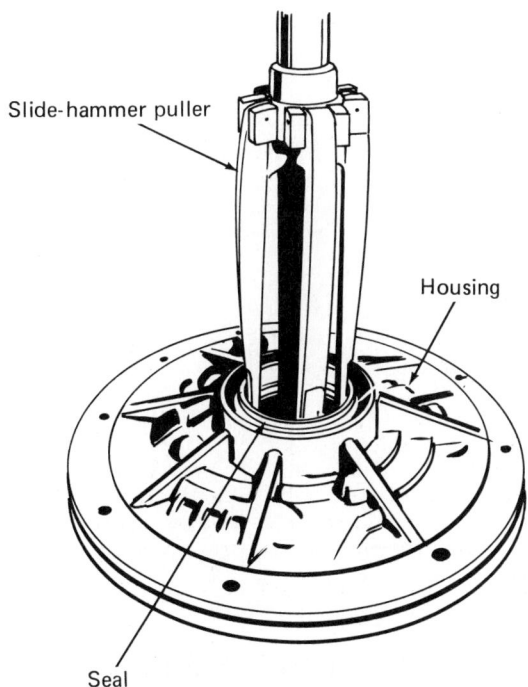

FIGURE 23-36 Using a puller to remove the pump seal.

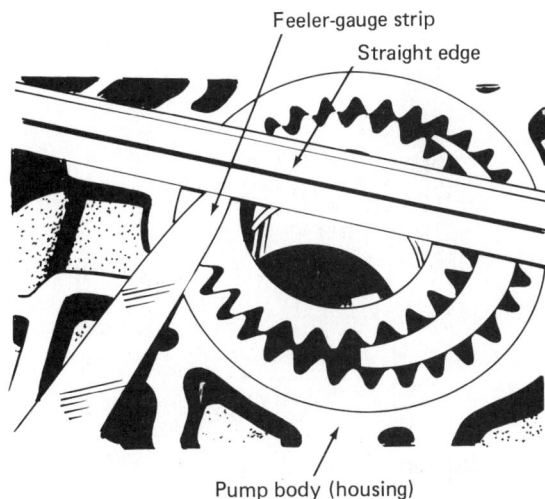

FIGURE 23-38 Measuring the pump housing-to-gear end clearance.

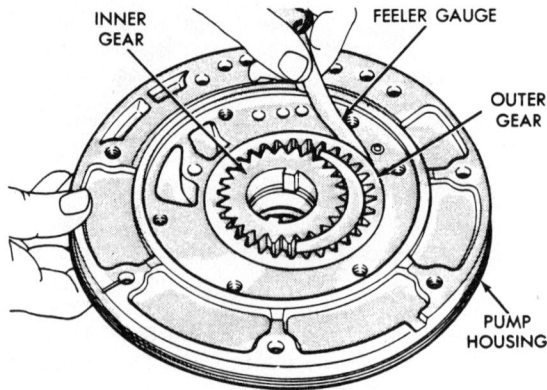

FIGURE 23-39 Measuring the clearance between the outer gear and the pump housing. (Courtesy of Chrysler Corp.)

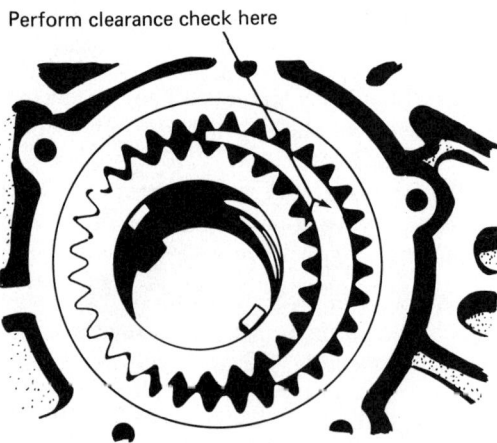

Perform clearance check here

FIGURE 23-40 Locations where you measure the clearance between both gears and the crescent.

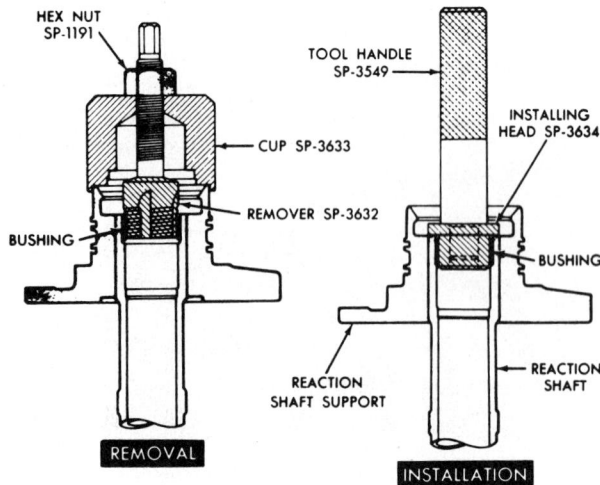

FIGURE 23-41 Removing and replacing the reaction shaft bushing. (Courtesy of Chrysler Corp.)

13. Using a feeler gauge of the specified thickness, measure the clearance between the outer gear and the pump housing bore (Fig. 23-39).

14. Using a feeler gauge of the specified thickness, check the clearance between the teeth of each gear and the pump crescent (Fig. 23-40). (NOTE: *If any of the above mentioned feeler gauge clearance checks are not to specifications, replace the pump assembly.*)

15. Remove all the sealing rings from the reaction shaft support. Inspect the reaction shaft support for worn or damaged ring grooves, clutch drum bearing surfaces, or splines. If wear or damage exists in any of these areas, replace the reaction shaft support or the pump assembly.

Service and Reassembly

To replace the pump bushings plus the seals and reassemble the unit, follow these procedures:

1. Again remove the gears from the pump housing.

2. With the tools shown in Fig. 23-41, remove and replace the reaction shaft bushings.

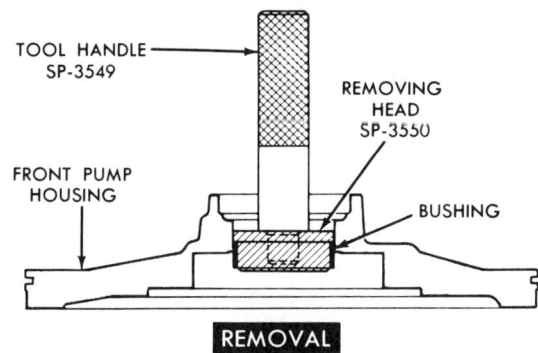

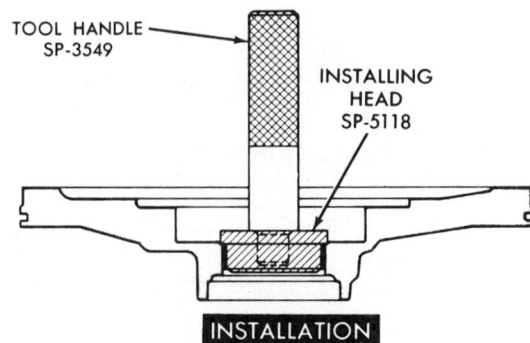

FIGURE 23-42 Replacing the pump bushing. (Courtesy of Chrysler Corp.)

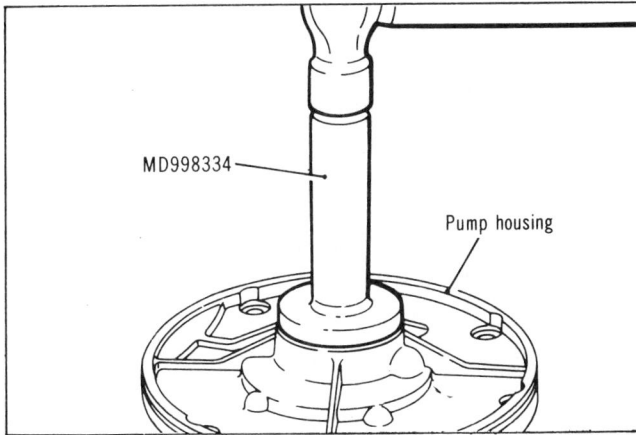

FIGURE 23-43   Installing a new pump seal. (Courtesy of Chrysler Corp.)

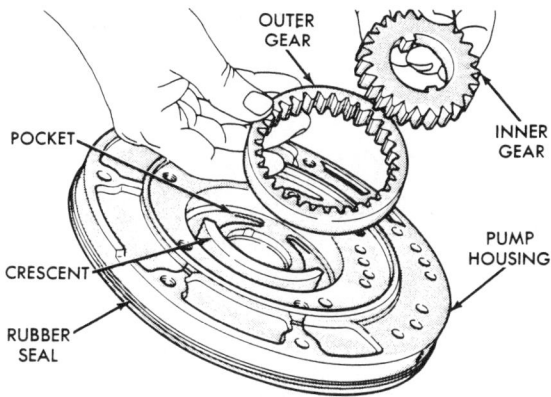

FIGURE 23-44   Installation of the pump gears into the housing. (Courtesy of Chrysler Corp.)

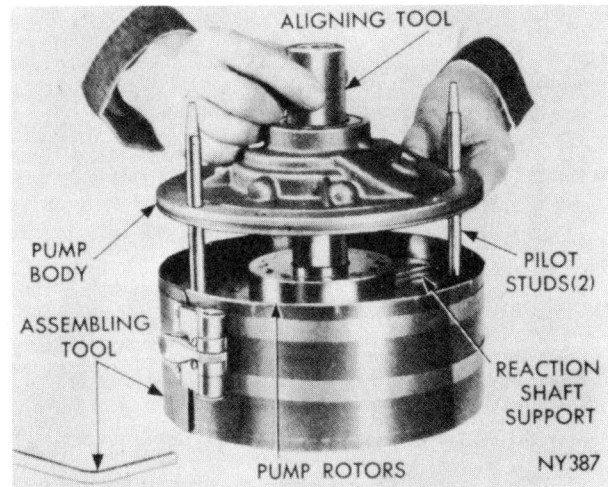

FIGURE 23-46   Aligning the pump housing and reaction shaft support before torquing their attaching bolts. (Courtesy of Chrysler Corp.)

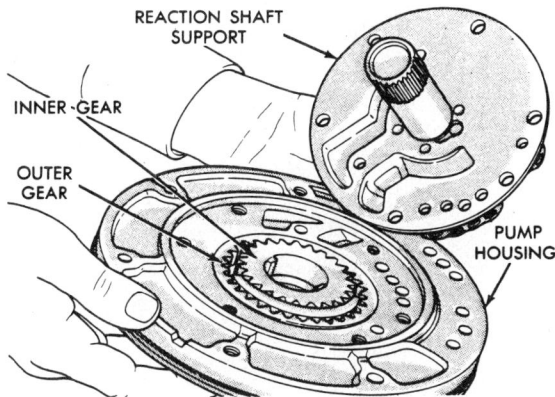

FIGURE 23-45   Reinstalling the reaction shaft support over the pump housing. (Courtesy of Chrysler Corp.)

3. Using the tool handle, removing and installing heads, remove and replace the converter support bushing in the pump housing (Fig. 23-42).

4. Install a new pump seal using the tool pictured in Fig. 23-43. (NOTE: If the new seal does not have a rubber or resin-coated backing, cover this area with a thin layer of nonhardening sealer before installation. Also, *make sure the sealing lip faces inward toward the gears.)*

5. Lubricate the gears with clean transmission fluid and reinstall them into the pump housing, aligning their locating marks (Fig. 23-44).

6. Position the reaction shaft support over the pump housing (Fig. 23-45), and install the attaching bolts. Tighten these bolts finger tight.

7. Install the alignment band over and around the pump housing and reaction shaft support as shown in Fig. 23-46. Also, insert the guide pin before torquing the reaction shaft support-to-housing bolts to specifications (See Table 23-1). Then, install a new rubber seal around the outer circumference of the housing.

8. Install the pump assembly over the converter hub, and *make sure the pump gears rotate freely.*

9. Replace the seal rings over the reaction shaft support (Fig. 23-47). After installation, check the rings for freedom of movement in their respective grooves.

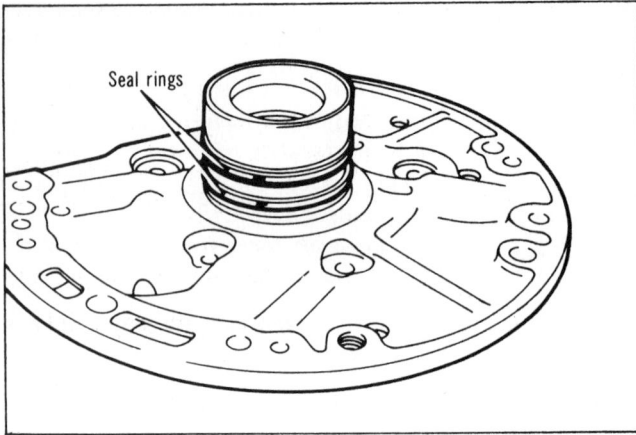

FIGURE 23-47   Replacing the seal rings over the reaction shaft support. (Courtesy of Chrysler Corp.)

FRONT CLUTCH RECONDITION

As part of the transaxle overhaul, the front clutch assembly must be reconditioned. This process involves disassembling the clutch, cleaning and inspecting its components, replacing the bushing, and reassembling the unit.

Disassembly

To tear down the front clutch, follow these steps:

1. Using a screwdriver, remove the waved snap ring from the clutch retainer (Fig. 23-48).

2. Remove the waved snap ring and thick steel pressure plate from the retainer (Fig. 23-49).

3. Take out the three clutch plates and three driving discs

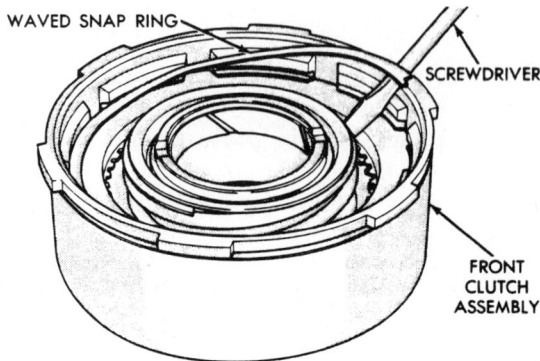

FIGURE 23-48   Removing the front clutch retainer snap ring. (Courtesy of Chrysler Corp.)

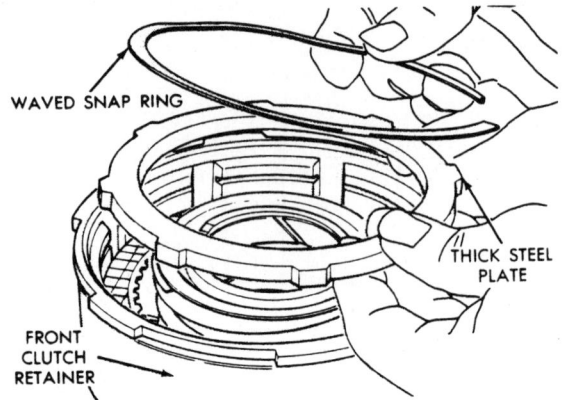

FIGURE 23-49   Removing the snap ring and steel pressure plate. (Courtesy of Chrysler Corp.)

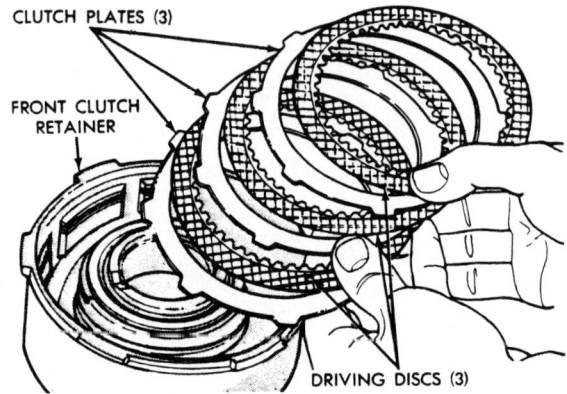

FIGURE 23-50   Lifting out the driving discs and clutch plates. (Courtesy of Chrysler Corp.)

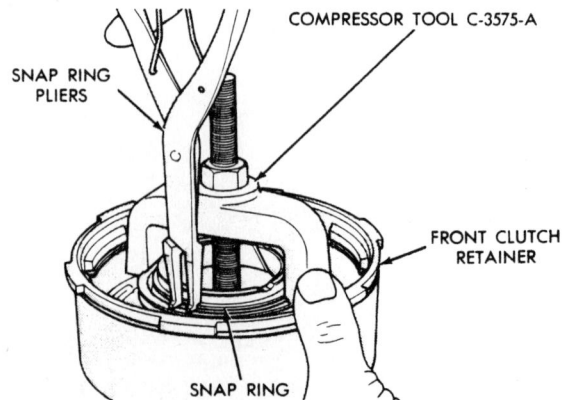

FIGURE 23-51   Compressing the return spring in order to remove the snap ring. (Courtesy of Chrysler Corp.)

from the retainer (Fig. 23-50). Tie all these parts plus the pressure plate and wavy spring together with a piece of fine wire as an aid to reassembly.

4. Install a compressor over the return spring retainer, and slowly compress the spring until you can easily remove the snap ring with pliers as shown in Fig. 23-51.

5. Remove the snap ring, retainer, and return spring (Fig. 23-52). Next, lift out the piston using a twisting motion to remove it from the retainer. (NOTE: It may be necessary to direct air pressure to the clutch apply port in order to force the piston from its bore.)

6. Remove the lip seal inside the retainer and the one located on the piston itself.

Clutch Component Cleaning

To prepare the clutch retainer and its various components for inspection, wash or clean all the metal parts in solvent and dry them all with low-pressure compressed air. CAUTION: *Do not wash or soak any rubber seals or driving discs in solvent or a varnish dissolving agent. These materials will deteriorate the rubber seals or lining on the discs.*

Clutch Component Inspection

With all the clutch parts clean and dry, inspect them as follows:

1. The retainer bushing for excessive wear or scoring.

2. The fit of the pressure plate in its retainer slots. There should be no appreciable radial play between the two components.

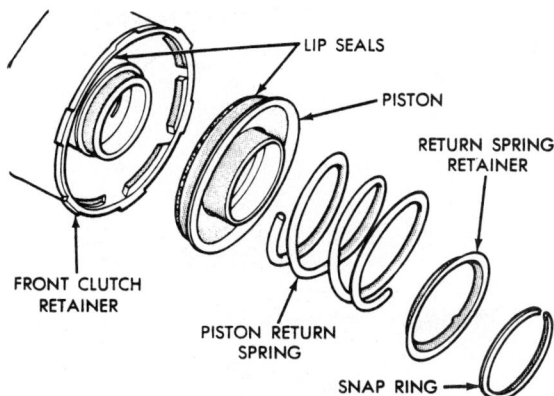

3. The retainer for burred, scored, or damaged thrust surfaces.

4. The inside of the retainer for rough or badly worn clutch plate slots or grooves.

5. The retainer for nicked or deeply worn sealing ring surfaces and for obstructed or restricted fluid passages.

6. The piston bore in the retainer for wear or scoring.

7. The surface of the retainer where the band rides for wear or scoring.

8. The piston for cracks, wear, or damage.

9. The steel residual check ball in the piston. Be sure that the ball is free to move in its bore and that the orifice leading to the ball is open.

10. The piston return spring for wear, damage, or signs of overheating.

11. The driving discs for excessive wear; cracked, charred, burned, glazed, or scored lining material; and distortion.

12. The steel clutch plates for distortion, surface scuffing or scoring, and damaged drive lugs.

13. If the clutch retainer or any of its various parts are defective, replace them as necessary.

Clutch Service and Reassembly

To service the clutch retainer plus the clutch plates and reassemble the unit, follow these steps:

1. With the tools shown in Fig. 23-53, remove and replace the front clutch retainer bushing.

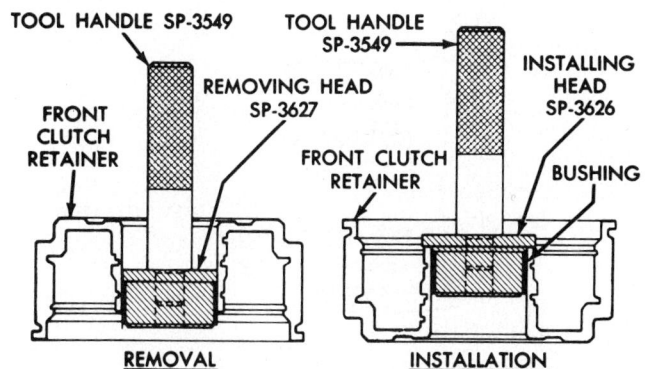

FIGURE 23-52 Removing the retainer, spring, and clutch piston. (Courtesy of Chrysler Corp.)

FIGURE 23-53 Removing the replacing the retainer bushing. (Courtesy of Chrysler Corp.)

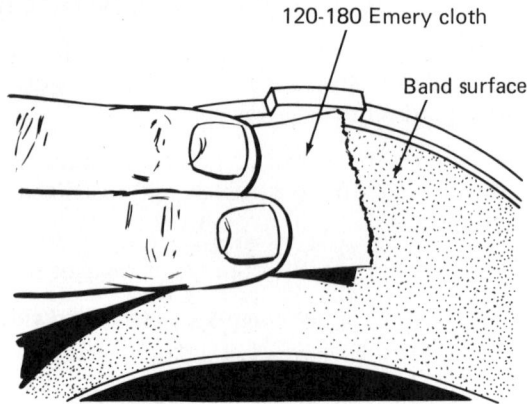

FIGURE 23-54 Deglazing the band surface with emery cloth.

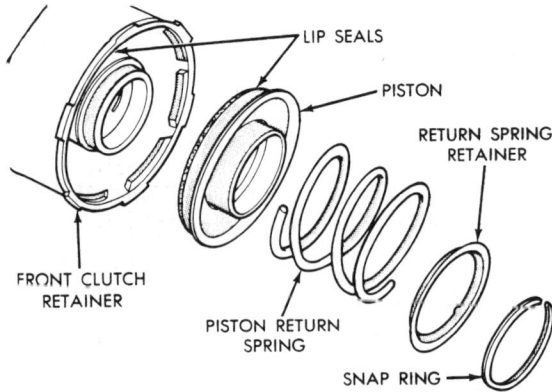

FIGURE 23-55 Installing the return spring and retainer. (Courtesy of Chrysler Corp.)

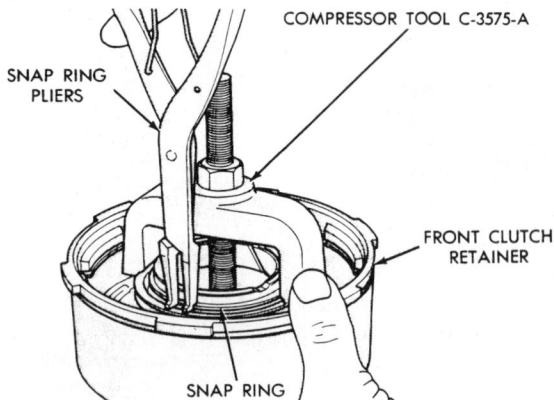

FIGURE 23-56 Compressing the return spring in order to install the snap ring. (Courtesy of Chrysler Corp.)

2. Deglaze the portion of the retainer where the band rides to restore its friction characteristic. To do this, sand the retainer surface with 120-180 grit sandpaper or emery cloth around the drum, but never in a front-to-back direction (Fig. 23-54).

3. Install a new lip seal over the hub inside the retainer; make sure its lip points toward the base of the bore.

4. Install a new lip seal in the groove on the outer circumference of the piston; be sure its lip faces toward the bottom of the retainer bore.

5. Lubricate both seals and the piston bore in the retainer generously with transmission fluid. Then, reinstall the piston into the retainer bore with a twisting motion. NOTE: It may be necessary to use a feeler gauge strip as an aid in installing the piston if its outer seal hangs up at the beginning of the retainer bore. If this is the case, slowly work the gauge strip around the circumference of the piston, while applying slight downward pressure on the piston. CAUTION: *Be careful not to tear or cut the seal during this process.*

6. Position the spring onto the piston. Next, place the retainer in position on top of the return spring (Fig. 23-55).

7. Using the compressor tool shown in Fig. 23-56, slowly compress the retainer and return spring far enough to permit the installation of the snap ring into its groove in the hub.

8. If the old steel clutch plates are still serviceable, bead blast or refinish them on both sides with 120-320 grit emery cloth or sandpaper (Fig. 23-57). When sanding

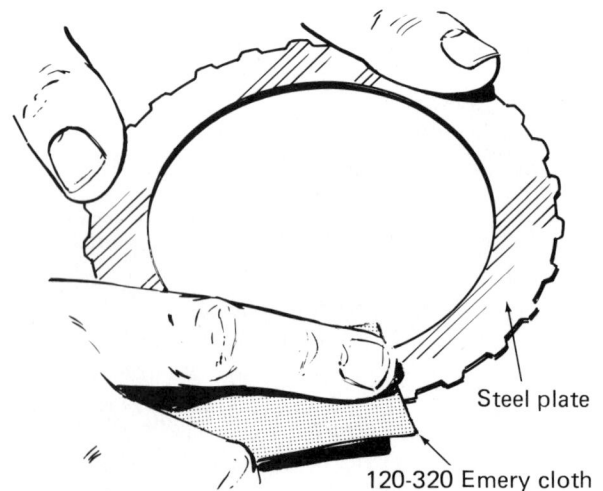

FIGURE 23-57 Deglazing a clutch plate with emery cloth.

**TABLE 23-2   Clearance and End-Play Specifications
for the A-417, A-413, and A-470 Automatic Transaxle**

TRANSAXLE SPECIFICATIONS

	(Millimeter)	(Inch)
Thrust Washers:		
Reaction Shaft Support (Phenolic) ....No. 1	1.55-1.60	.061-.063
Rear Clutch Retainer (Phenolic) ....No. 2	1.55-1.60	.061-.063
Output Shaft, Steel Backed Bronze ....(Select) No. 3	1.98-2.03	.077-.080
	2.15-2.22	.085-.087
	2.34-2.41	.092-.095
Front Annulus, Steel Backed Bronze ....No. 4	2.95-3.05	.116-.120
Front Carrier, Steel Backed Bronze .... Nos. 5, 6	1.22-1.28	.048-.050
Sun Gear (Front) ....No. 7	0.85-0.91	.033-.036
Sun Gear (Rear) ....No. 8	0.85-0.91	.033-.036
Rear Carrier, Steel Backed Bronze .... Nos. 9, 10	1.22-1.28	.048-.050
Rev. Drum, Phenolic ....No. 11	1.55-1.60	.061-.063

	(Millimeter)	(Inch)
Tapered Roller Bearing Settings:		
Output Shaft	0.0-0.07 Preload	0.0-.0028 Preload
Transfer Shaft	0.05-0.25 End Play	.002-.010 End Play
Differential	0.15-0.29 Preload	.006-.012 Preload

*S = Standard duty and W = Heavy duty.

	(Millimeter)	(Inch)
Pump Clearances:		
Outer Gear to Pocket	0.045-.141	.0018-.0056
Outer Gear I.D. to Crescent	0.150-.306	.0059-.012
Outer Gear Side Clearance	0.025-.050	.001-.002
Inner Gear O.D. to Crescent	.160-.316	.0063-.0124
Inner Gear Side Clearance	.025-.050	.001-.002

	(Millimeter)	(Inch)
End Play:		
Input Shaft	0.18-1.85	.007-.073
Front Clutch Retainer	0.76-2.69	.030-.106
Front Carrier	0.89-1.45	.007-.057
Front Annulus Gear	0.09-0.50	.0035-.020
Planet Pinion	0.15-0.59	.006-.023
Reverse Drum	0.76-3.36	.030-.132

		(Millimeter)	(Inch)
Clutch Clearance and Selective Snap Rings:			
Front Clutch (Non-Adjustable) Measured from			
Reaction Plate to "Farthest" Wave	2 Disc	1.94-2.95	.076-.116
	3 Disc	2.22-3.37	.087-.133
Rear Clutch (3 and 4 Disc)			
Adjustable	3 Disc	0.67-0.86	.026-.034
	4 Disc	0.67-1.10	.026-.043
Selective Snap Rings (4)		1.22-1.27	.048-.050
		1.52-1.57	.060-.062
		1.88-1.93	.074-.076
		2.21-2.26	.087-.089

Band Adjustment:
Kickdown, Backed off from 8 N·m (72 in. lbs.) ...... A-415   3 Turns,
A-413 & A-470   2-1/2 Turns

Low-Reverse........ A-415 .... Nonadjustable
A-413 & A-470 .... 3-1/2 Turns backed off from 5 N·m (41 in. lbs.)

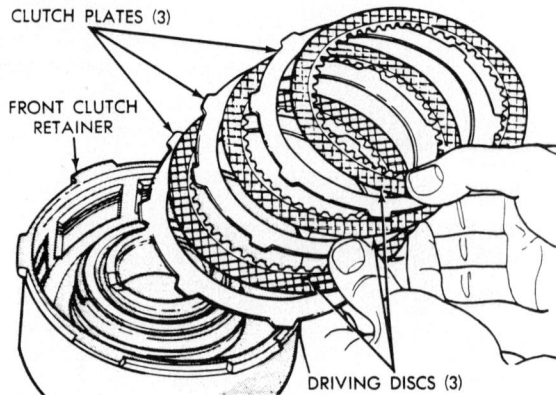

FIGURE 23-58 Installing the clutch plates and driving discs into the retainer. (Courtesy of Chrysler Corp.)

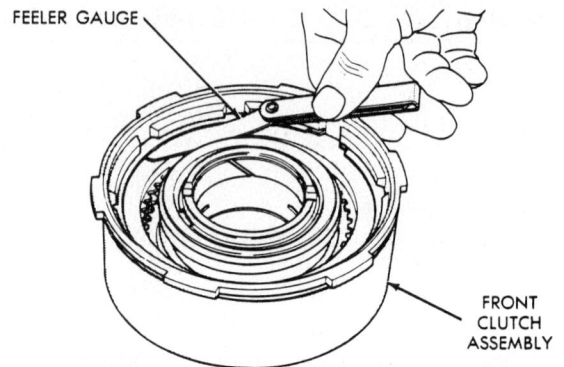

FIGURE 23-59 Installing the thick steel plate. (Courtesy of Chrysler Corp.)

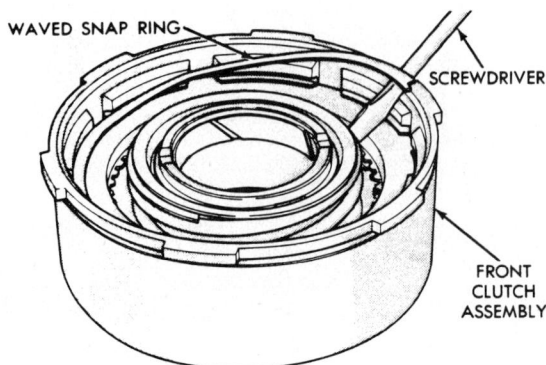

FIGURE 23-60 Installing the front clutch snap ring. (Courtesy of Chrysler Corp.)

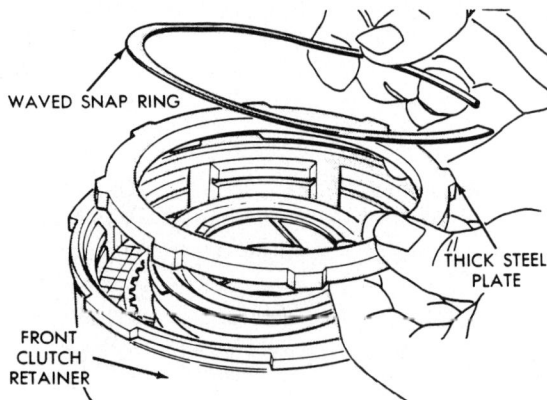

FIGURE 23-61 Checking front clutch pack clearance. (Courtesy of Chrysler Corp.)

the plates, rub until sanding marks are visible over the entire area on both sides of the plate. The bead blasting or sanding process removes polished surfaces and restores new roughness to the plate, which is necessary for proper clutch break in. After servicing the plates, clean and lubricate them with transmission fluid before installation.

9. Presoak all new driving discs in clean transmission fluid for no less than 30 minutes before installation.

10. Install into the retainer the clutch plates and the driving discs alternately, beginning with a steel plate (Fig. 23-58).

11. Reinstall the thick steel plate (Fig. 23-59) over the last driving disc.

12. Install the front clutch waved snap ring (Fig. 23-60).

13. With a feeler gauge, measure the front clutch pack clearance as shown in Fig. 23-61. It should be to specifications if new driving discs were installed in the assembly (see Table 23-2 on page 449).

REAR CLUTCH RECONDITION

During the overhaul of the transaxle, the rear clutch must be reconditioned before reuse. This procedure involves disassembling the clutch, cleaning and inspecting the components, servicing the clutch plates, and reassembling the unit.

Disassembly

To tear down the rear clutch, follow this procedure:

1. Using a screwdriver, remove the selective snap ring from its groove in the rear clutch retainer (Fig. 23-62).

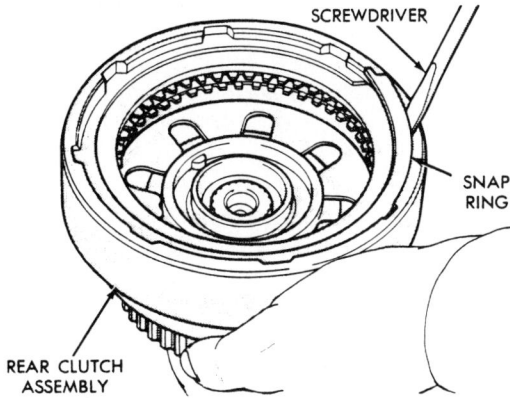

FIGURE 23-62   Removing the rear clutch snap ring. (Courtesy of Chrysler Corp.)

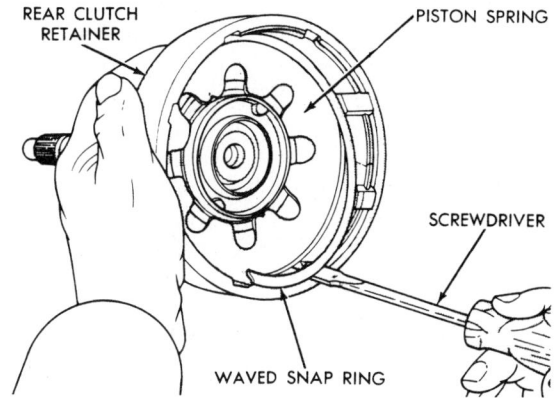

FIGURE 23-64   Removing the rear piston snap ring. (Courtesy of Chrysler Corp.)

2. From the retainer, take out the snap ring, thick steel pressure plate, driving discs, and clutch plates (Fig. 23-63). Tie all these parts together with a piece of fine wire as an aid during reassembly.

3. Using a screwdriver (Fig. 23-64), remove the wavy snap ring over the piston spring.

4. Remove the snap ring, piston spring, and piston from the rear clutch retainer (Fig. 23-65). As necessary, use air pressure to force the piston from its bore.

5. With the pliers as shown in Fig. 23-66, remove the input shaft snap ring.

6. As necessary, press out the input shaft from the rear clutch retainer.

7. Remove the inner and outer seals from the piston (see Fig. 23-65).

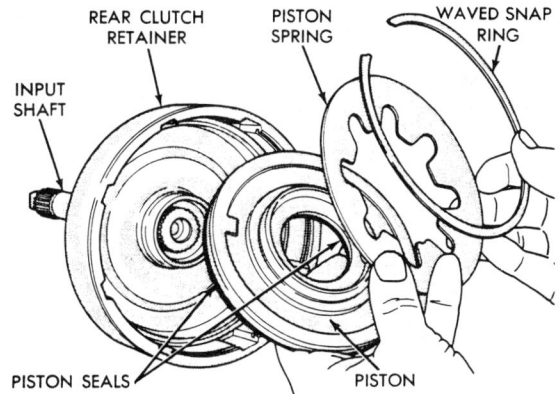

FIGURE 23-65   Removing the snap ring, piston spring, and piston from the rear retainer. (Courtesy of Chrysler Corp.)

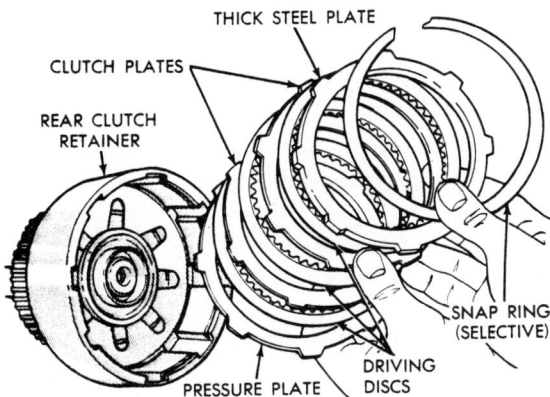

FIGURE 23-63   Taking out the snap ring, pressure plate, driving discs, and clutch plates. (Courtesy of Chrysler Corp.)

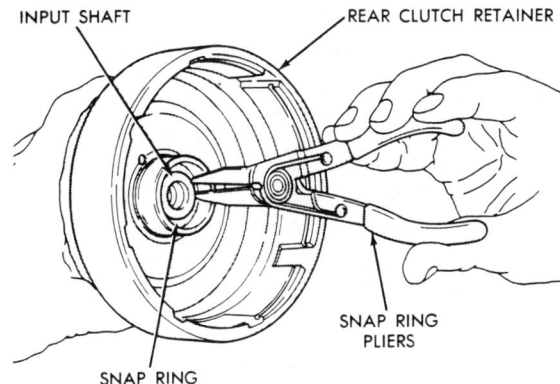

FIGURE 23-66   Taking off the input snap ring. (Courtesy of Chrysler Corp.)

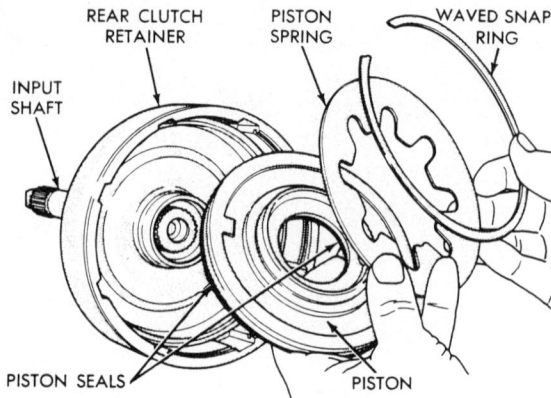

FIGURE 23-67 Installing the piston, spring, and waved snap ring. (Courtesy of Chrysler Corp.)

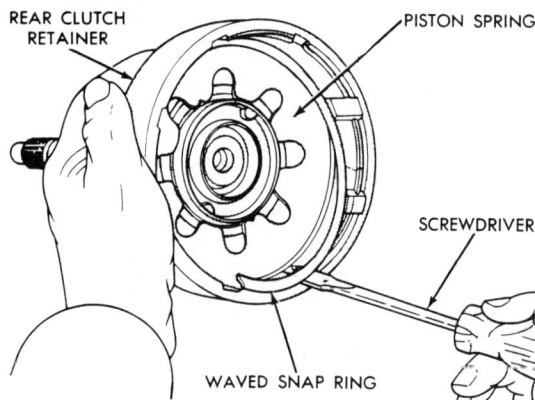

FIGURE 23-68 Inserting the waved snap ring into its groove with a screwdriver. (Courtesy of Chrysler Corp.)

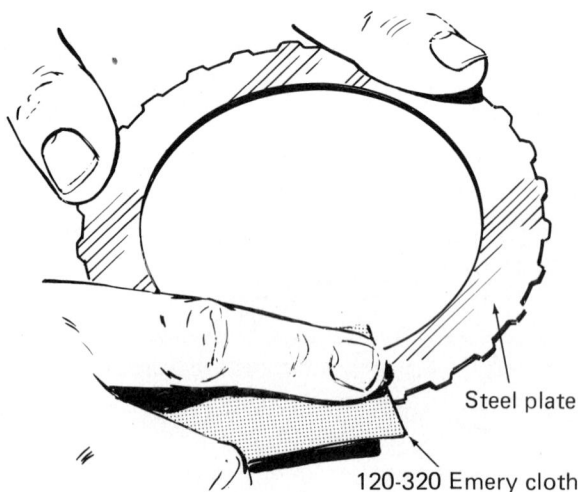

FIGURE 23-69 Sanding the steel clutch plates with emery cloth.

Clutch Component Cleaning

To prepare the rear clutch retainer and its various components for inspection, wash or clean all the metal parts in solvent, and blow dry with low-pressure compressed air. CAUTION: *Do not wash or submerge any rubber seals or driving discs in solvent or a varnish-dissolving cleaning agent. These materials will deteriorate the rubber seals and the lining on the driving discs.*

Component Inspection

With the rear clutch components clean and dry, inspect them all as follows:

1. The fit of the thick steel pressure plate in its retainer slots. There should be no appreciable radial play between the two components.

2. The input shaft splines for signs of wear or damage.

3. The inside of the retainer for rough or badly worn clutch plate slots or grooves.

4. The retainer for burred, scored, or damaged thrust surfaces.

5. The clutch driving disc splines for signs of excessive wear or damage.

6. The retainer for restricted or obstructed fluid passages.

7. The piston bore in the retainer for wear or scoring.

8. The piston for cracks, wear, or damage.

9. The residual check ball in the piston. Be sure that the ball is free to move in its bore and that the orifice leading to the ball is open.

10. The Belleville-type return spring for cracks, wear, damage, or signs of overheating.

11. The driving discs for excessive wear; cracked, charred, burned, glazed, pitted, or scored lining material; and distortion.

12. The steel clutch plates for distortion, surface scuffing or scoring, and damaged drive lugs.

13. If the clutch retainer or any of its various parts are defective, replace them as necessary.

Clutch Service and Reassembly

To service the clutch piston along with the clutch plates and reassemble the unit, follow these instructions:

1. Install two new seals in the grooves located on the inner and outer circumferences of the piston.

2. Lubricate both seals and the piston bore in the retainer generously with transmission fluid. Then, reinstall the piston into the retainer bore with a twisting motion.

3. Install the return spring and waved snap ring (Fig. 23-67). Make sure the inner lugs of the return spring bear against the piston.

4. Using a screwdriver, reinstall the waved snap ring into its groove in the retainer (Fig. 23-68).

5. If removed, press the input shaft back into the retainer and install its snap ring (See Fig. 23-66).

6. If the old steel clutch plates are still serviceable, bead blast or refinish them on both sides with 120-320 grit emery cloth or sandpaper (Fig. 23-69). When sanding the plates, rub until sanding marks are visible over the entire area on both sides of the plate. The bead blasting or sanding process removes polished surfaces and restores new roughness to the plate, which is necessary for proper clutch break in. After servicing the plates, clean and lubricate them with transmission fluid before installation.

7. Presoak all new driving discs in clean transmission fluid for no less than 30 minutes before installation.

8. Install the pressure plate into the clutch retainer. Then, insert the clutch plates and driving discs alternately, beginning with a driving disc next to the pressure plate (Fig. 23-70).

9. Install the thick steel plate over the last driving disc.

10. Install the selective snap ring into its groove in the retainer (Fig. 23-71).

11. With a feeler gauge, measure the rear clutch pack clearance as shown in Fig. 23-72. It should be to specifications if new driving discs were installed in

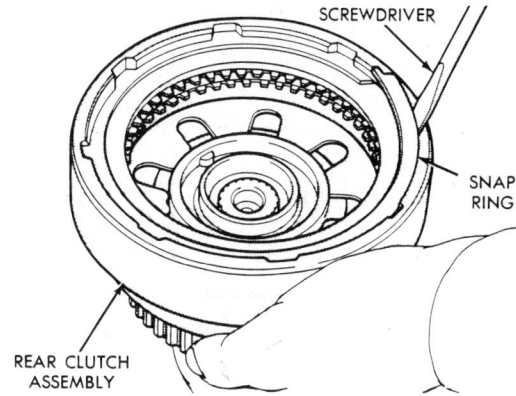

FIGURE 23-71 Inserting the snap ring into its groove in the rear clutch retainer. (Courtesy of Chrysler Corp.)

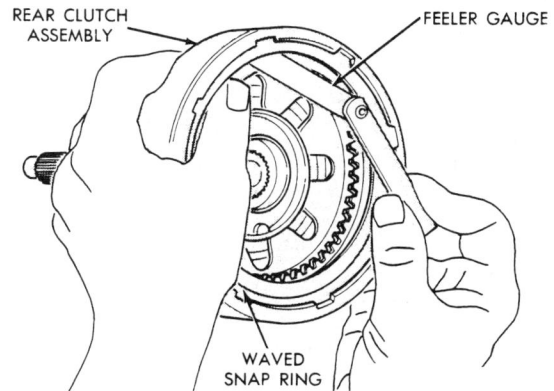

FIGURE 23-72 Measuring rear clutch pack operating clearance with a feeler gauge. (Courtesy of Chrysler Corp.)

the assembly (see Table 23-2). If not, install a smaller or larger thickness selective snap ring to correct the clearance problem.

FRONT PLANETARY AND ANNULUS GEAR RECONDITION

During the overhaul of the transaxle, the front planetary and annulus gear must be reconditioned. This process involves tearing down the unit, cleaning and inspecting all the components, and reassembling all the parts.

Disassembly

To tear down the front planetary and annulus gear, follow these directions:

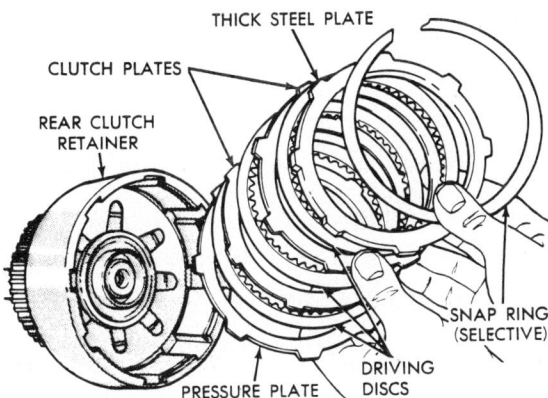

FIGURE 23-70 Installing the rear clutch plates, discs, and pressure plate. (Courtesy of Chrysler Corp.)

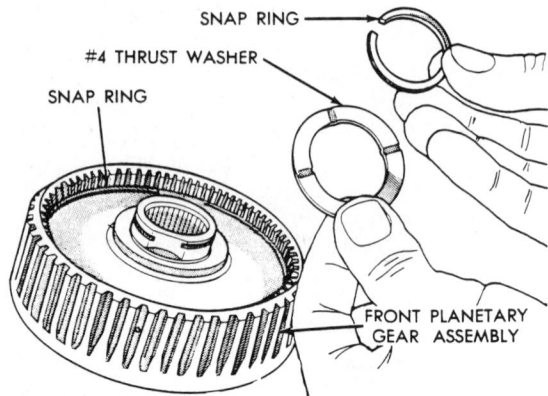

FIGURE 23-73   Removing the front planetary gear snap ring and No. 4 thrust washer. (Courtesy of Chrysler Corp.)

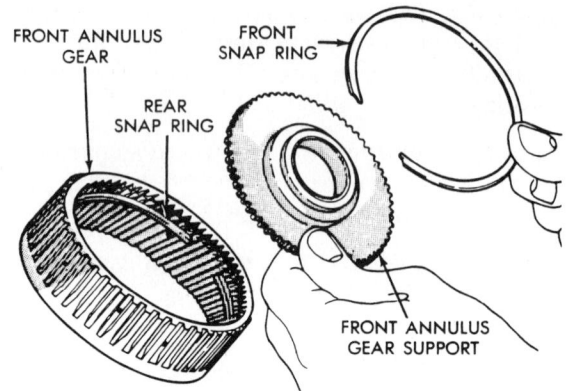

FIGURE 23-74   Separating the annulus gear and the No. 5 thrust washer from the planetary gear assembly. (Courtesy of Chrysler Corp.)

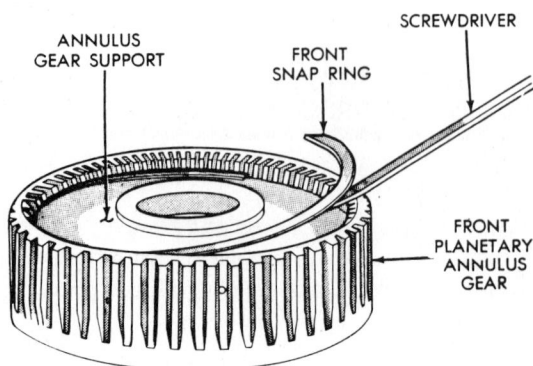

FIGURE 23-75   Taking out the front snap ring. (Courtesy of Chrysler Corp.)

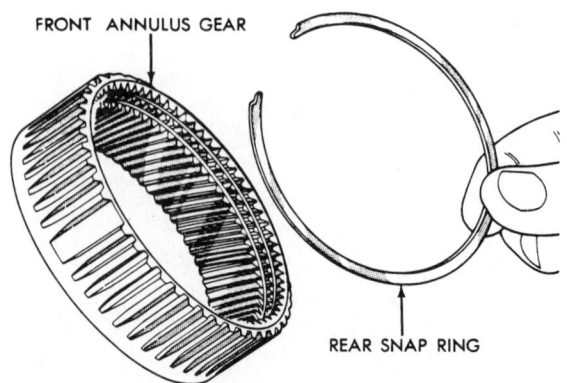

FIGURE 23-76   Separating the front snap ring and annulus gear support from the front annulus gear. (Courtesy of Chrysler Corp.)

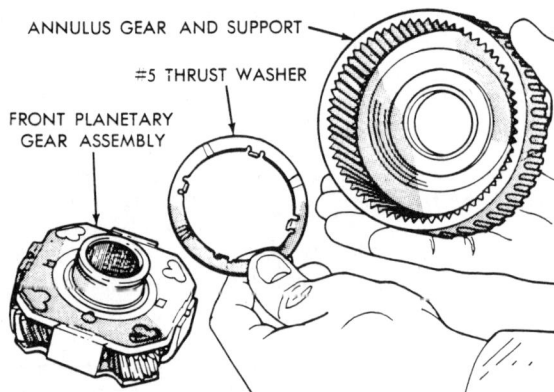

1. From the front planetary gear assembly, remove the snap ring and the No. 4 thrust washer (Fig. 23-73).

2. From the front planetary assembly, take off the annulus gear and support along with the No. 5 thrust washer (Fig. 23-74).

3. With a screwdriver, remove the front snap ring from above the annulus gear support (Fig. 23-75).

4. Separate the front snap ring and annulus gear support from the front annulus gear (Fig. 23-76).

5. Remove the front annulus gear support rear snap ring (Fig. 23-77).

Component Cleaning

Clean or wash all parts in solvent, and blow dry with low-pressure compressed air. CAUTION: *When drying the planetary gear assembly with air, do not permit the*

FIGURE 23-77   Taking out the rear snap ring from the front annulus gear. (Courtesy of Chrysler Corp.)

pinions to spin at high speeds. This will damage their support bearings.

Component Inspection

With the front planetary components clean and dry, check them as follows:

1. The front annulus support for burred, scored, or damaged support bearing surfaces.

2. The fit of the front annulus support splines in those within the annulus gear itself. There should be no appreciable radial play between the two.

3. The clutch hub slots on the outer circumference of the front annulus support for wear or damage.

4. The front annulus gear teeth for nicks, wear, or damage.

5. The Nos. 4 and 5 thrust washers for wear, scores, or damage.

6. The output shaft splines on the front planetary gear assembly for wear or damage.

7. All the front planet pinions for wear, nicks, tooth damage, and binding or looseness on their support bearings.

8. The end-play clearance of each pinion with a feeler gauge; it should be to specifications (see Table 23-2).

9. The front carrier for cracks or damage.

10. The front carrier for obstructed lubrication passages and worn or nicked bearing surfaces.

11. If any components are found defective, replace them before reassembling the planetary.

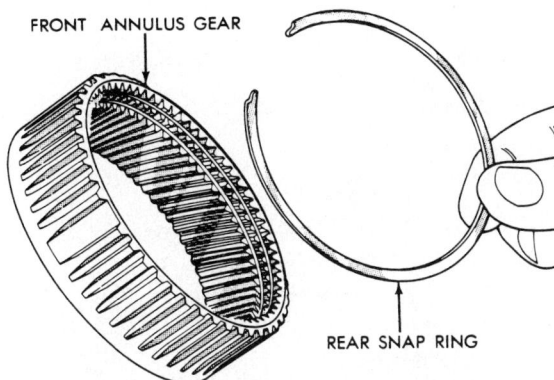

FIGURE 23-79 Installing the support and snap ring into the front annulus gear. (Courtesy of Chrysler Corp.)

Reassembly

To reassemble the front planetary and annulus gear, follow these instructions:

1. Reinstall the front annulus support rear snap ring (Fig. 23-78).

2. Insert the front annulus support into the gear and install the front snap ring (Fig. 23-79).

3. Coat the No. 5 thrust washer with petroleum jelly and position it in place over the front planetary gear assembly.

4. Lubricate the pinions with transmission fluid and install the front planetary gear assembly into the annulus gear (Fig. 23-80).

5. Coat the No. 4 thrust washer with petroleum jelly, and position it over the hub on the annulus gear support.

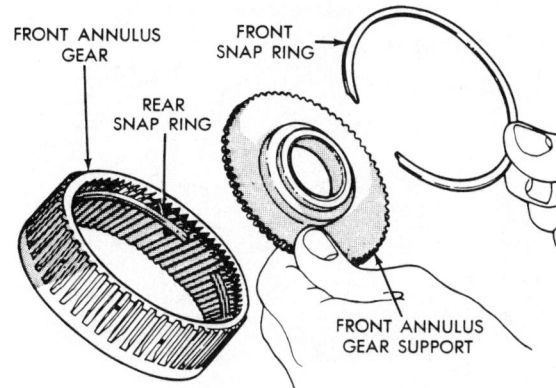

FIGURE 23-78 Installing the rear snap ring into the front annulus gear support. (Courtesy of Chrysler Corp.)

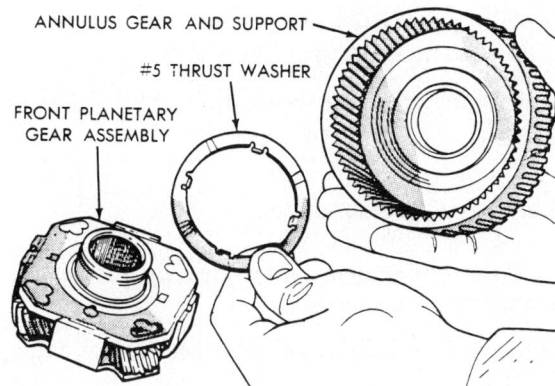

FIGURE 23-80 Reassembling the front planetary, No. 5 thrust washer, and the annulus gear. (Courtesy of Chrysler Corp.)

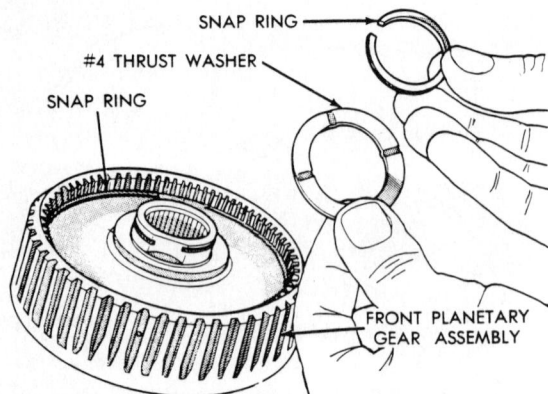

FIGURE 23-81 Installing the No. 4 thrust washer and the snap ring over the front annulus gear support. (Courtesy of Chrysler Corp.)

6. Install a new snap ring over the No. 4 thrust washer and into its groove in the annulus support (Fig. 23-81).

SUN GEAR DRIVING SHELL RECONDITION

When rebuilding the transaxle, the sun gear driving shell must be reconditioned. This procedure involves disassembling the unit, cleaning and inspecting its parts, and reassembling the shell and gear.

Disassembling and Cleaning

To tear down and clean this assembly, follow these steps:

1. Remove the snap ring from the sun gear driving shell (Fig. 23-82).

2. Carefully remove the sun gear and the Nos. 7 and 8 thrust washers from the driving shell.

3. Clean all the parts in solvent, and blow dry with low-pressure compressed air.

Inspection and Reassembly

With all the components clean and dry, inspect and reassemble them as follows:

1. The driving shell for cracks or other signs of damage.

2. The front clutch drive slots for wear.

3. The fit of the sun gear splines with those in the driving shell. There should be little or no radial play between them.

4. Both thrust washers for wear, burrs, or scoring.

5. The sun gear teeth for nicks, wear, or damage.

6. The sun gear bushing for wear or scoring.

7. If all the parts are serviceable, insert the sun gear through the No. 7 thrust washer and into the driving shell.

8. Install the No. 8 thrust washer over the exposed end of the sun gear, and reinstall the snap ring into its groove in the sun gear.

REAR PLANETARY ASSEMBLY AND ANNULUS GEAR RECONDITION

During the overhaul of the transaxle, the rear planetary gear assembly and annulus gear must be reconditioned before reuse. This process involves cleaning and inspecting the components.

Cleaning

Clean the components in solvent, and blow dry with low-pressure compressed air. CAUTION: *When drying the planetary gear assembly with air, do not permit the pinions to spin at high speed. This will damage their support bearings.*

Inspection

With the parts clean and dry, inspect them as follows:

1. The driving lugs on the outer circumference of the carrier for wear or damage.

2. The carrier for cracks, distortion, or signs of other damage.

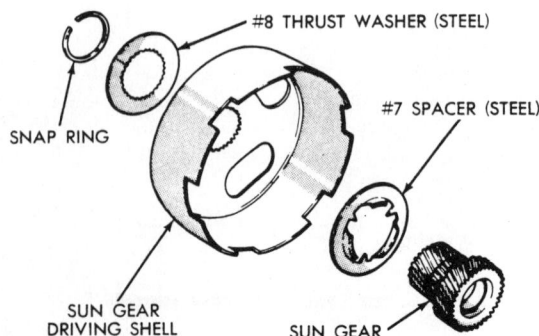

FIGURE 23-82 The sun gear driving shell disassembled. (Courtesy of Chrysler Corp.)

3. The rear carrier for obstructed lubrication passages and worn or nicked bearing surfaces.

4. The rear pinions for wear, nicks, tooth damage, and binding or looseness in their support bearings.

5. The end-play of each rear pinion with a feeler gauge; it should be to specifications (see Table 23-2).

6. The rear annulus gear for worn, nicked, or damaged teeth.

7. The output shaft for worn splines, obstructed oil passages, or excessive wear on its bearing surfaces.

8. If any of the rear planetary components are found defective, replace them before reassembling the transaxle.

KICKDOWN AND LOW-REVERSE BAND RECONDITION

When rebuilding the transaxle, both bands should be reconditioned. This involves cleaning and inspecting the bands, and if found defective, replacing either or both of them.

Cleaning and Inspection

To clean the bands, wash them in detergent and water solution and blow dry with low-pressure compressed air. CAUTION: *Do not soak the bands in solvent or a varnish dissolving solution because these chemicals can deteriorate the lining material.*

To inspect the bands, look for these conditions:

1. Distortion.

2. Cracked anchor ends.

3. Excessive or uneven lining wear.

4. Burned, charred, or glazed lining.

5. Loose lining.

6. Flaking or pitted lining.

7. If any of the above mentioned defects exist, replace the bands.

OVERRUNNING CLUTCH RECONDITION

During the rebuilding of the transaxle, the overrunning clutch must be reconditioned before reuse. This process involves disassembling the unit, cleaning and inspecting

its parts, replacing those that are defective, and reassembling the clutch.

Disassembly and Cleaning

To tear down and clean the assembly, follow these directions:

1. Push out the inner race from inside the clutch cam.

2. Remove the 8 overrunning clutch rollers and springs from the cam.

3. Wash all the parts in clean solvent, and blow dry with low-pressure compressed air.

Inspection

With all the components clean and dry, inspect them for the following defects:

1. The inner race for scored or damaged roller bearing surfaces or signs of brinelling.

2. The slots on the end of the inner race for wear and damage.

3. The clutch cam for scored or damaged roller ramp surfaces.

4. The clutch cam for worn, scored, or damaged band application surface.

5. Flat spots, chipped edges, or signs of excessive wear on each of the clutch rollers.

6. Worn, bent, or damaged roller springs.

7. If any part is found defective, replace it before assembling the clutch.

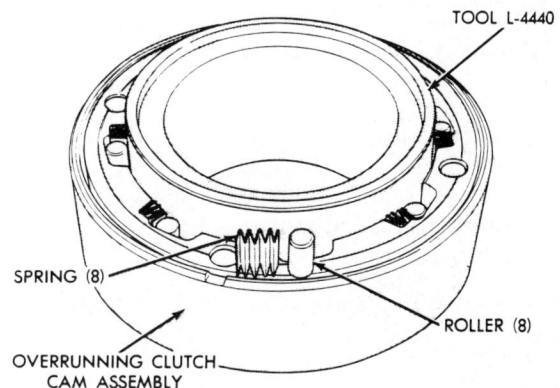

FIGURE 23-83   Reassembling the overrunning clutch. (Courtesy of Chrysler Corp.)

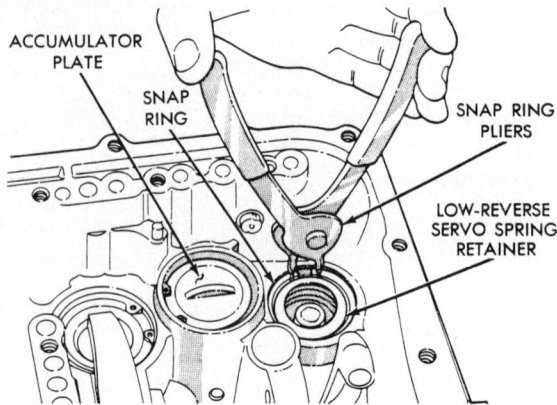

FIGURE 23-84 Removing the rear servo snap ring. (Courtesy of Chrysler Corp.)

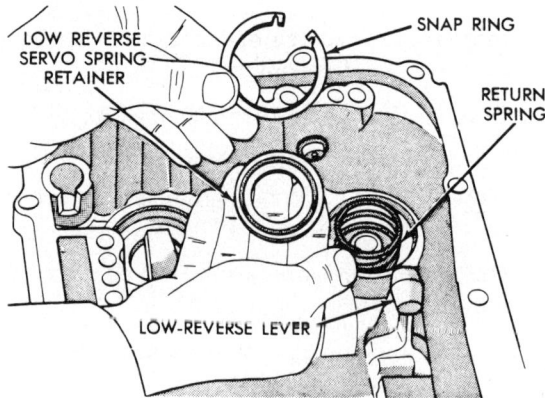

FIGURE 23-85 Taking out the low-reverse servo spring retainer and spring from the case. (Courtesy of Chrysler Corp.)

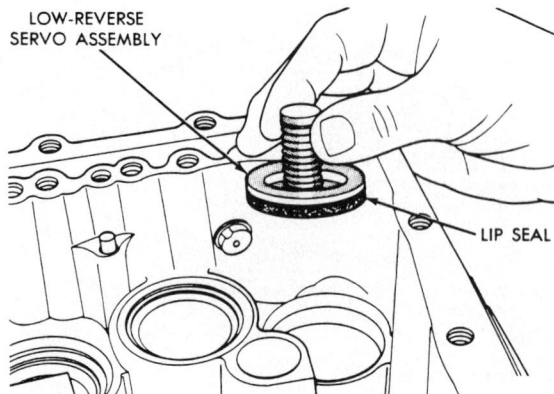

FIGURE 23-86 Lifting out the low-reverse servo assembly. (Courtesy of Chrysler Corp.)

Assembly

To reassemble the overrunning clutch, follow these steps:

1. As shown in Fig. 23-83, slide the special assembly tool inside the clutch cam.

2. Install the 8 springs and rollers into the cam.

3. Lightly coat the rollers and springs with clean ATF.

4. Push the special tool a small distance out from inside the cam assembly. Then, insert the inner race into the cam until it bottoms against the special tool. Continue to push in on the inner race until the tool passes completely out of the clutch cam.

5. Hold the cam assembly, and test the clutch by attempting to turn the inner race in both directions. It should rotate freely in one direction but lock up in the other.

LOW-REVERSE SERVO RECONDITION

Before reassembling the transaxle, the low-reverse (rear) servo must be reconditioned. This process involves removing the servo from the case, cleaning and inspecting the assembly, and reinstalling it.

Servo Removal

To remove the servo assembly from the case:

1. Push down slightly on the retainer, and with the pliers shown in Fig. 23-84, remove the servo snap ring.

2. Remove the low-reverse servo spring retainer and spring from the case (Fig. 23-85).

3. Lift out the low-reverse servo assembly (Fig. 23-86).

Cleaning

Remove the lip seal from the servo piston. Then wash all the servo components along with the case bore with clean solvent. Blow all the parts dry with low-pressure compressed air. Also, be sure to blow out the fluid passages in the case bore.

Inspection

With the components clean and dry, inspect them as follows:

1. The servo bore in the case for scores or signs of other damage.

2. The servo piston for cracks, wear, or signs of other damage.

3. The servo return spring for signs of overheating or distortion.

4. The spring retainer for damage or distortion.

5. If any of the parts are defective, replace them before installing the servo into the case.

Assembly

To reassemble the low-reverse servo into the case:

1. Lubricate the servo bore in the case with ATF.

2. Install the seal into its groove in the servo piston. CAUTION: *Make sure its lip faces toward the base of the servo bore.*

3. Install the servo assembly into the transaxle case (Fig. 23-87). As necessary, push the seal lip into the case bore with a small screwdriver to assist in installing the servo piston.

4. Reinstall the return spring (Fig. 23-88).

5. Position the retainer over the return spring (Fig. 23-89).

6. Push the retainer down slightly and install the snap ring with a pair of pliers as shown in Fig. 23-90.

ACCUMULATOR RECONDITION

Before reassembling the transaxle case, recondition the accumulator. This procedure involves removing the assembly, washing and inspecting its parts, and reassembling the accumulator into the case.

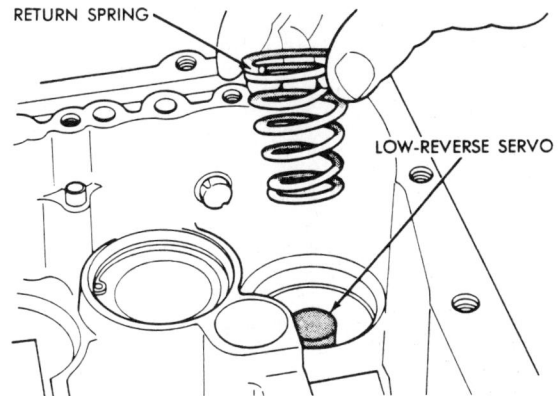

FIGURE 23-88   Installing the return spring. (Courtesy of Chrysler Corp.)

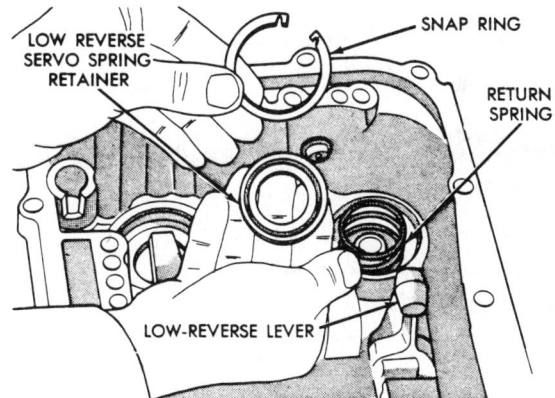

FIGURE 23-89   Positioning the retainer over the return spring. (Courtesy of Chrysler Corp.)

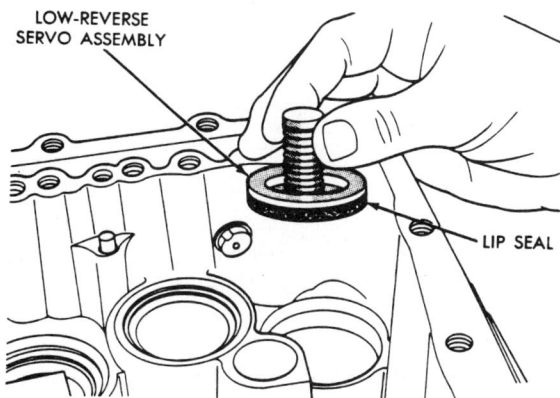

FIGURE 23-87   Installing the low-reverse servo assembly. (Courtesy of Chrysler Corp.)

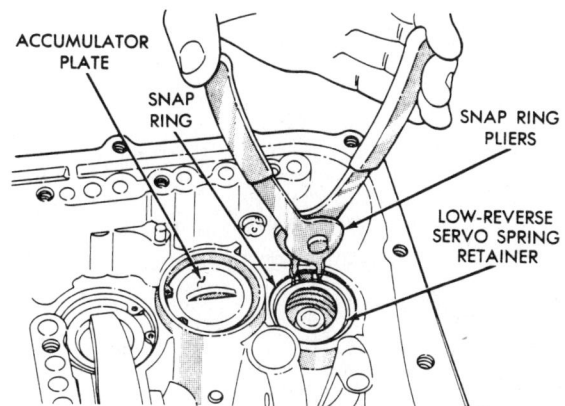

FIGURE 23-90   Reinstalling the snap ring. (Courtesy of Chrysler Corp.)

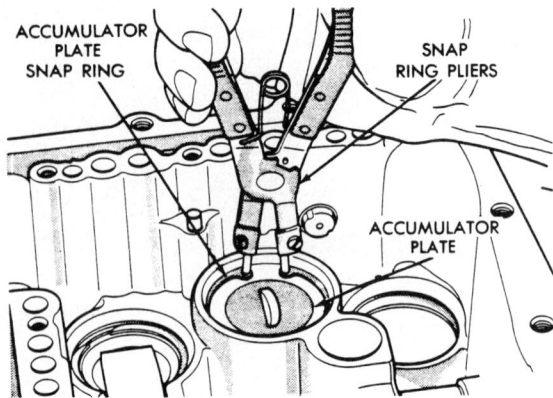

FIGURE 23-91 Removing the accumulator plate snap ring. (Courtesy of Chrysler Corp.)

Removal

To remove the accumulator assembly from the case:

1. Push down slightly on the accumulator plate and remove its snap ring with the pliers shown in Fig. 23-91.

2. Lift off the accumulator plate and snap ring (Fig. 23-92).

3. From the case bore, take out the accumulator spring and piston (Fig. 23-93).

Cleaning

Remove the two seal rings from the accumulator piston. Next, wash all the accumulator metal parts along with the case bore with clean solvent. Blow all the parts dry with low-pressure compressed air. Also, be sure to blow out the fluid passages in the case bore.

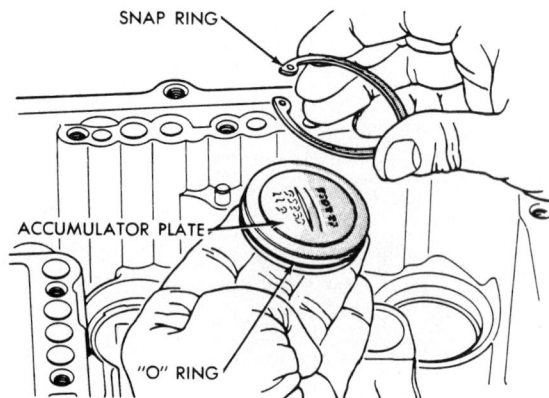

FIGURE 23-92 Removing the accumulator plate and snap ring. (Courtesy of Chrysler Corp.)

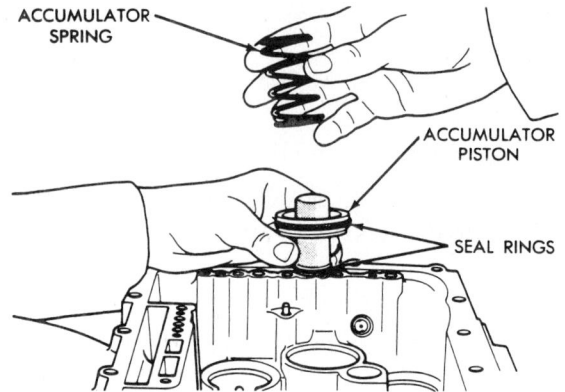

FIGURE 23-93 Taking out the accumulator spring and piston from the case. (Courtesy of Chrysler Corp.)

Inspection

With all the parts clean and dry, inspect them as follows:

1. The accumulator bore within the case for scores or signs of other damage.

2. The accumulator piston for cracks, wear, or signs of other damage.

3. The return spring for signs of overheating or distortion.

4. The accumulator plate for damage or distortion.

Assembly

To reassemble the accumulator into the case:

1. Lubricate the accumulator piston bore in the case with ATF.

2. Install two new seal rings into the grooves in the accumulator piston.

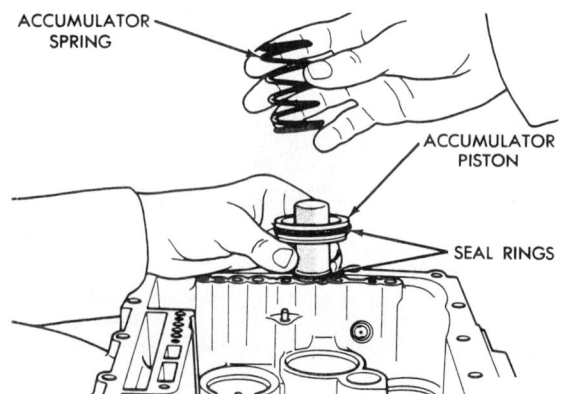

FIGURE 23-94 Installing the accumulator piston and spring. (Courtesy of Chrysler Corp.)

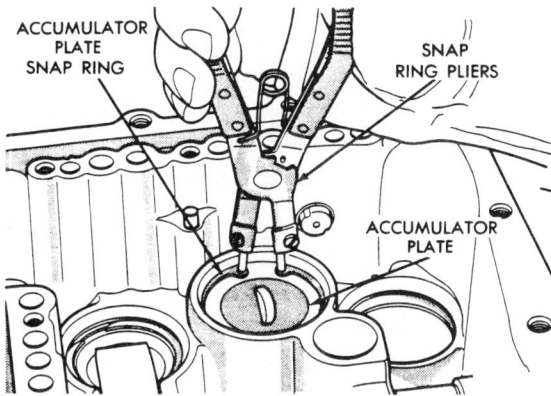

FIGURE 23-95 Installing the accumulator snap ring. (Courtesy of Chrysler Corp.)

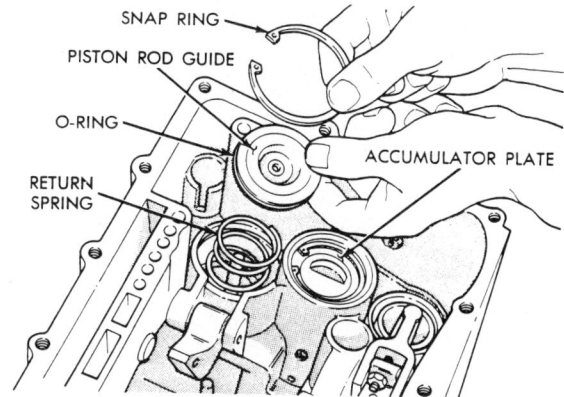

FIGURE 23-97 Removing the piston guide and snap ring. (Courtesy of Chrysler Corp.)

3. Install the accumulator piston and spring into the case bore (Fig. 23-94).

4. Position the accumulator plate over the return spring.

5. Push the accumulator plate down slightly and install the snap ring with a pair of pliers as shown in Fig. 23-95.

KICKDOWN SERVO RECONDITION

During the rebuilding of the transaxle, the kickdown servo must be reconditioned before the entire unit is reassembled. This involves removing the assembly, cleaning and inspecting its components, and reinstalling the servo into the case.

Removal and Tear Down

To remove the kickdown servo assembly from the case and prepare it for cleaning:

1. Slightly depress the servo piston guide, and remove the snap ring with the pliers as shown in Fig. 23-96.

2. Lift out the snap ring and piston guide (Fig. 23-97).

3. Remove the return spring and kickdown servo piston (Fig. 23-98).

4. Remove the snap ring that retains the piston rod into the kickdown piston (Fig. 23-99).

5. Remove the O-rings from the piston guide plus the piston rod and the seal rings from the kickdown piston.

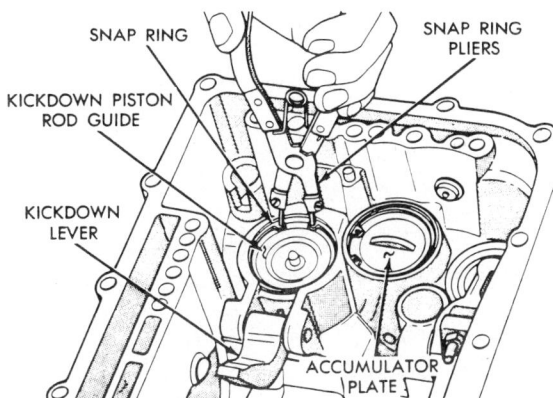

FIGURE 23-96 Taking out the kickdown servo snap ring. (Courtesy of Chrysler Corp.)

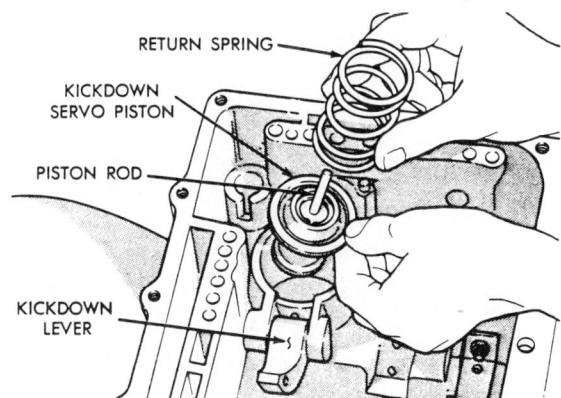

FIGURE 23-98 Lifting out the return spring and kickdown servo piston. (Courtesy of Chrysler Corp.)

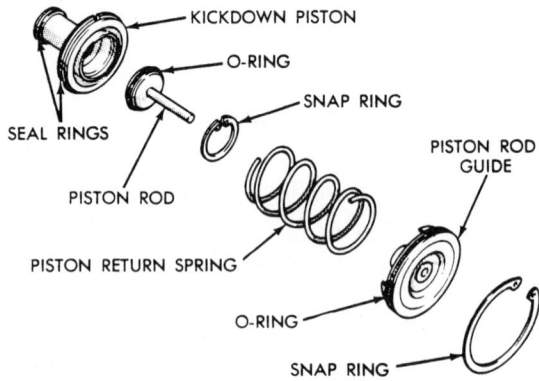

FIGURE 23-99 Removing the piston rod, seals, and rings from the kickdown servo assembly. (Courtesy of Chrysler Corp.)

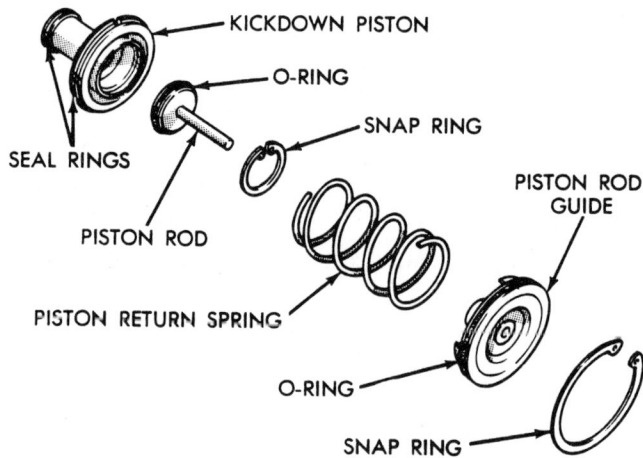

FIGURE 23-100 Installing the new seals and rings onto the servo assembly. (Courtesy of Chrysler Corp.)

Cleaning

Clean all the kickdown servo components along with the case bore with solvent. Then blow all the parts dry with low-pressure compressed air. Also, be sure to blow out the fluid passages within the case bore.

Inspection

With all the parts clean and dry, inspect them as follows:

1. The kickdown servo bore within the case for scores or wear.

2. The kickdown servo piston and rod for cracks, wear, or signs of other damage.

3. The return spring for signs of overheating or distortion.

4. The piston rod guide for cracks, wear, distortion, or signs of other damage.

Assembly

To reassemble the kickdown servo and install it into the case:

1. Install a new O-ring into the grooves located in the piston rod guide and the piston rod (Fig. 23-100).

2. Install two new seal rings onto the kickdown piston.

3. Lubricate all the O-rings and seal rings with ATF.

4. Carefully reinstall the piston rod into the kickdown piston and insert the snap ring.

5. Lubricate the servo bore with ATF, and reinstall the kickdown servo and return spring into the case (Fig. 23-101).

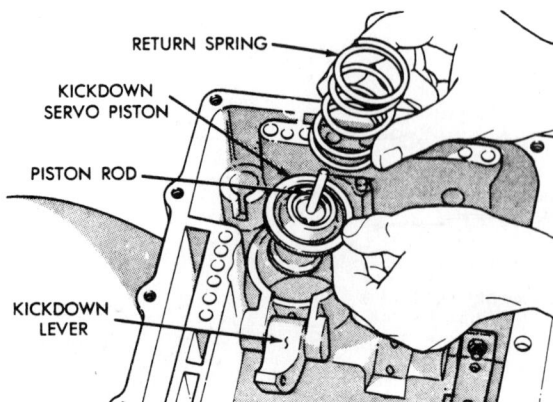

FIGURE 23-101 Installing the kickdown servo piston and return spring. (Courtesy of Chrysler Corp.)

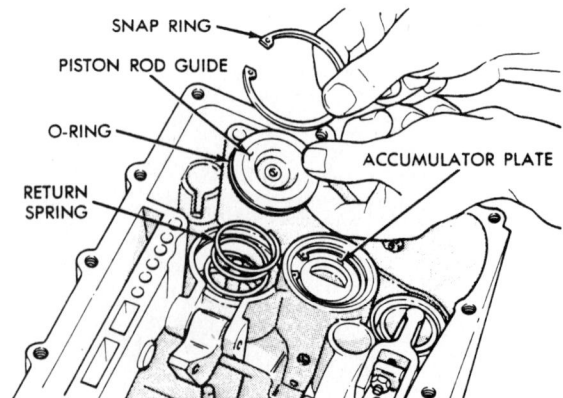

FIGURE 23-102 Installing the piston rod guide into the case. (Courtesy of Chrysler Corp.)

6. Install the piston rod guide over the return spring and into the case bore (Fig. 23-102).

7. While holding the piston rod guide slightly down, reinstall the snap ring (Fig. 23-103).

VALVE BODY RECONDITION

The inside of the valve body contains a great number of springs and valves. Each of these many parts performs a small but important function during the operation of the transaxle. When working together, they convert the valve body into a hydraulic computer that is responsible for upshifting or downshifting the transaxle according to the will of the driver, vehicle speed, and engine load.

General Instructions

Because of its many precision components and the very important job it performs, the mechanic must **carefully** and **thoroughly** clean and inspect the valve body as part of every transaxle overhaul, or a malfunction may develop. This procedure is usually the most difficult of all overhaul tasks because of the large number of small components (such as valves and springs), their delicacy, and the close tolerances under which the parts operate. For these reasons, the mechanic must follow these important general instructions when reconditioning the valve body:

1. Do not clamp any portion of the valve body or transfer plate in a vise. This can cause distortion of the aluminum body or transfer plate, which will result in sticking valves or excessive leakage.

2. When removing or installing valves or plugs, slide them in or out of their bores carefully. **Do not use force.**

3. Make certain your hands, tools, and the work bench are clean before working on the valve body.

4. Be careful never to drop or nick a valve spool during the service work. A nick can cause the valve to stick in its bore.

5. After cleaning, use low-pressure compressed air to dry a component. CAUTION: *Never use rags for this purpose. A single particle of lint can cause a valve to stick in its bore.*

6. When disassembling the valve body, lay the parts out on a clean surface in the order in which you removed them.

7. Using a parts breakdown, identify all the disassembled parts; number each spring on the diagram; and install it on a numbered peg of a spring holder.

8. Do not attempt to straighten any bent levers.

Disassembly

With these instructions in mind, tear down the valve body following these steps:

1. Using the special tool as shown, remove the detent spring attaching screw and spring (Fig. 23-104).

2. Loosen and remove the 16 valve body screws (Fig. 23-105).

3. Separate the valve body from the transfer plate (Fig. 23-106).

4. Loosen and then remove the separator plate screws. Next, lift the separator plate off the transfer plate.

5. Remove the 8 steel balls from the valve body (Fig. 23-107).

6. Remove the "E" clip with a screwdriver from the throttle shaft (Fig. 23-108).

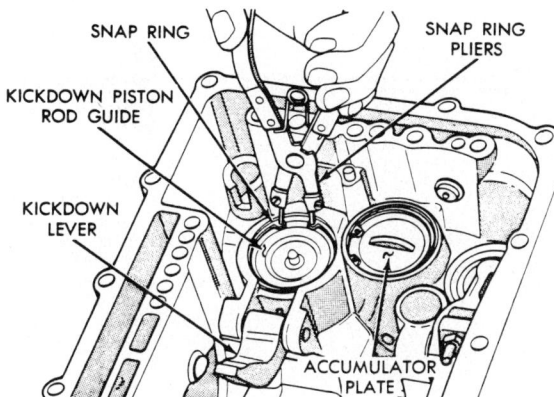

FIGURE 23-103   Reinstalling the kickdown servo snap ring. (Courtesy of Chrysler Corp.)

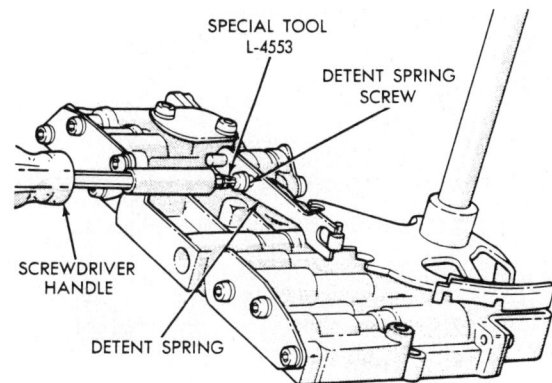

FIGURE 23-104   Removing the detent attaching spring and screw. (Courtesy of Chrysler Corp.)

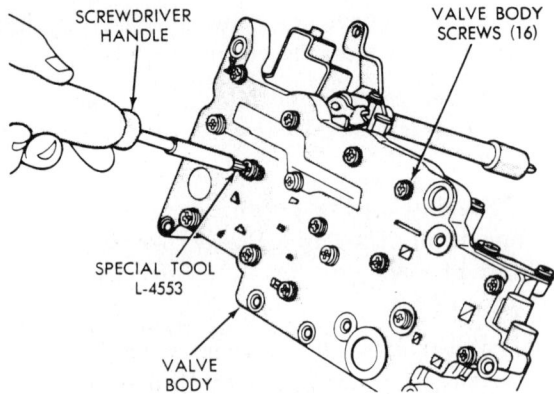

FIGURE 23-105   Taking out the valve body screws. (Courtesy of Chrysler Corp.)

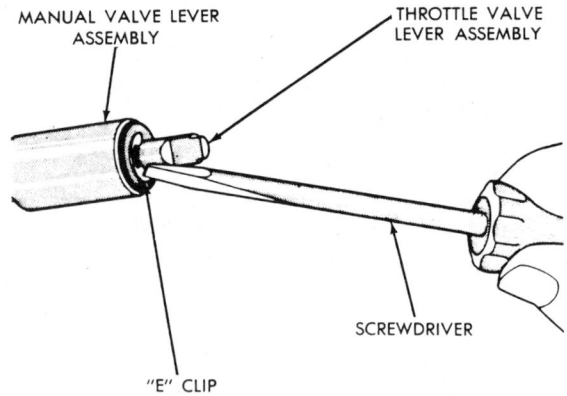

FIGURE 23-108   Removing the throttle shaft "E" clip. (Courtesy of Chrysler Corp.)

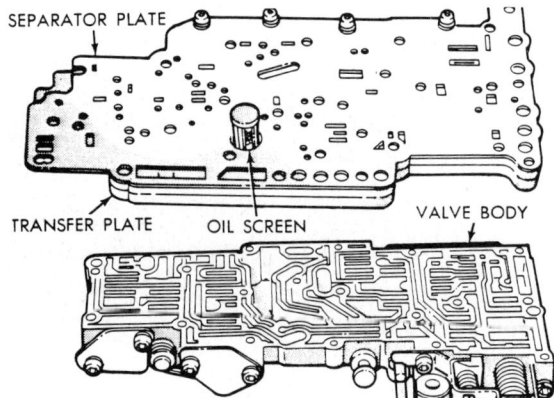

FIGURE 23-106   Separating the valve body from the transfer plate. (Courtesy of Chrysler Corp.)

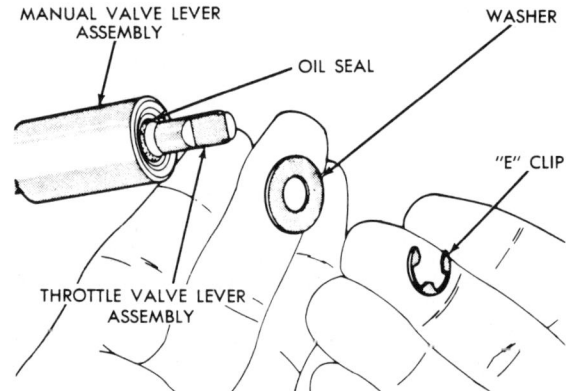

FIGURE 23-109   Removing the washer and oil seal. (Courtesy of Chrysler Corp.)

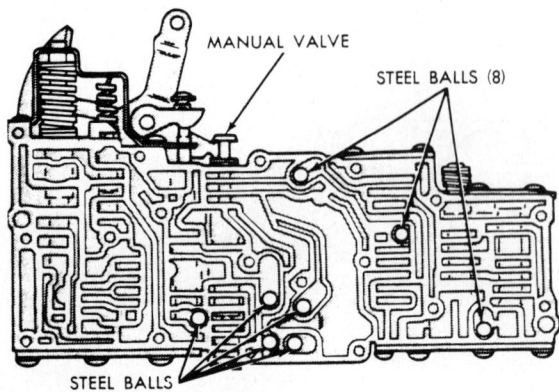

FIGURE 23-107   Steel ball locations in the valve body. (Courtesy of Chrysler Corp.)

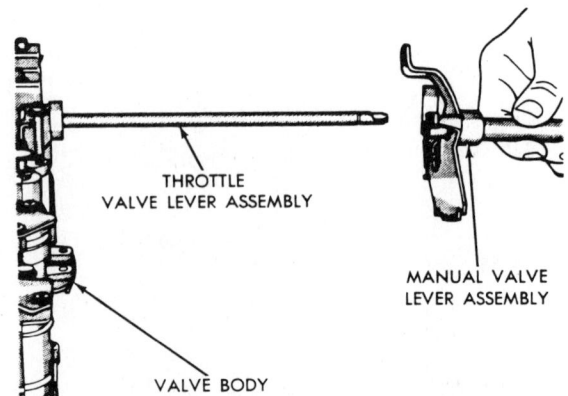

FIGURE 23-110   Removing the manual valve lever from the throttle valve lever assembly. (Courtesy of Chrysler Corp.)

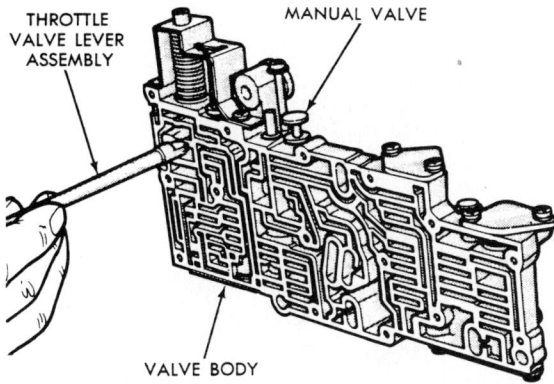

FIGURE 23-111   Removing the throttle valve lever assembly from the valve body. (Courtesy of Chrysler Corp.)

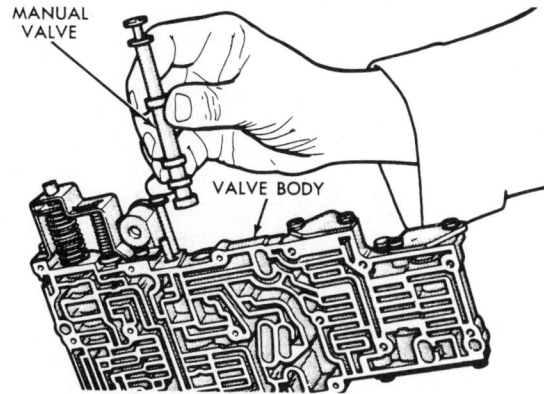

FIGURE 23-112   Taking out the manual valve from the valve body. (Courtesy of Chrysler Corp.)

7. Remove the washer and oil seal from the throttle shaft (Fig. 23-109).

8. Slip the manual valve assembly off the throttle valve lever (Fig. 23-110).

9. Take the throttle valve lever out of the valve body (Fig. 23-111).

10. Carefully remove the manual valve from the valve body (Fig. 23-112).

11. Loosen and remove the screws that hold the pressure regulator and adjusting screw bracket to the valve body (Fig. 23-113).

12. From under the adjusting screw bracket, remove the valves, springs, and the guide as shown in Fig. 23-114.

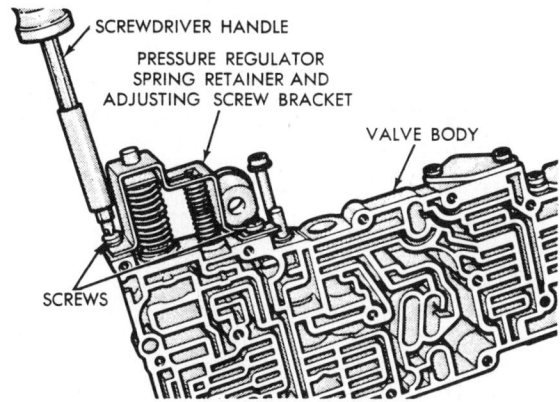

FIGURE 23-113   Removing the adjusting screw bracket screws. (Courtesy of Chrysler Corp.)

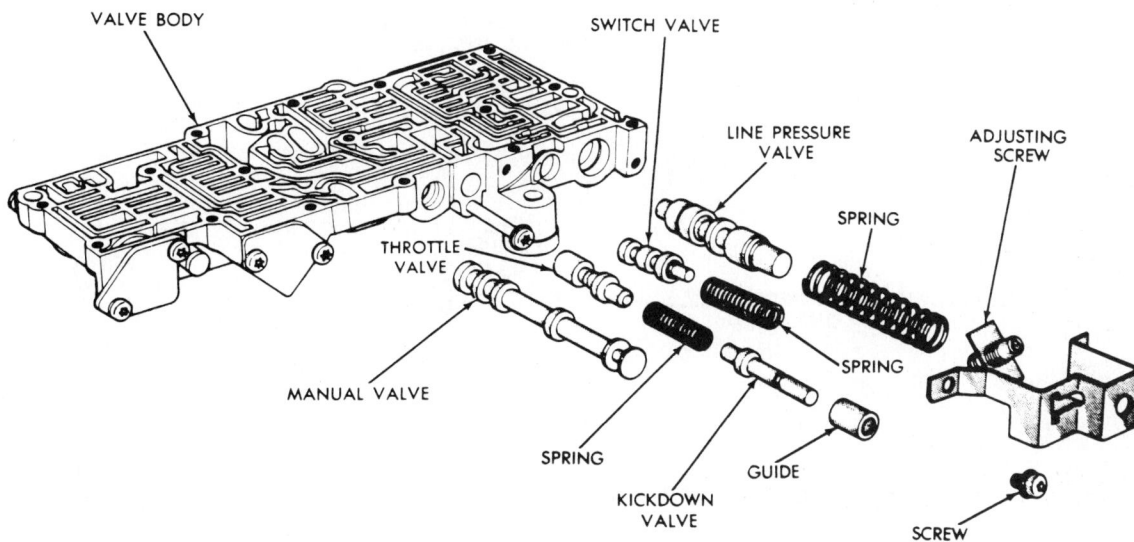

FIGURE 23-114   The pressure regulator and manual control components. (Courtesy of Chrysler Corp.)

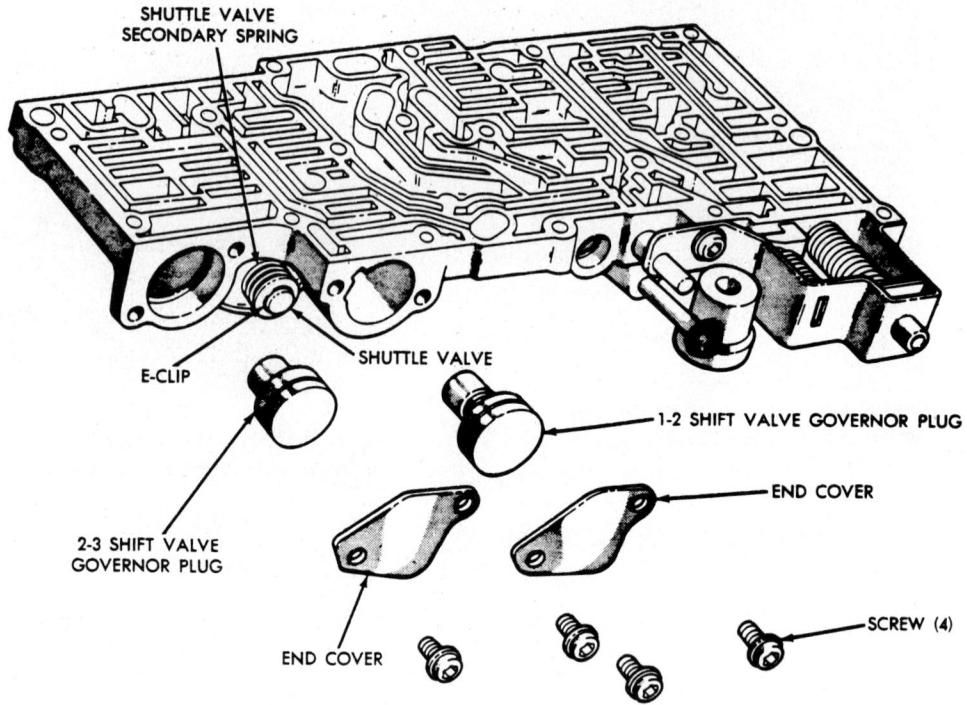

FIGURE 23-115 Removing the governor plugs. (Courtesy of Chrysler Corp.)

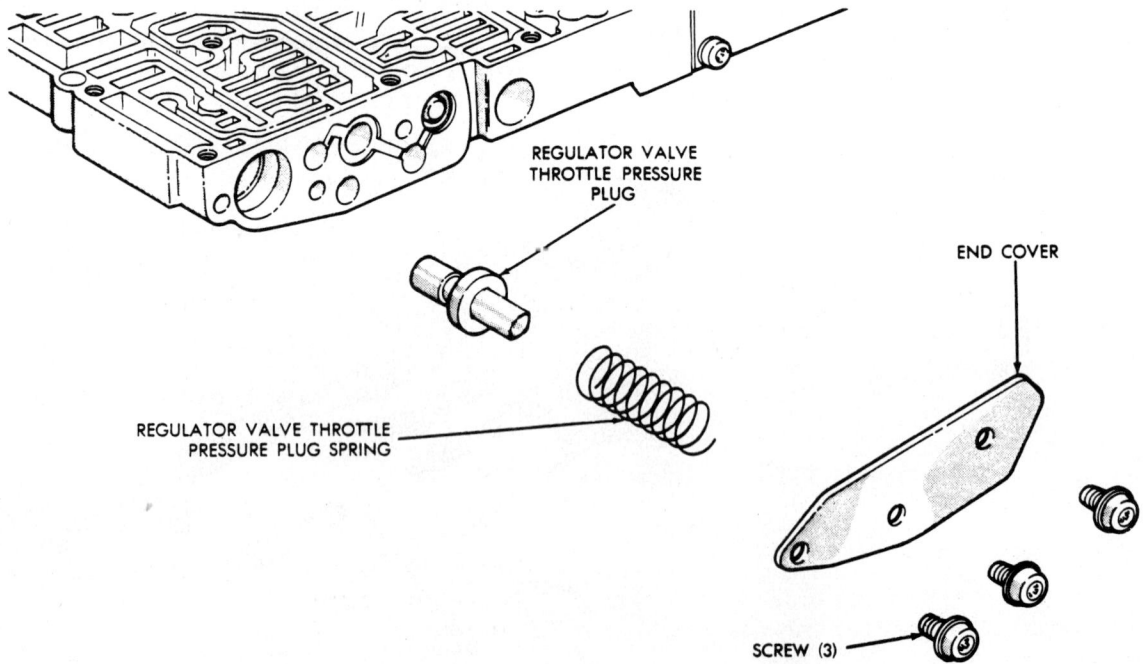

FIGURE 23-116 Removing the pressure regulator valve plug. (Courtesy of Chrysler Corp.)

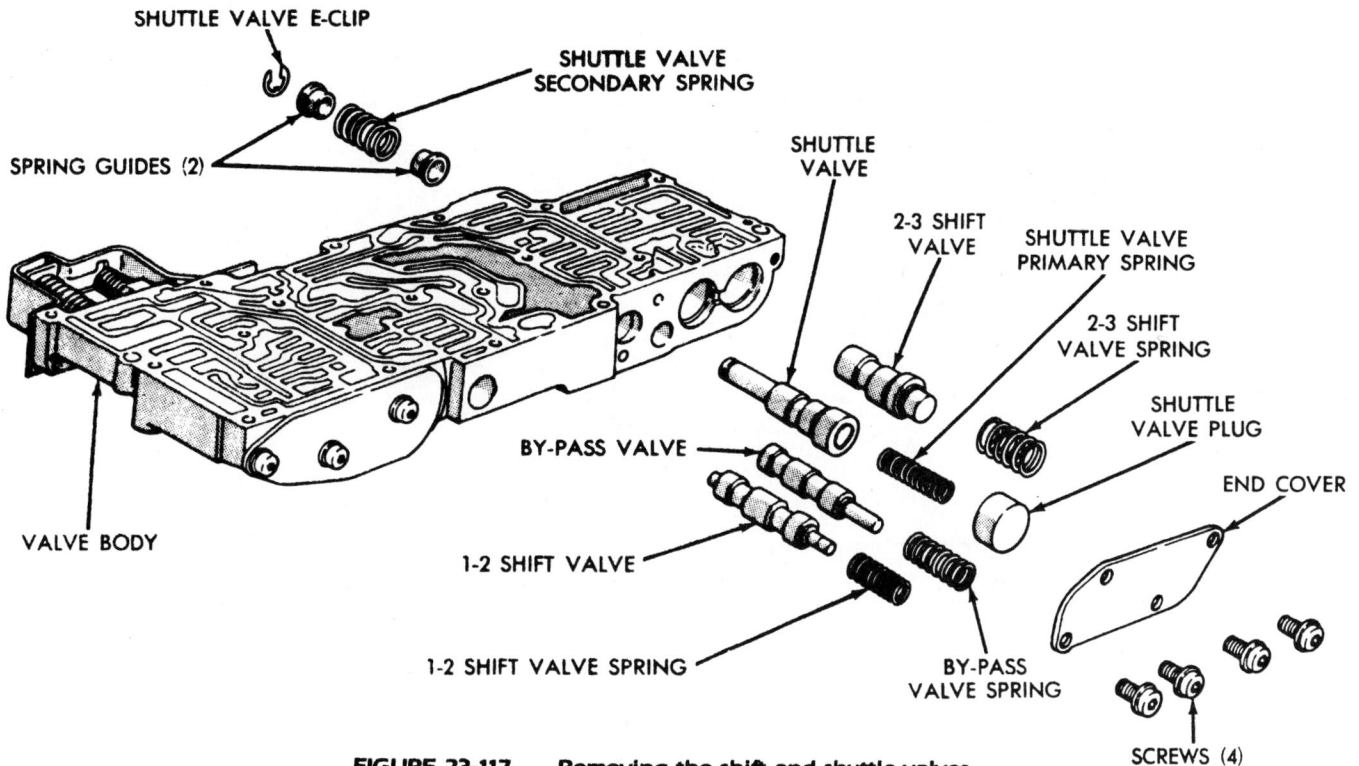

FIGURE 23-117   Removing the shift and shuttle valves. (Courtesy of Chrysler Corp.)

13. Loosen and remove the 4 screws that hold the end covers to the valve body. Then, remove the 1-2 and 2-3 shift valve governor plugs (Fig. 23-115).

14. Remove the "E" clip from the shuttle valve secondary spring guide; then take out the spring and two guides.

15. Loosen and remove the 3 screws from the regulator valve throttle pressure plug end cover (Fig. 23-116). Next, remove the spring and plug.

16. Loosen and take out the 4 screws from the shift valve end cover, and remove the valves (Fig. 23-117).

Cleaning

To clean the valve body components, follow these instructions:

1. Soak all valve body parts in a cold tank filled with a solution such as carburetor cleaner that dissolves varnish or shellac buildup.

2. Rinse off all parts in quantities of hot water.

3. After rinsing, immerse the parts in clean mineral spirits or solvent to separate the drops of water from the cleaned parts.

4. Blow all the parts dry with low-pressure compressed air.

Inspection

Inspect the valve body and its internal components for the following defects:

1. Bent manual valve.

2. Bent or damaged separator plate.

3. Cracked valve body casting and distorted or damaged mating surfaces.

4. Plugged or restricted fluid passages in the valve body castings.

5. Scored, scratched, or pitted valve bores.

6. Broken, bent, or worn valve springs.

7. Scored, cracked, or burred valves, guides, or plugs. Valve lands for shiny areas.

Service

Before reassembly, service the valve body parts as follows:

1. Shiny valve land areas indicate friction between the affected valve spool and its bore. To correct this problem, polish these areas away using 600-grit sandpaper or emery cloth. CAUTION: *Be careful not to*

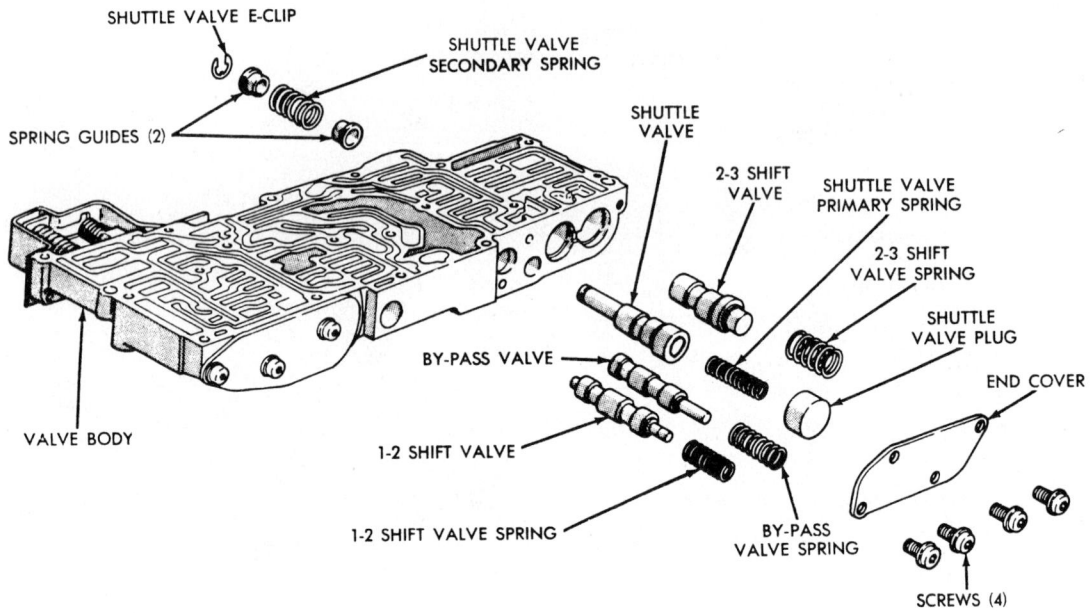

FIGURE 23-118 Installing the shift and shuttle valves. (Courtesy of Chrysler Corp.)

Labels in figure:
SHUTTLE VALVE E-CLIP
SHUTTLE VALVE SECONDARY SPRING
SPRING GUIDES (2)
VALVE BODY
SHUTTLE VALVE
2-3 SHIFT VALVE
SHUTTLE VALVE PRIMARY SPRING
2-3 SHIFT VALVE SPRING
SHUTTLE VALVE PLUG
END COVER
BY-PASS VALVE
1-2 SHIFT VALVE
1-2 SHIFT VALVE SPRING
BY-PASS VALVE SPRING
SCREWS (4)

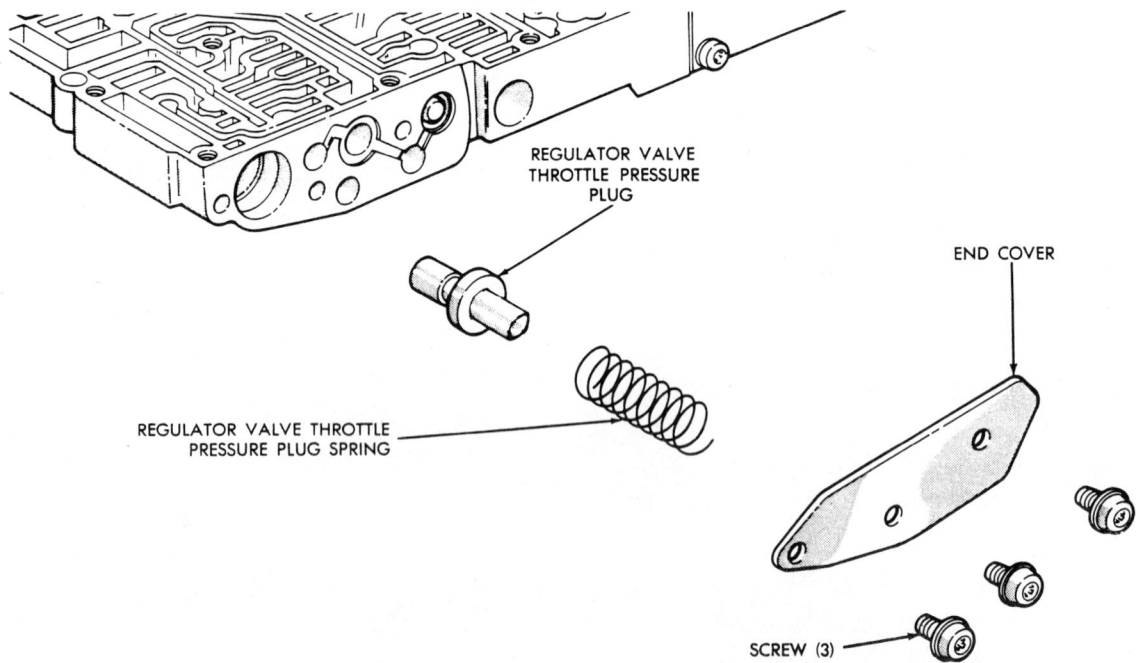

FIGURE 23-119 Installing the pressure valve throttle pressure plug and spring. (Courtesy of Chrysler Corp.)

Labels in figure:
REGULATOR VALVE THROTTLE PRESSURE PLUG
END COVER
REGULATOR VALVE THROTTLE PRESSURE PLUG SPRING
SCREW (3)

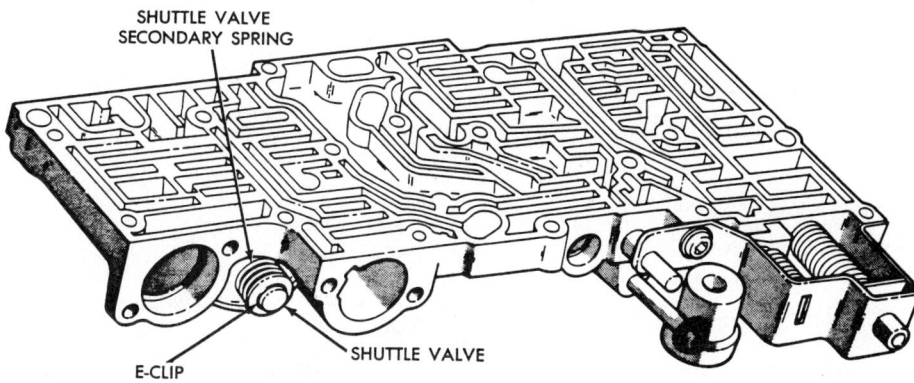

FIGURE 23-120 Installing the shuttle valve spring "E" clip. (Courtesy of Chrysler Corp.)

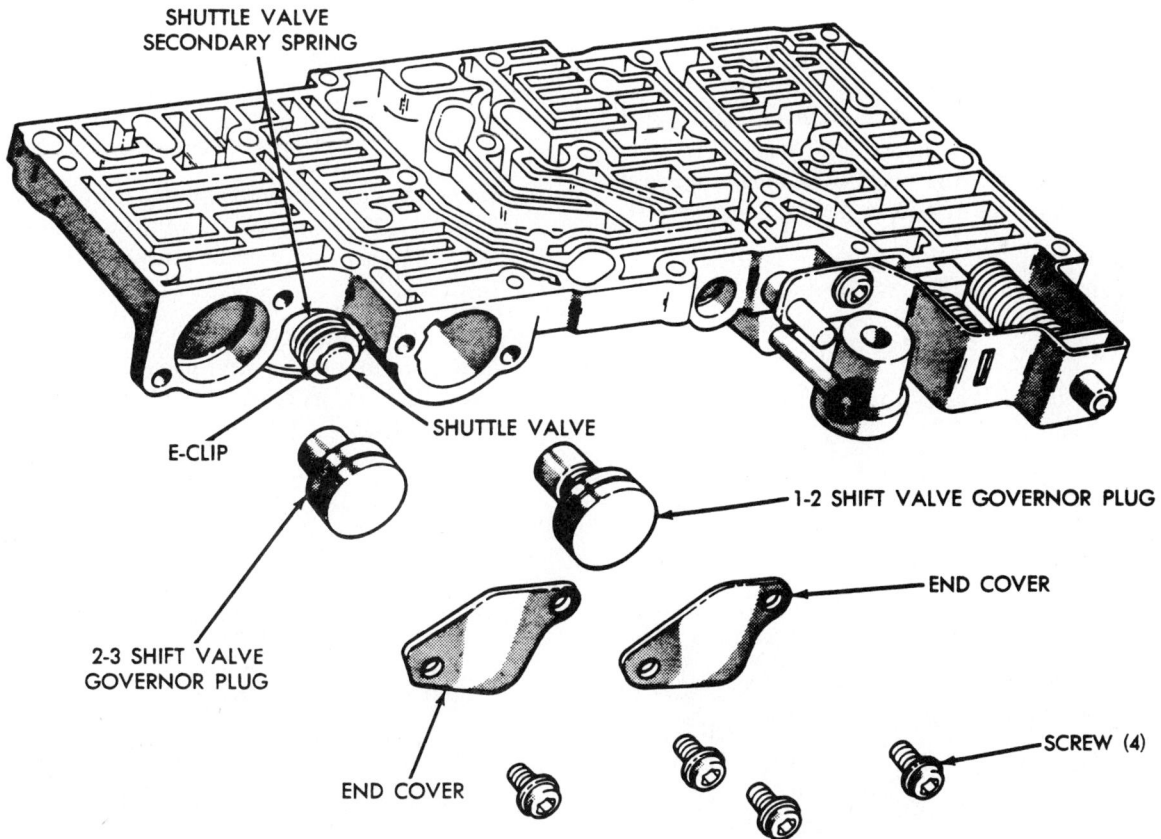

FIGURE 23-121 Installation of the governor plugs and end covers. (Courtesy of Chrysler Corp.)

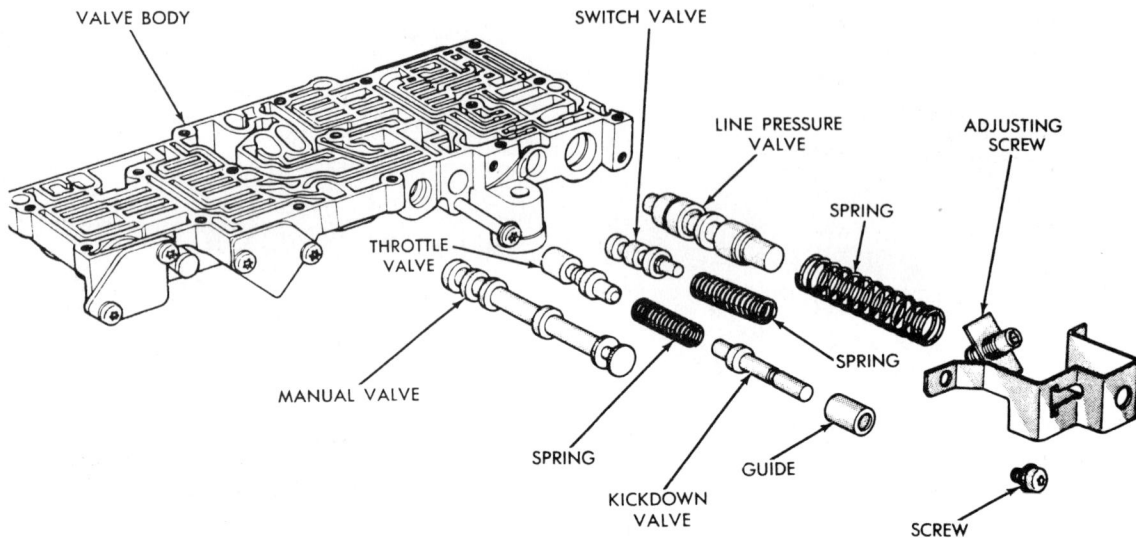

FIGURE 23-122 Installing the pressure regulator and manual control valves and springs. (Courtesy of Chrysler Corp.)

round off any edges of the valve when polishing. The sharpness of these edges is very important because they prevent foreign matter from lodging between the valve and its bore, thus reducing the possibility of sticking. Rewash and dry the polished valve.

2. Check each valve and plug for freedom of operation in its bore. Each valve or plug should freely slide back and forth in its bore due to its own weight without sticking or hesitating. If a valve or plug fails this test, polish or replace it.

3. Correct any slight valve body or transfer plate distortion by sliding the unit over a piece of crocus cloth fitted over a surface plate.

4. Before final installation, lubricate each valve with clean ATF.

5. Tighten all valve body screws to specifications (see Table 23-1).

Assembly

To reassemble the clean and serviced valve body:

1. Reinstall the valves, springs, and the plug as shown in Fig. 23-118. Then install the end cover along with its attaching screws; tighten these screws to specifications.

2. Slide the regulator throttle pressure plug and spring into their bore in the valve body. Next, install the end cover and screws (Fig. 23-119); tighten these screws to specifications.

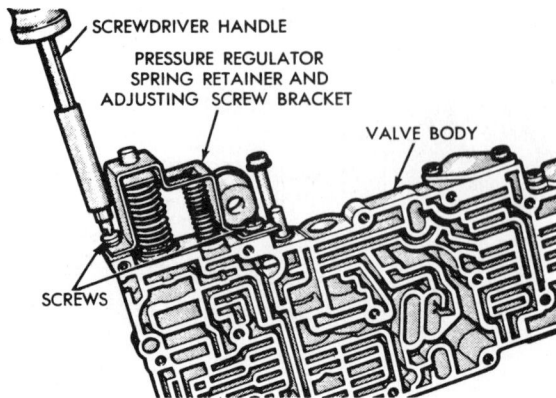

FIGURE 23-123 Installing the pressure regulator and adjusting screw bracket. (Courtesy of Chrysler Corp.)

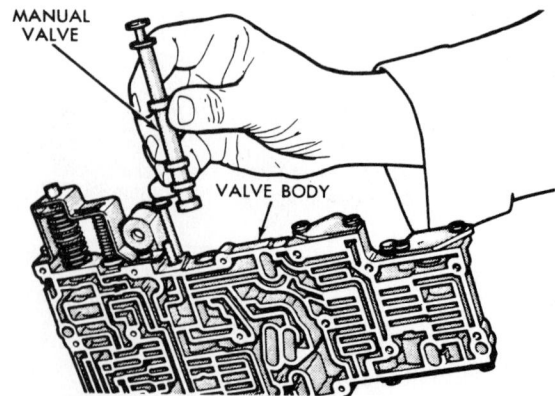

FIGURE 23-124 Installation of the manual valve. (Courtesy of Chrysler Corp.)

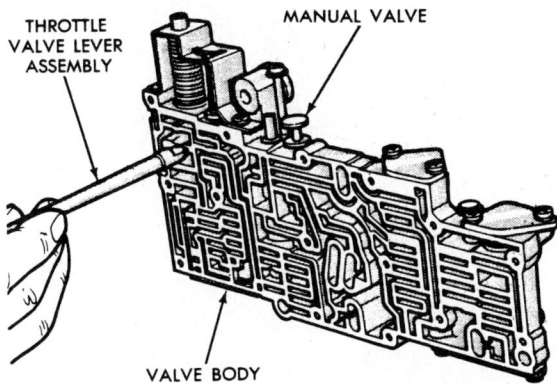

FIGURE 23-125   Inserting the throttle lever into the valve body. (Courtesy of Chrysler Corp.)

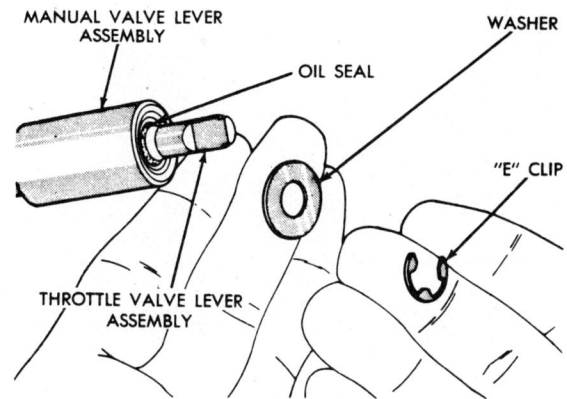

FIGURE 23-127   Installing the oil seal and washer into position. (Courtesy of Chrysler Corp.)

3. Insert the two guides and shuttle valve secondary spring into their bore and install the "E" clip (Fig. 23-120).

4. Carefully slide the 1-2 and 2-3 shift valve governor plugs into their respective bores. Then, install the two end covers along with their attaching screws (Fig. 23-121); torque the screws to specifications.

5. Install into their respective bores all the valves, springs, and the guide as shown in Fig. 23-122.

6. Position the adjusting screw bracket in place and install its screws (Fig. 23-123). Tighten these screws to specifications.

7. Carefully insert the manual valve into its bore in the valve body (Fig. 23-124).

8. Install the throttle valve lever assembly into the valve body (Fig. 23-125).

9. Slip the manual valve lever assembly over the throttle valve lever (Fig. 23-126).

10. Install the oil seal and washer as shown in Fig. 23-127.

11. Insert the "E" clip into its groove in the throttle valve lever assembly (Fig. 23-128).

12. Position the 8 steel balls in the valve body as shown in Fig. 23-129.

13. Position the separator plate over the transfer plate and install the attaching screws (Fig. 23-130); torque these screws to specifications.

14. Position the transfer plate over the valve body, and align all the screw holes.

15. Install and torque the 16 valve body screws (Fig. 23-131).

16. Install the detent spring and attaching screw as shown in Fig. 23-132.

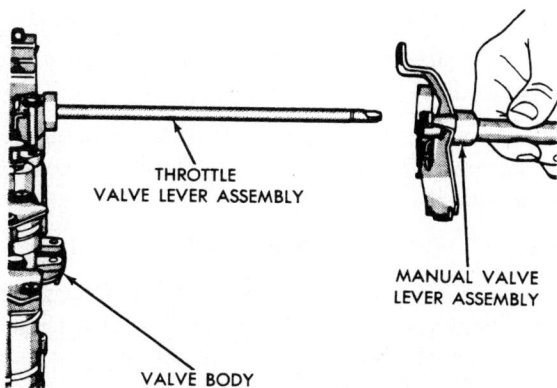

FIGURE 23-126   Installing the manual valve lever assembly. (Courtesy of Chrysler Corp.)

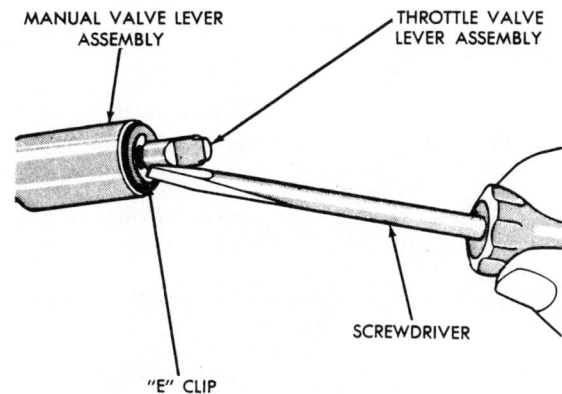

FIGURE 23-128   Installing the throttle lever "E" clip. (Courtesy of Chrysler Corp.)

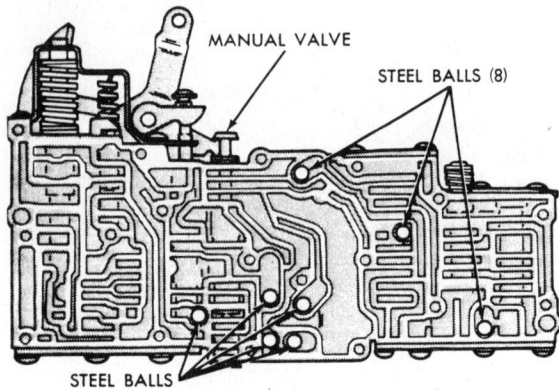

FIGURE 23-129 The correct steel ball locations in the valve body. (Courtesy of Chrysler Corp.)

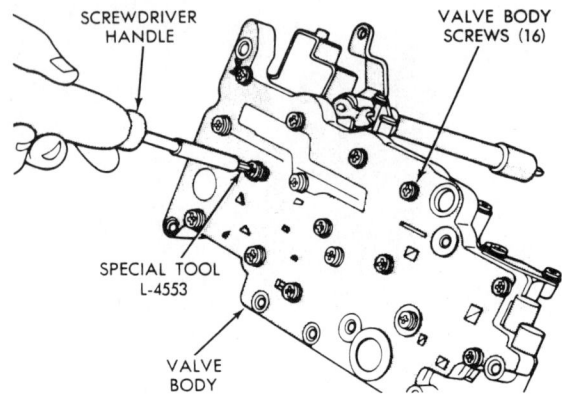

FIGURE 23-131 Installing the valve body screws. (Courtesy of Chrysler Corp.)

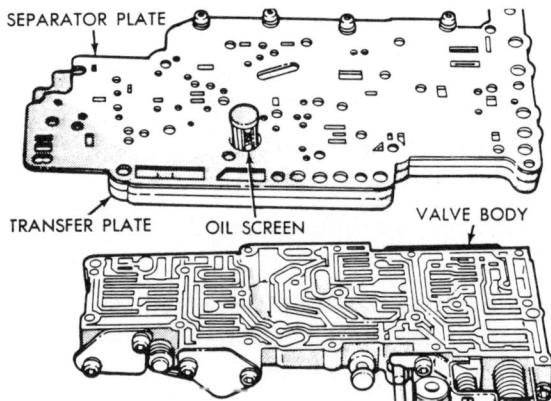

FIGURE 23-130 Assembling the separator plate, transfer plate, and valve body. (Courtesy of Chrysler Corp.)

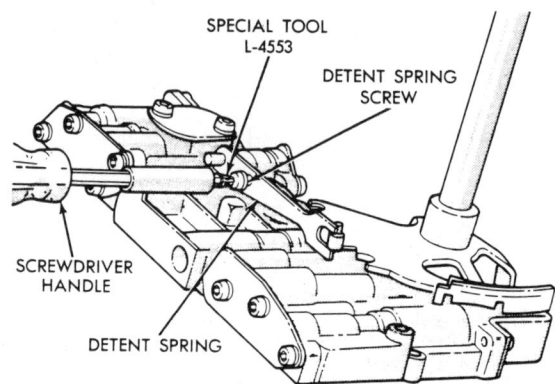

FIGURE 23-132 Installing the detent spring and attaching screw. (Courtesy of Chrysler Corp.)

TRANSAXLE REASSEMBLY

After reconditioning all the subassemblies, reassemble the transaxle following these steps:

1. Coat the No. 11 thrust washer with petroleum jelly and position it in place in the rear of the case (Fig. 23-133).

2. Install the low-reverse band and strut into the case (Fig. 23-134).

3. Reinstall the overrunning clutch cam assembly (Fig. 23-135).

4. Coat the No. 10 thrust washer with petroleum jelly and position it in place over the rear annulus gear (Fig. 23-136) or on the back of the rear planetary gear assembly.

5. Install the rear planetary gear assembly into the rear annulus gear (Fig. 23-137). Make sure the tabs on the No. 10 thrust washer index with the slots in the assembly and that the planetary gear assembly lugs enter the slots machined into the overrunning clutch inner race.

6. Coat the No. 9 thrust washer with petroleum jelly and place it in front of the rear planetary assembly (Fig. 23-138).

7. Install the sun gear driving shell (Fig. 23-139).

8. Coat the No. 6 thrust washer with petroleum jelly, and install it inside the sun gear driving shell (Fig. 23-140) or in back of the front planetary gear assembly.

9. Carefully install the front planetary gear assembly (Fig. 23-141). Make sure the tabs on the No. 6 thrust washer index with the slots in the gear assembly.

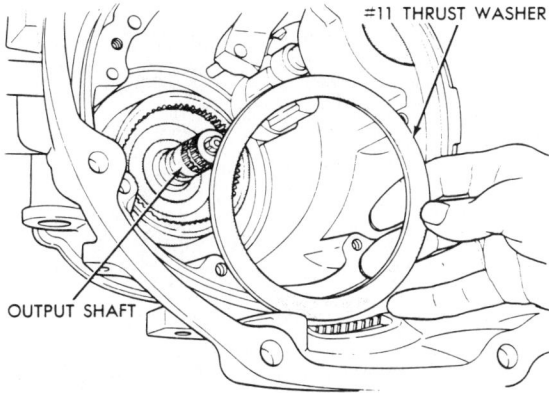

FIGURE 23-133 Installing the No. 11 thrust washer. (Courtesy of Chrysler Corp.)

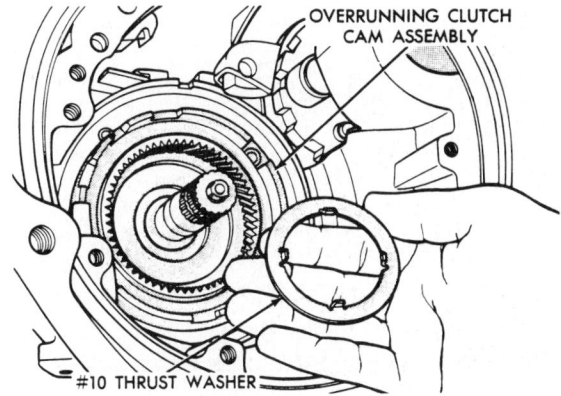

FIGURE 23-136 Installing the No. 10 thrust washer. (Courtesy of Chrysler Corp.)

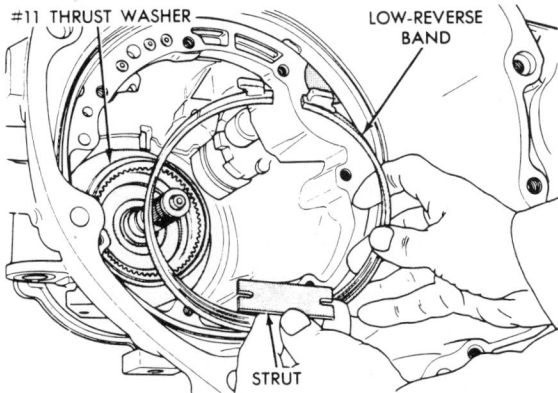

FIGURE 23-134 Installing the low-reverse band and strut. (Courtesy of Chrysler Corp.)

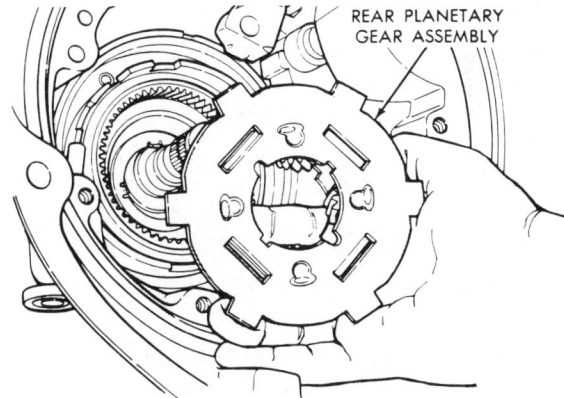

FIGURE 23-137 Installation of the rear planetary gear assembly. (Courtesy of Chrysler Corp.)

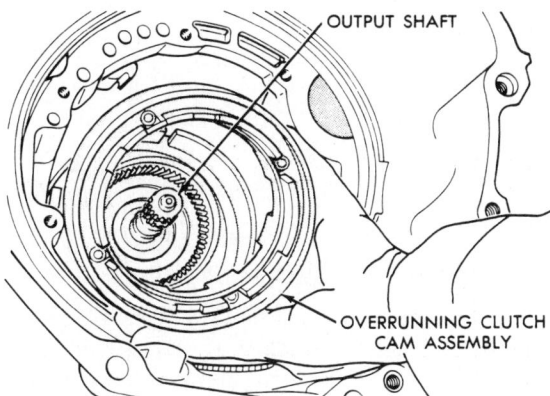

FIGURE 23-135 Installing the overrunning clutch cam assembly. (Courtesy of Chrysler Corp.)

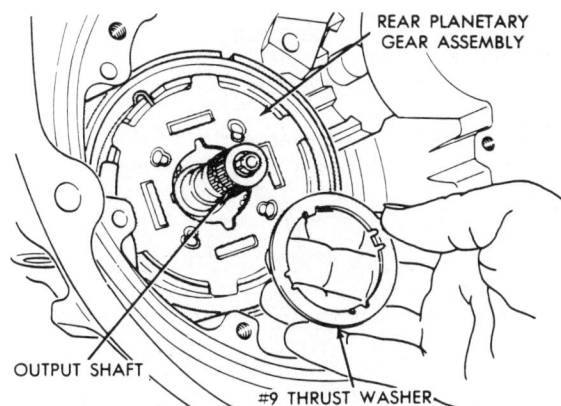

FIGURE 23-138 Installing the No. 9 thrust washer. (Courtesy of Chrysler Corp.)

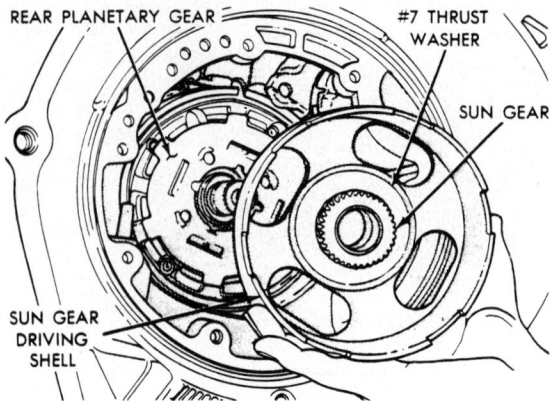

FIGURE 23-139   Installation of the sun gear driving shell. (Courtesy of Chrysler Corp.)

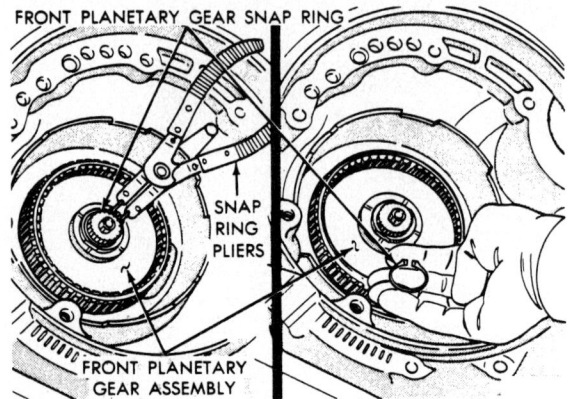

FIGURE 23-142   Installing the front planetary gear snap ring. (Courtesy of Chrysler Corp.)

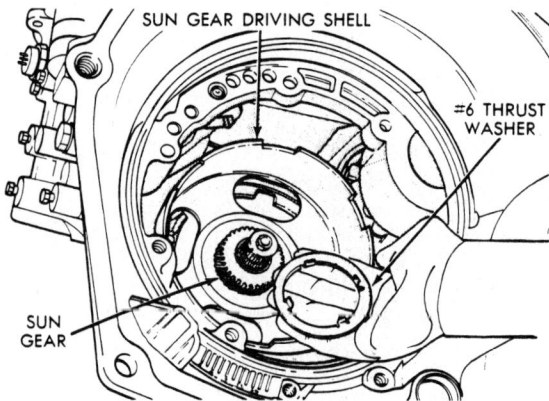

FIGURE 23-140   Installing the No. 6 thrust washer. (Courtesy of Chrysler Corp.)

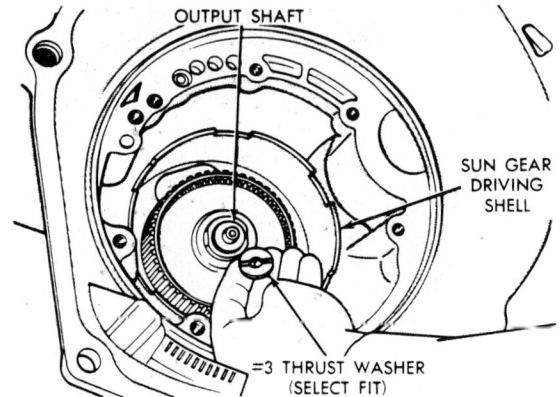

FIGURE 23-143   Installation of the No. 3 select fit thrust washer. (Courtesy of Chrysler Corp.)

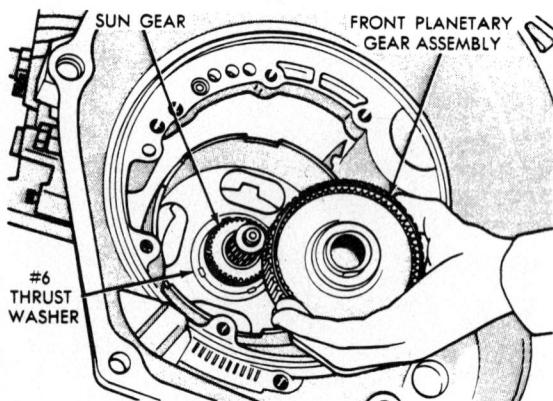

FIGURE 23-141   Installation of the front planetary gear assembly. (Courtesy of Chrysler Corp.)

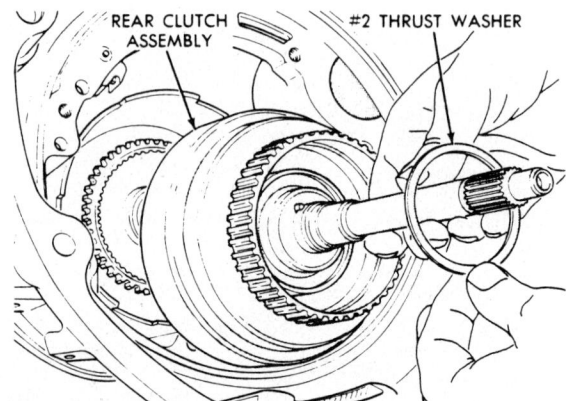

FIGURE 23-144   Installing the rear clutch assembly and the No. 2 thrust washer. (Courtesy of Chrysler Corp.)

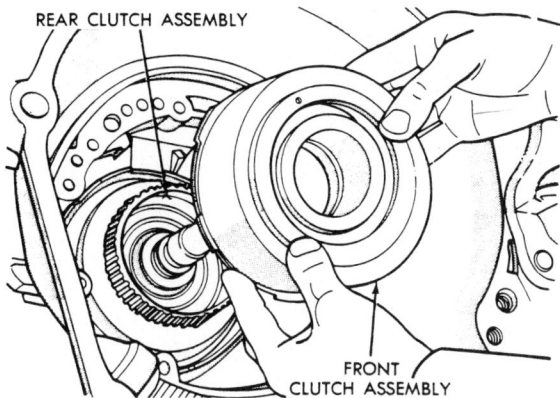

FIGURE 23-145   Installation of the front clutch assembly. (Courtesy of Chrysler Corp.)

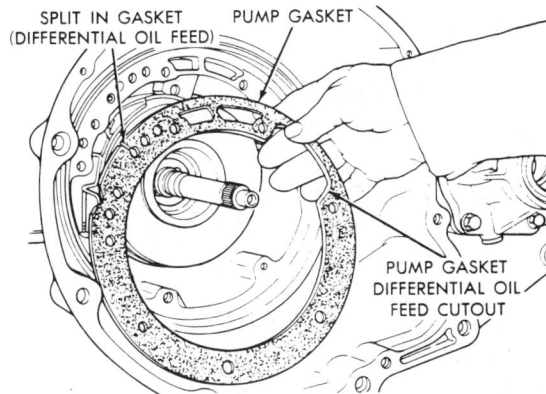

FIGURE 23-147   Installation of a new pump gasket. (Courtesy of Chrysler Corp.)

10. Install the front planetary gear snap ring (Fig. 23-142).

11. Install the correct size select fit thrust washer in place on the output shaft (Fig. 23-143). To determine the size of the washer needed, refer to the measurement taken during the input shaft end-play test.

12. Install the rear clutch assembly into the case (Fig. 23-144). Rotate the clutch sufficiently to index its driving discs with the front planetary clutch hub.

13. Coat the No. 2 thrust washer and position it in place on the rear clutch assembly.

14. Install the front clutch assembly (Fig. 23-145). Rotate the assembly until its driving discs index with the hub on the rear clutch.

15. Install the kickdown band and strut into the case (Fig. 23-146).

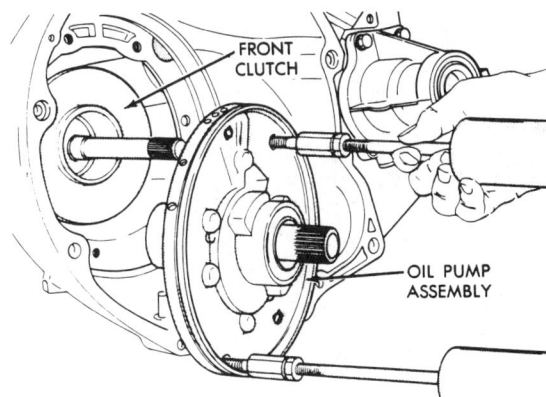

FIGURE 23-148   Installing the oil pump. (Courtesy of Chrysler Corp.)

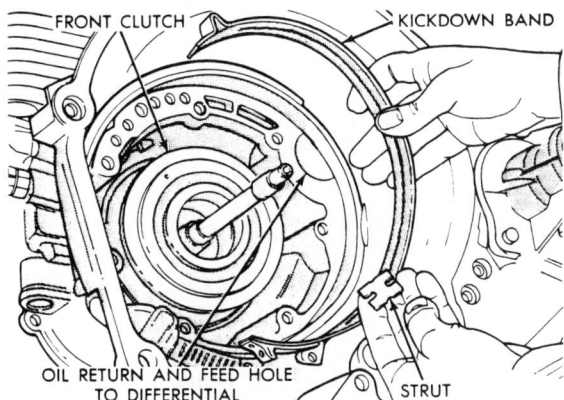

FIGURE 23-146   Installing the kickdown band and strut. (Courtesy of Chrysler Corp.)

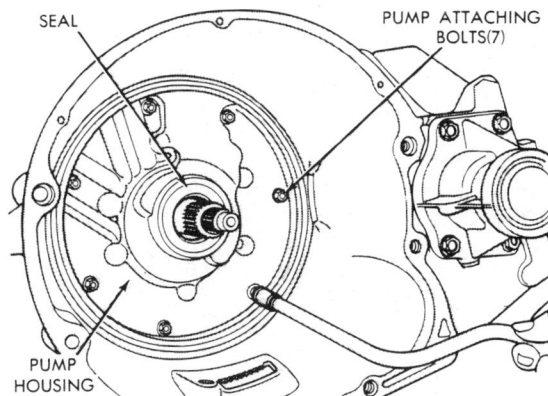

FIGURE 23-149   Installing the pump attaching bolts. (Courtesy of Chrysler Corp.)

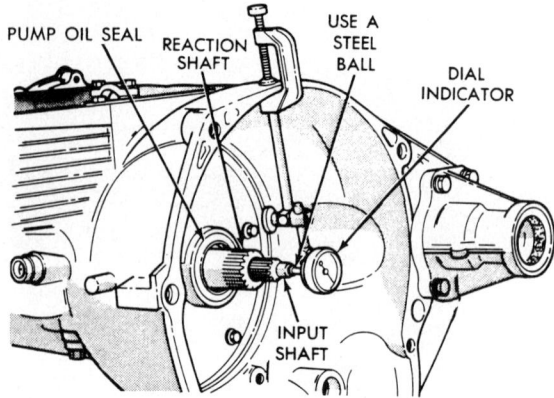

FIGURE 23-150   Performing an input shaft end-play test. (Courtesy of Chrysler Corp.)

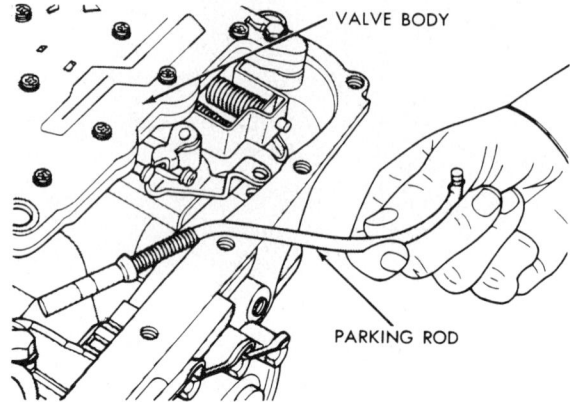

FIGURE 23-151   Installing the valve body and governor tubes. (Courtesy of Chrysler Corp.)

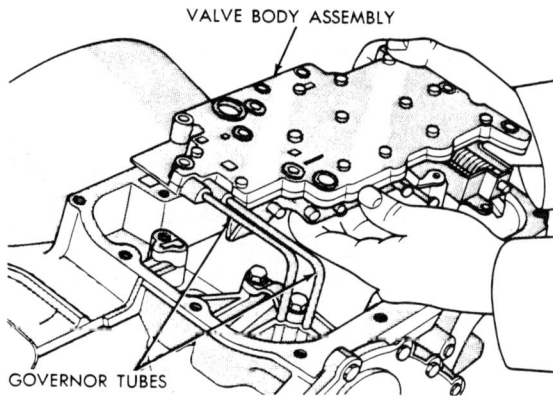

FIGURE 23-152   Installation and torquing of the valve body attaching bolts. (Courtesy of Chrysler Corp.)

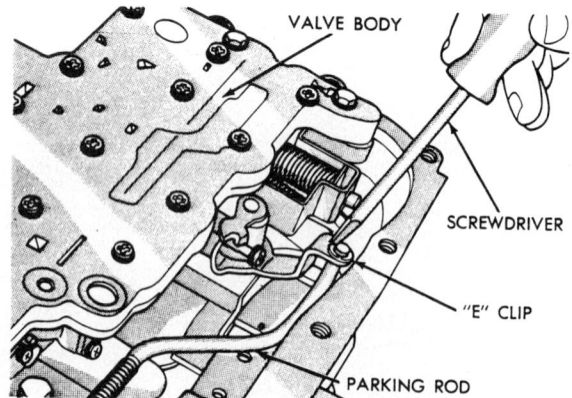

FIGURE 23-153   Installing the parking rod. (Courtesy of Chrysler Corp.)

16. Install the new pump gasket in place as shown in Fig. 23-147.

17. Reinstall the oil pump assembly and the No. 1 thrust washer as shown in Fig. 23-148.

18. Install the pump attaching bolts and torque to specifications (Fig. 23-149).

19. Attach a dial indicator onto the transaxle bell housing with its plunger resting against the end of the input shaft (Fig. 23-150).

20. Move the input shaft in and out to obtain an end-play reading, which should now be from 0.007 to 0.073 inch.

21. Carefully position the valve body and governor tubes in place (Fig. 23-151).

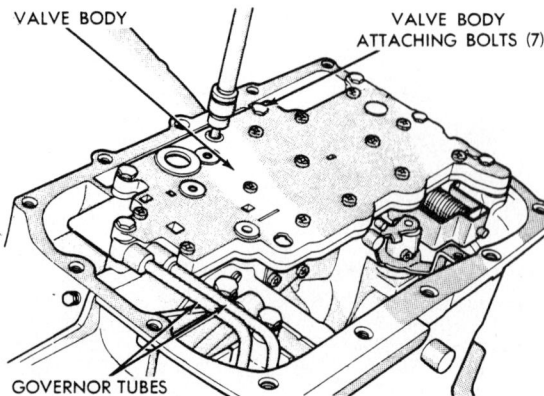

FIGURE 23-154   Installing the "E" clip in the groove on the end of the parking rod. (Courtesy of Chrysler Corp.)

22. Install the valve body attaching bolts and torque them to specifications (Fig. 23-152).

23. Install the parking rod into position (Fig. 23-153).

24. Insert the end of the parking rod into the lever, and install the "E" clip as shown in Fig. 23-154.

25. Install a new filter and its two attaching screws (Fig. 23-155); torque the screws to specifications.

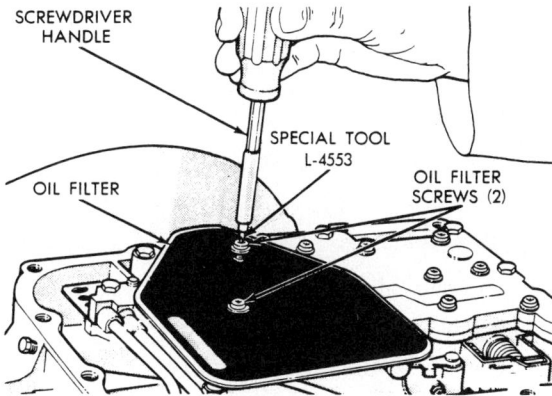

FIGURE 23-156 Installation of the oil pan and its attaching bolts. (Courtesy of Chrysler Corp.)

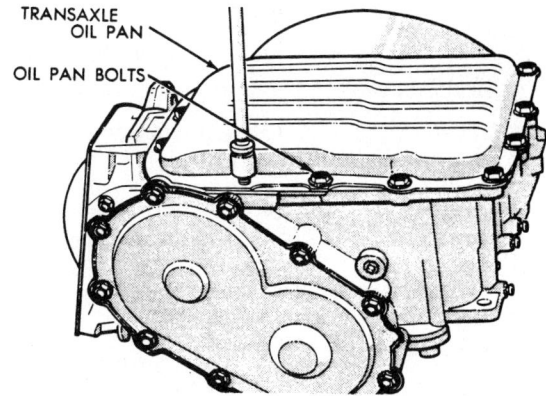

FIGURE 23-155 Installing the filter and its screws. (Courtesy of Chrysler Corp.)

26. Spread R.T.V. sealant onto the pan's gasket flange. Then position a new gasket over the flange, aligning all the bolt holes.

27. Install the pan onto the transaxle, and torque all its attaching bolts to specifications (Fig. 23-156).

28. Install the neutral-park, start, and backup lamp switch.

29. Adjust both bands to specifications (see Table 23-2).

REVIEW

This section will assist you in determining how well you remember the material contained in this chapter. Read each item carefully. If you can't complete the statement, review the portion of the chapter that covers the material.

1. Before disassembling the transaxle, the mechanic should _____ _____ _____ the case thoroughly.

2. Always install new _____, _____ _____, and _____ when rebuilding the transaxle.

3. Do not dry parts with a rag; instead, use _____ _____.

4. The end-play of the transaxle should be checked with a _____ _____.

5. The end-play of the A-415, A-413, or A-470 transaxle should be _____ inch.

6. No. _____ thrust washer controls the end-play of the transaxle.

7. Remove the kickdown band and strut after first taking off the _____ _____.

8. The No. 3 thrust washer is located on the _____ _____.

9. The location of the No. 7 thrust washer is on the _____ _____.

10. A _____ _____ secures the front planetary gear assembly to the output shaft.

11. A _____ _____ secures the kickdown servo in plate inside the case.

12. The location of the No. 10 thrust washer is on the back of the _____ _____ _____ .

13. The kickdown servo piston has _____ _____ to prevent fluid leakage.

14. The two oil pump gears require a mark on them as an aid to their proper _____ during pump reassembly.

15. Two large sealing rings fit in the grooves in the _____ _____ support.

16. To remove the front clutch spring retainer snap ring use a _____ tool.

17. Remove the front clutch retainer bushing with a _____ _____ and removing head.

18. Soak all the new driving discs in clean transmission fluid for _____ _____ before installation.

19. A three-disc rear clutch should operate with a clearance between _____ inch.

20. The front clutch clearance of a two-disc unit should be _____ inch.

21. Deglaze the band surface of the front retainer by sanding it with _____ grit sandpaper.

22. The overrunning clutch has _____ rollers and springs.

23. Use _____ _____ to hold all thrust washers in place.

24. When assembling the transaxle, _____ all fasteners to factory specifications.

25. To adjust the low-reverse band on the A-414 or A-470 transaxle, torque the adjusting screw to _____ inch-pounds and back it off _____ turns.

For the answers, turn to the Appendix.

Answers to Chapter 1 Review

1. a
2. c
3. a
4. d
5. b
6. c
7. d
8. b
9. a
10. d

Answers to Chapter 2 Review

1. c
2. a
3. a
4. b
5. d
6. a
7. c
8. b
9. b
10. a
11. d
12. c
13. b
14. a
15. d
16. a
17. c
18. b
19. c
20. a
21. b
22. a
23. d

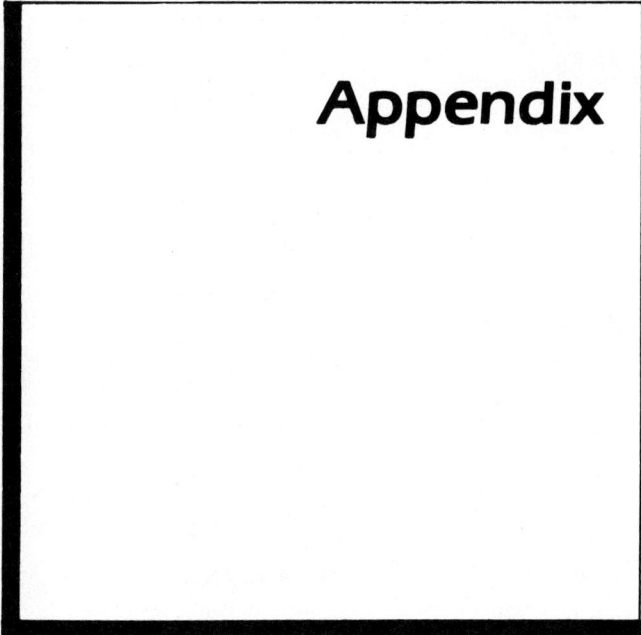

Appendix

Answers to Chapter 3 Review

1. c
2. a
3. c
4. b
5. d
6. c
7. c
8. a
9. c
10. d
11. a
12. c

6. a
7. b
8. d
9. d
10. b
11. c
12. a
13. c
14. b
15. c
16. a
17. c
18. a
19. b
20. c

Answers to Chapter 4 Review

1. d
2. a
3. b
4. c
5. b

Answers to Chapter 5 Review

1. c
2. a
3. b
4. b

5. d

6. a

7. d

8. c

9. b

10. d

Answers to Chapter 6 Review

1. b

2. a

3. d

4. c

5. d

6. d

7. a

8. c

9. d

10. a

11. a

12. b

Answers to Chapter 7 Review

1. d

2. c

3. a

4. b

5. b

6. c

7. c

8. a

9. b

10. b

11. a

12. c

Answers to Chapter 8 Review

1. c

2. a

3. b

4. a

5. c

6. a

7. a

8. a

9. b

10. c

11. a

12. c

13. a

Answers to Chapter 9 Review

1. b

2. a

3. b

4. a

5. d

6. b

7. a

8. b

9. d

10. a

11. b

12. a

13. b

14. d

Answers to Chapter 10 Review

1. c

2. a

3. d

4. a

5. b

6. d

7. a

8. b

9. a

10. a

11. c

12. b

13. c

14. c

Answers to Chapter 11 Review

1. c

2. a

3. b

4. d

5. b

6. a

7. d

8. a

9. b

10. b

11. a

12. c

13. d

14. a

Answers to Chapter 12 Review

1. d

2. c

3. a

4. b

5. c

6. b

7. a

8. a

9. d

10. c

11. c

12. d

13. c
14. d
15. b
16. a

Answers to Chapter 13 Review

1. c
2. a
3. b
4. a
5. d
6. a
7. b
8. c
9. b
10. d

Answers to Chapter 14 Review

1. b
2. a
3. d
4. c
5. a
6. c
7. d
8. b
9. c
10. a
11. d
12. a
13. c
14. d
15. c
16. b
17. a
18. d
19. c

20. b
21. a
22. c

Answers to Chapter 15 Review

1. recycle
2. Vortex
3. drain, plug
4. bench, checks
5. pulleys
6. pressure
7. electrical, kickdown
8. internal, pressures
9. four
10. steam
11. inside
12. air
13. diaphragm
14. inclined, plane
15. bearing
16. blind
17. single-post
18. two, four
19. telescoping, ratchet-type
20. wet
21. one
22. Heli-coil, repair, kit
23. smaller, lighter
24. air, ratchet
25. pulling-type, impact-type
26. snap, rings
27. pencil, magnet

Answers to Chapter 16 Review

1. torque, wrench
2. torquing
3. clicker, torque

4. torque, increased
5. micrometer
6. micrometer, 40
7. 0.001
8. ratchet, micrometer
9. bent, feeler, gauge
10. tapered, gauge
11. dial, indicator
12. dial
13. fastener
14. screw
15. pitch
16. bolt
17. flat
18. cotter, slotted
19. key, ball, pin
20. splines

Answers to Chapter 17 Review

1. logical, specific, thorough
2. level, condition
3. suction, gun
4. vacuum, modulator
5. cardboard, paper
6. black, light
7. cooler
8. manual, valve
9. shift, points, control, pressure
10. gauge, assembly
11. wide-open, throttle
12. road, test
13. band, clutch
14. clutch-, band-application
15. pressure
16. 1.5
17. high
18. modulator, selective

19. diagnosis
20. air
21. overhaul
22. rubber, metal
23. stethoscope
24. dynamometer
25. kickdown

Answers to Chapter 18 Review

1. oxidized
2. owner's, manual, service, manual
3. pan, converter
4. filler, tube
5. rear
6. wear
7. dimple
8. lint
9. paper
10. petroleum, jelly, grease
11. torque, wrench, factory, specifications
12. 6, o'clock
13. pump
14. stop
15. service, manual

Answers to Chapter 19 Review

1. planetary, member
2. tight
3. lining
4. external
5. selective
6. tightens or loads
7. Allen, eight-point
8. specifications
9. fluid, remove, pan
10. control, input
11. manual, valve
12. detent, pawl, shift, gate
13. throttle, kickdown or detent
14. service or transmission, manual
15. cable

Answers to Chapter 20 Review

1. linkages
2. drive, gear
3. inward
4. bushing
5. aligning
6. access, hole
7. seal
8. ball, joints
9. engine, support
10. remover

Answers to Chapter 21 Review

1. expensive
2. flush
3. before
4. true
5. two
6. fingernail
7. polish, crocus
8. replace
9. counterclockwise
10. thrust, washers, bearings
11. welds
12. flushing, machine
13. drain, plug
14. ring, gear
15. flow, pressure
16. layout, fluid, indelible, pen
17. passages
18. new, pump
19. straightedge

20. aligning, tool

Answers to Chapter 22 Review

1. housing or framework
2. varnish, shellac
3. low-pressure, compressed, air
4. replace
5. Heli-coil, insert
6. bushing
7. cut
8. shifter, shaft, seal
9. fluid, leak
10. epoxy, cement
11. extension, housing
12. bushing
13. inward
14. input, shaft
15. fine, wire
16. clutch, piston
17. drum, piston
18. feeler gauge
19. 40-60
20. one, direction
21. accumulator
22. governor
23. square, edges
24. shafts
25. thrust, washers
26. bearing
27. chisel, remover
28. valve, body
30. rag

31. 120-180
32. 30
33. scrape
34. Teflon
35. toward

Answers to Chapter 23 Review

1. steam, clean
2. gaskets, sealing, rings, seals
3. compressed, air
4. dial, indicator
5. 0.007-0.073
6. 3
7. oil, pump
8. output, shaft
9. driving, shell
10. snap, ring
11. snap, ring
12. rear, planetary, assembly
13. seal, rings
14. alignment
15. reaction, shaft
16. spring-compressor
17. tool, handle
18. 30, minutes
19. 0.026-0.034
20. 0.076-180
21. 120-180
22. 8
23. petroleum jelly
24. torque
25. 41, 3½

Index

Also from Reston by William Husselbee

Automotive Tune-up Procedures
Automotive Emission Control
Automotive Cooling, Exhaust, Fuel and Lubricating Systems